THIRD EDITION

HARTMANN'S PLANT SCIENCE

Growth, Development, and Utilization of Cultivated Plants

Margaret J. McMahon, Ph.D.
Department of Horticulture and Crop Science
The Ohio State University

Anton M. Kofranek, Ph.D.
Department of Environmental Horticulture
University of California at Davis

Vincent E. Rubatzky, Ph.D.
Department of Vegetable Crops
University of California at Davis

Prentice
Hall

Upper Saddle River, New Jersey 07458

Library of Congress Cataloging-in-Publication Data

McMahon, Margaret J.
  Hartmann's plant science: growth, development, and utilization of cultivated plants / Margaret
McMahon, Anton M. Kofranek, Vincent E. Rubatzky.—3rd ed.
    p. cm.
  Rev. ed. of: Plant science / Hudson T. Hartmann . . . [et al.].
  Includes bibliographical references (p.).
  ISBN 0-13-955477-7
    1. Plants, Cultivated. 2. Botany, Economic. I. Kofranek, Anton M. II. Rubatzky,
Vincent E. III. Plant science. IV. Title.

SB91 .P56 2002
631—dc21

2001036073

Editor-in-Chief: Stephen Helba
Executive Editor: Debbie Yarnell
Associate Editors: Kimberly Yehle, Michelle Churma
Managing Editor: Mary Carnis
Production Editor: Lori Dalberg, Carlisle Publishers Services
Project Coordinator: Eileen O'Sullivan
Director of Manufacturing and Production: Bruce Johnson
Manufacturing Buyer: Cathleen Petersen
Senior Design Coordinator: Miguel Ortiz
Cover Design: Carey Davies
Cover Photography: Frosted design: Richard Megna; Fundamental photographs: golf course, Tony
Stone Images; floral photographs and field scene, Margaret McMahon
Marketing Manager: Jimmy Stephens
Composition: Carlisle Communications, Ltd.
Interior Design: Carlisle Communications, Ltd.
Printer/Binder: Courier Westford

Pearson Education LTD.
Pearson Education Australia PTY, Limited
Pearson Education Singapore, Pte. Ltd.
Pearson Education North Asia Ltd.
Pearson Education Canada, Ltd.
Pearson Educación de Mexico, S.A. de C.V.
Pearson Education—Japan
Pearson Education Malaysia, Pte. Ltd.

10 9 8 7 6 5 4 3 2 1
ISBN 0-13-955477-7

Dedication

This third edition of *Hartmann's Plant Science: Growth, Development, and Utilization of Cultivated Plants* is dedicated to Dr. Hudson T. Hartmann. Dr. Hartmann died in 1994 and is remembered as a committed scientist and dedicated teacher. Dr. Hartmann's name has been added as a permanent part of the new title of this book in recognition of his long-lasting contributions to the field of plant science. We gratefully acknowledge those contributions.

About the Authors

Dr. Hudson T. Hartmann, December 6, 1914–March 2, 1994
Professor of Pomology, University of California, Davis

Dr. Hartmann was born in Kansas City, Kansas. After high school he entered the University of Missouri, earning a B.S. in Agriculture in 1939 and an M.S. in Horticulture in 1940. He then received a Teaching Assistantship in the Department of Pomology, University of California, which lead to graduate studies and achievement of his Ph.D. in Plant Physiology in 1947. An appointment as Assistant Professor at the University of California, Davis followed. At Davis he continued his research and teaching until retirement in 1980 as Emeritus Professor of Pomology.

Dr. Hartmann was a pioneer researcher in developing the use of plant growth regulators, intermittent mist, and other relevant practices for the rooting of cuttings, especially as they applied to fruit trees. Along with his research activities in plant propagation and fruit crop physiology he earned worldwide recognition for his expertise with olives.

Dr. Hartmann taught plant propagation at the University of California at Davis for more than 30 years until his retirement in 1980. Another of his proudest accomplishments was his activity with the International Plant Propagation Society. He served as Western Region Editor for the Society for 33 years, and International Editor for 21 years. Dr. Hartmann was elected as a Fellow of the American Society for Horticultural Science, and also received numerous other awards and honors.

During his career, he published over 40 research articles about plant propagation. Very notable was his collaboration with Dr. Dale Kester, in 1959 to prepare *Plant Propagation: Principles and Practices,* published by Prentice-Hall, which has since gone into a 7th edition, published in 2001. This textbook has been used by more plant propagation classes worldwide than any other propagation book, and was translated into Spanish, Italian, and Russian, as well as an English Economy Edition for Third World countries. Dr. Hartmann also was the senior author of the first and second edition of the college textbook, *Plant Science: Growth, Development, and Utilization of Cultivated Plants,* also published by Prentice-Hall in 1981 and 1988. Many of his contributions in those two editions continue to be relevant and useful in the present edition.

William J. Flocker, September 6, 1917–April 12, 1980
Professor of Vegetable Crops, University of California, Davis

William "Bill" Flocker was born and raised on a farm in Clinton, Indiana. After graduating from the University of Illinois in 1936, he enlisted in the U.S. Air Corps and served as a fighter pilot and squadron commander in World War II. However, his academic interests led him to earn his Ph.D. in Agricultural Chemistry and Soils at the University of Arizona in 1955. Shortly thereafter he joined the Vegetable Crops Department at the University of California at Davis.

Dr. Flocker is best remembered for his dedication to teaching and students, by whom he was revered. His interest in students is well demonstrated via his creation of the course, Plant Science 2, in 1965. He early recognized the potential for new approaches for teaching in order to orient students in plant science to develop an appreciation for the significance of agriculture. His pioneer work in audio-tutorial teaching techniques utilized in that heavily enrolled course was an outstanding example of his innovation.

Through his research programs he became an authority in soil physics, soil physical properties, and soil-water movement. His research interests included procedures for maintaining and improving soil structure for maximum crop production. A valued contribution was his research and demonstration of the importance of dealing with soil compaction problems. Concurrent with his concern for environmental quality he did significant research for incorporating solid wastes, particularly cannery wastes, into soil to improve marginal land. For this and similar soil amendment research he was recognized and presented the Environmental Quality Award by the American Society for Horticultural Science.

Dr. Flocker's desire for students to become better aware about the principles of plant science and the utilization of cultivated plants directed him to join with Drs. Hudson T. Hartmann and Anton M. Kofranek in the preparation of the first edition of *Plant Science: Growth, Development, and Utilization of Cultivated Plants.* Many of his contributions to that publication continue to be useful and appreciated in this edition.

Dr. Margaret (Peg) J. McMahon, Born April 8, 1948, in Cleveland, Ohio
Associate Professor of Horticulture and Crop Science,
The Ohio State University

Peg is a fourth generation horticulturist and grew up working on the family farm and in the ornamental and vegetable greenhouses owned by her family. She majored in Horticulture at The Ohio State University, earning her B.S. in Agriculture in 1970. After graduation she worked for fifteen years as a grower and propagator for Yoder Brothers, Inc., a multinational greenhouse company, starting at their Barberton, Ohio location, transferring to operations in Salinas, CA and then Pendleton, SC. She was responsible at one time or another for the production or propagation of nearly all of the company's crops.

While in South Carolina she enrolled in the graduate school at Clemson University. She earned an M.S. in Horticulture and a Ph.D. in Plant Physiology in 1988 and 1992, respectively. Peg's masters project focused on the early detection of chilling injury in tropical and subtropical foliage plants. Her Ph.D. research was in photomorphogenesis, specifically the development and use of far-red absorbing filters as a technology to reduce unwanted stem elongation in greenhouse crop production. She continues her research in far-red absorbing filters. Far-red absorbing filters are now coming into the product line of several agricultural plastics manufacturers partly as a result of her research.

In 1994 she accepted a position as an Assistant Professor in the Department of Horticulture and Crop Science (H&CS) at The Ohio State University. Her responsibilities are in teaching and research. She teaches several H&CS courses including the introductory and senior capstone crop science classes as well as greenhouse production courses. The introductory and capstone classes are required for all students with majors and minors in Turfgrass Science and Crop (agronomic and horticultural) Science. In 1999 she was promoted to Associate Professor.

Dr. McMahon has published over 20 peer-reviewed articles on photomorphogenesis, teaching methods, and other subjects.

Anton M. Kofranek, Born February 5, 1921, in Chicago, Illinois
Retired as Professor Emeritus from University of California at Davis in 1987

Dr. Kofranek obtained his B.S. in Horticulture at the University of Minnesota in 1947, an M.S. in Plant Physiology at Cornell University in 1949, and his Ph.D. in Plant Physiology at Cornell University in 1950. His first academic appointment was as Instructor in the Department of Horticulture at the University of California at Los Angeles in September 1950. In 1968 when the department was transferred to the University of California at Davis and renamed as the Department of Environmental Science he also relocated and continued his floricultural research and teaching at Davis until retirement in 1987.

His research specializations dealt with photoperiod research for numerous floricultural crops. Other interests included postharvest physiology and nutritional investigations of flower crops. Dr. Kofranek produced over 200 publications. He co-authored *Plant Science: Growth, Development, and Utilization of Cultivated Plants,* and *The Azalea Manual.*

Throughout his career, Dr. Kofranek was highly respected and appreciated by persons in all segments of the floriculture trade, and he received numerous awards and recognition from producers, wholesalers, landscape architects, and other allied industry segments. He has been repeatedly recognized for his efforts by the California Flower Growers. He also received considerable academic recognition, and in 1979 was installed as a Fellow of the American Society for Horticultural Sciences. Dr. Kofranek carried out sabbatical study programs in The Netherlands, Israel, England, South Africa, and at Cornell University. He was a consultant with USAID programs with Egypt and India. His extensive travels led to collaboration with scientists in several countries.

Vincent E. Rubatzky, Born October 24, 1932, in New York
Retired as University of California Extension Vegetable Specialist,
Emeritus in 1995

Dr. Rubatzky obtained a B.S. in Vegetable Crops at Cornell University in 1956, a M.S. in Plant Physiology at Virginia Polytechnic Institute in 1958, and the Ph.D. in Plant Physiology and Horticulture at Rutgers University in 1964. His first academic position was as Extension Vegetable Specialist at the University of California at Davis in 1964, and where he conducted his extension outreach and research program until retirement.

His research involvement dealt with variety development and evaluation, harvest mechanization, crop physiological disorders, and crop scheduling and cultural practices. He also served in a liaison capacity with several California vegetable industry organizations and groups. For several years Dr. Rubatzky taught a class entitled Evolution, Biology, and Systematics of Vegetables, and a Field Study of the California Vegetable Industry course. Sabbatical studies and consultations within Poland, France, United Kingdom, Italy, The Netherlands, Peru, and Ecuador provided many travel experiences. In 1995, he was elected as an American Society for Horticultural Sciences Fellow, Man of The Year by the Pacific Seedman's Association in 1987, and received additional recognition from other organizations. Following retirement Dr. Rubatzky continued to serve on a voluntary basis as Specialist, Emeritus until mid-1999. Presently residing in New Jersey, he collaborates with colleagues in vegetable production research in California and New Jersey.

Dr. Rubatzky produced many applied research and extension publications during his University service. He co-authored *Plant Science: Growth, Development, and Utilization of Cultivated Plants; World Vegetables: Principles, Production, and Nutritive Values,* and *Carrots and Related Vegetable Umbelliferae,* and was an editor for the *Third International Symposium for Diversification of Vegetable Crops,* the *Atlas of the Traditional Vegetables in China,* and several other vegetable publications.

Contents

CHAPTER 5

Propagation of Plants   68

CHAPTER 6

Vegetative and Reproductive Growth and Development   96

CHAPTER 7

Photosynthesis, Respiration, and Translocation   121

CHAPTER 12

Harvest, Preservation, Transportation, Storage, and Marketing   231

UNIT II

AN OVERVIEW OF THE FRUIT CROPS AND ORNAMENTAL PLANTS   249

CHAPTER 13

Cultural Practices in Orchards and Vineyards   251

CHAPTER 14

Flowering and Fruiting in Fruit Crops   275

CHAPTER 15

Nursery Production: Field, Above-Ground Container, and Pot-In-Pot Cultures 288

CHAPTER 16

Landscape Trees: Deciduous, Broad- and Narrow-Leaved Evergreens 302

CHAPTER 17

Ornamental Shrubs: Deciduous, Broad- and Narrow-Leaved Evergreens 324

CHAPTER 18

Floriculture 337

Preface

Enter the wonderful world of plant science. Discover why we depend on plants and the people who understand them for our survival. Find out how plants keep us alive and allow us to enjoy our lives. Learn about growing, maintaining, and utilizing plants for the benefit of ourselves and the environment in which we live. No matter if your interests range from the family farm to a tournament golf course to the boardroom of an international business, rewarding, challenging, and fulfilling careers are open to anyone skilled in plant science.

Human survival absolutely depends on the ability of plants to capture solar energy and convert that energy to a form that can be used as food. The captured energy stored in plant tissues also provides fiber and oil for fuel, clothing, and shelter. The production of plants that meet our needs for survival is an important application of the knowledge of plant science.

However, the essentials for nutrition and shelter can be provided by relatively few plant species. Life would be very boring if those few were the only species produced for our needs. Fortunately, for those who dislike boredom, thousands of species of plants are grown that add enjoyment to life by providing a variety of flavors and textures in food and fiber. Other plants brighten our lives when used in landscaping and interior decoration. Recently, the importance of turfgrass in athletic and outdoor recreation sites has risen dramatically around the world.

Animal feeds are another important use of plants. Animals provide nutrition and variation in our diets and materials for clothing and shelter, reduce labor in some areas of the world, and add pleasure in our lives through recreation or as pets. As a result of our need for plants to survive and enjoy our lives, plants have a tremendous economic impact in developed and developing nations. The career opportunities created by the need for people with an understanding of plant growth are unlimited.

Plant Science is written for anyone with an interest in how plants are grown and utilized for maintaining and adding enjoyment to human life. The beginning sections of the text are designed to provide the fundamentals of botany, plant physiology, and environmental factors affecting plant growth in a way intended to be easily comprehended by students familiar or unfamiliar with plant science. The later sections are the integration of the aforementioned topics into strategies of producing plants for human use as food, fiber, and recreation.

The third edition of *Plant Science* has been updated to include the most recent statistics, production methods, and issues concerning the production and utilization of plants. To aid the student and instructor, key learning concepts and summaries of those key learning concepts have been added to the chapters in the first two sections. An appendix listing Web sites that provide useful crop production information has been added. A chapter on nursery production has also been added. The chapters concerning herbaceous ornamental production have been consolidated into one chapter.

The contribution of the writers and reviewers of the material in the first and second editions is greatly appreciated. The authors have called upon many scientific workers in the United States and other countries to review chapters and parts of chapters in this book. They gave their time most generously and greatly assisted us in assuring accuracy in the subject matter presented. The responsibility, however, for the final version of the text is the authors'. The following persons reviewed chapters or parts of chapters for either the first or second edition, or both: Curtis Alley, A. H. Allison, Victor Ball, Robert F. Becker, Brian L. Benson, Alison Berry, Itzhak Biran, Arnold J. Bloom, James W. Boodley, Robert A. Brendler, Royce Bringhurst, Thomas G. Byrne, Jack Canny, Will Carlson, Kenneth Cockshull, Charles A. Conover, Beecher Crampton, Began Degan, Frank Dennis, Francis DeVos, Dominick Durkin, Roy Ellerbrock, James E. Ellis, Clyde Elmore, Thomas W. Embledon, Harley English, Elmer E. Ewing, Michael Farhoomand, Gene Galleta, Melvin R. George, Marvin Gerdts, Ernest M. Gifford, Victor L. Guzman, L. L. Haardman, Abraham H. Halevy, Richard W. Harris, R. J. Henning, Charles Heuser, James E. Hill, Jackson F. Hills, Karl H. Ingebretsen, Lee F. Jackson, Subodh Jain, Merle H. Jensen, Robert F. Kasmire, Thomas A. Kerby, Dale Kester, Paul F. Knowles, C. Koehler, Harry C. Kohl, Dale W. Kretchman, Harry Lagerstedt, Pierre Lamattre, Robert W. Langhans, Roy A. Larson, Robert A. H. Legro, Andrew T. Leiser, Gil Linson, Warner L. Lipton, Oscar A. Lorenz, James Lyons, Harry J. Mack, John H. Madison, Vern L. Marble, George Martin, Shimon Mayak, Keith S. Mayberry, Richard Mayer, Donald N. Maynard, Arthur McCain, Charles A. McClurg, Harry A. Mills, Franklin D. Moore III, Yoram Mor, Julia Morton, James L. Ozbun, Jack

L. Paul, William S. Peavy, Nathan H. Peck, Jack Rabin, Allan R. Rees, Michael S. Reid, Charles Rick, Frank E. Robinson, J. Michael Robinson, Norman Ross, Edward J. Ryder, Kay Ryugo, Roy M. Sachs, John G. Seeley, Robert W. Scheuerman, Art Schroeder, William Sims, Donald Smith, L. Arthur Spomer, George L. Staby, W. B. Storey, Vernon T. Stoutemyer, Walter Stracke, Mervin L. Swearingin, Herman Tiessen, Edward C. Tigchelaar, Ronald Tukey, Benigno Villalon, Stephen Weinbaum, Ortho S. Wells, Bernard H. Zandstra, Naftaly Zieslin, and Frank W. Zink. The following are appreciated for their contributions in writing or updating chapters in the third edition: Dr. Hannah Mathers wrote Chapter 15. Dr. Jeff Kuehny wrote Chapter 18. Dr. John Streeter, aided by Dr. Edward McCoy, updated Chapters 7, 8, and 9. Dr. David Ferree, aided by Dr. Joseph Scheerens, updated Chapters 13 and 14. Dr. Martin Quigley updated Chapter 20. Drs. Peter Thomison and James Beuerlein updated Chapter 21. Dr. Matthew Kleinhenz updated Chapters 23, 24, and 25. Dr. Charles Graham updated Chapters 26, 27, and 28.

Margaret J. McMahon
Anton M. Kofranek
Vincent E. Rubatzky

UNIT

I

Plants: Structure, Classification, Growth, Reproduction, and Utilization

CHAPTER

The Role
of Cultivated Plants
in the Living World

KEY LEARNING CONCEPTS

After reading the chapter you should be able to:

- ♦ Understand the role of cultivated plants as food sources.
- ♦ Understand the benefits of cultivated plants in addition to supplying food.
- ♦ Understand the challenges to plant scientists as they try to increase or improve food and other plant benefits to the peoples of the world.

If we could look back some 150 million years to the middle of the geologic period known as the Mesozoic era, the world would not appear entirely strange to us. Among plants, we might recognize some members of the angiosperms,[1] to which belong our present-day grasses, flowers, vegetables, shrubs, and most of our modern trees. In the animal kingdom, by contrast, we would confront those ancient reptiles known as dinosaurs. The careful observer also might note the existence of some small primitive mammals— to which we human beings can trace our own beginnings. Thus, both forms of life dominant today originated during the height of the dinosaur reign. It is still a matter of conjecture why the dinosaurs became extinct, but we do know that mammals flourished during the subsequent Cenozoic era, beginning about 65 million years ago. The human race can be traced back some 3 million years. However,

---

[1] Derived from the Greek words for "vessel" and "seed."

the modern human species, *Homo sapiens (archaic)*, came much later, toward the end of the Pleistocene epoch, some 250,000 years ago, with *H. sapiens* (modern) about 28,000 years ago.

Plants evolved much earlier than humans. The gymnosperms, which include the cone-bearing timber and ornamental trees of today, had their beginnings 300 million years ago. The angiosperms, which humans and other animals depend on for much of their food and other uses, appeared about 90 million years ago. Fossil records show that plants such as hickories, oaks, beans, grapes, and other species were well established by the time even the most ancient forms of humankind appeared. These species would have provided shelter and food then as they do today. However, the plants would have been gathered from the wild, not grown specifically for use. The growing or **cultivation** of plants for use came about 10,000 years after the appearance of modern humans. With the development of cultivation came the need to understand how plants grow. Thus, plant science was germinated and began to grow. The first plant scientists were those who observed how plants grew, developed ideas about the process and how to improve it, tested those ideas, and then came to conclusion.

Today the growing and cultivation of plants for use requires the same understanding of basic plant science. Anyone who makes decisions regarding cultivated crops, their development, and their use—whether it be in the lab or in the field—is a plant scientist and must use sound, scientific-based information when making decisions.

The first cultivated food crops apparently were cereals such as wheat and barley. Several grains of barley have been recovered from archeological sites in Egypt, indicating that ground grain was an important economic resource as early as 18,000 years ago. An agricultural type of existence spread to the European continent some 6,000 years ago, replacing a nomadic hunting existence. There is evidence that corn was first cultivated in what is now Mexico about 5,000 years ago, followed about 1,500 years later by potatoes in South America and rice in the Far East. Some of our present-day fruit crops—grapes, figs, olives, dates, pomegranates, and mulberries—were being cultivated in the eastern Mediterranean when history was first being recorded.

CULTIVATED PLANTS

It is startling to realize that if animals ceased to exist on earth, plants would continue not only to survive but also to thrive very well. However, if plants ceased to exist, all forms of animal life, including human life, would soon disappear. They would simply starve to death. It is true, however, that without humans our highly prized cultivated plants—most of them developed and maintained by our interest in them— would disappear within several crop generations. Replacing the beautiful ornamentals, the tasty fruits and vegetables, and

the highly productive cereal grains would be wild grasses, thistles, and certain shrubs and trees. They would be able to survive in the natural environment without human aid by their production of seeds and their tolerance of prevailing insects and diseases. The variability that would appear in surviving seedlings would enable some of them to grow in a given, perhaps harsh, environment, while others would perish.

Food Sources

To understand the importance of plants to the continued existence of animal life on earth, it is necessary to examine the fundamental chemical reaction known as *photosynthesis.* Photosynthesis takes place in the leaves and other green parts of plants, specifically in the chloroplasts within the green cells. Carbon dioxide absorbed from the air combines with water taken in by the plants' roots from the soil to form organic food materials known, appropriately, as carbohydrates. An important byproduct of this reaction is the free oxygen the plant releases into the atmosphere. The basic energy source for this process is the visible spectrum from the sun.

Once carbohydrates are formed, the plant is able to add nitrogen and other elements absorbed by the roots from the soil to form proteins and, subsequently, to produce fats and oils. All these nutrients—carbohydrates, proteins, fats, and oils—are used by plants to sustain their own growth. Scientists refer to living things that produce their own food as *autotrophic.* However, these nutritive materials can also be utilized as food by humans and other animals, which are *heterotrophic*—dependent upon plants for their food. Without nutritive materials from plants, such as cereal grains, seeds, fruits, leafy plant parts, and tuberous roots and stems, animals would have no way to sustain life.

Plants are thus essential to humans and other animals because they produce food through the photosynthetic process. Selection of the kinds of plants for the human food supply becomes a history of agricultural development, going back into prehistoric times. Obviously the particular kinds of plants selected vary among the agricultural regions of the world. They are the ones best suited to a given set of climatic conditions and are the most efficient in producing food materials.

Only a few species of plants feed the world's peoples either directly or indirectly through animals. These plants are:

1. Cereal crops—wheat, maize (corn), rice, barley, oats, sorghum, rye, and millet. (Over half the world's food supply comes from the photosynthetic activity of these crops.)
2. Roots and tubers—potatoes, sweet potatoes, and cassavas.
3. Oil crops—soybeans, corn, peanuts, palm, coconuts, sunflowers, olive, and safflower.
4. Sugar—sugar cane and sugar beets.

5. Fruit crops—bananas, oranges, apples, pears, and many others.
6. Vegetable crops—tomatoes, lettuce, carrots, melons, asparagus, and so forth. (Fruits and vegetables add to the variety and palatability of our daily meals and supply much-needed vitamins and minerals.

Table 1–1 shows some of the common crops ranked in relation to the calories and proteins produced per unit of land area. Not all of the total production of food materials becomes available for human consumption. Much is lost during harvesting, transportation, and marketing, primarily from attacks by insects, diseases, birds, and rodents. Also, some of the production is saved to be used as seed for future plantings.

A different kind of energy loss occurs when plants are used to produce human food in the form of animal products. For example, it takes about 10 kilograms (22 pounds) of grain (which could be consumed by humans directly) to produce 1 kilogram (2.2 pounds) of beef. But some of the kinds of feed consumed by beef cattle is ordinarily not eaten by humans. A bushel of corn consumed as whole corn meal would meet the daily energy and protein requirements of 23 people, but when this same bushel is fed to chickens and consumed as eggs, it meets the energy requirements of only two persons and the protein requirements of eight. Nevertheless, meat, milk, and eggs are important in the human diet since they contain proteins of the proper quality (balanced quantities of essential amino acids), as well as some of the necessary minerals and vitamins.

Assessing the world's food situation involves other factors besides the utilitarian one of meeting minimal food needs. When the people of a country become more affluent, they want and can afford to purchase a greater proportion of their protein requirements in the form of the more palatable animal products—steaks, chops, eggs, processed meats, and

dairy products. This shift in food consumption patterns, part and parcel of the modern world's "rising expectations," coupled with the tremendous increase in world population, especially in developing regions (see Fig. 1–1), requires continuing increases in the world's food-producing capability. Much of the world's best agricultural land is already under cultivation, although there is still unused productive land awaiting development in Argentina, Brazil, Canada, Sudan, and Australia. However, in most if not all developed countries such as the United States (see Fig. 1–2), Japan, and those in Europe, farmland is being lost forever to industrial, residential, or recreational development. In addition, in the United States, even though the number of farms is decreasing, the

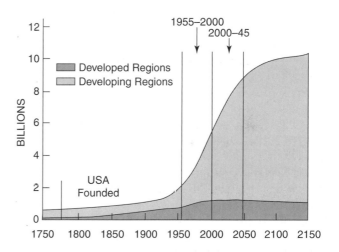

Figure 1–1 Trends and Projections (UN Medium Variant) in World Population Growth, 1750–2150. *Source:* World Resources Institute, 1996, p. 174, from United Nations, Population Division, 1992.

NUMBER OF FARMS IN THE UNITED STATES:

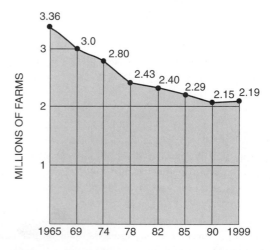

Figure 1–2 The steady decline in the number of farms in the United States is predicted to continue. *Source:* USDA and Office of Technology Assessment, U.S. Congress.

	Some Important Food Crops Ranked According to Calorie and Protein Production per Unit of Land Area	
TABLE 1–1		

Rank	Calories Produced per Unit Area	Protein Produced per Unit Area
1	Sugar cane	Soybeans
2	Potato	Potato
3	Sugar beets	Corn
4	Corn	Peanuts
5	Rice	Sorghum
6	Sorghum	Peas
7	Sweet potato	Beans
8	Barley	Rice
9	Peanuts	Barley
10	Winter wheat	Winter wheat

Source: USDA, IR-1 Potato Introduction Station, Sturgeon Bay, Wisconsin.

size of farms is increasing. The loss of the small family farm and the increase in corporate "megafarms" have become a serious social and political issue. Plant scientists are being asked to not only increase productivity and profitability of existing crops, but also to find and develop new crops or new uses for old crops.

In the mid-1970s various projections implied that the world was on the brink of famine or ecological disasters due to desperate food needs. But this assessment changed in the 1980s, especially in the less developed countries, by improved production technology and greater incentives to use it. Agricultural research made new **cultivars**[2] of high-yielding wheat, rice, corn, and other crops available to highly populated developing countries. Much of this improved technology can be attributed to assistance from agricultural researchers in the United States and other developed countries working with less developed countries.

But there are still areas where people starve or are badly malnourished. These underfed populations are mostly in the sub-Saharan regions of Africa, where drought, wars, and political instability are big problems, along with high human fertility rates, low per capita income, and insufficient monetary investment in agricultural production. In addition to the above-mentioned factors, poor distribution of available food contributes to starvation in many areas. The United States Department of Agriculture estimates that by 2009 food consumption in 30 of 67 developing countries is likely to lag behind minimal nutritional standards. While food distribution is expected to improve in many developing regions, poor distribution of food to these 30 countries is expected to intensify by 2009. Improving the nutritional value of food, so that the food distributed provides better nourishment, is a challenge to current and future plant scientists.

OTHER BENEFITS FROM CULTIVATED PLANTS

Imagine for a moment torrential rains falling on a landscape of rolling hills completely denuded of plants. It is easy to visualize the increasing worthlessness of such a landscape as it slowly erodes into the rivers, lakes, and oceans. The fertile topsoil disappears, uncovering a rocky base of little value to humans or animals. However, plants help prevent such erosion because their roots act as webs to hold the soil in place. In addition, their leaves and branches slow the force of the falling water, permitting it to trickle down and become stored in the depths of the soil, gravel, and rocks rather than running in torrents over the surface. Also, as the grasses, shrubs, and trees age and die, their decomposing roots, stems, and leaves add to the soil's mass, forming a humus material that constitutes a favorable environment for the germination of seeds

[2]Cultivar = *culti*vated *vari*ety.

and growth of seedlings. For eons the presence of plants on the earth's surface has protected the fragile topsoil layer, which is the natural habitat for the roots of both naturally occurring and cultivated plants, from the wearing effects of rain and winds. Erosion-controlling techniques are a necessary part of any cultivation system.

Wood and Wood Products from Timber Trees

The world's forests have provided wood for human shelter since the dawn of civilization. For centuries wood has also served as a source of fuel for warmth and for cooking. Fortunate indeed are those countries endowed by nature with luxuriant stands of tall timber trees. Unlike coal, gas, and oil, wood is a renewable natural resource. With proper management, forests can be enlarged and made more productive by proper tree selection and care. It is essential that the importance of wood for a nation's future be realized and that forests not be exploited for the immediate profits of the day.

Wood and wood products provide material for construction and other utilitarian uses, such as home furnishings. The ornamental value of trees used for landscape and other aesthetic purposes such as carving is nearly inestimable. The biochemicals found in some wood species provide products for industry and medicine. Industrial products include latex, pitch, and resin. Wood, of course, is used for purposes other than building homes and other structures. For example, in the United States, billions of cubic feet of wood per year are used in the manufacture of paper.

Textiles from Fiber-Producing Crops

Fiber crops such as cotton, flax, and hemp supply much of the fabric that clothe us and shelter us. Cotton, hemp, and also jute are used to make string, twine, rope, and burlap. Hemp (*cannabis sativa*) is a crop that has a deep root system, which can aid in re-conditioning a compacted soil. Unfortunately, hemp grown for fiber is the same species as marijuana. Marijuana plants produce the drug tetrahydro-cannabinol (THC). As a result, the production of hemp for fiber is illegal in many areas of the world. Although closely related, the two types are very different in morphological characteristics. Fiber hemp has the long, unbranched, and sturdy stems required for fiber production. Hemp leaves are relatively small with low THC content. In contrast, the marijuana-type plant is highly branched with larger leaves that have high THC levels. Morphological identification of one plant from the other is not difficult. Because the hemp plant has environmental benefit as a field crop and because the fiber from hemp produces an excellent clothing fabric, many countries, such as Canada, are once again allowing it to be grown. Currently, hemp production in the United States is illegal but increasing pressure on government agencies is forcing a reevaluation of that policy.

Drugs and Medicines

Plants are the source of many drugs and medicines. From the earliest history of modern man, the shaman and wise elders of a tribe used herbs to treat diseases and disorders among tribal members. Much of this wealth of plant lore is in danger of being lost as indigenous cultures are absorbed into modern civilizations. Ethanobotany, the study of plant usage by indigenous cultures and the preservation of that knowledge about plants, has become an important branch of plant science. One example of a modern-day medicine that can trace its roots to usage by Native Americans is aspirin. Salicylic acid, the pain-relieving component of aspirin, is found in the bark of willow trees (*Salix*). Native Americans chewed willow bark to relieve pain.

Many other modern medicines are directly derived from plant materials. Examples include quinine, which is used to treat malaria, and the heart medication, digitalis. Morphine, opium, codeine, and heroin are all pain-relieving extracts of the opium poppy (*Papaver somniferum*). Cocaine, another painkiller, is derived from the coca plant (*Erythroxylum coca*). However, each of these pain-relieving drugs also has addictive properties. Addiction to these chemicals has become a major problem worldwide.

Tobacco (*Nicotiana tabacum*) is a crop grown for recreational use. However, because of the connection of the drug nicotine to cancer and other health problems, the demand for the crop has dropped. Tobacco farmers are now looking for alternate crops to grow.

One of the newest plants to be recognized for its medicinal benefits is a yew (*Taxus*) that produces taxol, a chemical that shows promise in the treatment of cancer. The taxol-producing species grows wild in the Pacific Northwest and is now being evaluated as a potential crop plant for areas where it would grow well.

Plants also provide latex, pitch, waxes, essential oils, perfumes, spices, and other products. Although relatively small crops, each one can provide a significant portion of a region's or a farmer's income.

Latex, Pitch, Waxes, Essential Oils, Perfumes, and Spices

The vegetative parts of living plants can be thought of as chemical factories, using as their raw materials water and mineral elements taken in by the roots, carbon dioxide absorbed by the leaves, plus energy from the sun. From these materials many products are produced, some in copious amounts, that can be extracted or drained off to provide us with some extremely useful materials. Some important examples of these products are described below.

Latex from the Rubber Tree

It is said that Columbus, in his voyages to the New World during the fifteenth century, saw the native people throwing about resilient balls that had been made from the extract of a tree. Later, during the explorations of the sixteenth, seventeenth, and eighteenth centuries, this gummy material produced from an exudate of a native tropical South American tree (*Hevea brasiliensis*) was brought back to Europe and became familiar to Europeans. Not much was done with the latex material until about the middle of the nineteenth century, when it was discovered that by heating it and adding sulfur, many useful products—including the pneumatic tire—could be fashioned from it. Rubber soon became one of the indispensable products of modern civilization. In the early 1900s, huge rubber plantations were developed, largely through the efforts of the British and Dutch in Ceylon (now Sri Lanka), Dutch East Indies (now Indonesia), and Malaya, also in Central America and the West Indies. Still later, American companies developed rubber plantations in West Africa.

The inner bark of the rubber tree produces in latex vessels a material containing about 30 percent rubber, the rest mostly water. Cutting through these latex vessels to the right depth—not deep enough to injure the growing layer (the cambium)—and then attaching a container to the tree allows the latex to flow and be caught for processing. This latex base, from which many rubber products have been made, has been of inestimable value to civilization.

Before World War II most applications that called for a resilient material used natural rubber. During the war, however, the Allies lost access to most of the world's rubber-producing regions. This stimulated the consuming countries to manufacture synthetic rubber, mainly from petroleum and coal, in great quantities. By the mid-1960s the synthetic product had largely replaced natural rubber.

Pitch, Turpentine, and Resin

From the resin canals in the trunks of various cone-bearing trees, such as longleaf pine (*Pinus palustris*) and slash pine (*Pinus elliottii*), comes an exudate known in the crude state as "pitch," obtained by tapping or chipping into the tree's trunk. Pitch was used in earlier days to caulk wooden ships. Distillation separates pitch into resin (often called rosin) and turpentine, both of which are widely used in the manufacture of varnishes. In colonial times in the United States, both materials were an important source of income for the residents of the Carolinas and later for those of Florida and Georgia.

Essential Oils, Perfumes, and Spices

The practice of obtaining plant extracts with a pleasing odor—to improve the taste of foods, to aid in food preservation, and to adorn the body—dates to antiquity. When the pharoahs were ruling ancient Egypt, perfumes were commonly used. Cleopatra undoubtedly used perfumes to help charm Marc Antony and Julius Caesar. In the early days of colonial expansion by the English, Spanish, Dutch, and Portuguese, empires rose and fell on the search for spices and perfumes produced by plants in faraway countries. These extracts were highly prized symbols of wealth and prosperity among the ruling classes, and no effort or outlay of money was spared to obtain them.

Perfume-producing materials in great demand included rose oil from the flowers of *Rosa damascena* in the Balkans, Asia Minor, and India; jasmine from the flowers of *Jasminum grandiflorum* in southern France and India; geranium oil from the foliage of the *Pelargonium* species; citronella from the leaves of *Cymbopogon nardus* in Java; and patchouli oil from *Pogostemon cablin* in southeast Asia and the Philippines.

Valued spices were black pepper from the dried fruits of *Piper nigrum* in China and India; vanilla from the pods of the climbing orchid, *Vanilla planifolia,* in Aztec Mexico and the islands off the southeast coast of Africa; cloves from the dried flower buds of *Caryophyllus aromaticus* in the East Indies; nutmeg from the seeds of *Myristica frangrans* grown in the West Indies, Sri Lanka, and Indonesia; and cinnamon from the dried inner bark of *Cinnamomum zeylanicum* found in Sri Lanka and the Malabar coast of India.

Plants for Aesthetic Purposes

Modern civilization, with its needs for food, shelter, and clothing satisfied and with ever-increasing leisure time, takes great satisfaction in the aesthetic value of plants. Many people greatly enjoy tending and watching their houseplants, working with the flowers and vegetables in their gardens, and observing the year-by-year increase in the size of their shrubs and fruit trees. The therapeutic value of working among plants surrounded by a restful garden and peaceful green lawns has no doubt permitted many a person to survive the rigors of today's hectic, high-speed urban existence. Plants in their natural habitat—the forest trees, the native shrubs, and the wildflowers—preserved for public enjoyment in innumerable county, state, and national parks and forests provide great pleasure and relaxation.

CHALLENGES

At the start of the new millennium, plant scientists face not only traditional but new challenges, many of them more social than production-improvement driven. Improvements in production efficiency have increased yield per acre dramatically in developed countries, and currently enough food is produced to feed the world's population. However, starvation exists in both developed and developing countries, often because of social and political issues. Providing food to these areas will require creative thinking by those developing and producing food crops.

In the farming communities of developed countries, increased costs, low profit margins, and surpluses have decreased farm income. As a result, many small family farms have gone out of business, with the land sold for development or to become part of a megafarm, especially in the United States. The loss of family farms and the farming community is itself a serious social problem. Crops and production strategies will have to be developed that allow the family farm to remain viable. The solutions will no longer lie primarily in increasing productivity.

Many of the past improvements in production were developed to not only reduce labor and increase productivity and profit, but to allow farmers to be better stewards of the land and the environment. High-oil corn has been bred to yield a product that can be used to replace petrochemicals in some industrial uses. No- or low-till farming reduces labor costs and is less detrimental to the soil than traditional cultivation practices. However, it requires the use of herbicides to control weeds. To make herbicide use more effective and reduce the amount of herbicide used, genetically modified organisms/crops (GMOs) were developed that are resistant to a very effective herbicide, Round-up®. Weeds are susceptible to the chemical, but crops are resistant. As a result only one or two applications of a single herbicide is required in a season to get the same or better weed control compared with multiple applications of several herbicides in nonresistant fields.

A genetically modified corn (Bt®) synthesizes a naturally occurring protein that is lethal to the larva of many species of the order Lepidoptera, such as corn borer. The production of a natural larval toxin in corn has reduced the need to spray insecticides for control of a very destructive pest.

Rice has been genetically modified to produce β-carotene in the grain. β-carotene (vitamin A) provides a critical nutritional element that plant scientists predict will save the eyesight of millions of children in areas where vitamin A deficiency causes blindness.

Genetically modified crops such as those mentioned previously could not have been developed without the technology of gene transfer from one organism to another. Traditional breeding could not impart the traits because the trait does not exist in the gene pool of those crops or species with which the crops are breeding compatible. Not all GMOs are the result of gene transfer; sometimes an existing gene is altered. In the case of Flavr Savr tomatoes, the gene that produced the enzyme that softened the ripening fruit was scrambled so that the enzyme produced was dysfunctional. The fruit ripened but softened much slower than unmodified fruit. The shelf life of the harvested fruit was increased while the traditional tomato flavor was maintained.

Although plant scientists have been working to reduce the impact of farming on the environment and increase the productivity and profitability of farming, a large segment of the public has not shared the same viewpoint. This is the case in developed countries where agricultural practices, which were developed not only to increase profitability of farming but also to protect the environment and improve human life, are now perceived to harm the environment or put humans at great risk. As a result, plant scientists need to give serious consideration to the power of public opinion. Today's plant scientists must find ways to assure the public that what is being done is in the public's best interest and will not cause undo risk to the environment and living creatures.

SUMMARY AND REVIEW

Although plants have been used by humans for food since humans first appeared, deliberately growing (cultivating) plants for human consumption did not occur until about 10,000 years ago. The earliest crops were species native to a region. Through the ensuing millennia plant scientists learned how to produce more food in an area, different kinds of food, and ways to preserve and transport food, allowing foods from any region to be consumed in any other region. As a result, the selection of plant-based foods now include everything from the oldest of cultivated crops to new and/or exotic crops from local and remote growing areas of the world.

Plants provide humans with much more than a direct source of food. Plants provide nutrition for animals that humans use for food, labor, and recreation. Our buildings are constructed from materials derived from plants. Many industrial, cosmetic, and pharmaceutical chemicals are derived from plants. Our lives are improved by the use of ornamental plants for decoration and recreation.

The challenge for plant scientists continues to be providing sufficient and nutritious foods for the developed countries. A relatively new challenge in developing countries is finding ways to preserve the family farm and the contribution, including independence from foreign or large corporate food suppliers, that way of life makes to society.

The problems facing developing countries are also a great concern for plant scientists. Starvation is endemic in many areas. Although enough food is currently produced globally to feed the world's population, food is not easily distributed to the areas where starvation occurs. Often the hindrances to distribution are social and political rather than physical. Plant scientists are working to improve the productivity or nutritional quality of the foods that can be produced locally in these areas, thus circumventing or reducing the problem of poor distribution.

EXERCISES FOR UNDERSTANDING

1–1. In a two-minute time period, make a list of all the objects within view that are directly or indirectly derived from plants. Choose two or three of the objects and determine where the plant material from which it was derived was produced. What were the steps in the process getting the plant into its present form?

1–2. Describe what you see as the most difficult challenge facing plant scientists today. How would overcoming the challenge benefit society? What would you do to overcome the challenge?

REFERENCES AND SUPPLEMENTARY READING

BALIK, M. J. and P. A. COX. 1996. *Plants, People, and Culture: The Science of Ethanobotany.* Scientific American Library. New York. Freeman.

DAILY, G. C. and P. R. EHRLICH. 1992. Population, Sustainability, and the Earth's Carrying Capacity. *Bioscience,* 42:761–771.

SMITH, B. D. 1995. *The Emergence of Agriculture.* Scientific American Library. New York. Freeman.

VAN DER KAMP, J. W. 2000. Genetically modified foods in Europe. *Cereal Foods World.* 45:19–20.

CHAPTER

Structure of Higher Plants

KEY LEARNING CONCEPTS

After reading the chapter you should be able to:

♦ Describe and recognize parts of the plant cell, plant tissues, and plant organs.
♦ Understand the basic functions of cells, tissues, and organs.

Every day we can identify with familiar plants in our immediate surroundings. Some plants and their features can be identified and appreciated from their external structure, but their internal structure and function are often overlooked. The beauty of an orchid blossom is greatly admired, but just as impressive are the parts of a cell as recorded with a scanning electron microscope. The purpose of this chapter is to develop an understanding of the internal and external structures of the higher plants.

An approach that will capitalize on what is already known is to follow a plant from seed germination to full size and then to observe the formation of fruits and seeds. We can then appreciate how the plant grows and develops and, at the same time, acquire the vocabulary necessary to understand the growth processes of plants. This approach will allow us to study both the external form of the plant, or its **morphology,** and its internal structure, or **anatomy** and **histology** (microscopic features).

Our major food, fiber, wood, and ornamental plants belong to two main classes—the **gymnosperms,** represented mainly by the narrow-leaved, evergreen trees; and the **angiosperms,** usually broad-leaved, flowering plants. In the temperate zone, angiosperms are by far the most common in everyday life. Angiosperms are divided into two subclasses: the **monocotyledons,** which have an embryo with one cotyledon, and the **dicotyledons,** which have an embryo with two cotyledons. These names are often shortened to **monocot** and **dicot.** We begin by examining the life cycle of a common monocot, corn (Fig. 2–1), and a common dicot, the bean (Fig. 2–4).

THE LIFE CYCLE OF A CORN PLANT (A MONOCOT)

When a corn seed is planted in moist soil, it imbibes (absorbs) water from the soil. Germination begins with the emergence of the **radicle** (the primary root) and the **plumule** (the primary shoot). These two enlarging axes form the primary body of the plant.

The radicle grows downward through a protective sheath, the **coleorhiza,** from which the primary root devel-

ops and the secondary roots branch. A mature corn plant can develop roots 2 m (6.1 ft) long. **Adventitious roots** (roots other than those that develop from the radicle) grow from the shoot axis just at or above the soil surface (Fig. 2–2): These roots, also called anchor, brace, or prop roots, branch out in the soil to give added support to the plant.

The emerging plumule is protected by a sheathlike leaf, the **coleoptile,** that envelops the main stem as it grows upward through the soil. As the true foliage leaves develop, the main stem continues to produce sheathing leaves that encircle the stems at each node. The growth of the stem is complex and is discussed in detail in Chapter 6.

When the corn plant has reached a given size, producing a set number of leaves, female flowers, known as **pistil-**

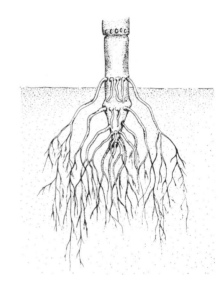

Figure 2–2   Adventitious or anchor roots develop above the soil line on the lower stem of a corn plant. These add support to the plant.

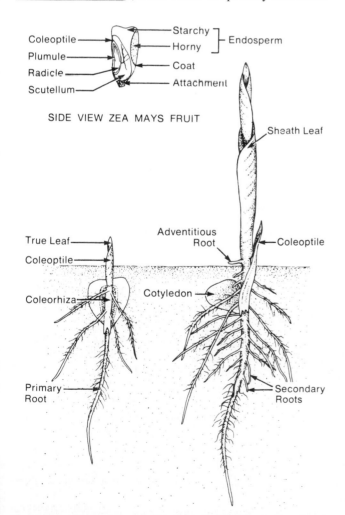

Figure 2–1   Structure of the seed and seedling of corn (*Zea mays*), a monocotyledonous plant.

Figure 2–3   *Left:* A pollen-bearing corn tassel (staminate flower). *Right:* The ear (pistillate flower), showing the "silks" that intercept the wind-blown pollen grains.

late flowers or ears, appear at the base (axil) of one or more sheath leaves. Later, the male flowers, known as **staminate flowers** or tassels, develop at the top of the plant. Figure 2–3 shows both kinds of flowers. Blown by the wind, pollen grains from the tassels fall upon and pollinate the long pistillate filaments (silks) and subsequently fertilize the ovaries, which become the individual corn kernels borne on a stalk (cob) (see Fig. 5–1). Each ovary develops into a fruit, called a **caryopsis,** that encloses the true seed. After the kernels mature and dry, the fruits (containing the seeds) are harvested and stored over the winter. The seeds can be sown when weather conditions are favorable for germination, and the life cycle repeats itself.

THE LIFE CYCLE OF A BEAN PLANT (A DICOT)

After a bean seed has been sown in moist soil, it imbibes water and swells. The seed coat bursts and the radicle emerges (Fig. 2–4). The radicle grows downward and the hook of the bean, known as the **hypocotyl,** emerges above the soil, carrying the two cotyledons with it. Between the cotyledons lies a growing point (**apical** or **shoot meristem**) flanked by two opposite primary foliage leaves. The stem region just above the cotyledons and the first trifoliate leaves is called the **epicotyl** (Fig. 2–4). Under favorable conditions the shoot apical meristem rapidly produces two trifoliate leaves opposite

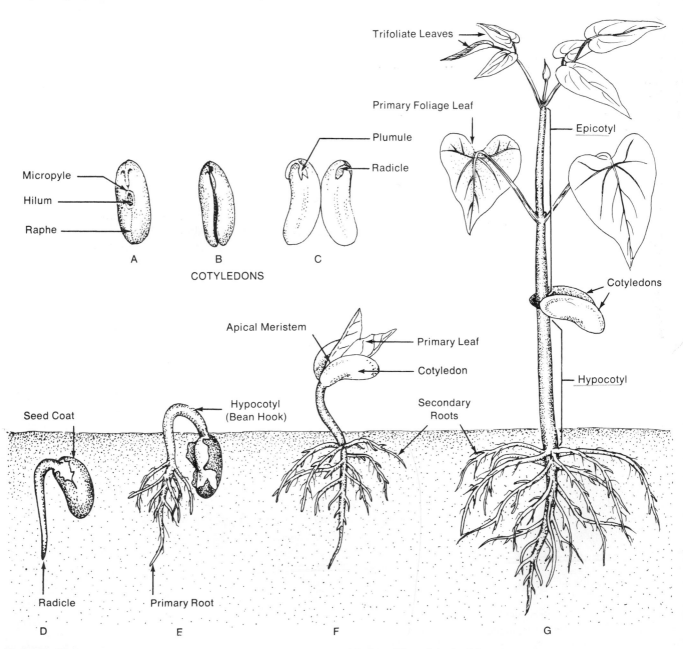

Figure 2–4   Structure of the seed and seedling—in several growth stages—of the bean (*Phaseolus vulgaris*), a dicotyledonous plant.

each other on the stem. These cotyledons, which have been supplying much of the reserve food for this initial growth, shrivel and abscise (drop off). The plant's green leaves are now capable of manufacturing food for future growth of the seedling. The bean plant produces trifoliate leaves, and flowers begin to develop in the axils of about the fourth set of leaves and in each succeeding set. These flowers are self-pollinated; thus fruits (pods) develop as long as environmental conditions are favorable. The seeds mature and dry within the pod, and they can be sown at once to produce another generation of bean plants. The difference in emergence of the growing points of beans and corn from beneath the soil affects the tolerance of each crop to light frosts. A late frost would be more likely to severely damage a newly emerged bean seedling than a newly emerged corn seedling because the growing point of corn is below the soil and protected. The leaves of the corn plant may be damaged, but the growing point will survive to generate new leaves. However, the bean plant's growing point would be severely injured or killed and unable to generate new growth.

The life cycles of plants like the corn and bean are more or less familiar to most of us from our own observations, but to enlarge our knowledge about higher plants, we must consider the largely unfamiliar areas of plant anatomy and histology.

THE CELL

The plant **cell** is the basic structural and physiological unit of plants, in which most reactions characteristic of life occur. The tissues of the plant develop through an orderly process of cell division and differentiation. **Cytology** is the branch of biology involved in the study of the components of cells and their functions.

Cells vary greatly in size. The smallest must be measured in micrometers (1/1000 of a millimeter), but some wood fiber cells are several centimeters long.

Early cytological studies were conducted with light microscopes, which can demonstrate general cellular features, but which cannot resolve all the fine details within cells. Electron microscopes and enhanced light microscopes have revealed that living cells are not empty chambers but highly organized complexes of subcellular compartments with specialized metabolic functions. In the living cell, these complexes are distributed through a dynamic and orderly flow of materials within the **cytoplasm.**

CELL STRUCTURE

There are two types of cells. **Prokaryotic cells** have no separate subcellular units; for example, nuclear material is not enclosed in a membrane. These cells, considered primitive, are found in bacteria and blue-green algae. **Eukaryotic cells**

are made up of compartments bounded by membranes, with specialized structures and functions. These units, called **organelles,** include the **nucleus, mitochondria, plastids,** microbodies, **vacuoles,** dictyosomes, and **endoplasmic reticulum** (Fig. 2–5). Plant cells are eukaryotic cells.

The Protoplast

The organelles of the plant cell are contained within a membrane-bounded **protoplast,** which in turn is encased within a cell wall. The major features of the protoplast are the outer membrane or plasma membrane, the cytoplasm, the nucleus, and the vacuole.

Plasma Membrane

The plasma membrane, also called the **plasmalemma,** is a lipid bilayer surrounding the cytoplasm. This membrane is important in maintaining a surface area for selective absorption and secretion by the cell, and plays a role in generating energy as well. Proteins embedded within the bilayer can function as enzymes; as surface receptors, both on the inside surface and in communication with the cell environment; or as channels for the uptake and efflux of ions.

Cytoplasm

The **cytoplasm** is a viscous fluid composed of matrix proteins, bounded by the semipermeable plasma membrane. The flow of organelles within the cytoplasmic matrix, called *cytoplasmic streaming,* is clearly visible in active leaf cells under a light microscope. Also within the cytoplasm is a very important network of membranes, the **endoplasmic reticulum** (ER). **Proteins** are synthesized on the surfaces of the ER throughout the cell, on small discrete structures called **ribosomes.** Proteins may be further processed inside the ER and transported to destinations which will be sites of activity within the cell.

Plastids of several types are located within the cytoplasm. The colorless leucoplasts serve as storage bodies for oil, starch, and proteins. Chromoplasts contain the various plant pigments, including **chlorophyll.** Chromoplasts with chlorophyll are called **chloroplasts** and are responsible for photosynthesis in leaves and in some stems. Enclosed by a double membrane, most chloroplasts also contain other **pigments,** large quantities of proteins and **lipids,** and some stored **starch.** Within the chloroplast, light energy is first harvested by pigments bound to stacked membranes called grana and then converted into chemical energy in the form of sugars.

Mitochondria are cytoplasmic bodies that are smaller than plastids. Like the chloroplasts, they are surrounded by a double membrane, and contain a specialized inner membrane system. The mitochondria are sites of respiration and are also involved in protein synthesis. They produce energy-rich compounds such as **adenosine triphosphate** (ATP).

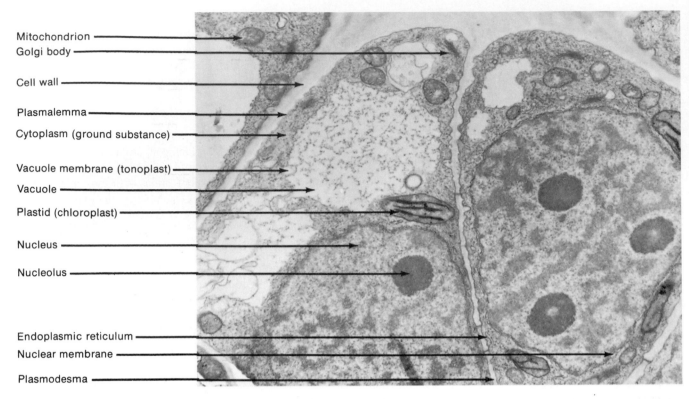

Mitochondrion

Golgi body

Cell wall

Plasmalemma

Cytoplasm (ground substance)

Vacuole membrane (tonoplast)

Vacuole

Plastid (chloroplast)

Nucleus

Nucleolus

Endoplasmic reticulum

Nuclear membrane

Plasmodesma

Figure 2–5 Photomicrograph of a plant cell showing the various parts and organelles, × 6000. *Source:* Keith Weinstock.

The Nucleus

The **nucleus** is a prominent organelle within the cell, enclosed by a double membrane known as the **nuclear envelope** and containing one or more bodies called nucleoli. Within the nucleus are the **chromosomes,** long lengths of DNA (**deoxyribonucleic acid**) and associated proteins which contain the genetic information coding for all cell functions, for differentiation of the organism, and for reproduction. During cell division, the chromosomes replicate and condense into discrete rod-shaped bodies; a set of chromosomes is passed on to each of the daughter cells (see Chapter 4), thus assuring continuity of genetic information from old to new cells. **Genetic codes** are transcribed from the DNA in the nucleus and translated into proteins on the **ribosomes.**

DNA is also found outside the nucleus in the mitochondria and in the chloroplasts, thereby giving these bodies a role in heredity independent of the nucleus. Unlike nuclear DNA, mitochondrial DNA and chloroplastic DNA are inherited only from the female parent. There is no sexual segregation of genetic traits. This has proven useful in determining the relationship between species or individuals.

Vacuoles

Vacuoles may occupy a major portion of the interior of plant cells. In actively dividing cells, vacuoles are very small, but they can account for up to 90 percent of the volume of ma-

ture cells. The membrane surrounding the vacuole is called the **tonoplast,** and serves an important role in regulating ion flow within the cell, maintaining cell turgor, and other functions. Vacuoles contain a watery solution of dissolved materials, including inorganic salts, blue or red pigments (**anthocyanins**), sugars, organic acids, and various inclusions of crystals. The vacuole serves as a storage reserve for water and salts, as well as for toxic products.

The Cell Wall

The **cell wall** protects the protoplast, provides an external structure, and in some tissues (e.g., bark, wood) may act as a strong support for the plant. The cell wall is nonliving, made up of **cellulose, pectic substances,** and lignins. Between cells lies an intercellular layer called the **middle lamella,** which contains many of the mucilaginous pectic compounds that hold adjacent cell walls together. Adjacent to the middle lamella is the primary wall, which is mostly composed of cellulose. This elastic but strong material is the chief constituent of most plant cell walls.

The secondary wall layer, which lies within the primary wall and is laid down only after the primary wall is complete, is usually thicker than the primary wall when fully developed. The secondary wall is also composed of cellulose, but in some cells and tissues it may contain **lignins, suberins,** or **cutins.** Lignins are closely associated with the cellulose and give it added strength, as well demonstrated by wood fibers. Large quantities of water are contained and transferred in cellulosic walls, which act as wicks. In some specialized cells, (for example, cork), water loss or flow is prevented by the presence of the waxy material, suberin, in the walls.

Individual cells in a tissue are connected to one another via strands of cytoplasmic material, called **plasmodesmata,** which extend through the plasma membrane. The surrounding cell wall forms channels around the plasmodesmata, called **pits.** As the wall grows thicker, these **pits** are preserved and can become quite long, clearly visible in the light microscope. Water and dissolved materials can move from cell to cell through these connections.

PLANT TISSUES

Large tracts of organized cells of similar structure that perform a collective function are referred to as **tissues.** Tissues of various types combine to form complex plant organs such as leaves, flowers, fruits, stems, and roots.

In all plants, both young and mature, two basic kinds of tissues can be distinguished. One kind is the **meristem,** or **meristematic tissue.** This is comprised of actively dividing cells that develop and differentiate into yet other tissues and organs. Cells in the meristematic tissues have thin walls and dense protoplasts. Meristematic tissues are found in the root and shoot tips, just above the nodes (**intercalary meristems**) and in woody perennials, as cylinders in the shoots and roots (the **cambium layer**).

The second kind of tissue is that which develops from the meristems and has differentiated fully. This is the **permanent tissue** of which there are two kinds: the **simple,** which includes the epidermis, parenchyma, schlerenchyma, and collenchyma; and the **complex,** which includes the xylem and phloem.

Meristematic Tissues

The common categories of meristematic tissues are:

Apical meristems *FiG 2-6*
  Shoot *Tips only*
  Root *Tips only*
Subapical meristems *—SIDE LEAVES*
Intercalary meristems *— MONOCOTS ONLY*
Lateral meristems *—ONLY IN PLANTS W/ANNUAL GROWTH (GIRTH)*
  Vascular cambium *— PROD XYLEM & PHLOEM*
  Cork cambium

Apical Meristems

Shoot meristems, frequently referred to as shoot apical meristems, are the termini of the above-ground portions of

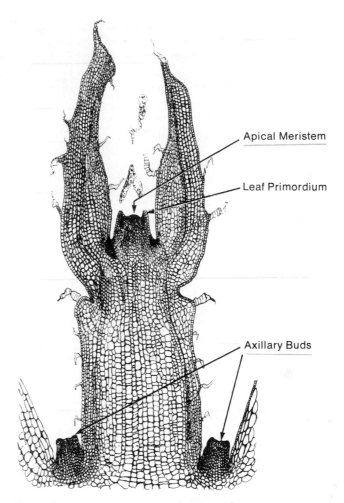

Figure 2–6 Longitudinal section of a shoot tip showing the apical meristem. Cell division in this region, along with cell elongation, is responsible for shoot growth. Buds in the axils of leaves also have meristematic regions at their tips that can develop into shoots. Should the apical meristem die or be broken off, an axillary bud can become the dominant shoot tip and continue shoot growth.

the plant (see Fig. 2–6). They are responsible for producing new buds and leaves in a uniform pattern at the terminus of the stem and laterally along stems. The pattern of leaves and lateral buds that form from the shoot meristems vary with the species of plant. For example, in the maples (*Acer*), ashes (*Fraxinus*), and in members of the mint (*LABIATAE*) and olive (*OLEACEAE*) families, the leaves and buds are opposite—at a 180° angle—to one another. On the other hand, in the oaks (*Quercus*) and walnuts (*Juglans*), for example, the leaves and buds alternate from one side of the stem to the other, and in the pines (*Pinus*) they form a spiral pattern.

The shoot apical meristem produces epidermis, cortex, primary xylem and phloem, and the central pith, tissues that form the primary structure of the stem. The shoot apex may eventually develop terminal inflorescences (floral groupings) instead of continuing to produce leaves and lateral buds as, for example, in the chrysanthemum, poinsettia, and sunflower.

Some shoot meristems always remain vegetative and continue to produce leaves and lateral buds, as in many vines such as the grape. In such cases, flowers or inflorescences are borne in the axils of lateral leaves somewhat behind the terminal growing point. Many trees also have this growth pattern; the dominant meristem (central leader) enables the plant to grow upright and gain height. In this case individual flowers or inflorescences are usually borne on side branches in the axils of leaves at some distance below the apex. These growth and flowering characteristics are known for most cultivated plants. Gardeners, orchardists, and foresters use that knowledge to prune trees and other plants to direct the growth in the manner that they desire.

Root meristems, located at the various termini of the roots, are the growing points for the root system. Some plants have a dominant **tap root,** which develops downward, together with limited lateral root growth (see Fig. 2–7). Examples of plants with tap roots are carrots, beets, and turnips, all well-known root crops. Other species with tap roots include oaks, pecans, alfalfa, and cotton. Many plants, however, do not have a dominant tap root. Instead, the roots branch in many directions creating a **fibrous root** system (Fig. 2–7). Examples of these are the grasses, grain crops, and many kinds of shallow-rooted trees.

The root meristem lies just behind the root cap, which protects the meristem as the root grows through the soil. These root cap cells are constantly being destroyed, but the apical root meristem produces more to replace them. The root meristem produces the primary tissues—such as protoderm, ground meristem, and procambium—that later become the epidermis, cortex, and vascular cylinder of the mature root.

Subapical Meristems

The **subapical meristem** produces new cells in the region a few micrometers behind an active shoot or apical meristem. The subapical meristematic region has long been thought of as a region where cells only elongate and expand. Cells do, indeed, expand in this region and thus increase internode length, thus adding to the growth in height of the plant. However, since new cells also form in this subapical region, it is a true meristem. The activity of the subapical meristem can particularly be seen in certain plants that lack tall stems when they are first producing leaves and that grow as a rosette. Examples are beets, carrots, China asters, lettuce, mustard, and turnips. These plants form rosettes of leaves on very short internodes. Later, when the shoot apical meristem initiates flowers, the stem below the flower elongates rapidly (bolts) because of the activity of the subapical meristem. During the period of fast growth, cells divide as well as elongate. The division and elongation together account for the rapid stem growth below the terminal flower buds.

Intercalary Meristems

The **intercalary meristems** are active tissues that have been separated from the shoot terminal meristem by regions of

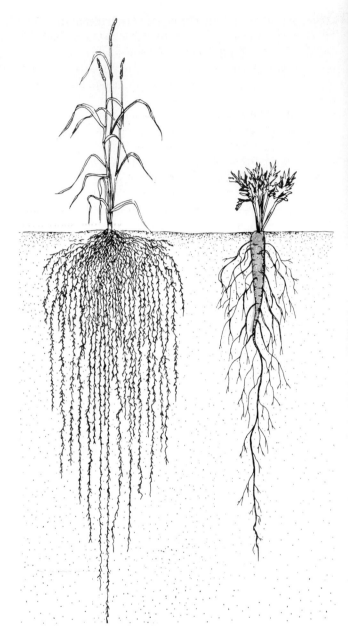

Figure 2–7   Two types of root systems. *Left:* Fibrous root system of a cereal plant. *Right:* Tap root system as developed by the carrot plant (*Daucus carota*).

NODE : LEAF MEETS STEM

more mature or developed tissues. The separation occurs at an early stage of development, and therefore the separated cells retain their ability to divide. The best examples of intercalary meristems are found in monocots and especially in the grasses. The active meristematic cells just above the nodes in the lower region of the leaf sheath divide, and those cells develop (expand and elongate) rapidly. Grass leaves elongate at the base in a like manner (see Chapter 19).

Lateral Meristems

GIRTH/WIDTH

The **lateral meristems,** which produce secondary growth, are cylinders of actively dividing cells starting somewhat below the apical or subapical meristems and continuing through the plant axis. These meristems are the **vascular cambium**—which produces new xylem (water and mineral conducting elements) and new phloem (photosynthate conducting elements)—and the **cork cambium**—which chiefly produces bark, the protective covering of old stems and roots (Fig. 2–8). Stem girth of woody perennial plants and trees increases mainly by the activity of these lateral meristems. The number of growth rings indicates the tree's age (Fig. 2–9). Measuring the width of the annual growth rings in the stems is one way to determine the rapidity of lateral growth of a tree. Long-lived trees, such as the redwood (*Sequoia*), and certain pines, such as the bristle cone pine (*Pinus aristata*), increase greatly in size by lateral growth, but short-lived summer plants (annuals) such as marigolds (*Tagetes*), tomatoes (*Lycopersicon*), and peppers (*Capsicum*) develop only a limited girth of the lower stem before the frost destroys them. Tomatoes, however, are perennials if they are grown in the frostfree tropics. Vascular and cork cambial growth turns the lower stem into a small trunk if growth continues.

MERISTEMS DVLOP INTO PERMANENT TISSUES.

Permanent Tissues

Permanent tissues can be classified into simple and complex tissues. The **simple tissues** are uniform, composed of only one type of cell. Examples are epidermis, parenchyma, sclerenchyma, collenchyma, and cork. **Complex tissues** are mixed, containing different kinds of cells. Examples are xylem and phloem.

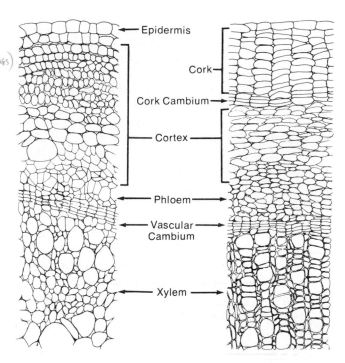

(RINGS)

Figure 2–8 *Left:* Cross section of the stem of a young dicotyledonous plant showing the tissues of the meristematic region, the vascular cambium. The epidermis lies on the outside. The dividing cells of the vascular cambium layer, producing phloem to the outside and xylem to the inside, account for the thickening of the stem as it grows older. *Right:* Cross section of an older dicot stem where a cork cambium has developed. This meristematic layer produces the cork cells (bark), which protects the inner, more tender tissues.

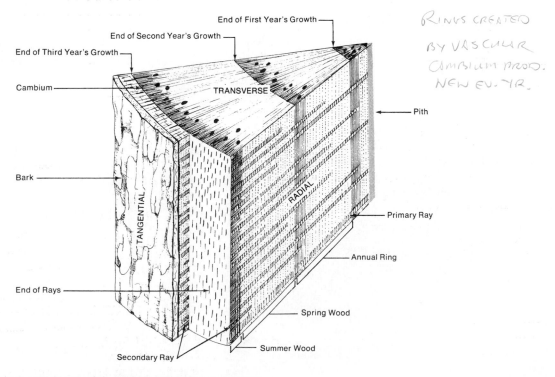

RINGS CREATED BY VASCULAR CAMBIUM PROD. NEW EV. YR.

Figure 2–9 Section of a three-year-old stem of pine (*Pinus*) showing the annual rings by the end of the third summer. The porous, fast-growing spring wood is followed by the more dense, slower-growing summer wood.

Simple Tissues

The **epidermis** is a single exterior layer of cells that protects stems, leaves, flowers, and roots. The outside surface of epidermal cells is usually covered with a waxy substance called cutin, which reduces water loss. The epidermis of leaves is usually colorless except for the **guard cells** of the **stomata** (Fig. 2–10), which contain **chlorophyll** and are green. Some leaf epidermal cells are elongated into hairs and are called trichomes (Fig. 2–11). The root epidermis lacks cutin. It develops protuberances called root hairs, which actively absorb water from the soil.

Parenchyma tissue is made up of living thin-walled cells with large vacuoles and many flattened sides. This is the principal tissue of the cylindrical zone under the epidermis extending inward to the phloem. This region is called the **cortex.** Parenchyma tissue, however, is not confined to stems but can be found in all plant parts. Parenchyma in leaves is active in photosynthesis. Parenchyma cells, when wounded, are capable of becoming meristematic and then proliferating to heal wounds and to regenerate other kinds of tissues.

Sclerenchyma tissue is composed of thick-walled cells found throughout the plant as fibers or **sclereids.** The protoplasts eventually die in these lignified cells, and they are nonliving. Sclerenchyma cells are common in stems and bark and are also found as stone cells in pear fruits and walnut shells.

Collenchyma tissue gives support to young stems, petioles, and the veins of leaves. The walls and corners of the cells are thickened, primarily by cellulose, to provide reinforcement.

Cork tissue occurs commonly in the bark of maturing stems, the trunks of trees, and potato skins. The cell walls are waterproofed with a waxy material called suberin. Cork cells soon lose their protoplasts and die but continue to retain their shape.

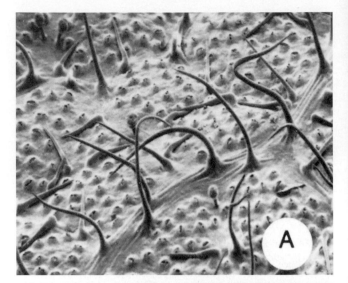

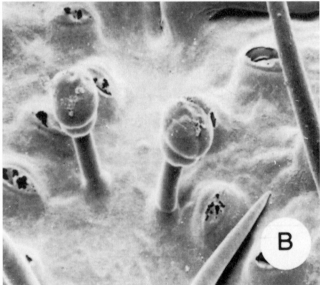

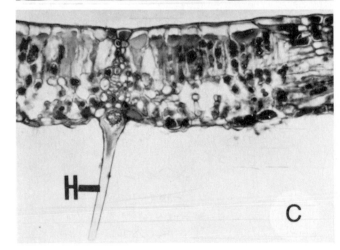

Figure 2–11 Leaf surfaces of *Ulmus elata:* (A) Lower surface (× 110). (B) Lower surface stomata and trichomes (× 550). (C) Leaf lamina transection (× 230). H = hair. *Source:* Meyer, R. E., and S. M. Meola. 1978. Morphological characteristics of leaves and stems of selected Texas woody plants. USDA Tech. Bull. 1564.

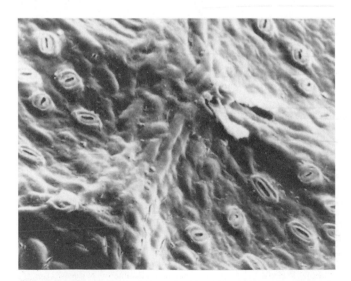

Figure 2–10 Stomates on the underside of a leaf. *Source:* Dr. Irving B. Sachs, Forest Products Laboratory, USDA Forest Service.

Complex Tissues

Xylem is a structurally complex tissue that conducts water and dissolved minerals from the roots to all parts of the plant. The cells found in the xylem may be vessels, tracheids, fibers, and parenchyma. **Vessels** are long tubes made up of short vessel members (Fig. 2–12) that are united after the end walls of the cells have dissolved (Fig. 2–13). **Tracheids** are long, tapered, dead cells that conduct water through pits (Fig. 2–14). Tracheids contribute significant strength and support to the stems of gymnosperms. **Fibers** are thick-walled sclerenchyma cells that provide support. The parenchyma cells in xylem are arranged in vertical files (Fig. 2–15) and act as food storage sites. Not all these cell types occur in the xylem tissue of any one plant species; usually one or two are absent.

Phloem conducts food and metabolites from the leaves to the stem, flowers, roots, and storage organs. A complex tissue, phloem comprises sieve tubes, sieve tube members, companion cells, fibers, and parenchyma. **Sieve-tube members** are long slender cells with porous ends called **sieve plates** (Fig. 2–16). Sieve tube members occur only in angiosperms. The equivalent cell in gymnosperms is the **sieve cell,** which is like the sieve-tube element except that it lacks a sieve plate. **Companion cells** aid in metabolite conduction and are closely associated with sieve-tube members (Fig. 2–16). **Phloem fibers** are thick-walled cells that provide stem support (Fig. 2–17). The parenchyma cells in the phloem serve as storage sites.

THE PLANT BODY

The various tissues are united in a structured and organized pattern to form organs such as roots, stems, leaves, flowers, fruits, and seeds. These make up the plant body. When a plant

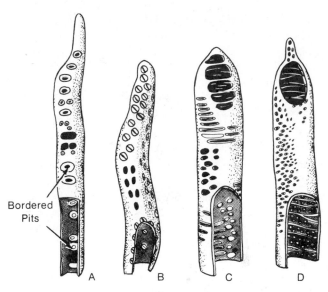

Bordered Pits

Figure 2–13 Types of tracheids and vessel elements found in xylem tissue. *Left to right·* (A) tracheid in pine (*Pinus*); (B) tracheid in oak (*Quercus*); (C) vessel element in magnolia (*Magnolia*); (D) vessel element in basswood (*Tilia*). *Source:* Weier, T. E., C. R. Stocking, M. G. Barbour, and T. L. Rost. 1982. *Botany, an introduction to plant biology.* 6th ed. New York: John Wiley. State University of New York College of Environmental Science and Forestry.

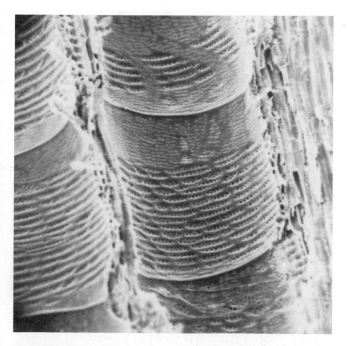

Figure 2–12 A section of the xylem of *Quercus rubra,* red oak, showing the barrel-shaped vessel members. The pitting in the lateral walls allows the movement of water to adjacent cells. Scanning electron microscope (SEM) photo provided by Dr. Irving B. Sachs, Forest Products Laboratory, USDA Forest Service.

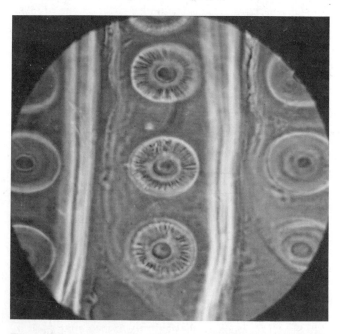

Figure 2–14 The tracheids with bordered pits of Larch (*Larix lyalli*) (×800). The central part is the torus, and surrounding is the thin margo through which liquids diffuse. *Source:* Dr. Irving B. Sachs, Forest Products Laboratory, USDA Forest Service.

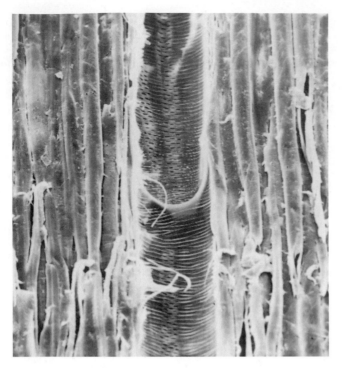

Figure 2–15   Long files of parenchyma cells surround the vessel. Pitting in lateral vessel wall in red maple *(Acer rubrum). Source:* Dr. Irving B. Sachs, Forest Products Laboratory, USDA Forest Service.

first begins to grow from seed, the original organs are the radicle and plumule. These organs form the primary plant body. As the plant continues to grow, the primary organs develop into mature organs made up of permanent tissues.

Roots

Roots are responsible for absorbing and conducting water and mineral nutrients and for anchoring and supporting the plant. In addition, some roots, as in sugar beets and carrots, act as storage organs for photosynthesized food. The roots of some plants develop secondary xylem from cambial activity and an abundance of parenchyma cells, which are able to store photosynthates and water. Dissolved mineral nutrients and water required for growth are absorbed by the root hairs, which are extensions of the epidermal cells (Fig. 2–18).

A few kinds of trees develop aerial roots from the underside of branches. Once these roots reach the soil and penetrate it, they become functional as ground roots. Good examples of this are the strangler fig *(Ficus aurea)* and the banyan tree *(Ficus benghalensis)* of the tropics. Some of the strangest roots are those in certain tropical orchids *(Cattleya, Phalaenopsis, Aerides,* and *Vanda).* The roots contain chlorophyll for photosynthesizing food; they cling to rocks or tree surfaces and are fully exposed to receive light. Frequent rains and mist supply the moisture and nutrients necessary for growth.

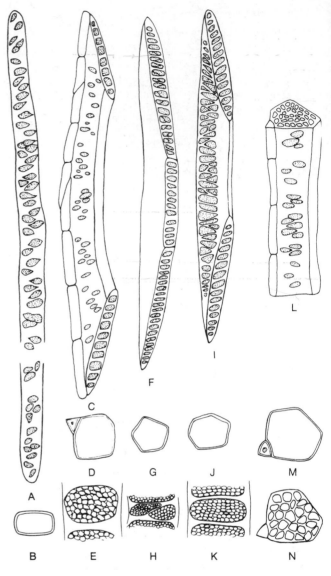

Figure 2–16   Sieve cells, sieve-tube elements, and companion cells in side view and cross section, showing detail structure of sieve plates.

A, B: From Canadian hemlock *(Tsuga canadensis)*; only one-third of cell shown.

C, D, E: From tulip tree *(Liriodendron tulipifera).* C, D, with companion cells attached; E, detail of sieve plate.

F, G, H: From apple *(Malus pumila)*; H, detail of sieve plate.

I, J, K: From black walnut *(Juglans nigra)*; K, part of sieve plate in detail.

L, M, N: From black locust *(Robinia pseudo-acacia)*, with companion cells attached; N, detail of sieve plate.

Source: Eames, A. J., and L. H. McDaniels. 1947. *An introduction to plant anatomy.* New York: McGraw-Hill.

The root system is a significant portion of the entire dry weight of any plant, about one-quarter to one-third of the total, depending on the storage or fibrous nature of the root. Measuring the total root system of a single mature rye plant showed that the plant had about 600 km (380 miles) of roots! Many of the functional roots of woody plants extend only into the top 1 m (3 ft) of soil. The depth that tree roots

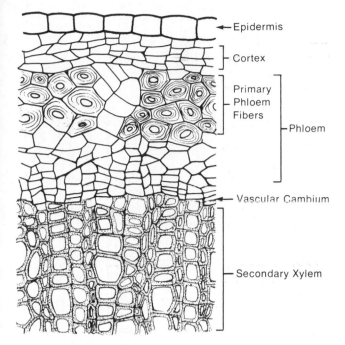

Figure 2–17   Cross section through stem of flax (*Linum*) showing thick-walled strengthening phloem fibers. *Source:* Esau, K. 1965. *Plant anatomy.* 2nd ed. New York: John Wiley.

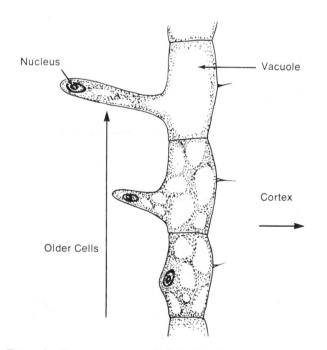

Figure 2–18   Section of epidermis of a young root showing three stages (bottom to top) in the development of root hairs.

penetrate depends largely on the species of tree and on the structure and water status of the soil.

After the radicle or primary root emerges from the seed, it can continue to grow principally as a tap root, or it may develop branch roots and form a fibrous root system (Fig. 2–7). The tap root usually grows downward, and the

branch roots grow downward or horizontally. The tap root can be encouraged to branch at an early stage by removing or breaking the apical root meristem. This often happens when some seedlings are transplanted into the garden; for example, the tap roots of tomatoes are broken when the young plants are removed from the container, and the roots become fibrous. When young trees are transplanted from the nursery, the tap root is cut and the roots branch after the tree is transplanted into the home garden or commercial orchard.

The meristematic region of a root is composed of small, thin-walled cells with dense protoplasm that produce primary tissue at a rapid rate. Behind this active meristematic region lies the zone of elongation. Here the cells expand, especially in length, new protoplasm forms, and the size of the vacuoles increases. The apical meristem and the region of elongation take up only a few millimeters of each root. Behind the region of elongation is the region of maturation, where the enlarged cells differentiate into the tissues of the primary body. In the epidermis of this young region the cells protrude and elongate and begin to form root hairs (Fig. 2–19). New root hairs arise in the newly developed region to replace old root hairs destroyed as the roots penetrate the soil.

The root apical meristem produces tissues different from those produced by the shoot apical meristem. The root meristem gives rise to the **root cap, epidermis, cortex,** and **central vascular cylinder.** The root cap is a thimble-shaped group of cells that protect the actively dividing meristem as it penetrates the soil (Fig. 2–19). These moist cells are sloughed off as the root comes in contact with sharp soil particles; the meristem forms new cells on the inner part of the cap to replace the damaged or lost cells. The meristem produces long rows of cells under the root cap. One of these becomes the **protoderm** that gives rise to the epidermis, or outer layer, of the root. The **ground meristem,** the tissue layer that gives rise to the cortex just below the epidermis, is usually thicker than the ground meristem found in stems. The cortical region is mainly composed of storage parenchyma cells, which have large intercellular spaces. A single layer of inner cortical cells forms the **endodermis,** a tissue found only in the root and not the stem. Each thin-walled endodermal cell is completely encircled by a narrow, thickened band of waterproofed material known as the **Casparian strip.** The solution of water and nutrients entering the root from the soil cannot penetrate the Casparian strip. For the soil solution to enter the inner tissue (pericycle) of the root, it must pass through the permeable endodermal cell walls and the protoplast (Fig. 2–20).

The Procambium Layer

The **procambium layer** gives rise to various tissues of the vascular cylinder. These include the **pericycle,** which is the outermost layer of cells of the central core and lies just inside the endodermis. The pericycle develops from a single parenchyma cell layer on the outer portion of the

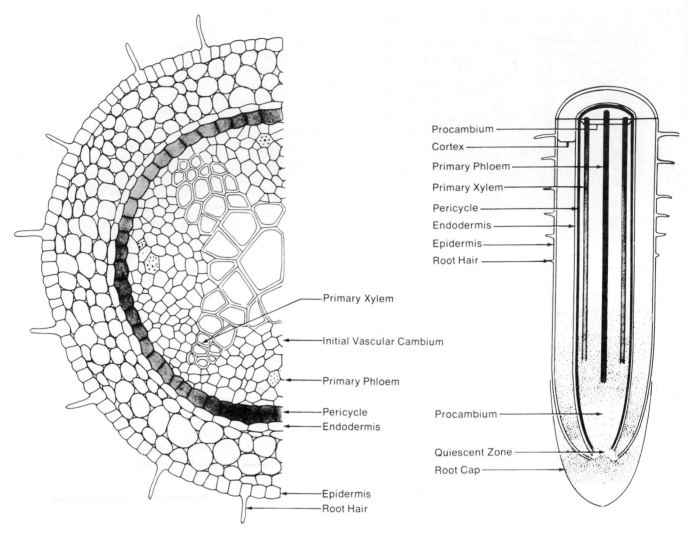

Figure 2–19 *Left:* Cross section of a young root showing the parts of the primary plant body and their location. *Right:* Developmental occurrences in the root tip, showing the various components and their relative location. *Source:* Weier, T. E., C. R. Stocking, M. G. Barbour, and T. L. Rost. 1982. *Botany, an introduction to plant biology.* 6th ed. New York: John Wiley.

procambium. The pericycle is a meristematic region producing lateral (branch) roots that grow outwardly through the cortex and epidermis. It may also give rise to vascular and cork cambium. The procambium layer also produces a vascular cylinder of primary phloem and xylem, vascular cambium, and, in some species, pith. The pericycle and the vascular cylinder are collectively called the **stele.** The primary xylem is a central mass of tissue that may extend as arms beyond the primary phloem (Fig. 2–19). A layer of procambium cells separates these two primary tissues; this layer is the meristematic region for any new vascular tissue that subsequently forms.

As the root grows in girth and the plant matures, a continuous ring of secondary phloem forms outside the vascular cambium and the primary phloem becomes less important than at earlier growth stages. The cambium layer develops

from the procambium and from pericycle cells outside the primary xylem. The primary xylem with its extending arms remains, but it is encircled by secondary xylem formed by the adjacent vascular cambium. Annual rings develop in the secondary plant body of the root as it grows in girth, much as in stems except for the star-shaped primary xylem at the core of the root. The cork cambium formed from pericycle produces a corky layer outwardly from the vascular system.

Adventitious roots form at any place on plant tissue other than the radicle of a germinating seed and its extensions. Adventitious roots arise from meristematic cells adjacent to vascular bundles (in herbaceous dicot stems) or from cambium or young phloem cells in young stems of woody perennials. This production of adventitious roots is the basis for propagation by stem cuttings (see Chapter 5). Adventitious roots can arise from plant parts other than stems, such

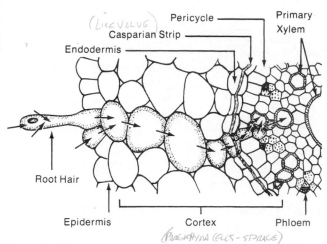

Figure 2–20 (LIKEVALUE) labels: Pericycle, Casparian Strip, Endodermis, Primary Xylem, Root Hair, Epidermis, Cortex, Phloem, (PARENCHYMA (ELLS - SPRACE))

Figure 2–20   Cross section of a young wheat (*Triticum*) root showing the path of entrance of soil solution into the root from a root hair to the tracheary elements of the xylem. *Source:* Esau, K. 1965. *Plant anatomy.* 2nd ed. New York: John Wiley.

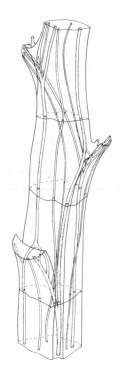

Figure 2–21   Three-dimensional skeletal view through a potato stem showing the primary vascular system extending through the stem with branches into the cutaway leaf petioles. *Source:* Eames, A. J., and L. H. McDaniels. *An introduction to plant anatomy.* 1947. New York: McGraw-Hill.

as from leaf petioles or leaf blades or even from old root pieces. Plant parts to be rooted are usually detached from the parent plant and placed under favorable environmental conditions for rooting. In the propagation procedure known as layering, adventitious roots are induced to form on plant parts still attached to the parent plant (see Chapter 5). Adventitious roots also develop on intact plants, as in the corn plant (see Fig. 2–2).

The ability of some plants to readily form adventitious roots allows seedlings and transplants to be planted deep in the ground if the seedling or transplant has become tall and spindly. By planting deep and having adventitious roots form, the young plant is much less likely to be damaged by wind or rain. For example, tomato transplants that have gotten tall and spindly are often planted deep in the field or home garden.

Stems

The main stem and its branches are the scaffold of the plant, supporting the leaves, flowers, and fruits. The leaves and herbaceous green stems manufacture food, which is transported to the roots, flowers, and fruits through the phloem. Figure 2–21 illustrates the complexity of the primary vascular system. The greater part of the vascular system consists of xylem and phloem. The secondary xylem also serves as the major structural support in woody perennial plants.

The stem develops from three primary tissues produced by the apical meristem: the protoderm, the ground meristem, and the procambium. These give rise to the epidermis, cortex, and vascular cambium, respectively.

The **epidermis,** which is usually a single layer of surface cells, protects the stem. Epidermal cells are usually cutinized on their outer surface to retard desiccation. The epidermis of leaves and young stems has pores, the stomata (Fig. 2–10), that allow for the exchange of gases.

The **cortex** lies just beneath the epidermis and encircles the inner core of the vascular tissue. The cortex comprises **parenchyma, collenchyma, sclerenchyma,** and secretory cells, with parenchyma cells the most numerous. Some parenchyma cells have chloroplasts and are called chlorenchyma. Parenchyma cells have the ability to divide and form new tissue when wounded, thus providing a protective mechanism for the stem. Collenchyma is the outer cell layer of the cortex adjacent to the epidermal layer (Fig. 2–22). These cells may be thickened at the corners, and their walls contain cellulose, hemicellulose, and pectin. This tissue, therefore, adds strength to the stem. Sclerenchyma cells have thick lignified walls. They can form long fibers, which are the source of strength in mature stems. Secretory cells produce resinous substances and are commonly found, for example, in the resin ducts of pine trees.

The **vascular system** of seed-bearing plants consists of the pericycle, phloem, vascular cambium, xylem, pith rays, and pith. The arrangement of these complex tissues in the vascular system differs among three broad groups of plants: (1) the gymnosperms and the woody dicotyledonous angiosperm perennials (which live for long periods), such as trees and shrubs; (2) the herbaceous dicotyledonous plants, such as potato, petunia, and phlox; and (3) the monocotyledonous plants, such as corn and date palms. These three groups are discussed separately to distinguish among their different internal stem structures and growth patterns.

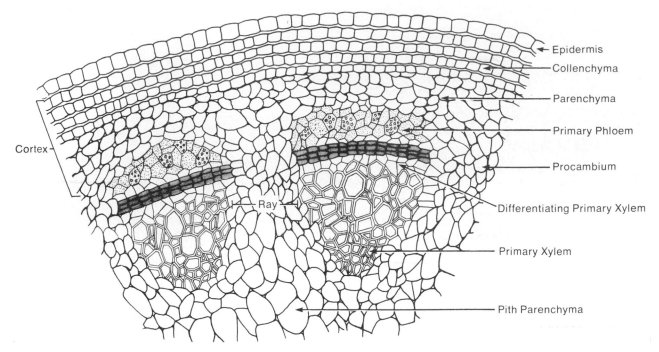

Figure 2–22 Cross section of a young woody plant stem toward the end of primary growth, showing various tissues present.

Woody Perennials (Dicotyledonous Angiosperms and the Gymnosperms)

All the cells and tissues originate from a terminal shoot meristem that forms protophloem and protoxylem (primary tissue) (Fig. 2–22). As the stem grows in length, the secondary tissues form from the vascular cambium. The secondary phloem develops toward the outside of the stem and the secondary xylem forms inwardly from the vascular cambium. These growing secondary and permanent tissues crush the primary tissues until they become difficult to see. Secondary xylem is actively produced by the vascular cambium in the early spring and less actively in late summer. This xylem tissue becomes the early (porous) and late (dense) wood that form the annual growth rings in trees (Figs. 2–9 and 2–23). The vessels or tracheids formed during the spring flush of growth are larger than those formed during the summer (Figs. 2–21, 2–23, and 2–24). The narrow-leaved evergreen trees belonging to the gymnosperms are usually referred to as the softwoods or nonporous wood trees (Figs. 2–24 and 2–25). The xylem of gymnosperms consists mainly of tracheids (Fig. 2–13). The broad-leaved angiosperm trees are called hardwoods or porous wood trees (Fig. 2–26); the xylem tissue is made up mostly of vessel elements (Figs. 2–12 and 2–13).

Both the gymnosperms and the woody perennial angiosperms grow in girth each year when the cells of the vascular cambium divide, forming annual rings of xylem. A stem nearing the end of its first season of growth is mostly xylem. The phloem, as it is crushed by the expanding xylem, is constantly being renewed by the vascular cambium. The

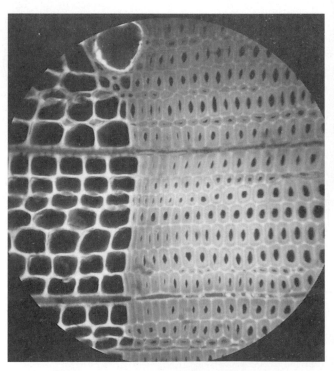

Figure 2–23 A cross section through the xylem of Douglas fir (*Pseudotsuga menziesii*) (× 200). The large pores to the left are the spring wood and the more dense cells were formed in summer of the same year. The very large pore at the top is a resin duct. *Source:* Dr. Irving B. Sachs, Forest Products Laboratory, USDA Forest Service.

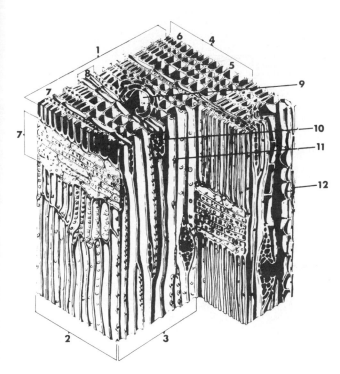

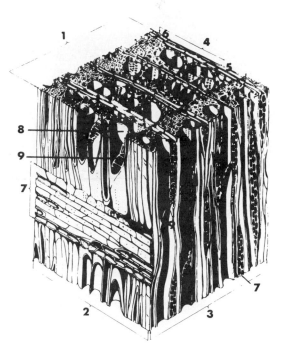

Figure 2–24 Three dimensional view of the wood of a softwood forest species: (1) cross-sectional face; (2) radial face; (3) tangential face; (4) annual ring; (5) early wood; (6) late wood; (7) wood ray; (8) fusiform ray; (9) vertical resin duct; (10) horizontal resin duct; (11) bordered pit; (12) simple pit. *Source:* U.S. Forest Products Laboratory, Madison, Wis.

Figure 2–26 Three-dimensional view of the wood of a hardwood forest species: (1) cross-sectional face; (2) radial face; (3) tangential face; (4) annual ring; (5) early wood; (6) late wood; (7) wood ray; (8) vessel; (9) perforation plate. *Source:* U.S. Forest Products Laboratory, Madison, Wis.

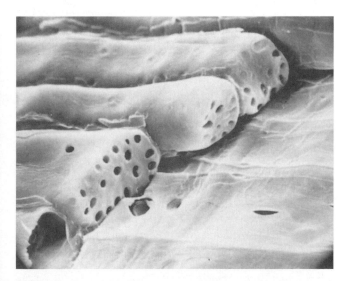

Figure 2–25 End-wall perforations in the ray cells of Douglas fir (*Pseudotsuga menziesii*). *Source:* Dr. Irving B. Sachs, Forest Products Laboratory, USDA Forest Service.

Figure 2–27 The rough, thick bark of the cork oak (*Quercus suber*), which is commercially stripped for cork products.

phloem is a relatively thin layer of complex tissue protected by the bark or cork layer.

The cork cambium (**phellogen**), which is a meristematic tissue, provides cells that grow both outward and inward. The outward cells become cork cells; the inward, phelloderm. The cork cells become suberized and are, therefore, resistant to entry or loss of water. The cork cells soon die but

retain their ability to resist desiccation, disease, insects, and extreme temperatures. The unusually thick bark of the cork oak (*Quercus suber*) is stripped for a multitude of commercial uses such as corks and insulation (Fig. 2–27).

In young twigs and small trunks of many kinds of trees and shrubs, pore openings (**lenticels**) allow the inward and outward diffusion of gases (Fig. 2–28).

Figure 2–28   The bark of a birch tree (*Betula verrucosa*) showing the lenticels.

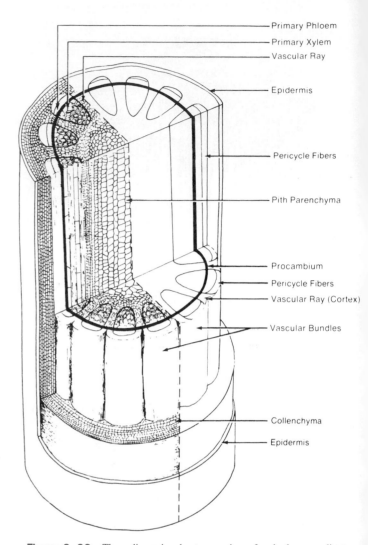

Figure 2–29   Three-dimensional cutaway view of an herbaceous dicot stem, showing the vascular bundles.

Herbaceous Dicotyledonous Plants

The early stem growth of plants in this category is much like the early growth of woody dicot stems. The vascular bundles (fascicles) of an herbaceous dicot usually remain separated and distinct; they are arranged in a single circle in the stem (Fig. 2–29). A larger proportion of the herbaceous stem is cortex and pith than xylem or phloem. Stem strength comes from the pericycle fibers adjacent to the phloem or from collenchyma or sclerenchyma tissue just beneath the epidermis (Fig. 2–29).

Herbaceous Monocotyledonous Plants

Stem growth originates from an apical meristem that produces vascular bundles scattered throughout the parenchyma (Fig. 2–30). The vascular bundles form most frequently near the epidermis. The sclerenchyma cells near the epidermis and thick-walled cells surrounding the bundles provide the principal support in monocot stems. Monocots (see Fig. 2–30) have no continuous cambium and, therefore, lack secondary growth. Stem diameter from the base to the apex is usually more uniform in monocot stems than in dicot stems with secondary vascular growth.

Woody Perennial Monocotyledonous Plants

In trees such as date and coconut palms (PALMAE), the thickness at the stem apex increases by the activity of a primary thickening meristem. In the trunk below the terminal growing point, parenchyma cells continue to divide and enlarge allowing for lateral stem enlargement. This is termed diffuse secondary growth, since no actual lateral meristem is involved.

Stem Forms

When most people think of the stem of a plant, they envision the upright portion that bears branches, leaves, flowers, and fruits. Stems come in other forms, too. For example, certain fruit trees, such as apples, cherries, plums, and pears, bear flowers and fruits each spring on persistent shortened stems called spurs (Fig. 13–5). Stems can also grow horizontally, as in a pumpkin or cucumber vine. Some species of plants have underground stems; only a small portion of the stem shows above ground for a relatively short period in the spring. These are the so-called bulbous plants (Chapter 20). The white (Irish) potato plant (*Solanum tuberosum*), ready for harvest, exemplifies two kinds of stems: the aboveground stem that bears the leaves and flowers, and an underground stem whose terminal portion swells into a tuber as it accumulates starches and sugars from photosynthesis in the leaves (Fig. 2–31). Just like other stems, the white potato tuber has buds (eyes) that sprout, when planted, to form new above-ground stems.

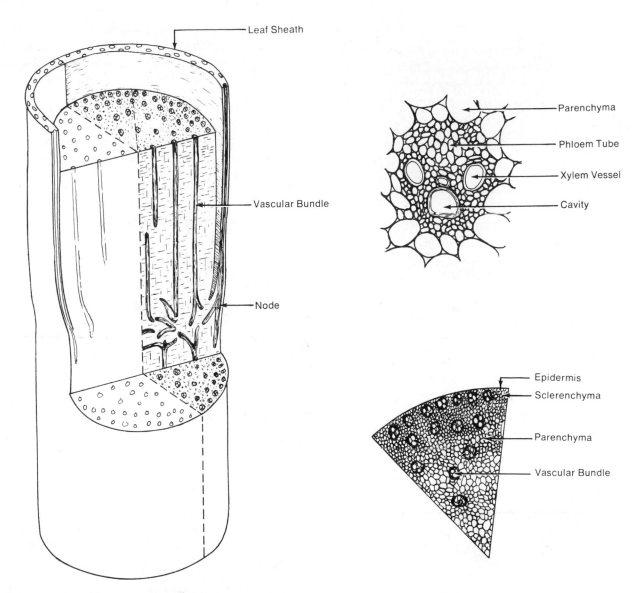

Figure 2–30 *Left:* Three-dimensional cutaway view of the stem of an herbaceous mono-cot (solid stem), showing the scattered vascular bundles. *Upper right:* Enlarged view of vas-cular bundle. *Lower right:* Enlarged cross-section area.

A **rhizome** is an underground stem that grows hori-zontally. Examples of plants with rhizomatous stems are ba-nanas, cannas, certain irises, certain bamboos (Fig. 5–21), and some grasses, such as quack grass, Johnson grass, and bermuda grass.

Stolons are stems that grow horizontally above ground. Sometimes called runners, stolons can develop roots in the soil at every node or at every other node (strawberry). Examples of species with stolons are ajuga, bermuda grass, and some ferns.

Corms are thickened compressed stems that grow un-derground (Fig. 2–32). Buds on corms sprout to produce up-right stems, which bear leaves and flowers. Gladiolus, crocus, freesia, and ixia are some examples.

Bulbs are highly compressed underground stems to which numerous storage leaves (scales) are attached. These highly developed stems provide a means for some species to survive the cold of winter and the dry soil of summer. One or more buds on the bulb sprout in the spring to produce an elongated stem with leaves and flowers. Through photosyn-thesis, the leaves produce carbohydrates, which are translo-cated to the bulb for storage. The storage of food in bulbs is a mechanism to enable the species to survive through periods of unfavorable climatic conditions. Hyacinths, lilies, onions, and tulips are examples of bulbous plants.

Stem **tubers** are the enlarged, fleshy, terminal portions of underground stems. The white potato, as discussed previ-ously, is a good example (Fig. 2–31).

Figure 2–31 A potato plant at time of harvest. The tubers are stem structures growing below ground whose terminals have stored sufficient photosynthate to swell into tubers.

Figure 2–32 A gladiolus corm, which is thickened compressed stem tissue, toward the end of the growing season. The spring-planted corm (below) has shrivelled with a new corm (above) forming above the old corm. New small cormels are forming at the base of the new corm.

Leaves

Leaves develop in a complex series of events closely associated with stem development. Some of these developmental events are still not clearly understood. Leaves are initiated by the apical shoot meristem. Their prescribed pattern, position, and shape are influenced to some extent by their environment. For example, the leaves of cacti are adapted to growth in the desert, whereas leaves of ferns are adapted for growth in a rain forest. Most monocots, such as the grasses and palm trees, have strap-shaped leaves with parallel veins and inter-

veinous connections between major veins. The veins contain sheaths of vascular bundles including xylem and phloem elements. **Palisade** and **spongy mesophyll parenchyma cells,** containing chlorophyll for photosynthesis, surround these veins (Figs. 2–33 and 2–34).

The leaves of dicotyledonous plants vary considerably in size and shape (Fig. 2–35). Practically all have veins arranged in the shape of nets. The large primary veins divide into smaller secondary veins. The veins are made up of xylem and phloem connected to all segments of the leaf. The spongy **mesophyll** parenchyma (Figs. 2–33 and 2–34) contains the intercellular spaces through which carbon dioxide, oxygen, and water pass. The outside layer or skin of the leaf is largely made up of epidermal cells. This epidermal layer contains openings or pores called **stomata,** each surrounded by two **guard cells.** There are generally more stomata in the lower epidermal layer of the leaf than in the upper epidermal layer. Water lilies, however, are an exception to this pattern.

The primary function of leaves is photosynthesis (see Chapter 7); a secondary function is transpiration. The **guard cells,** which occur in pairs on both sides of the stomata, control the opening and closing of the stomata through which carbon dioxide, one of the raw materials for photosynthesis, enters the plant, and oxygen, a product of photosynthesis, is released. Water also enters or escapes through the stomata. The loss of water from the leaf by evaporation is a process called **transpiration.** Transpiration helps regulate leaf temperature and provides the force that draws water into and through the xylem. Some plants have modified leaf surfaces that affect the rate of transpiration. The leaves of some plants, such as cabbage, have a thick waxy surface (**cuticle**) that greatly reduces water loss. In leaves of other plants the epidermal cells produce elongated hairs that reduce the wind velocity at the leaf surface, thus reducing the transpiration rate. Some kinds of plants, especially those native to hot dry climates, such as the olive tree, minimize water loss by having stomata sunken deep in the epidermal layer.

Plants often have leaves modified to perform functions other than photosynthesis and transpiration. For example, the leaf stipules of the black locust (*Robinia pseudoacacia*) are modified to become thorns that aid in protecting the plant. Some viny plants, like the grape, have leaves modified in the form of tendrils that help support the vine when trained on trellises.

In most dicotyledonous plants, the leaf is made up of the **blade,** the flat thin part; the stemlike **petiole,** which attaches the blade to the stem; and, in some plants, the **stipules** at the base of the petiole. Some leaf blades are attached directly to the stem and lack a petiole or stipules. These are termed **sessile** leaves.

Leaf shapes among the many plant species vary greatly, and a special morphological terminology describes the leaf shape, margin, tip, and base. Leaves are usually classified (see Fig. 2–35) as **simple** (a single leaf) or **compound**

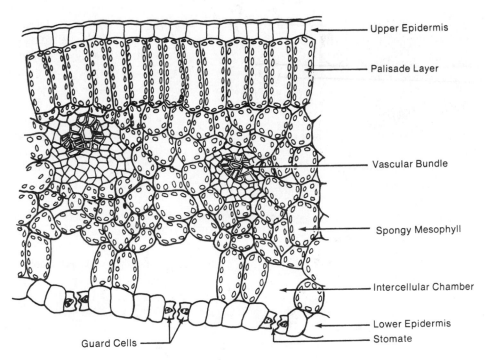

Figure 2–33   Cross section through a lily leaf showing the tissues involved in photosynthesis, transpiration, and translocation.

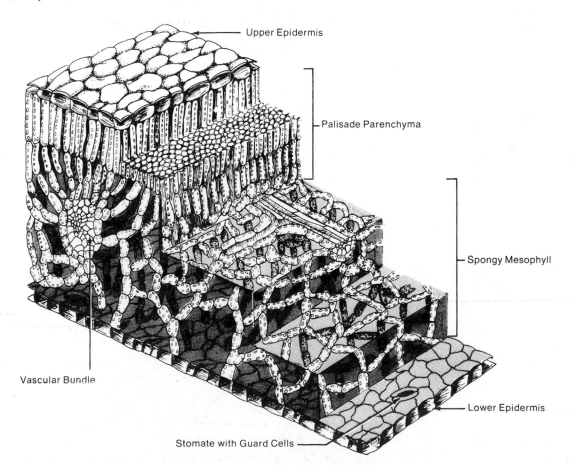

Figure 2–34   Three-dimensional cutaway view of an apple leaf showing the relation of cells in the various tissues. *Source:* Eames, A. J., and L. H. McDaniels. 1947. *An introduction to plant anatomy.* 2nd ed. New York: McGraw-Hill.

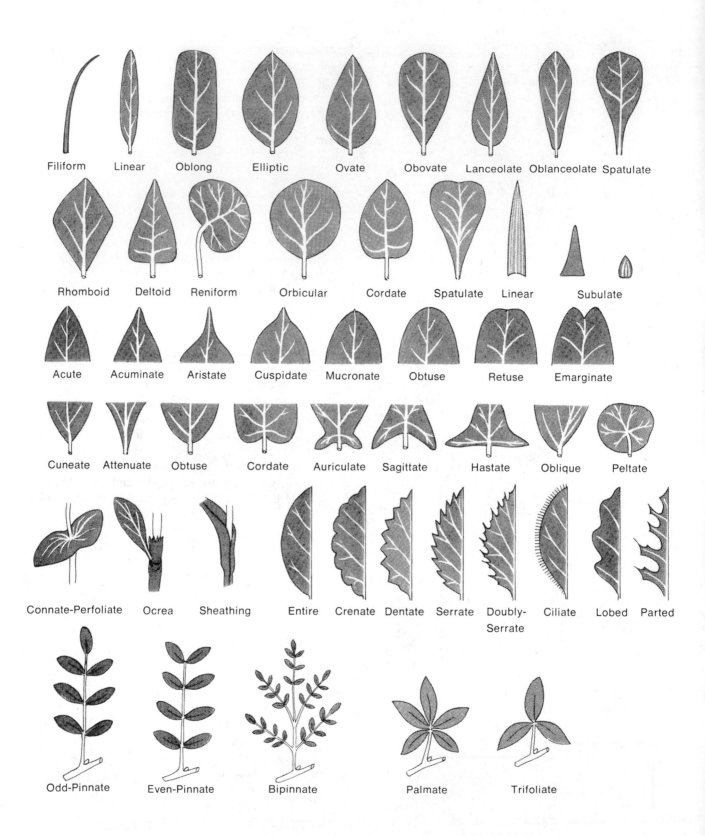

Figure 2–35 Examples of leaf patterns, with descriptive names. The top two rows are the shapes of simple leaves; the center rows are simple leaves cut to depict leaf tips or bases; one row up from the bottom illustrates examples of leaf attachments or leaf edges; the bottom row shows examples of compound leaves. These characteristics are used to describe leaves in plant identification keys (Chapter 3).

(one with three or more leaflets). Distinguishing between simple and compound leaves can be difficult. The best test is to examine the base of the petiole. A true leaf has a bud in this location; a leaflet does not. A compound leaf resembling a feather is termed **pinnate;** one resembling the palm of a hand, **palmate.** A **trifoliate** compound leaf has three leaflets, as in the bean plant.

The shapes of simple leaves are described as linear, oblong, elliptical, lanceolate, deltoid, and so forth. The shapes of the tips and bases of simple leaves are also categorized as cordate (heart-shaped), sagittate (arrow-shaped), auriculate (ear-lobed), and sheathing, as in the grasses (Fig. 19–2). Leaf edges or margins range from entire (smooth), dentate (tooth-like), serrate (saw-like) to lobed (rounded edges). These leaf characteristics aid in plant identification, as we will see in Chapter 3.

All gymnosperms native to North America and Europe have needlelike leaves and are evergreen. Some gymnosperms native to Australia, South Africa, and other parts of the world are broad-leaved plants.

Buds

Plant stems generally produce buds in the axils of leaves at the nodes or terminally on shoots. Buds usually do not occur on roots. A **bud** can be defined as an undeveloped shoot or flower, largely composed of meristematic tissue, and generally protected by modified leaf scales. These are: (1) vegetative buds, which develop into a shoot; (2) flower buds, which open to produce a flower or flowers; and (3) mixed buds, which open to produce both shoots and flowers. Cutting through a bud longitudinally reveals the miniature parts of either a stem growing point or, in a flower bud, all the miniaturized parts of a flower.

Buds are especially prominent in winter on deciduous plants when the leaves have fallen. A leaf scar, where the leaf petiole was attached, is visible just below each bud. Buds may occur opposite each other on a stem or in an alternate arrangement around the stem.

Buds are initiated by terminal growing points as shoots elongate during the growing season. Some buds continue to grow after they are formed, developing into shoots. The growing points in other buds remain dormant until the following spring. Some buds may remain latent for long periods of time and become embedded in enlarging stem tissue; these become latent buds.

Adventitious buds can develop in places where buds generally do not form, such as buds arising on root pieces when root cuttings are made.

Buds of deciduous woody species usually go into a physiological resting or quiescent state shortly after they are formed in the summer and stay that way until they are subjected to low-temperature winter chilling, which overcomes their resting condition and enables them to resume growth the following spring. Buds of tropical and subtropical plants, however, generally do not develop such a resting condition.

Flower buds form by the differentiation of vegetative buds into flower parts. This change is shown microscopically for the cherry (see Fig. 14–1) later in the text.

Flowers

In the angiosperms, specialized floral leaves borne and arranged on the stem are adapted for sexual reproduction; these are the **flowers.** After fertilization (Chapter 6) portions of the flower develop into a **fruit,** which bears the **seed(s).**

The flowers may be borne at the apex of a stem, as in the sunflower, rose, and poinsettia, or in the axils of the leaves lower down on the stem, as in the tomato, fuchsia, and many of the temperate zone fruit trees. Flowers or **inflorescences** vary in shape and form among the species, a fact that aids in identifying a plant's **species, genus,** and **family.** As with stems, botanists classify flowers in a specialized morphological terminology (see Figs. 2–36 and 2–37).

Complete Flowers

Complete flowers usually have four parts—sepals, petals, stamens, and pistil—which are usually borne on a receptacle (Fig. 2–36). The **sepals** are the leaf-like scales that encircle the other flower parts, as in the carnation and rose. Most often the sepals are green, but sometimes they are the same color as the petals, as in tulips and lilies. The sepals may fold back (rose) or remain upright (carnation) as the petals grow and emerge. The sepals collectively are called the **calyx.** The **petals** are the next whorl of floral leaves inward from the sepals. The collective term for petals is **corolla.** The petals are usually brightly colored with some yellow, and they often contain **nectaries** that secrete nectar to attract insects, which pollinate the flowers. Collectively sepals and petals are called the **perianth** as in bulbous plants.

The next whorl of floral organs in a complete flower is the male part, or **stamen.** Each stamen consists of a **filament** and an **anther;** the anther produces the **pollen.** Collectively a group or whorl of stamens is the **androecium.**

The **pistil,** the central female component of the flower, is composed of three parts: the **stigma,** the receptive surface that receives the pollen; the **style,** a tube connected to the stigma; and the **ovary,** attached to the lower end of the style. The ovary contains undeveloped **ovules** that are attached to a **placenta;** the ovules develop into **seeds** after pollination and fertilization (Chapter 6). The pistil can be simple (i.e., has but one carpel) or compound (i.e., has two or more fused carpels). Collectively, the carpels are known as the **gynoecium.** The apricot and the apple (Figs. 2–38 and 2–39) are examples of complete flowers with a simple and a compound pistil, respectively.

The stamens and pistils are considered the essential parts of the complete flower for sexual reproduction. The

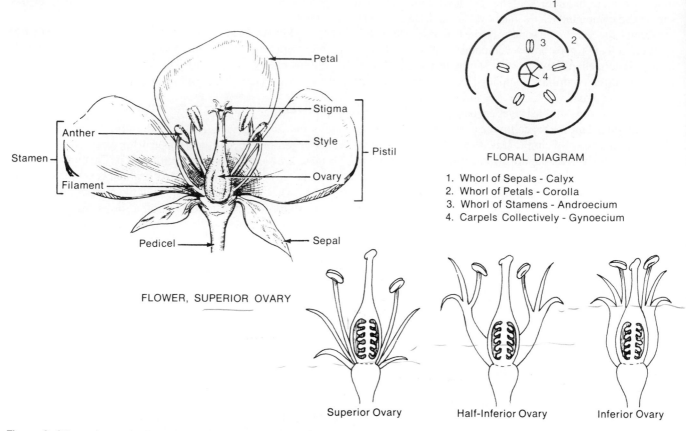

FLORAL DIAGRAM

1. Whorl of Sepals - Calyx
2. Whorl of Petals - Corolla
3. Whorl of Stamens - Androecium
4. Carpels Collectively - Gynoecium

FLOWER, SUPERIOR OVARY

Superior Ovary   Half-Inferior Ovary   Inferior Ovary

Figure 2–36 Types of flower patterns in angiosperms, showing the relative position of the ovary in relation to accessory structures.

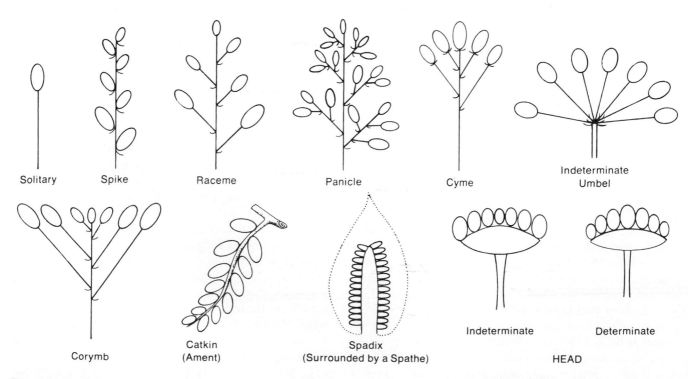

Solitary   Spike   Raceme   Panicle   Cyme   Indeterminate Umbel

Corymb   Catkin (Ament)   Spadix (Surrounded by a Spathe)   Indeterminate   Determinate   HEAD

Figure 2–37 Some typical stylized inflorescences used to describe plants.

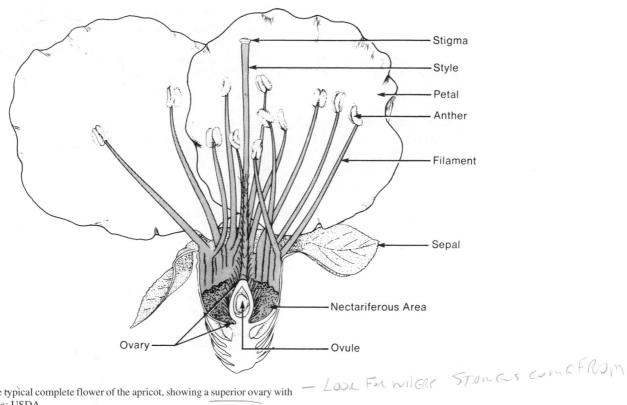

Stigma

Style

Petal

Anther

Filament

Sepal

Nectariferous Area

Ovary

Ovule

Figure 2–38 The typical complete flower of the apricot, showing a superior ovary with a simple pistil. *Source:* USDA.

— LOOK FOR WHERE STAMES COMEFROM

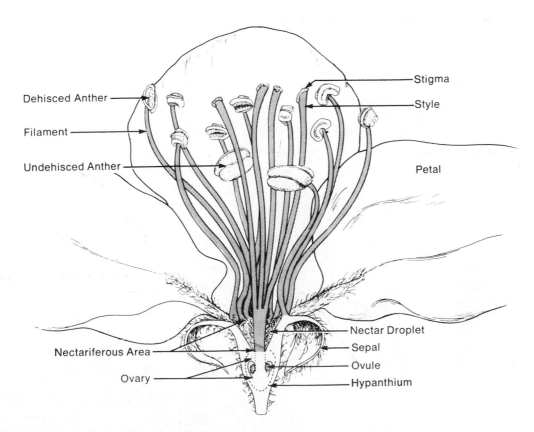

Dehisced Anther

Filament

Undehisced Anther

Stigma

Style

Petal

Nectar Droplet

Nectariferous Area

Sepal

Ovule

Ovary

Hypanthium

Figure 2–39 The typical complete flower of the apple, showing an inferior ovary with a compound pistil. *Source:* USDA.

sepals (calyx) and the petals (corolla) are accessory flower parts.

Incomplete Flowers

Incomplete flowers lack one or more of the four parts: sepals, petals, stamens, or pistil. Flowers with both stamens and pistils are called **perfect flowers.** Flowers with stamens only and no pistils are called **staminate** flowers; those with pistils but no stamens are called **pistillate** flowers. Staminate or pistillate flowers are by definition **imperfect** flowers. Plants having both staminate and pistillate flowers borne on the same plant are termed **monoecious** (e.g., alder, corn, walnut) (Fig. 2–40). If the pistillate and staminate flowers are borne on separate individual plants (male and female plants), the species is called **dioecious.** Examples are date palm, kiwifruit, gingko, pistachio, and asparagus.

Some flowers, like the tulip, are borne singly on a stalk and are called **solitary,** but others are arranged in multiples or in clusters known as **inflorescences** (see Fig. 2–37). The **corymb** is a short, flat-topped flower with an indeterminate cluster that continues to produce flowers until conditions become unfavorable; the lower flowers open first. An example is the cherry. The **cyme** resembles the corymb, except that the central or topmost flower is the first to open. Examples are chickweed and strawberry. The **raceme** is a single elongated indeterminate arrangement of stalked flowers, found in the mustard and cole crops (CRUCIFERAE), for example. The **spike** is an elongated, simple, indeterminate inflorescence with sessile (no stalk) flowers,

as in wheat, oats, and gladiolus. The **catkin** is a spike with only pistillate or staminate flowers exemplified by alder, poplar, walnut, and willow. The **panicle** is an indeterminate branching raceme found in many of the grasses. The **umbel** is an indeterminate, often flat-topped, cluster of flowers that are of equal length and arise from a common point, as in carrots, dill, and onions. A **head** is a short dense spike; daisies and sunflowers have heads. A **spadix** is a complete densely flowered structure surrounded by a spathe (calla lily in the ARACEAE family).

Fruits

A fruit is a matured ovary plus associated parts; it is generally a seed-bearing organ, but there are parthenocarpic fruits which are seedless (see p. 131). The fruit protects the seed in some plants and helps disseminate it. For example, the seeds of fleshy fruits like peaches or plums occur inside their pits, which are discarded by animals or humans eating the fruit. The seeds in the discarded pits may eventually germinate and grow. One might tend to think of all fruits as fleshy organs, but there are many dry fruits such as nuts, capsules, legume pods, and grains. Fruits develop after pollination and fertilization (Chapter 6). Flowers are self-pollinated or cross-pollinated by wind or insects. The pollen grows from a pollen grain on the stigma through the style and fertilizes the egg, causing the fruit to develop. Fruits may consist of a single carpel, as in beans and peas, or a combination of several carpels, as in apples (Fig. 2–39) and tomatoes. The fruit matures quickly, in a matter of weeks in the case of some summer annuals and strawberries, or it requires as long as nine months, as with oranges. The ovary wall, which is called the **pericarp,** can develop into different structures. The peel of an orange is part of the pericarp. The pod of a pea, or the shell of a sunflower seed, and the skin, flesh, and pit of a peach are all derived from the ovary wall or pericarp.

Simple Fruits

Simple fruits have a single ovary formed from one flower. The most common classification of simple fruits categorizes them as fleshy, semifleshy, or dry by the texture of the mature pericarp.

Fleshy Fruits The entire pericarp and accessory parts develop into succulent tissue.

> *Berry.* A pulpy fruit from one or more carpels that develops few to many seeds. Examples are bananas, dates, grapes, peppers, tomatoes, and papayas.
>
> *Hesperidium.* A fruit with several carpels with inner pulp juice sacs or vesicles enclosed in a leathery rind; for example, orange, lemon, lime, and grapefruit.
>
> *Pepo.* A fruit formed from an **inferior ovary** that develops from multiple carpels bearing many seeds. The pericarp is a thick and usually hard rind.

Figure 2–40 Flowers of a monoecious species, the walnut. *Left:* Female flowers and *Right:* male flowers or catkins are borne in separate structures on the same plant. The female flowers are wind pollinated.

Cucumbers, melons, squashes, and watermelons are examples.

Dry-Fleshy Fruits   Some parts of the pericarp become dry and the other portions remain succulent.

Drupe.   This is a simple fruit derived from a single carpel. The **exocarp** (the outer layer) becomes the thin skin; the **mesocarp** (the middle layer) becomes thick and fleshy; the **endocarp** (the inner layer) becomes hard and stony and is often referred to as the pit (and often erroneously as the "seed"). Peaches, plums, cherries, apricots (Fig. 2–38), almonds, and olives are examples of drupe fruits.

Pome.   This is a simple fruit made up of several carpels. The outer (and edible) portion forms from an accessory structure, the hypanthium of the flower, which surrounds the multiple **carpels.** Apple (Fig. 2–39), pear, and quince are examples.

Dry Fruits (Papery or Stony)   The entire pericarp is dry at maturity.

Dehiscent fruits.   These fruits split at maturity to expose the seeds.

Legume or *pod.*   A fruit from a single carpel which usually dehisces along both sutures. This fruit is typical of the pea family (FABACEAE), such as peas and beans.

Capsule.   The fruits form from two or more carpels, each of which produces many seeds. Splitting can occur in several different ways. Iris, poppy, and jimson weed are examples.

Follicle.   The fruits form from a single carpel that splits along one suture. *Delphinium* and *Helleborus* are examples.

Silique.   Fruits form from two carpels with a septum between. The two halves separate longitudinally, exposing the seeds on a central membrane. Mustard, *Lunaria,* and stocks are examples.

Indehiscent fruits.   These fruits do not split open when mature.

Achene.   Simple, one-seeded, thin-walled fruit attached to an ovary wall. Very often achenes are mistaken for seeds as in the case of strawberry "fruits" (Fig. 2–41), the so-called seeds of the rose-hip, and sunflower fruits.

Caryopsis (grain).   A one-seeded fruit with a thin pericarp surrounding and adhering tightly to the true seed. Corn, rice, wheat, and barley are examples.

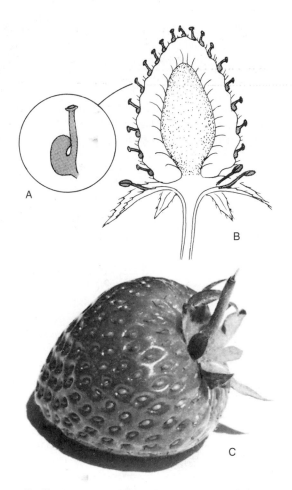

Figure 2–41   Strawberry "fruit." The true strawberry fruits on the left (A) are the small achenes borne on the surface of the large fleshy receptacle (B). The complete edible part is shown in (C).

Nut.   A one-seeded fruit with a thick, hard, stony pericarp. Oak (acorn), chestnut, filbert, walnut, and hickory are examples.

Samara.   A one-seeded (elm) or a two-seeded (maple) fruit with a winglike structure formed from the ovary wall. Ash is another example.

Schizocarp.   A fruit formed from two or more carpels that split at maturity to yield two one-seeded halves. Carrots, dill, caraway, and parsley are examples.

Aggregate and Multiple Fruits

Aggregate and **multiple fruits** form from several ovaries. The true fruits are attached to, or contained within, a receptacle or an accessory structure.

Aggregate fruits develop from many ovaries on a single flower. The strawberry, for example, has many achenes (true fruits), each attached to a single fleshy receptacle (Fig. 2–41). The many achenes of the rose-hip develop inside the receptacle. The blackberry and raspberry are similar to the strawberry except that individual small drupes, instead of achenes, are attached to the fleshy receptacle.

Multiple fruits develop from many individual ovaries fused into a single structure borne on a common stalk. The fig "fruit" we eat is made up of small drupes (the true fruits) contained inside a fleshy receptacle. The whole structure is termed a syconium. The pineapple is a large accessory structure covered with seedless (parthenocarpic) berries. Mulberries are multiple drupelets borne on a fleshy receptacle.

Seeds

Seeds vary considerably in size, shape, structure, and mode of dissemination. The seeds contained in samaras take to the air on their small wings. The downy tufts of milkweed and dandelion seeds enable them to be carried great distances on wind currents. The coconut is known to have floated to new land masses on ocean currents. Some seeds are attached to barbs or hooks or are contained within burrs or siliques that can catch in clothing or animal fur and be transported great distances. Squirrels and other rodents bury nuts, many of which germinate later. Birds and other animals disseminate seeds by eating the fruit and passing resistant seeds through their digestive tract. Some kinds of seeds are forcefully expelled from dehiscing dried fruits as they mature. These varied means of seed dissemination aid in species survival by promoting plant growth in new locations with possibly different environments.

A seed is a mature ovule. The three basic parts are the embryo, the food storage tissue (endosperm, cotyledons, or perisperm), and the seed coats. Chapter 6 covers the development of the seed in detail.

The **embryo** is a miniature plantlet formed within the seed from the union of the male and female gametes during fertilization. Basically, the embryo has two growing points: the **radicle,** which is the embryonic root, and the **plumule,** which is the embryonic shoot. One or two **cotyledons** are located between these two growing points on the root-shoot axis.

Food can be stored in the endosperm, cotyledons, or perisperm in the form of starch, fats, or proteins. Stored food in the cereal grains is largely endosperm. Seeds having a large portion of their food stored as endosperm are called **albuminous** seeds. Those seeds with no endosperm or only a thin layer surrounding the embryo are called **exalbuminous** seeds. Such seeds store food either in fleshy cotyledons, as in beans, or occasionally in the perisperm (developed nucellus), as in beets.

The **seed coverings** are usually tough, preventing damage to the enclosed embryo. They are also relatively impervious to water to save the embryo from desiccation, but again there are exceptions. There may be one or two seed coats (**testa**), which form from the **integuments,** the outer layers of the ovule. In such dry fruits as achenes, samaras, and schizocarps, the fruit adheres closely to the seed coat; consequently the true fruit is usually called a "seed." The pits or stones of such drupe fruits as peaches and apricots are often called "seeds"; however, they are actually the hardened inner wall (endocarp) of the pericarp that surrounds the seed.

The scar that remains after breaking the seed from the stalk is called the **hilum,** and the small opening near the hilum is the **micropyle.** The ridge on the seed is the **raphe** (Fig. 2–4).

SUMMARY AND REVIEW

Plants are made up of cells and subcellular organelles including, but not limited to, the nucleus, mitochondria, chloroplasts, vacuoles, and endoplasmic reticulum. The organelles of the plant cell and the surrounding fluid matrix called the cytoplasm are contained within the plasmalemma. Surrounding the plasmalemma is the cell wall.

Groups of cells that perform a collective function are called tissues. Some of the major tissues that make up a plant include meristems, pith, epidermis, cambium, xylem, phloem, palisade layer, and spongy mesophyll. Tissues can be grouped together to form complex systems of organs such as roots, stems, leaves, flowers, and seeds.

Each cell, tissue, and organ has a specific function. Cells are the site of DNA and RNA synthesis, respiration, photosynthesis, protein synthesis, and other biochemical processes.

Tissues are a collection of cells that have a specific function. Tissues can be simple or complex. Simple tissues

are the meristems, epidermis, and parenchyma. Meristematic tissues, including apical, subapical, intercalary, and vascular cambium, are the regions of cell division (mitosis) and contribute to the growth of the plant by adding more cells. The epidermis is the exterior layer of living cells that protects all parts of the plant and includes the guard cells of the stomata. Some epidermal cells produce hair-like projections called trichomes. Parenchyma tissue is made up of thin-walled cells that are found in all parts of the plant where they perform different functions depending on their location. These cells can become meristematic and regenerate the same or different tissues when a plant is injured.

Complex tissues are the vascular tissues, xylem and phloem, which move liquids and dissolved solids throughout the plant. Xylem tissue is nonliving and can be composed of vessels, tracheids, and fibers. Phloem tissue is living and is composed of sieve tubes, sieve tube members, companion cells, fibers, and parenchyma.

Plant organs are groups of tissues organized to perform very complex functions in the plant. They are the roots, stems, leaves, flowers, fruits, and seeds.

Roots can be tap or fibrous in form. They anchor the plant and provide the mechanism to bring water and nutrients into the plant. They are often the storage site of photosynthesized food. Parts of the root include the root cap, epidermis, root tip, cortex, endodermis, procambium layer, and stele (comprised of the vascular cylinder and pericycle). *branch roots arise*

The stem makes up the support system of the plant supporting the leaves, flowers, and fruit. Stem tissues vary between plants but most, if not all, stems have an epidermis, cortex, vascular system of xylem and phloem and other tissues, and vascular cambium (not found in monocots). Stems can have different forms besides that usually associated with the aerial portion of a plant. These forms include rhizomes, stolons, corms, bulbs, and stem tubers.

Leaves are the site for most of the photosynthetic activity in a plant. Leaves also transpire large quantities of water to help regulate plant temperature and provide the force for carrying xylem contents through the plant. Leaves are comprised of epidermis, vascular bundles, palisade, and spongy mesophyll cells. Quite often, leaf shape is a distinguishing characteristic of a plant.

Flowers are the sexual reproductive structures of angiosperms. Flower parts include sepals, petals, stamens, and pistils. When a flower has all four parts it is called complete. Flowers missing one or more parts are called incomplete. The stamens and pistils are the male and female reproductive organs, respectively. Stamens are composed of anthers (site of pollen production) and filaments. Pistils are composed of stigma, style, and ovary (site of ovule production). Perfect flowers have both stamens and pistils. Imperfect flowers lack one or the other. Monoecious plants have both male and female flowers on the same plant. Dioecious plants have female flowers on one plant and male flowers on another. Flowers can occur singly on a stalk (solitary) or there can be multiple flowers per stem (inflorescence).

Fruits are matured ovaries plus associated parts. Generally fruits are seed-bearing, but some fruits are parthenocarpic and are seedless. The ovary wall (pericarp) can develop in many different ways and often is a significant portion of the fruit. Fruits can be simple (developing from one ovary), such as berries (grapes and tomatoes), drupes (peaches and cherries), pomes (apples and pears), pods (beans), achenes (strawberries), caryopsis (grains), nuts, and others. Multiple and aggregate fruits form from several ovaries.

EXERCISES FOR UNDERSTANDING

2–1. As you learn the various parts of a plant, think about how each part contributes to the economic value of a crop. For example, how does the condition of a leaf and its components affect the value of a hay crop, of a carrot crop, of a bunch of cut roses, or of a golf course green?

2–2. As you look at the diagrams of plant cells, tissues, and organs, notice how they fit together to make the plant function as a whole. Consider what would happen if the plant lost the function of one of its tissues or organs. For example, what do you think would happen to the rest of the plant if a plant would lose its root system because of field flooding?

2–3. Think of cases where the knowledge of a plant's anatomy and structure would help you with crop management decisions. For example, would knowing that the apical meristem of a newly emerged corn seedling is below ground affect your decision to replant or not replant a 500-acre cornfield after a light frost damaged the leaves of the seedlings?

REFERENCES AND SUPPLEMENTARY READING

RAVEN, P.H., R.F. EVERT, and S.E. EICHORN. 1999. *Biology of Plants,* Sixth Edition. New York: Freeman.

TAIZ, L. and E. ZEIGER. 1998. *Plant Physiology.* Second Edition. Sunderland, Massachusetts: Sinauer.

CHAPTER

Naming and Classifying Plants

KEY LEARNING CONCEPTS

After reading the chapter you should be able to:

♦ Understand how plants are named and classified.
♦ Use the nomenclature and system of taxonomic classification to identify plants and their relationship to each other.

There are over 500,000 different kinds of plants, and for man to be able to communicate about them, some method of classifying and naming them had to be developed. The process of developing plant names started over 4,000 years ago, but the first recorded names were attributed to Theophrastus (370–285 B.C.). Naming plants, no doubt, began with simple names that referred to the plant's use, growing habit, or other visible attribute. One example is the milkweed—so named because its sap (latex) is milky in appearance. One difficulty with such names is that often they are only used locally. People in one place know the plant by one name, while elsewhere the same plant is known by a different name. The weed *Tribulus terrestris* is known as the "puncture vine" in some areas because the seeds have sharp spines that puncture tires or bare feet; another common name for the same plant in other areas is "goat-head" because the shape of the seed resembles a goat's head. As plant knowledge expanded and the exchange of this knowledge became desirable, it was obvious that a uniform and internationally acceptable system was needed to name and classify plants.

There are many ways to classify plants, and any system depends on how the classification is to be used. Some classifications relate directly to specific environmental requirements of the plant for satisfactory growth. For example, such a classification could categorize plants as to their climatic requirements.

CLIMATIC AND RELATED CLASSIFICATIONS

Farmers obviously have to be able to identify and name crops. But this alone is not enough. They also have to distinguish which of the many crops suit their climate. Most farmers in the United States are working in the temperate zone. For example, some fruit and nut crops grown in the temperate zone are almond, apple, apricot, cherry, peach, pear, pecan, and plum. Fruit growers in a tropical region would have an entirely different choice of crops, such as cocao, cashew and macadamia nuts, banana, mango, papaya, and pineapple. The fruits of the subtropical region, between the temperate zone and the tropics, cannot withstand the severe winters of the temperate zone but need some winter chilling. Some of these subtropical plants are citrus, date, fig, olive, and pomegranate. Thus, by using climate as a criterion, plants can be classified into distinct groups.

Agronomic crops like grains, forage, fiber, and oil crops and the vegetable and ornamental plants can also be classified by their temperature requirements. For example, some annuals have specific climatic requirements for growth and flowering and are distinguished as winter or summer annuals. Winter annuals are planted in the fall and bloom early the following spring. Summer annuals are planted in the spring and bloom through the summer and fall.

Some crops grow best in certain seasons and are thus classified. For example, warm-season plants such as corn, beans, tomatoes, peppers, watermelons, petunias, marigolds, zinnias, and bermuda grass grow best where monthly temperatures average 18°C to 27°C (65°F to 80°F), while broccoli, cabbage, lettuce, peas, flowering bulbous plants, snapdragons, cyclamen, and bluegrass are cool-season crops growing best at average monthly temperatures of 15°C to 18°C (60°F to 65°F). Plants can be classified by the seasons in which they are most likely to flower and fruit or when the quality of the product can be expected to be at its maximum. Numerous flower and vegetable **cultivars** can be classified as early, midseason, or late maturing.

Vegetables are classified into groups according to their edible parts. Some are grown for fruits and seeds, such as the tomato, bell pepper, string bean, pea, and corn. Many are grown for their shoots or leafy parts, such as asparagus, celery, spinach, lettuce, and cabbage. Others are grown for their underground parts (either roots or tubers), such as the carrot, beet, turnip, and potato.

Ornamentals are sometimes classified by use—that is, as houseplants, greenhouse plants, garden plants, street trees,

Figure 3–1 Jeffrey pine (*Pinus jeffreyi* Grev. and Balf.) tree growing in the Sierra Nevada mountains of northern California near Lake Tahoe. This tree is about 2 m (6.5 ft) in diameter at breast height (DBH), the point where lumber trees are measured. *Source:* Robert A. H. Legro.

and various classes of landscape plants. Houseplants, which have become very popular, are often classified according to their foliage, flowers, or growth habits. Common foliage plants, which stay green all year, are the philodendrons, dieffenbachias, and ferns, to name a few. Blooming of some flowering plants may be only seasonal, as in the lily and poinsettia, but the African violet and chrysanthemum are available in flower the year round. Plants outside the home are typically classified by use; for example, bedding plants such as petunias, marigolds, and zinnias, and landscape plants like trees and shrubs.

The forester classifies trees into two broad groups: the hardwoods and softwoods. Some hardwood types are oaks (*Quercus*), maples (*Acer*), birch (*Betula*), and beech (*Fagus*). Some softwood trees are pines (*Pinus*) (Fig. 3–1), firs (*Abies*), redwood (*Sequoia*) (Fig. 3–2), cedars (*Cedrus*), and spruce (*Picea*). Trees are also classified according to the hardiness zones in which they can survive. Some are able to withstand very low temperatures during the winter, whereas others are subject to frost damage and therefore must be grown in a subtropical climate.

COMMON AND BOTANICAL NAMES

Most plants are generally known by their **common names** because common names are often easier to remember, pronounce, and use. Maple or elm trees growing along the streets are referred to by common names in everyday conversation. Common names often evolve because of certain plant characteristics.

Figure 3–2 A grove of coast redwoods (*Sequoia sempervirens* D. Don, Endl.) along Highway 101 on "The Avenue of the Giants" in northern California. The camper truck is dwarfed by these majestic trees.

A common name has value in conversation only if both persons know exactly what plant is being discussed. This is most likely when persons are from the same community and the common name of the plant cannot be mistaken for another. But take the case of the jasmine, a plant known all over the world and prized for its fragrance, flavoring (tea), and landscaping. Many common plant names contain the word *jasmine,* but such plants do not resemble one another and may not even be closely related botanically. Some of the so-called jasmines are listed below, with the botanical (italicized) name following the common name:

     Star jasmine   (*Jasminum gracillimum*)

     Star jasmine   (*Trachelospermum jasminoides*)

     Blue jasmine   (*Clematis crispa*)

     Cape jasmine   (*Gardenia jasminoides*)

     Crape jasmine   (*Tabernaemontana divaricata*)

     Night jasmine   (*Cestrum nocturnum*)

     Night jasmine   (*Nyctanthes arbor-tristis*)

It is obvious from these examples that common names have their limitations for universal written or verbal communication. There are too many and they are too variable to serve most scientific purposes.

DEVELOPMENT OF BOTANICAL CLASSIFICATIONS

Theophrastus (370–285 B.C.), a student of Aristotle, classified plants by their texture or form. He also classified many as herbs, shrubs, and trees. He noted the annual, biennial, and perennial growth habits of certain plants, and described differences in flower parts that enabled him to group plants for purposes of discussion. He is known as the Father of Botany for these significant contributions.

Carl von Linné (1707–1778), better known as Carolus Linnaeus, devised a system of categorizing plants that led to the modern taxonomy or nomenclature of plants.

Scientific Classification

The scientific system of classification has all living things divided into groups called **taxa** (sing. taxon) based on physical characteristics. The first taxon, called **Domain,** divides all living things into two Domains: **Prokaryotes** (cells having no separate subcellular units) and **Eukaryotes** (cells having subcellular units). The Eukaryote Domain is divided into the four Kingdoms of Fungi, Protista, **Plantae,** and Animalia. The Plantae Kingdom is divided into two groups: **bryophytes** (includes mosses and liverworts) and **vascular plants.** The vascular plants are divided into two subgroups: seedless and seeded. Seedless and seeded plants are further classified by **Phyla.** Seedless phyla include the Pterophyta (ferns). Seeded phyla include Cycadophyta (cycads), Ginkgophyta (ginkgo), Coniferophyta (conifers), and Anthophyta (angiosperms, which are subdivided into monocotyledon and dicotyledon). Almost all commercially important crop plants are in the seeded group. After phylum, plants are classified in descending rank by **class, order, family, genus,** and **species.** Each descending rank more closely defines the physical characteristics common to members of that rank.

PLANT IDENTIFICATION AND NOMENCLATURE

The family is usually the highest taxon commonly included in plant identification or study. Students of plant science are usually required to learn the family, genus, and species of some plants as well as their common names.

Since the early Christian era, naturalists wrote their books in Latin, which was the language of all educated people in Europe. Thus, Linnaeus used names of Latin form. Most of the names he gave, which describe morphological characteristics of the plants, came from Latin words, although some were derived from Greek and Arabic.

Figure 3–3 The southern magnolia (*Magnolia grandiflora* L.). The name *grandiflora* is more than justified; the flower measures 15 cm (6 in.) across.

The names are usually phonetic and often give a clue to the plant's characteristics, its native habitat, or for whom it is named. Such derivatives are numerous. Names that refer to leaves include *folius, phyllon,* or *phylla,* usually as suffixes. The names can also have prefixes, such as *macro* or *micro.* Thus words are created, such as *macrophylla* (large leaf), *microfolius* or *microphylla* (small leaf), *illicifolius* (holly leaf), and *salicifolius* (willow leaf). The Latin for flower is *flora;* add the prefix *grand* and it becomes *grandiflora* (large flower), as in *Magnolia grandiflora* L., the southern magnolia (Fig. 3–3). Shapes or growing habits of plants can be described with *altus* or *alta* (tall), *arboreus* (treelike), *compactus* (dense), *nanus* or *pumilus* (dwarf), *repens* or *reptans* (creeping), and *scandens* (climbing). Names based on flower or foliage color include *albus* or *leuco* (white), *argentus* (silver), *aureus* or *chryso* (gold), *rubra, rubens,* or *coccineus* (red), and *croceus, flavus,* or *luteus* (yellow). Species names sometimes reflect the plant's place of origin. Examples are *australis* (southern), *borealis* (northern), *canadensis* (from Canada), *chinensis* or *sinensis* (from China), *chilensis* or *chileonsis* (from Chile), *japonica, nipponica,* or *nipponicus* (from Japan), *campestris* (field), *insularis* (island), and *montanus* (mountain). Sometimes Linnaeus took names, like *Narcissus,* from classical mythology, or devised names to honor other scientists, like *Rudbeckia,* honoring Linnaeus' botany professor, Rudbeck.

Each plant has a two-word, or **binomial,** name[1] given in Latin. The first name refers to the plant's **genus,** the sec-

ond to its **species.** The Latin binomial name is international and understood universally.

Complete Linnaean names have a third element—the **authority,** or the abbreviated name of the scientist who named the species. Consider the name for the common white (Irish) potato, *Solanum tuberosum* L. The "L." means Linnaeus, and his initial appears commonly, because Linnaeus named so many species. Books and journals often omit the authority for brevity and simplicity, but it is important in determining which taxon is being referred to in situations where different botanists have used different binomials for the same plant. In such cases, the botanical or scientific name that was published first takes precedence.

Wild or naturally occurring plants are named under the rules of the *International Code of Botanical Nomenclature.* Cultivated plants are named according to the same principles but are covered by the *International Code of Nomenclature Cultivated Plants.* Some of the basic rules of nomenclature follow.

The generic name always begins with a capital letter; it is underlined when written by hand or typewriter and italicized in print. Thus, the genus name for potato is *Solanum.* The specific epithet *tuberosum* is likewise underlined or italicized. The specific epithet usually begins with a lowercase letter, but it *may be* capitalized if it is a person's name; it is always correct if written entirely in lowercase. Many of the original species names are frequently capitalized (e.g., *Pinus Jeffreyi* Grev. and Balf. or *Pinus jeffreyi* Grev. and Balf.).

To complete the binomial name, the authority for describing and naming the plant is given after the genus and species; thus, *Solanum tuberosum* L. In this text, unless otherwise specified, both the binomials and the authorships agree with those given in *Hortus Third* which is a widely recognized compilation of the cultivated plants for the United States and Canada. These authority names are often abbreviated. Each taxonomy book has a list of the full names of these authorities.

When several plants in the same genus are listed, the genus name is given in full for the first plant, then shortened to the first initial (which is always capitalized) for the other plants in the list. As an example, the apricot, the European plum, and the peach are all members of the genus *Prunus.* Listed as scientific binomials, they would be *Prunus armeniaca, P. domestica,* and *P. persica,* respectively. This procedure should not be used if there is any chance of confusion with another genus with the same first initial.

The singular and plural spelling of *species* is the same. Occasionally the plant genus is known but the exact species is not known because it is difficult or impossible to identify. In such a case, the genus name is given and followed by the lowercase letters "sp." for species (singular) and "spp." for species (plural). An example is *Prunus* spp. The "spp." usually refers to all of the many species in the genus. However, the "sp." refers to a definite plant whose specific epithet is not known. The "sp." or "spp." is never underlined or italicized.

---

[1] The binomial is a binary combination of the name of the genus followed by a single specific epithet. If an epithet consists of two or more words, these are to be united or hyphenated [Article 23 *International Code of Botanical Nomenclature* (2)]. Although the second word of the binomial is technically the "specific epithet," many persons refer to that second name as the "species." This shorter and therefore more convenient form is the one used in this text.

Subspecific Categories

Sometimes a botanical binomial is not sufficient to identify a species—wild or cultivated. Botanists and horticulturists may form subspecific categories, such as **botanical variety, cultivar,** and **group.**

Botanical Variety

A plant group can be so different in the wild from the general species described originally that it warrants a **botanical variety** classification below that of species. An example of this is *Buxus microphylla* Sieb. and Zucc. var. *japonica* Rehd. and Wils., which is native to Japan. The "var." stands for *varietas,* Latin for "variety." Another botanical variety originated in Korea; it is *Buxus microphylla* var. *koreana* Nakai. These botanical varieties are sufficiently different to warrant unique names and authorities to distinguish them from one another. In this case the name of the variety *japonica* or *koreana* is underlined or italicized. When a varietas epithet is formed from a surname, it may or may not be capitalized depending on the personal preference of the author. However, the trend is to not capitalize them, as recommended by the International Codes.

Cultivar

Many kinds of plants that are valuable in agriculture must be propagated with little or no genetic change in the offspring. In agriculture and horticulture there are cultivated varieties that remain genetically true. These cultivated varieties may be different from botanical varieties and are called **cultivars,** a contraction of *culti*vated *vari*ety. There are two main categories of cultivars—the clones and the lines. If propagated by vegetative methods, they are called *clones;* if by seeds (under certain specified conditions), they are called *lines.* The word *cultivar* is abbreviated "cv." and the plural is "cvs." A cultivar is often a distinct variant selected by someone who believed it was uniquely different from any plant already in cultivation. The flower color may have changed from red to white because of a mutation, as in some carnations. Perhaps a plant has fewer spines or thorns than does the ordinary species; an example is a Chinese holly (*Ilex cornuta* Lindl. and Paxt.) found to have few or no spines. It was named *Ilex cornuta* cv. Burfordii. The cultivar name is always capitalized but never underlined or italicized. The term "cv." after *cornuta* may be dropped in favor of single quotes around the cultivar name. Either way of expressing the cultivar name is acceptable, and both can be used in the same article. Either single quotes or the term "cv." are used, but *never both.* Tables or lists usually use "cultivar" or "cv." in the heading to avoid single quotes around each cultivar name.

Many annual flowers, vegetables, grains, and forage crops are cultivars that are propagated by seed. Others are F$_1$ hybrids, uniform and nonuniform assemblages. An example is *Petunia* $\times$ *hybrida* Hort. Vilm.-Andr.—the hybrid garden petunia. A breeder may develop a new strain that is believed to warrant a cultivar name such as 'Fire Chief' or 'Pink Cascade'. The parent plants can be maintained and crossed to produce the same F$_1$ hybrid cultivar year after year. Many vegetables and flowering annuals are maintained as cultivars in this manner, with the parents maintained to produce new crops of seed each year for planting. Cultivars of fruit trees, grapes, and woody ornamentals are usually maintained as true to type **clones** by **vegetative** propagation methods.

Group

The group category is used for some vegetables and some ornamentals such as lilies, orchids, roses, and tulips. It is a category below the species and not used as frequently as the cultivar category. A group includes more than one cultivar of a particular kind of plant. For example, when there are evident differences among plants of the same species, they can be further categorized by a group name. When a species has many cultivars, cultivars that are similar are categorized into groups. For example, cultivars of *Brassica oleraceae* can be grouped into the Acephala Group, the Alboglabra Group, the Botrytis Group, or the Capitata Group, depending upon their morphological characteristics. These groups have the same botanical name—*Brassica oleracea.* The name of the group is written within parentheses between the species name and the cultivar name, as *Brassica oleraceae* (Capitata) 'King Cole', and the group name is always capitalized but not enclosed in single quotes.

Family

The family is a group of closely related genera. The relationship can be based on certain plant structures or on chemical characteristics, such as the presence of latex in the milkweed family ASCELPIADACEAE, but flower structure is the usual basis for association. The nightshade family SOLANACEAE contains not only *Solanum* (potato) but also *Lycopersicon* (tomato), *Capsicum* (pepper), *Nicotiana* (tobacco), *Datura* (deadly nightshade), *Petunia,* and many others. This is a large family (about 90 genera and more than 2,000 species), most of which are native to the tropics. All species in this family have similar flower structures; the similarities between a tobacco, a tomato, and a potato flower, for instance (see Fig. 3–4), are readily seen.

The first letter of family names is always capitalized and the names are sometimes underlined or italicized. The family names may be written entirely in capital (uppercase) letters, the method used in this text. Most families' names end with -*aceae* (pronounced ace-ay-ee) attached to a genus name; for example, SOLANACEAE, ROSACEAE, AMARYLLIDACEAE, LILIACEAE, and MAGNOLIACEAE. Eight families, however, do not follow this standard rule. For the sake of uniformity, new names have been

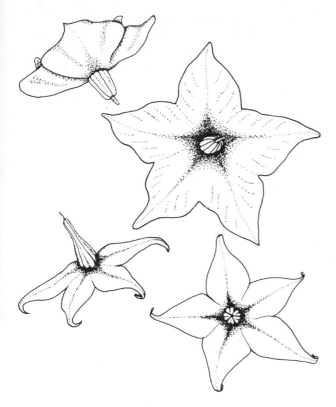

proposed for these families. Either name may be used. The new names appear in parentheses following the old names:

ASTERACEAE (COMPOSITAE)
BRASSICACEAE (CRUCIFERAE)
POACEAE (GRAMINAE)
CLUSIACEAE (GUTTIFEREAE)
LAMIACIACE (LABIATAE)
FABACEAE (LEGUMINOSAE)
ARECACEAE (PALMAE)
APIACEAE (UMBELLIFERAE)

Plant classifications can be studied in detail in various plant biology or taxonomy books and references.

PLANT IDENTIFICATION KEY

See Table 3–1 for a simplified yoked (indented) key used to identify some commonly known seed-bearing plants (Spermatophyta).

To use a key one needs to know the vocabulary of plant structure (see Chapter 2). Examine the plant in question to decide if its characteristics fit in one category or the other offered by the key. Keying plants is a process of elimination by making yes or no decisions to characteristics offered in the

Figure 3–4 Side and front views of potato (*upper*) and tomato (*lower*) flowers. These views show the similarities of the flowers of plants in the same family, SOLANACEAE. *Source:* Moira Tanaka.

TABLE 3–1	A Simplified Yoked (indented) Key for Identifying Some Seed-Bearing Plants (Spermatophyta)

1. Ovules and seeds borne naked on scales in cones without typical flowers (Fig. 3–5); trees or shrubs, often evergreen. Gymnospermae
  2. Plant foliage palmlike. CYCADACEAE (Fig. 3–7)
  2. Plant foliate not palmlike
    3. One seed in a cup-shaped, drupelike fruit. TAXACEAE (Fig. 3–8)
    3. Many seeds in a dry woody cone
      4. Leaves alternate and single
      4. Leaves alternate and in clusters; needle-shaped
        5. Cone-scale without bracts with two to nine seeds
        5. Cone-scales in axils of bracts, flattened, with two seeds. PINACEAE
          6. Cones upright on top of branchlets. *Abies*
          6. Cones not upright on branchlets. *Pinus*
            7. Twigs not grooved. White pines or soft pines
            7. Twigs grooved. Pitch or hard pines
1. Plants with seeds borne in an ovary (base of pistil) with typical flowers. Herbs, trees, and shrubs. Angiospermae
  8. Leaves usually parallel veined, flower parts usually in multiples of three. One seed-leaf or cotyledon. Do not form annual rings when increasing in stem girth. Monocotyledonae
  9. Plant with palmlike leaves. PALMAE
    10. Leaves fanlike
    10. Leaves featherlike. Feather and fishtail palms
      11. Lower feathery leaves not spinelike
      11. Lower feathery leaves spinelike, fruit fleshy with long grooved seed. *Phoenix* spp.
        12. Plant is a tree with shoots at base, trunk about 50 cm (20 in.) in diameter. Fruit edible. *Phoenix dactylifera*, date palm
        12. Plants not as above. Other *Phoenix* spp.
  9. Plants without palmlike leaves
    13. Perianth none or rudimentary
      14. Stems solid. CYPERACEAE
      14. Stems mostly hollow. GRAMINEAE

15. Plants woody, bamboolike. Bamboos
15. Plants herbaceous, not bamboolike
  16. Grasses that produce sugar. *Saccharum* spp.
  16. Grasses that produce little sugar
    17. Small grains and their kin (rice, wheat, etc.)
    17. Cornlike plants and their kin
      18. Plants monoecious. *Zea mays,* corn
      18. Plants not monoecious. *Sorghum* spp.
13. Perianth present
  19. Pistils several, not united. APONOGETONACEAE
  19. Pistils one, carpets united, ovary and fruit superior. AMARYLLIDACEAE
    20. Anthers six, stem a fibrous rhizome. *Agapanthus* spp.
    20. Anthers six, stem a corm or bulb *Allium* spp.
      21. Leaves large, usually hollow and cylindrical. Bulb rounded and large. *Allium cepa,* onion
      21. Leaves large, usually hollow and cylindrical. Bulb slightly thicker than neck. *Allium fistulosum*
8. Leaves usually without parallel venation, two cotyledons. Herbs, trees, and shrubs with stems increasing in thickness with cambium cells, which form annual rings in woody plants. Dicotyledonae
22. Corolla absent or not apparent, calyx present or lacking
22. Corolla present, calyx usually forming two series of calyxlike bracts
  23. Petals united
  23. Petals separate
    24. Ovary inferior or partly so
    24. Ovary superior
      25. Stamens few, not more than twice as many as petals
      25. Stamens numerous, more than twice as many as petals
        26. Habit aquatic. NYMPHAEACEAE, water lilies
        26. Habit terrestrial
          27. Pistils more than one, filaments of stamens united into a tube. MALVACEAE
            28. Style—branches slender, spreading at maturity, seeds kidney-shaped. *Hibiscus* spp.
            28. Styles united, ovary several carpels, calyx deciduous, seed angular. *Gossypium* spp.
              29. Staminal column long, anthers compactly arranged on short filaments. *G. barbadense,* sea-island cotton
              29. Staminal column short, anthers loosely arranged and of varying lengths. *G. hirsutum,* upland cotton (Fig. 3–9)
          27. Pistil more than one, filaments not united into a tube. ROSACEAE
            30. Ovaries superior (Fig. 2–38), fruit not a pome
              31. Pistils one, leaves simple and entire. *Prunus* spp.
                32. Fruit soft and pulpy. *P. armeniaca,* apricot
                32. Fruit dry and hard. *P. mume,* Japanese apricot
              31. Pistils two to many, leaves compound (at least basal leaves)
                33. Plans woody shrubs. *Rosa* spp.
                  34. Styles not extended beyond mouth of hip. Stamens about one-half as long as styles. *R. odorata*
                  34. Styles extend beyond mouth of hip, stamens about as long as styles. *R. multiflora*
                33. Plants herbaceous. *Fragaria* spp.
                  35. Underside of leaves are bluish white. *F. chiloensis,* wild strawberry (fig. 3–10)
                  35. Underside of leaves are green. Other *Fragaria* spp.
            30. Ovaries inferior, fruit a pome
              36. Fruit with stone cells. *Pyrus* spp., pears
              36. Fruit without stone cells. *Malus* spp., apples

key—rejecting those that do not apply (dichotomy). Keys must be written to include the unknown plant to assure accuracy, so that there is no question about whether one should pursue one branch of the dichotomous key or its alternative.

To use a simplified key (Table 3–1), eliminate the alternative that does not pertain to the plant in question, and then proceed to the next pair of numbers directly under the proper choice. As an example, the first choice in this key is to determine whether the seeds of the plant being identified

are borne naked, as in the cone-bearing plants of the gymnosperms (Fig. 3–5), or enclosed in an ovary, as in the angiosperms. Once this yes or no decision has been made, the next step is to compare the next pair of descriptions directly under the previously chosen characteristic. If the seeds are enclosed within an ovary, the plant in question is an angiosperm and the next pair of numbers to compare is 8. Note that the two 8s are separated by several pairs of subsidiary descriptions. A choice between the two 8s must be made

Figure 3–5   Cones of the sugar pine (*Pinus lambertiana* Dougl.) in closed and open conditions. When the cones dry, the winged seeds, shown on the 17.8-cm (7-in.) ruler, are released and can be disseminated by the wind.

Figure 3–7   The inflorescence of a cycad (*Dioon edule* Lindl.) native to Mexico. The inflorescence is about 23 cm (9 in.) tall. The pinnate leaves might lead one to erroneously call this plant a palm.

Figure 3–6   An inflorescence of the pear (*Pyrus communis* L.) showing the five petals and the many stamens typical of the ROSACEAE. Mature fruits do not develop for another four months.

before proceeding further. If the plant in question has parallel-veined leaves and the seeds have one cotyledon, the plant belongs to the monocotyledon subclass. This procedure is followed until the plant to be identified "fits" a given set of plant characteristics.

A slight variation of this yoked key appears in Chapter 19 for the identification of turfgrasses; the number at the end of the phrase in that case indicates the next number to pursue.

Certain plant parts might not be available because they may not be in season—when needed to identify the plant. An example of this occurs in the simplified key in Table 3–1. In the family ROSACEAE (the second item), both flowers and fruit are needed to be certain of making the correct choice. In the case of strawberries, both fresh flowers and fruits can be obtained at the same time; however, it is not possible to have flowers and fruits at the same time for the apricot or the pear (Fig. 3–6). In such cases it might be necessary to preserve

Figure 3–8   The fruits of *Taxus baccata* L. 'Lutea'. The single seed (arrow) is within the cup-like fruit, which is poisonous. *Source:* Robert A. H. Legro.

flowers for future examination or simply describe them thoroughly in the spring when they are abundant, and then wait for the fruit to develop in order to accurately determine its characteristics.

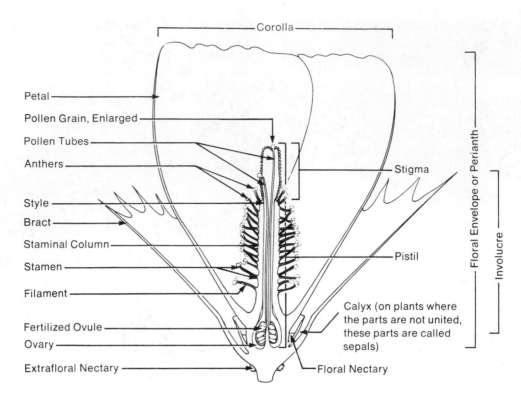

Figure 3–9 Cotton (*Gossypium hirsutum* L.) flower in a longitudinal section. *Source:* USDA.

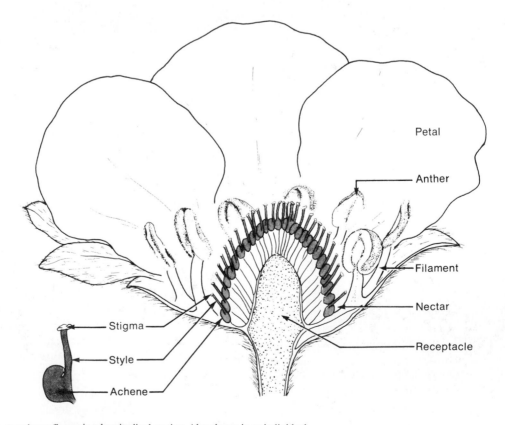

Figure 3–10 A strawberry flower in a longitudinal section. Also shown is an individual achene. *Source:* USDA.

This simple key illustrates how choices are made between two characteristics—eliminating alternatives that are not pertinent in order to finally identify the plant in question. Much of the upper portion of the key can be ignored by a trained taxonomist or a good botany student because, to them, it is easy to identify a gymnosperm or an angiosperm. The family characteristics are determined by observing and studying certain flower characteristics. Sometimes it is difficult but necessary to distinguish between an inferior or superior ovary (the characteristic that divides the LILIACEAE from the AMARYLLIDACEAE families; see Fig. 2–36), but one finds these determinations easier after some experience is gained. Once the family is known, the correct genus and species are identified by referring to taxonomic books with detailed keys.

Obviously one must be familiar with plant parts and structures (Chapter 2) to use a botanical key and determine the identity of the plant. Various plant parts and shapes are useful in illustrating plant identification (see Figs. 2–35, 2–36, and 2–37).

SUMMARY AND REVIEW

Plants are classified in many different ways. They may be classified by taxonomy (common physical characteristics), climate (temperate, tropical), or season (warm, cool), or by temperature requirements, edible or usable parts, or other criteria.

Taxonomic ranking goes from general to very specific common characteristics. The most general category is Domain. The most specific is species.

Plants usually have two different names: common and scientific. The common name is usually in the language of the region where the plant is growing. The same plant can have many different common names. Likewise, different plants can have the same common name. To avoid confusion, every species of plant is now given a scientific name based on its taxonomic ranking. The scientific name of a species is composed of the genus and specific epithet. Assigning a plant a scientific name is a very formal process designed to ensure that each species has a name that is distinct from any other species. Most scientific names are in Latin although Greek and Arabic words are sometimes used.

EXERCISES FOR UNDERSTANDING

3–1. Using a seed catalog for home gardeners or professional growers, identify the advertised plants by their common and scientific names. Look for variety or cultivar names. Note how many varieties or cultivars exist for each species. Also look for the same common name appearing for different plants or different names with the same plant.

3–2. Obtain a taxonomic key for weeds or wildflowers growing in your state or region. Use the key to identify a weed or wildflower that you know. Then use it to identify one that you do not know. Note: sometimes weeds and wildflowers are the same species, it depends on the frame of reference.

REFERENCES AND SUPPLEMENTARY READINGS

BAILEY, L. H., E. Z. BAILEY, and staff of L. H. BAILEY HORTORIUM. 1976. *Hortus third.* New York: Macmillan.

BRICKELL, C. D., ed. 1980. International code of nomenclature for cultivated plants—1980. *Regnum Vegetabile* 104:7–32.

GRIFFITHS, M. 1994. *Index of garden plants.* Portland, Ore.: Timber Press.

LANJOUW, J., ed. 1966. *International code of botanical nomenclature.* Utrecht, Netherlands: International

Bureau for Plant Taxonomy and Nomenclature of the International Association for Plant Taxonomy.

LAWRENCE, G. H. M. 1951. *Taxonomy of vascular plants.* New York: Macmillan.

RAVEN, P. H., R. F. EVERT, and S. E. EICHHORN. 1999. *Biology of plants.* New York: Freeman.

WOODLAND, D. W. 2000. *Contemporary Plant Systematics,* Third Edition. Berrien Springs, Maryland: Andrews University Press.

CHAPTER

4

Origin, Domestication, and Improvement of Cultivated Plants

KEY LEARNING CONCEPTS

After reading the chapter you should be able to:

♦ Understand how several crops originated and where they were domesticated.
♦ Know how crops can be domesticated and improved.
♦ Understand the importance of saving germplasm from extinction and the government system to preserve germplasm.
♦ Understand the genetic principles of crop improvement and how those principles are used to improve crops.
♦ Recognize the potential of biotechnology and the social controversy surrounding some biotechnology techniques.

ORIGIN OF CULTIVATED PLANTS

Most of the crop plants important today were cultivated in a primitive way long before recorded history. Agriculture began some 10,000 years ago when ancient peoples selected certain plant types they found growing about them and thus enlarged their food sources beyond hunting and fishing. Most of these early food plants are still cultivated but in much improved forms. Others that were unknown to early humans have been added over the centuries to make up our present day selection of food plants. Then, much later, as people became more concerned about the aesthetics of their environment, they also added shade trees, shrubs, flowering plants, and lawn grasses to the list of cultivated plants.

Several botanists have brought together fascinating information on the subject of where and when the crops that now feed the world's peoples, their livestock, and other domestic animals originated. The Swiss botanist Alphonse DeCandolle wrote *Origin of Cultivated Plants*, first published in 1833. Later, the famous Russian plant geneticist Nikolai Vavilov also studied the origin of cultivated plants. The results of his studies were published in *The Origin, Variation, Immunity, and Breeding of Cultivated Plants,* translated from the Russian and published in English in 1951. Vavilov concluded from his studies that the various cultivated plants originated in eight independent centers: (1) central China, (2) India, (3) Indochina and the Malay Archipelago, (4) the Turkey-Iran region, (5) the Mediterranean area, (6) the Ethiopia-Somaliland area of east Africa, (7) Mexico and Central America, and (8) the Peru-Ecuador-Bolivia and the Brazil-Paraguay area of South America. Many archeological studies carried out in later years generally confirm Vavilov's locations for these centers.

More recent studies and theories on the origins and movements of the world's agricultural crops have been summarized by Carl Sauer in *Agricultural Origins and Dispersals*, by Jack Harlan in *Crops and Man*, by Jack Hawkes in *The Diversity of Crop Plants*, and by Barbara Bender in *Farming in Prehistory*.

Sauer proposed that the cradle of world agriculture was in Southeast Asia, in the area of present-day Thailand, some 10,000 years ago. This region at 10°N latitude probably would have had abundant rainfall and a warm to hot climate, which were good conditions for plant cultivation. He believed, too, that the founders of agriculture were a sedentary, well-fed people with some skills that would lead them to try new practices in growing plants. According to Sauer, agriculture must have started in woody areas with a diversity of plant materials rather than in large river valleys subject to periodic flooding. He proposed that a fair amount of propagation of plants in early agriculture was done by vegetative methods, which immediately fix superior plant forms, rather than by seed propagation where domestication might be slow.

Harlan disagrees with Vavilov's concept that agriculture originated in definite centers. Even if some centers are identified for selected crops, many plant species originated at the same time over wide geographical areas (see Fig. 4–1)— what Harlan calls "noncenters". Harlan concluded that these noncenters were based on three large independent systems: one in the Near East and Africa, one in China and Southeast Asia, and one in southern Mexico and South America.

Bender focused on the important period some 11,000 years ago when humans made the gradual transition from hunting and gathering to producing food, which profoundly changed human culture. Various ideas on the climatic and cultural factors that might have motivated man to "invent agriculture" are discussed by Harlan and Hawkes. Clearly, this invention marks a momentous step toward the evolution of early human societies.

DOMESTICATION OF PLANTS

Toward the end of the Ice Age, technically the Pleistocene epoch of the Cenozoic era, some 11,000 to 15,000 years ago, when the glaciers were in full retreat and early humans were wandering about the earth, the stage was being set for the initiation of food production (Table 4–1). Before this time, tool-using hunter-gatherers had been on earth for some 4 million years. Then—only about 400 generations ago—there was a gradual transition to food-producing activities. This transition phase suggests several interesting questions, such as why it took so long for humans to become food producers and why such activities began in various parts of the world— Southwest and Southeast Asia, middle America, and western South America—at about the same time.

There is definite evidence from archeological sites that agricultural villages existed about 8000 to 9000 B.C. in the area of Southwest Asia known as the Fertile Crescent. It extends from the alluvial plains of Mesopotamia (now Iraq) across Syria and down the eastern coast of the Mediterranean sea to the Nile Valley of Egypt. Radiocarbon dating suggests that plants and animals were being domesticated at several places in this area by at least 5000 to 6000 B.C. There is evidence from site excavations that einkorn wheat, emmer wheat, barley, lentil, chickpeas, oats, and vetch were being cultivated, as well as dates, grapes, olives, almonds, figs, and pomegranates. Excavations at Jarmo in Iraqi Kurdistan, at ancient Jericho in the Jordan Valley, and many other sites in the Fertile Crescent have supplied much of the evidence to support these conclusions.

An indigenous savanna type of agriculture apparently was developing from domesticated native plants about 4000 B.C. in a belt across central Africa (see Fig. 4–1). This area was the first home of the human race, as we understand it now. The genus *Homo* originated here, where most of human evolution subsequently occurred. Some important world crops brought under cultivation in this area include coffee, sorghum, millet, cowpeas, yams, and oil palm.

The Chinese center of agricultural origins became important about 4000 B.C. according to radiocarbon dating. Crops domesticated include millet, chestnuts, hazelnuts, peaches, apricots, mulberries, soybeans, and rice. The West did not learn of the all-important rice plant until about 350 B.C., and the peach was unknown outside China until about 200 A.D.

A farming culture in Southeast Asia and what is now Indonesia apparently had domesticated rice around 6000 B.C. Other important crops appearing later under cultivation in this area were sugar cane, coconut, banana, mango, citrus, breadfruit, yams, and taro.

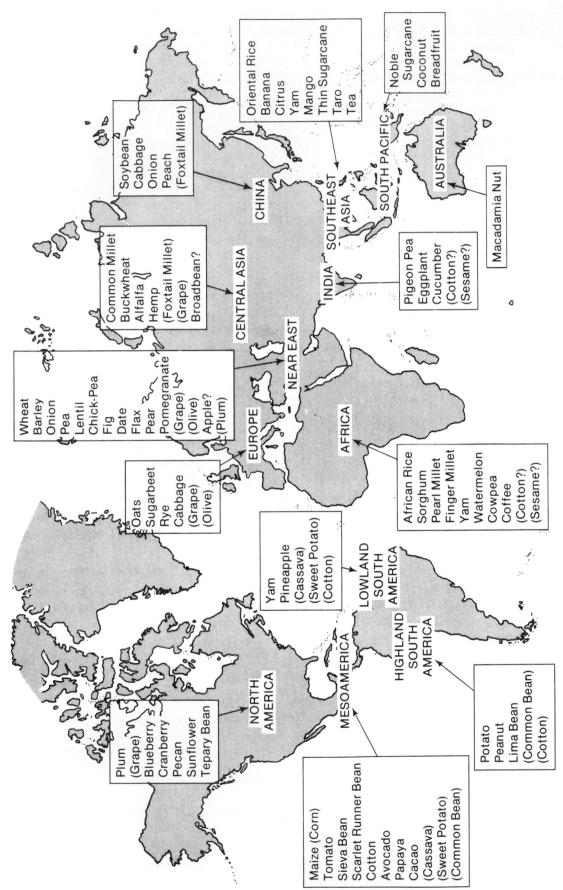

Figure 4–1 Regions of the world where major food crops were domesticated. Crops that apparently originated in several different areas are shown in parentheses. Question marks after the name indicate doubt about the location of origin. *Source:* Harlan, J. R. 1976. The plants and animals that nourish man. *Sci. Am.* 235(3):88–97. © September 1976 by Scientific American, Inc. All rights reserved.

50

TABLE 4–1 Beginnings of Agricultural Development in the Old and New World

The Postglacial Time Scale / Date	Climatic Changes in Northern Europe	Old World Cultural Stages	New World Cultural Stages
10000 B.C.	Last glacial stage (Würm-Wisconsin ice)	Late Paleolithic hunting cultures (Cro-Magnon, etc.)	
9000 B.C.	Retreat of the glaciers (Pre-boreal period, cold dry)	Mesolithic fishing, hunting, collecting cultures	
8000 B.C.			Hunting cultures established (Folsom Man, etc.)
7000 B.C.		Agriculural beginnings	
6000 B.C.			
5000 B.C.	Boreal period (Warm dry)		
	Atlantic period (warm moist)	Neolithic agriculture established and spreading	
4000 B.C.		Beginnings of civilization (Egyptian-Sumerian)	
3000 B.C.		Neolithic agriculture in northern Europe	
2000 B.C.			American agricultural beginnings
	Subboreal period (colder dry)	Babylonian Empire Invasions: Aryans to India; Medes and Persians to S.W. Asia and Mesopotamia	
1000 B.C.		Rise and flowering of Greek civilization	
B.C.–A.D.			Early Mexican and Mayan civilizations
	Sub-Atlantic period (cool moist)	Roman empire	Decline of Mayans
		Invasions: Goths, Huns	
		Rise of Islam	
1000 A.D.		Norsemen to America	
		Mongol and Tartar invasions	Aztecs and Incas
		Voyages of discovery and colonization by Europe	
		Industrial revolution and modern period	
2000 A.D.			

Source: Dasmann, R. F. 1976. *Environmental conservation.* 4th ed. New York: Wiley.

In the New World, evidence from archeological sites shows agricultural beginnings in two areas. One is present-day southern Mexico and Central America, where plant cultivation began about 5000 to 7000 B.C. The plants grown were early forms of maize (corn), sweet potato, tomato, cotton, pumpkin, peppers, squash, runner beans, papaya, avocado, and pineapple.

The second American region is a broad "noncenter" of agricultural origins stretching from Chile northward to the Atlantic Ocean and eastward into Brazil. There is evidence that both the snap and lima beans were cultivated here by about 6000 B.C. or earlier. Other important cultivated crops from this region are the potato, peanut, cacao, pineapple, cashew, papaya, avocado, Brazil nut, peppers, tobacco, guava, tomato, yam, manioc, and squash.

No major cultivated crop originated in the area of the present-day United States. Agriculture here relies in a large measure on introduced crops. There are, however, many minor native American fruit and nut crops, such as the American grapes and plums, the pecan, chestnut, hickory nut, hazelnut, black walnut, persimmon, blueberry, raspberry, blackberry, and cranberry. The sunflower (which has long been an important oil crop in the former Soviet Union and Eastern Europe) originated in the United States, as did hops, the tepary bean, Jerusalem artichoke, and some grasses. Many plants now used as ornamentals, mostly in improved forms, and many of the world's great timber tree species also originated in the United States.

The only food crop to originate in the entire continent of Australia was the macadamia, or Queensland nut.

Methods of Plant Domestication

When ancient peoples started to domesticate plants—that is, propagating and growing plants under their own control—they would have had the choice of vegetative propagation or seed propagation. While both methods have advantages and disadvantages in crop improvement, because of genetic segregation between generations the offspring from sexual propagation are often different (not true-to-type) from parents and desired traits may not be present in the offspring. For this reason, vegetative propagation is the more immediate and direct approach to selecting and maintaining superior plants.

Vegetative or Asexual Propagation Methods

It is probably no coincidence that some of the oldest cultivated woody plants are the easiest to propagate by such vegetative methods as hardwood cuttings. These plants include the grape, fig, olive, mulberry, pomegranate, and quince.

An observant person in prehistoric days might have noticed among a wild stand of seedling grapevines, for example, one single plant standing out from the others in its heavy production of large, high-quality fruits. From past ex-

perience, too, the person might have known that a piece of the grape cane stuck into the ground in early spring would take root and produce a new plant just like the original. The next logical step, of course, would be to break off and plant many cane pieces (cuttings) from this one very desirable grape plant, then to protect them against animals and the competition of weeds. This action would establish a vineyard based on cultivating a superior seedling and disregarding all seedlings with inferior characteristics.

Many of the tree fruit species—fig, almond, quince, apple, pear, cherry, pomegranate, and walnut—are native to areas of the Near East on the southern slopes of the Caucasus Mountains. Today, there are still forests in the area consisting entirely of seedling wild trees of these species, showing great variability in many characteristics. One may also see trees in which the local farmers have grafted the superior wild types onto inferior seedlings, making use of them only as rootstocks. Also, while clearing the brush and forests to make space for grain fields, the farmers have left superior individual wild apple, pear, and cherry seedlings in the fields for harvesting.

Many types of ancient plants brought under domestication by early people could have been—and probably were—easily maintained and increased by unsophisticated vegetative methods. Such plants would have included the potato and sweet potato, propagated by dividing the tubers; banana, bamboo, and ginger, propagated by cutting up rhizomes; pomegranate, quince, fig, olive, and grape, propagated by stem cuttings; filbert, propagated by layering; and pineapple and date, propagated by a form of suckering.

Seed or Sexual Propagation Methods

Domestication of seed-propagated plants, such as the cereal crops, by the ancient agriculturists probably began as the purposeful harvesting of wild grass seeds, some of which were sown to produce the next year's crop. This established two population types for the next harvest—one that was not harvested and reseeded itself, and one that was sown from the harvested seed. In cereals this procedure immediately starts to separate the "shattering" types (the seed separates from the head) from the "nonshattering" types. Most of the seeds that shatter fall to the ground, while nonshattering seeds are harvested and can be resown. Thus, seed with the desirable nonshattering trait is easily obtained without intentional selection. The mere practice of harvesting sets up a selection pressure for improved forms.

Planting the harvested seeds close together in a cultivated plot kept free of competing weeds automatically selects for the stronger, more vigorous plants. In some species, seedlings derived from the larger seeds—those having a considerable amount of stored foods—are likely to germinate and grow rapidly, thus suppressing seedlings arising from the smaller seeds with low vitality. The harvested crop would, therefore, come from the more vigorous plants producing the desirable large grains, and the plants producing the smaller, inferior grains would be eliminated. Thus seedling competition sets up an automatic selection pressure for an improved form.

By these procedures, early peoples unconsciously developed superior forms of their cereal crops. Other desirable characteristics arising from selection pressures were loss of seed dormancy, increased flower numbers and larger inflorescences, and a trend toward determinate rather than indeterminate growth.[1] These superior characteristics in the cultivated races over the wild types can usually be obtained with only small genetic differences between the two.

Plant size, productivity, seed output, and ecological adaptations are often complex quantitative traits and involve many small genetic changes. Their evolution is a slow, continuous process. Often some other characteristics, such as nonshattering seeds, resistance to natural enemies, or novel fruit or seed color and shape might be controlled by one or very few genes. Such traits can evolve rapidly by early man's observation and by harvesting separately.

EXAMPLES OF IMPROVEMENT IN SOME IMPORTANT CROP PLANTS[2]

The following examples illustrate patterns in plant improvement, beginning with: (1) the harvest of crops from wild plants by primitive humans, followed by (2) selection of superior types from prehistoric eras to the present time, and going on to (3) modern methods of plant breeding that can dramatically increase crop yield and quality by applying genetic principles and gene transfer to developing improved cultivars. It should be realized that the present, improved cultivars of such important crops as wheat, corn, rice, potato, sweet potato, and all fruit crops have been so adapted to conform to human cultural practices that they all now completely depend upon our care for their continued existence.

Grains and Vegetable Crops

Wheat (Triticum aestivum *and* T. turgidum Durum group)

Wheat is the most widely cultivated plant in the world today, the chief cereal, and is used worldwide for making bread. Present wheats evolved from wild wheatlike grass plants found and cultivated by ancient humans in the Near East region about 7000 B.C. Native wheatlike plants can still be found in this area. Even in early prehistoric times, wheat probably improved naturally by spontaneous hybridizations, by chromosome doubling, and by mutations to increase fer-

tility. Wheat species occur in a series with increasing chromosome numbers (polyploidy): First is the small primitive diploid einkorn wheat (7 pairs of chromosomes); second is the much larger tetraploid emmer wheat (14 pairs of chromosomes); and third is the hexaploid bread wheats (21 pairs of chromosomes), the ones grown today. Humans probably had no role in this early advancement of wheat. From the Near East these early wheat forms were taken into ancient Egypt, the Balkans, and central Europe. The Spanish brought wheat to the Americas; eventually, the United States, Canada, and Argentina became the world's largest wheat producers. A chance introduction of 'Turkey Red' wheat into central Kansas by a small group of Mennonite immigrants from Russia in 1873 established the basis for the tremendous hard red winter wheat industry of the central Great Plains area of the United States.

The two major wheat species today are *Triticum aestivum,* used for flour in making breads and pastries, and *T. turgidum* (Durum group), used for such products as macaroni, spaghetti, and noodles. In the twentieth century, plant breeders, through hybridization and selection, have produced perhaps a thousand cultivars of bread wheat alone designed for certain climates, for high productivity, special milling properties, and particularly for disease resistance. Wheat cultivars resistant to the devastating stem rust disease must continually be developed to cope with the mutating stem rust pathogen.

Today's plant breeders, too, have developed dwarf forms of wheat, which can produce high seed yields without falling over (lodging) when heavily fertilized. Some of these new dwarf cultivars were developed by Norman E. Borlaug, working at the Rockefeller Foundation's International Maize and Wheat Improvement Center in Mexico. Borlaug was awarded the Nobel Peace Prize in 1970 for his work. Certain of these cultivars have been so successful with fertilizers and irrigation in Pakistan and India that both countries have become exporters rather than importers of wheat, a result of the so-called "green revolution."

Wheat plants are self-pollinated, allowing farmers to save their seeds for future planting. F_1 hybrid wheat for increased plant vigor and yields has not been developed to the extent it has with corn, owing to difficulties arising from the wheat's flower structure. Wheat has perfect (bisexual) flowers, making cross-pollination difficult, whereas corn has the male and female flowers separate on the same plant and is cross-pollinated easily.

Corn (Zea mays)

Corn (or maize) originated in the New World about 5000 to 6000 B.C., but its earliest history is still a mystery. Maize is known only as a domesticated plant. There is no wild form except, apparently, teosinte, a close relative. The economic life of the ancient American civilizations—the Aztecs, Mayas, and Incas—depended on corn. At the time of Columbus' expeditions to the New World, corn was being grown by

---

[1]In determinate growth, shoot elongation stops when flowers form on the shoot terminals. In indeterminate growth, elongation continues after flowering.

[2]For a discussion of the genetic terminology used in this section, see the section "Some Basic Genetic Concepts in Plant Improvement" on page 60.

Figure 4–2 Improvement in corn (maize) from a primitive type (*left*) still growing on the eastern slope of the Andes mountains in South America, to modern hybrid corn (*right*). The primitive type corn contains a valuable trait—multiple-aleurone layers in the grain—that is being transferred by plant breeders to modern dent corn. This example illustrates the need for maintaining the germplasm of seemingly worthless plant types. *Source:* USDA.

types. The Indians of America must have made considerable conscious selection of corn for so many types to be introduced to European agriculture following the New World explorations.

The development of hybrid corn in the 1930s is one of the outstanding achievements of modern agriculture (see Fig. 5–1). High-yielding F_1 corn hybrids were developed for different climatic zones. In 1935, 1 percent of the corn planted in the United States was of the hybrid type. By 1970 virtually all corn produced in the United States was of the hybrid type. The planting of hybrid cultivars, along with fertilizers, irrigation, and mechanization, dramatically improved production. Corn is now one of the world's chief food crops for both humans and domesticated animals, but it would not survive in nature without man's attention and cultivation. The latest development in corn is high-oil corn.

Rice (*Oryza sativa*)

Rice is the basic food for more than half the world's population and one of the oldest cultivated crops. It is believed to have originated in Southeast Asia about 5000 years ago, or even earlier, and it spread to Europe and Japan by the second century B.C.

There are about 25 species of *Oryza*, but *O. perennis,* widely distributed throughout the tropics, is probably the one from which cultivated rice was developed. Primitive humans most likely collected seeds and cultivated the wild types. During cultivation of rice, mutations, plus hybridization with other *Oryza* species, probably occurred, leading to improved forms with larger grains and nonshattering fruit stalks. Rice was first cultivated in America along the coast of South Carolina about 1685. Four rice experiment stations were established by the U.S. government in the early 1900s. Breeding work at U.S. government rice experiment stations has resulted in many superior cultivars being introduced to the rice-growing areas of the United States.

Much of the rice harvested in Asia was from old, unproductive types grown under primitive conditions. The Ford and Rockefeller Foundations established the International Rice Research Institution (IRRI) in the Philippines in 1962, bringing together scientists from eastern and western nations for the purpose of improving rice culture. This group of researchers developed new high-yielding dwarf cultivars by hybridization that did not fall over (lodge) when heavily fertilized. These cultivars dramatically increased yields, and some are highly resistant to insects and disease pathogens native to the Far East, South America, and Africa.

Today, a genetically modified rice produces β-carotene in its kernels, giving the rice a yellow color. Adding β-carotene, a precursor for vitamin A (which is essential for human health), to rice increases the nutritional value of the rice significantly. It is predicted that the modified rice will help prevent blindness in millions of children in developing countries where rice is a dietary staple. However, with public sentiment being strongly against genetically modified

Indian tribes from Canada to Chile. Corn probably originated in several places in both Mexico and South America. An early form of corn is still growing in South America (see Fig. 4–2).

A number of hypotheses have been advanced to explain the origin of corn: (1) it developed from "pod corn," a type in which each individual kernel is enclosed in floral bracts—as in the other cereals; (2) it originated from teosinte, corn's closest relative, by gradual selection under the influence of harvesting by humans; (3) corn, teosinte, and *Tripsacum* (a perennial herb) descended along independent lines directly from a common ancestor; or (4) there is the tripartite theory that (a) cultivated corn originated from pod corn, (b) teosinte is a derivative of a hybrid of corn and *Tripsacum,* (c) the majority of modern corn cultivars are the product of an admixture with teosinte or *Tripsacum* or both.

Corn, even today, is an extremely variable species, from the color of the grains to the size and shape of the grains and ears (Fig. 4–2). Corn mutates easily, forming new

foodstuffs, it is not known if the rice will ever be available to these children.

Soybean (Glycine max)

Sometimes called the "Cinderella crop," the soybean has risen spectacularly to prominence in the United States in recent years with, for example, an increase in production from 135,000 metric tons (5 million bushels) in 1925 to about 75 million tons (3 billion bushels) in 2000. This great increase is due, in part, to the availability of more productive, disease-resistant cultivars. From about 1910 to 1950, large numbers of new strains and seed lots of soybeans were introduced into the United States from the Orient, its native home, largely by USDA plant explorers. From these diverse sources of germplasm, hybridization programs developed many superior cultivars. Much of this work was done at the USDA Regional Soybean Laboratory at Urbana, Illinois, in cooperation with various Midwest agricultural experiment stations. These cultivars give high yields, proper bean maturity for the particular area, strong erect plants that hold their seeds until harvest, and high disease resistance and bean quality. The soybean is particularly well adapted to the United States' Midwest "corn belt" and the southeastern states, which together account for more than 26 percent of the world's soybean production.

Sugar Beet (Beta vulgaris)

The modern sugar beet is a plant developed entirely by human efforts in plant breeding. It is our only major food crop that was not grown in some primitive form in ancient times. It was developed only about 250 years ago in Europe as a source of sugar to compete with the then very expensive cane sugar. The German chemist Andreas Marggraf found in 1747 that the kind of sugar in cultivated garden beets was the same chemically as that in cane sugar. Breeding and selection increased the sugar content in the root from about 2 percent to about 16 to 20 percent.

In the early part of the nineteenth century, Napoleon encouraged the development and production of the sugar beet industry in France to free that country from the British monopoly on cane sugar. By the end of the nineteenth century, sugar beets were being grown in North America, and they have now become an important temperate zone crop in many areas of the United States and southern Canada, and even more so in other parts of the world. All production phases are now completely mechanized, permitting the sugar beet to compete favorably with the tropical sugar cane plant. To maintain profitable production, however, plant breeders have had to continue to develop sugar beet cultivars resistant to virus and fungus diseases.

Potato (Solanum tuberosum)

Potatoes are one of the big four crops that feed the world's population. Wild potato species are widespread in South America, particularly in the Andes Mountain region. Potatoes were probably cultivated by primitive peoples in this area more than 4,000 years ago, but they became a more productive crop over the years with selection of superior types. The potato was cultivated over the length of the Andes Mountain area at the time the Spanish explorers arrived. They carried it back to the European continent about 1575 where today 90 percent of the world's potato crop is grown.

The potato plant produces pink, white, or blue flowers that develop into small green berries containing seeds. The seeds, when planted, produce new types of potato plants much different from the parent plant and from each other in many respects. In the early days of potato culture, the South American Indians undoubtedly selected superior plant types resulting from natural crosses. Once one single such superior plant was obtained, it could be perpetuated and increased in great numbers as a clone by tuber division. Some of these superior selections propagated by vegetative methods over the years became commercially successful cultivars. It was noted, however, that certain of these cultivars would degenerate after many generations of such asexual propagation and yield only weak unproductive plants. It was observed, too, that sowing seeds from such plants gave progeny plants with changed characteristics, including renewed vigor and productivity. It is now known that these clonal cultivars had become infected with viruses that passed along through the tubers to the new plants generation after generation. Such viruses did not pass through the seed to the new seedling plant, so that its growth was no longer inhibited by the virus.

In modern commercial potato growing, the planting stock ("seed" tubers) is produced under carefully observed conditions in regions where any viruses can readily be detected and the infected stock discarded. The grower who plants only "certified" stock can be reasonably sure that his fields will not be infected with viruses or other pathogens and will be true to type. In the United States, certified "seed" potatoes are produced by commercial growers in the northern states and strictly inspected by state government agencies. This is also true for many other, but not all, countries.

Many potato cultivars now being grown have been developed by plant breeders who have introduced superior characteristics into an existing cultivar. An example is resistance to *Phytophthora infestans* (late blight)—from *Solanum demissum*, a wild potato species, obtained from collections in Mexico. Many cultivars are being grown today for different purposes and for different climatic zones. New cultivars are constantly being introduced by plant breeders and older ones discarded.

Tomato (Lycopersicon esculentum)

The cultivated tomato originated from wild forms in the Peru-Ecuador-Bolivia area of the Andes Mountains in South America. Prehistoric Indians carried it to Central America and Mexico. Early explorers introduced the tomato to Europe about 1550, and it was brought back west, to the

Carolinas in North America, about 1710. Thomas Jefferson grew tomatoes on his plantation around 1780. In those days, most people considered the tomato poisonous, but in the United States it started gaining acceptance as a food plant about 1825. The early Indians undoubtedly improved the tomato by planting seeds taken only from the best fruits on the most productive plants. This selection process continued in the early days of tomato culture in Europe and the United States. Early cultivars introduced by U.S. seed companies were Stone and Globe in 1870. In the early 1900s, the USDA and the state experiment stations began breeding tomato cultivars to include specific characteristics. Marglobe was introduced by the USDA in 1925 and Rutgers by New Jersey in 1934. Some tomato cultivars, more recently introduced, carry resistance to fusarium wilt, verticillium wilt, and nematodes. Cultivars developed especially for machine harvesting have firm-fleshed fruits that all ripen at one time. Vigorous, highly productive F_1 hybrids, marketed both as seeds and as bedding plants, are recent developments by seed companies.

Fruit Crops

The major fruit crops are all heterozygous. They do not "come true" when propagated by seed, so vegetative propagation must be used to maintain an improved seedling selection. In ancient times, most kinds of fruit—other than those very easily propagated vegetatively, such as bananas, grapes, figs, pomegranates, and olives—were probably seed propagated and the variability in offspring accepted by farmers. Later, however, as more sophisticated methods of vegetative propagation were developed, such as budding and grafting, it would have been found that certain superior individual fruit plants could be maintained and increased by these methods. Nevertheless, considerable seed propagation of fruit species were undoubtedly practiced in the early days of fruit growing.

Apple (Malus pumila)

The early U.S. colonists planted many seedling apple trees, probably because it was easier to bring seeds taken from their favorite apple trees in their native homes rather than material for grafting. As a result, they continued to increase their apple orchards by planting seedlings. In these early days, much of the apple crop was preserved as cider, for which fruit from seedling trees was quite satisfactory. Certain individual seedling trees, no doubt, were much superior to the others and formed the starting point for vegetative propagation and the origin of the many hundreds of apple cultivars grown in the United States up to the early part of the twentieth century. These numbers dwindled, however, until the 1980s, when only 15 cultivars accounted for over 90 percent of the apples produced in the United States. Important cultivars are Delicious, Golden Delicious, McIntosh, Rome Beauty, Jonathan, Winesap, York, Stayman, Cortland, and Granny Smith. Al-

though apple breeding programs have been conducted by the USDA and by some state agricultural experiment stations, all the major apple cultivars now being grown originated as chance seedlings many years ago. For example, the 'McIntosh' apple was found growing as a seedling tree near Dundela, Ontario, Canada by John McIntosh in 1796. The 'Delicious' apple started as a single chance seedling near Peru, Iowa, about 1870. The 'Golden Delicious' also originated as a chance seedling in Cass County, West Virginia, about 1910. However, more recent cultivars released from breeding programs at the New York Agricultural Experiment Station—such as Empire, Macoun, and Jonagold—are starting to find grower and consumer acceptance.

Pear (Pyrus communis)

In pears, too, cultivars originating as chance seedlings have dominated the markets. The Bartlett (Williams Bon Chretien) originated in England as a chance seedling in 1796 and has been the world's leading pear cultivar ever since. Other leading pear cultivars in the United States, such as Beurre d'Anjou and Beurre Bosc, originated in Belgium as open-pollinated seedlings of cultivars then grown there. No pear cultivar from a controlled breeding program in the United States has become commercially important, although European peer breeders have developed several superior cultivars that produce well in France, Italy, and Belgium.

Peach and Nectarine (Prunis persica)

In contrast to the apples and pears, the important peach and nectarine cultivars grown today are the products of public and private plant breeding programs. Plant breeders have provided the consuming public with truly outstanding peaches and nectarines, in contrast to the small-fruited, nonproductive cultivars of earlier days that got their start as chance seedlings.

Strawberry (Fragaria × Ananassa)

Today's garden strawberry first originated in France about 1720 as a natural hybrid between two native American *Fragaria* species. From this and subsequent hybridizations, a number of cultivar selections were made and maintained vegetatively by runners in the early days of strawberry culture in Europe. However, many of these early cultivars were susceptible to viruses, verticillium wilt, and other diseases and were low in productivity, so that around 1945 the entire U.S. strawberry industry was falling into a precarious position.

Since World War II, a parade of new strawberry cultivars has been replacing older ones, coming from USDA and several state strawberry breeding programs and from similar programs in other countries. New cultivars have been developed for specific climatic regions and for such characteristics as adaptability of fruits to be used for freezing or for fresh shipping, resistance of plants to viruses and fungi and to winter cold, and fruit appearance and flavor.

PLANT IMPROVEMENT PROGRAMS

From the time of Neolithic man up to the early 1900s, sexual plant improvement mostly consisted of selecting seeds from those individual plants in a mixed population that had the desired characteristics. The seed was planted and from that population seeds were again taken from the most desirable plants, and so on. While improvements resulted, this method offered no way to transfer desirable characteristics from one line to another.

Evolution and Darwinism

Charles Darwin, the great English naturalist, working in the middle of the nineteenth century, provided scientific explanations of how evolution occurred, published in his monumental work (1859), *On the Origin of Species by Means of Natural Selection, or the Preservation of Favoured Races in the Struggle for Life*. Darwin's concept of evolution, generally accepted today, is:

1. Variation exists in an initial population of plants or animals.
2. Environmental stresses place certain individuals at an advantage.
3. Because certain individuals survive and reproduce more successfully, they leave more offspring, which then carry the same genetic traits.
4. The abundance of the advantageous traits increases in this way in every generation, but variation still persists.

In Darwin's time, nothing was scientifically known about heredity. Darwin, in his proposals, had great difficulty in accounting for a sufficient supply of variation. But the discoveries by the Austrian monk Gregor Mendel, in the 1860s, demonstrated the genetic mode of plant inheritance and developed the foundation for the science of genetics. His published works lay unappreciated in an obscure journal for 34 years, but in 1900 his papers were discovered independently by three European botanists. The discoveries of Mendel provided exactly the mechanisms needed by Darwin, and removed his difficulties in explaining variation. The integration of Darwinian selection and Mendelian genetics are now generally accepted as the proper explanation of evolution.

Many crop scientists have discovered patterns of evolution in various crop genera, primarily from two points of view: (1) relationships between crops and their related wild and weedy species, and (2) adaptive variation in geographical races. Plant breeders have made fascinating and useful applications of this knowledge in terms of collecting and utilizing many germplasm resources. In fact, all of the processes of evolution in native plants have direct analogues in breeding methods. For example, induced mutations and hybridization follow natural sources of novel variation; selection and cycles of recombination in hybrid materials provide the genetic changes in both natural and breeders' populations.

THE MECHANISM OF EVOLUTION

Evolution, as accepted today, can be explained as follows:

1. Genes, in the chromosomes, are largely responsible for the structure, metabolism, and development of plants and animals.
2. The complement of genes does not remain constant, since mutations occur which modify the metabolism and structure of the individuals that contain them.
3. Mutations, which may cause considerable change, may kill or greatly weaken the plant, since they can upset the equilibriums that exist between the plant and its environment.
4. Hybrids differ from their parents because of the resulting new combinations of genes.
5. If the variants, as developing from mutations or by hybridization, are better adapted to the existing environment than the parent plants, then the parent plants may be replaced by the new forms.
6. As the earth's surface and climate change, or as plants' habitat may be changed in other ways, those plants best adapted to the new environment replace those poorly adapted.
7. Evolution thus results from slow changes in the environment, variations occurring in plants and animals, and adjustments taking place between the changes in the environment and changes in the living organisms.

Based on Mendel's work and utilizing the expertise of trained geneticists, modern breeding programs have produced an array of new cultivars for many crops. These have been bred for such characteristics as resistance to disease, insects, and cold, and for productivity, flavor, and nutritive value. Today's breeding programs have been called "directed and accelerated evolution."

A product of the twentieth century, public and private plant breeding programs have had a tremendous beneficial impact on our food supply and range of ornamental plants. These programs have produced a great many new superior cultivars for almost all cereal, vegetable, forage, fruit, and ornamental crops. The USDA and most state agricultural experiment stations in the United States maintain such programs, often with a separate department of plant breeding. Most other countries also have plant breeding programs, often specializing in certain crops. Private plant breeding has mostly been done by seed companies producing new agronomic, vegetable, and flower cultivars.

Innovations by plant breeders include the development of F_1 hybrid corn and new vegetable and flower cultivars. Most of the new vegetable and flower lines are far superior to previous cultivars in vigor and in insect and disease resistance; the new vegetable cultivars are also superior in flavor, appearance, and productivity.

By developing plants that show strong resistance to insects and disease, plant breeders are reducing the need for insecticides and fungicides. This, in the long run, would be the best method of pest control. For example, potato cultivars have been developed that are resistant to the late blight disease (*Phytophthora infestans*) which was responsible for the nineteenth-century Irish potato famine. Other potato cultivars have been developed that are resistant to the golden nematode, permitting potatoes to be produced in soils infested by these worms without expensive soil fumigation. Wheat breeders must continually develop new wheat cultivars resistant to stem rust (*Puccinia graminis tritici*) because this fungus continually changes to attack formerly resistant cultivars.

Plant breeders have a useful procedure for obtaining improved plant forms by spontaneous or induced mutations resulting from chromosome or gene changes. These changes can be induced by chemical treatment with colchicine or by irradiation with gamma or x-rays.

In one instance, plant breeders have gone beyond just improving native plants. They have created a new man-made cereal, triticale, by hybridizing the ancient grains, wheat and rye. The name *triticale* derives from the generic names of these two grains: *Tritium* and *Secale*. The hybrid combines the high yield and protein content of wheat with the winter hardiness of rye. The triticale plant is disease resistant and thrives in some unfavorable soils and climates.

SEARCHING FOR AND MAINTAINING NEW GERMPLASM[3]

There is no assurance that we are at present cultivating all the useful food and ornamental plants in existence on earth, or that all the germplasm containing useful genes has been found. Plant explorers have long roamed the world and they continue their searches. They have found many plants that have subsequently made a major impact on the world's agriculture, often in different parts of the world from the plants' native regions. Accessible plants have always been moved about over the world by explorers on land and sea, armies, immigrants, and travelers. Plants moved into a new region often perform much better than they did in their original home. For example, the coffee plant (*Coffea arabica*) is native to the Ethiopian area of eastern Africa. It never developed into much of a crop there, but when moved to Brazil and Colombia in South America, the coffee tree prospered so well that these countries now produce the bulk of the world's coffee supply.

A number of prominent plant explorers have contributed much to the wealth of available plant materials.

Many of the early plant collecting trips were less than successful because the plants did not survive the long trip back. A London physician who was an amateur horticulturist, Nathanial Ward, invented the Wardian case early in the 1800s. The case was a small glass-enclosed box containing soil in the bottom. Plants kept in the Wardian case could survive long sea voyages, permitting the importation of species never before received alive. Large, magnificent, ornate glasshouses were built in England about this time to house the many tropical and subtropical plants brought back by the plant hunters.

Many plant explorers from the United States brought back plants that would later be the foundation for several crops grown in the United States. Colonel Agoston Haraszthy (wine grapes), N. E. Hansen (cold-resistant fruit and cereal plants), Mark Carleton (hard red winter wheat), David Fairchild, Frank Meyer, and others brought back many different species that have become the genetic basis for many of our edible and ornamental crops.

Plant collecting trips to the native home of certain desirable plant types are still being made by plant explorers. Plant explorers may be looking for an entirely new plant species to serve as a crop in a particular climatic region of their own country. Safflower came to the northern Great Plains of the United States this way, and it has proven most profitable. Plant explorers may also be searching for new germplasm for existing crops. Closely related types can be used by plant breeders to introduce, for example, genes for insect or disease resistance or improved vigor or quality into cultivars already being grown.

To eliminate duplication of effort among countries engaged in plant exploration and to maintain some record of what is being collected and where it is being maintained, the Food and Agriculture Organization of the United Nations (FAO) has set up a clearinghouse procedure.

Moving plant materials about the world can also introduce devastating insect or disease pests into a country where they have never before appeared. For example, the chestnut blight fungus (*Endothia parasitica*) was inadvertently introduced on imported plant material to the New York area from the Orient in the late 1800s. By about 1935 this fungus had practically eliminated all varieties of the beautiful native American chestnut trees (*Castanea dentata*) from the eastern United States. To guard against the introduction of such pests, most countries have set up elaborate inspection, fumigation, and quarantine procedures. Plant material can be introduced as seeds, bulbs or corms, rooted cuttings, or as scions or budwood for grafting or budding onto related growing plants. Seeds are the easiest to ship and pose the least danger of carrying pathogens. Rooted plant parts with soil particles around the roots are particularly suspect since the soil may contain nematodes or soil-borne diseases. Vegetative plant material coming into the United States is usually fumigated, then held under postentry quarantine for as long as two years before distribution to nurseries is permitted.

---

[3]The protoplasm of the sexual reproductive cells containing the units of heredity (chromosomes and genes).

Shipment of plant material is now easy and highly successful because of specialized packing materials, refrigeration, and frequent worldwide air flights. This is in contrast to the earlier days when only slow ship transport, often through hot tropical seas, was possible.

Preservation of Desirable Germplasm

A concerted worldwide effort is needed to ensure the survival of the earth's endangered plant species so as to preserve genetic diversity. There is this type of an international germplasm network, known as the Consultative Group on International Agricultural Research (CGIAR). This network interacts with, and is supplemented by, many national seed banks and agricultural centers. Such gene pools are needed for developing improved crops in the future, for introducing beneficial genes into existing crops from close relatives, for maintaining attractive plants and trees valued for their aesthetic purposes, and for keeping plants intact as part of ecosystems where their presence is necessary to the survival of other plant and animal species. The genetic diversity of plants, as well as animals and microbes, is of fundamental importance to our survival on earth. Food and other agricultural crops are derived from the genetic diversity of natural plant populations.

To help prevent eradication of many plant species, the U.S. Congress in 1973 passed the Endangered Species Act, which directed the Smithsonian Institution to prepare a list of endangered plant species and to recommend measures for saving them. As many as one in ten plant species is now extinct or endangered because of the encroachment of agricultural operations, the removal of rare plants by plant collectors, and the general destruction of vegetation from various causes. Such endangered germplasm can be saved and stored for future use as seeds or as living plants in special protected locations.

In the United States, the National Plant Germplasm System is already established and functioning. The system is a coordinated network of institutions, agencies, and research units working cooperatively to introduce, maintain, evaluate, catalog, and distribute all types of plant germplasm. Financial support comes from the USDA and from state agricultural experiment stations as well as from commercial plant breeding and seed trade organizations. The general mission of the system is to provide plant scientists with the germplasm needed to carry out their work, for example, in breeding new cultivars resistant to certain insects, diseases, smog, or high soil salinity.

The activities of the National Plant Germplasm System are:

1. **Introduction** of plant materials into useful scientific channels is done by planned foreign and domestic exploration trips, by exchanges with foreign agencies, or by traveling scientists. There may also be useful domestic germplasm that should be maintained—for example, mutations, species hybrids, or germplasm resistant to a certain insect or disease—which may be valuable for future crop development. These, along with introduced foreign material, are eligible to enter the National Germplasm System.

2. **Maintenance** of this potentially valuable germplasm for future research programs is the responsibility of the regional and interregional plant introduction stations, the National Seed Storage Laboratory, and the curators of collections of specific crops.

3. **Evaluation** of the plant genetic resources is done by initial screening and subsequent tests in the field, greenhouse, and laboratory by cooperating state, federal, and private scientists.

4. **Distribution** of plant germplasm is made free of charge to all qualified scientists and institutions requesting it, in sufficient amounts to enable them to initiate their research program.

A National Plant Germplasm Committee represents the elements in the System, advising on policy and coordinating activities to meet the immediate and long-term national goals of agriculture in the United States.

BROADENING THE BASE OF AGRICULTURAL PRODUCTION

The world's peoples are largely fed today by only about 20 crops. Reliance on so few crops could lead to a catastrophic famine if but a few of them were obliterated by insect or disease attacks or by climatic changes. The ravage of U.S. corn plantings by the corn blight disease in 1970 is an example of such a possibility.

In an attempt to broaden the base of agricultural plants in the tropics and to promote interest in neglected but seemingly useful tropical plants with economic potential, the U.S. National Academy of Science promoted a compilation of plants nominated by plant scientists around the world and published an account of 36 plants selected from the 400 proposed. Each plant was described along with its special requirements, research needs, selected readings, research contacts, and germplasm sources.

All 36 plants were thought to have considerable potential, but most have not been cultivated out of their own limited region of origin. Among the cereals, for example, the report cited an almost completely neglected grain species in the genus *Amaranthus,* native to Central America, that has very high levels of protein and the essential amino acid lysine, which is usually deficient in plant proteins. Among the vegetables studied, the wax gourd (*Benicasa hispida*) gives three crops a year of a large melonlike fruit that can be stored for 12 months without refrigeration. The mangosteen (*Garcinia mangostana*) is perhaps the world's best-tasting fruit, but it is little known outside the very humid tropics of Southeast Asia.

Since 1990 crop scientists have met regularly in the United States to share information on new crops and the potential for new crop development. These scientists are looking for solutions to global problems such as hunger, malnutrition, deforestation, desertization, and agricultural sustainability.

Other crops, native to the Americas, which have undeveloped but great potential are jojoba (*Simmondsia chinensis*), guayule (*Parthenium argentatum*), groundnut (*Apios americana*), and the tepary bean (*Phaseolus acutifolius*), as well as *Cuphea* species, whose seeds contain valuable oils. The leguminous tree *Leucaena leucocephala,* native to Central America, already has its own success story. It is fast growing, with high-protein leaves, and is used in India for feeding dairy cattle. In Africa, the leaves are used as a much-needed fertilizer, while in Sumatra and the Philippines it is used as a commercial regenerative timber source.

A number of exotic tropical fruits are found in America's supermarkets, many being grown in southern Florida, but others arriving by plane from various tropical regions.

SOME BASIC GENETIC CONCEPTS IN PLANT IMPROVEMENT

To understand why the characteristics of parent plants do or do not appear in the seedling offspring and to see how genetic manipulation can improve plants, you need some knowledge of genetic processes and the structures controlling inheritance in plants.

Chromosomes

Some plants consist entirely of living cells, whereas others, such as woody perennials, are made up of both living and dead cells. The living cell consists of a **cell wall** containing a fluid, the **cytoplasm,** in which the **nucleus** is suspended, along with many other structures. The nucleus contains the **chromosomes,** which carry most of the genetic information and transmit that information from one generation of cells to the next. The number of chromosomes is the same in all vegetative cells of an entire plant species. This is usually the **2n,** or **diploid,** chromosome number (although higher numbers—tetraploid or octaploid, for example—also occur). In the sex cells—the egg and sperm—the number is reduced by half and is termed the **haploid,** or **1n,** chromosome number.

Although usually constant for a given species, the number, size, and appearance of chromosomes vary considerably between different plant species. The chromosomes can be counted by microscopic examination of the nucleus at a stage just before cell division. The chromosome numbers are known for most plant species. For example, the diploid chromosome number for alfalfa (*Medicago sativa*) is 32; for barley (*Hordeum vulgare*), 14; for corn (*Zea mays*), 20; and for sugar beet (*Beta vulgaris*), 18.

Chromosomes change in appearance during cellular development and division, but basically each is a long, thread-like structure consisting of **deoxyribonucleic acid (DNA)** plus associated **ribonucleic acids (RNA)** and various proteins. DNA can replicate itself and it can transmit information to other parts of the cell. DNA is a polymer—a very large molecule made up of many repeating units, but the repeating units called **nucleotides** can vary. DNA, the active genetic material, is a very large molecule composed of two spiral strands (see Fig. 4–3). The "backbone" of these strands is composed of **sugar residues** (S) linked by **phosphates** (P) on each side. A sugar residue on one strand is connected with a sugar residue on the other by two **bases** that are linked to each other by **hydrogen bonds** (Fig. 4–3). These bases are **cytosine** (C), **guanine** (G), **adenine** (A), and **thymine** (T). The sugar residues in each strand are held tightly together by the phosphate radicals, but the two spiral strands are bound together more loosely by the hydrogen bonds.

One of the characteristics of DNA that makes it possible for the chromosomes to transmit genetic information from one cell generation to the next is its ability to replicate itself. This ability arises from the double-strand structure of the molecule and the properties of the four bases. Adenine and thymine are held together by two hydrogen bonds (A = T). Their molecular structure is such that only they can join together. In the same manner, only cytosine and guanine can join together, and they are held by three hydrogen bonds (C = G). When the chromosome divides during cell division, the two spiral strands of the DNA molecule unravel and separate at the position of these hydrogen bonds. Every base (cytosine, guanine, adenine, or thymine) attached to each strand attracts its complementary base, so that each single strand immediately becomes a new double strand exactly the same as the original double strand (see Fig. 4–3).

Although DNA includes the self-perpetuating **genetic code,** a related substance—ribonucleic acid (RNA)—actually controls the growth processes in the cell. DNA and RNA have certain differences. DNA is double-stranded whereas RNA is a single strand. RNA sugars have one more oxygen atom than DNA sugars. RNA has uracil (U) as a base in place of thymine (T). The DNA molecule acts as a template from which a complementary strand of RNA is formed (in the same manner by which a complementary strand of DNA forms from a single strand).

The form of RNA that carries the genetic instructions as a complementary copy of the DNA series of bases is called **messenger RNA.** Still another form of RNA—**transfer RNA**—brings the amino acids to the ribosomes to construct the proteins.

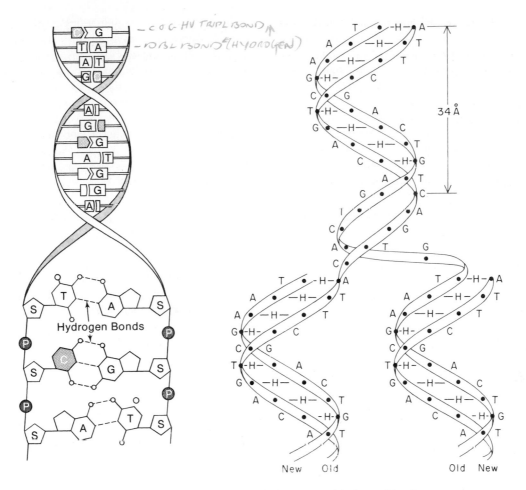

Figure 4–3   *Left.* Schematic diagram of the DNA molecule showing the helical structure and the base pairing. Lower detailed view shows the alternate attachment of sugars (S) and phosphates (P) on each strand. Connecting the two strands are the base pairs adenine (A) –thymine (T) guanine (G) –cytosine (C), which are held together by hydrogen bonds and attached to the sugars on each strand. *Right:* During cell division and chromosome splitting, the DNA molecule replicates itself by unraveling, separating at the hydrogen bonds. Then each strand quickly becomes a new double strand just like the original, with each base attracting to itself its complementary base (A = T and G = C). *Source:* (left) from McElroy and Swanson. 1976. *Modern cell biology.* 2nd ed. Englewood Cliffs, N.J.: Prentice-Hall. (right) Wright, J. W. 1976. *Introduction to forest genetics.* New York: Academic Press.

Genes

Structurally, a gene[4] is a sequence of triplet organic bases (cytosine, guanine, adenine, and thymine) along a DNA molecule. The gene is the ultimate hereditary unit that functions as a certain part of a chromosome determining the development of a particular characteristic in an organism. Thus, one gene (or several interacting genes) may determine plant height, leaf shape, flower color, or fruit size. Genes are much too small to be seen, even with an electron microscope. There are, of course, a great many genes in each cell of the higher plants; they number in the thousands. Any individual

gene may have a large effect or a small effect. Some genes act independently, whereas others act only in conjunction with other genes.

Since genes are arranged along the chromosome, genes on the same chromosome are **linked**—that is, genes on the same chromosome move from one cell generation to the next as a unit. Linkage is not perfect, however; sometimes during meiosis (reduction cell division), chromosomes break and exchange parts, as we will soon study in more detail.

Homologous Chromosomes

In each vegetative cell there are pairs of each individual chromosome. These are called **homologous chromosomes** and are properly defined as such if they each have the same gene or genes affecting the same traits at corresponding

---

[4]Virus particles are similar to genes. Since viruses can be isolated and easily studied, much knowledge of gene structure is obtained from the study of viruses.

positions. Genes are termed **alleles** to each other if they occupy the same position on homologous chromosomes and affect the same trait. If a plant had the genetic constitution ABCDEFG/abcdefg on a certain pair of homologous chromosomes, the genes A and a would be alleles, B and b would be alleles, and so forth.

Allelic genes can be dominant or recessive to each other. A **dominant gene,** A, is one that causes a certain characteristic to be expressed whether the plant is **homozygous,** AA—both alleles the same—or **heterozygous,** Aa—the two alleles different. A **recessive gene** causes the character it controls to be expressed only if both alleles are recessive, aa. (By convention, capital letters express dominant genes, while lowercase express recessive genes.)

Mitosis

Cell division in the shoot and root tips, axillary buds, leaf primordia, and the vascular cambium—all of which increases plant size—is called **mitosis.** These vegetative cells usually contain two sets of homologous chromosomes—the 2n or diploid number. During cell division, the chromosomes split longitudinally, replicating to produce two chromosomes that are identical to each other. One of each pair goes to one daughter cell, and one to the other. An equatorial plate, as it is called, develops between them to form a new cell wall and thus two new cells, each with its full complement of chromosomes—and genes. So each daughter cell has a genotype identical to that of the mother cell.

Meiosis and Fertilization

Meiosis refers to the type of cell division, sometimes called **reduction division,** that occurs in the flower to form—in the angiosperms—the cells from which the pollen grains and the embryo sac (which contains the egg) develop (see Figs. 4–4 and 4–5). In this type of cell division, the homologous chromosomes separate from each other without replicating, one going to one daughter cell and one to the other, thus reducing the number of chromosomes—the 1n or haploid number. During pairing of the two sets of homologous chromosomes, **crossing over** can occur. The chromosomes may break at the same locus on each so that they may rejoin after exchanging segments. Thus, if the original gene sequence on the homologous chromosomes was ABCDEFGHI/abcdefghi and crossing over occurred between E and F, then the gene sequence for this particular chromosome appearing in the pollen grain or egg cell would be ABCDEfghi or abcdeFGHI. This gene alteration would, of course, be expressed in altered characteristics of the new plants.

In **fertilization** in angiosperms, one male gamete (1n) from the pollen grain unites with a female gamete—the egg (1n)—to form the zygote (2n), which develops into the embryo and finally the new plant (see Figs. 4–4, 4–5, and 4–6). The manner in which these gametes can segregate and re-

unite to form new combinations is illustrated for both a monohybrid and a dihybrid cross (Fig. 4–6). Also during fertilization in the angiosperms, one male gamete (1n) unites with the two polar nuclei (1n each) in the embryo sac to form a food storage tissue, the endosperm (3n), that can serve as a nutritive material for the developing embryo. The seed coats, developing by mitosis from the female parent, cover both the endosperm and the embryo. In gymnosperms, the endosperm tissue—more correctly termed the female gametophyte—is 1n tissue, rather than 3n as in the angiosperms (see Fig. 4–5).

Mutations

As plants grow, hundreds of thousands of cells may be dividing constantly. At each cell division, DNA must replicate itself—the double strand must untwist, the millions of nucleotide triplets must be reproduced exactly on new single strands, and the single strands must then twist around each other to form new double strands. It is a marvel that this complicated process takes place on such a grand scale. With so many cells involved, however, errors can and do occur during replication. When they do they are called **mutations,** and the altered genes may possibly result in changes in the characteristics of the plant—although most mutations have such slight effects that they go unnoticed. The great majority of mutations are deleterious, but some are not and they provide a source of variability that aids the plant breeder in developing new cultivars. Mutation rates can be vastly increased by treatment with ionizing radiation and by certain chemicals. Mutations that occur during the formation of pollen grains and egg cells and appear in the plant's seedling offspring are particularly important to the plant breeder.

In addition to gene mutations, hereditary modifications can also be caused by gross changes in chromosome number or structure. This can involve the doubling of chromosome numbers, the addition or subtraction of an entire chromosome, or some other structural change in a chromosome. Such gross chromosome changes may cause pronounced changes in the plant's characteristics.

Polyploidy

Polyploidy is a condition in which individual plants have more than two sets of homologous chromosomes in their somatic (vegetative) cells. Beyond the normal diploid (2n) number, plants may be triploid (3n), tetraploid (4n), pentaploid (5n), hexaploid (6n), and so forth. Polyploid plants may arise by duplication of the chromosome sets from a single species—**autoploidy**—or by a combination of chromosome sets from two or more species—**alloploidy.** The latter is the more common type of polyploidy in nature. Many of the cultivated crop species evolved in nature as polyploids, as shown in Table 4–2 for oats, wheat, and tobacco.

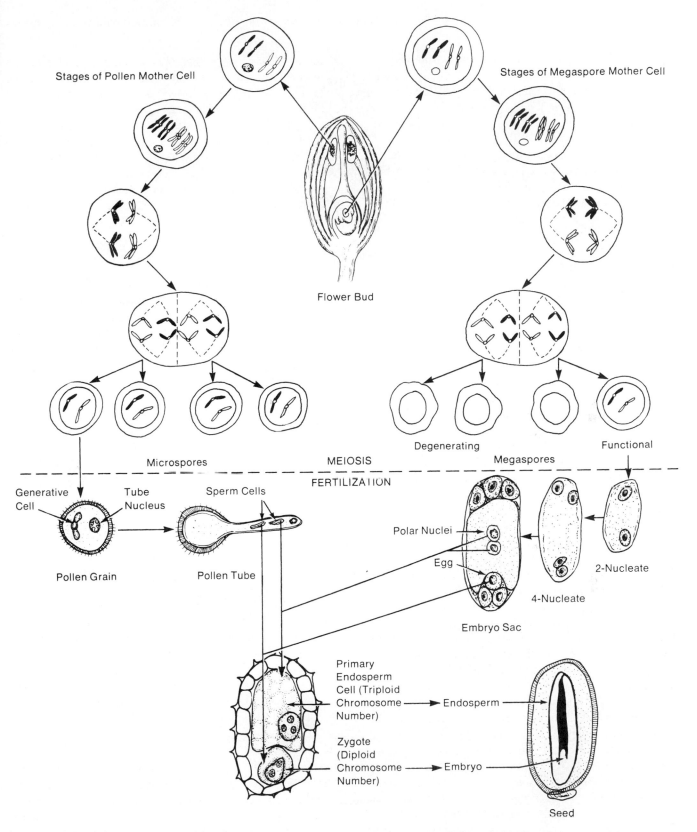

Stages of Pollen Mother Cell

Stages of Megaspore Mother Cell

Flower Bud

Microspores MEIOSIS Degenerating Megaspores Functional

FERTILIZATION

Generative Cell Tube Nucleus Sperm Cells

Polar Nuclei

Egg

2-Nucleate

Pollen Grain Pollen Tube

4-Nucleate

Embryo Sac

Primary Endosperm Cell (Triploid Chromosome Number) ——→ Endosperm

Zygote (Diploid Chromosome Number) ——→ Embryo

Seed

Figure 4–4 Diagrammatical representation of the sexual cycle in angiosperms. Meiosis occurs in the flower bud in the anther (male) and the pistil (female) during the bud stage. During this process the pollen mother cells and the megaspore mother cells, both diploid, undergo a reduction division in which homologous chromosomes segregate to different cells. This is immediately followed by a mitotic division which produces four daughter cells, each with half the chromosomes of the mother cells. In fertilization a male gamete unites with the egg to produce a zygote, in which the diploid chromosome number is restored. A second male gamete unites with the polar nuclei to produce the endosperm. *Source:* Hartmann, H. T. and D. E. Kester. 1983. *Plant propagation.* 4th ed. Englewood Cliffs, N.J.: Prentice-Hall.

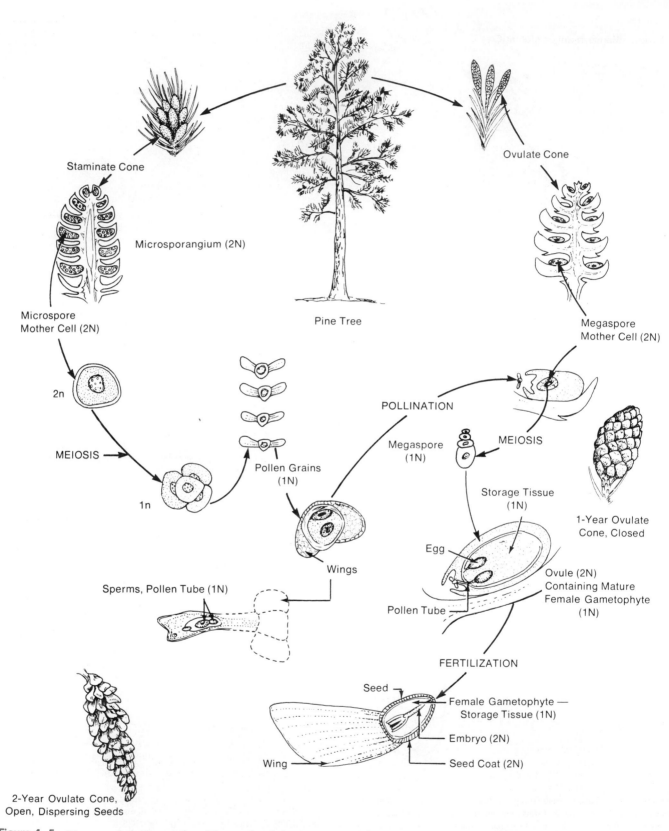

Staminate Cone

Microsporangium (2N)

Microspore
Mother Cell (2N)

2n

MEIOSIS

1n

Pollen Grains
(1N)

Pine Tree

Ovulate Cone

Megaspore
Mother Cell (2N)

POLLINATION

Megaspore
(1N)

MEIOSIS

Storage Tissue
(1N)

1-Year Ovulate
Cone, Closed

Egg

Wings

Sperms, Pollen Tube (1N)

Pollen Tube

Ovule (2N)
Containing Mature
Female Gametophyte
(1N)

FERTILIZATION

Seed

Female Gametophyte —
Storage Tissue (1N)

Embryo (2N)

Wing

Seed Coat (2N)

2-Year Ovulate Cone,
Open, Dispersing Seeds

Figure 4–5 Diagrammatical representation of the sexual cycle in a gymnosperm (pine), showing meiosis and fertilization.
Source: Hartmann, H. T. and D. E. Kester. 1983. *Plant propagation.* 4th ed. Englewood Cliffs, N.J.: Prentice-Hall.

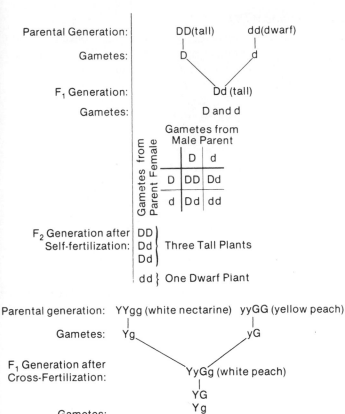

TABLE 4–2	Polyploidism in Oats, Wheat, and Tobacco	
Common Name	**Species**	**Somatic Chromosome Number**
Sand oats	*Avena strigosa*	14
Slender wild oats	*Avena barbata*	28
Cultivated oats	*Avena sativa*	42
Einkorn	*Triticum monococcum*	14
Emmer	*Triticum dicoccum*	28
Common wheat	*Triticum aestivum*	42
Wild tobacco	*Nicotiana sylvestris*	24
Cultivated tobacco	*Nicotiana tabacum*	48

Cytoplasmic Inheritance

While most inherited characteristics are transmitted by the genes in the nucleus, certain characteristics in some herbaceous plants can be controlled by cytoplasmic factors, which are contributed only by the female parent. Male sterility in corn is due, in part, to a cytoplasmic factor and is used to produce hybrid seed without the laborious procedure of hand detasseling.

Genotype and Phenotype

The term **genotype** refers to the genetic makeup of the plant—its genetic constitution—the kinds of genes it has on the chromosomes and the order in which they are situated. **Phenotype** refers to the plant's appearance, behavior, and chemical and physical properties. The phenotype expressed is also influenced by the environment. A plant may have genes for very vigorous growth but when grown under a deficiency of soil nitrogen, for example, its inherent vigor is not expressed. However, the genes still control the plant's characteristics even if an environmental factor drastically alters the phenotype. Therefore, if seeds are taken from stunted plants grown under deficient soil nitrogen conditions and planted in rich fertile soil, the plants again show the strong, vigorous growth called for by their genotype.

BIOTECHNOLOGY

Biotechnology, in a broad sense, can be defined as "the management of biological systems for the benefit of humanity." By this definition, it is obvious that farmers have been practicing biotechnology for many years in the production of their crops. But the word biotechnology, as generally used in the late twentieth century, means the *new* biotechnology, developed by means of molecular biology and molecular genetics using such advanced techniques as protoplast, cell, and tissue cultures, somatic hybridization, embryo manipulation, and such recombinant DNA techniques as gene splicing, replication, and transfer to other organisms.

Figure 4–6 *Above:* Inheritance involving a single gene pair in a monohybrid cross (one characteristic involved). In garden peas, tallness (D) is dominant over dwarfness (d). A tall pea plant is either homozygous (DD) or heterozygous (Dd). Segregation occurs in the F_2 generation to produce three genotypes (DD, Dd, dd) and two phenotypes (tall and dwarf). *Below:* Inheritance in a dihybrid cross (two characteristics involved) involving peach (*Prunus persica*). Fuzzy skin (G) of a peach is dominant over the glabrous (i.e., smooth) skin of the nectarine (g). White flesh color (Y) is dominant over yellow flesh color (y). In the example shown, the phenotype of the F_1 generation is different from either parent. Segregation in the F_2 generation produces nine genotypes and four phenotypes. *Source:* Hartmann, H. T. and D. E. Kester. 1983. *Plant propagation.* 4th ed. Englewood Cliffs, N.J.: Prentice-Hall.

Micropropagation

For many years propagation of both herbaceous and woody plants has been done by using tiny explants under sterile conditions. The medium is controlled so as to induce shoot development on the explants by the use of cytokinins; then roots are induced to form on the new shoots by changing the medium to include auxins. One of the early examples of micropropagation is the mass asexual reproduction of orchids by the use of protocorms.

Genetic Engineering

Genetic engineering to produce plants designed at the gene level is an interdisciplinary effort by cell biologists, biochemists, and geneticists. The techniques of *in vitro* cell and tissue culture are well developed, and they offer great promise for crop improvement. These techniques are already being used to free plants of viruses by shoot tip cultures, to propagate plants rapidly by inducing shoot and root formation on tiny tissue explants, to produce plants genetically altered so as to be resistant to damage by herbicides, using somaclonal variation[5] to improve soluble solids in tomatoes, to produce high-quality disease-tested potato cultivars, and to develop synthetic seed (somatic embryos encapsulated in a hydrated polymer gel).

Genetic engineering in plants should be viewed as a technique to complement rather than replace traditional plant breeding procedures. The patent protection offered living organisms developed by genetic engineering procedures due to U.S. Government decisions in 1980 and in 1985 has encouraged the establishment of many biotechnology companies with the promise of high monetary rewards.

[5]Somaclonal variation is existing genetic variation which may not be seen until after plant cells have been through aseptic culture—or the culture may force the change.

Recombinant DNA

The techniques involved, gene splicing, replication, and transfers to other organisms, were basically developed in the 1970s, although it was realized that many more refinements would be made in the future. Genes can be cut out of one organism with a restriction enzyme that behaves like a scalpel. Such genes can then be inserted into a cell or cells of the plant that is to be genetically altered. These new cells develop into new plants. Plants and other organisms that have been genetically engineered to carry the DNA of other species are known as genetically modified organisms (GMOs). Genetically modified plants have been developed with pest or disease resistance to reduce the use of chemicals in the field or with herbicide resistance to allow more environmentally sound tillage practices. Other GMOs have increased nutritional value. However, at the time this book is being revised, public sentiment in some areas of the world is strongly against the production and utilization of GMOs. The power of public perception and pressure is forcing food-processing companies to eliminate the use of GMOs in many of their products. Farmers are also reducing the percentage of GMO crops they grow. The agricultural community is facing the difficult task of handling strong public sentiment that is often emotionally, but not necessarily scientifically, driven.

Protoplast Fusion

By the mid-1970s, work in various cell biology laboratories showed that new or improved plants could be developed without hybridization through the mechanisms of the flower. Fusing protoplasts from cells of different plants (cell hybridization) under aseptic conditions in a test tube may develop entirely new plants with new combinations of chromosomes. An example of this might be the fusion of barley and soybean protoplasts to produce a leguminous grain crop that could supply its own nitrogen, extracted from the air, by the bacteria-containing nodules on the roots.

SUMMARY AND REVIEW

Agriculture began approximately 10,000 years ago when primitive humans began using certain plant types growing in the regions they inhabited for food. Several theories exist on the specifics of how and where plants came to be cultivated, but in general everyone agrees that certain regions gave rise to crops that developed from native plants in those regions. Most major geographic land areas of the globe have crops that can be traced back to that area. Sometimes the same crop developed in more than one area. Although few major agricultural crops originated in the United States, several minor ones can be traced to the United States.

Crops are domesticated by selecting and propagating plants with superior characteristics. The propagation can be by sexual (seed) or asexual (vegetative) means. In many cases, vegetative or asexual propagation allows desirable traits to be easily passed from one generation to the next. Care must be taken to ensure that over time the daughter material has not mutated or become infected with pathogens. In some cases asexual propagation is not practical, as with most agronomic and forage crops. Therefore, propagation by seed is necessary. As long as the seeds produce plants that maintain the desired characteristics of the parent(s) (true-to-type), the seeds can be saved from one crop to produce the next. When seeds do not produce true-to-type, then seeds that are the result of specific parental crosses must be obtained for each crop. Crop improvement has progressed from its early stages when superior plants were merely selected by a farmer and the seeds of those plants used for the next crop to modern-day methods that involve very rigid and competitive breeding and biotechnology programs throughout the world.

When plants are domesticated, the result is often a loss of genetic traits that, at the time of original domestication, were not considered important or may not have been noticed. Many times these traits are what allowed the plant to survive in the wild. With time, quite often these traits do become important but are lost because the alleles that carried those traits are no longer found in domesticated cultivars. Recently humans have come to understand the value of genetic diversity and the importance of the wild ancestors of many of our crops. Plant collectors today often search for the ancestors of domesticated crops as well as for new species that have crop potential. To facilitate the collection and preservation of germplasm, centers around the world have been established, including the United States National Germplasm System.

The basis for both natural selection and directed plant breeding is founded on basic cell biology and Mendel's principles of heredity. DNA carries the genetic code for each plant. During meiosis, DNA segregates and reproductive cells may differ from each other and the parent from which they generate. When combined to make a new plant, the recombination of genes can result in offspring different from either parent. The offspring may have traits that are similar, inferior, or superior to the parent. In natural selection the pressures of the environment will determine if the traits of the offspring will favor its survival or not. In directed plant breeding programs, humans make the determination of trait favorability, which has more to do with appearance or flavor than survival. As a result, many crops are no longer able to survive without the aid of intense cultivation practices. Hence, the need to locate and reintroduce survival traits is sometimes necessary.

At the start of the new millennium, the potential for crop improvement through the use of biotechnology and genetic engineering is inestimable. No longer is the genetic information carried in a plant limited to what can be obtained through compatible sexual crossing. The ability to impart resistance, productivity, or nutrition in plants appears to be limitless from a scientific basis. However, today's public sentiment does not support genetically modified organisms (GMOs). The agricultural community is being forced to find ways to show the public that genetic engineering is scientifically sound and environmentally safe.

EXERCISES FOR UNDERSTANDING

4–1. Choose your favorite food or flower and trace it back to its place of origin. If possible, find pictures or samples of its wild ancestors to see what similarities still exist.

4–2. In seed catalogs, locate those varieties or cultivars that are referred to as heirlooms or have similar designations. These designations indicate varieties or cultivars that have been around for a long time. Find out what characteristics make them more desirable than more modern cultivars, which in many cases may produce a better plant or fruit than the heirloom.

4–3. Check the daily newspapers or weekly magazines for articles on public reaction to GMOs. Determine how much scientific data are presented by both those attacking and those defending the use of GMOs. Evaluate the validity of the data presented.

REFERENCES AND SUPPLEMENTARY READING

BAKER, H. G. 1978. *Plants and civilization.* Third Edition. Belmont, Calif.: Wadsworth.

BENDER, B. 1975. *Farming in prehistory.* London: John Baker.

DeCANDOLLE, A. 1886. *Origin of cultivated plants.* Second Edition. Reprinted 1959. New York: Hafner.

DUNN, L. C. 1965. Mendel, his work and his place in history. *Proc. Amer. Phil. Soc.* 109:189–98.

FAIRCHILD, D. 1938. *The world was my garden.* New York: Scribner's.

HARLAN, J. R. 1971. Agricultural origins: centers and noncenters. *Science* 174:468–74.

HARLAN, J. R. 1975. *Crops and man.* Madison, Wisc.: American Society of Agronomy and Crop Science Society of America.

HARLAN, J. R. 1975. Our vanishing genetic resources. In P. H. Abelson, ed., *Food, politics, economics, nutrition, research.* Washington, D.C.: American Association for the Advancement of Science.

HAWKES, J. G. 1983. *The diversity of crop plants.* Cambridge: Harvard University Press.

JANICK, J., ed. 1996. *Progress in new crops.* Alexandria, Va.: American Society of Horticultural Science Press.

KOHN, D., ed. 1986. *The Darwinian heritage.* Lawrenceville, N.J.: Princeton Univ. Press.

LEMMON, K. 1969. *The golden age of plant hunters.* Cranbury, N.J.: A. S. Barnes.

MENDEL, GREGOR. 1866. Versuche über Pflanzenhybriden (Experiments with plant hybrids). *Trans. Brünn Soc. for the Study of Natural Science.*

SAUER, C. O. 1969. *Agricultural origins and dispersals.* Cambridge, Mass.: M.I.T. Press.

UCKO, P. J., and G. W. DIMBLEBY, eds. 1969. *The domestication and exploitation of plants and animals.* Chicago: Aldine.

VIETMEYER, N. 1981. Rediscovering America's forgotten crops. *Nat. Geo. Mag.* 159(5):702–12.

CHAPTER

5

Propagation of Plants

KEY LEARNING CONCEPTS

After reading the chapter you should be able to:

♦ Recognize several techniques used to propagate plants.
♦ Understand the principles of seed production, testing, and germinating.
♦ Understand the principles of the asexual propagation techniques of cutting, grafting, layering, micropropagation (tissue culture), and others.

Plants are propagated by either sexual (seed) or asexual (vegetative) methods. Some kinds of plants are almost always propagated by one method or the other; other kinds can be propagated successfully either way. The various ways of propagating plants are outlined in Table 5–1.

CHOICE OF PROPAGATION METHODS

In propagating plants we are often interested in starting with one or a few plants and producing many more—maybe hundreds of thousands more—all just like the original. A successful propagation method is one that will transmit all the desirable characteristics of the original plant to all the progeny. If the characteristics of the original plant are lost or changed during the propagation procedures, that particular method is unsatisfactory for that type of plant. It is then necessary to use another propagation procedure that will preserve these characteristics.

TABLE 5–1	Methods of Propagating Plants, with Typical Examples

I. Sexual
   A. Propagation by seed
II. Asexual (vegetative)
   A. Apomictic embryos—citrus mango
   B. Cuttings
   C. Grafting
   D. Budding
   E. Layering
   F. Runners—strawberry
   G. Suckers—red raspberry, blackberry
   H. Separation
   I. Division
      1. Stem tubers—white potato
      2. Tuberous roots—sweet potato, dahlia
      3. Rhizomes—iris, canna
   J. Micropropagation

It is important to realize that the kinds of agricultural and ornamental plants being grown today are ones having particularly desirable characteristics. As Chapter 4 pointed out, such plants originated from a plant or plants found growing in the wild, in cultivated plant populations, from mutations, or from breeding programs conducted by government agencies or private plant breeders. With the appropriate propagation procedures, such plants can become the starting point for populations of many millions of individuals, all just like the original mother plant. As explained in Chapter 3, the groups of plants that people have developed and cultivated have been given cultivar names; for example, Redhaven peach, Bing cherry, Pawnee wheat, Imperial Blue pansy, Peace rose, Ranger alfalfa, etc. The propagation methods used for increasing the populations of these groups of plants must do so without changing their characteristics.

There are several kinds of cultivars. If the plant group will reproduce "true" by seeds—with no characteristics changed—the cultivar is termed a **line.** A line is homozygous[1] and, if self-pollinated (or if cross-pollination is prevented), seed propagation will give progeny like the original plant. Many of the world's leading economic plants—the cereals, the vegetables, and the garden flowers—are made up of groups of these lines. They are seed propagated and will faithfully maintain their characteristics when propagated in this manner. Many forest species, although seed propagated, are not considered lines because of their higher level of variability. They are known by their species names, such as Douglas fir (*Pseudotsuga menziesii*), ponderosa pine (*Pinus ponderosa*), or coast redwood (*Sequoia sempervirens*).

In addition to lines, there are other types of seed-propagated cultivars, such as **inbred lines**—used to produce hybrid cultivars—and **hybrids**—as in hybrid corn (see Fig. 5–1).

Many groups of plants are heterozygous[2] rather than homozygous. These include fruit and nut species, many forage crops, and woody ornamentals. These plants have many dissimilar genes controlling their characteristics. During meiosis and fertilization leading to embryo formation in the seed, these genes segregate and recombine in a great many different ways so that the resulting plants differ from each other and from their parent. With these plants, seed propagation cannot maintain the characteristics of the female parent and cannot be used as a successful propagation procedure. With such heterozygous plants, vegetative (asexual) propagation is usually used. A piece of vegetative tissue—a section of a stem, root, or leaf—is placed in a suitable environment, such as a warm, humid rooting frame in the greenhouse. In time, the piece of tissue may regenerate the missing part: a stem piece forms roots, a root piece forms shoots, or a leaf forms both shoots and roots. By these means, new plants form that are exactly the same genetically as the plant from which the piece of vegetative tissue was taken; therefore, the new plant has all the same characteristics as the parent. The flower is not involved in vegetative propagation, allowing no opportunity for genetic change (unless, perhaps, a mutation has occurred, which does happen, but rarely). When such vegetative propagation is used, cultivars, even though heterozygous, can be perpetuated generation after generation, involving hundreds of thousands or more individual plants. Cultivars originating from a single plant or plant part and maintained in this manner by vegetative propagation are called **clones.**

The vegetative procedures just described—where a piece of tissue regenerates the missing part—is termed **cutting propagation.** There are a number of other, somewhat more complicated vegetative propagation methods, such as grafting, budding, layering, and runners.

SEXUAL PROPAGATION

In the sexual reproduction of plants, a seed must be produced in a flower. Seed formation is preceded by a type of cell division termed **meiosis** or reduction division, in which the number of chromosomes in the cells is reduced by half to form the male sperm cell and the female egg. The egg and sperm combine during fertilization in the ovule to form the zygote that develops into the embryo.

Seed Production

If a cultivar—Great Lakes head lettuce, Calrose rice, or Kombar barley, for example—is to be maintained by seed propagation, careful control of the seed source is essential. If

---

[1]Having similar genes of a Mendelian pair present in the same cell as, for example, a dwarf pea plant with genes (tt) for dwarfness only.

[2]Having different genes of a Mendelian pair in the same cell as, for example, a tall pea plant with genes for tallness (T) and genes for dwarfness (t).

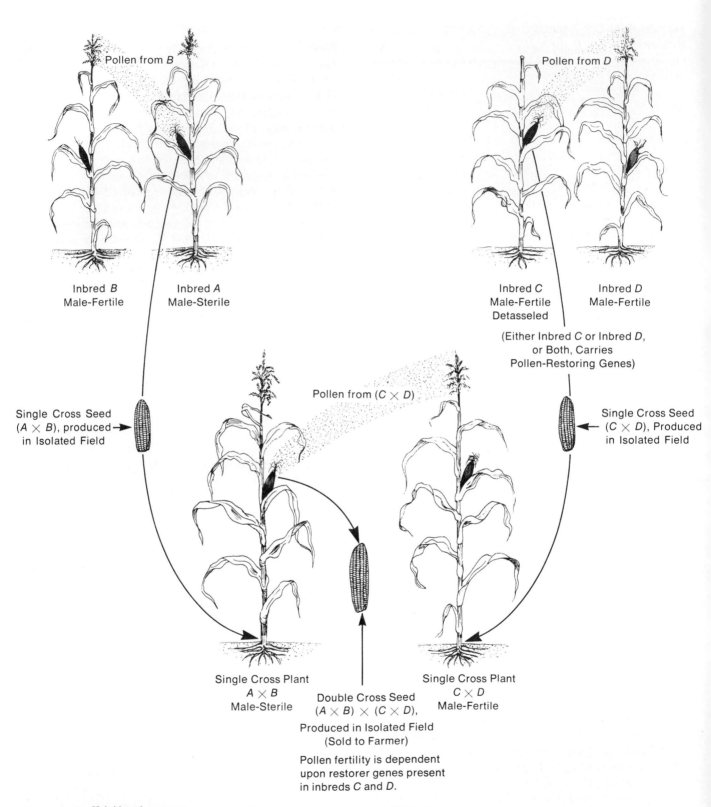

Pollen from B

Pollen from D

Inbred B
Male-Fertile

Inbred A
Male-Sterile

Inbred C
Male-Fertile
Detasseled

Inbred D
Male-Fertile

(Either Inbred C or Inbred D,
or Both, Carries
Pollen-Restoring Genes)

Single Cross Seed
($A \times B$), produced →
in Isolated Field

Pollen from ($C \times D$)

Single Cross Seed
← ($C \times D$), Produced
in Isolated Field

Single Cross Plant
$A \times B$
Male-Sterile

Double Cross Seed
($A \times B$) $\times$ ($C \times D$),

Produced in Isolated Field
(Sold to Farmer)

Pollen fertility is dependent
upon restorer genes present
in inbreds C and D.

Single Cross Plant
$C \times D$
Male-Fertile

Figure 5−1 Hybrid seed corn production by the utilization of cytoplasmic male sterility in the production of single cross and double cross. In this example only one inbred, A, is male sterile. The cytoplasmic male sterility is transmitted to the single cross $A \times B$. Pollen-restoring genes carried by the inbreds C or D give pollen fertility to the double cross $(A \times B) \times (C \times D)$ also. *Source:* USDA.

the cultivar is homozygous and self-pollinated, seed purity is generally assured. If the cultivar is homozygous but will cross-pollinate with other cultivars or with other species, then the plants used to produce the seeds must be separated by considerable distances from other plants with which they are likely to cross to prevent pollen contamination and loss of genetic purity.

With certain plants propagated by seed, some variability may be tolerated, or it may even be desirable. Examples are reforestation projects, tree and shrub plantings in shelter belts, plantings for wildlife cover, and plant breeding projects.

Most states in the United States and many other countries have established seed certification programs to protect and maintain the genetic quality of cultivars. Government agencies set standards for seed, particularly for such field and forage crops as soybeans, rice, wheat, oats, alfalfa, and clovers. Isolation standards for the seed-producing fields are established, as are requirements for roguing or culling out off-types, diseased plants, and weeds. Field inspections and final seed testing are included in such programs. Thorough cleaning of seed harvesting equipment between seed lots is required.

Seed certification programs in the United States recognize four classes of seeds in agronomic crops, such as cotton, alfalfa, soybeans, and cereal grains:

1. *Breeder seed.* This is produced only in small amounts and is under the control of the plant breeder. It is planted to produce foundation seed. Breeder seed is labeled with a white tag.
2. *Foundation seed.* This is multiplied from breeder seed; it is available only in limited amounts and is planted to produce registered seed. It is controlled by public or private foundation seed stock organizations. Foundation seed is also labeled with a white tag.
3. *Registered seed.* This is the seed source for growers of certified seed and is under the control of the registered seed producers. It is the progeny of either breeder or foundation seed. It is labeled with a purple tag.
4. *Certified seed.* This seed is available in large quantities and is sold to farmers for general crop production. It is labeled with a blue tag. Certified seed is of known genetic identity and purity.

Genetic quality of vegetable and flower seeds, however, is largely regulated by the seed companies, which maintain careful control over their seed plantings and continually test seed purity in special test gardens.

The production of plants from **hybrid seeds** has been one of the outstanding scientific breakthroughs in agricultural history. The use of hybrid seed has more than doubled the yield of both sweet corn and field corn. In the United States, virtually all corn is now produced from hybrid seed. This method of producing hybrid corn seed was developed in 1918, but it was not used commercially until about 1935 (Fig. 5–1). Hybrid barley and alfalfa cultivars have also been released, and almost all grain sorghum produced in the

United States comes from hybrid seed. Considerable research and millions of dollars have been expended by public and private agencies in trying to develop practical methods of producing hybrid seeds of wheat to increase yields, but these are still without the spectacular success achieved with corn. Wheat flowers are not structured for easy cross-pollination, as are the flowers of corn.

The use of F_1[3] hybrid seeds of several of the herbaceous bedding plants—petunias, zinnias, pansies, and marigolds—has made possible superior flower types and increased plant vigor, and has given a great boost to the bedding plant industry. Hybrid seeds also give greatly increased yield and quality of such vegetable crops as tomato, squash, cucumber, muskmelon, cauliflower, and broccoli.

A seed production industry has developed since about 1935, and it has made outstanding contributions to the increased production of world crops. Working with government and private plant breeders and seed certification agencies, the seed industry has supplied the channels for rapid seed increase and distribution while safeguarding genetic cultivar purity. Many progressive farmers in North America have become accustomed to using certified seed obtained from recognized seed producers rather than relying on their own or a neighbor's "bin-run" seed of dubious quality.

Seed Formation

Seeds originate as the final product of a plant's sexual reproduction system.

A seed has three essential parts:

1. The **embryo**, which develops into the new plant.
2. **Food storage material,** which is available to nourish the embryonic plant. This may be either endosperm tissue or the fleshly cotyledon(s), a part of the embryo itself.
3. **Seed coverings,** which are usually the two seed coats but may include other parts of the ovary wall.

The parts of the seed as they develop from the parts of the flower are:

Ovary → **Fruit** (sometimes composed of more than one ovary, plus additional tissues)

Ovule → **Seed** (sometimes coalesces with the fruit)

Integuments → **Seed coats** (two)

Nucellus → **Perisperm** (usually absent or reduced but sometimes develops into storage tissue)

2 polar nuclei + 1 sperm nucleus → **Endosperm** (triploid—3n)

[3]F_1 hybrid is the first generation offspring of a cross between two individuals differing in one or more genes.

Egg nucleus
+ 1 sperm nucleus → Zygote → Embryo
(diploid—2n)

After pollination and fertilization are completed, many changes occur in the flower to produce the fruit and the seed. Fruits and seeds appear in innumerable forms, depending upon the species, but they all contain the same three essential parts listed above.

Seed Storage and Viability Testing

For a seed to germinate, the embryo must be alive. In some plants, such as the willow, maple, and elm, the embryos are very short-lived, remaining viable for only a few days or months. In others, such as the hard-seeded legumes, the embryo generally remains alive for a great many years. Seeds of other kinds of plants range between these extremes, the length of embryo viability often depending upon seed storage conditions. For example, seeds stored in a sealed container under refrigeration at 0°C to 4°C (32°F to 40°F) and at low relative humidity (e.g., 15 percent, which would give a seed moisture level of about 4 to 7 percent) generally retain viability considerably longer than seeds sorted at room temperature and high relative humidity. Seeds of certain plants, however, soon become desiccated if stored under dry conditions, and the embryos die. Examples are citrus, maple, oak, hickory, walnut, and most tropical species.

It is often advisable before planting seeds, especially those of woody perennials and those which may have been stored for several years, to test the viability of a representative sample of the seed lot to be planted. There are several **seed viability tests.**

One easy means of determining the possible germinability of a seed lot is a **cut test.** Seeds of a representative sample are simply cut in half to see whether there is an embryo inside. Often the embryo has aborted or has been eaten by insects and, of course, the seed would not germinate. The mere presence of an embryo, however, does not mean it is alive.

Another simple test is to **float** the seeds in water. Quite often the "floaters" are empty seeds and can be skimmed off. The sound, full seeds sink and are the ones to be planted.

X-ray photographs of seeds do essentially the same thing as a cut test and are used in some seed laboratories to determine if the seed is empty or the embryo is shrunken. X-ray tests can also be used to determine the optimum time to harvest seeds by observing when the embryos have completely filled the seed.

These tests are not, strictly speaking, viability tests but are useful to rule out seeds that have no possibility of germinating. They still do not give the viability status of seeds with full-sized, apparently sound embryos.

Germination Test

The germination test is useful for seeds that have no dormancy problems, such as flower, vegetable, and grain seeds.

They can be tested for germinability by several methods, such as rolling them in several layers of moist paper toweling, or actually planting the seeds in flats of a suitable growing medium in a greenhouse. Seed-testing laboratories use elaborate seed germination boxes with controlled lighting and temperature. After several days or weeks, viability is calculated as the percentage of seedlings developing from the number of seeds planted.

Tetrazolium Test

The chemical 2,3,5-triphenyl tetrazolium chloride, which is colorless when dissolved in water, changes to the red-colored chemical, triphenyl formazan, whenever it contacts living, respiring tissue. In living tissues, enzymes change the tetrazolium salt to formazan. In dead, nonrespiring tissues, these enzymes are not active. In this test, the seeds are usually soaked in water to allow them to become completely hydrated, then cut in half lengthwise to expose the embryo. They are then placed in a 0.1 to 0.5 percent tetrazolium solution and held at room temperature or somewhat higher for several hours to several days, the exact time depending upon the species of seed. The parts of the seed that are living (and respiring) will become red; the nonliving parts remain white. If the embryo turns red, the seeds are viable; if the embryo remains white, the seeds are nonviable. Sometimes only a portion of the embryo becomes red, making it difficult to interpret the results of the test. This difficulty in interpretation is the principal weakness of this test, which, nevertheless, is widely used.

Excised Embryo Test

The embryos in the seeds of many woody plant species have profound dormancy conditions (see below) and do not respond in a direct germination test. However, if the embryos are carefully excised from the seed and placed on moist paper in a covered dish, viable embryos will show some activity—possibly a greening and separation of the cotyledons with definite indications of life. Nonviable embryos remain white and succumb quickly to fungus and bacterial attacks. Although this method takes time and requires skill in removing the embryo, it is routinely used by some seed laboratories with good results.

Seed Dormancy

Seeds of many plant species, especially woody perennials, do not germinate when extracted from the mature fruit and planted, even though all temperature, light, and moisture conditions favor germination. This is an important survival mechanism for the species and a result of evolutionary development. These species have survived because their seeds have not germinated just before adverse weather conditions that would kill the young, tender seedlings. Thus, in nature, these dormancy factors prevent seed germination of woody perennials in the autumn, allowing the embryonic plant within the seed to overwinter in a very cold-resistant form.

Any plant species whose seeds did germinate in the fall in an area with severe winters would likely not survive in that region. Often, the causes for dormancy can persist indefinitely in the seed and require specific treatments to overcome them before germination will take place. This poses problems for the propagator and requires a knowledge of seed dormancy and how to overcome it. Seed dormancy can result from structural or physiological conditions in the seed coverings, particularly the seed coats, or in the embryo itself, or both.

Seed Coat Dormancy

Seed coats or other tissues covering the embryo may be impermeable to water and gases, particularly oxygen, which therefore cannot penetrate to the embryo and initiate the physiological processes of germination. This situation usually occurs in species whose seeds have hard seed coats, such as alfalfa, clover, and other legumes as well as in some pine, birch, and ash species. In nature, continued weathering, the action of microorganisms, or passage through the digestive tract of animals can soften the seed coverings sufficiently so that they do become permeable and germination can proceed. In some species, the seed coats are apparently permeable to water and gases but have such high mechanical resistance to embryo expansion that germination does not occur unless the seed coats are softened in some manner.

Various artificial methods of softening seed coats are widely used to enhance germination. Three principal procedures are:

Scarification The surface of the seed is mechanically scratched or ruptured. This is often done by rubbing the seed between sheets of sandpaper.

Heat treatment In many kinds of seed, exposure to heat, usually boiling water, for a short time will sufficiently disrupt the seed coat to permit passage of water and gases.

Acid scarification Soaking seeds with impervious coverings in concentrated sulfuric acid for the proper length of time will etch their coats enough for germination.

Embryo Dormancy

Embryo dormancy is very common in seeds of woody perennial plants. It is due to physiological conditions or germination blocks in the embryo itself that prevent it from resuming active growth even though all environmental conditions (temperature, water, oxygen, light) are favorable.

It has been known for hundreds of years that dormant seeds, if allowed to winter outdoors in regions with cold climates so that they received some chilling while being kept moist, will germinate readily in the spring. From this arose the practice known as **stratification,** in which boxes are filled with alternate layers of moist sand and seed and set out-

doors in a protected shady place to overwinter. The following spring, the seeds are removed from the box and planted. The critical conditions in seed stratification are:

1. *Chilling temperatures*—from about 1°C to 7°C (34°F to 45°F).
2. *Moisture.* The seeds should be soaked in water to start, then kept moist.
3. *Adequate oxygen.* The seeds should have adequate air and not be kept in an airtight container.
4. *Period of time.* The optimum stratification time varies considerably among species. Seeds of the American plum (*Prunus americana*), for example, require at least 90 days chilling while, at the other extreme, apricot seeds (*prunus armeniaca*) require only 20 to 30 days.

More precise stratification treatments can be given if controlled refrigeration is used rather than natural outdoor winter cold, which can fluctuate considerably. Polyethylene plastic bags are suitable containers for stratifying seeds. The seeds are mixed with a slightly moist medium—sand, vermiculite, or peat moss—and placed in the bag. Polyethylene allows sufficient oxygen to pass through for the seed's requirements but slows water loss. It is advisable to soak seeds in water for 24 hours to thoroughly saturate the tissues before the chilling treatment begins.

Seeds of some species will germinate better if they are given a warm (24°C; 75°F) moist stratification period for several weeks just before the cold stratification period.

In nursery practice many propagators sow seeds having embryo dormancy in outdoor nursery beds in the fall, allowing the natural winter chilling to satisfy the embryo's chilling requirement.

There is evidence that during the stratification treatment, growth-promoting hormones (e.g., gibberellins and cytokinins) in the seeds increase while the level of growth-inhibiting hormones (e.g., abscisic acid) decreases, thus permitting germination.

The term **after-ripening** is often used to describe the physiological changes in the seed that allow germination to take place.

Double Dormancy

Seeds of some species have both seed coat and embryo dormancy. An example is redbud (*Cercis occidentalis*). To obtain good germination of such seeds, they first should be treated in some manner as described above to soften the seed coats, then given a cold stratification treatment to overcome the embryo dormancy.

Rudimentary Embryos

Some plants shed their fruits before the embryo within the seed has matured enough to germinate. A period of time— several weeks to several months—after harvest is required for the embryo to develop to the point where it can continue growth. This process can take place either while the seed is in storage or after planting. Seed dormancy due to rudimentary embryos occurs in such genera as *Fraxinus, Ilex, Magnolia, Pinus, Ranunculus,* and *Viburnum.*

Chemical Inhibitors

In many species, the seeds contain one or more chemicals that can block essential steps in the germination process. Sometimes—for example, in iris seeds—the inhibitor is not in the embryo but in the endosperm tissue. If the embryo is excised from the seed, it will start to grow readily in a sterile nutrient culture. Chemical inhibitors can also occur in seed coats or in the pericarp (ovary wall). Often, leaching such seeds in running water for several hours removes the inhibitors and permits germination. Seeds of some desert plants contain chemical germination inhibitors that are leached out by heavy soaking rains (but not by light showers). The heavy rains soak the soil sufficiently so that the seedlings can become established before the soil dries out. This interesting evolutionary phenomenon permits the continuation of these species by allowing survival of the seedling offspring in a difficult environment.

Some of these germination inhibitors are well-known chemicals such as coumarin and caffeic acid. Seeds of certain fleshy fruits—tomatoes, lemons, strawberries—do not germinate while still attached to the fruit because of certain of these germination inhibitors in the fruits.

Secondary Dormancy

Seeds that are ready to germinate after all germination blocks are removed can become dormant again because of exposure to some environmental condition. For example, seeds of some woody perennial plants, after undergoing stratification to overcome embryo dormancy, become dormant again if the germination temperatures reach 26°C to 32°C (79°F to 90°F). Exposing winter barley or spring wheat seeds to certain unfavorable conditions, such as high temperatures or high moisture levels, can also induce a secondary dormancy.

Seed Germination

If the seeds have viable embryos, have all germination blocks removed, and are placed under proper environmental conditions of moisture, temperature, and (sometimes) light, the quiescent embryos in the seeds will resume their growth. The nutrients stored in the endosperm or cotyledons of the seed nourish the developing embryo until the new shoot rises above ground, develops leaves, and produces its own food by photosynthesis.

Germination can proceed in several ways, depending upon the species (see Figs. 5–2 and 5–3). Sometimes the cotyledons are pushed above ground (**epigeous germination**) and sometimes they remain below ground (**hypogeous germination**). The sequence of events during seed germination is:

1. *Imbibition of water by the seeds.* The colloidal properties of seed tissues give them great water-absorbing properties. Moist seeds may swell to a size much larger than the dry seeds. The cells become turgid and seed coverings soften and rupture, permitting easy passage of oxygen and carbon dioxide.

2. *Activation of hormones and enzymes.* After water is absorbed, various enzyme systems are activated or synthesized, often as a result of stimulation by hormones. The enzymes convert complex food storage molecules into simpler food materials that can be

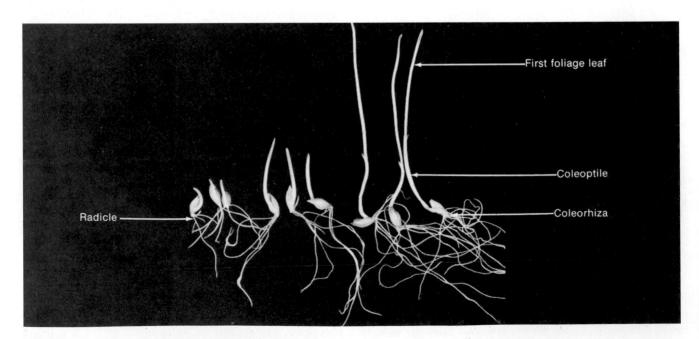

Figure 5–2   Seed germination in a monocotyledonous plant, barley.

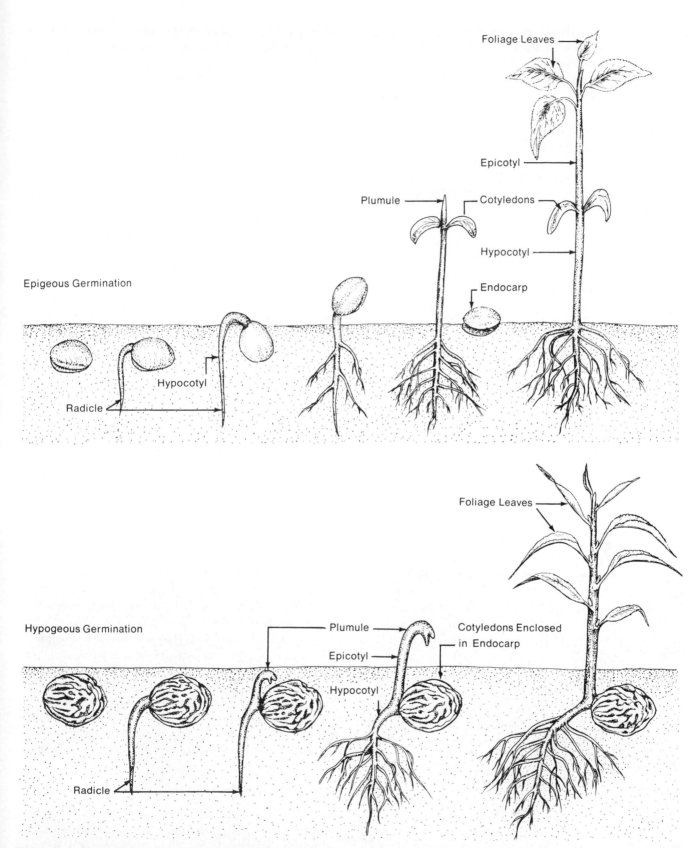

Figure 5–3 Seed germination in dicotyledonous plants. *Above:* Epigeous germination as shown in cherry. The cotyledons are above ground. *Below:* Hypogeous germination as shown in peach. The cotyledons remain below ground. *Source:* Hartmann, H. T., and D. E. Kester. 1983. *Plant propagation.* 4th ed. Englewood Cliffs, N.J.: Prentice-Hall.

readily translocated and used for growth. Other enzymes are involved in the respiratory processes, which release energy for cell division and growth. Food materials are translocated to root and shoot growing points.

3. *Embryo growth and development.* The root-shoot axis (plumule, epicotyl, hypocotyl, and radicle) grows by cell division and enlargement. At the same time, food materials translocate to the growing points from the storage tissues, which gradually become depleted. The seed coats rupture and photosynthetic tissue (green leaves and shoots) emerge into the light to carry on photosynthesis. In addition, the embryonic root (radicle) emerges and grows into moist soil to supply the newly developed leafy tissues with water—which will be lost through transpiration. By this time, if no unfavorable environmental influences interfere, the seedling has become established and can exist as an independent plant.

Environmental Factors Influencing Seed Germination

For successful seed germination and seedling growth, certain environmental conditions are required. These are:

1. Adequate moisture
2. Proper temperature
3. Good aeration
4. Light (in some cases)
5. Freedom from pathogenic organisms
6. Freedom from toxic amounts of salts

Moisture   It is essential that water be available in adequate amounts to initiate the physiological and biochemical processes in the seed that result in reactivation of embryo growth. Germination usually takes place satisfactorily at moisture levels between field capacity and permanent wilting percentage, although seeds of some species (lettuce, peas, rice, beets, celery) germinate best at high soil-moisture levels, whereas those of others (spinach) do best with low moisture.

Temperature   The temperature can strongly influence the percentage and rate of seed germination, varying with the kind of seed. Seeds of the cool-season crops germinate best at relatively low temperatures of 0°C to 10°C (32°F to 50°F). Examples are peas, lettuce, and celery. Seeds of warm-season crops germinate best at temperatures ranging from 15°C to 26°C (59°F to 79°F); examples are soybeans, beans, squash, and cotton. Seeds of many other species will germinate over a wide temperature range. In addition, many kinds of seeds germinate much better when the temperature fluctuates daily about 10°C (18°F) between maximum and minimum.

Aeration   Respiration rates are high in germinating seeds, which thus require adequate oxygen. The usual amount in the air is 20 percent. If this concentration is decreased, germination rate and percentage germination of most kinds of seeds will be retarded.

In seedbeds that are overwatered and poorly drained, the soil pore spaces may be so filled with water that the amount of oxygen available to the seeds becomes limiting and germination of most kinds of seeds is retarded or prevented.

Light   Light is essential to the germination of some kinds of seeds, such as lettuce, celery, most grasses, and many herbaceous garden flower plants. Such seeds should be planted very shallow for good germination. However, seeds of other plants—onion, amaranth, nigella, and phlox—are inhibited by light and will not germinate unless planted deep enough to avoid light. The light requirement for seed germination is very complex, depending on the age of seed, degree of seed imbibition with water, temperature, day length, and certain germination-stimulating chemicals. A pigment in seeds—phytochrome—is involved in the controlling mechanism.

Pathogenic Organisms   **Damping off** describes the situation in which the seedlings die during or shortly after germination. Damping off is caused primarily by attacks of certain universally present and very destructive fungi—*Pythium ultimum* and *Rhizoctonia solani,* and to a lesser extent, *Botrytis* spp. and *Phytophthora* spp. Mycelia of these organisms and the spores of *Pythium* and *Phytophthora* are often found in the germination medium, on the seed surfaces, in the water, or on tools. The best control methods are fumigation or heat pasteurization of the germination medium, surface treatment of the seeds with fungicides before planting, and good sanitation procedures.

Salinity Problems   If the germination medium is watered lightly but frequently after the seeds have been planted, evaporation of water from the surface leaves salt deposits. If this situation continues, the salinity can increase to such a level that it will injure or kill the seedlings as they germinate. This is a particular problem with small, shallow-planted seeds that may dry out quickly, and in areas having high salt concentrations in the water. Such salts may originate in the germination medium, the irrigation water, or added fertilizers. Salinity damage to seedlings often looks like damping-off injury. This problem can be prevented by using soil mixtures and water low in salts, withholding fertilizers, and by irrigating more copiously but less frequently so that excess salts are leached out.

VEGETATIVE PROPAGATION

Vegetative or asexual propagation is accomplished entirely through **mitosis,** the same nonreductive cell division process by which the plant grows. Each daughter cell is an exact replica of its mother cell. Chromosome numbers and composition do not change during cell division. Mitotic cell division, as illustrated in Figure 5–4, produces the adventitious

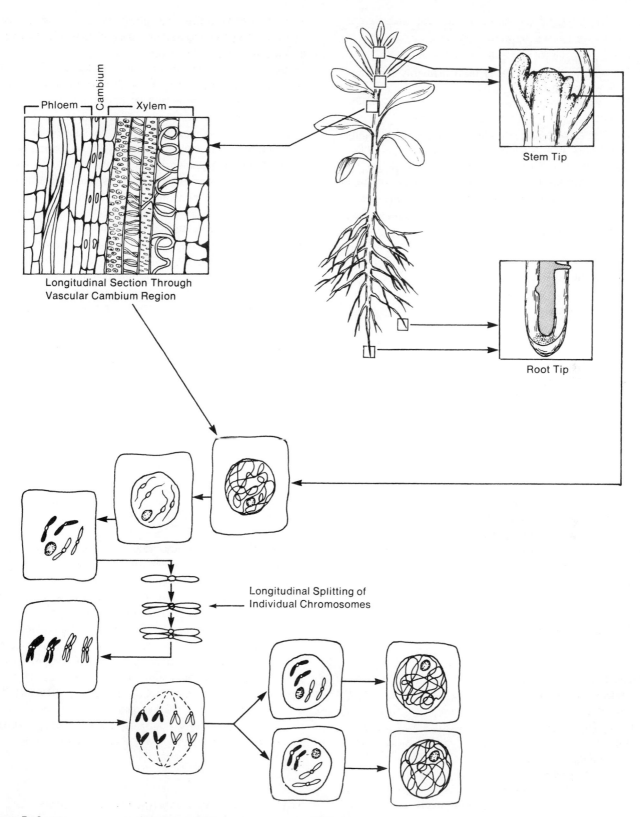

Figure 5–4 The process of growth and asexual reproduction in a dicotyledonous plant. Mitosis occurs in three principal growing regions of the plant: the stem tip, the root tip of primary and secondary roots, and the cambium. A meristematic cell is shown dividing into two daughter cells whose chromosomes will (usually) be identical with those of the original cell. *Source:* Hartmann, H. T., and D. E. Kester. 1983. *Plant propagation.* 4th ed. Englewood Cliffs, N.J.: Prentice-Hall.

roots and shoots[4] as well as the callus[5] (parenchyma cells required for healing of a graft union) necessary for successful vegetative propagation (see Fig. 5–5).

Vegetative propagation is used primarily for woody perennial plants that are highly heterozygous; that is, ones that do not "breed true" from seed. In these plants, the mother plant's desirable characteristics will be lost if seed propagation is used. To maintain the genetic identity of the mother plant in such cases, it is necessary to avoid use of the flower and seed altogether in reproducing the plant. Vegetative tissues (stem, root, or leaf) are used to develop new plants by inducing the formation of adventitious shoots, roots, or both. If pieces of stem tissue will not produce adventitious organs (roots, shoots, or both), or if a particular kind of rootstock is required, it is necessary to graft or bud two pieces of tissue together, one to become the top part of the plant and one to become the root system.

[4]**Adventitious shoots** are those appearing any place on the plant other than from the shoot terminals or in the axils of leaves.
Adventitious roots rise any place on the plant other than from the radicle (root tip) of the seed or its branches.
[5]**Callus** is a mass of undifferentiated and proliferating parenchyma cells.

As noted previously, a cultivar that must be reproduced by asexual methods to maintain its characteristics is termed a *clone,* as distinguished from a *line,* which will maintain its characteristics without change by seed propagation. Almost all fruit and nut cultivars and many ornamental cultivars are clones. All plants that are members of the same clone have the same genetic makeup and are, in reality, exact descendants of the mother plant from which the clone originated, although the original mother plant may have died many years earlier. Some clones, such as the 'Thompson Seedless' grape, have been in existence since ancient times and consist of many millions of individual plants scattered all over the world.

Cultivated clones originate in two ways. The first, and usual, way is as seedling plants that some person recognizes as having some superior qualities and proceeds to propagate vegetatively. For example, the world-famed 'Golden Delicious' apple originated as a seedling tree that A. H. Mullins found growing on his farm in Clay County, West Virginia. Mullins recognized this apple's value. Later it came to the attention of Stark Brothers Nurseries of Louisiana, Missouri, who in 1912 paid Mullins for the propagation rights to the tree. They named it 'Golden Delicious' and introduced nursery trees for orchard planting in 1916. Although the original tree died long ago, millions of 'Golden Delicious' apple trees

 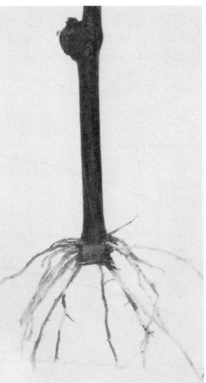 

Figure 5–5   Regeneration in asexual propagation. *Left:* Adventitious shoots growing from a root cutting. *Center:* Adventitious roots developing from the base of a stem cutting. *Right:* Callus tissue produced to give healing of a graft union. *Source:* Hartmann, H. T., and D. E. Kester. 1983. *Plant propagation.* 4th ed. Englewood Cliffs, N.J.: Prentice-Hall.

are now growing throughout the world, all with the same genetic makeup as the one original tree and all producing the same kind of apples. Although many clones have originated as chance seedlings found growing in the wild, they can also originate from controlled crosses made by plant breeders.

A second way clones originate is from mutations (bud sports). A single bud on a plant may have its genetic makeup altered during cell division so that, as the bud grows and develops into a branch, one or more of its characteristics differs from those of the rest of the plant. Many mutations are minor or inferior or go unnoticed. But occasionally some strikingly superior characteristic appears; someone notices it and propagates new plants taken from shoots on this mutated branch. This, then, becomes the start of a new clone. The desirable pink-fleshed 'Ruby' grapefruit originated in 1929 as a mutated branch on the 'Thompson Pink' grapefruit which, in turn, had originated in 1913 as a mutated branch on a white 'Marsh' grapefruit tree.

The primary advantage of clones is the uniformity of the member plants. All members have the same genetic makeup (**genotype**) and potentially they can all be exactly alike. However, environmental factors can so modify the appearance and behavior of the plant (**phenotype**) that individual plants can differ strikingly. For example, a vineyard of 'Concord' grapevines, pruned, irrigated, sprayed, and fertilized properly, would appear totally different from an adjacent abandoned vineyard of the same cultivar, yet the vines in both would be genetically identical.

The mutation process can also cause undesirable genetic changes in clones. This happens frequently in many species and must be guarded against to prevent deterioration of the clone.

Some mutations, called **chimeras,** genetically change only a portion rather than the entire shoot. Some of the variegated leaf patterns found in the foliage of certain plants such as *Pelargonium* or citrus are due to chimeras (Fig. 5–6).

Disease Problems in Clones

Propagating plants vegetatively with clonal material has one great disadvantage: clones can become infected with systemic viruses and mycoplasma-like organisms that are passed along to the daughter plants during asexual propagation procedures. In time, all clonal members may become infected with viruses. Some viruses are latent in particular nonsusceptible clonal material, but if this material is used in a graft combination where the virus can move through the graft union to the graft partner—which is susceptible—then the entire grafted plant will be killed by the virus. On the other hand, virus-free seedlings can be obtained in many species by seed propagation because the virus is not transmitted through the embryo.

Some viruses can be removed from clonal material by **heat treatment.** The virus-infected plant material, perhaps a small nursery tree growing in a container, is held at 37°C to 38°C (98°F to 100°F) for two to four weeks or longer. This

APOMIXIS—MAINTAINING A CLONE BY THE USE OF SEEDS

Apomixis is an interesting phenomenon in which the genetic identity of the mother plant is transmitted to daughter plants that develop by seed formation and germination. Apomixis is a form of asexual propagation because there is no union of male and female gametes before seedling production. There are several types but a common one is that found in citrus seeds where, in addition to the sexual embryo formed through the usual pollination and fertilization processes, embryos also arise in the nucellar tissue (**nucellar budding**). The nucellar tissue enclosing the embryo sac has not undergone reduction division and has the same genetic makeup as the female parent. So the nucellar embryos, although developing in a seed, are exactly the same genetically as the mother plant and thus maintain the clone.

Such seeds can contain several nucellar embryos in addition to the sexual embryo. Thus, several seedlings are obtained from one seed, a situation known as **polyembryony.**

Even though plants arising by apomixis from nucellar embryos maintain the clone, they go through the juvenile to mature transition stages just as any woody plant seedling would, taking a number of years to flower and fruit.

Figure 5–6 A variegated pink 'Eureka' lemon, a type of chimera.

combination of time and temperature eliminates the virus. After treatment, cuttings can be taken for rooting, or buds may be taken for budding into virus-clean seedling rootstocks.

Another procedure to eliminate viruses from clones is **shoot-tip culture.** In virus-infected plants the terminal growing point is often free of the virus. By excising this shoot apex aseptically and growing it on a sterile medium, roots will often develop, producing a new plant free of the virus. Here again, a starting point becomes available for continued propagation of the clone but without the virus. This method has succeeded with many herbaceous plants such as carnation, chrysanthemum, hops, garlic, rhubarb, orchid, and strawberry. Sometimes, in strawberry plants for example, both procedures are required—heat treatment of the plant, followed by excision and culture of the shoot tip—to free the plant of known viruses.

In recent years certification programs have been established by government agencies in many states in the United States and in other countries to provide nurseries with sources of true-to-name, pathogen-free propagation material. Elaborate programs, for example, have been established for citrus in Florida and California and for deciduous tree fruits, grapes, strawberries, potatoes, and certain ornamentals in many states.

In such programs, mature flowering or fruiting plants known to be true-to-name and true-to-type are selected as mother plants. These are "indexed" by certain grafting procedures to be sure that no known viruses or other diseases are present. If no pathogen-free source plants can be located, then procedures like those described above for eliminating viruses must be used to obtain pathogen-free plants. Once a "clean" source is obtained, it must then be maintained under isolated and sanitary conditions, with periodic inspection and testing to ensure that it does not again become infected. Sometimes it is necessary to grow the plants in insect-proof screenhouses or greenhouses or in isolated areas far from commercial production fields.

Distribution systems from these foundation plantings are necessary, sometimes requiring plots of land, "increase blocks," to grow a greater amount of propagating material that may be needed. This material can be termed "certified stock" if it is grown under the supervision of a legally designated agency with prescribed regulations designed to maintain certain standards of cleanliness and clonal identity.

Many of the major agricultural industries, based on crops that are susceptible to various pathogens, particularly viruses, could not exist without such certification programs. These crops include citrus, grapes, potatoes, roses, cherries, peaches, and strawberries.

Propagation by Cuttings

Cutting propagation is a vegetative method widely used for propagating herbaceous and woody ornamental plants and, to a much lesser extent, fruit species. A cutting is essentially a piece of vegetative tissue that, when placed under the proper environmental conditions, will regenerate the missing parts—roots, shoots, or both—and develop into a self-sustaining plant.

Cuttings can be classified according to the part of the plant from which they are obtained:

Stem cuttings
    Hardwood
        Deciduous
        Narrow-leaved evergreen
    Semihardwood
    Softwood
    Herbaceous
Leaf cuttings
Leaf-bud cuttings
Root cuttings

Stem cuttings already have terminal or axillary buds (potentially a new shoot system), but new roots must develop at the base of the cutting before a new plant will be formed. Stem cuttings can be prepared to include the shoot tip, or cuttings can be made from the more basal parts of the shoot that have only axillary buds. **Leaf cuttings** have neither buds nor roots, so both must form. **Leaf-bud cuttings** have a bud at the base of the petiole—for the new shoot system—so only new roots must form. **Root cuttings** must produce a new adventitious shoot and continue growth of the existing root piece, or develop roots from the base of the new shoot. Figure 5–7 illustrates types of cuttings.

Plant species and cultivars vary markedly in their ability to develop adventitious roots. Cuttings from some kinds of plants root easily even when the simplest procedures are used. Cuttings of others root only if the influencing rooting factors, as described starting on page 86, are carefully observed. Cuttings of still other kinds of plants have never been rooted, or rooted only rarely and in meager numbers, despite great efforts and much research. The basic reasons for such differences in rooting ability among different kinds of plants are not well understood.

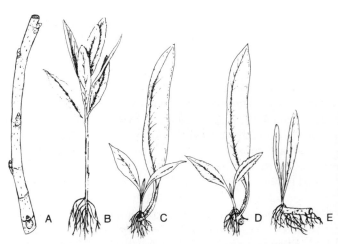

Figure 5–7   Types of cuttings. (A) Hardwood stem cutting. (B) Leafy stem cutting. (C) Leaf cutting. (D) Leaf-bud cutting. (E) Root cutting.

Stem Cuttings

Deciduous Hardwood Stem Cuttings Such cuttings are made in late winter or early spring, using leafless shoots that grew the previous summer. The shoots are cut into lengths of 15 to 30 cm (6 to 12 in.) and propagated outdoors in nursery beds by planting vertically with just the top bud showing. Figure 5–8 shows quince cuttings prepared for spring planting and the resulting plants after they rooted and grew through the summer. A porous, sandy loam soil is best for rooting hardwood cuttings. Many plants can be propagated in this manner, including some deciduous trees, such as willow and poplar; many deciduous shrubs, such as forsythia, weigela, privet, spiraea, honeysuckle, and roses; and several fruit species, such as grape, mulberry, currant, gooseberry, quince, olive, fig, and pomegranate.

Narrow-Leaved Evergreen Hardwood Stem Cuttings
Many conifers can be propagated by stem cuttings made in various lengths of about 7 to 15 cm (3 to 6 in.). Cuttings are best made in early winter. Needles are removed from the lower half of the cutting but left on the upper. The cuttings are usually inserted into flats or directly into greenhouse benches filled with a rooting medium of sand or of equal parts perlite and peat moss. The cuttings are best rooted in a cool, humid greenhouse or cold frame, preferably with high light intensity, and kept

lightly watered or misted until they root. Rooting may take several weeks or even months, depending upon the species.

Semihardwood Stem Cuttings Most broad-leaved evergreen ornamentals—rhododendron, camellia, pittosporum, holly, evergreen azalea, escallonia, euonymus, and boxwood—as well as some fruit species—citrus and olive, for example—can be propagated by this type of cutting. Cuttings are best taken in midsummer, following the flush of spring growth. They are made about 10 to 15 cm (4 to 6 in.) long. Four or five leaves should be retained on the upper portion of the cutting and all lower leaves removed, as shown in Figure 5–9.

The cuttings are inserted into flats or greenhouse benches. A porous rooting medium, such as equal parts of perlite and peatmoss or of perlite and vermiculite, is used. The cuttings are placed in a well-lighted, humid location to minimize water loss from the leaves. Closed rooting frames or flats covered with polyethylene plastic sheeting are suitable for keeping the humidity around the cuttings high.

Most commercial nurseries root this type of cutting in **mist propagating beds,** where a fine mist is sprayed over the cuttings intermittently during the day.

Softwood Stem Cuttings These are similar to semihardwood cuttings except they are prepared from young leafy shoots arising in the spring from deciduous trees or shrubs. Such plants as crape myrtle, pyracantha, mock orange, forsythia, weigela, roses, pomegranates, and plums are easily

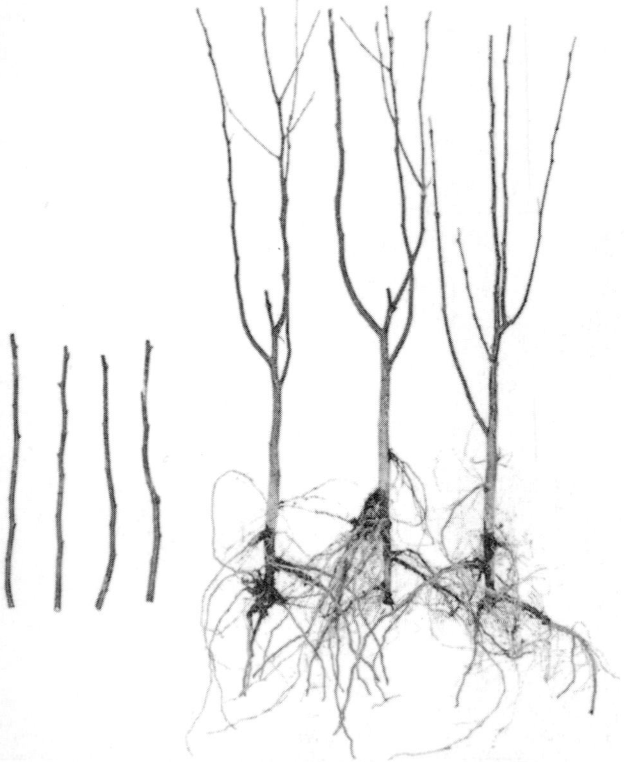

Figure 5–8 *Left:* Hardwood cuttings of quince (*Cydonia oblonga*) prepared and ready for planting in early spring. *Right:* Rooted cuttings after one year in the nursery. *Source:* Univ. of Calif. Div. Agr. Sci. Leaflet 21103, p. 55. 1979.

Figure 5–9 Semihardwood leafy cuttings of escallonia showing typical method of preparing this type of cutting. Cuttings were treated with a rooting hormone—indolebutyric acid at 0, 1,000, 3,000, and 8,000 parts per million before they were stuck in the rooting bed.

propagated by this type of cutting. Cuttings are prepared in the spring in the same manner as semihardwood cuttings and are rooted under similar conditions. All leafy cuttings must be rooted under high humidity conditions to reduce water loss from the leaves.

Herbaceous Stem Cuttings   This type of cutting is used in propagating such plants as coleus, carnation, geranium, chrysanthemum, and many tropical house plants, all of which root easily. Typically a stem is harvested and the basal end is placed in a rooting medium in a warm (75°F to 80°F), draft-free environment and misted (see Fig. 5–10) until new roots develop. Rooting time depends on species and cultivar, often taking several days to several weeks. Heating the root zone area to 75°F will decrease the time of root formation and make rooting more uniform. Misting frequency should decrease as callus and roots form to reduce the risk of disease and increase hardiness of the cutting.

Leaf Cuttings

There are various types of leaf cuttings, as shown in Figure 5–11. A common one consists of a single leaf blade and petiole, as might be taken from an African violet. The petiole is inserted into the rooting medium to a depth of about 2.5 cm (1 in.). As with other types of leafy cuttings, the humidity must be kept very high, preferably by using a closed rooting frame, as for the herbaceous cuttings, or by the use of a humid greenhouse. Roots and shoots generally develop from the same point at the base of the petiole and grow into a plant independent of the leaf blade, which functions to nourish the new plant. African violet, peperomia, begonia, and sansevieria are examples of plants commonly started by leaf cuttings.

Leaf-Bud Cuttings

Leaf cuttings of some species will form roots at the base of the petiole but do not develop a shoot, resulting only in a rooted leaf that may stay alive for months (or years). To avoid this, a leaf-bud cutting can be prepared. This cutting consists of a short piece of stem with an attached leaf and a bud in the axil of the leaf, as shown in Figure 5–12. Such cuttings are rooted under high humidity, as described for semihardwood or herbaceous cuttings. The axillary bud develops into the new shoot system. Leaf-bud cuttings are useful as substitutes for stem cuttings in obtaining as many plants as possible from scarce propagating material. Leaf-bud cuttings give one and perhaps two (if the plant has opposite leaves) new plants from each node, whereas each stem cutting generally requires a minimum of two nodes.

Root Cuttings

Many plant species can be propagated by cutting the small, young roots into pieces about 2.5 cm (1 in.) long and planting them horizontally in soil about 1.3 cm (0.5 in.) deep or vertically with the upper end (nearest the crown of the plant) just below the soil level. One or more new adventitious

Figure 5–10   Mist propagation bed operating in a greenhouse. The mist is on only a few seconds—just long enough to wet the leaves—then turns off. When the leaves start to dry the mist is turned on again. The mist is generally off all night.

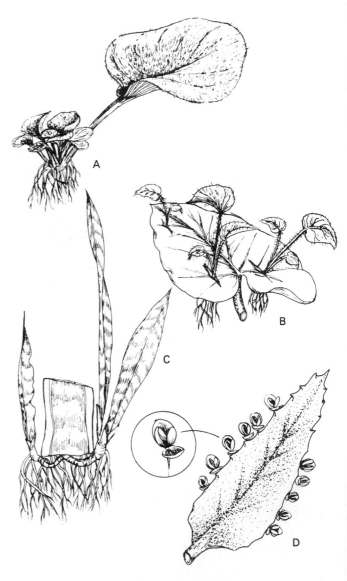

Figure 5–11   Types of leaf cuttings. (A) New plants arising from base of petiole in African violet (*Saintpaulia*). (B) New plants arising from cuts in veins of a begonia leaf. (C) New plant arising from base of leaf blade in sansevieria. (D) New plants growing from notches of leaf in *Kalanchoe (Bryophyllum)*.

Figure 5–12   Leaf-bud cuttings of peperomia. Each cutting consists of a leaf blade, petiole, axillary bud, and a piece of stem. Arrows show axillary buds starting to grow.

shoots form along the root piece, and either this shoot forms roots or the root piece itself develops new branch roots, thus producing a new plant. The best time to obtain the root pieces from the mother plant and to prepare and plant the root cut-

tings is late winter or early spring. The roots contain the highest quantity of stored foods at that time and the cuttings will start to grow at the beginning of the growing season. Cuttings can be planted in an outdoor nursery or in flats of soil in a greenhouse or cold frame.

Origin of Adventitious Roots in Stem Cuttings

In stem cuttings of herbaceous plants, adventitious roots generally originate laterally and adjacent to the vascular bundles, whereas in cuttings of woody perennials the roots originate in the region of the vascular cambium, often in young phloem parenchyma (Fig. 5–13). In each case the new roots are in position to establish a vascular connection with the conducting tissues of the xylem and phloem in the cutting.

Much study has been given to the mechanisms that lead to the initiation of adventitious roots in cuttings. There is convincing evidence that auxin, one of the natural growth hormones, is essential. It has long been known that the leaves on cuttings strongly promote root initiation. Materials

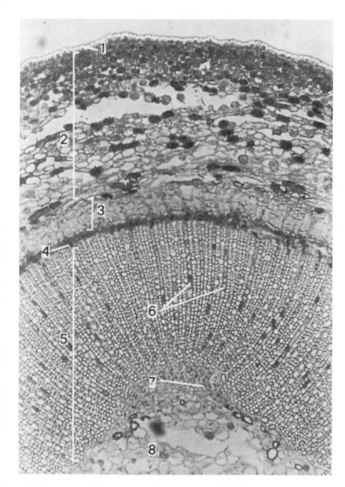

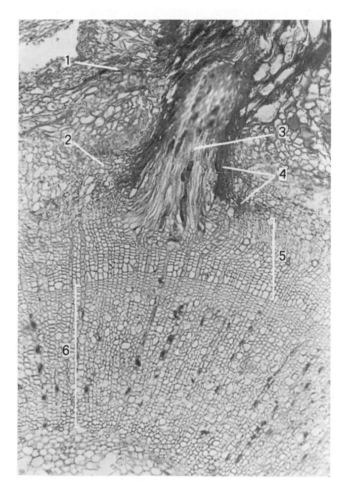

Figure 5–13   *Left:* Cross section of euonymus shoot before rooting: (1) epidermis, (2) primary bark, (3) phloem, (4) cambium, (5) xylem, (6) medullary rays, (7) perimeduler zone, (8) pith. *Right:* Cross section of euonymus shoot (cutting) after an adventitious root has formed: (1) primary bark, (2) phloem, (3) newly-formed adventitious root, (4) cambium, (5) xylem formed after rooting, (6) primary xylem. *Source:* Prof. Bojinov Bogdanov and *Proceedings International Plant Propagators' Society,* Vol. 35, pp. 449–53. 1985.

originating in the leaves, called rooting cofactors, are believed to be essential to rooting; they combine with auxin to form a complex that directs RNA to activate enzymes that cause root initials to form. The composition of these cofactors is not clear but some, at least, are likely to be phenolic compounds. Other natural hormones, such as gibberellins and abscisic acid, also influence adventitious root formation. Controlled studies under aseptic conditions have shown, too, that sugar, nitrogen, calcium, and other mineral nutrients must be present for roots to form.

Factors Influencing the Rooting of Cuttings

There is a large group of plants whose cuttings root only with considerable difficulty. It is necessary to carefully consider the factors described below in order to satisfactorily root cuttings in this group.

Source of Cutting Material   Generally, the cuttings most likely to root come from stock plants that are growing in full sun at only a moderate rate and that have thus accumulated carbohydrates in their tissues. If the cutting material for woody plant species can be taken from the young, nonflowering plants only a few years away from a germinated seed, rooting will be much better than when the cuttings are taken from old, mature flowering and fruiting plants. The **juvenility influence** in the young material can sometimes be retained by keeping the stock plants cut back heavily each year to force new shoot growth out from the lower part of the plants near the ground level.

Time of Year the Cutting Material Is Taken   In woody perennial plants, cutting material can be taken at any time of the year. In some species the time the material is taken can dramatically influence rooting. Hardwood cuttings often root best if the material is gathered in late winter, while softwood cuttings usually root best if taken in the spring shortly after new shoot growth has attained a length of 10 to 15 cm (4 to 6 in.). Semihardwood cuttings are best taken in midsummer after the spring flush of growth has matured somewhat. Herbaceous cuttings can be easily rooted any time of the year, especially when succulent.

Etiolation   It has long been known that stem tissue developing in complete darkness is more likely to initiate adventitious roots than tissue exposed to light. Thus, if the basal part of shoots that are later to be made into cuttings can be kept in darkness, they are likely to form roots. Such techniques are used successfully in rooting cuttings of difficult species, such as the avocado. The mechanism for etiolation[6] in promotion of rooting is not clear, but the harmful effects of light on rooting may be due to photoinactivation of one or more natural rooting factors in the stem tissues.

---

[6]Etiolation is the growth of shoots in the absence of light or in low light, causing them to be abnormally elongated and colored yellow or white due to the absence of chlorophyll.

Treatment of Cuttings with Auxins   In the mid-1930s it was discovered that one of the natural plant hormones, auxin (indoleacetic acid, or IAA), stimulated the initiation of adventitious roots on stem cuttings. Synthetic IAA was just as effective as the natural material. But it was soon discovered that other closely related synthetic auxins—indolebutyric acid (IBA) and naphthaleneacetic acid (NAA)—were even more effective. This knowledge was quickly picked up by plant propagators, who now routinely treat the base of cuttings with one of these materials, particularly IBA, just before the cuttings are stuck in the rooting medium. Commercial preparations of IBA dispersed in talc are available; the lower end of the cutting is swirled around in the mixture to coat it with the powder. The propagator can also prepare solutions from the pure chemicals, then dip the base of the cuttings in the solution for about five seconds just before the cuttings are stuck in the rooting medium. The optimum concentration to promote rooting varies with the species but ranges from about 2,000 to 10,000 parts per million (ppm).

Misting   Misting is an essential factor in the propagation of most cuttings, especially herbaceous cuttings. The frequency of misting is most critical during the early stages of rooting, before callus formation. As rooting takes place frequency tapers off. The primary purpose of misting is to prevent cuttings from dehydrating by keeping relative humidity near 100 percent around the cutting. Misting also helps to keep the cutting cool, which reduces transpirational loss of water. Frequency is determined not only by root development, but by environmental factors such as light intensity, relative humidity, and air temperature. Low light, high relative humidity, and cool temperatures reduce the need for mist. Timers can be used to control the frequency of misting but are not the best mechanism. Timers do not sense changes in environment and do not make adjustments as conditions change. Using light or relative humidity sensors to set misting frequency allows misting frequency to change as environmental conditions that affect the need for mist change.

Bottom Heat in the Cutting Beds   To force rooting at the base of cuttings before shoot growth starts, it is advisable to maintain temperatures at the base of the cuttings at about 24°C (75°F)—or about 6°C (10°F) higher than that at the tops of the cuttings, 18.5°C (65°F). This is best done by providing bottom heat from thermostatically controlled electric heating cables or hot water pipes under the rooting frames. Bottom heat under the cuttings often greatly stimulates rooting.

Propagation by Grafting and Budding

Grafting and budding are vegetative methods used to propagate plants of a clone whose cuttings are difficult to root. These methods are also used to make use of a particular rootstock rather than having the plant on its own roots. Certain rootstocks are often utilized to obtain a dwarfed or invigorated plant or to give resistance to soil-borne pests.

Grafting

Grafting can be defined as the art of joining parts of plants together so that they will unite and continue their growth as one plant. The **scion** is that part of the graft combination that is to become the upper or top portion of the plant. Usually the scion is a piece of stem tissue several inches long with two to four buds. If this piece is reduced in size so there is just one bud, with a thin layer of bark and wood under it, then the operation is termed **budding.** The **rootstock** (or **understock** or **stock**) is the lower part of the graft combination, the part that is to become the root system. **Root grafting** is a common method of propagation in which a scion is grafted directly onto a short piece of root and the combination is then planted. Sometimes grafting is used to change the fruiting cultivar in a fruit tree or grapevine to a different one (**top-grafting** or **top-working**). Grafting may be used to repair the damaged trunk of a tree (**bridge-grafting**) or to replace an injured root system (**inarching**). Grafting, or budding, is sometimes used to study the transmission of viral diseases (**indexing**). The indexing test involves inserting a bud from a plant suspected of carrying a virus into another indicator plant by T-budding. A definite visual response results from the presence of a virus, such as gumming around the inserted bud.

Several standard methods of grafting and budding have been widely used over the world for a great many years.

Whip Grafting   The **whip graft** (Fig. 5–14) is useful in grafting together material about 0.6 to 1.3 cm (0.25 to 0.5 in.) in diameter. It is often used to make **root grafts** in late winter. A small stem piece of the scion cultivar is grafted onto a root piece. The completed grafts are buried for two or three weeks in a moist material, such as wood shavings, at about 21°C (70°F) to encourage callusing or healing of the union. After the union has healed, the graft can be planted in the nursery.

Various **grafting machines** can be used as a substitute for the whip graft and are considerably faster than hand grafting. Such machines have been used mostly in grape grafting but are also satisfactory for making fruit tree root grafts.

> In all types of grafting and budding the two parts must be held together very tightly and securely by wedging, tying, nailing, or wrapping with string, rubber bands, or with plastic or cloth tape. The graft union must also be completely covered by grafting wax to prevent the cut surfaces from drying out. Polarity must be observed: the scions must be inserted so that the buds point upward.

Cleft Grafting   The **cleft graft** (Fig. 5–15) is mostly used in top-grafting, where scions 0.6 to 1.3 cm (0.25 to 0.5 in.) in di-

ameter are inserted into stubs of older limbs that are 8 to 10 cm (3 to 4 in.) in diameter after they have been cut off a foot or so out from the trunk of the tree. It is very important in cleft grafting to match the cambium layer of the scion as closely as possible with the cambium layer of the stock so that the two pieces will heal together. Cleft grafting is usually done in late winter or early spring.

Bark Grafting   The **bark graft** is also used for top-grafting. It can be done only when the bark separates easily from the wood along the cambium layer. Thus, bark grafting is usually done in early-to-mid spring after new growth is well underway. Bark grafting is easy to do and is widely used, especially for species considered somewhat difficult to graft successfully.

> The proper selection of scionwood and budwood is very important in all types of grafting and budding. Scionwood and budwood should be taken from source trees true-to-type for the cultivar to be propagated. They should be free of known viruses and any other diseases. Some American states and some other countries have programs to supply propagating material of many fruit and woody ornamental species certified as produced under disease-free conditions.
>
> For all types of grafting and budding, the buds on the scionwood and budwood must be dormant when the grafting or budding is done. For the whip and cleft graft, where the grafting is done during the dormant season, the scionwood can be collected from dormant source trees and used immediately in the grafting operation. For the bark graft, made later in the spring after vegetative growth has started, it is necessary to gather the scionwood earlier during the dormant season, wrap it in polyethylene bags with some slightly damp peatmoss, then hold it under refrigeration, preferably at about 0°C (32°F), until grafting. For broad-leaved evergreens, such as citrus or olives, scionwood can be obtained from the tree at the time the bark grafting is to be done, using two-year-old shoots having dormant latent buds.

Budding

T-budding   This technique is widely used in propagating fruit trees and roses. Buds taken from budsticks are inserted under the bark of small seedling rootstock plants a few inches above ground level. Buds are then inserted and tied into place with budding rubbers, but the tops of the seedling rootstocks are not cut off above the inserted cultivar bud until just before growth starts the next spring. For fruit trees, budding is usually done in late summer. Roses are usually budded in the spring, with the top of the rootstock broken over above the bud after about two weeks following budding. The break forces the bud to grow, and after the shoot has grown 10 to 20 cm (4 to 8 in.) from the bud, the rootstock is completely cut off above the inserted bud. By late fall, a sizable rose plant is ready to dig.

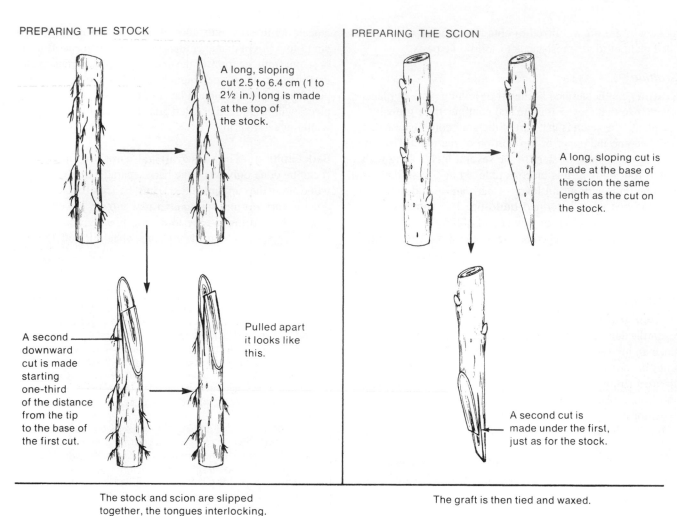

A long, sloping cut 2.5 to 6.4 cm (1 to 2½ in.) long is made at the top of the stock.

A long, sloping cut is made at the base of the scion the same length as the cut on the stock.

A second downward cut is made starting one-third of the distance from the tip to the base of the first cut.

Pulled apart it looks like this.

A second cut is made under the first, just as for the stock.

The stock and scion are slipped together, the tongues interlocking.

The graft is then tied and waxed.

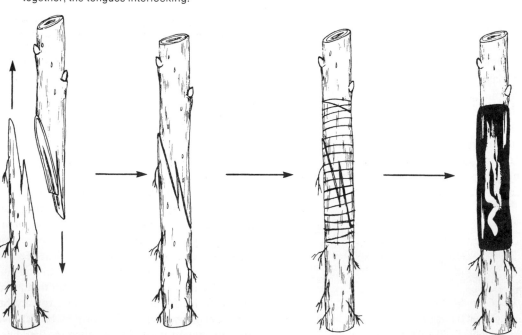

Figure 5–14 The whip, or tongue, graft. This method is widely used in grafting small plant material and is especially valuable in making root grafts, as illustrated here. *Source:* Hartmann, H. T., and D. E. Kester. 1983. *Plant propagation.* 4th ed. Englewood Cliffs, N.J.: Prentice Hall.

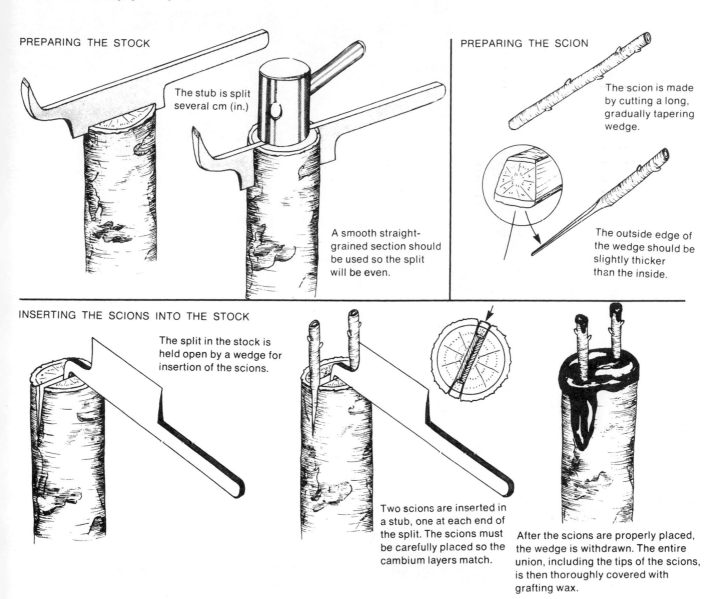

PREPARING THE STOCK

The stub is split several cm (in.)

A smooth straight-grained section should be used so the split will be even.

PREPARING THE SCION

The scion is made by cutting a long, gradually tapering wedge.

The outside edge of the wedge should be slightly thicker than the inside.

INSERTING THE SCIONS INTO THE STOCK

The split in the stock is held open by a wedge for insertion of the scions.

Two scions are inserted in a stub, one at each end of the split. The scions must be carefully placed so the cambium layers match.

After the scions are properly placed, the wedge is withdrawn. The entire union, including the tips of the scions, is then thoroughly covered with grafting wax.

Figure 5–15 Steps in making the cleft graft. This method is very widely used and is quite successful if the scions are inserted so that the cambium layers of stock and scion match properly. *Source:* Hartmann, H. T., and D. E. Kester. 1983. *Plant propagation.* 4th ed. Englewood Cliffs, N.J.: Prentice-Hall.

Patch-budding Patch-budding is mostly used for such plants as walnuts, pecans, and other species that are difficult to T-bud because of their thick bark. Patch-budding must be done during the growing season when the bark on both the budstick and the rootstock is "slipping" easily (separating from the wood along the cambium layer). For propagating nursery trees, patch-budding is generally done in mid-to-late summer; the subsequent handling is the same as for T-budding.

Healing of the Graft and Bud Union

In preparing a graft or bud combination, the two parts are joined by one of the methods just described so that the cambial layers of stock and scion exposed by the grafting cuts are brought into intimate contact. They are held in place by wedging, nailing, or wrapping so that the parts cannot move

about or become dislodged. Then the graft union is thoroughly covered with plastic or cloth tape or, better, by grafting wax to keep out air. The union heals by callus production from young tissues near the cambium layers of both stock and scion. Temperature levels must be conducive to cellular activity—generally from about 10°C to 30°C (50°F to 86°F)—and no dry air must contact the cut surfaces because it would desiccate the tissues.

The steps in healing of a graft union are illustrated in Figure 5–16. Healing usually takes about two weeks and must be completed, with vascular connections made for translocation of water through the xylem, before the buds on the scion start to grow and develop leaves. The transpiring leaves would soon desiccate the scion unless the graft union has healed by this time.

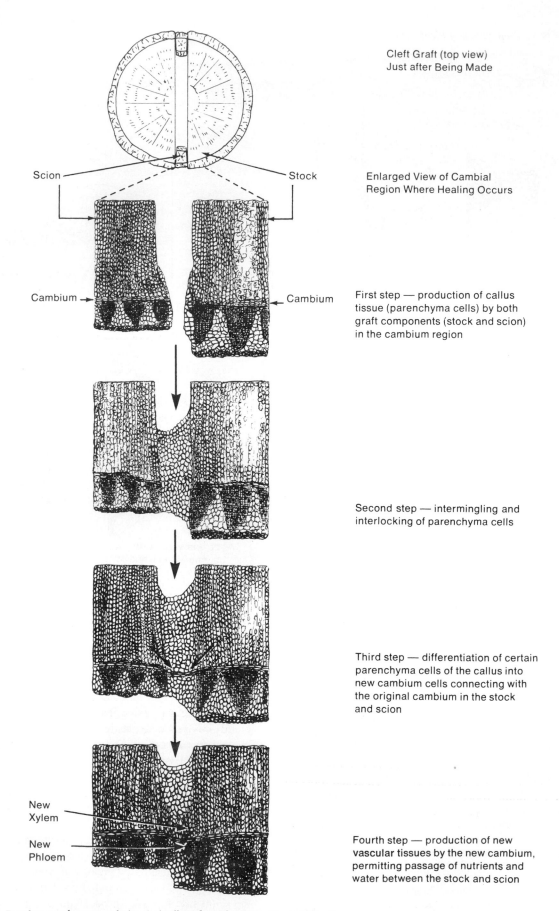

Cleft Graft (top view)
Just after Being Made

Scion

Stock

Enlarged View of Cambial
Region Where Healing Occurs

Cambium

Cambium

First step — production of callus
tissue (parenchyma cells) by both
graft components (stock and scion)
in the cambium region

Second step — intermingling and
interlocking of parenchyma cells

Third step — differentiation of certain
parenchyma cells of the callus into
new cambium cells connecting with
the original cambium in the stock
and scion

New
Xylem

New
Phloem

Fourth step — production of new
vascular tissues by the new cambium,
permitting passage of nutrients and
water between the stock and scion

Figure 5–16 Developmental sequence during the healing of a graft union as illustrated by the cleft graft. *Source:* Hartmann, H. T., and D. E. Kester. 1983. *Plant propagation.* 4th ed. Englewood Cliffs, N.J.: Prentice-Hall.

Limits of Grafting

There are certain limits to the combinations that can be successfully established by grafting or budding. The partners (stock and scion) in the combination must have some degree of botanical relationship—the closer the better. For instance, woody perennial plants in different botanical families have never, as far as is known, been successfully grafted together. For example, it is useless to attempt to graft a scion taken from a grapevine (VITACEAE) onto an apple tree (ROSACEAE).

In only a few cases have completely successful graft combinations been made between plants in the same family but different genera. For example, the thorny, deciduous large shrub trifoliate orange (*Poncirus trifoliata*) is widely used commercially as a rootstock for the common sweet orange (*Citrus sinesis*), a large evergreen tree. Both are in the family RUTACEAE, but different genera—*Poncirus* and *Citrus.*

If the two graft partners are in the same genus but different species, then the chances of success are greatly improved. Still, many such graft combinations will not unite. Plant propagation books should be consulted for information that has been accumulated by trial and error over the years. For example, in the family ROSACEAE and the genus *Prunus,* it is well established that almonds (*P. dulcis*), apricots (*P. armeniaca*), European plums (*P. domestica*), and Japanese plums (*P. salicina*) can all be successfully grafted onto peach seedlings (*P. persica*) as a rootstock. But two other members of the genus *Prunus,* sweet cherry (*P. avium*) and sour cherry (*P. cerasus*) fail when grafted onto peach roots.

If the two partners are different cultivars (clones) within a species, the chances are almost 100 percent that the graft combination will succeed. For instance, if you have a 'Jonathan' apple tree (*Malus pumila*), you could successfully top-graft on it any other apple cultivar, for example, the 'Golden Delicious' (*Malus pumila*).

Graft Incompatibility

There are many puzzling, unexplained situations among various graft combinations. Some pear cultivars are commercially grafted onto quince roots (an intergeneric combination), but scions of other pear cultivars grafted on quince roots soon die. The reverse, quince on pear roots, always fails. Plums can be successfully grafted on peach roots but peaches on plum roots are a failure. Japanese plums (*Prunus salicina*) can be grafted on European plum (*P. domestica*) roots, but in the reverse combination, the trees soon die.

There are many examples of incompatible graft combinations. **Compatibility** in grafting is the ability of two different plants, grafted together, to produce a successful union and to develop satisfactorily into one composite plant. The causes for **graft incompatibility** are little understood in spite of many years of research into the problem. There is some evidence, however, that in certain graft combinations one partner (scion or stock) produces chemicals that are toxic to the other, killing the entire plant.

Effect of Rootstock on Growth and Development of the Scion Cultivar

Tree fruit growers often select a certain rootstock for a particular fruiting cultivar because it will dwarf the tree to some extent and thus make harvesting easier. This is particularly true in apples, where an entire series of clonal rootstocks is available to produce apple trees with any desired degree of dwarfness. Quince roots will dwarf pear trees. Trifoliate orange roots will dwarf sweet orange trees.

These dwarfing influences extend to the tree only—not to the fruits produced by the trees. But in some species, particularly citrus, the kind of rootstock used can also strongly influence the quality of fruit produced by the scion cultivar. For instance, when sweet orange seedlings are used as the rootstock for orange trees, the fruits will be of much higher quality than when 'Rough' lemon is used as the rootstock.

What mechanisms are involved in such influences? Why are trees dwarfed by certain rootstocks? Considerable research has been conducted by plant physiologists and horticulturists over the years to try to explain these results, but convincing arguments are few. Probably the best hypothesis is that certain dwarfing rootstocks produce relatively large amounts of growth inhibitors, such as abscisic acid (see p. 118) that translocate through the graft union to the top fruiting cultivar and slow its growth. In addition, such dwarfing stocks produce low amounts of growth promoters, such as the gibberellins.

Layering

A simple and highly successful method of propagating plants is layering. Layering is similar to propagation by cuttings except that, instead of severing the part to be rooted from the mother plant, it is left attached and receives water and nutrients from the mother plant. After the stem piece (layer) has rooted—no matter how long it takes—it is cut from the mother plant and transplanted to grow independently. Various layering procedures can be used for many different kinds of plants (Fig. 5–17).

Tip Layering

This is a natural means of vegetative propagation, shown by black raspberries and the trailing blackberries. The tips of the long canes, if buried a few inches deep in the soil toward the end of their first summer's growth, will root and produce a shoot that grows upward through the soil, forming a new plant. The cane can be cut from the parent bush and the new plant dug and replanted.

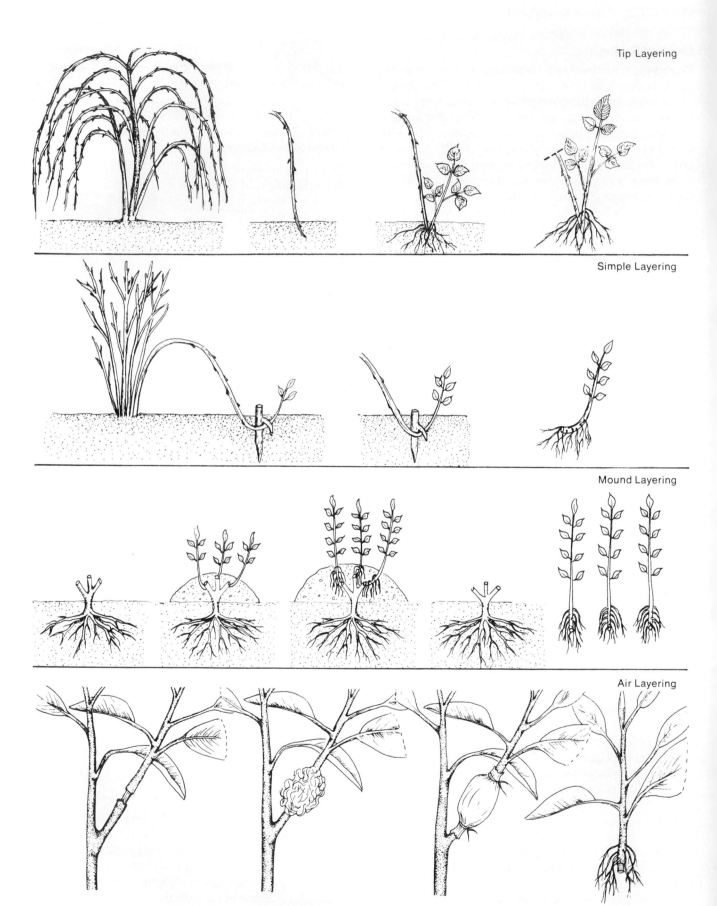

Tip Layering

Simple Layering

Mound Layering

Air Layering

Figure 5–17 Steps in preparing four kinds of layers. See text for details.

Simple Layering

This method can be used with plants that produce long shoots arising from the plant at ground level. In early spring the ends of these shoots can be bent over, placed in a hole in the soil several inches deep, and recurved so that the shoot tip is exposed above ground. The curved section to be buried should be nicked or twisted slightly to retard translocation of food materials through the stem; this promotes rooting. The hole is then filled in with loose, moist soil tamped firmly in place. Sometimes it is necessary to drive a stake with a hook on it into the soil beside the layer to hold it in place. After the first summer's growth, the layer will usually have rooted and can be cut from the parent plant, dug, and transplanted. Simple layering is used to propagate filberts commercially.

Mound Layering

This method, also called **stooling,** is widely used commercially to propagate apple and plum clonal rootstocks as well as currant and gooseberry cultivars. In late winter, the mother plants in the stool bed are cut back almost to the ground. As the new shoots start to grow in the spring, moist soil—or sometimes a mixture of soil and wood shavings or sawdust—is placed around the base of the shoots. As the shoots continue to grow, the mixture is mounded higher. The stool beds are kept continually moist, usually by overhead sprinklers. Roots develop from the base of these shoots, which by the end of the summer are well rooted. In early spring the soil is pulled away, and the rooted layers are cut off and transplanted to the nursery row for another season's growth. The stool bed is then handled again in the same manner during the next season.

Air Layering

In this method the rooting medium is brought up to the stem to be rooted, rather than bringing the stem down to the soil. It is used for propagating the India "rubber" plant (*Ficus elastica*) or for *Dieffenbachia,* as well as many stiff-stemmed tropical plants, such as the litchi (*Litchi chinensis*) and Jackfruit (*Artocarpus heterophyllus*). The leaves of the branch to be air-layered are removed just below a good clump of foliage. The stem is girdled by cutting the bark away down to the wood for a width of 2.5 cm (1 in.) and a root-promoting auxin powder is rubbed into this cut. A ball (about two handfuls) of moist (not wet) sphagnum moss is wrapped around the girdling cut, then a sheet of polyethylene plastic or aluminum foil is wrapped snugly around the sphagnum moss and tied tightly above and below the ball. No further watering is needed because the moss absorbs moisture from the plant itself and the covering retards water loss. After several weeks, roots will start developing. When a heavy root system has formed, the layer can be cut off just below the root ball, removed from the plastic, and, without disturbing the roots, planted in a large pot of soil. A few leaves should be removed to reduce water loss and the plant should be put in a cool, humid, and shady place until it becomes well established on its own new roots system.

Other Plant Structures Providing Natural Propagation Methods

Runners

Some plants, such as the strawberry (Fig. 5–18), grow as a rosette crown, with runners (stolons) arising from the crown.

Figure 5–18 Strawberry plant with new plants developing from nodes on the runners (stolons).

New plants arise from nodes at intervals along these runners. From these runner plants additional new runners arise, thus developing a natural clonal multiplication system. The runner plants must have favorable moist soil conditions to root. The strawberry produces runners in the summer in response to long days, and stops producing runners as the days shorten in the fall. Then the strawberry runner plants can be dug, packed in polyethylene-lined boxes, and placed in cold storage ($-2°C$; $28°F$) for planting later, usually the following spring. Other plants with this natural asexual reproduction system are the ground cover, *Duchesnea indica,* and the strawberry geranium (*Saxifraga stolonifera*).

Suckers

Some plants, such as the blackberry and red raspberry, produce **adventitious** shoots—or suckers—from their horizontal root system, which eventually spread to form a dense thicket of new plants. These individual shoots with a piece of the old root attached can be dug and replanted. This is a simple, highly successful asexual propagation method.

Crowns

Many perennial plants exist as a single unit, becoming larger each year as new shoots arise from the crown (root-shoot junction) of the plant. Vegetative propagation of such plants consists of **crown division**—cutting the crown into pieces, each having roots and shoots, and transplanting to a new location (Fig. 5–19).

Propagation Using Specialized Stems and Roots

A number of herbaceous perennial plants have structures such as **bulbs, corms, tubers, tuberous roots,** or **rhizomes.** These structures function as food storage organs during the plant's annual dormant period and also as vegetative propagation structures. In bulbs and corms, the newly formed plants break away from the mother plant naturally. This type of propagation is termed **separation.** The remaining structures—tubers, tuberous roots, and rhizomes—must be cut apart; this is termed propagation by **division.**

Bulbs

These are short, underground organs having a basal plate of stem tissue with fleshy leaf scales surrounding a growing point or flower primordium (see Fig. 20–8). In the axils of some of these leaf scales, new miniature bulbs develop that eventually grow and split off from the parent bulb to form a new plant. There are two types of bulbs: (1) the **tunicate bulb** has a solid tight structure with fleshy scales arranged in concentric layers and covered by a dry membranous protective layer; examples are onion, daffodil, tulip, and hyacinth; (2) the **scaly bulb,** such as the lily, has loose, separate scales with no protective layer to prevent desiccation. The scales can be removed from the bulb; placing them under moist,

Figure 5–19 Crown of an herbaceous perennial, the Stasta daisy. Lateral shoots develop from the underground portion of older stems and root. These rooted shoots can be cut from the mother plant and replanted. Underneath is a rooted shoot cut from the mother plant.

humid conditions will promote the development of one or more new miniature bulbs at the base of each scale.

Corms

A **corm** is the swollen underground base of a stem axis; it has nodes and internodes and is enclosed by dry scalelike leaves. The gladiolus, freesia, and crocus are examples of plants having a corm structure. Following bloom in gladiolus, one or more new corms develop just above the old corm, which disintegrates. In addition, several new small corms called cormels are produced just below each new corm. These may be detached and grown separately for one or two years to reach flowering stage.

Stem Tubers

The edible Irish (white) potato is a good example of a plant having a tuber structure. The underground tuber consists of swollen stem tissue with nodes and internodes. When a potato tuber is cut into sections and planted, shoots arise from the buds (eyes) on the potato piece, which itself serves as a food supply for the developing shoot. From the lower portion of such shoots, adventitious roots develop, along with several underground horizontal shoots that are stem tissue. The terminal portions of these horizontal shoots enlarge greatly to form the fleshy potato tuber. Potato cultivars are clones and are propagated by dividing each tuber into several sections. The crop eventually developing from each section usually consists of three to six new tubers.

Tuberous Roots

Tuberous roots are root tissue. The dahlia, tuberous-rooted begonia, and the sweet potato (Fig. 5–20) are examples.

Sweet potatoes are propagated vegetatively by placing the tuberous roots in beds so they are covered with about 5 cm (2 in.) of soil. Adventitious shoots (slips) develop from these root tubers; adventitious roots form from the base of the shoots. These rooted slips are then pulled from the tuberous

root and planted. As the sweet potato vines grow, some of their roots swell to form the familiar edible sweet potato.

Rhizomes

Certain plants, such as the German iris and bamboo (Fig. 5–21), have the main stem axis growing horizontally just at or slightly below the soil surface. This stem is a rhizome. Some plants have thick and fleshy rhizomes while others have thin, slender rhizomes. Like any stem, rhizomes have nodes and internodes. Leaves and flower stalks and adventitious roots develop from the nodes. Propagating plants with rhizomes is easy. At a time of year when the plant is not actively growing, the rhizome can be cut into pieces several inches in length and transplanted. Noxious weeds, such as Johnson grass, that have a rhizome structure cannot be controlled by cultivation because it merely breaks up the rhizomes and spreads the pieces about, each piece developing a new plant. Many important economic plants are propagated from rhizomes; examples are banana, ferns, ginger, and many grasses. Sometimes the above-ground horizontal stems are termed stolons, particularly in such grasses as Bermuda grass.

MICROPROPAGATION (TISSUE CULTURE)

A major advance in plant propagation involves the use of very small pieces of plant tissue grown on sterile nutrient media under aseptic conditions in small glass containers. These small pieces of tissue, called **explants,** are used to regenerate new shoot systems, which can be separated for rooting and growing into full-size plants. Some of the pieces of tissue are retained for further regeneration. The increase is geometrical, giving rise to fantastically large numbers of new individual plants in a short time. Some nurseries pro-

Figure 5–20 *Left:* Sweet potato tuberous root, producing rooted adventitious shoots. *Right:* Two detached shoots (slips) ready for planting.

Figure 5–21 Rhizome of bamboo with lateral buds and adventitious roots arising at nodes. Plants with rhizomes are propagated simply by cutting the rhizome into pieces and planting them.

duce millions of plantlets a year by tissue culture methods, which require the use of highly trained technicians working under sterile conditions.

Micropropagation was used at first with herbaceous plants, such as ferns, orchids, gerberas, carnations, tobacco, chrysanthemum, asparagus, gladiolus, gloxinia, strawberry, and many others. Later, with modifications of the media, it was found that many woody plants, such as rhododendrons, kalmias, deciduous azaleas, roses, plums, apples, and many others could also be successfully propagated by tissue culture methods (Fig. 5–22).

Different parts of the plant can be taken as the explant. Entire seeds themselves can be used; for example, very tiny orchid seeds have been commercially germinated in sterile culture for many years. Embryos can be extracted from seeds and grown on a sterile nutrient medium. Shoot-tip culture involves excision of the growing point, which increases in size when kept in a nutrient medium and is divided over and over, greatly increasing plant numbers. This method has been successful with orchids, ferns, apples, and carnations. In other cases, tissue culture involves, for example, the excision of a piece of stem tissue and placing it on a nutrient medium. The explant then develops masses of **callus** by continuous cell division. From these callus clumps, roots and shoots may differentiate to form new plants. This method has been used with carrot, tobacco, asparagus, endive, aspen, Dutch iris, and citrus.

Single pollen grains of some plants, such as tobacco, have been germinated in sterile culture, when taken at just the right stage, to form **haploid** (1n) plants. Doubling the chromosome number with colchicine treatments yields diploid homozygous plants.

In addition to propagation, these aseptic tissue cultures procedures are used for pathogen elimination and germplasm preservation. The nutrient media used for micropropagation are also favorable substrates for the growth of bacteria, fungi, and yeasts, so the prepared nutrient medium and its containers must be sterilized. The sealed containers are placed in an autoclave or pressure cooker for 20 to 30 minutes at 120°C (248°F). The plant tissue itself underneath the epidermis is sterile, but surface sterilization of the explants with a material such as a 10 percent Clorox solution is necessary, followed by rinsing with sterile water. Excision and insertion of the plant tissues onto the nutrient medium in the sterilized containers must be done with laboratory skill and procedures to prevent recontamination.

The nutrient media upon which the excised plant parts are grown include mineral salts, sugar, vitamins, growth regulators (auxins and cytokinins), and sometimes certain organic complexes such as coconut milk, yeast extract, or banana puree. Many different media have been developed for obtaining shoot and root proliferation on explants of various species. The constituents of one widely used medium are given in Table 5–2. A mixture of micronutrient elements, including iron, is often added.

Figure 5–22 Small plantlets of plum trees being propagated in a sterile nutrient medium in test tubes (10 cm rule). *Source:* F. Loreti and S. Morini.

TABLE 5–2 Components of the Murashige and Skoog Medium for Growing Tissue Explants Under Sterile Culture

NH$_4$NO$_3$	400 mg/l	indoleacetic acid	2.0 mg/l
Ca(NO$_3$)$_2$·4H$_2$O	144	kinetin	0.04-0.2
KNO$_3$	80	thiamin	0.1
KH$_2$PO$_4$	12.5	nicotinic acid	0.5
MgSO$_4$·7H$_2$O	72	pyridoxine	0.5
KCl	65	glycine	2.0
NaFe—EDTA	25	myo-inositol	100
H$_3$BO$_3$	1.6	casein	
MnSO$_4$·4H$_2$O	6.5	hydrolysate	1,000
ZnSO$_4$.7H$_2$O	2.7	sucrose	2%
KI	0.75	powdered purified agar	1%

Micronutrient elements may or may not be required but are usually added routinely. The following stock solution will provide the required materials. One milliliter of this solution is added per liter of culture medium.

MnSO$_4$·4H$_2$O	1.81 g	CuSO$_4$·5H$_2$O	0.08 g
H$_3$BO$_3$	2.86 g	(NH$_4$)$_2$MoO$_4$	0.09 g
ZnSO$_4$·7H$_2$O	0.22 g	Distilled water	995 ml

Iron is usually essential and can be supplied in several ways: iron tartrate (1 ml of a 1 percent stock solution), inorganic iron (FeCl$_3$.6H$_2$O, 1 mg/l) or FeSO$_4$, 2.5 mg/l; or chelated iron. Chelated iron can be supplied as NaFeEDTA, 25 mg/l, or by mixing Na$_2$EDTA and FeSO$_4$·7H$_2$O in equimolar concentrations to give 0.1 mM Fe.

SUMMARY AND REVIEW

Plants can be propagated sexually by seed or asexually by techniques that include cutting, grafting, and layering, and by the use of plant structures such as runners, suckers, crowns, bulbs, corms, rhizomes, and so on. Micropropagation involves the use of tissue culture.

Proprogation of varieties and cultivars by seed requires careful control of the seed source. The genetic complement of the seed must remain the same as that of previous generations of the variety or cultivar. Seed certification programs are designed to maintain the genetic quality of each generation of seed. Most agronomic (field) crops have seed certification programs. In the case of horticultural crops (vegetables and ornamentals), seed genetic quality is maintained by the seed companies producing the seed. The development of hybrid seed has been one of the more important breakthroughs in crop improvement and seed production.

Seed production starts with seed formation either through natural fertilization or hybridization and subsequent maturation of the seed and embryo. After harvesting, the seeds are stored in the appropriate environment (usually dry and cold) until planted. Before planting, the stored seeds are sampled and tested for viability (the ability to germinate) by a test such as cut, float, X-ray, germination, tetrazolium, or excised embryo.

In many species at the time of seed maturation and harvest the embryo is dormant and requires some change in environmental factor to break dormancy. Stratification (exposure to cold temperatures) will break dormancy of some seeds. Other seeds require soaking or leaching to remove a chemical inhibitor present in the seed.

Seed germination is a series of events that include imbibition of water, activation of hormones and enzymes, then embryo growth and development. Environmental factors that affect seed germination include water, temperature, aeration, light (for some seeds), pathogens, and salts.

Asexual propagation involves reproducing an entire plant from a part of another plant. Except in some cases such as mutation or chimera, the new plant has exactly the same genetic makeup as the first plant. The technique is very useful when genetic integrity cannot be maintained by seed production or when seed production is inefficient.

Cutting production involves removing a part of a stem or root of a plant and regenerating the missing organs and tissues by providing an environment that favors the regeneration process.

Grafting is the attaching of a twig called the scion from a plant with one genetic complement to the rootstock from a plant with a different genetic complement. The scion and rootstock must be closely enough related for there to be grafting compatibility. There are several different kinds of grafts, including whip and cleft. Budding is similar to grafting except the scion is a bud.

Layering is similar to cutting except the part to be propagated is not removed from the parent plant until the missing stem or roots have regenerated. Layering can be simple, mound, or air.

Other plant structures that can be used for propagation are plantlets that can be separated from the parent plant or crowns, bulbs, corms, and tubers that can be divided. Plantlets can come from runners, which extend from the main portion of the plant or form naturally in some species along leaf margins and at leaf bases.

EXERCISES FOR UNDERSTANDING

5–1. Using seed obtained from a local garden center or feed and seed store, test for viability and germination rate using some of the techniques mentioned. Check the label to see if viability and germination data are provided. If they are, compare your results to those on the label. If germination rate is low, try different techniques such as stratification, soaking in water, or placing the seeds in light or dark to improve germination rate.

5–2. With permission of the owner, try to air-layer a tropical houseplant such as dieffenbachia (dumb cane) or dracaena. Likewise, try to propagate root cuttings from herbaceous garden or houseplants by placing the cuttings in a pot of moist medium and covering with a clear plastic bag sealed to the pot. Do multiple pots with some in high light, some in low light, some on or near a heater, and others away from a heat vent. Warning: do not try to sell the new plants you may produce. Many plants sold commercially are protected by patent laws and sale without permission of the patent holder is illegal.

REFERENCES AND SUPPLEMENTARY READING

HARTMANN, H. T., D. E. KESTER, F. T. DAVIES, and R. L. GENEVE. (1997). *Plant propagation: principles and practices.* Upper Saddle River, N. J.: Prentice-Hall.

RICE, L. W., and R. P. RICE, JR. 2000. *Practical horticulture.* Upper Saddle River, N. J.: Prentice-Hall.

CHAPTER

6

Vegetative and Reproductive Growth and Development

KEY LEARNING CONCEPTS

After reading the chapter you should be able to:

♦ Recognize definitions and measurements of plant growth and development.
♦ Understand factors that affect plant growth.
♦ Recognize the categories of plant hormones and understand the role of hormones in plant growth and development.

VEGETATIVE GROWTH AND DEVELOPMENT

Shoot and Root Systems

In living plants we primarily see the shoot system, in all its diverse patterns—from the mosses to the magnificent towering redwoods, oaks, and pines. In many crop plants grown for livestock forage—alfalfa, corn for silage, and pasture grasses—the entire shoot system is harvested. A beautiful lawn results from the shoot and leaf growth of millions of small grass plants. A timber tree cut for lumber is the product of many years of shoot growth. The much-admired potted foliage plants in our homes are the shoot and leaf systems of particular kinds of plants brought into cultivation many years ago after their discovery by plant explorers in tropical lands.

In growing most plants—with the exceptions, perhaps, of bonsai[1] plants and container-grown ornamentals—we are interested in obtaining as much vigorous **vegetative** growth as we can as quickly as we can. This is particularly true for crop plants, where the difference between a profitable year and a loss for the farmer can be the amount of shoot growth harvested or the amount of leaf area produced to nourish the developing flowers, fruits, grains, or seeds.

The roots of plants are largely unseen and tend to be forgotten, but in the higher plants they are essential for growth. The four principal functions of roots in higher plants are:

1. anchoring plants in the soil
2. absorbing water and mineral nutrients
3. conducting water and dissolved minerals, as well as organic materials
4. storing food materials; for example, in plants such as sweet potatoes, sugar beets, and carrots

The roots of some plants can also function in vegetative reproduction, as in root cuttings.

The root system and the shoot system tend to maintain a balance. As the top of the plant grows larger and larger, the leaf area increases and water loss through transpiration increases. This increased water loss is made up by water absorption from an increasing root system. The enlarging shoot system also requires greater amounts of minerals that are absorbed by the increasing root system.

Definitions and Measurements

What is plant growth and how is it measured? We generally think of growth as an irreversible increase in volume or dry weight. The swelling of wood after it becomes wet is not growth because the wood will shrink upon drying. Growth can be measured as increases in fresh weight or dry weight, or in volume, length, height, or surface area. As its gross size increases, a plant's form and shape change as directed by genetic factors. Plant growth is a product of living cells, with all their myriad metabolic processes. A definition of plant growth is as follows: size increase by cell division and enlargement, including synthesis of new cellular material and organization of subcellular organelles.

Plant shoot growth can be classed as **determinate** or **indeterminate.** In the case of determinate growth, after a certain period of vegetative growth, flower bud clusters form at the shoot terminals so that most shoot elongation stops. Many vegetable species and cultivars grow this way, remaining "dwarfed" bush plants; bush snap beans are an example of a determinate type. Plants with indeterminate growth bear the flower clusters laterally along the stems in

the axils of the leaves so that the shoot terminals remain vegetative and the shoot continues to grow until it is stopped by senescence or some environmental influence. Trailing pole beans, grapevines, and forest trees are examples of plants with an indeterminate growth pattern. The determinate, bush-type plants produce much less vegetative growth than do the indeterminate type.

Shoot Growth Patterns: Annuals, Biennials, and Perennials

Annuals, which are herbaceous (nonwoody) plants, complete their life cycle (seed to seed) in one growing season. Shoot growth commences after seed germination and continues in a fairly uniform pattern—provided no environmental influences are limiting—until growth is stopped by frost or some senescence-inducing factor. Flowering, followed by fruit and seed production, occurs at intervals through the summer. General growth curves for the annuals are shown in Figure 6–1; a detailed growth curve for barley, an herbaceous annual, is shown in Figure 6–2.

Figure 6–3 shows various events in the life cycle of a typical **angiosperm** annual plant. All these events occur during a single summer growing season.

The **biennials,** which are herbaceous plants, require two growing seasons (not necessarily two years) to complete their life cycle (seed to seed). As shown in Figure 6–4, stem growth is limited during the first growing season, the shoot

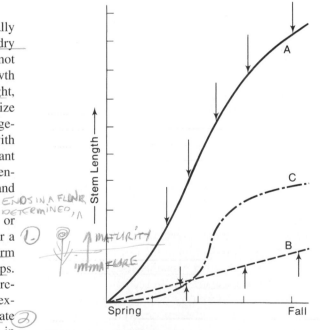

Figure 6–1 Vegetative growth patterns of annula plants. (A) Indeterminate vine-type plants. (B) Determinate, bush-type plants. (C) Terminal-flowering plants, such as cereals and grasses. Arrows indicate times of flower initiation. *Source:* Adapted from Rappaport, L., and R. M. Sachs. 1976. *Physiology of cultivated plants.* Davis, Calif.: UCD Bookstore.

[1]Bonsai is a Japanese term describing the art of dwarfing and shaping trees, shrubs, or vines grown in containers by careful pruning of roots and tops.

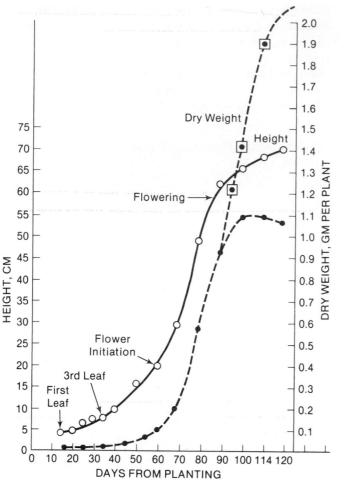

Figure 6–2 Growth curve of a field-grown barley plant from leaf emergence to grain maturity. O = plant height. ● = dry weight of plant minus grain weight. ◉ = dry weight of plant plus grain weight. *Source:* Adapted from Noggle, G. R., and G. J. Fritz. 1976. *Introductory plant physiology.* Englewood Cliffs, N.J.: Prentice-Hall.

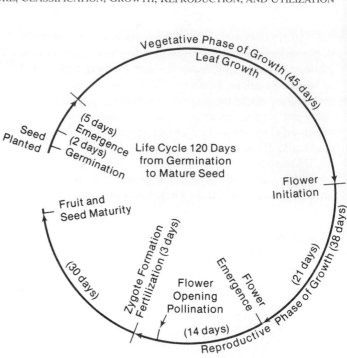

Figure 6–3 Events in the life cycle of a typical annual plant—from seed planting to seed maturity—accomplished in four months. *Source:* Adapted from Noggle, G. R., and G. J. Fritz. 1976. *Introductory plant physiology.* Englewood Cliffs, N.J.: Prentice-Hall.

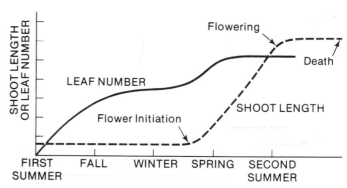

Figure 6–4 Growth curves of biennial plants during the first growing season (vegetative growth only), a required winter chilling-period, and a second growing season (flowering, fruiting, and seed production). *Source:* Adapted from Rappaport, L., and R. M. Sachs. 1976. *Physiology of cultivated plants.* Davis, Calif.: UCD Bookstore.

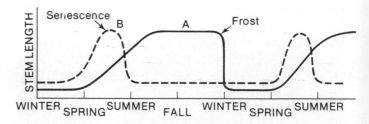

Figure 6–5 Shoot growth patterns of herbaceous perennials over a two-year period. (A) Plants such as garden (outdoor) chrysanthemums, peony, and phlox whose growth is stopped by cold weather in the fall. (B) bulbous plants such as tulips, narcissus, and hyacinths, whose growth is terminated after spring flowering. *Source:* Adapted from Rappaport, L., and R.M. Sachs. 1976. *Physiology of cultivated plants.* Davis, Calif.: UCD Bookstore.

system remaining mostly as a rosette. The plants remain alive but dormant through the winter. Exposure to chilling temperatures triggers hormonal changes leading to stem elongation, flowering, fruit formation, and seed set during the second growing season. Senescence and death of the plant follows shortly thereafter, Examples of herbaceous biennials are celery, Swiss chard, beets, and such cole crops as cabbage and Brussels sprouts.

Most annual and biennial plants flower and fruit only once before dying. The production of the flowers and fruits or, perhaps, just the flowering stimulus itself apparently causes the plants to senesce and die. In such plants, continued removal of flowers and fruits often delays senescence.

Perennials are either herbaceous or woody. In herbaceous perennials the roots and shoots can remain alive indefinitely but the shoot system may be killed by frosts in cold-winter regions or by senescence-inducing factors. Shoot growth resumes each spring from latent or **adventitious** buds at the crown of the plant. Figure 6–5 shows typical growth patterns for two types of herbaceous perennials. Many perennial ornamental plants (e.g., pelargoniums) are grown as annuals in areas with severe winters such as the Midwest region of the United States. When grown in areas with mild winters such as parts of the South and West (United States), these plants exhibit their perennial characteristics.

In woody perennial plants both the shoot and root system remain alive indefinitely, each growing to the ultimate size for the particular plant as programmed by its gene complement. Shoot growth of temperate zone plants takes place annually during the growing season, as indicated by Figure 6–6, adding to the growth accumulated in previous seasons. The magnitude of growth can vary considerably from season to season under the influence of several environmental factors.

In the tropics some of the evergreen tree species continually grow vegetatively throughout the year, but other species exhibit intermittent growth, sometimes correlated with changes in weather, especially rainfall. Vegetative growth of coffee trees, for example, is much greater in the wet season than in the dry.

In the temperate zones all trees show intermittent annual vegetative growth. These growth patterns can be placed into four categories: NB TO KNOW WHEN PRUNING ✱

1. A single flush of terminal shoot growth during the growing season followed by the formation of a resting terminal bud; oaks, hickories, many conifers, and most fruit tree species are examples.
2. Recurrent flushes of terminal growth with terminal bud formation at the end of each flush; examples include the pines of the southeastern United States (*Pinus taeda* and *P. virginiana*) and certain subtropical evergreen and deciduous tree species such as citrus and Persian walnut.
3. A flush of growth followed by shoot tip abortion at the end of the season, with the shoot the following season starting up from a lateral bud. This gives rise to a

zigzag pattern of shoot development, exemplified by the elm, birch, willow, beech, and honey locust.
4. A sustained growth flush for varying periods, producing new growing points that develop as late-season leaves, and ending the season with the formation of a distinct terminal bud; examples are the tulip tree (*Liriodendron tulipifera*) and sweet gum (*Liquidambar styraciflua*).

There is a pronounced diurnal variation in shoot growth of woody perennials. Shoots tend to grow more rapidly at night than in the day provided temperatures are favorable. This difference is probably due to a lower water stress at night than during the day. Some species show twice the growth at night compared with day growth.

Root Growth Patterns

Various studies of root growth patterns conclude that the roots of mature deciduous woody perennials start to grow in the spring before bud break and, in some species, stop growth at various times in the summer, but in others continue to grow until after leaf drop in the autumn. Growth peaks in the spring and again in late summer or early autumn. Roots of some species, however, continue to grow even during the winter months whenever soil temperatures and moisture are favorable. Not all roots of a tree may be growing at any one time. While some are actively growing, other may be quiescent. The spring flush of root growth results from the accumulated foods stored in the tree the previous year. When this source is depleted, root growth slows but, following gradual accumulation of carbohydrates from photosynthesis through the summer, root growth again increases in the autumn.

It does not appear that the growing points of the roots are under the same control as the shoot buds which in many woody perennials go into a resting period in late summer and require chilling through the winter to be reactivated.

How the Plant Grows

The importance of meristems in the growth of plants must be clearly understood. In dicotyledonous plants, vegetative buds at the shoot tips (**apical meristems**) and in the axils of leaves contain meristematic cells that are capable of dividing and redividing, by mitosis and cytokinesis,[2] producing millions of cells along a longitudinal axis. Cell division, together with cell elongation, causes the shoots to grow. A growing point at a shoot tip, with its meristematic region pro-

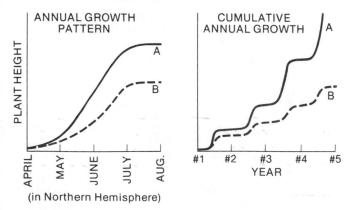

Figure 6–6   *Left:* Growth patterns for temperate zone woody perennials in the northern hemisphere during one growing season, and *Right:* over a period of several years. (A) Curve for a rapid-growing species such as poplar. (B) Curve for a slow-growing species such as oaks. *Source:* Adapted from Rappaport, L., and R.M. Sachs. 1976. *Physiology of cultivated plants.* Davis, Calif.: UCD Bookstore.

[2]Cell division occurs in two parts: (1) *mitosis,* where one nucleus is divided into two nuclei, and (2) *cytokinesis,* which is the division of the resulting binucleate cell into two uninucleate cells. The latter is accomplished by the formation of the cell plate, a thin layer of polysaccharide materials that develops across the spindle between the daughter nuclei toward the end of mitosis, thus dividing the cytoplasmic constituents of the mother cell.

ducing new cells year after year for many hundreds of years, can account for the towering height of the redwood, Douglas fir, and other forest trees. Similar meristematic cells are also located in the root tip, just behind the root cap.

The thickening of the trunks of growing trees and other woody perennials is due to secondary growth produced by another meristematic region, the vascular cambium, a cell layer that lies between the xylem and the phloem and encircles the tree from the roots to almost the top of the shoot. Cells of this vascular cambium also have the capability of dividing, producing both the permanent woody xylem tissues toward the inside that give the tree its girth and mechanical strength and the more fragile transitory phloem cells to the outside. External to the vascular cambium is another meristematic region, the cork cambium, which produces the cork in the bark layer.

Genetic Factors Affecting Plant Growth and Development[3]

One of the marvels of biology is the manner in which the offspring of plants and animals resemble, yet differ from, their parents. A peach tree, not a cherry tree, grows from a germinated peach seed. In a field planted with 'Marquis' wheat seed, 'Marquis' wheat plants develop, not oats or barley. The organism developing by cell division and elongation from the fertilized egg—the zygote—in every case is under the genetic control of the genes inherited from the parents at the time of fertilization. The genes, which direct the form and shape of the organism, are passed on to the daughter cells. All of them are on the chromosomes in the nucleus of each cell of the organism.

As the plant enlarges from the fertilized egg (zygote) to its mature size, many developmental processes take place. Certain segments of DNA (genes) direct the synthesis of enzymes that catalyze specific biochemical reactions required for growth and differentiation. The genes involved in protein synthesis are referred to as **structural genes.** Genes involved in regulating the activity of the structural genes are referred to as **regulatory** and **operator genes.** During a particular stage of development, not all the genetic information stored in the chromosomes is necessarily active. As development proceeds, some genes become activated—switched on—whereas other genes are inactivated—switched off. The regulatory genes are involved in activating or deactivating operator genes that, in turn, control the activity of structural genes.

What are the signals that trigger the action of the regulatory genes? Although not clearly understood, they are believed to include growth regulators, certain inorganic ions, coenzymes, and other metabolites. Environmental factors such as temperature or light can also function as signals during certain developmental stages. Thus, the particular combination of genes directs the form and size each plant is

finally to assume, as altered by stresses from environmental influences, either beneficial or deleterious.

Environmental Factors Influencing Plant Growth and Development

Light　The sun is the source of energy for photosynthesis and other plant processes but only a small amount of the sun's radiation reaches the earth's surface. The atmosphere is composed of several gases, water vapor, and dust that filter out much of the radiation. Ozone absorbs the long wavelengths, keeping the earth's surface from becoming too warm. Thus the atmosphere is a gaseous envelope that filters out many of the sun's rays but allows enough of the visible and some of the invisible rays (both long and short wavelengths) to pass to maintain the earth at the moderate temperature both plant and animal life needs. These rays of light strike the earth's surface, heating the soil and large bodies of water which, in turn, give off heat (reradiating long wavelengths) that is trapped by the atmosphere and warms it.

Light intensity is high where there are no clouds or where there is little moisture in the air, as in deserts. Light intensity is lower in humid or cloudy regions because the water vapor absorbs much of the radiation. Since the earth is tilted on its axis in relation to the sun (at an angle of 23 1/2°), light intensity varies greatly with the season in the temperate zones. In the summer the sun's rays strike the earth's surface more directly than in the winter. Day length varies greatly near the poles: from 24 hours of light in summer to 24 hours of darkness in winter. These variations become less extreme toward the equator, where the day lengths range from 12 to 13 hours all year.

The narrow band of light that affects plant photoreaction processes ranges from about 300 nanometers (nm) to about 800 nm. The most important process, **photosynthesis,** is very efficient in the orange-red (650 nm) and in the blue range (440 nm) of the visible spectrum. The stomates, which have chlorophyll in their guard cells and allow carbon dioxide to enter, open in the light and are generally closed in the dark. They respond effectively to red light (660 nm) and blue light (440 nm); thus stomatal opening and closing is very closely correlated with photosynthesis. Stomates may close about midday, as the plants wilt slightly because of excessive transpiration.

Light also affects plant growth and development through a process called **photomorphogenesis**, meaning shape determined by light. Most photomorphogenic responses are controlled by the **phytochrome** pigment system. Phytochrome is a pigment that allows the plant to detect changes in the environment by sensing changes in light quality, specifically the shift in red to far-red light. (Far-red light is considered to be the region of light between 700 and 800 nm with the region between 730 and 740 nm being the most effective.) Phytochrome has the capability of absorbing both red and far-red light, but not at the same time as shown in Figure 6–7. If the pigment has been exposed to red light, it

---

[3]For a discussion of some basic genetic concepts, see Chapter 4.

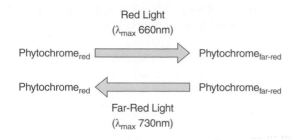

Red Light
(λ_{max} 660nm)

Phytochrome$_{red}$ ➡️ Phytochrome$_{far-red}$

Phytochrome$_{red}$ ⬅️ Phytochrome$_{far-red}$

Far-Red Light
(λ_{max} 730nm)

Figure 6–7 How red and far-red light affect phytochrome. Phytochrome$_{red}$ can only absorb red light, while phytochrome$_{far-red}$ can only absorb far-red light. The wavelength of light that is most effectively absorbed by the phytochrome molecule is signified by λ_{max}. Slightly longer and shorter wavelengths are also absorbed, but less effectively.

changes form and can then absorb far-red light when available and vice versa. Through this mechanism the plant detects the ratio of red: far-red (R:FR) light in the environment. The shifting ratios of light indicate to the plant when neighboring (competing) plants are present and when the day has begun and ended. Some forms of phytochrome degrade slowly in the dark, which also affects the status of the phytochrome pool and may give plants a method to sense the length of day and night. Although the mechanism of how the signal received by phytochrome is translated to a growth response is unknown (it presumably affects the hormonal balance in the plant), the responses elicited by phytochrome have tremendous impact on the production of many crops.

Plant architecture in an environment where there is more red than far-red light, such as in full sun, is very distinguishable from one where far-red dominates, such as in crowded or shaded field or greenhouse bench conditions. Plants grown in environments where red light dominates are short and compact with dark green leaves. These plants are sturdy and stand up well to wind, rain, and harvesters. The number of branches (tillers) multiplies, increasing the number of flowers or seed heads per plant. The compact, highly branched, dark green form is very desirable for most ornamental crops produced in greenhouses and nurseries. Plants grown in environments where far-red dominates are tall, spindly, and weak with few branches and leaves that are large, thin, and light green. These plants lodge (fall over) in the field and are unsightly as ornamentals.

Other plant processes strongly influenced by light are phototropism and photoperiodism. **Phototropism** is the movement or bending of stems, leaves, and flowers toward the light. This bending results from more cell growth on the side away from the light source than on the side toward it. It is believed that auxin, a natural hormone, accumulates and is transported to the shaded side, promoting cell expansion there. Blue light is the most active in this process, but the exact pigment involved is unknown.

Another photomorphogenic response is **photoperiodism**, the physiological responses of plants to variations in length of light and darkness. Plants have many such responses. As the day length gradually shortens in late summer and fall, a 'signal' sent from the leaves starts certain processes that help the plant survive the winter. The onset of dormancy in some plants begins when days shorten in the fall, resulting in growth cessation and enabling the plant to survive the winter cold. Development of fall leaf color in certain trees is also attributed to the shortening day. Flower initiation begins in such plants as the strawberry, poinsettia, and chrysanthemum with the onset of short days in the fall. Tubers of potatoes, dahlias, and tuberous-rooted begonias begin to form when the days start to shorten after midsummer; onions, on the other hand, form bulbs when the days begin to lengthen in late spring. Photoperiodism in flower induction is discussed later in this chapter.

Temperature The seasonal variation of light intensity is responsible for the temperature changes from summer to winter in the various temperature zones. Farmers depend on climatic records in their areas to predict the last day of frost in the spring and the number of available "growing days" before the first killing frost in the fall. The greater the distance from the equator, the fewer the number of available growing days to mature crops. It is possible to grow temperature-sensitive tomatoes in Alaska or northern Europe, but precautions must be taken to protect the seedlings from frost in early spring. But once summer arrives in these northern latitudes, the days are so long and the temperatures are warm enough that plants grow, flower, and set fruit rapidly. The growing season may be short, but plants develop quickly.

All plants have optimal temperatures for maximum vegetative growth and flowering, as noted for plants discussed as crops in subsequent chapters. Most temperate-region plants grow between temperatures of 4°C (39°F) and 50°C (122°F), but these are generally the limits of plant growth. The high temperatures destroy the protoplasm of most cells; however, some spores and seeds can withstand the temperature of boiling water for short periods. At the low temperatures, most plants just fail to grow owing to lack of cell activity. However, there are some arctic or mountain plants that function near freezing, but these are rare exceptions.

Plant parts are injured by very high temperatures, even if the exposure is short. Leaves may be **solarized** or sunburned when exposed to high light intensities. In the leaf, light energy converts to heat, which destroys the cells. Young trees in orchards are prone to sunscald, which kills the cambium layer just under the thin bark of the trunk and limbs. Injury can be prevented by whitewashing the bark to reflect the heat. Occasionally, plants suffer heat damage when the relative humidity drops suddenly because of hot, desiccating winds. Heat damage in greenhouse-grown plants can be prevented if the relative humidity is increased by misting in the immediate vicinity of the plant.

The most common low temperature injury is evident after the night of the first killing frost in the fall. Plant surfaces may be coated with frost depending on the dew-point temperature, but soon after the sun shines on the leaves, the damage is evident in blackened leaves. The contents of the cells are damaged by the ice and, upon thawing, the water in the protoplasm moves to the intercellular spaces, the protoplasm dehydrates, and death occurs. Frost damage to plants due to heat loss at night by radiation can be avoided to some extent by placing an opaque shield between the plant and the clear sky on a calm night. This shield prevents radiation from the leaves to the cold sky. Transparent polyethylene plastics allow heat from longwave radiation to escape at night, and are therefore useless as a radiation shield. Citrus and grape growers sometimes use smudge pots burning oil to create heat that radiates to the trees. In locations where there are temperature inversions above the plants (slightly warmer temperatures in a layer some distance above the soil than at ground level), wind machines may be useful. The power-driven propellers mix the warm air above with the cool air below, thus preventing low temperature injury.

Many woody perennials grown in locations with severe winters enter into a rest period—brought on by shortening days in fall—and become resistant to low winter temperatures. A covering of snow helps the plants withstand the winter since it acts as a good insulator at extremely low temperatures.

Some plants of tropical origin may be injured at temperatures just above the freezing point. This is referred to as a chilling injury. The leaves may wilt and never recover and developing fruits may not mature; some examples are avocado, banana, mango, okra, and tomato. Unripe tomatoes, which show pink color, will be injured and will not finish the ripening process if refrigerated at 4°C (39°F) or less. Many ornamental foliage plants are susceptible to chilling injury. These plants need protection when being shipped during the winter from warm production areas such as Florida to colder areas. They should also be kept away from cold areas in homes and office buildings.

Water   Most growing plants contain about 90 percent water. It is stored in various plant tissues and is used as one of the raw materials for photosynthesis. A corn plant near harvest may contain as much as two liters (2.1 qt) of water, but during its growth from a seedling to a mature plant with ears, it may extract 100 times this amount from the soil. Some of this water is used in growth and development, but most of it is lost through the leaf stomates in the transpiration process. Therefore, in order to produce a mature crop of corn or tomatoes, the equivalent of 30 to 60 cm (12 to 24 in.) of water must be applied to the soil surface as rain or irrigation. The quantity of water necessary depends on the crop (some plants use much more than others) as well as the available sunshine during the season. When the light from the sun is strong, the leaves tend to lose large quantities of water by transpiration and, additionally, the soil loses water by evaporation. Substantially more soil water is lost from transpiration by plants than by evaporation from the soil. The total soil water loss by both means is called **evapotranspiration.** The rate of evapotranspiration is known in many farming regions and is a guide to the farmer concerned with how much irrigation water must be applied to compensate for the soil water loss.

The plant obtains water from the soil by forces in the transpiration process. The mesophyll cells of the leaves, filled with water, are connected to the intercellular spaces that lead to the stomates; the stomates, when open, lose moisture to the air. Thus, water is "pulled" through the plant via the conducting tissue (xylem). If more water is lost through the stomates than the roots can supply, the plant wilts. Herbaceous plants may wilt slightly at midday or later on a bright sunny day but usually recover during the night. Deciduous fruit plants often fail to recover from this water loss. Wilted plants eventually die if they cannot recover enough soil water to regain their turgidity. While the plant is wilted, the stomates are closed, cutting off the intake of CO_2 for photosynthesis, and thus reducing carbohydrate manufacture.

The quality of soil water, as determined by the quantity of minerals and salts dissolved in the water, is very important to the growth of plants. If there is too much dissolved salt, or if one salt, such as sodium chloride, predominates, the roots and shoots become damaged and the top of the plant suffers from lack of water. The plant may suffer from both lack of water and toxic effects of specific ions, e.g., sodium (Na) and chloride (C1). Some desert plants are resistant to high salts in water and manage to survive. In contrast, most domesticated plants require good quality water.

Agricultural production is influenced by both the quality and quantity of water available in a region of favorable temperatures. The quality of water is no problem when there is adequate rainfall. However, if the rain must be stored as runoff water in reservoirs, poor quality may result if the water accumulates too much salt before it is collected and used in irrigation.

Gases   The two gases most important to the growth of green plants are oxygen (O_2) and carbon dioxide (CO_2). Green land plants and the extensive phytoplankton of the oceans help keep a balance of O_2 and CO_2 in the atmosphere throughout the world. Green plants use CO_2 for photosynthesis and return O_2 to the atmosphere. Plants and animals use O_2 and give off CO_2 during respiration. Ocean and freshwater algae account for about 90 percent of the photosynthesis in the world! These plants do most of the work to keep out atmosphere in a favorable balance. Of the land plants, the forests of the world account for most photosynthetic activity.

According to one theory, stomatal opening and closing is regulated by the CO_2 level which is influenced strongly by photosynthesis. When the CO_2 concentration in the leaf cells behind the stomata is lower than that found in the atmosphere, stomata open. The use of CO_2 during photosynthesis keeps the concentration lower than atmospheric concentration and the stomata remain open if other conditions are favorable. (However, the whole process regulating stomatal opening and closing is more complex than just CO_2 level.)

Oxygen is important in the respiration of all plant parts. Respiration is the release of energy captured and stored in the carbohydrates (sugars) synthesized during photosynthesis. The released energy is used to drive the complex biochemical reactions needed for the growth and development of not only plants but also all living organisms.

Soils saturated or waterlogged for long periods do not contain adequate O_2 for root respiration activity. Roots cannot absorb minerals from the soil if the soil is lacking O_2 due to low respiration rates. When plants are grown with their roots in water (hydroponics), air is bubbled through the water to supply O_2 in constant small quantities so that the roots can respire.

Atmospheric pollution is a by-product of our technical society. We burn fossil fuels at a high rate for heat, transportation, and manufacturing. The burning releases the energy and CO_2 that plants (millions of years ago) stored as carbohydrates through photosynthesis. While increasing CO_2 may increase the photosynthetic productivity of crops, more CO_2 is being released than can be absorbed by plants, resulting in an increase in atmospheric CO_2, which has been associated with global warming. The loss of millions of acres of forestland has significantly reduced the amount of CO_2 that land plants can absorb. Reforestation on a large scale is being considered as a means to reduce CO_2 buildup in the atmosphere.

Other atmospheric pollutant by-products of burning include ozone, nitrogen oxide, sulfur dioxide, fluorides, and ethylene, all of which can be harmful to plants. Ozone is formed in a photooxidation process in the presence of strong sunlight. It is the major pollutant in smog and is usually found in large cities that have many automobiles. Exhausts from cars, in the presence of nitrogen oxide and abundant sunlight, produce ozone and peroxyacetyl nitrates (PAN), which harm both plants and animals. Plants affected by pollutants may have a leaf margin burn, as in some grasses, or the leaves—of petunias, lettuce, and beans—may lose chlorophyll in the center. This area of the leaf then becomes bleached or dies. Automobile exhausts also emit ethylene into the air. Ethylene is a growth regulator that can cause unusual plant growth patterns, such as leaf distortions. Sulfur dioxide results from the combustion of fossil fuels (except most natural gases) and is particularly evident in smelting processes using sulfide ores. Sulfur dioxide can cause pronounced areas of damage on leaves of sensitive plants. Fluorides can be emitted into the atmosphere by the manufacture of cement, glass, ceramics, bricks, aluminum, and phosphate fertilizers. Fluoride damage is usually seen along the margins or tips of leaves, especially in monocots.

Phase Change: Juvenility, Maturation, Senescence

Seedling plants undergo a phasic development throughout their lives that is essentially the same as in animals. They pass through embryonic growth, juvenility, a transition stage, maturity, senescence, and death.

During maturation, seedlings of many woody perennial species differ strikingly in appearance at various stages of development. When eucalyptus trees are very young they have opposite, broad leaves with no petiole. As the tree gets older and taller, the leaves in the upper portion of the tree assume a different appearance: they become alternate, are narrower, and have a distinct petiole, and the leaf color changes from bluish-green to a deep green. Young citrus seedlings are very thorny but as they continue their growth to full-size trees, thorns are no longer produced on the shoots. Figure 6–8 shows the transition stages in leaf patterns of an *Acacia melanoxylon* seedling as the shoot elongates and ages.

The familiar English ivy (*Hedera helix*) in its juvenile stage has three-or five-lobed palmate alternate leaves and hairy vinelike stems with no flowers. On the same plant can be found branches in the adult stage that have nonlobed ovate opposite leaves and smooth nonclimbing stems with prominent flower stalks and flowers.

The morphological changes of age are accompanied by important physiological changes. A primary criterion for judging that a plant is in the juvenile stage is its inability to form flowers even though all environmental conditions are conducive to flowering. The onset of flowering and fruiting indicates the termination of the juvenile phase. Another important indication of phase change from juvenility to maturity is a loss or reduction in the ability of cuttings to form adventitious roots.

Seedlings of many woody perennial species, including conifers, undergo this juvenile-to-mature phase change. The juvenile stage in certain forest species can last as long as 30 to 40 years and during this time, flowering does not take place.

Carbon dioxide is the third most abundant gas in the atmosphere, behind N_2 and O_2. In relative amounts, CO_2 is very small. Nitrogen is approximately 78 percent ($\approx$780,000 ppm), O_2 is approximately 21 percent ($\approx$210,000 ppm), and CO_2 is approximately 0.035 percent ($\approx$350 ppm). Atmospheric CO_2 rose 20 to 30 ppm during the 1900s. This elevation is believed to be enough to contribute significantly to global warming. Global warming results from atmospheric gases, especially CO_2, which trap heat at the earth's surface. The phenomenon is sometimes referred to as the "greenhouse effect." Gases that contribute to global warming are often misleadingly referred to as greenhouse gases. As a result, greenhouses are often looked upon by the public as being contributors to global warming. Ironically, because of the plants growing in them, greenhouses most likely reduce the greenhouse effect.

Figure 6–8 *Acacia melanoxylon* seedling showing phase change from juvenile to mature form. Lower, juvenile leaves have compound bipinnate structure. Upper, mature "leaves" are actually expanded petioles (phyllodia). Transition stages are evident in between. *Source:* Hartmann, H.T., and D.E. Kester. 1983. *Plant propagation.* 4th ed. Englewood Cliffs, N.J.: Prentice-Hall.

The phase changes are not genetic but result from little-understood epigenetic changes—altered patterns of gene expression, leading to biochemical and physiological alterations. These changes apparently originate in the apical meristems of the plant. Strangely, the lower part of many woody plants—the oldest in terms of calendar days—remains physiologically the youngest, least likely to flower, and most likely to produce adventitious roots on its cuttings. On the other hand, the top of the plant—the youngest in time—develops into the part that is the most mature, most likely to flower, and the most unlikely to form roots when cuttings are taken.

Once the mature phase is attained, a plant is relatively stable and does not revert easily to the juvenile state, although treatments with plant hormones such as gibberellins or grafting tissues from mature plant parts back onto juvenile plants can cause the mature tissue to revert to the juvenile stage.

Breeders of fruit or nut crops are eager to bring the seedlings from their controlled crosses through the juvenile stage as rapidly as possible so that they will flower and fruit and the breeders can see the results of their work. Plant propagators, on the other hand, would like to maintain the juvenile stage in stock plants as long as possible so that cuttings taken from them will root rapidly and in high percentages.

Not every young woody perennial plant is necessarily in the juvenile growth stage. It may have been propagated vegetatively—perhaps as a rooted cutting taken from a plant in the adult phase. On the other hand, if it was propagated from a germinated seed it would be in a juvenile growth phase. Young nursery stock of all cultivars of fruit trees is propagated asexually from material taken from mature trees, and such stock is usually many generations from the original seedling mother tree for the cultivar—which itself did go through the transitory phases from a juvenile to an adult form.

Aging and Senescence

The life spans of the different kinds of flowering plants differ greatly, ranging from a few months to thousands of years. Some of the coniferous evergreen forest trees are the earth's oldest living organisms. Living bristlecone pines (*Pinus aristata*) said to be over 4,000 years old are growing in California's Sierra Nevada Mountains (Fig. 6–9). Some of the giant California redwoods (*Sequoia sempervirens*) are known to be over 3,000 years old. Olive trees with huge trunks found in the eastern Mediterranean area are believed to be several thousand years old (see Fig. 6–9).

There seems to be no reason why a clone[4] could not continue to exist indefinitely if it was protected from diseases (especially viruses) and other environmental stresses and repropagated frequently. Some clones are known to be very old—the 'Winter Pearmain' apple, cultivated in England, was being grown there about 1200 A.D. The 'Reine Claude' plum was also grown in England as long ago as 1500 A.D. The 'Black Corinth' grape has been grown in Greece for thousands of years. Other present-day clones of plants that are easily propagated by vegetative means, such as the fig, olive, and date, were being grown in Biblical times.

Senescence is considered to be a terminal, irreversible deteriorative change in living organisms, leading to cellular and tissue breakdown and death. It is a conspicuous period of physical decline, particularly evident toward the end of the life cycle of annual plants (population senescence) and of individual plants (whole plant senescence), but it can also occur in leaves, seeds, flowers, or fruits (organ senescence). Plants exhibit senescence in different ways. In annuals the entire plant dies at the end of one growing season, after and probably because of fruit and seed production. In herbaceous perennials, the tops of the plants die at the end of the growing season, perhaps killed by frost, but the shoot grows again in the spring and the roots can live for many years. In deciduous woody perennials, the leaves senesce, die, and fall off each year but the shoot and root systems remain alive, often for a great many years.

---

[4]A clone is genetically uniform material derived from a single individual and propagated exclusively by vegetative means, such as cuttings, divisions, or grafts.

Figure 6–9 *Above:* Ancient olive tree *(Olea europaea)* growing on the Mount of Olives in Jerusalem. Trees of this species are known to live for several thousand years. *Below:* Bristlecone pine *(Pinus aristata),* which grows in the high mountains from California to Colorado. These trees are among the earth's oldest living plants, some reaching the age of 4,000 years. *Source of P. aristata:* E. Memmler.

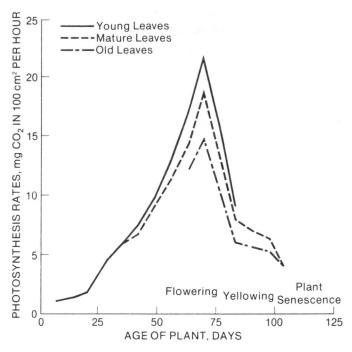

Figure 6–10 Photosynthetic rate in the wheat plant increases sharply as it matures, until flowering, when it decreases rapidly. Young, mature, and old leaves all exhibit the same behavior. *Source:* Leopold, A.C., and P.E. Kriedmann. *Plant growth and development.* 2nd ed. New York: McGraw-Hill. Originally appeared in Singh, B.N., and K.L. Lal. 1935. Investigations of the effect of age on assimilation of leaves. *Ann. Bot.* 49:291-307.

Senescence is usually considered to be due to inherent physiological changes in the plant, but it can also be caused by pathogenic attack or environmental stress. As individual trees age, for example, they are more and more vulnerable to lethal attacks by fungi, bacteria, and viruses. The long-lived trees mentioned above characteristically have very durable heartwood, containing high levels of resins and phenolic compounds that resist decay. As large trees get older, the ratio of foliage area to the surface area of roots, trunk, and limbs often diminishes progressively, thus lessening the amount of food the tree produces in relation to the amount it

consumes in respiration. Eventually the leaves are unable to supply adequate amounts of food to nourish the tree. The reduced foliage cover for the trunk and large limbs as the trees age can also lead to sunburn damage and the death of exposed tissues.

Considerable study has been given to senescence in plants, particularly in regard to leaves and their abscission. During leaf senescence DNA, RNA, proteins, chlorophyll, photosynthesis, starch, auxins, and gibberellins decrease, sometimes drastically. Senescence is not entirely degradation, however; particular mRNAs and proteins are synthesized only in senescing tissues.

Figure 6–10 shows the decline in photosynthetic activity of wheat plants following flowering; the decline in photosynthesis, of course, soon leads to senescence and death. Plant senescence is hastened, too, by the transfer of stored nutrients to the reproductive parts—the flowers, fruits, and seeds—as they develop and mature, at the expense of the root and shoot systems. As a result, senescence can be postponed in many plants by picking off the flowers before seeds start to form. In sweet peas, for example, removing the flowers once they start to wither and before seeds form prolongs the blooming period. Just as plant hormones are involved in many plant functions they are involved, too, in senescence. For example, ethylene plays a major role in fruit ripening and deterioration.

REPRODUCTIVE GROWTH AND DEVELOPMENT

Fruit and seed production involves several phases:

1. Flower induction and initiation
2. Flower differentiation and development
3. Pollination
4. Fertilization
5. Fruit set and seed formation
6. Growth and maturation of fruit and seed
7. Fruit senescence

Flower Induction and Initiation

Some annuals mature and can flower in only a few days or weeks after the seeds sprout; some forest and fruit trees require years before flowering. Once mature, the plant can be induced to flower by becoming sensitive to the conditions of its environment. What brings about the formation of flowers? In some species it is daylength (Photoperiodic effect) and/or low temperatures (vernalization), although in most trees neither daylength nor cold temperatures induces flowering.

Photoperiodism (Daylength)

The influence of the photoperiod on the flowering of several plant species was first studied in detail by the USDA at Beltsville, Maryland, and the results were published by W. W. Garner and H. A. Allard in 1920. They grew plants in containers that could be wheeled into dark sheds at the end of the work day and returned to the sunlight in the morning. They found that 'Maryland Mammoth' tobacco and certain cultivars of soybeans and cosmos required short days for flower induction. A set number of successive short days was required to complete differentiation (change from vegetative to reproductive growing points or shoot terminals). Once induced by short days, the plants could be moved to a long daylength without interfering with the flowering process. These plants, such as strawberry, poinsettia, and soybean, were called **short-day plants.** Later studies by many other workers found that long days were necessary to induce flowering of some plants, such as spinach, sugar beets, and winter barley. These were classified as **long-day plants.** A third group of plants, including tomato, corn, fruit trees, and cucumber, were those in which flowering was not affected by daylength and were called **day-neutral plants.** This phenomenon whereby daylength controls certain plant processes, as noted above, was termed photoperiodism. Other workers in later experiments found this phenomenon to be more complicated than this and that many plants did not fit nicely into these three categories because of interactions of daylength with temperatures.

Further studies showed that the length of night rather than day was actually the critical factor. However, the term *photoperiodism* still remains.

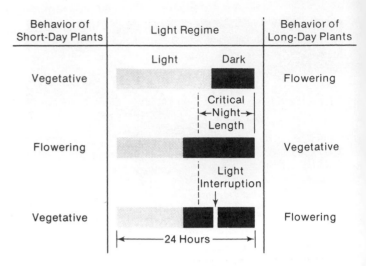

Figure 6–11   The effect of light interruption of the dark period on flowering in short-day and long-day plants. *Source:* Adapted from Galston, A.W., and P.J. Davies. 1970. *Control mechanisms in plant development.* Englewood Cliffs, N.J.: Prentice-Hall.

The short-day plant (SDP) is, in fact, a long-night plant requiring a certain period of darkness not interrupted by light. This is illustrated in Figure 6–11, which demonstrates how a flash of light (or night break) of sufficient intensity or duration will inhibit flowering of a short-day plant (long-night plant) but may induce flowering of a long-day plant (LDP). This information has been useful to commercial chrysanthemum growers who grow these short-day plants on a year-round schedule. When they want the young plants to reach a size adequate for flowering, the growers use incandescent or fluorescent lamps over the chrysanthemum plants (near midnight), each night for one to four hours, depending on time of year and latitude. This "night break" inhibits flowering until the plants reach the desired height. Conversely, when the natural daylight of summer is too long for chrysanthemum plants to flower, they cover the plants of proper size with black cloth or plastic each evening about 6 P.M. and remove it in the morning at about 8 A.M. This shortens the plant's day (lengthens the night) enough to induce and fully develop the flowers. These manipulations enable chrysanthemum plants to be flowered every day of the year. Poinsettias flower the same way and could be produced year-round. However, because of the association with Christmas, the public will only purchase poinsettias for Christmas.

The stimulus, or stimuli, to flowering appears to be transported from exposed leaves to the apical meristems. This was discovered by partial leaf removal when plants were placed under the inductive photoperiod for flower formation, then failed to flower. The SDP *Xanthium* (cocklebur) exposed to long noninductive days would initiate flowers if light was blocked from a single leaf. Some experiments in which flowering plants (donor) were grafted to nonflowering plants (receptor) caused the latter group to flower. These and other experiments gave rise to the theory that a flowering hormone, which might be similar in all plants, was responsible for flower induction. However, the hormone will remain a hypothetical compound unless it can be isolated and characterized.

In the 1950s, USDA researchers studied how interrupting the night with short lightbreaks affected cocklebur flowering. They found that red light, 660 nm, was the most effective part of the action spectrum for the inhibition of flow-

ering. This could be nullified by subjecting the plants to a series of exposures of light at 730 nm. This led to the discovery of phytochrome, which was described earlier in this chapter.

There has been considerable documentation to categorize plants as short-day, long-day, or day-neutral. Although not complete, one such list appears in Table 6–1. The critical daylength of many of the species shown in the table may be

TABLE 6–1	A Partial List of Long-Day, Short-Day and Day-Neutral Plants		
Long-Day Plants	**Length of Daily Light Period Necessary for Flowering**	**Short-Day Plants**	**Length of Daily Light Period Necessary for Flowering**
Althea (*Hibiscus syriacus*)	More than 12 hours	Bryophyllum (*Bryophyllum pinnatum*)	Less than 12 hours
Baby's breath (*Gypsophila paniculata*)	16 hours	Chrysanthemum (*Chrysanthemum × morifolium*)	15 hours
Barley, winter (*Hordeum vulgare*)	12 hours	Cocklebur (*Xanthium strumarium*)	15.6 hours
Bentgrass (*Agrostis palustris*)	16 hours	Cosmos, Klondyke (*Cosmus sulphureus*)	14 hours
Canary-grass (*Phalaris arundinacea*)	12.5 hours	Cotton, Upland (*Gossypium hirsutum*)	14 hours
Chrysanthemum frutescens	12 hours	Kalanchoe (*Kalanchoe blossfeldiana*)	12 hours
Chrysanthemum maximum	12 hours	Orchid (*Cattleya trianae*)	9 hours
Clover, red (*Trifolium pratense*)	12 hours	Perilla, Common (*Perilla crispa*)	14 hours
Coneflower (*Rudbeckia bicolor*)	10 hours	Poinsettia (*Euphorbia pulcherrima*)	12.5 hours
Dill (*Anethum graveolens*)	11 hours	Rice, winter (*Oryza sativa*)	12 hours
Fuchsia hybrida	12 hours	Soybean (*Glycine max*)	12 hours
Henbane, annula (*Hyoscyamus niger*)	10 hours	Strawberry (*Fragaria × Ananasia*)	10 hours
Oat (*Avena sativa*)	9 hours	Tobacco, Maryland Mammoth (*Nicotiana tabacum*)	14 hours
Orchardgrass (*Dactylis glomerata*)	12 hours	Violet (*Viola papilionacea*)	11 hours
Ryegrass, early perennial (*Lolium perenne*)	9 hours		
Ryegrass, Italian (*Lolium italicum*)	11 hours		
Ryegrass, late perennial (*Lolium perenne*)	13 hours		
Sedum (*Sedum spectabile*)	13 hours		
Spinach (*Spinacia oleracea*)	13 hours		
Timothy, hay (*Phleum pratensis*)	12 hours		
Timothy, pasture (*Phleum nodosum*)	14.5 hours		
Wheatgrass (*Agropyron smithii*)	10 hours		
Wheat, winter (*Triticum aestivum*)	12 hours		

Day-Neutral Plants			

Balsam (*Impatiens balsamina*)
Bluegrass, annual (*Poa annua*)
Buckwheat (*Fagopyrum tataricum*)
Cape jasmine (*Gardenia jasminoides*)
Corn (maize) (*Zea mays*)
Cucumber (*Cucumis sativus*)
English holly (*Ilex aquifolium*)
Euphorbia (*Euphorbia peplus*)
Fruit and nut tree species
Globe-amaranth (*Gomphrina globosa*)
Grapes
Honesty (*Lunaria annua*)
Kidney bean (*Phaseolus vulgaris*)
Pea (*Pisum sativum*)
Scrofularia (*Scrofularia peregrina*)
Senecio (*Senecio vulgaris*)
Strawberry, everbearing (*Fragaria × Ananasia*)
Tomato (*Lycopersicon lycopersicum*)
Viburnum (*Viburnum* spp.)

Source: Modified from F.B. Salisbury and C. Ross. 1985. *Plant physiology.* 3rd ed. Belmont Calif.: Wadsworth.

changed by a slight shift in temperature above or below the optimum for that species.

Understanding photoperiodic response allows crop producers to select species and cultivars that will flower and seed at the right time for their geographic location or market window.

Low Temperature Induction

Some plants, including many of the biennials, require low temperature for flower induction. The term for this is **vernalization,** which means "making ready for spring." It was first observed in winter wheat over a century ago. Vernalization is any temperature treatment that induces or promotes flowering. The temperatures required to vernalize a given plant and the length of the vernalization period vary among species and may even differ among cultivars of the same species. Broadly speaking, however, vernalization temperatures range between 0° and 10° C (32° and 50°F). Some of the biennials that require vernalization are beets, Brussels sprouts, carrots, celery, and some garden flowers such as Canterbury bells and foxglove. Winter annuals—such as the cereal crops, barley, oats, rye, and wheat—also respond to cold by flowering. Some plants, such as lettuce, peas, and spinach, can be induced to flower earlier with vernalization, but vernalization is not an absolute requirement; they will eventually flower without it. Some species can be vernalized as seeds (beet and kohlrabi), but most plants must reach a minimum size or produce a certain number of leaves to be sensitized by the cold.

Garden perennials, plants with corms or tubers, and many flowering shrubs and fruit trees require low temperatures to overcome the rest period, but few require low temperatures for flower induction. The true bulb plants, such as the hyacinth, narcissus, and tulip, do not require a vernalization period to break rest, but low temperatures are required to promote flower development once the flower has formed within the bulb. The olive tree (*Olea europaea*) does need chilling temperatures for induction of flower parts. In the kiwifruit *(Actinidia deliciosa),* too, there is no evidence of reproductive structures in the bud until after chilling temperatures occur.

Experiments have shown that gibberellins (discussed later in the chapter) applied to the meristems of some long-day plants replace their low temperature requirement for flowering. The gibberellins stimulate cell division and activate the flowering stimulus.

It is important to note that many plants do not respond with flower induction to changed daylength or low temperature. In fact, the majority of agricultural plants are self-inductive for flowering; that is, they initiate or form flowers at almost any photoperiod and without vernalization. Many of the garden annuals are good examples of plants that flower when they reach a certain morphological maturity. Most fruit trees, shrubs, woody plants, garden perennials (roses, carnations, gerbera) and vegetable crops (beans, peas, tomatoes, peppers, cucumbers) have self-induced flowering.

Flower Development

When an apparent floral stimulus is transmitted from the leaves to the apical meristems, a change from a vegetative to a flowering state takes place. In the case of some short-day plants, such as *Xanthium,* only one inductive short day (long night) is required, whereas chrysanthemums, poinsettias, and kalanchoe require three to four consecutive short days for induction. Once the apex has changed to a flower primordium, the process is not reversible. The floral apex may abort, however, if the subsequent environmental conditions are not favorable for full flower development. In such a case the axillary buds below the aborted floral apex will usually grow vegetatively until daylength or temperature conditions once again are favorable for flower induction.

The number of days from flower initiation to anthesis (time of flower opening) depends on the species and the cultivar. These periods of development can be modified somewhat by raising the temperatures slightly in the last third to half of the development phase. Very high temperatures in the early developmental stages, however, may cause flower abortion.

Pollination

In the production of most floral crops and flowering shrubs— for example, carnations, petunias, chrysanthemums, roses, and camellias—the flower itself is the desired product. There is little interest in any resulting fruits and seed, except in the case of the plant breeder working with such species. But in the food crops—the cereals, fruits, and many vegetable species—the postflowering structures are the desired products. It is the grains, fruits, and seeds that are harvested.

In angiosperms, **pollination** is defined as the transfer of pollen from an anther to a stigma. The anther and stigma may be in the same flower (self-pollination), in different flowers on the same plant (self-pollination), in different flowers on different plants of the same cultivar (self-pollination), or in different flowers on plants of different cultivars (cross-pollination).

Pollen grains come in many sizes and shapes and, while essential for sexual plant reproduction, can be devastating to many people as allergy producers. See Figure 6–12.

Figure 6–13 shows the various parts of a simple flower dependent upon pollination for fruit set.

If a plant is **self-fertile,** it produces fruit and seed with its own pollen, without the transfer of pollen from another cultivar. If it is **self-sterile,** it cannot set fruit and seed with its own pollen, but instead requires pollen from another cultivar. Often this is due to **incompatibility,** where a plant's own pollen will not grow through the style into its embryo sac (see Figs. 6–13 and 6–15). Sometimes, too, cross-pollination between two particular cultivars is ineffective because of incompatibility, which is believed to be due to factors that inhibit pollen tube germination or elongation.

Figure 6–12   Scanning electron micrograph of pollen grains produced in a staminate cone of red pine *(Pinus resinosa). Source:* USDA.

Pollen transfer from the anthers to the stigmas is principally by:

1. Insects, chiefly honeybees (Fig. 6–14). Insect pollination is common among cultivars with white or brightly colored flower parts and attractive nectar. Most fruit crops, many vegetables, and legume forage crops are pollinated by insects.

   Adequate pollination is so important in many crops that considerable efforts are made to aid the bees in their pollen distribution. Commercially prepared pollen is collected and tested for vigor and compatibility and applied to the crop flowers by various wind-generating devices. Also pollen inserts are placed at the entrances of hives to coat bees with pollen as they enter and exit.

2. Wind. This is the main pollinating agent for plants with inconspicuous flowers—the grasses, cereal grain crops, and forest tree species, as well as some fruit and nut crops such as the olive, walnut, pistachio, and pecan.

Other minor pollinating agents are water, snails, slugs, birds, and bats.

Figure 6–15 shows a longitudinal section through the pistil of a flower following pollination. Note the elongated pollen tube. A pollen grain that germinated the sticky surface of the stigma has grown down through the style carrying the male gametes to the embryo sac in the ovary.

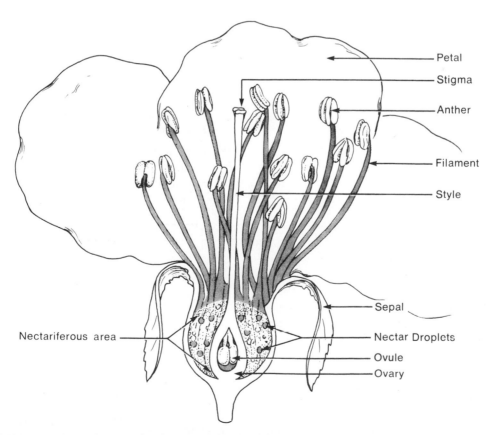

Petal
Stigma
Anther
Filament
Style
Sepal
Nectariferous area
Nectar Droplets
Ovule
Ovary

Figure 6–13   Longitudinal section of a cherry flower showing the structures involved in transfer of pollen from anthers to stigma (pollination). *Source:* USDA.

Figure 6–14 Honeybees, collecting nectar from the flowers, also cause pollination by distributing pollen from the anthers to the stigma. Bees perform a great service in the culture of many crops by their pollination activities. *Source:* USDA.

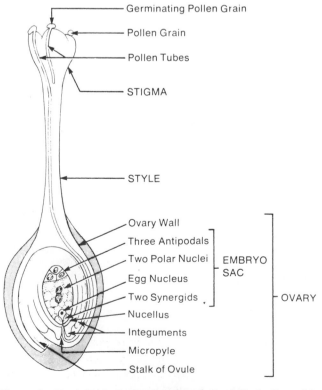

- Germinating Pollen Grain
- Pollen Grain
- Pollen Tubes
- STIGMA
- STYLE
- Ovary Wall
- Three Antipodals
- Two Polar Nuclei
- Egg Nucleus — EMBRYO SAC
- Two Synergids
- Nucellus
- Integuments
- Micropyle
- Stalk of Ovule

OVARY

Figure 6–15 A longitudinal section through the pistil of a flower following pollination and just before fertilization.

Parthenocarpy

If pollination—and subsequent fertilization—do not occur, fruit and seed rarely develop. One important exception, however, is fruit that sets parthenocarpically. **Parthenocarpy** is the formation of fruit without the stimulation of pollination and fertilization. Without fertilization, no seeds are produced; therefore, parthenocarpic fruits are seedless. There are many examples of parthenocarpic fruits such as the 'Washington Navel' orange, the 'Cavendish' banana, the oriental persimmon, and many fig cultivars. (Not all seedless fruits are parthenocarpic, however; sometimes, as in certain seedless grapes, pollination and fertilization occur and the fruit forms but the embryo aborts, thus no viable seed is produced.)

Fertilization

In the angiosperms the pollen tube grows through the micropyle opening in the ovule into the embryo sac and discharges two sperm nuclei (1N each). One unites with the egg (1N) to form the zygote (2N), which will become the embryo and eventually the new plant. The other sperm nucleus unites with the two polar nuclei (1N each) in the embryo sac to form the endosperm (3N), which develops into food storage tissue. This process is termed **double fertilization.** The elapsed time between pollination and fertilization in most angiosperms is about 24 to 48 hours.

In the cone-bearing plants of the gymnosperms, the entire process of pollination and fertilization is different than in the angiosperms. Staminate, pollen-producing cones are produced on the tree separately from the ovulate cones. There is no wall enclosing the ovaries as in the angiosperms, thus producing "naked" seeds on the cone scales.

Fruit Setting

Following formation of the zygote—at which time the genetic makeup (the genotype) of the new seedling plant is determined—many significant changes occur, leading to the formation of the fruit and (usually) seeds within the fruit.

Accessory tissues in the flower are often involved in fruit formation, such as the enlarged fleshy receptacle surrounding the ovary wall in the apple and pear. Botanically, however, the true fruit is just the enlarged ovary.

Figure 6–16 shows various stages in the development of the different tissues in a lettuce fruit.

In many plants only a small percentage of the flowers develop into fruits. This is particularly true in fruit crops where a tree could not possibly mature as many fruits as there are flowers. Many of the flowers drop without fertilization of the egg, and many of the flowers with a fertilized egg abort at the zygote stage or later. When the zygote fails to develop and no seed forms, the immature fruit usually drops. In some seedless grapes, such as 'Thompson Seedless', the embryo starts to develop, then aborts; the seed fails to develop, and the fruit does not abscise but grows to full size.

ENDOSPERM TISSUE IS TRIPLOID (3N) ~VIRICH NUTRITION (LIVE ABLEGG)

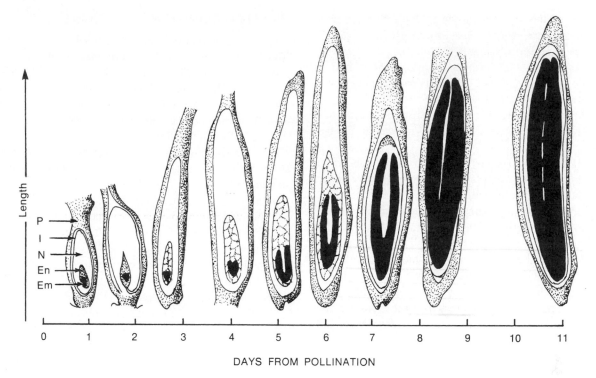

Figure 6–16 Developmental pattern of the tissues in a lettuce fruit, from the fertilized egg to the mature fruit. The ovary wall (the pericarp) is firmly attached to the seed coat (integument), so the structure is correctly considered a fruit and not a seed. P = pericarp; I = integument; N = nucellus; En = endosperm; Em = embryo. *Source:* Adapted by Hartmann, H.T., and D. E. Kester. 1983. *Plant propagation.* 4th ed. Englewood Cliffs, N.J.: Prentice-Hall. From Jones, H. A. 1927. Pollination and life history studies of the lettuce *(Latuca sativa). Hilgardia* 2:425-79.

Certain of the plant hormones appear to be involved in fruit setting, but the actual physiological mechanisms are largely unknown. In fruits of some species—tomatoes, peppers, eggplants, and figs—applied auxin can replace the stimulus of pollination and/or fertilization. Fruit set can also be induced in grapes, certain stone fruits—apricots, for example—and apples and pears by gibberellin sprays. Cytokinins also stimulate fruit set in grapes.

One of the chief problems in fruit production is obtaining the optimal level of fruit setting. Too low a fruit set gives a light, unprofitable crop. Too heavy a set leads to undesirable small, poor-quality fruits that mature late, possibly exhausting the tree's food supply and often resulting in little or no crop the following year. To overcome excessive fruit set, half—or more—of the fruits are removed at a very early stage, either by hand thinning, machine shaking, or chemical sprays. Interestingly, fruits of some species (the Washington Navel orange, for example) are self-thinning. Most of the tiny fruits originally forming drop, leaving an optimum number to develop to maturity.

As might be expected, temperature strongly influences fruit set. Temperatures that are too low or too high at this critical period are often responsible for crop failures. Low light intensity and lack of adequate soil moisture can also adversely affect fruit set.

Fruit Growth and Development

Once fruit has set, the true fruit and, sometimes, various associated tissues begin to grow. Food materials move from other parts of the plant into these developing tissues. Hormonal substances, such as the auxins, gibberellins, ethylene, and cytokinins, may be involved in some phases of fruit growth just as they are in fruit set. These materials originate in both the developing seeds and fruit, although it is significant to note the parthenocarpic fruits (without seeds) continue to grow to full size.

An interesting relationship between fruit growth and the presence of auxin has been observed in strawberry fruits. Removing some of the achenes ("seeds") from the surface of the strawberry at an early growth stage causes it to be lopsided; the strawberry fails to develop under the section where the achenes were removed. The stimulatory effect of some mobile material originating in the achenes is lost. Presumably this material is an auxin, because application of auxin paste to the area where the achenes were removed allows the strawberry to develop normally.

Evidence of the participation of gibberellins in fruit growth has been shown in the grape. Application of gibberellin to 'Thompson Seedless' grape clusters at an early stage of berry development markedly increases the ultimate fruit size. The size increase is so pronounced that virtually all table grapes of this cultivar grown in California are now treated with gibberellin. This effect on size also holds true for certain other grape cultivars (see Fig. 6–22).

While various plant hormones may be involved in fruit growth, the basic mechanisms are still barely understood. During flower development and in the early stages of fruit growth, there is considerable cell division. Following this period of intense cell division, most fruits increase in size because of cell enlargement.

Fruits have two basic patterns of growth, as shown in Figure 6–17. One is the simple sigmoid growth curve—typical of such fruits as the orange, apple, pear, pineapple, olive, almond, tomato, and strawberry—in which there is a slow start followed by a period of rapid size increase, then a decrease in growth rate near fruit maturity. The second pattern is a double sigmoid growth curve, in which the single sigmoid growth curve is repeated. Near the center of the growth period, the growth curve is flat; the fruit increases little, if at all, in size. The stone fruits—peach, apricot, plum, and cherry—as well as the grape and fig show a double sigmoid growth pattern. In the stone fruits, which have a hard endocarp or pit, the pit hardens during the second phase of fruit development. In addition, some important changes take place in seed development within the pit, as illustrated in Figure 6–18.

PLANT GROWTH REGULATORS

In plants, as in animals, many of the behavioral patterns and functions are controlled by hormones. Hormones are produced in minute amounts at one site in the plant and translocated to other sites where they can alter growth and development. The natural hormones and other materials are essentially "chemical messengers," influencing the many patterns of plant development.

A distinction must be made between the terms *plant hormone* and *plant growth regulator*. A **plant hormone** is a natural substance (produced by the plant itself) that acts to control plant activities. Plant hormones that are chemically synthesized can initiate reactions in the plant similar to those caused by the natural hormones. **Plant growth regulators,** on the other hand, include plant hormones—natural and synthetic—but also other, nonnutrient chemicals not found naturally in plants but that, when applied to plants, influence their growth and development.

There are five recognized groups of natural plant hormones: **auxins, gibberellins, cytokinins, ethylene,** and **abscisic acid** (Fig. 6–19). The discovery and subsequent study of these plant hormones is one of the most exciting and fascinating chapters in the history of plant physiology. Despite considerable study of hormones, however, the mechanism of their actions in the plant is still not completely understood.

Auxins

Auxins were the first group of plant hormones to be discovered. The discovery came in the mid-1930s, and for many years thereafter auxins and their activities in plants were studied intensely throughout the world.

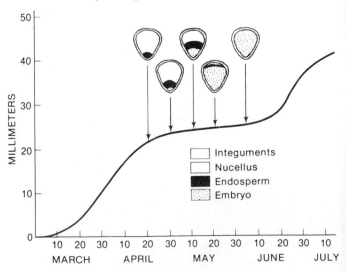

Figure 6–18   Growth curve of an apricot fruit through the growing season. During the second growth period the pit (endocarp) hardens and the seed within the pit develops mostly from nutritive tissue (nucellus and endosperm) to finally consist entirely of the embryo. *Source:* Adapted from Crane, J. C., and P. Punsri. Comparative growth of the endosperm and the embryo in unsprayed and 2,4,5-trichlorophenoxyacetic acid sprayed royal and Tilton apricots. *Proc. Amer. Soc. Hort. Sci.* 68:96-104. 1956.

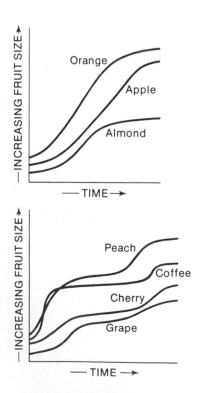

Figure 6–17   Growth curves of representative kinds of fruits showing the two characteristic types. *Top:* The sigmoid growth curve. *Bottom:* The double sigmoid growth curve.   (STONE FRUIT)

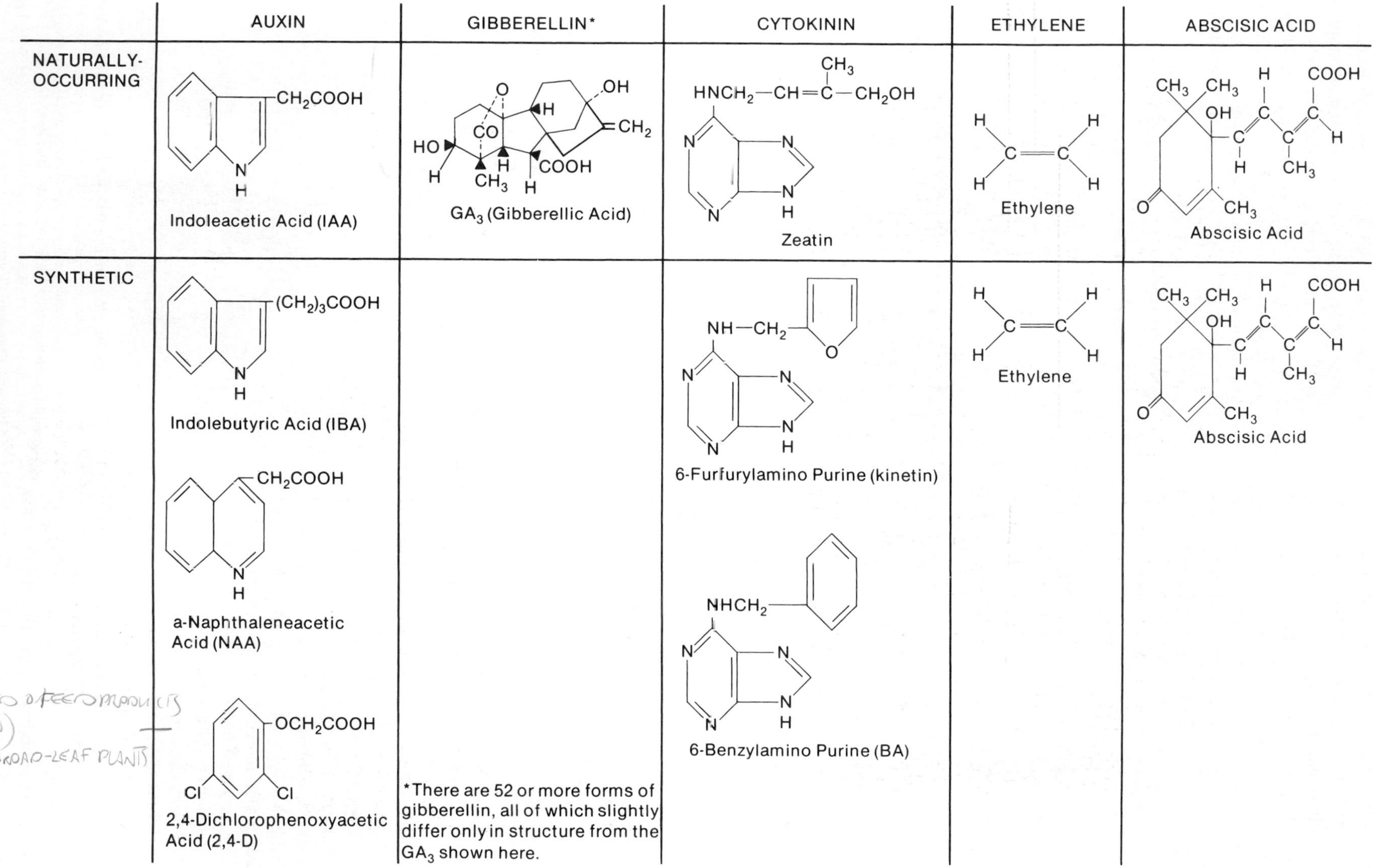

Figure 6–19 Structural formulas of some natural and some synthetic plant growth regulators.

113

The auxins, both natural and synthetic, influence plant growth in many ways, including cell enlargement or elongation, photo- and geotropism, apical dominance, abscission of plant parts, flower initiation and development, root initiation, fruit set and growth, cambial activity, tuber and bulb formation, and seed germination. Auxins operate at the cellular level, affecting such activities as protoplasmic streaming and enzyme activity. Auxins are related to many other chemical control mechanisms and are readily transported throughout the plant, principally in an apex-to-base direction (basipetally).

The natural auxins originate in meristems and enlarging tissues, such as actively growing terminal and lateral buds, lengthening internodes, and developing embryos in the seed. Auxins are produced in relatively high amounts in the shoot tip or terminal growing point of the plant and move down the plant through the vascular tissues, causing the phenomenon known as **apical dominance** (blockage of growth of lateral buds by presence of terminal buds). High levels of auxin in the stem just above the lateral buds block their growth. If the shoot tip supplying this auxin is broken or cut off, the auxin level behind the lateral buds is reduced and the lateral buds begin to grow. This is part of the reason why, when a shoot tip is removed, many new shoots arise from buds down along the stem.

One of the most widespread auxins that occurs naturally in plants is indoleacetic acid (IAA) (see Fig. 6–19). Several other natural auxins have also been identified, and there are others whose chemical structure is yet unknown. There are also many synthetic auxins which induce the same effects as natural auxin. Some of these are indolebutyric acid (IBA), naphthaleneacetic acid (NAA), and 2,4-dichlorophenoxyacetic acid (2,4-D). (See Fig. 6–19).

Some important commercial uses of these synthetic auxins are:

1. *Adventitious root initiation.* One of the first responses attributed to auxins was the stimulation of root formation in stem cuttings. Two synthetic auxins, indolebutyric acid and naphthaleneacetic acid, are now widely used commercially in treating the bases of stem cutting to stimulate the initiation of adventitious roots.
2. *Weed control.* The synthetic auxin, 2,4-dichlorophenoxyacetic acid, is in widespread commercial use as a selective weedkiller that eliminates broad-leaved weeds in grass or cereal fields.
3. *Inhibition of stem sprouting,* Many kinds of woody ornamental trees produce masses of vigorous sprouts from the base of the trunk that, if not removed, would transform the tree into a bush. Continual removal of these sprouts by hand is costly and time consuming. It has been found that treatment of the tree trunks with the auxin naphthaleneacetic acid at about 10,000 ppm (1.0 percent) strongly inhibits the development of such sprouts.

4. *Tissue culture.* The initiation of roots and shoots on small pieces of plant tissue cultured under aseptic conditions has become a standard method of micropropagation of some plant species. Often an auxin, such as IAA or 2,4-D, had to be included in the culture medium for roots to initiate.

Gibberellins (GA)

The gibberellins are a group of natural plant hormones with many powerful regulatory functions. The most obvious is to stimulate stem growth dramatically, far more than auxins can. Gibberellins may stimulate cell division, cell elongation, or both, and they can control enzyme secretion.

In some plants, GA is involved in flower initiation and sex expression (male or female flower parts). Fruit set as well as fruit growth, maturation, and ripening seem to be controlled by gibberellin in some species. Senescence of plant parts, particularly leaves, is also affected by GA. Certain dwarf cultivars of peas and corn, if treated with GA, grow to a normal height, indicating that the dwarfed plants lack a normal level of gibberellin (Fig. 6–20).

Gibberellins are also involved in overcoming dormancy in seeds and in buds. Their role in the germination of barley seed has received much study (see Fig. 6–2l). After the seed has been moistened and placed at room temperature, a natural gibberellin produced in the embryo translocates to the aleurone layer surrounding the endosperm. Triggered by the GA, cells in the aleurone layer synthesize such enzymes as amylases, proteases, and lipases. These enzymes then diffuse throughout the endosperm, hydrolyzing starches and proteins into sugars and amino acids that then become available to the embryo for its growth and development.

Figure 6–20 Overcoming dwarfness in corn by spraying with gibberellin. *Left:* Untreated, genetically dwarf corn plants. *Center:* Nondwarf corn sprayed with gibberellin. *Right:* Genetic dwarf corn sprayed with gibberellin. Photographs taken six weeks after spraying. *Source:* From *Plant Growth Substances in Agriculture* by Robert J. Weaver. W.H. Freeman and Company. Copyright © 1972.

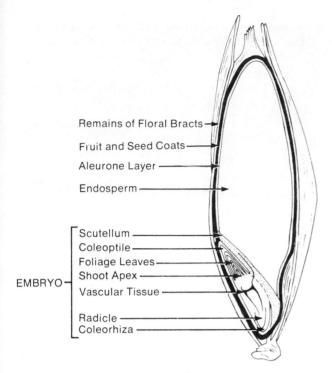

Figure 6–21 Longitudinal section of a barley seed. *Source:* From a drawing by Peter J. Davies, In: *The Life of the Green Plant* (3rd edition), by Arthur W. Galston, Peter J. Davies, and Ruth L. Satter. Prentice-Hall. Copyright 1970 and 1980.

Figure 6–22 Effect of gibberellin sprays on growth of 'Black Corinth' grapes. (A) Untreated control. (B) Stem girdling control. (C) Plants sprayed at an early growth stage with gibberellin at 5 ppm. (D) At 20 ppm. Photos taken 59 days after spraying. *Source:* From *Plant Growth Substances in Agriculture* by Robert J. Weaver, W.H. Freeman and Company. Copyright © 1972.

The molecular structure of the gibberellins is well known; a typical one is shown in Figure 6–19. By 2000 more than 100 different gibberellins had been discovered in tissues of various plants. Some common ones are GA_1, GA_3 (gibberellic acid), GA_4, and GA_7.

Gibberellins were first discovered in 1926 by Japanese researchers studying a disease of rice plants caused by the fungus *Gibberella fujikuroi.* Plants infected with the fungus grew excessively and abnormally. Extracts from this fungus applied to noninfected plants stimulated the same abnormal growth. By 1939 the active ingredient was extracted from the fungus, crystallized, and named gibberellin. This early work with gibberellin in Japan went unnoticed in the Western world until the early 1950s when a great surge of gibberellin research began, particularly in the United States and England. This research led to the isolation of many different forms of gibberellin extracted from the *Gibberella* fungus and from higher plants.

Gibberellins are synthesized in the shoot apex of the plant, particularly in new leaf primordial. They are also found in embryos and cotyledons of immature seeds and in fruit tissue. In addition, the root system synthesizes large quantities of gibberellin, which moves upward throughout the plant. GA translocates easily in the plant in both directions, unlike auxin, which moves largely in an apex-to-base direction.

Pharmaceutical companies produce crystalline GA_3 as the acid or the potassium salt for research studies and certain commercial applications. These preparations are all obtained from growth of the *Gibberella* fungus in a process similar to that used to produce antibiotics.

Even though it has been demonstrated that gibberellins occur naturally in many of the higher plants, little is known of the physiological mechanisms of gibberellin action or transport.

Although gibberellins are a powerful and important group of plant hormones involved in many of the plant functions, only a few agricultural uses have been found for them. Some are:

1. *Increasing fruit size of seedless grapes.* This is the principal commercial application of gibberellin. Practically all vines of the 'Thompson Seedless' grape grown for table use in California are sprayed each year. Berry size of other grape cultivars, such as 'Black Corinth', is also increased by gibberellin sprays, as shown in Figure 6–22.
2. *Stimulating seed germination and seedling growth.* A number of cases have been reported where soaking seeds in solutions of gibberellic acid before germination greatly stimulates seedling emergence and growth. Such responses have been obtained with barley, rice, peas, beans, avocado, orange, grape, camellia, apple, peach, and cherry. Figure 6–23 shows the stimulation obtained with grape seedlings by gibberellin treatment.
3. *Promoting male flowers in cucumbers.* When pollen is wanted for hybrid seed production, a single application of GA_3 to the leaves stimulates maleness of a cucumber. This has proved an important discovery for hybridizers.
4. *Overcoming the cold requirement for some plants.* Azalea plants require six weeks of cool temperatures (8°C or 46°F) to develop flower buds. Several leaf

NO DIFF. (handwritten)

Figure 6–23 Effect of gibberellin on germination and growth of 'Tokay' grape seeds. Seeds soaked (before planting) at 0, 100, 1,000, or 8,000 ppm in potassium gibberellate solution for 20 hr. *Source: From Plant Growth Substances in Agriculture* by Robert J. Weaver. W.H. Freeman and Company. Copyright © 1972.

applications of 1,000 ppm gibberellin will completely or partially replace this cold requirement for flower bud development. It has been shown experimentally that gibberellins applied to biennial plants that require a cold period before they flower causes early flowering. Most of these treatments, however, have limited commercial value.

Cytokinins

STIMULATES mitosis (CELL DIVISION + SHOOT GR) (handwritten left margin)

Prod in Roots TIPS → move up to shoot tips counteracts apical dominance (ie: vs Auxins) (handwritten)

This group of plant hormones primarily promotes cell division but they also participate in a great many aspects of plant growth and development, such as cell enlargement, tissue differentiation, dormancy, different phases of flowering and fruiting, and in retardation of leaf senescence.

Cytokinins interact with auxins to influence differentiation of tissues. As shown in Figure 6–24, externally applied cytokinin alone stimulates bud formation in tobacco stem segments; auxin applied alone causes roots to develop; but when cytokinin plus auxin are applied together, there is a canceling effect—only masses of undifferentiated callus form.

There are both natural cytokinins, such as zeatin, and synthetic forms, such as kinetin and benzyladenine (BA) (see Fig. 6–19). There are over 100 known natural and synthetic cytokinins. Cytokinins occur in many plant tissues as both the free hormonal material and as a component of transfer RNA. They are found in abundance in embryos and germinating seeds and in young developing fruits—all tissues with considerable cell division. Roots supply cytokinins upward to the shoots.

The mechanism of cytokinin action in the plant is not clear. Cytokinins indirectly increase enzyme activity and increase the DNA produced in some tissues. Their regulatory effects seem to result from interactions with other hormones in the plant.

Cytokinins were discovered when scientists at the University of Wisconsin in 1955 used a synthetic material—kinetin, later named a "cytokinin"—to cause cell division of tobacco stem pith. After many interesting physiological activities of kinetin became apparent, plants were examined for

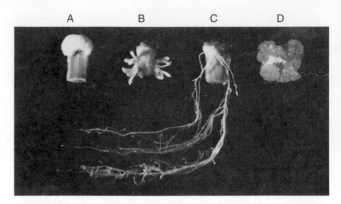

Figure 6–24 Effects of a cytokinin and an auxin on growth and organ formation in tobacco stem segments. (A) Control, no treatment. (B) Cytokinin—bud formation but no root formation. (C) Auxin—root formation with prevention of bud development. (D) Cytokinin plus auxin—stimulation of callus growth but no organ formation. *Source:* Adapted from Hartmann, H. T., and D. E. Kester. 1983. *Plant propagation.* 4th ed. Englewood Cliffs, N.J.: Prentice-Hall. Originally in Skoog, F., and D. J. Armstrong. 1970. Cytokinins. *Ann. Rev. Plant Physiol.* 21:359-84.

possible natural similar materials. In 1964 such a material was isolated from young corn seeds by researchers in New Zealand and was named zeatin. An active promoter of cell division known to exist in coconut milk was finally determined to be a zeatin-riboside.

Even though cytokinins are strongly involved in plant growth regulation, no important agricultural uses have been developed for them. In media for aseptic tissue culture, however, cytokinin usually must be added to induce shoot development. Applications of cytokinins to green tissue have been shown to delay senescence. These materials have also been used experimentally, and in limited commercial applications, on greenhouse roses and potted chrysanthemum plants to stimulate growth of axillary buds by overcoming natural bud inhibitors.

Ethylene

It has been well established for many years that ethylene gas (see Fig. 6–19) evokes many varied responses in plants. As long ago as 1924 it was found the ethylene could induce fruit ripening; in 1925, it was determined that ethylene could overcome bud dormancy in potato tubers; in 1931, that it could induce leaf abscission; in 1932, that it could induce flowering in pineapple plants *(BROMELIADS)* (handwritten); in 1933, that it could cause roots to form on stem cuttings. By the mid-1930s it was determined that ethylene was itself a plant product, and arguments arose among plant scientists about whether ethylene, a gas, should be considered a plant hormone.

Little further attention was paid to possible roles of ethylene as a natural growth regulator until the development in the 1960s of gas chromatographic techniques that permit the detection of ethylene in concentrations as low as one part per billion. Vast amounts of research of ethylene physiology in the 1960s and 1970s established ethylene as a plant hormone.

Ethylene itself is a tiny molecule ($CH_2 = CH_2$), compared with the other plant hormones. The pathway for ethylene biosynthesis in plants has been fairly well elucidated. Ethylene, as a gas, diffuses readily throughout the plant, moving much like carbon dioxide, and it can exert its influence in minute quantities. Its solubility in water also enhances its movement through the plant. The cuticular coatings on external cell surfaces tend to prevent losses from the plant. Ethylene is apparently produced in actively growing meristems of the plant, in ripening and senescing fruits, in senescing flowers, in germinating seeds, and in certain plant tissues as a response to bending, wounding, or bruising. Synthetic ethylene, from ethephon, applied to plant tissues can cause a great burst of natural ethylene production—an autocatalytic effect.

Just how ethylene exerts its regulatory effects is no better known than the basic mechanisms involved in the action of the other plant hormones. One theory is that ethylene regulates some aspect of DNA transcription or RNA translation, thus changing RNA-directed protein synthesis and, consequently, enzyme patterns. But many other mechanisms are also likely to be in operation.

The possible commercial uses of ethylene were greatly increased with the development in the 1960s of ethylene-releasing compounds such as ethephon (2-chloroethyl) phosphonic acid. This compound applied as an agricultural spray gradually releases ethylene into plant tissues. In contrast to some of the other plant hormones, ethylene and ethylene-releasing chemical have several valuable commercial applications:

1. *Fruit ripening.* Ethylene gas, injected into airtight storage rooms, is used commercially to ripen bananas, honeydew melons, and tomatoes. The ethylene-releasing chemical ethephon will also ripen tomatoes that are green but horticulturally mature. To harvest canning tomatoes by machine harvesting equipment, where the entire crop is picked at one time, it is important that most of the fruits be ripe and fully colored at the time of harvest. Spraying the field with an ethylene-releasing material before harvest promotes uniform red color development of the green fruits. Ethephon is used as a preharvest spray to promote uniform ripening of apples, cherries, figs, blueberries, coffee, and pineapple.

2. *Flower initiation.* Ethylene gas released from ethephon has initiated flowers in several ornamental bromeliad species, including *Ananas* spp., *Aechmea fasciata, Neoregelia* spp., *Billbergia* spp., and *Vriesia splendens.* Ethephon has been widely used to promote uniform flowering in the cultivated banana.

3. *Changing sex expression.* Ethylene application to certain plants, such as cucumbers and pumpkins, can dramatically increase the production of female flowers. Some cucumber cultivars produce both female and nonfruiting male flowers on the same plant. Spraying the vines with ethephon causes all flowers to be female, which develop into fruits and thus increase yields. This practice gives results similar to previous studies where auxins were used.

4. *Degreening oranges, lemons, and grapefruit.* Sometimes the rind of maturing oranges and grapefruits remains green owing to high chlorophyll levels, even though the eating quality, juice content, and ratios of soluble solids to acid are high enough to meet grade standards for harvest. Citrus packers can treat such fruits with ethylene at about 20 ppm for 12 to 72 hours. This breaks down the chlorophyll and allows the orange and yellow carotenoid pigments to show.

5. *Harvest aids.* Certain fruit and nut crops, such as sour cherries and walnuts, are harvested by mechanical tree shakers that shake the trees until the crop falls into catching frames or onto the ground to be picked up later. Often this practice is not completely successful because the fruits or nuts are so tightly attached that the tree shakers do not remove them. However, by spraying the trees about a week before harvest with an ethylene-releasing compound—ethephon, for example—the abscission-inducing effects of the ethylene result in a much higher percentage of crops removed.

Ethylene can also harm plants. It can cause unwanted leaf abscission and can hasten senescence of most flowers. The introduction of a few parts per billion of ethylene into the surrounding air will cause carnation flowers to close, rose buds to expand prematurely, and orchid flower petals and sepals to develop a water-soaked appearance. The pollination of an orchid flower can generate sufficient ethylene to cause injury to the flower parts. Ethylene can cause flower bud abortion of bulbs during shipment. A few diseased tulip bulbs give off enough ethylene in a packing crate to stop further development of the flower buds within the bulbs. In certain commercial flowers such as carnations and geraniums, the deleterious action of ethylene may be blocked with the application of silver thiosulfate.

Inhibitors

Some plant growth regulators inhibit rather than stimulate plant growth. One of these—abscisic acid—is a natural plant hormone. Other compounds found in plants, principally phenolic materials, inhibit growth but are not classed as plant hormones. A third group consists of synthetic chemicals, not found in plants, that have important and commercially useful growth-inhibiting or retarding properties.

Abscisic Acid (ABA)

Abscisic acid tends to interact with other hormones in the plant, counteracting the growth-promoting effects of auxins and gibberellins. ABA is involved in leaf and fruit abscission, as well as in the onset of dormancy in seeds and in the early stages of the rest period in vegetative and flower buds of woody perennial shrubs and trees. Dormancy induced by applications of synthetic ABA is overcome by gibberellin treatment—an example of the antagonistic effects of these two hormones. ABA is very effective in inducing closure of the stomata in leaves, indicating a possible role in the stress physiology in plants. Large increases in the ABA content of leaves have been noted following water stress in the plant.

Abscisic acid is widespread in the plant body and can be readily obtained by alcoholic extraction of many plant tissues. Chemically ABA is a sesquiterpene; its structural formula is shown in Figure 6–19. Several bioassays are used to detect abscisic acid in plant material; one is based on the closure response of stomata mentioned above.

Studies conducted in California in 1964 on the development of maturing cotton fruits revealed an unknown inhibitory material whose concentration changed dramatically, reaching its highest levels during fruit abscission. The material was isolated from cotton fruits, crystallized, and named abscisin II for its abscission activity. About the same time, in England, an inhibitory material was found in the leaves of sycamore trees, particularly in the autumn. This material, called dormin, was subsequently found to be identical to the abscisin II isolated from cotton. Then, by mutual consent, a decision was made to rename them both "abscisic acid." For many years before abscisic acid was identified, a widespread and powerful growth inhibitor or mixture of inhibitors in plants was known to exist and was referred to as the β inhibitor complex. Subsequently, one of the most important components was shown to be abscisic acid.

Abscisic acid moves readily through the plant. Fruits, especially rose fruits, are high in ABA. Also, ABA appears to be synthesized by leaves.

The mechanism of ABA action in plants is believed by some to involve an alteration in the nucleic acid and protein synthesis systems, although the rapid action of ABA when it is applied to plants implies a more direct effect.

Abscisic acid is difficult and expensive to synthesize, so it is usually used only in experimental studies that require small amounts. No commercial applications have been developed, owing partly to the lack of adequate amounts of test material.

Other Inhibitors

Plants accumulate a wide range of materials that seem to have no defined metabolic roles, although they can act as growth inhibitors and, in some cases, are toxic to feeding insects or animals. These inhibitors are largely phenolic compounds such as benzoic acid, cinnamic acid, caffeic acid, and coumarin. In some plants they occur in very high concentrations, second only to carbohydrates.

Synthetic Growth Retardants A rather diverse group of growth retardants developed since about 1950 has several important commercial uses regarding ornamental plants, principally in obtaining compact, dwarf-type plants. These materials generally act by slowing, but not stopping, cell division and elongation in subapical meristems, usually without causing stem or leaf malformations. The primary effect of these materials is the opposite of gibberellin, often converting a tall-growing plant into a rosette. Most act by blocking gibberellin synthesis. Plants treated with these growth retardants have a compact scaled-down appearance, which is often more attractive than larger, untreated plants with a loose, open growth. The treated plants also often have darker, more attractive foliage and more flowers than untreated plants (see Fig. 6–25). Some of the better known synthetic growth retardants are described below.

Daminozide (succinic acid-2, 2-dimenthyl hydrazide, Alar, B-Nine). Tests have shown that daminozide effectively retards growth and stimulates flowering of several kinds of herbaceous and woody ornamental plants and will enhance the size and color of various fruit species. Those that respond well include chrysanthemums (Fig. 6–25), various bedding plants (2,500 to 5,000 ppm), and azaleas (2,500 ppm). Sometimes two applications two or three weeks apart are required to maintain the desired dwarf form.

Chlormequat [(2-chloroethyl) trimethylammonium chloride, Cycocel, CCC]. Chlormequat is effective in retarding the height of some ornamental plants. The height of poinsettias may be controlled if chlormequat is applied as a drench to the soil or as a spray to the stems and foliage.

Ancymidol (A-Rest), a-cyclopropyl, a-4-methoxypropyl, α-5-pyrmidie methanol. This growth retardant is very effective for reducing the height of some bulbous and other potted ornamental crops.

Figure 6–25 Growth retardation in chrysanthemum plants when treated with a growth retardant, *Chrysanthemum x morifolium.* 'Circus' plants sprayed with daminozide (B-Nine) to retard shoot growth; *left:* Control, no daminozide; *center:* 2,500 ppm; *right:* 2,500 ppm sprayed Aug. 14 and again on Aug. 21.

Five cuttings were planted in 6-in. (15 cm) pots on July 31, pinched (apex removed) Aug. 7, sprayed Aug. 14, given short days under black cloth on Aug.18 to initiate flowers; the plants flowered about Oct. 15, when the photo was taken. Average heights above the soil at flowering time were: *left:* 37 cm; *center,* 31 cm; *right,* 26 cm.

Paclobutrazol (ICI-PP333). [2RS, 3RS]-1-[4-chloro-phenyl] 4-4-dimethyl-2-1, 4-triazol-yl-pentan-3-ol] Bonzi®. Paclobutrazol® and its close chemical relative uniconazole (Sumagic®) are very potent growth-retarding chemicals that effectively control the height of many herbaceous and woody ornamentals. The rates used for these chemicals are much lower than for other growth retardants.

> **Caution: Chemical use on plants must be in accordance with the law.** Before using any chemical check the label to be certain that the crop you are treating, the rate you are using, and the intended use, along with any other considerations, are in agreement with the label.

> In the United States the application of chemicals to plants for commercial use is strictly controlled by Environmental Protection Agency (EPA) regulations. Before the chemicals can be used legally, an EPA registration must be obtained for each crop, stating the dosage allowable and the time of year application is permissible. Application for registration is usually made by the chemical company manufacturing the material after a patent has been obtained. Regulations for obtaining a registration for use of chemicals to be applied to food crop plants are much stricter than those for ornamentals.

SUMMARY AND REVIEW

Growth is the increase in size of a plant by cell division and enlargement. Shoot growth can be determinate (shoot elongation ceases with the formation of reproductive structures) and indeterminate (bearing clusters of fruits and flowers along the stem). Plant growth patterns include annual (complete life cycle in one growing season), biennial (life cycle covers two growing seasons), and perennial (growth resumes each growing season for several years). Shoot growth often occurs in "flushes," which are periods of growth followed by inactive periods during a growing season. Vegetative growth is the growth of the roots and nonflowering or fruiting top growth. Reproductive growth is the growth and development of flowers, fruit, and seed.

Factors that affect plant growth can be genetic, environmental, and developmental. Genetic factors include the overall genetic (DNA) composition of the plant and the active or inactive state of genes at any particular time. Environmental factors include light, heat, water, and dissolved nutritional minerals, and atmospheric gases. Light provides the energy for photosynthesis and the accumulation of carbohydrates needed for growth. Changes in light quality or duration direct the shape of the growing plant, including the flowering of photoperiodic plants. Heat determines how fast most plants grow and develop. Most plants have an optimum growing temperature. Temperatures well below or above the optimum will slow growth rate and may be detrimental or even lethal. Water is stored in cells and used in cell processes. Water and solutes are carried throughout the plant through the vascular system. Most water entering the plant leaves the plant through the transpirational stream. Water evaporating out of the leaves cools the plant through the process of evapotranspiration. The most important atmospheric gases for plants are CO_2, needed for photosynthesis, and O_2, needed for respiration. Stages of development include juvenile, mature, and senescence. During each period specific types of growth and development can occur that may not occur in other stages.

There are five classes of plant hormones: auxins, gibberellins (GAs), cytokinins, ethylene, and abscisic acid (ABA). Endogenous auxins influence cell enlargement, photo- and geotropism, apical dominance, and other growth traits. Agriculturally, auxins and related compounds are used to promote rooting of cuttings, as weed killers, and to promote rooting in tissue culture systems. Gibberellins can promote flower initiation, stem elongation, and overcome dormancy in seeds and buds. Agricultural uses of GAs include increasing the size of seedless grapes and overcoming or reducing cold requirements of some plants. Most chemical growth retardants act by blocking synthesis of GAs in the plant. Cytokinins promote cell division, a slow leaf senescence, and influence other physiological processes. They are used agriculturally to promote shoot development in tissue culture. Ethylene is the only known gaseous hormone. It promotes senescence and fruit ripening in many species. Agriculturally, ethylene is used to ripen fruit on the field or after harvest, to defoliate cotton and tomatoes to aid in harvest, to initiate flowering in some species, and to increase the number of female flowers, and subsequently fruit number, in cucumbers, melons, and pumpkins. Unwanted exposure to ethylene can cause severe damage to plants. Some fruits give off ethylene as they ripen. The ethylene can harm any ethylene-sensitive produce, including flowers, stored in the same area. Ethylene can build up to toxic levels in greenhouses as a result of incomplete combustion and improper venting of flame-burning heaters. Some fungal pathogens, especially those found in bulb crops, generate ethylene, causing damage to infected plants and nearby plants. Abscisic acid promotes the closing of stomates when plants are stressed by low water availability. In some plants, ABA also promotes leaf abscission and/or dormancy in some plants.

EXERCISES FOR UNDERSTANDING

6–1. Sow some fast-germinating seeds such as marigold and corn side by side on moist paper and observe the order in which plant parts appear and how each one develops. Also observe how the two species differ from each other in appearance and development. What is controlling any differences between species? At the same time, plant some seeds in 4-in. or 6-in. pots (one species and plant per pot) and observe their growth and development over a longer period of time. Describe what plant hormones are likely to be active during the different stages of growth.

6–2. Allow some plants growing in pots to dry out on a sunny day. Observe and describe the plants' response to low water as the stress progresses. Plants growing in a greenhouse are often subjected to very high temperatures (>100°F) for several hours on sunny afternoons and the pots become dry faster than water can be applied. A similar situation happens to field crops, orchards, and in turfgrass and landscapes if there is no rain or irrigation. What hormone is activated to reduce the loss of water from the plant?

6–3. Using marigolds growing in pots as described in exercise 6–1, treat half of the plants in pots with a chemical growth regulator at the recommended rate (be sure the chemical is labeled for the species you are growing). Observe how the chemical affects the appearance of the plant and how long the chemical appears to affect the plant. What plant hormone is most likely being affected by the chemical? What do you think would happen to plant growth if that hormone is applied to the plant after the chemical is applied? You can try two different chemical classes, such as daminozide versus paclobutrazol, and observe any similarities and differences in plant response.

REFERENCES AND SUPPLEMENTARY READING

HALEVY, A. H., ed. 1985–1987. *Handbook of flowering.* Six vols. Boca Raton, Fla.: CRC Press.

SALISURY, F. B., and C.W. ROSS. 1985. *Plant physiology.* Third Edition. Belmont, Ca.: Wadsworth.

TAIZ, L., and ZEIGER, E. 1998. *Plant physiology.* Second Edition. Sunderland, Mass.: Sinauer.

WILKINSON, R. E., ed. 1994. *Plant-environment interactions.* New York, N.Y.: Dekker.

CHAPTER

7

Photosynthesis, Respiration, and Translocation

KEY LEARNING CONCEPTS

After reading the chapter you should be able to:

♦ Understand the complex process of photosynthesis including carbon fixation and carbohydrate synthesis.
♦ Understand how environmental and physiological factors influence photosynthesis in plants.
♦ Understand the process of respiration and the factors that affect it.
♦ Understand the process of translocation and the factors that affect it.
♦ Understand the relationship and interdependency among the processes of photosynthesis, respiration, and translocation.
♦ Understand why the processes are important in crop production.

PHOTOSYNTHESIS

It is principally through the plant kingdom that solar energy is converted to chemical energy and hence made available for use by the animal kingdom. The sun is the original source of all important fuels, including coal, oil, gas, wood, and alcohol. Plants are the source of all food, either directly as grain, fruits, vegetables, and nuts or indirectly through animals as meat, eggs, and dairy products. An exception to the rule may be the mineral content of animals (e.g., bones); but minerals are not usually considered "food." Clothing is another

121

example of an essential human requirement that is derived from plants.

This chapter describes how the sun's energy, in the form of light, is transformed by plants into usable chemical energy. The conversion of the sun's energy into chemical energy is the single most significant reaction on earth and could easily be called "the reaction of life." The conversion is the total of many reactions, which in a generalized form is called **photosynthesis**.

Total worldwide productive capacity, including steel mills, automobile factories, shipyards, oil production, milling, and so on, is almost insignificant when compared to the quiet but continuing productive capacity of the earth's green plants. It has been estimated that about 1.4×10^{14} kg (3.1×10^{14} lb) of carbon from carbon dioxide in the air is converted to carbohydrates each year by the green plants that live on the land and in the ocean, seas, and lakes. A number of this magnitude is beyond our comprehension. To put it another way, assume that the 1.4×10^{14} kg of carbon is converted entirely to an equivalent amount of coal, which would be 1.41×10^{11} MT (1.55×10^{11} t). Assume further that a standard-size railroad car holds 45.5 MT (50 t); then the carbon fixed annually by plants would yield enough coal to fill 97 cars every second of every hour of every day all year long.

The light energy from the sun is captured by chlorophyll in the green parts of plants, mainly the leaves, and is used to convert carbon dioxide from the air and water from the soil (plus some water vapor from the air) into carbohydrates, which are stored in the plant for future use. In some superficial ways, photosynthesis and respiration can be thought of as the same reaction, with respiration being the reverse of photosynthesis. Thus the products of photosynthesis are the raw materials for respiration, and the fuels for photosynthesis are the products of respiration (Equation [2]). Photosynthesis is a synthesizing (building up) reaction. In contrast, respiration is a degrading reaction in which carbohydrates are ultimately broken down to carbon dioxide and water with the release of the stored energy. Plants perform both reactions and are said to be autotrophic. Human beings and animals, on the other hand, are heterotrophic: they can carry on respiration only. Hence, all living organisms obtain their energy directly or indirectly from the photosynthetic process—the basis of our dependence on plants.

Properties of Photosynthesis in the Leaves of Higher Plants

As we previously noted, the process of photosynthesis includes the absorption of atmospheric CO_2 as well as its subsequent incorporation and metabolism into carbohydrate compounds within the leaf (see Fig. 7–1 and its legend). Photosynthetic CO_2 assimilation takes place in the chloroplasts of leaf mesophyll cells of all higher plants. It is within the chloroplast that there are organized membranes containing chlorophyll as well as protein and lipid components. These membrane systems, in conjunction with chlorophyll molecules in these membranes, serve to trap energy from the sunlight and convert it into the stable energy of chemical compounds needed for the processes of incorporation of CO_2 into leaves. Also, it is the soluble or aqueous phase of the leaf chloroplast that is the actual site of incorporation of CO_2 into organic compounds. This process will be described later in this chapter.

Photosynthesis occurs mainly in the leaves in most plants, but also to some extent in stems and green fruit. Many epiphytic (nonparasitic plants that grow on other plants) orchids (*Cattlyea, Phalaenopsis, Vanda, Aerides*) have chlorophyll in their root tips; thus, the roots of these plants also photosynthesize.

All the photosynthetic reactions are summed up in a chemical equation:

$$nCO_2 + nH_2O \xrightarrow[\text{light}]{\text{green plant}} (CH_2O)n + nO_2 \quad [1]$$

In a general way, the equation says that a green plant receives light, uses it to join n number of carbon atoms together, and combines two oxygen atoms to form one molecule of oxygen gas (O_2). The new compound $(CH_2O)_n$, made up of carbon, hydrogen, and oxygen, is a carbohydrate. This general chemical formula represents any number of carbohydrate compounds.

Note that the generation of molecular oxygen is extremely important. The primitive earth was essentially anaerobic and the beginning of photosynthesis changed the course of evolution in favor of organisms that depend on oxygen for their survival. Even today, the existence of much of the fauna we see around us is dependent on the oxygen generated by plants.

Energy Requirements for Photosynthesis: The Photolysis[1] of Water

While Equation [1] indicates the raw ingredients and the final products, it neither shows the reaction quantitatively nor indicates the complexity of the mechanisms. It was recognized long ago that photosynthesis probably occurs in a series of numerous reactions, which can be summarized thus:

$$6CO_2 + 12H_2O \xrightarrow[\text{light energy}]{\text{green plant}} C_6H_{12}O_6 + 6H_2O + 6O_2 \quad [2]$$

Equation [2] shows that 6 molecules of carbon dioxide react with 12 molecules of water to produce one molecule of glucose, 6 molecules of water, and 6 molecules of

---

[1]"Photolysis" is simply the light-dependent splitting of water to yield its components, hydrogen and oxygen. This process requires substantial energy which, in photosynthesis, is supplied by light.

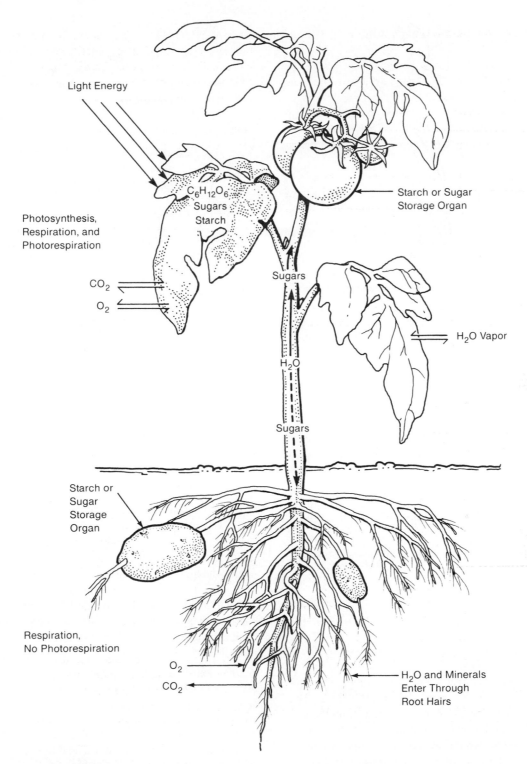

Light Energy

$C_6H_{12}O_6$
Sugars
Starch

Photosynthesis,
Respiration, and
Photorespiration

CO_2

O_2

Starch or Sugar
Storage Organ

Sugars

H_2O Vapor

H_2O

Sugars

Starch or
Sugar
Storage
Organ

Respiration,
No Photorespiration

O_2

CO_2

H_2O and Minerals
Enter Through
Root Hairs

Figure 7–1 Schematic representation of photosynthesis, respiration, leaf water exchange, and translocation of sugar (photosynthate) in a plant. In the green leaves, CO_2 from air, water from roots, plus light energy interact through photosynthesis to produce intermediates which eventually are metabolized to form sugars, and O_2 is evolved as a by-product. In most plants, sucrose is the most common sugar that is synthesized in the light, and this compound is translocated to growing points throughout the plant. Sucrose is also stored in leaf cell vacuoles, but it is mobilized from the vacuoles during dark periods. In addition, starch is formed in the leaf chloroplasts during photosynthesis. The sucrose that is translocated may be converted into starch in such storage organs as fruits or tubers. Respiration takes place in all living cells of the plant, both above and below ground. Stored carbohydrates are metabolized, during O_2-dependent respiration, and CO_2 is released as a by-product.

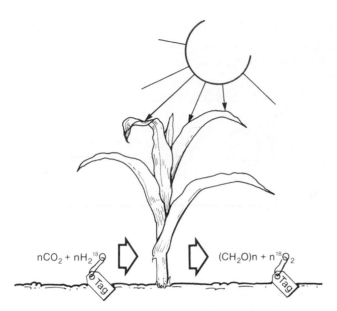

Figure 7–2   C. B. Van Neil's work in the 1930s showed conclusively that water is the source of oxygen given off by green plants. The experiment used the heavy isotope of oxygen (^{18}O) as a radioactive tag to identify certain oxygen atoms.

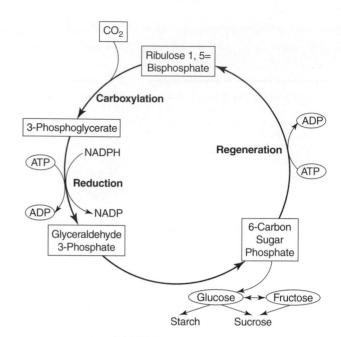

Figure 7–3   Simplified portrayal of the CO_2 assimilation cycle involved in photosynthesis. The carboxylation step is the point at which new carbon enters the metabolism of the cells. The reduction step requires both energy in the form of adenosine triphosphate (ATP) and electrons, which are derived from water and enter the pathway via pyridine nucleotides (NADPH). Sucrose is "drawn off" of the cycle and some of the carbon is recycled to regenerate the CO_2 acceptor—ribulose 1,5-bisphosphate.

oxygen[2]. The reason that water appears on both sides of the equation is discussed on page 133, footnote 6.

The source of the released oxygen puzzled plant scientists for years. It was thought that carbon dioxide was the most likely source. The discovery and use of radioactive tracer elements made possible the study of the pathways of the atoms during the synthesis of compounds, which had previously been impossible. By using water enriched with the heavy isotope of oxygen (^{18}O),[3] scientists found that water was the source of the oxygen gas evolved in photosynthesis; that is, when ^{18}O-labeled water was used, the labeled ^{18}O atom was found in the molecular O_2 released (Fig. 7–2). This discovery was one of the great breakthroughs in the study of photosynthesis. It also became evident that light provides the energy necessary to break apart the water molecules to produce hydrogen ions (with associated electrons) plus oxygen gas.

During photosynthesis, water is separated into hydrogen and oxygen ions, as already noted. In addition, the oxygen in carbon dioxide is separated from carbon. The hydrogen atoms with their associated electrons from the water recombine with the carbon to form stable phosphorylated, 3 carbon compounds, which are converted into 6 carbon sugar phosphates. These 6 carbon sugar phosphates, in turn, are metabolized to sugars, for example, sucrose (Fig. 7–3).

For CO_2 to be converted to carbohydrate, the carbon atom must gain electrons—i.e., be converted from an oxidized state to a reduced state. The electrons required for this process come from the oxygen atoms in the water during the photolysis step.

The overall photosynthetic reaction is strongly endothermic,[4] requiring about 114,000 calories of energy for each 44 grams (1 mole) of carbon dioxide changed to carbohydrate. In nature, the light from the sun provides the necessary energy.

It is interesting to note that only a very small amount of the total energy received on earth from the sun is actually captured by plants. Efficiency of solar energy conversion has been calculated for a number of different crops and environmental conditions. The range of conversion of solar energy to chemical energy by plants varies from 0.1 percent for poor growing conditions on cloudy days and 3 percent for intensive cropping in good light to 25 percent for plants grown in controlled laboratory conditions. The average solar energy conversion efficiency for normal agricultural cropping systems is about 2.0 to 2.5 percent. The remaining 98 percent or so of the solar energy reaching the earth is lost by reflection and reradiation into the atmosphere; is used to evaporate water from oceans, lakes, and soil surfaces; and increases soil and plant temperatures. The following discussion deals with how plants absorb light only at certain wavelengths, thus further reducing their efficiency in the use of light energy.

---

[2]Some bacteria (purple bacteria) have the ability to use hydrogen sulfide (H_2S) as green plants use H_2O to convert CO_2 into bacterial protoplasm. They have a pigment similar to chlorophyll. The overall reaction is:

$$6CO_2 + 12H_2S \rightarrow 6CH_2O + 6H_2O + 12S$$

[3]The heavy isotope of oxygen is ^{18}O. The most abundant isotope found in the atmosphere is ^{16}O.

[4]Endothermic reactions require energy in order to proceed. Exothermic reactions, by contrast, give off energy as they proceed.

THE NATURE OF LIGHT

The sun's energy is derived from thermonuclear reactions that convert hydrogen atoms to helium atoms with the release of tremendous amounts of energy. In the sun, four hydrogen atoms each weighing 1.008 units of atomic mass combine to form one helium atom weighing 4.003 units of atomic mass. Thus 4.032 atomic mass units of hydrogen are converted to 4.003 atomic mass units of helium with the conversion of 0.029 units of atomic mass to energy. The amount of energy is calculated from the equation $E = mc^2$ where the mass (m, 0.029 atomic mass units in this case) is multiplied times c^2 (c is the speed of light, 3×10^{10} cm/sec). The product of these equals 8.7 $\times 10^8$ ergs. Researchers have estimated that the sun's mass is being coverted to energy at the rate of 120 million tons per minute. This energy travels through space to the earth as electromagnetic radiation waves at the speed of light, about 300,000 Km/secs (186,000 mi/sec). The wavelengths of this radiation vary from very long radio waves of more than one kilometer (0.62 mi) to very short cosmic rays of less than 10^{-4} hm (3.9×10^{-10} in). Visible light is that portion of the electromagnetic spectrum between 400 and 760 mn; however, plants respond to the somewhat wider spectrum of about 300 to 800 nm (Fig. 7–4). Light is transmitted to the earth in discreet bundles of energy called photons or quanta. The number of waves passing a given point in space per second is called its frequency. Light traveling in shorter wavelengths at higher frequencies is more energetic than light at longer wavelengths and lower frequencies. Figure 7–5 shows the relationship between wavelength and frequency.

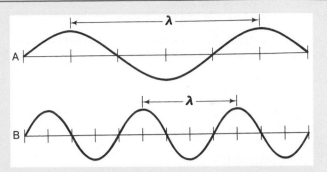

Figure 7–5 The wave nature of light. The wavelength (A) is twice that of (B) and the wave frequency in (A) is half that in (B). The short wave- length and high frequency make (B) more energetic than the long wave- length and low frequency of (A).

The rate of photosynthesis can be determined by measuring the rate at which carbon dioxide is absorbed by a leaf. A leaf is placed in a sealed glass chamber in the light, and the loss of carbon dioxide from inside the chamber is measured. But while carbon dioxide is being used by photosynthesis, some is also being evolved by the leaf in either light or darkness because of cell respiration. The above method of analysis measures **net photosynthesis** because some of the photosynthesis taking place is reusing the carbon dioxide evolved from respiration. **Total photosynthesis** is determined by measuring the carbon dioxide absorbed in photosynthesis and adding to this the carbon dioxide evolved during respiration in the light. The relationship is: rate of total photosynthesis = rate of net photosynthesis + rate of respiration in light.

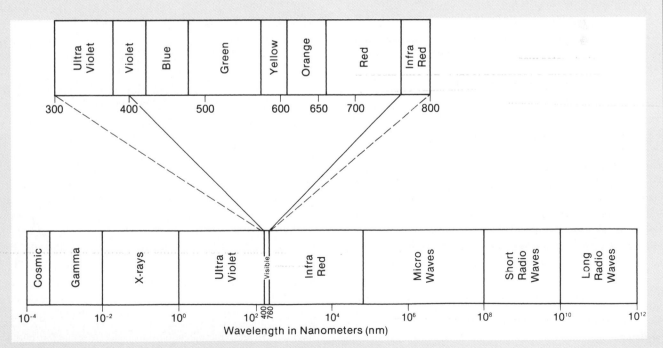

Figure 7–4 The electromagnetic spectrum of radiant energy with the expansion of the photosynthetic active radiation (PAR) spectrum.

Factors Affecting the Rate of Photosynthesis in Higher Plants

Environmental and plant growth factors that affect the rate of photosynthesis are:

1. light quality (wavelength)
2. light intensity (the amount of incident light energy absorbed by the leaf)
3. carbon dioxide concentration
4. heat
5. water availability
6. plant development and source-sink relationships

Light Quality

The energy level of different colors of light differs. Blue light with its shorter wavelength and higher frequency is about 1.8 times more energetic than the same number of photons of red light.

The plant must absorb light to keep the photosynthetic machinery running. Photosynthesis occurs in special structures, or organelles, called chloroplasts found within some plant cells. The chloroplasts contain pigments capable of intercepting light and converting electromagnetic energy into the chemical energy necessary to drive the photosynthetic processes. When these pigments (chlorophyll a, chlorophyll b, and some carotenoids) are irradiated with light containing all visible wavelengths, they absorb mostly from the red and blue portions of the spectrum and reflect the green portion. This can be demonstrated by growing groups of uniform plants at the same light intensity (energy) but at different wavelengths, then measuring the photosynthetic activity ($CO_2 \rightarrow O_2$ gas exchange) of a leaf (see box) and plotting the activity against the different wavelengths. This kind of graph is called an **action spectrum.** The example given in Figure 7–6 shows peaks of activity in the blue and red wavelength range, indicating that these are the wavelengths of light most responsible for photosynthesis.

Light quality is particularly important when plants are grown under artificial light. Ample radiation from the red and blue wavelengths must be assured for photosynthesis. When the action spectrum for photosynthesis is compared with a growth curve for plants, the similarity between the two curves is obvious. In one experiment, groups of green bean plants (*Phaseolus vulgaris*) were grown for six weeks under lights of similar intensities but different wavelengths—violet, blue, green, yellow, and red. The heights of the plants were measured and the averages plotted, as shown in Figure 7–6. The graph shows peaks of growth in the blue and red wavelength ranges similar to the action spectrum for photosynthesis.

Light Intensity

Comparing Figures 7–4 and 7–6, we can see that the wavelengths of light that are active in photosynthesis are not uni-

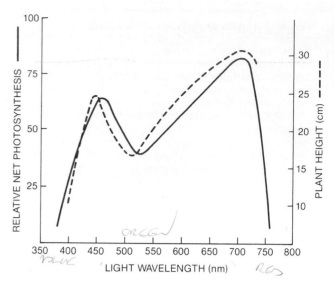

Figure 7–6 Action spectrum of a plant leaf (solid line), relating wavelength of light to relative net photosynthetic activity. The broken line shows the effect of light quality on the plant height growth of bean plants. Note the similarity of this curve to the action spectrum.

form across the entire visible spectrum. That is, the wavelengths below 600 nm are relatively unimportant. The light that is active in photosynthesis is referred to as photosynthetic photon flux PPF. PPF is usually expressed as "photon flux density" which has units of micromoles of photons per m^2 per second.[5] Sensors which accurately measure PPF over the 400 to 700 nm range are used in plant science research. Until recently their cost prohibited their use by growers; however, models are now available that growers can afford. The PPF of full sunlight is approximately 2,000 μmol $\times$ $m^{-2} \times sec^{-2}$.

Light intensity affects plant growth, in part by influencing the rate of photosynthetic activity. The effect varies with different plants. Some species require high light intensities to grow well and are often termed "sun-loving" plants; examples are corn, potatoes, sugarcane, many turf grasses, and some fruit trees. On the other hand, plant species that do not grow well in high light intensities are sometimes refereed to as "shade-loving" plants. Many such plants grow in the dense shade of the forest floor, and some of them are useful as house ornamentals. Other plant species are intermediate between the sun-loving and shade-loving types and grow well in moderately intense light.

Light intensity distinctly affects the size and shape of leaves (Table 7–1). Generally, the leaves of a given plant

[5] Units of footcandles and lux are sometimes used to express light "intensity," but these measurements and units refer to the entire visible spectrum. Although these light measurements are useful for some activities (e.g., photography), they are not very relevant to plant growth. These units of light are no longer accepted in the scientific literature relating to photosynthesis.

TABLE 7–1	Effects of Light Intensity on Bean (*Phaseolus vulgaris*) Seedlings	

Total Darkness (Etiolated)	**High Light Intensity**
No photosynthesis	High photosynthesis
Yellow to white in color	Green
Long internodes	Shorter internodes
None or tiny leaves	Large normal leaves
Spindly stem	Stout normal stem
Fine hairlike roots	Large normal roots

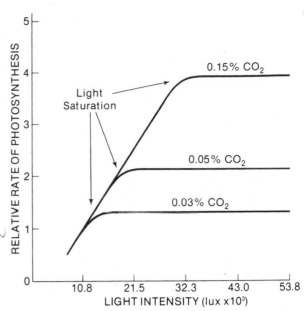

Figure 7–8 Light saturation at different CO_2 concentrations, and at constant temperature of 25°C (77°F). Increasing CO_2 concentration increases photosynthetic activity. Note that the point of light saturation is also increased at higher CO_2 concentration.

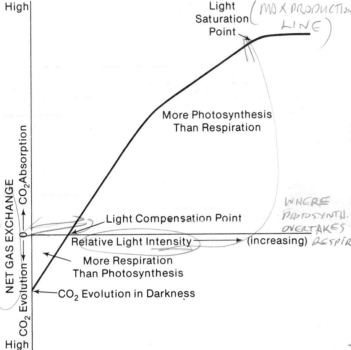

Figure 7–7 The effect of light intensity on net photosynthetic activity.

species will grow longer and larger in area at the low light intensity than leaves grown at the high light intensity. Also, leaves of plants grown in high light intensities are thicker and broader than those grown in low light intensities. These responses highlight the role of leaf blades in collecting solar energy.

In a given plant, there is a light intensity at which photosynthesis and respiration rates are equal and net gas exchange is zero. This intensity is the **light compensation point** (Fig. 7–7). At the light compensation point, steady state equilibrium exists between respiration and photosynthesis but CO_2 is still exchanged. The gas is evolved in respiration at the same rate that it is used in photosynthesis. The plant is said to be **light saturated** when further increases in light intensity increase photosynthesis little or not at all. At very high light intensities, the rate at which CO_2 is available

to the plant could limit the photosynthetic rate. The light intensity at which saturation occurs increases as the CO_2 concentration surrounding the plant rises (Fig. 7–8).

Carbon Dioxide

When a plant is placed in a closed system, CO_2 will be assimilated until the CO_2 concentration reaches equilibrium at some low value. The carbon dioxide concentration at equilibrium is such that the amount of carbon dioxide evolved in respiration exactly equals the amount consumed in photosynthesis. This CO_2 concentration is called the **carbon dioxide compensation points.** It is reached when a plant is growing at a constant light intensity higher than the light compensation point. At the CO_2 compensation point, no further assimilation (photosynthesis) occurs until an equal amount of CO_2 is evolved by respiration.

The concentration of CO_2 in the air surrounding the leaves markedly affects photosynthesis (Fig. 7–8). Normally the atmosphere contains an average of about 0.03 percent CO_2 and 21 percent O_2. Plant physiologists have found that increasing the CO_2 concentration in a closed system, such as a sealed greenhouse or an acrylic plastic chamber, to about 0.10 percent approximately doubles the photosynthetic rate of certain crops such as wheat, rice, soybeans, some vegetables, and fruits. Many greenhouse crops, such as carnations, orchids, and roses are grown commercially in a CO_2-rich atmosphere. Increasing the CO_2 concentration is feasible in the greenhouse or laboratory, but it is not possible to markedly increase the CO_2 concentration in the air above a corn or wheat field.

However, there is now good evidence to indicate that the CO_2 concentration in the atmosphere is steadily increasing due to the enormous consumption of fossil fuels related to human activity on the earth. In fact, there is presently considerable research on the impact of higher CO_2 concentrations on the growth and reproduction of horticultural and agronomic crops.

A few practices can have an impact on CO_2 contraction in some field situations. For example, the density of the crop and the height of the crop canopy can sometimes be altered to increase the diffusion rate of CO_2 which, in turn, increases its concentration in the vicinity of the leaves. Applications of organic matter in the form of crop residues or green manure crops to the soil tends to increase CO_2 levels in the atmosphere above the soil. On a warm, sunny day after a rain and with no air movement, some plants grow so rapidly that CO_2 availability at the leaf surfaces become limiting. Under such conditions, wind machines can increase the available CO_2 by circulating and mixing the air. In greenhouses, the use of horizontal airflow fans (HAFs) increases airflow around plants and exposes the leaves to a constant supply of CO_2. The use of HAFs may eliminate the need for supplemental CO_2 in most greenhouses.

Heat

At low light levels, temperature has little effect on the rate of photosynthesis because light is the limiting factor. However, as a general rule, if light is not limiting, the rate of photosynthetic activity approximately doubles for each 10°C (18°F) increase in temperature for many plant species in the temperate climates. The effect of temperature varies with species; plants adapted to tropical conditions require a higher temperature for maximum photosynthesis than those adapted to colder regions. However, excessively high temperatures will reduce the photosynthetic rate of some plants not accustomed to such high temperatures by causing the stomata of the leaves to close. A reduced rate of photosynthesis, together with the increased respiration rate at high temperatures, lowers the sugar content of some fruits (for example, the cantaloupe) grown under these conditions.

Water

Under conditions of drought (low soil moisture and hot, drying winds), plants often lose water through transpiration faster than their roots can absorb it. The rapid loss of water causes the stomata to close and the leaves to wilt temporarily. When this occurs, the exchange of CO_2 and O_2 is restricted, resulting in a dramatic drop in photosynthesis. Thus, water deficit reduces photosynthesis, in part, by markedly reducing gas exchange.

Excessive soil moisture sometimes creates an anaerobic condition (lack of oxygen) around the roots, reducing root respiration and mineral uptake and transport of water and minerals to the leaves—thus indirectly depressing photosynthesis in the leaves.

Plant Development and the Source-Sink Relationship

Provided that all other environmental factors are constant, the growth of the plant also significantly influences the rate of net and total photosynthesis, both in single leaves and in the total canopy of leaves on a plant. As the leaves expand and grow larger, chloroplast development and replication proceed in the new cells until the leaf has fully expanded and has reached maturity.

As single leaves on a plant develop, their photosynthetic rate rises in step with the expansion of the leaf until the leaf has matured. For example, when a soybean or spinach plant leaf begins expansion, its chlorophyll content per square decimeter of leaf area and its rate of net photosynthesis per square decimeter of leaf area both rise to maximum values just after the leaf reaches full expansion. When the leaf reaches full expansion, it is often called a **source leaf** because the carbohydrate synthesized in that leaf is in excess of "local" requirements and is exported to other parts of the plant that are actively growing (Fig. 7–1). These sites of active growth and metabolism are often called **sink tissues,** because it is in these growing tissues where the imported carbohydrate is metabolized to compounds employing information of new structures or in storage. Examples of sinks are roots and reproductive organs (i.e., fruits and seeds). Note that juvenile leaves are also sinks until their photosynthetic capacity provides carbohydrate above amounts required for local growth.

Considerable research has established that sink tissues can place a "demand" upon source leaves such that there can be an actual increase in the activity of photosynthetic CO_2 assimilatory enzymes. The result is reflected in a sustained increase in photosynthetic rate in the source leaves.

In many plant species, the photosynthesis rates in all the mature leaves begin to decline drastically when the plants flower and envelop fruits and seeds. Actually, proteins in the leaves, and specially photosynthetic enzyme proteins in the chloroplasts, are degraded to amino acids as the leaves begin the process called **senescence**. The amino acids in the leaves are transported to the developing seeds, and subsequently seed storage protein is synthesized from the amino acids derived from protein degradation in the leaves. In the case of plants such as soybean and wheat, when the seeds are totally mature, the foliar photosynthetic activity declines to zero, followed by death of the foliage.

The Plant as a Metabolic Machine

As indicated at the opening of this chapter, not all reactions in the plant are synthesizing reactions. Some, as will be shown later, are destructive transformations that break carbohydrates and other substances down to simpler compounds. The constructive reactions (photosynthesis) and the destructive reactions (respiration) are collectively called

Enzymes are organic catalysts that facilitate the thousands of reactions that make up the metabolism of living cells. They are proteins consisting of chains of amino acids of varying length. These molecules are involved in one way or another in many photosynthetic or respiratory reactions and vary considerably in size, solubility, and function. It must be emphasized that while the enzymes are involved in these reactions, they themselves are not reactants and, thus, are not "consumed" or permanently transformed in the reaction.

Amino acids are fundamental building blocks of proteins. Although plant cells contain dozens or even hundreds of amino acids, only 20 common amino acids are incorporated into proteins. A typical enzymatic protein will contain hundreds of amino acids. Because of this high number, the 20 amino acid types can be linked in an enormous variety of combinations, thus providing a basis for the widely different properties of different enzymes. Many enzymes also contain mineral elements such as iron or molybdenum and these mineral atoms are usually involved in electron transfer properties of the enzymes.

In some reactions, enzymes require a **cofactor**—a nonprotein compound that participates as a reactant and is subsequently regenerated so that it is again available to react with the main substrate. Examples of cofactors are the pyridine nucleotides—compounds often involved in electron transfer in an oxidation/reduction reaction. NAD^+ and $NADP^+$ are the oxidized forms and NADH and NADPH are the reduced forms; we will often see these compounds as participates in metabolic reactions, including reactions associated with photosynthesis.

Adenosine triphosphate (ATP) is also involved in many enzymatic reactions, including those of photosynthesis. ATP is not involved in electron transfer. Instead it is an energy-rich compound; when one of the phosphate groups is hydrolyzed from ATP to form adenosine diphosphate (ADP), a large amount of energy is made available to drive the reaction.

metabolism; individually, they are termed **anabolism** and **catabolism.** Anabolic processes occur in the plant only if energy is available. The atoms are joined together with chemical bonds that utilize and store the energy. In the catabolic processes, these chemical bonds are broken and the compounds degraded to simpler compounds, molecules, or atoms. These reactions release stored energy.

For convenience, the overall photosynthetic process is often separated into two parts. First, in the light reaction, energy in the form of light is received by the plant and used to reduce nicotinamide adenine dinucleotide phosphate ($NADP^+$) to NADPH (see preceding box). Second, in the Calvin cycle, carbon taken into the plant as CO_2 from the air is reduced to form carbohydrate. We will now consider these two components of photosynthesis in more detail.

The Photolysis of H_2O and Photosynthetic Electron Transport

Photolysis occurs in the membranes within chloroplasts when the rays of white light fall upon a plant leaf. The chloroplasts contain molecules of chlorophyll a and chlorophyll b (pigments), both of which absorb red light (600 to 700 nm) and blue light (400 to 500 nm); see Figure 7–6. Chloroplasts of higher plants also contain another group of pigments, the carotenoids, which absorb blue light but not red. A photon of light interacts with an electron in the chlorophyll molecule. The electron accepts the light energy and becomes excited; that is, it is raised to a higher energy level. The electron returns to its ground (before excitement) state by different means, depending on the nature of the environment. It returns to its original state by heat emission, by chemical reaction with other molecules, by remitting the energy as fluorescence, or by becoming involved in the photochemical reactions of photosynthesis.

The most important result of the electron transport reactions is the reduction of $NADP^+$ to form NADPH. The source of the electrons required for NADPH synthesis come from the splitting of water molecules (Equation [3]). Light energy is collected by the chlorophyll molecules until the electrons from water are raised to an energy level sufficient to reduce the $NADP^+$ (Equation [4]).

$$H_2O \rightarrow 2H^+ + 2e^- + 1/2\ O_2 \qquad [3]$$

$$NADP^+ + 2H^+ + 2e^- \rightarrow NADPH + H^+ \qquad [4]$$

The Calvin Cycle

In the Calvin cycle, the electrons and energy captured in the light reaction are used to convert carbon dioxide to carbohydrates and water. The overall reaction is:

$$6CO_2 + 24e^- + 24H^+ \rightarrow C_6H_{12}O_6 + 6H_2O \qquad [5]$$

A simplified version of the pentose cycle is shown in Figure 7–9. In the *first step,* one molecule of CO_2 from the air enters the cycle. Bear in mind that there is no beginning or ending of the cycle; that is, it is continuous. Also, it should be pointed out that while only one molecule of CO_2 enters at a time, six carbon atoms are required to synthesize the final glucose sugar ($C_6H_{12}O_6$); therefore, this cycle (Fig. 7–9) revolves six times to form one glucose molecule. The CO_2 molecule joins a five-carbon sugar (ribulose 1,5-bisphosphate) that was synthesized earlier in the dark reaction. Some authors claim that an unstable six-carbon sugar is temporarily formed in the *second step,* which immediately enters the *third step* and splits into two three-carbon compounds (3-phosphoglyceric acid, or PGA). The plants that form these initial three-carbon compounds are called C_3 plants. As we will see later in this section (p.131), some plants form

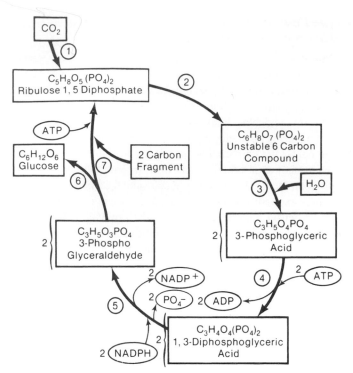

Figure 7–9　A schematic representation of the Calvin cycle associated with CO_2 assimilation.

The dark reaction is historically known as the Calvin cycle or the reductive pentose phosphate cycle. It is sometimes referred to as the Blackman or enzymatic reaction. It was first assumed that sugars were formed in the dark reaction from the stepwise synthesis of a number of intermediate products. Because of formaldehyde's formula CH_2O, biochemists logically suspected formaldehyde as the first product of photosynthesis. These early attempts to discover the stepwise synthesis of sugars ended in failure because formaldehyde was never isolated. In fact, they showed that formaldehyde kills plant cells. Thus, the products and the steps involved in the dark reactions of photosynthesis remained a secret until radioisotopes became available. It was shown that after feeding plants radioactive carbon dioxide for five seconds, most of the radioactivity was located in a three-carbon compound, 3-phosphoglyceric acid. Following this discovery, many of the intermediate steps were then determined.

Although the Calvin cycle was originally thought of as a series of dark reactions, this is not a correct view. Actually, the Calvin cycle involves a series of 12 enzymes (protein catalysts), and at least five enzymes are rate-limiting to the assimilation of CO_2. These same five enzymes are activated in the light, and this means that they become less rate-limiting so that CO_2 assimilation can proceed at the highest rates. Indeed, in many higher plants, there is only negligible dark CO_2 fixation, because plastid enzymes are not activated.

Furthermore, within the chloroplast, there are several other processes which are not associated directly with CO_2 assimilation but which are driven in the light through reductant and ATP derived from water photolysis. For example, in the plant cell cytoplasm is reduced to the level of amino acids employing reductant (a protein called reduced ferredoxin), directly derived from water splitting and subsequent electron transport. Sulfate is reduced to the sulfhydryl level in the chloroplast in another sequence of reactions dependent upon photolytically derived ATP and reductant. Also, a recent study has made it clear that both carbon and nitrogen photoassimilation in higher plants can proceed simultaneously without internal competition for reductant and ATP. Indeed, it is becoming clear to present-day researchers that the term photosynthesis, applied to all higher plants and algae, must be broadened to include all of the processes inside and outside the chloroplast that are driven by reductant and ATP that are derived directly or indirectly from the photolysis of water.

four-carbon compounds first and are called C_4 plants. In the C_3 cycle, each of the two 3-phosphoglyceric acid molecules contains one phosphate group. In the *fourth step,* a phosphate group from each of two ATP molecules from the light reaction joins each of the two 3-phosphoglyceric acid molecules, forming two molecules of 1,3-diphosphoglyceric acid and changing each ATP to ADP. In the *fifth step,* each of the two 1,3-diphosphoglyceric acid molecules is reduced by reacting with two NADPH molecules from the light reaction. This forms two molecules of a three-carbon phosphorylated compound, called glyceraldehyde-3-phosphate, and two NADP+ molecules. In the *sixth step,* portions of the glyceraldehydes-3-phosphate molecules join together to form a six-carbon diphospho-sugar that is enzymatically transformed into glucose ($C_6H_{12}O_6$). In the *seventh step,* the remainder of the glyceraldehyde-3-phosphate carbon is recycled through a number of reactions, called the pentose phosphate shunt, to form more ribulose bisphosphate by joining some two-carbon fragments with three-carbon fragments. Note that for every six molecules of carbon dioxide that enter the cycle and yield one glucose molecule (or two triose phosphate molecules), 18 molecules of ATP and 12 molecules of NADPH are used.

C_3, C_4, and CAM Type Plants

Three groups of plants can be distinguished by certain specific physiological, morphological, and biochemical characteristics—C_3, C_4, and Crassulacean acid metabolism (CAM)

plants. One of the distinctions for these three types of plants is that the pathway for photosynthetic carbon fixation is different for each group.

The C_3 plants fix carbon dioxide by combining it with ribulose 1,5-bisphosphate to form two molecules of 3-phosphoglyceric acid, as just noted. This reaction is catalyzed by the enzyme ribulose bisphosphate carboxylase (Fig. 7–10).

In contrast, the C_4 plants fix carbon during photosynthesis by combining CO_2 with phosphoenolpyruvic acid in the presence of the enzyme phosphoenolpyruvate carboxylase to

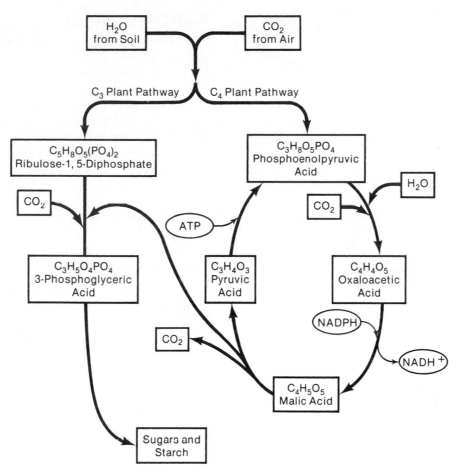

Figure 7–10   Carbon dioxide fixation (photosynthesis) pathways for C_3 and C_4 plants. In the C_4 pathway, CO_2 and H_2O react in the presence of an enzyme to form a four-carbon compound oxaloacetic acid. This compound is transformed to malic acid, which is converted to pyruvic acid to regenerate more phosphoenolpyruvic acid. The CO_2 generated in the conversion of malic pyruvic enters the Calvin (C_3) cycle via the pentose phosphate pathway.

produce oxaloacetic acid (a four-carbon acid). The CAM-type plants often display a diurnal pattern of organic acid formation and fix CO_2 in a modified C_4 pathway. Some of these plants have large succulent leaf cells, with stomata that open at night, allowing carboxylase enzymes to fix CO_2 into four-carbon acids. Certain members of the cactus, orchid, and pineapple families are examples of C_4 plants. The biochemistry of the different pathways of fixation all ultimately return to pathways that are parts of the Calvin cycle (Fig. 7–10).

Another important difference between C_3 and C_4 plants is their differential response in net photosynthesis to various light intensities (Fig. 7–11). The C_3 plants have low net photosynthetic rates, high carbon dioxide compensation points (50 to 150 ppm CO_2; 150 ppm = 0.015 percent, and high photorespiration rates). Plants of this type include cereal grains (barley, oats, rice, rye, wheat), peanuts, soybeans, cotton, sugar beets, tobacco, and spinach, as well as some evergreen and deciduous trees.

The C_4 plants, on the other hand, have high net photosynthetic rates, low carbon dioxide compensation points (0 to 10 ppm CO_2), and low photorespiration rates. This group includes about half of the species of the grass family of plants (e.g., corn, sorghum, sugarcane, millet, crabgrass, bermuda grass) as well as many of the broad-leaf plants such as pigweed *(Amaranthus)* and *Atriplex*. The C_4 plants are the more efficient users of CO_2 (Fig. 7–11). C_4 plants have the additional advantage of greater water use efficiency than the C_3 plants, and the CAM plants are even more water efficient than the C_4 plants. Another characteristic difference between the C_3 and C_4 groups is that at the normal CO_2 concentration (0.03 percent), light saturation is difficult to attain for C_4 plants, but light saturation is easily reached at low light intensities for C_3 plants.

Photosynthetic Intermediate Flow: Chloroplast to Cytoplasm and Sucrose Synthesis

In almost all vegetable crop plant leaves, whether they display C_3, C_4, or CAM type photosynthesis, sucrose and certain amino acids are considered to be the major primary final

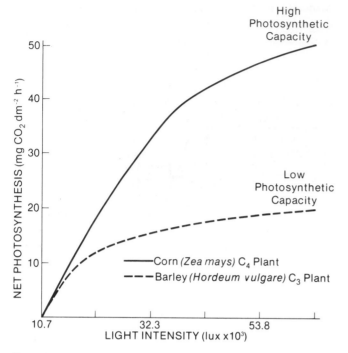

Figure 7–11   Effect of light intensity on net photosynthetic rate of a C_4 plant (corn) and a C_3 plant (barley). The C_4 plant is more efficient in using the sun's energy.

products of chloroplast metabolism (Fig. 7–3). The synthesis of these compounds actually begins in the leaf chloroplasts. As previously described, the three-carbon, phosphorylated intermediate, 3-phosphoglyceric acid (3-PGA), 3-phosphoglyceraldehyde (GAP), and dihydroxyacetone phosphate (DHAP) are the first measurable products of carbon assimilation in the leaf chloroplast. As previously noted, the latter two intermediates combine in the chloroplast to form fructose-1,6-diphosphate (FDP), which is further metabolized to fructose-6-phosphate (F6P), glucose-6-phosphate (G6P), glucose-1-phosphate (G1P), and starch. Also, some molecules of F6P are recycled in the PPRC to pentose phosphates in order to regenerate ribulose-1,5-diphosphate, the acceptor for atmospheric CO_2 (Figs. 7–9 and 7–10).

Additionally, molecules of 3-PGA, GAP, and DHAP are also exported from the chloroplast, across the chloroplast outer membrane, and into the leaf cell cytoplasm. There 3-PGA is metabolized in a series of reactions which subsequently give rise to amino acids. At the same time, in the cytoplasm, GAP and DHAP are reconverted to FBP. Then FBP is phosphated to F6P, some of which is converted, in turn, to G1P. Through a complex series of reactions F6P and G1P (hexose phosphates) are combined to form sucrose phosphate, a phosphorylated, 12-carbon compound (Fig. 7–3, in the cytoplasm). The phosphate group is removed, and sucrose is then translocated to active "tissue sinks," where it is metabolized to important primary and secondary compounds used in tissue development. Alternatively, in the light, sucrose is stored in special cell organelles which are called **vacuoles.**

PLANT RESPIRATION

The stepwise release of the energy captured and stored in photosynthesis is called **respiration.** Respiration reconverts the carbohydrates synthesized in photosynthesis to carbon dioxide, water, and energy.[6] The biological processes responsible for these changes are of prime concern because the energy released by respiration is the source of the energy for life processes. Sugars are oxidized in stepwise fashion, and phosphorylated compounds and organic acids are common intermediates. A greatly simplified reaction to represent the process is as follows:

$$C_6H_{12}O_6 + 6\ H_2O + 6\ O_2 \rightarrow 6\ CO_2 + 12\ H_2O + \text{energy} \quad [6]$$

From Equation [6], it is obvious that respiration resembles the burning of sugar. The principal difference between respiration and combustion is the rate of decomposition. Respiration is a stepwise process of degradation that releases a small amount of energy with each step; combustion releases the energy quickly.

Factors Affecting Rates of Respiration

Temperature

Temperature strongly affects respiration rates. Within the range of 0°C to 35°C (32° to 95°F), the rate increases about two times for each 10°C rise. The effect of temperature on respiration is an important factor in the storage of some crops. A harvested plant part that is stored or preserved is often living tissue and unless the product has been cooked or processed, the enzymes are active and vital processes continue. Since respiration is a degradation process, it should be retarded as quickly and completely as possible to prolong storage life. One way to retard respiration is to refrigerate the product. Temperatures must be carefully controlled, however. In some crops—apples, for instance—optimum storage temperature in 0°C (32°F); lower temperatures can severely damage the fruit. Some fruits such as bananas and tomatoes and certain flowers suffer tissue damage at storage temperatures below about 10°C (50°F), causing them to turn brown and soft.

Most plants grow best when nighttime temperatures are about 5°C (9°F) lower than daytime temperatures. The higher daytime temperatures favor photosynthesis, thereby resulting in the production of more carbohydrate for storage and growth. Lower nighttime temperatures reduce the rate of respiration, again allowing more plant growth or storage of photosynthates produced during the daytime.

[6]The overall reaction for respiration is sometimes written with no water on the left side of the equation and only six molecules of water on the right side. Omitting water on the left side obscures the fact that oxygen does not react directly with glucose; it is the water molecules as intermediate products that do. The oxygen given off in photosynthesis came from the water. In respiration the hydrogen atoms from the breakdown of sugar are used to reduce the oxygen gas to water.

Oxygen Concentration

Oxygen is an essential ingredient of respiration. With other factors being constant and not limiting, the rate of respiration decreases as oxygen concentration decreases. Lowering the oxygen concentration—accomplished by increasing the concentration of either carbon dioxide or nitrogen—is useful in storing certain fruit and vegetable crops. This modified storage atmosphere discourages rapid respiration.

Soil Conditions

Compacted or water-logged soils generally are poorly aerated. This condition reduces respiration in the roots, resulting in poor plant growth. Mineral nutrient deficiencies affect the respiratory enzymes, indirectly causing a reduction in respiration.

Respiration Pathways

A sugar molecule is degraded during respiration by a series of reactions. The first series is commonly referred to as **glycolysis.** Glucose (a six-carbon sugar) is chemically split into two three-carbon molecules in a series of stepwise reactions. The degradation of glucose to pyruvic acid, catalyzed by cytoplasmic enzymes, proceeds either aerobically or anaerobi-

cally. The degradation of pyruvic acid to carbon dioxide and water is referred to as the Krebs cycle, citric acid cycle, or the **tricarboxylic acid (TCA) cycle.**

For the discussion here, glucose is considered the substrate for respiration. However, other metabolites can also be respired; that is, broken down to small compounds, principally carbon dioxide. Examples of other potential substrates of respiration would include organic acids, amino acids, and lipids.

Glycolysis

In glycolysis, glucose in the presence of the enzyme hexokinase accepts a phosphate group from adenosine triphosphate (ATP), yielding glucose 6-phosphate (phosphorylated glucose) and converting the ATP to adenosine diphosphate (ADP). The stepwise degradation of glucose to pyruvic acid is schematically represented in Fig. 7–12. Each molecule of glucose yields two molecules of pyruvic acid and, in the process, two molecules of NAD^+ are reduced to NADH. The process consumes two molecules of ATP but generates four ATP molecules as well, resulting in a net gain of two ATPs.

It would be a mistake to consider the pathways of glycolysis as so many one-way streets. Most are reversible; that is, under certain conditions the plant cell is able to synthesize

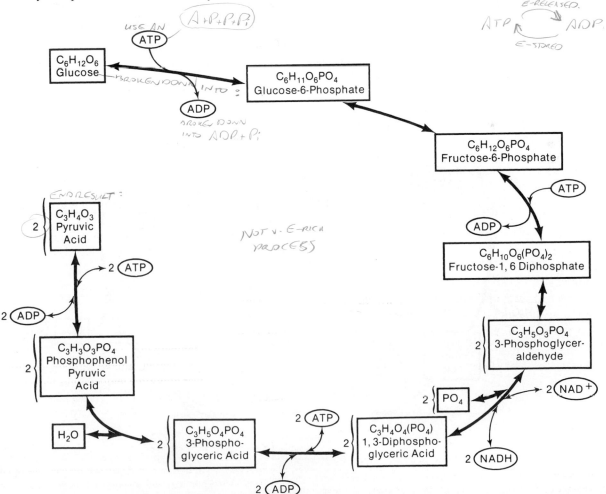

Figure 7–12 Glycolysis. Numerals in front of the molecules indicates the number of molecules in the reaction.

fructose 1,6-diphosphate from pyruvic acid or glucose from fructose 1,6-diphosphate that has been dephosphorylated. Thus, in some environments, plants not only consume carbohydrates during glycolysis but also are capable of synthesizing them with reverse glycolysis.

The Tricarboxylic Acid (TCA) Cycle

The stepwise degradation of pyruvic acid begins the tricarboxylic acid or TCA cycle. These respiration reactions occur in the cell's mitochondria. Under normal aerobic conditions, the final products of the TCA cycle are CO_2 and H_2O. For simplicity we will consider only the main steps, omitting some of the intermediate products. There are specific enzymes that catalyze each reaction. The principal reactions are shown in Fig. 7–13. Coenzyme A and NAD^+ (1 in Fig. 7–13) are necessary for the removal of one carbon dioxide molecule (A in Fig. 7–13) from pyruvic acid, generating a two-carbon fragment and one molecule of NADH. The two-

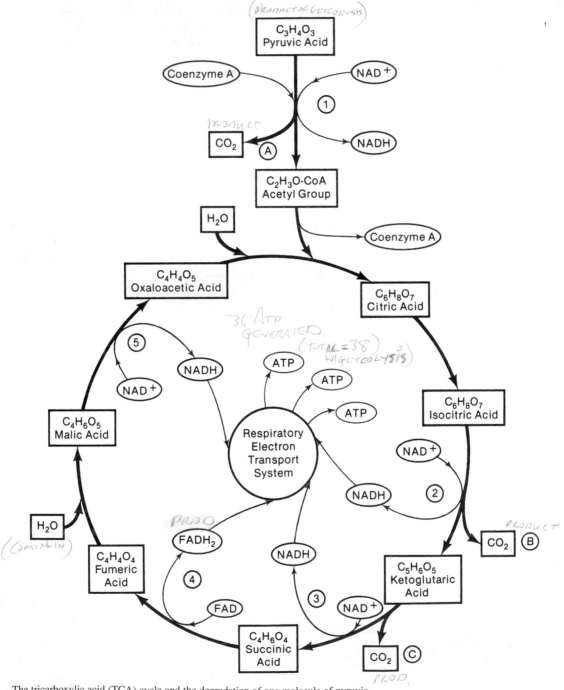

Figure 7–13 The tricarboxylic acid (TCA) cycle and the degradation of one molecule of pyruvic acid. Two molecules of pyruvic acid were produced from one glucose molecule; therefore, six molecules of CO_2 are released from one glucose molecule. Circled numbers and letters indicate places where NADH or FADH and CO_2 are generated in the cycle, respectively.

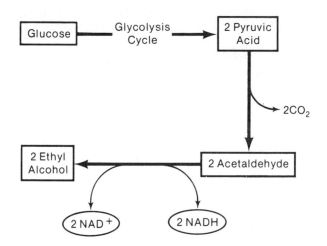

Figure 7–14 Anaerobic respiration (fermentation). Plant tissue has much less anaerobic respiration than aerobic respiration.

carbon fragment joins with a four-carbon oxaloacetic acid (left over from the cycle) to form a six-carbon citric acid molecule. This molecule is enzymatically rearranged in two steps to form isocitric acid, which reacts with NAD^+ (2 in Fig. 7–13) to form α ketoglutaric acid (a five-carbon) plus one carbon dioxide molecule (B in Fig. 7–13). NAD^+ is changed to NADH. In another series of reactions, catalyzed by several different enzymes, ketoglutaric acid is converted to succinic acid (four-carbon), and NAD^+ loses another carbon dioxide molecule (C and 3 in Fig. 7–13). Succinic acid is acted upon by FAD^7 (4 in Fig. 7–13) and changed to fumaric acid, which is enzymatically converted to malic acid by the addition of a water molecule. To complete the cycle, malic acid is oxidized to oxaloacetic acid, and this reaction is coupled to the formation of another NADH (5 in Fig. 7–13).

Fermentation

Under anaerobic conditions (i.e., where no molecular oxygen is available), pyruvic acid is also broken down to products such as acetaldehyde (a two-carbon compound) and ethyl alcohol. The latter is a product of fermentation used in the manufacture of wines and other alcoholic beverages (Fig. 7–14).

ELECTRON TRANSPORT SYSTEM

The last portion of the respiration process consists of a series of electron transfer reactions illustrated in Figure 7–15. The multiple enzymes involved are located in the membranes of

[7]Flavin-adenine dinucleotide (FAD), an electron carrier, acts in a way similar to NAD^+ in biological oxidation reactions. $NAD^+ + 2e^- + H^+ \rightarrow NADH$, and $FAD + 2e^- + 2H^+ \rightarrow FADH2$.

mitochondria. Electrons present in the NADH generated in the TCA cycle (Fig. 7–13) are channeled through a series of oxidation-reduction steps, eventually leading to the reduction of molecular oxygen to form water. The most important product of this cycling of electrons is the generation of high-energy ATP; namely, three ATP molecules are generated for each pair of electrons passed through the electron transport system. The ATP generated in the electron transfer portion of the respiration process can then be employed in a wide variety of reactions required for building and/or maintaining other metabolites or structures, for example, protein synthesis.

It is important to reemphasize that respiration can occur in any plant tissue or cell, wherever a supply of carbohydrate is available to fuel the process. For example, in the production of fruit and seed, carbohydrate is used for "localized" production of NADH and ATP; that is, carbohydrate, not NADH and ATP, are transported long distances in the plant. In contrast to the ubiquitous occurrence of respiration, photosynthesis occurs only in green plant tissues and mainly in the leaf blades.

Also, respiration is thought to have two purposes. One role of respiration is to generate "local" supplies of ATP required for a variety of metabolic processes, as already noted. Examples would include the synthesis of membranes and the active transport of minerals and metabolites across the membranes. The other role of respiratory generation of ATP is the maintenance of existing processes or structures. For example, it is well known that many enzymatic proteins are constantly degraded and resynthesized, thus requiring a supply of energy just to keep the metabolic machinery up and running smoothly. This latter portion of respiration is usually referred to as "maintenance respiration" and is a relatively small portion of the overall respiratory requirement of the growing plant.

To summarize respiration—and to account for the ATP molecules produced—assume the completed degradation of glucose to carbon dioxide and water. As each glucose molecule is consumed in glycolysis, two molecules of ATP and two molecules of pyruvic acid form (Fig. 7–12). The two pyruvic acid molecules from the glycolysis of one glucose molecule enter the TCA cycle and yield six molecules of CO_2 (Fig. 7–13). Eight NAD^+ molecules accept 16 electrons, and 2 FAD molecules accept 4 electrons (a total of 20 electrons). Four electrons were accepted by two NAD^+ molecules during the glycolysis of glucose to pyruvic acid (Fig. 7–12). All 24 of the electrons involved in the photolysis of water are accounted for (Equation [3]). To complete the cycle, the 24 electrons and 12 hydrogen ions are carried by enzyme carriers and finally return to form 6 water molecules and 6 oxygen molecules (Equation [3] and Fig. 7–15). Each of the 8 NAD^+ molecules plus the 4 FAD molecules (total, 12) provide sufficient energy to convert 3 ADP molecules to ATP (Fig. 7–15). This makes a total of 36 molecules of ATP formed. Remember also that during the respiration of glucose to pyruvic acid, 2 molecules of ATP were formed. Thus, the grand total is 38 molecules of ATP.

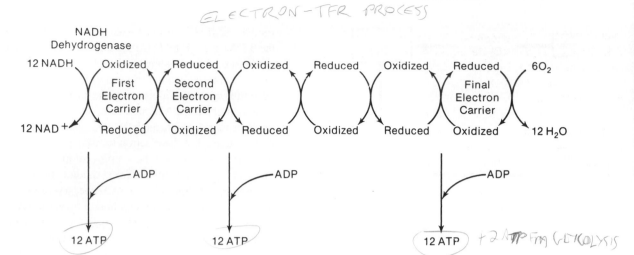

Figure 7–15   The essential features of the system are that electrons from NADH are passed through various electron carriers to the ultimate electron acceptor—oxygen—which is reduced to water. The protons required for water formation come from the NADA and from free protons in the water of the cells.

TABLE 7–2	Summarized Comparison of Photosynthesis and Respiration

Photosynthesis	*Respiration*
Requires CO_2 and water.	Requires O_2 and carbon compounds.
Produces O_2 and $C(H_2O)_n$.	Produces CO_2 and H_2O.
Light energy trapped by chlorophyll.	Energy released.
Occurs in light.	Occurs in both light and darkness.
Only cells with chlorophyll photosynthesize.	All living cells respire.
ATP produced (photophosphorylation).	ATP produced (oxidative phosphorylation).
Hydrogen from water used to reduce $NADP^+$ to NADPH.	Hydrogen from $C(H_2O)_n$ used to oxidize NADPH to $NADP^+$ or reduce $NADP^+$ to NADPH.

The principal characteristics of photosynthesis and respiration are summarized and compared in Table 7–2.

PHOTORESPIRATION

To this point in the discussion of respiration, we have considered only mitochondrial respiration—also sometimes called "dark respiration." This latter designation distinguishes mitochondrial respiration from another type of respiration termed **photorespiration.** Photorespiration is a process that occurs in the light and begins with the *oxygenation* of ribulose bisphosphate; that is, the enzyme that normally catalyzes the carboxylation of ribulose bisphosphate also catalyzes the oxygenation of the substrate (refer to Fig. 7–10).

The reactions that follow the oxygenation step are complex and are beyond the scope of this text. These reactions lead to the formation of 3-phosphoglycerate—a product of the "normal" pathway of photosynthesis. A negative aspect of the photorespiration pathway is that additional energy (ATP) and reductant (NADPH) are consumed in the process of forming 3-phosphoglycerate. Thus, photorespiration has a net negative impact on the efficiency of C_3 photosynthesis.

Photorespiration is strongly dependent on the ratio of CO_2 to O_2 in the chloroplast. Some plants have mechanisms for concentrating CO_2, and these plants have low or nil rates of photorespiration. An example of plants able to concentrate CO_2 are the C_4 plants already discussed; the CO_2 assimilation pathway for these plants was illustrated in Figure 7–10. However, most crop plants are C_3 plants and, for this group, photorespiration is a process that negatively influences the capture of light energy in the reduced carbon of carbohydrates.

ABSORPTION, TRANSLOCATION, AND ASSIMILATION

The absorption and transport of raw materials used for photosynthesis in the green parts of the plant and the translocation of the products of photosynthesis to areas of storage or consumption are important to an understanding of plant growth and development.

Absorption and Conduction of Water

Water is absorbed by the roots from the soil and distributed throughout the plant even to its highest leaves—perhaps 60 m (200 ft) or more above the source. This formidable task requires considerable energy, estimated to equal about 16 kg/cm^2 (225 lb/in^2) of pressure. The energy to move the water is supplied by pressure developed in the roots and by the negative pressure resulting from transpiration (water loss) in the leaves. In tall plants, this latter force is the principal force driving water movement.

The upward movement of water occurs mainly in the xylem. Minerals dissolved in the water are also carried to leaves, stems and fruits via the xylem stream. Examples of important minerals are nitrogen used for protein synthesis, phosphates used to make ATP during photosynthesis, and magnesium that becomes a part of the chlorophyll molecule. Another conducting system is the phloem, through which the manufactured sugars move from the leaves to storage organs and various tissues for utilization.

Water absorption and translocation by the plant are of primary importance. Of all materials taken in by plants, water is absorbed in the largest quantities, but only a fraction of that taken in is used in metabolic processes. Most of the water is lost from the leaves by transpiration. Water loss in the plant is regulated to a certain extent by the opening and closing of stomata on the leaf surfaces.

The upward movement of water from the roots through the xylem in the stems to the uppermost leaves is sometimes called the **transpiration stream,** because transpiration is the primary cause of this movement. The upward transpiration pull in the leaves is started by the evaporation of water molecules from the outer surfaces of the mesophyll cells. As this water is lost, the mesophyll cells become water deficient and a potential difference is created between the dry mesophyll cells and the walls of adjacent moist cells. Because of water's cohesive properties, water from the wetter adjacent cell walls begins to diffuse into the less hydrated cells, thus relieving the pressure difference. Continued loss of water molecules at the leaf surface establishes a flow of water throughout the plant from cell to cell, from the roots to the leaves. In addi-

tion, a pressure difference exists between the outer cells of the root tops and the soil moisture, creating a driving force that moves water into the roots. Part of the water is absorbed into the roots by osmosis.[8] The proportion taken in by osmosis compared with the amount taken in by the transpiration stream depends upon the rate of transpiration. Normally, however, the amount entering by osmosis is relatively small.

Absorption and Transport of Mineral Nutrients

In higher plants, the minerals initially needed to start growth of a new plant are normally provided by the seed or stored in the propagating tissue. But as these are used, additional nutrients for the plant's continued growth must be absorbed from the soil.

Chemical analyses of the sap in root cells for various mineral nutrients reveal that these cells have concentrations 500 to 10,000 times higher than those of the same element in the soil solution. If simple diffusion was the only mechanism involved in taking up soil nutrients, the mineral nutrients could not move into the roots against such a high concentration gradient. In addition, the plasma membranes of cells are largely impermeable to the movement of ions. Thus, energy is required to move ions against the concentration gradient and through the impermeable membranes. In the root cells, this energy is obtained from the respiration of starches and sugars that originated in the photosynthetic processes.

---

[8]Osmosis is the flow or diffusion that takes place through a semipermeable membrane (as in a living cell) that typically separates either a solvent (water) and a solution, or a dilute solution and concentrated solution. These concentration gradient differences create flow conditions through the semipermeable membrane until the membrane are equal and equilibrium is established. At this equilibrium concentration osmosis ceases.

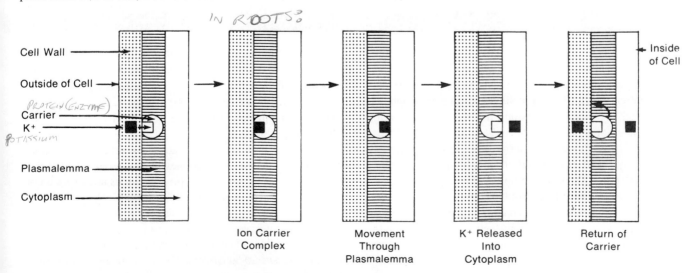

Figure 7–16 A schematic diagram of carrier ion transport. *Source:* Adapted from Weier, T. E., C. R. Stocking, M. G. Barbour, and T. Rost. 1982 *Botany: An introduction to plant biology.* 6th ed. New York: John Wiley.

A plant absorbs mineral nutrients against a high concentration gradient by a process called **active transport.** Active transport is facilitated by special proteins present in the plasma membranes of cells. Various models for these "carrier proteins" have been proposed, and one model is shown in Figure 7–16. In this model, a potassium ion, which has moved through the cell wall, is bound by the carrier at the outer surface of the plasma membrane. In the energy-requiring step, the protein moves or changes shape so that K^+ ion is moved to the inside of the membrane. Following release of the ion, the protein resumes its original shape or position.

Translocation of Sugars

Sugars which are synthesized during photosynthesis move throughout the plant, primarily in the phloem tissues. The movement is mostly downward from leaves to roots, but lateral or even upward movement from leaves to fruits or buds or other storage organs also occurs. Translocation takes place in the long sieve elements connected end to end to form sieve tubes. In woody perennials, the phloem tissue is just underneath the bark following development of secondary vascular tissue. Thus, when this phloem tissue is severed by "girdling," the tissue above the cut proliferates whereas the tissue below the cut is starved for photosynthate (Fig. 7–17).

The rate of translocation of sugars in the phloem is, in some instances, more than a thousand times faster than simple diffusion of sugar through water. The rate of translocation has been measured in many plants, and average values of 1 to 6 $g/cm^2/hr$ have been found in developing fruits and tubers.

To conclude the discussion of transport, it is useful to note that metabolites, ions, and water move through the plant

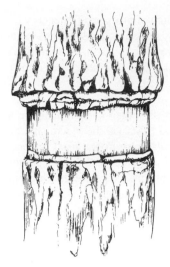

Figure 7–17　Grape shoot girdled by removing the bark and the phloem and leaving the xylem intact causes sugar and other carbohydrates to concentrate above the girdle. Grape shoots are often girdled commercially to increase fruit size.

by a variety of processes, some relatively slow and some relatively fast:

1. diffusion, which transports ions and molecules slowly
2. cytoplasmic streaming, which transports molecules and ions within the cytoplasm at a considerably faster rate than diffusion
3. downward mass flow translocation of material in the phloem
4. very rapid upward movement of water and mineral nutrients through the xylem
5. lateral transport of materials along the vascular rays radially from sieve tubes into the cambium tissue and xylem

SUMMARY AND REVIEW

Photosynthesis uses solar energy to provide the only source of food and oxygen for animals, including humans. The essential requirements for photosynthesis are carbon dioxide, water, light energy, and living cells containing chlorophyll. The products are carbohydrates and oxygen. There are two reactions in photosynthesis, the light reaction (which requires light) occurs when the energy from light is harvested by chlorophyll and transferred to the high-energy compounds ATP and NADPH. In the process, water is split and oxygen is released into the atmosphere.

The dark reaction (which can proceed in dark or light) uses the energy stored in NADPH and ATP to fix atmospheric carbon from CO_2 into the carbohydrate glucose, a six-carbon sugar. The Calvin cycle is the biochemical pathway that leads to the formation of glucose. It takes six revolutions of the cycle to form one glucose molecule.

Photosynthetic rate and efficiency depend on many environmental factors. The concentration of CO_2 in the atmos-

phere and within the leaf, light intensity and quality, the availability of water to the plant, and heat are important environmental factors. When any of these factors are out of balance, such as too much or too little heat, or in an inappropriate form, such as light of the wrong wavelength (color), photosynthesis is inhibited.

Within the plant, the physiological state of the plant affects photosynthesis. The biochemical systems that are a part of photosynthesis are regulated by enzymes and other metabolic activity. The plant must be in good physiological condition for the necessary enzymes to be synthesized and active and for other processes such as transpiration and active transport to take place. The age of a plant often determines photosynthetic activity and efficiency. Many plants, for example wheat and soybeans, exhibit a reduction or even cessation of photosynthetic activity when the seed has been formed.

Respiration is the release of energy stored in carbohydrates. Its end products are CO_2 water and energy available in the form of ATP. Respiration occurs in all living organisms. The energy released through respiration is used to drive all the biochemical reactions in nonphotosynthesizing organisms and nearly all the reactions in photosynthesizing organisms. Respiration can be divided into three steps. The first is glycolysis or the degradation of glucose (6c) into two molecules of pyruvate (3C). The second is the TCA cycle which further degrades pyruvate, releasing 3 molecules of CO_2 for each pyruvate molecule degraded. The third reaction is the formation of ATP from the FADH and NADH produced as a part of the TCA cycle. As a result of the third step, O_2 is combined with H^+ and water is formed.

Transpiration is the movement of water from the soil through the plant and into the atmosphere. Water moves through plants primarily through the xylem. Negative water pressure develops in the xylem because of water pressure in the roots, and loss of water through the leaves creates water pressure differences with the greatest pressure being in the roots and the least in the leaves. Water moves from high to low pressure so water movement is from the roots to the leaves. In the water are dissolved minerals that were transported into the xylem along with the water. These move up into the plant where they can be used metabolically for plant growth.

The sugars synthesized in photosynthesis and other dissolved substances move through the plant mainly through the phloem. Unlike xylem which is made of dead cells that form long, nearly hollow cylinders, phloem cells are living. Water and solutes move through phloem by active transport. Movement can be very selective with specific contents of the water and water itself directed to specific sites.

Photosynthesis, respiration, and translocation are not independent processes. Translocation systems provide the water necessary for photosynthesis and remove the synthesized sugars. Without the removal of the sugars, feedback mechanisms would shut down photosynthesis when the sugars accumulated. Respiration recovers the energy that was stored by photosynthesis and provides the energy for all plant processes other than photosynthesis, including the active transport of photosynthesized sugars. When the energy output of photosynthesis is less than the energy utilization of respiration, there is a loss of carbohydrates in the plant. Compensation points are the level at which either CO_2 or light is so low that respiration equals photosynthesis and carbohydrates are not being stored. The processes must work in a coordinated way or plant growth and development will be abnormal.

Growing and utilizing crops requires constant decision making about the managing of the crop. These decisions often have to be made swiftly, and even more often have to be modified as changes in environment affect the development of a crop. When the growing environment cannot be manipulated, an understanding of the impact of environment on key plant processes is needed to allow wise decision making in the management of the harvest.

In situations where the environment can be manipulated, an agronomist, horticulturist, or turf grass specialist has to understand how manipulations will affect photosynthesis, respiration, and translocation. For instance, in prolonged periods of cloudy weather, the development of flowers in a greenhouse, growth of grass on a golf green, or the filling of soybean pods may be slowed or halted because the light level is below the compensation point. In cases of low light, a decision may have to be made regarding reducing temperatures, withholding irrigation, or reducing fertilization to slow plant growth enough to prevent carbohydrate loss to respiration and to maintain crop quality. When light conditions improve, the decision of when to resume normal growing practices has to be made. A grower also has to keep in mind how the manipulations themselves may affect such things as condition or timing of the harvest.

EXERCISES FOR UNDERSTANDING

7–1. Observe a crop growing in a field or greenhouse, a flower garden, or the turf on a sports or recreation field. Make a list of all the environmental factors that you can see or think of that are affecting the plants as you look at them. Assuming these factors will remain fairly constant for at least a week, predict how each one is affecting photosynthesis, respiration, and translocation in the plant and how the plant may look in a week's time in response to these factors.

7–2. If you could manipulate the environment in the previous growing situation, what could you do to change or improve the growth of the plants you are observing? How would your manipulations affect photosynthesis, respiration, and translocation?

REFERENCES AND SUPPLEMENTARY READING

BUCHANAN, B. B., W. GRUISSEM, and R. L. JONES. 2000. *Biochemistry and molecular biology of plants.* Rockville, Md.: American Society of Plant Physiologists.

LAMBERS, H., F. S. CHAPIN, and T. PONS. 1998. *Plant physiological ecology.* New York: Springer-Verlag.

SALISBURY, F. B., and C. W. ROSS. 1992. *Plant physiology.* Fourth Edition. Belmont, Calif.: Wadsworth.

TAIZ, L., and E. ZEIGLER. *1998. Plant physiology.* Second edition. Sunderland, Mass.: Sinauer Associates.

CHAPTER

8

Soil and Soil Water

KEY LEARNING CONCEPTS

After reading the chapter you should be able to:

♦ Describe soil and its components.
♦ Understand the factors involved in soil formation and how those factors influence soil characteristics.
♦ Understand the physical properties of soil and the influence those properties have on plant growth.
♦ Understand the chemical properties of soil and the influence those properties have on plant growth.
♦ Understand the nature of soil water and how its properties affect plant growth.

Plants require oxygen, carbon dioxide, water, and certain mineral elements in order to grow and develop. As noted in Ch. 7, some of the required oxygen and carbon dioxide is exchanged through the leaves and roots; oxygen and mineral elements are absorbed by the roots from the soil solution. This chapter discusses the role of soil, soil water, and soil atmosphere in providing plants with anchorage, essential mineral nutrients, and water.

DEFINITION OF SOIL

Soil can be defined as that portion of the earth's crust that has formed from the decomposition of rocks and minerals through physical, chemical, and biotic forces, in which the roots of plants grow. Soil is a very complex and dynamic system of many interacting factors that affect and are affected by plants. Soil consists of:

1. the solid fraction; that is, rock fragments and minerals
2. the organic fraction; the decayed and decaying residues of plants, microbes and soil animals
3. the liquid fraction and its dissolved minerals
4. the soil atmosphere or soil air.

The voids found between the solid fraction are called **pore spaces.** Pore spaces vary in size and continuity and are an important physical property of soils.

The kind of soil and its ability to permit plant growth is influenced by the relative amounts of each of these four components. Disregarding soil texture, an "ideal" soil by volume would contain about 25 percent air, 25 percent water, 40 percent mineral material, and 10 percent organic matter. In reality, the idealistic percentages seldom occur; furthermore, water and air contents can vary greatly depending on when observations are made.

FACTORS INVOLVED IN SOIL FORMATION

Soil is derived from parent rocks, minerals, and decaying organic matter. The two major processes in soil formation are (1) accumulation and (2) differentiation of the parent material. **Parent material** accumulates from the breakdown of parent rocks by weathering. This process must occur before soil can begin to form. The parent material accumulates as an unconsolidated mass that later differentiates into characteristic layers called **horizons.**[1] Differentiation occurs by mechanical separation and/or dissolution of the parent material. As the process continues, the horizons generally become more distinguishable and finally develop into a **soil profile.**[2]

The factors responsible for soil formation are (1) parent material, (2) climate, (3) all forms of organic matter, (4) topography, and (5) time.

Parent Material

The formation and accumulation of material by chemical and physical weathering of parent rocks is the first step in the development of soil.

---

[1]Horizon—a distinct layer of soil having physical and/or chemical differences resulting from soil-forming processes as seen in a vertical cross section.
[2]Soil profile—a vertical section of soil extending through all its horizons from the surface to the parent material.

Chemical Weathering

This entails four distinct processes.

Carbonation is the reaction between carbonates and other minerals (such as feldspars) and carbonic acid (H_2CO_3) formed when carbon dioxide from the air dissolves in water. Carbonation reactions produce compounds that are less resistant to decomposition. For example, limestone ($CaCO_3$) reacts with water (H_2O) and carbon dioxide (CO_2) to form calcium bicarbonate [$Ca(HCO_3)_2$], a compound more soluble than the limestone:

$$CaCO_3 + H_2O + CO_2 \rightarrow Ca(HCO_3)_2$$

Hydration adds molecular water to form a hydrated material more vulnerable to pulverization. An example is calcium sulfate ($CaSO_4$) absorbing water to form gypsum ($CaSO_4 \cdot 2\,H_2O$), a hydrated calcium sulfate:

$$CaSO_4 + 2\,H_2O \rightarrow CaSO_4 \cdot 2\,H_2O$$

Hydrolysis is the reaction between the parent material and water to form a more soluble product. In the following hydrolysis reaction, potassium ions (K^+) are made more available to plants by the reaction of the slowly soluble feldspar mineral ($KAlSi_3O_8$) with water (H_2O) to form soluble potassium hydroxide (KOH) plus aluminum silicate (Al_2SiO_3):

$$KAlSi_3O_8 + H_2O \rightarrow HAlSi_3O_8 + KOH$$

$$2\,HAlSi_3O_8 + H_2O \rightarrow Al_2O_3 \cdot H_2O + 6\,H_2SiO_3$$

Oxidation reactions form oxides of parent material by reaction with oxygen. For example, ferrous oxide (FeO) reacts with oxygen (O_2) to yield ferric oxide (Fe_2O_3), a product more oxidized than the reactant:

$$4\,FeO + O_2 \rightarrow 2\,Fe_2O_3$$

Physical Weathering

Changes in temperature strongly affect the rate of physical weathering. Differential rates of contraction and expansion caused by temperature changes bring about cracking and peeling of the outer layers of rocks by a process called **exfoliation.** A second process is caused by the presence of different materials within a rock, each with its own characteristic coefficient of expansion. Because of these differences, sudden large temperature changes cause uneven expansion or contraction, cracking the rocks. A third process is the cracking of rocks caused by the expansion of water as it freezes in rock fissures.

The mechanical action of glaciers causes rocks embedded in the ice to scrape against other rocks as the glacier moves. This action grinds the rocks into increasingly smaller rock fragments. This is a powerful physical process, which reached a tremendous magnitude during the Ice Ages

Figure 8–1 Massive gullies formed by severe stream erosion during periods of heavy rainfall. These gullies are beyond reclamation by practicable methods. *Source:* USDA Soil Conservation Service.

(Pleistocene epoch). The product of glacial weathering is called **glacial till,** and it comprises rock particles ranging in size from clay to boulders. This material is deposited in various ways beneath, beside, and at the terminus of the melting glacier.

Physical weathering is also caused by moving water, as stream erosion, sheet erosion, rill erosion, or wave action (Fig. 8–1). The action is similar to that of glaciers. Water from spring rains and melting snow moves rapidly down mountain streambeds, carrying parent rock fragments of varying sizes. As these fragments move along, they are gradually worn down to smaller and smaller particles, eventually forming parent material. In arid regions, wind action is similar to water. Coarse sand particles (parent material) are swept along the ground with sandblasting action wearing away other larger parent rocks. The action of plant roots can sometimes physically break down parent rocks. For example, a tree root growing into a crack in a rock can ultimately fracture the rock. While this is not considered weathering, it is a physical soil-forming process. In this case, some chemical weathering must occur first to provide nutrients for the plants.

Kinds of Rocks

Parent rocks contain the nutrients that will be found in parent material, and later in the soil. Rocks are made up of consolidated material, unconsolidated material, or both (Fig. 8–2).

1. **Igneous** rocks—for example, granite formed ages ago or recently (as in Hawaii)—are consolidated, hard, and generally weathered in place. They are formed from the hardening of various kinds of molten rock material and are composed of the minerals quartz and feldspar, among others.
2. **Sedimentary** rocks are generally unconsolidated, composed of fragmental rock material that has been

Figure 8–2 Parent rocks from which soil can be formed. (A) Igneous. (B) Sedimentary. (C) Metamorphic. *Source:* Rosa Maria Marquez.

transported and deposited by wind, water, or glaciers. Limestone, sandstone, and shale are examples.
3. **Metamorphic** rocks form from igneous or sedimentary rocks that have been subjected to sufficiently high pressures and temperatures to change their structure and composition. Slate, gneiss, schist, and marble are examples of metamorphic rocks.

Climate

The climate affects soil formation. In areas of high rainfall, soils are highly leached and acid in reaction. Chemical weathering proceeds at a rapid rate, especially if high rainfall is coupled with high temperature. The fertility level of soils formed under high rainfall is generally low because many of the plant nutrients are leached from the root zone. Many of these soils are red or yellow in color, indicating a relatively high percentage of iron oxide, which remains after other elements have been removed.

On the other hand, soils developed in arid climates are not highly leached. Calcium and magnesium carbonates tend to accumulate, and chemical weathering proceeds at a much slower rate. Soils formed under arid conditions often contain excessive quantities of other salts besides carbonates, and are not productive until the amounts of salts are reduced. Land can be desalinated by flooding with water and leaching the salts downward through the soil profile.

Organic Fraction

The organic components of the soil are residues of plants and animals living in the soil. The amount and kind of these organisms are influenced by the climate. For example, the climate of an area determines whether grasses or trees predominate which, in turn, influences the soil-forming processes.

Vegetation aids soil formation by supplying organic matter in the form of dying and decomposing plants. Grasses decompose into a different kind of organic residue than do conifer trees. Also, the amount of organic matter varies according to the type of vegetation (Fig. 8–3). Peat soils form where reeds, sphagnum moss, and grasses grow and where decomposition of the organic matter is minimized. Soils in arid regions normally contain low amounts of organic matter because of the limited growth of desert grasses, shrubs, and cacti in these climates.

The amount of organic matter in soil is influenced to a large extent by the difference between the accumulation and decomposition of such material. In cases where decomposition rates are very high, the organic fraction accumulates very slowly. Thus the amount of organic matter remains low. Consider the case where temperatures are high and rainfall or irrigation keeps the soil moist. Under these conditions, the rate of decomposition nearly equals the rate of accumulation, and the organic matter content changes little over time. On the other hand, in cold areas or sometimes under anaerobic conditions, decomposition is inhibited and the organic matter accumulates. Finally, when a soil is tilled, the decomposition rate can exceed the accumulation rate and organic matter is lost from the soil. Many factors affect the rates of accumulation or decomposition, all of which play a role in determining the amount of organic matter present in a given soil at a particular time and location.

The type of root system produced by the different plant species also influences soil formation. Dense fibrous root

Figure 8–3 The organic matter produced from the decomposition of forest litter is different from that produced from prairie grasses. In this example, the forest is not so dense that it excludes grasses, and both types of vegetation are seen and produce both types of organic matter. *Source:* USDA Soil Conservation Service.

systems of grasses often lead to permeable soils, such as some of the Mollisols (prairie soils) formed in the central states of the United States and in some areas of the USSR.

Topography

Topography influences drainage and runoff. Steep slopes are subject to erosion because water flows faster downhill, with less percolating into the soil. Erosion in turn affects the speed of the soil-forming process. Leaching is also either speeded up or retarded by the topography. Soils developed in humid regions of high rainfall and flat topography are often highly leached or waterlogged, depending upon internal drainage. Gentle slopes that are heavily covered with vegetation slow the water flow and allow more time for water to percolate into the soil, permitting the development of a well-defined profile. Rapid surface runoff causes more erosion, and if the vegetation is removed or absent, even deeper gullies are cut into the sloping land (Fig. 8–4).

Topography has a marked effect on climate. High altitudes mean lower soil and air temperatures, which influence the amount and type of vegetation. On the average, air temperature decreases 1°C for each 300 meters (2°F/1,000 ft) of elevation for the first 10 to 15 km (6 to 9 mi) of altitude. Thus

Figure 8–4   Sloping topography accelerates rill erosion, which removes topsoil. Constant removal of topsoil exposes the parent material to weathering and more rapid soil formation. *Source:* USDA Soil Conservation Service.

an increase in elevation changes the kind and type of vegetation. Grasses and deciduous trees grow at lower elevations, with coniferous evergreens at higher elevations. Above the timber line, due to the lower temperature and humidity, little or no vegetation exists.

Time

Hard-to-decompose rocks such as granite require millions of years before parent material can accumulate; softer rocks such as limestone require less time. Interactions between biological and chemical agents reacting with parent material over long periods of time differentiate the soil into horizons as aging takes place. Soils without well-developed horizons are classified as young soils even though the parent material may have been present for a great many centuries. For example, some desert soils, constantly shifted and transported by wind, are young soils because they lack profile development. Also, silt is often moved along in rivers, deposited, and then moved again, leaving little chance for profile development. Such soils remain both pedologically and geologically young no matter how many years they may have been in existence.

As they develop, mature soils attain well-defined profiles consisting of three principal layers or zones (Fig. 8–5). The surface layer, the **A horizon,** is often referred to as the **zone of leaching.** It varies in depth and contains most of the plant roots. This leached zone often lacks some of the important mineral nutrients, but it does contain the largest

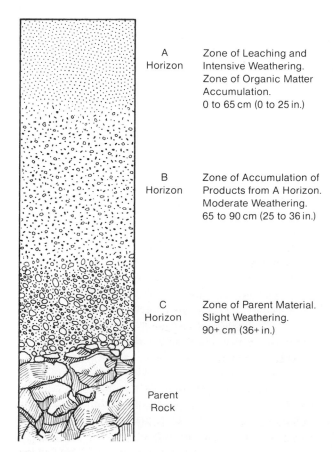

A Horizon — Zone of Leaching and Intensive Weathering. Zone of Organic Matter Accumulation. 0 to 65 cm (0 to 25 in.)

B Horizon — Zone of Accumulation of Products from A Horizon. Moderate Weathering. 65 to 90 cm (25 to 36 in.)

C Horizon — Zone of Parent Material. Slight Weathering. 90+ cm (36+ in.)

Parent Rock

Figure 8–5   An idealized soil profile developed by weathering.

amount of organic matter. The organic matter makes the A horizon permeable and dark-colored.

Below the A horizon is the **B horizon,** known as the **zone of accumulation.** Plant nutrients, silts, clays, and other materials from the upper layer are leached into and accumulate in this horizon. The color is generally lighter than that of the A horizon, and less organic matter is present (although the roots of deep-rooted plants do reach into the B horizon).

The **C horizon** consists of unweathered to slightly weathered material like or unlike that from which the A and B horizons are formed. It can also include accumulated calcium carbonates or other salts.

The factors affecting soil formation are obviously interrelated. Parent material affects, along with other factors, the capacity of the soil to support plant life, which in turn influences the kind of vegetation. Topography and temperature also influence vegetative growth. Temperature interacts with organic matter. Topography, temperature, and time influence parent rock conversion into parent material, each interacting so as to yield different soil formation and soil characteristics. Thus each factor affects and is affected by all the others. It would be difficult to say that any one is more important than another in soil formation.

PHYSICAL PROPERTIES OF SOIL

Soil Texture

An important physical property of soil is its texture. **Soil texture** is defined as the percentage of sand, silt, and clay particles in a soil. Soil particles vary in size from coarse rock fragments (>2 mm) to those so small (<0.002 mm) that an electron microscope is needed to observe them (Table 8–1). To measure soil texture, the soil particles are separated into their respective sizes and the percentage in each size category is calculated. A textural classification is then made with the aid of a soil textural triangle (Fig. 8–6).

Soil texture influences many of the soil's properties pertaining to crop production (Table 8–2). The distribution of different particle sizes determine the ability of soils to hold and transmit water. Soil with a high percentage of sand loses water quickly, retaining little for plant use. Plants growing in these soils will experience water deficits sooner after wetting than those grown in loam or clay soils. Texture also influences soil aeration. In soils largely composed of very fine clay particles, movement of both air and water can be limited. Plant roots need oxygen for respiration, and soils with low rates of gaseous diffusion restrict respiration and plant growth. Also, many beneficial soil microorganisms require well-aerated soils.

A soil consisting of a mixture of 40 percent sand, 40 percent silt, and 20 percent clay produces a loam soil that retains sufficient water for good plant growth and permits its movement without restricting aeration. From its texture alone, such a soil would approach an ideal soil.

TABLE 8–1 Dimensions of Soil Particle Size Classes

Diameter of Particles			
Millimeters 2.0	0.05		0.002
Inches 0.080	0.002		0.00008
gravel, stones	sand	silt	clay
Particles visible with the naked eye		Particles visible under microscope	Particles visible under electron microscope

TABLE 8–2 Some Soil Properties Influenced by Soil Texture

Soil Property	Textural Class		
	Sand	Silt Loam	Clay
Aeration	excellent	good	poor
Cation Exchange	low	medium	high
Drainage	excellent	good	poor
Erodibility[a]	easy	moderate	difficult
Permeability[a]	fast	moderate	slow
Temperature (spring)	warms fast	warms moderately	warms slowly
Tillage	easy	moderate	difficult
Water-holding Capacity	low	moderate	high

[a]By water.

Soil texture can affect soil fertility. A soil high in clay has a higher cation exchange capacity. This reduces leaching losses of positively charged cations.

The ease of tillage (plowing, disking, cultivating) is influenced by soil texture. Sandy or loam soils at the proper moisture content are easier to till than clay soils. Root penetration is sometimes restricted in soils of high clay content. Other factors being equal, most crops grow better in loam soils than in either sandy or clay soils.

Soil Structure

Soil structure is defined as the arrangement of primary soil particles into secondary units; that is, the manner in which individual primary particles clump and hold together. The secondary unit (aggregate) is a clump of soil particles held together which acts as an individual larger particle with specific characteristics. The kind of soil structure is determined not only by the relative amounts of each primary particle but also by the manner in which these particles are arranged into aggregates. The size and form of aggregation is known as the structure of soil.

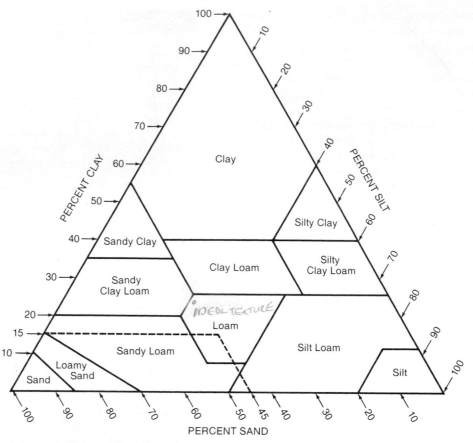

Figure 8–6   A soil textural triangle. To illustrate how the triangle works, assume that a sample of soil has been analyzed and found to contain 45 percent sand, 15 percent clay and 40 percent silt. On the triangle, locate the 45 percent value
on the sand (bottom) axis and draw a line parallel to the lines in the direction indicated by the arrow. Next, locate either
the 15 percent value for clay on the clay axis (left side of triangle) or the 40 percent silt on the silt axis (right side of triangle), and draw another parallel line along either of the two axes in the direction indicated by the respective arrows. The
two lines intersect in the loam area of the triangle; thus, this soil is classified as a loam. *Source:* USDA.

Descriptive words are used to classify soil structure:
for example, prismatic, subangular blocky, blocky, columnar, platy, and granular. These words are based upon the size,
shape, character, and appearance of the aggregates (Fig.
8–7). Aggregates may vary from a fraction of a centimeter to
several centimeters in diameter and may be held together
strongly or weakly.

Structure can be changed by mechanical operations.
Tilling the soil when its moisture level is higher than **field capacity**[3] can destroy good structure. Tillage at a moisture
level slightly below field capacity is less likely to damage
soil structure. Soil with crusts and compacted layers can have
their structure improved through tillage.

A soil with good structure has soil pores that are large
enough to transmit water and air without restriction, yet
small enough to retain some water against the pull of gravity. To form pores of this size requires soil aggregates about
1.5 to 6 mm (0.06 to 0.24 in.) in diameter. Smaller aggregates

form pores that are too small to permit adequate drainage and
soil aeration, whereas larger aggregates form pores that are
too large to hold much water, even though some is held
within the aggregate itself. Good soil structure provides a favorable air and water relationship required for plant growth
and microbial activity.

Aeration

Soil pore spaces serve as passageways for the transmission
of oxygen needed for respiration by plant roots and microorganisms, and for the escape of waste carbon dioxide. Soil
aeration can become limiting when soil pores are filled with
water, when the soil is compacted, or when a crust forms at
the surface. The relative amounts of oxygen, nitrogen, and
carbon dioxide vary in the soil. Generally, soil air contains
less oxygen but more carbon dioxide than the above-ground
atmosphere. Concentrations of 5 percent carbon dioxide
have been found in soil air, whereas the above-ground
atmosphere contains about 0.03 percent. Under flooded
conditions, oxygen content in the soil is often low. Many
plants suffer when oxygen concentrations in the soil atmos-

[3]**Field capacity** is that amount of water retained in the soil after gravitational drainage of excess water.

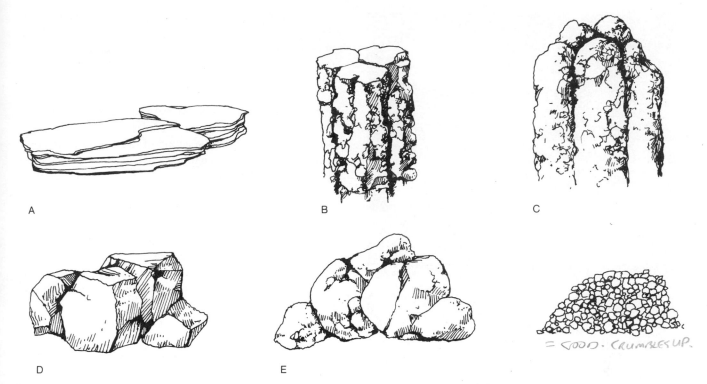

Figure 8–7 Several kinds of soil structure. (A) Platy. (B) Prismatic. (C) Columnar. (D) Angular block. (E) Subangular blocky. (F) Granular.

Golf course greens and athletic fields often have sand as the primary component supporting the turf. The use of sand allows golf course and athletic field superintendents to manage fertilization and irrigation programs with less difficulty than when finer-textured soils are used. However, without the finer particles of clay and silt, the soil matrix is too coarse to provide stability. One of the challenges facing turf grass scientists is to develop materials that hold the sand in place, providing stability and allowing a vigorous root system to develop.

Figure 8–8 Heavy earthmoving equipment such as this or other heavy machinery can contribute to the serious problem of soil compaction and especially so when the soil is moist. This equipment, which is used to level land, can easily carry 20 MT (19.5 t). *Source: The Daily Democrat,* Woodland-Davis, California.

phere drop below 10 to 12 percent or if carbon dioxide concentrations rise above 5 percent. Studies have shown that oxygen deficiency in soils causes poor plant growth more frequently than does carbon dioxide excess.

Soil Compaction

The use and size of farm equipment has increased considerably in recent years, and it is common to see heavy tillage equipment make many trips across the fields. Trucks heavily loaded with harvested crops travel over the land. Any of these can seriously compact soil (Figure 8–8). Grazing animals can also compact soils when pasture soils are wet.

Soil compaction is a serious problem, directly related to land preparation, tillage, and particularly to harvesting operations (Fig. 8–9). The term **soil compaction** is difficult to define precisely and often means different conditions to dif-

ferent people. To growers, a compacted soil is best described as one with abnormally high bulk density with very small pore spaces. Soil is not compacted all at once. The problem develops slowly, worsening each time heavy equipment is used. Often years pass before the problem reveals itself in declining crop yield or quality. Research has found no easy or simple solution to the problem. The problem can be recognized and alleviated somewhat through proper soil management and organic matter additions. Soil compaction is especially serious for the fresh market vegetable farmer, because vegetables are often harvested when the soil is wet. Farmers in irrigated regions sometimes irrigate up to the time of harvest to keep plants fresh and turgid, then quickly

CRUMB/
GRANULAR

Figure 8–9   Seedbed prepared on soils having different degrees of compaction, using normal tillage operations such as plowing, double disking, and harrowing with spike-toothed harrow. Note excessive cloddiness on the moderately and severely compacted soils.

harvest the crop, often using heavy equipment. Also, wholesale market prices for fresh vegetables frequently change dramatically from day to day. Thus, growers may harvest the crop even though the soil moisture is high and the danger of soil compaction great (Fig. 8–10).

Whenever a soil becomes more compacted, the process is accompanied, among other changes, by an increase in **bulk density.** In simple terms, an increase in bulk density means that the soil has become more dense through a loss of pore space.

Compacting the soil causes the particles to be pressed together, reducing pore space. This reduction in pore space, however, occurs to a greater extent for the larger soil pores and to a lesser extent for progressively smaller pores. Consequently, the selective reduction of the larger-sized pores leads to reduced water infiltration and gas exchange with the atmosphere (Figs. 8–11 through 8–13).

Soil compaction impedes root penetration and thereby limits the volume of soil from which plant roots can extract nutrients and water. Indirectly, this can result in nutrient deficiencies, particularly of phosphorus. The reduction in porosity can result in a greater accumulation of CO_2 in the soil.

Soil crusting, similar to soil compaction, appears as a dense, hard layer of varying thickness on the soil surface. Soil crusts result from the action of raindrops or irrigation, which disperse surface aggregates. Crusting is particularly severe when the amount of water applied exceeds the infiltration rate. Soil crusts restrict seedling emergence, gas exchange, and water infiltration (Fig. 8–14).

In some instances a certain degree of soil compaction is beneficial. Some compaction after seeding assures intimate contact between soil and seeds, enhancing imbibition of soil water by the planted seeds and thus increasing chances of germination. Also, some soil fumigation procedures entail purposeful soil crusting. The crust slows the leaking of the injected toxic gases into the atmosphere and thus makes the fumigation more efficient.

Soil compaction is easier to prevent than to correct. The risk of compaction can be reduced by good soil management. This includes using tillage implements of the lightest possible weight at the proper soil moisture content, avoiding unnecessary tillage, keeping equipment off the land when the soil is wet, and incorporating green manure crops or crop residues, preferably fibrous, into the soil.

CHEMICAL PROPERTIES OF SOIL

Effect of Climate

The type of parent material predominately influences the chemical characteristics of young soils. As weathering proceeds, soils tend to show the effects of climate, and the resemblance to parent material lessens or disappears. The chemical properties of the soil are determined largely by the colloid-sized (not visible with an ordinary microscope) silicate and aluminosilicate clay minerals.

In the temperate zones, chemical weathering is less intense in arid than humid regions. Soluble salts released by weathering are not lost by leaching from soils in arid regions, and the soil becomes more alkaline—that is, it has a higher pH.[4]

In tropical zones, with higher temperature and more rainfall than temperate regions, weathering and leaching is greater. The silicate and aluminosilicate minerals are more

[4]$pH = -\log_{10}H_a^+$. In soils, only the active hydrogen ions—those dissociated from the colloids—determine soil pH; hydrogen ions on exchange sites do not. Therefore, it is more descriptive to refer to H_a^+ (hydrogen ion activity) than to $[H^+]$ (hydrogen ion concentration).

Figure 8–10 The upper-portion of a soil profile showing a compacted layer about 25 to 30 cm (10 to 12 in.) below the soil surface. This layer was caused by repeated traffic over the field when soil was wet. Photo was taken soon after a crop of tomatoes was harvested from the field. The previous fall and early winter, the field had been planted to spinach, which was harvested from excessively wet soil during the rainy season.

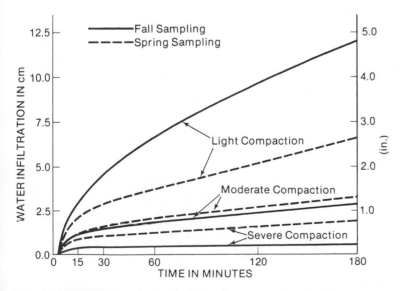

Figure 8–11 Water infiltration rates were measured on each of three plots mechanically compacted to three densities. Tests were made in the fall after compaction and then repeated the following spring on the lightly, moderately, and severely compacted plots. Obviously, severe compaction decreases the water infiltration rate.

149

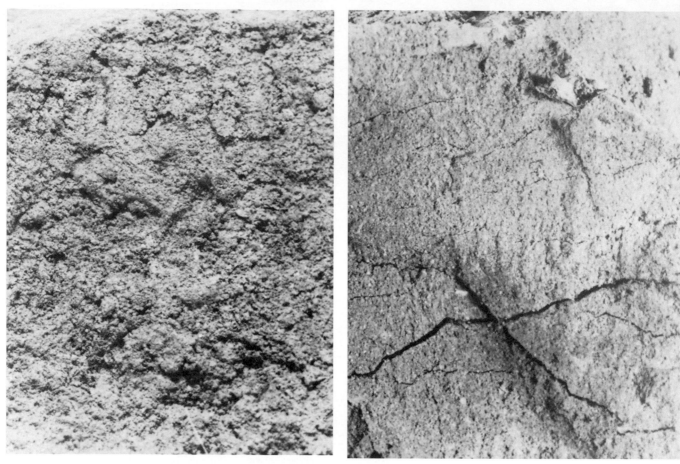

Figure 8–12   Soil samples taken from the top 10 cm (4 in.) of adjacent test plots of a non-compacted (*left*) and compacted (*right*) soil. The photographs are magnified 40 times. Note the greater porosity in noncompacted soil samples.

decomposed, resulting in soils known as Oxisols. These soils contain high concentrations of iron and aluminum, are generally red to reddish brown in color, and are low in cation exchange capacity, fertility, and organic matter. Even though large amounts of vegetation are produced, organic matter is low because dead plant matter is rapidly decomposed by high microbial activity.

In the cold humid regions, forest vegetation, mainly conifer trees, combine with climatic factors to produce Spodosol soil groups. The silica content of these soils is high in the surface layer in contrast to the iron and aluminum content in the Oxisols. These soils are highly leached and inherently low in plant nutrients. Organic compounds produced by decaying conifer needles form acid solutions that dissolve iron and alumina oxides and basic compounds (Ca and Mg salts). The ions are leached to lower depths in the soil profile and redeposited there together with some aluminum and dissolved organic matter to produce a dense layer 60 to 90 cm (2 to 3 ft) below the surface. These soils can be identified by an ashy, bleached white layer immediately above the dense layer.

Cation Exchange Capacity

An important property of clay and of the organic humus fraction of the soil is its ability to attract and hold cations—positively charged ions, some of which are essential plant nutrients. The ability of one cation to replace or be replaced by another is called **cation exchange.** Clay colloids carry thousands of negative charges in some cases throughout the clay particle and in other cases concentrated at the broken edges of the clay's alumina and silica layers. Thus, a clay colloid acts as a large, highly negatively charged particle (anion) (Fig. 8–15). A soil's capacity to hold cations is called its **cation exchange capacity** (CEC). Organic matter has a greater net negativity than clay and has a high CEC. The CEC is measured in terms of the number of meq[5] per 100

[5]meq = milliequivalent = 1/1000 of an equivalent. An equivalent is defined as 1 gm atomic weight of hydrogen or any other ion that will react with or replace this amount of hydrogen. When used with CEC, a more descriptive definition of meq is 6.023×10^{23} (Avogadro's number) charges in 1 gm atomic weight of a given cation.

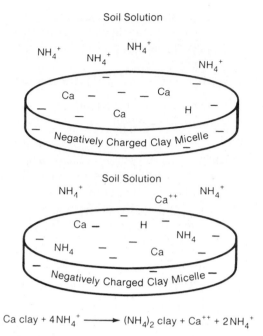

Soil Solution

Ca clay + 4 NH$_4^+$ $\longrightarrow$ (NH$_4$)$_2$ clay + Ca^{++} + 2 NH$_4^+$

Figure 8–15 If a soil colloid made up of numerous negatively charged clay micelles is treated with fertilizer containing ammonium cations, the added ammonium cations will replace an equivalent amount of another cation, such as calcium, on the clay colloid and force Ca^{++} ions into the soil solution.

Figure 8–13 Soil samples from noncompacted and compacted fields were air-dried and screened to simulate tillage. Water was allowed to infiltrate into each for four hours. Note that much more water infiltrated into the noncompacted soil, showing that the ill effects of previous compaction still remained.

Figure 8–14 This 25 percent stand of potato plants (*left*) resulted from severe soil crusting that prevented and delayed shoot emergence. A loss of this magnitude necessitates redisking and replanting the field. The partly excavated cross section of a raised bed (*right*) exposes the strength and thickness of a soil crust. It is apparent why potato shoots would have difficulty penetrating this barrier. *Source:* Herman Timm.

grams of soil. Not all cations are attracted or held with equal energy. The strength of attraction for some cations when present in equivalent amounts is:

$$\text{hydrogen } (H^+) > \text{calcium } (Ca^{++}) > \text{magnesium}$$

$$(Mg^{++}) > \text{potassium } (K^+) > \text{ammonium } (NH_4^+)$$

$$> \text{sodium } (Na^+)$$

Most plant nutrients are cations, which can be held by soil particles. The most notable exception is nitrate (NO_3^-), which is not held and is readily leached out of the root zone. The cation exchange capacity depends somewhat on the soil pH. Other things being equal, the cation exchange capacity is lower in acid soils and higher in alkaline soils.

Soil Acidity and Alkalinity

The acidity or alkalinity of the soil is defined by its pH. Unlike cation exchange capacity, pH is not a fixed characteristic of soil and, depending on a number of conditions, varies over time.

Values for pH vary considerably among soils, ranging from about 4.0 for an acid soil to 10.0 for an alkaline soil (Fig. 8–16).

Plants do not grow well in highly acid or highly alkaline soils, although within these extremes plants differ in their pH tolerance. Some grow well in acid soil, others grow better in alkaline soil, others in neutral soil (pH 7.0), and still others grow well over a wider pH range (Table 8–3).

The availability of certain plant nutrients is regulated by the acidity or alkalinity of the soil (Fig. 9–31). In certain acid soils, aluminum solubility increases to toxic levels at low pH values, but its availability decreases to deficient levels at high pH values. Iron and zinc become less available to plants as the pH increases, but molybdenum is more available at higher pH levels. Phosphorus is more available at a soil pH of about 6.5 to 7.0 than at either higher or lower values. Calcium and magnesium applied as carbonate salts not only increase their availability, but also will decrease soil acidity, raising the soil pH.

Soils in climates with high rainfall and humidity generally tend to be acid, while those found in arid climates tend to be alkaline. In wet climates the base elements (sodium,

TABLE 8–3	Soil pH Range for Optimum Growth of Selected Plants	
4.5 to 5.5	**5.5 to 6.5**	**6.5 to 7.5**
azalea	barley	alfalfa
bent grass	bean (snap, lima)	apple
blueberry	carrot	asparagus
camellia	chrysanthemum	beets (sugar, table)
chicory	corn (field, sweet)	broccoli
cranberry	cucumber	cabbage
dandelion	eggplant	cauliflower
endive	fescue	celery
fennel	garlic	chard
fescue	oats	hydrangea
gardenia	peas	(sepals become
hydrangea	pepper	pink)
(sepals become	poinsettia	leek
blue)	pumpkin	lettuce (head, Cos)
potato	radish	muskmelon
poverty grass	rye	onion
red top	squash	parsnip
rhododendron	strawberry	soybean
rhubarb	timothy	spinach
shallot	tobacco	sweet clover
sorrel	tomato	
sweet potato	turnip	
watermelon	wheat	

potassium, calcium, and magnesium) are removed from the soil by leaching as well as by the harvested crops that have absorbed them. As the base elements are lost, the exchange sites on the clay colloids become occupied with hydrogen ions, making the soil acid.

Normally, sudden and large changes in soil pH do not occur. Change is generally gradual, especially with fine-textured soils. These and other soils resist such change due to **buffering.**

When necessary or desirable, soil pH can be altered by adding certain materials. The pH of an acid soil can be increased by adding basic amendments or fertilizers containing such elements as calcium, potassium, sodium, or magnesium (Ch. 9). Most often, calcium carbonate ($CaCO_3$), or agricultural lime, is used. This finely ground limestone is spread evenly over the surface and tilled into the soil. Calcium carbonate is effective in reducing soil acidity by its ability to provide calcium (Ca^{++}) and hydroxyl (OH^-) ions in the soil

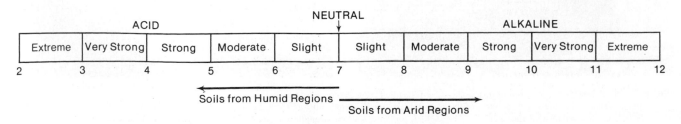

	ACID				NEUTRAL		ALKALINE			
Extreme	Very Strong	Strong	Moderate	Slight	Slight	Moderate	Strong	Very Strong	Extreme
2	3	4	5	6	7	8	9	10	11	12

← Soils from Humid Regions →

Soils from Arid Regions →

Figure 8–16 The pH for many mineral soils ranges from about 4 to 10. The pH range for most agricultural soils lies between 5 and 8.5

solution. These subsequently displace and react with hydrogen (H^+) ions in the soil. Estimating the amount of lime required involves several factors that include the soil pH, texture, organic matter content, structure, crops to be grown (Table 8–3), and the fineness of grind (mesh).

In arid regions the soil pH is usually neutral or alkaline. Calcium is abundant, making liming undesirable. More likely, the pH has to be lowered to make iron, manganese, and zinc more available to the plants.

Lowering the soil pH can be performed by adding acid-forming chemicals such as ferrous sulfate ($FeSO_4$) and elemental sulfur (S). These sulfur-containing compounds lead to the formation of sulfuric acid (H_2SO_4) and acidify the soil. Alternatively, though less effective, acidic organic materials—such as peat and decomposed plant materials—can also be mixed with the soil to lower pH.

The prolonged use of chemical fertilizers that are residually acid such as ammonium sulfate [$(NH_4)_2SO_4$], ammonium nitrate (NH_4NO_3), and ferrous sulfate ($FeSO_4$) tend to make the soil acid.

Residually basic (alkaline) fertilizers make the soil reaction more alkaline. Examples include sodium nitrate ($NaNO_3$), potassium nitrate (KNO_3), calcium nitrate [$Ca(NO_3)_2$], and calcium carbonate ($CaCO_3$).

Saline and Sodic Soils (WHEN TOO ALKALINE)

Saline soils contain unusually large quantities of soluble salts. The soluble salts are typically chlorides and sulfates of calcium (Ca^{++}), magnesium (Mg^{++}), and sodium (Na^+), although other soluble cationic salts may contribute as well. Sodic soils differ from saline soils in that a large percentage (over 15 percent) of the total cation exchange sites of the soil are occupied specifically by sodium ions (Table 8–4). Displacement of sodium (Na^+) ion is the main objective in reclamation of a sodic soil. Agriculturally, saline and sodic soils are problem soils that require special handling for successful farming. Excessive amounts of soluble salts are harmful to plants and, when cations are predominantly monovalent (with a single charge), they have adverse effects on soil structure. Soils can be classified on the basis of the kind and amount of salts present, as shown in Table 8–4.

TABLE 8–4	Characteristics of Saline and Sodic Soils[a]		
Soil	**Electrical Conductivity**	**Exchangeable Sodium**	**pH**
saline	>4 mmhos/cm[b]	<15%	<8.5
saline-sodic	>4 mmhos/cm	>15%	<8.5
nonsaline-sodic	<4 mmhos/cm	>15%	>8.5

[a]Specific thresholds are approximate because many factors affect the measured characteristics.
[b]mmhos = millimhos = 1/1000 mho. The mho is the reciprocal of ohm, a unit of resistance.

SOIL ORGANISMS

A microscopic examination of a soil sample reveals a wide variety of animal and plant life, some beneficial and essential to human well-being, and some harmful, often causing problems for people, their livestock, or crop plants. The animals include earthworms, gophers, insects, mice, millipedes, mites, moles, nematodes, slugs, snails, sowbugs, and spiders. Plant organisms in the soil include actinomycetes, algae, bacteria, and fungi. Certain beneficial fungi live in symbiotic associations with plant roots; the association is called **mycorrhiza.**

Soil organisms act both chemically and physically. They digest crop residues and other organic matter enzymatically, and may physically move the residues from one place to another, mixing it with the soil. Earthworms and burrowing animals mix large quantities of material with the soil mass. The kind and amount of soil organisms depend upon several factors, including climate, vegetation, soil pH, fertility level, temperature, and soil moisture.

Consider the ways that soil organisms increase crop productivity. Roots of higher plants are a good source of organic matter. Their decomposition produces organic acids and gluelike materials that bind soil particles together to form the aggregates necessary for good soil structure. Following decomposition of roots, open channels are left in the soil, improving drainage and aeration. Organisms also decompose stems, leaves, and other crop residues.

Nitrogen fixation, sulfur oxidation, and nitrification are processes carried on by soil bacteria essential to higher plants. Legume crops inoculated with *Rhizobium,* a symbiotic nitrogen-fixing bacterium, are frequently grown as a nitrogen source. As the legume crop grows, the bacteria converts unavailable atmospheric nitrogen (N_2) into nitrogenous compounds that the legumes use while the plant furnishes energy to the bacteria. Such a relationship between two dissimilar organisms living together for mutual benefit is called **symbiosis.** When the legume crop dies or is plowed under and decomposes, the nitrogen compounds become a source of nitrogen for a succeeding crop.

Elemental sulfur is not immediately available to higher plants; it must first be oxidized to the sulfate form. Autotrophic *Thiobacillus* bacteria bring about this transformation through a complicated series of reactions. Under certain conditions, autotrophic bacteria oxidize iron and manganese to compounds that are less soluble and thus less available to plants. The action of these bacteria helps prevent toxic amounts of iron and manganese from being taken up by the plants.

Not all soil organisms are beneficial. Some of the most injurious plant pests are soil borne. For example, nematodes attack and destroy plants in a wide range of species. *Phylloxera,* an aphid that attacks grape roots, devastated large vineyard areas until resistant rootstock cultivars were developed. Pathogenic soil-borne bacteria and fungi are also responsible for significant crop losses.

SOIL ORGANIC MATTER

Soil organic matter content has a profound effect on its biological, chemical, and physical properties. Through the decomposition of organic matter, nutrients become available to crop plants. Organic matter provides food and energy for soil organisms. Most organic matter, except for a small animal fraction, comes from plants. By weight, about 90 percent is made up of carbon, hydrogen, and oxygen. The remainder is usually sulfur, phosphorus, nitrogen, potassium, calcium, and magnesium plus a minute amount of microelements.

The speed of organic matter decomposition varies according to its chemical composition. It is rapid for simple carbohydrates and slow for fats and lignins. Essentially the decomposition reaction is the oxidation of carbon compounds to carbon dioxide, water, and energy (see p. 00)

$$(CH_2O)_n + O_2 \rightarrow CO_2 + H_2O + energy$$

The proteins, fats, and other complex compounds decompose in a multitude of reactions to form amino acids, ammonia, nitrates, phosphates, carbon dioxide, and others. After complete decomposition, a complex, amorphous, colloidal substance called **humus** remains that is resistant to further decomposition. This is the material that helps improve soil structure, imparts the dark color to the soil mineral fraction, and increases the soil's water-holding and cation exchange capabilities. For example, the cation exchange capacity of a mineral soil ranges from about 10 to 100 meq per 100 gm while the capacity of humus ranges from about 100 to 300 meq per 100 gm. Being colloidal in size, humus acts similarly to clay colloids in cation exchange reactions, but it is composed chiefly of carbon, hydrogen, and oxygen with small amounts of other elements, while clay colloids largely consist of aluminum, silicon, and oxygen.

Carbon: Nitrogen Ratio

A widely used practice for improving the physical condition of the soil is to incorporate crop residues, green manure crops, or other organic matter. This is a beneficial practice, but one must be aware of the resulting effect it has on the carbon:nitrogen (C:N) ratio. Under natural conditions, there is a close relationship between the amount of carbon and the amount of nitrogen in the soil. This ratio is nearly constant at about 12 parts of carbon to one part nitrogen. Variations when present seem to correlate with climate, especially temperature and rainfall. For instance, the C:N ratio tends to be smaller (less carbon, more nitrogen) in arid and warmer regions than in humid and cooler regions. Incorporation of the straw residue from a high-yielding grain crop having a high C:N ratio (50:1) will result in a change in the soil's C:N ratio. Soil organisms, primarily fungi and bacteria, rapidly multiply because of the large source of added carbon food (Fig. 8–17).

These organisms also require large amounts of nitrogen for their own growth. Consequently, they tie up the soil

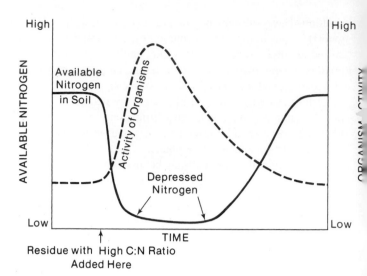

Figure 8–17 The relationship between the available nitrogen in a soil and the activity of soil microorganisms after a heavy application of organic matter with a high C:N ratio.

nitrogen, causing a temporary nitrogen deficiency for the growing crop.

With time, the organic residues continue to decompose until the soil organisms exhaust their supply of food and begin to die. The nitrogen from the decomposing organisms is then returned to the soil and made available to crops again at about the original nitrogen level.

SOIL WATER

Characteristics of Water

Water is the universal solvent; it dissolves more substances than any other liquid. It is one of our renewable natural resources; the world's supply of water has not changed but is constantly being recycled.

The characteristics of water arise from its unique structure, which also accounts for its remarkable stability. Water is one of nature's most stable compounds, so much so that for centuries it was considered a single element, not the compound it is. The water molecule in fact comprises two hydrogen atoms attached to an oxygen. The water molecule is not symmetrical. This lack of symmetry causes one end of the molecule to have a more positive electrical charge and the opposite end a more negative charge (Fig. 8–18). This phenomenon of polarity creates an attraction between water molecules: the positive end of one molecule attracts the negative end of an adjacent molecule. Water molecules can also attract or be attracted by cations, such as Na^+, K^+, and Ca^{++}, or by anions or clay colloids in the soil. Although we write the chemical designation for ions as in the last sentence, they are, in fact, hydrated in the "real world," giving them unique chemical and physical properties.

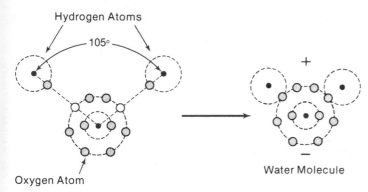

Figure 8–18 Two hydrogen atoms join one oxygen atom at an angle of 105°, giving the water molecule an asymmetrical configuration that accounts for many of its unique properties. Note the electrical polarity of the molecule.

Surface tension is that physical property of water in the liquid state that is due to the intermolecular attraction (hydrogen bonding) between the water molecules. The molecules in the surface film in water are inwardly attracted resulting in a strong surface tension. Were the water's surface tension not so strong, soil would hold little water, water could not reach the top of a tall tree, and blood would not flow through our bodies. This strength of attraction between water molecules is illustrated by the fact that a steel needle can be floated on the surface of water.

In the liquid state, the attraction among water molecules is chaotic and random, but as the liquid freezes, a rigid symmetrical lattice with an open porous structure forms. This arrangement accounts for the reduction in density as the water solidifies. Water also has an unusually high **specific heat.**[6]

Uses of Water in the Plant

Water must be absorbed by seeds to initiate the enzyme activity necessary for germination. In addition, water is used in photosynthesis and all other metabolic processes associated with plant growth and development. Plants absorb more water than any other material, most of it entering the plant via the roots from the soil. Much of this water is not retained, but is released by transpiration to the air as vapor. Water transpired through leaf openings (stomata) requires heat to evaporate. Most of the heat comes from sunlight intercepted by the plant leaves and, when this heat dissipates as a result of water evaporation, the temperature in and around the leaves is lowered.

Before nutrient elements can enter the plant roots, they need to be dissolved in water. Water also functions as a transport system within the plant, moving nutrient materials to the

sites where they are converted into products of photosynthesis. It then transports the synthesized materials to sites of storage or use in the plant.

Soil Water

Even though water is present in the soil, it sometimes is not available to the plant for reasons that will be discussed later in this chapter. Water held by and moving within the soil supplies the plant with mineral nutrients and oxygen, as well as water. It moves through the soil pore spaces in the liquid or gaseous state. The pore spaces are always filled with water, air, or a mixture of both. When the pore spaces are filled with water, the soil is said to be **saturated.** Saturation is an unhealthy condition for plants if it lasts too long because the oxygen needed for respiration is missing. On the other hand, when the pore spaces are filled mostly with air, the soil is too dry for good plant growth. The number and size of the soil pores vary with the soil's texture and structure. Clay soils have smaller but more numerous pores than sandy soils. Thus, an equal volume of clay soil holds more water than a sandy soil when the pores are filled (Fig. 8–19). The ability of the soil to retain water is called its **water-holding capacity.** The capacities of different textured soils to hold water are plotted in Fig. 8–20.

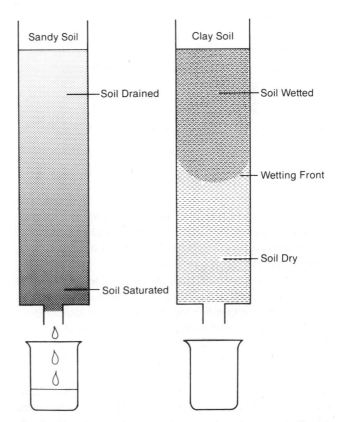

Figure 8–19 Fifty ml of water were added to each column. After one hour, water has drained to the bottom and some dripped out of the sandy soil (*left*), while the water remained in the top half of the clay soil (*right*). Obviously, clay soil can hold more water.

---

[6]Specific heat is the ratio of the quantity of heat required to raise a substance one degree to that required to raise an equal mass of water one degree.

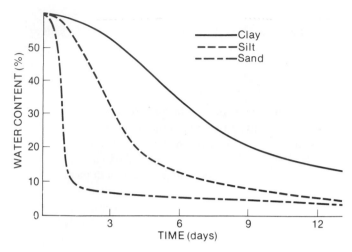

Figure 8–20 The speed with which soil water moves into or out of a soil is determined by the soil's texture. For example, if a sandy soil (coarse-texture), a silt soil (medium texture), and a clay soil (fine texture) are saturated at the same time (t = 0) and if the moisture contents of the three soils are determined and plotted with time, the graph shows that the sandy soil loses moisture very quickly, the silt loses water moderately, and the clay soil loses it slowly. As this indicates, sandy soils have to be irrigated more frequently than silt or clay soils.

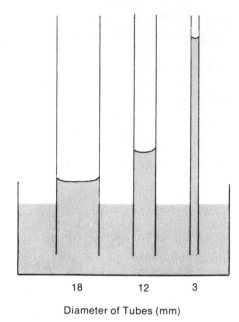

Diameter of Tubes (mm)

Figure 8–21 Capillary rise. Water ascends in capillaries, reaching higher levels in smaller tubes. Doubling the diameter of the tube doubles the area for the water molecules to adhere to—but it also quadruples the weight of the water to be pulled up. Therefore, water does not rise as high in the columns with larger diameters.

Three forces—gravity, adhesion, and cohesion—are responsible for water movement within the soil. **Gravity** causes water to move downward under no tension, and is the principal force when a soil is saturated. **Adhesion** is the force of attraction between unlike molecules (soil particles and water). **Cohesion** is the force of attraction between like molecules (water and water). The latter two forces can cause water to move by capillarity in any direction—upward, downward, or laterally—and are the principal forces that move water in an unsaturated soil.

The upward movement of water, called **capillary rise,** is responsible for the loss of water from the soil surface by evaporation. Capillary rise can be demonstrated with one end of a strip of blotting paper inserted partway into water or with capillary tubes (Fig. 8–21).

As soil dries, the water film surrounding each soil particle thins. Consequently, the adhesive and cohesive forces of attraction increase rapidly, making it more difficult for the plant to extract water from the soil.

There are two particularly important aspects to consider regarding soil moisture: (1) the quantity of water in the soil in terms of volume or weight and (2) the availability of the water in terms of moisture tension. The greater the force of attraction between soil particles and water molecules, the greater the moisture tension and the lower the availability of the water (Fig. 8–22).

The movement and retention of water in the soil, its uptake and movement within the plant, and its transpiration into the atmosphere are responses to changes in energy levels. The energy comes from three sources: (1) **gravitation,** due to the earth's gravity; (2) **matric,** from the adhesion and cohesion between soil particles and water, resulting in adsorbed water and capillary water; and (3) **osmotic,** caused by dissolved salts in the soil water.

Consider a box with one transparent side, filled with dry soil. A small V-shaped channel simulates an irrigation canal in the soil surface. Water is applied at a constant rate to the soil. The water immediately begins to wet the soil by moving downward through the profile as a result of the force of gravity. As water is added to the channel, it also moves slowly both laterally and downward. Gravity does not move water laterally, so a different energy source must be responsible. The lateral movement of water is under the influence of matric energy resulting from the attraction between soil particles and water molecules (cohesive forces). Salts are present in soil water. The salts create osmotic energy, and if the salts are present in a sufficiently high concentration, the osmotic energy prevents water movement into the plant (Fig. 8–23).

Instruments called tensiometers are used to measure the tension in the soil, providing information useful in determining when to apply irrigation water to a crop. Tensiometers in theory can operate between 0.0 and 1.0 atm, but in practical use they operate between 0.0 and about 0.8 atm of soil moisture tension (Fig. 8–24).

After a prolonged rain or irrigation, the air in the soil pores is displaced with water. In this condition the soil is saturated, the soil moisture tension or suction is zero, and no energy is required to remove water from the soil particles (Fig. 8–25). This state will prevail as long as water is ap-

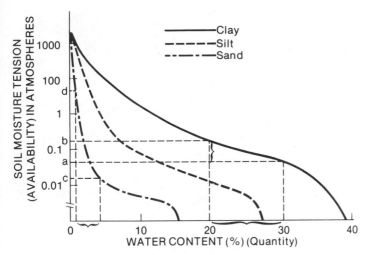

Figure 8–22 Soil moisture release curves showing relationship between water content and soil moisture tension with three different soil textures. For example, compare the 10 percent loss of water from the clay soil (30 percent–20 percent) and the accompanying slight increase in soil moisture tension (from *a* to *b*), with the 3 percent loss of water from the sandy soil (4 percent—1 percent) and the accompanying large increase in soil moisture tension (from *c* to *d*). A similar analysis can be made on this graph with the same textured soil in different soil moisture ranges. In the wet range (high soil moisture percentage) a large loss of water produces a small increase in soil moisture tension, but in the dry soil moisture range, a small loss of water produces an increase in soil moisture tension that could easily go from wet enough to sustain plant growth to the permanent wilting point very quickly.

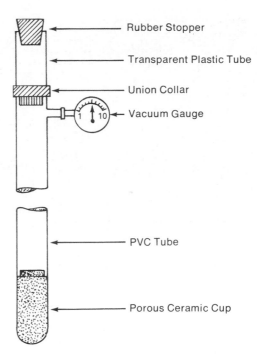

Figure 8–24 Principal parts of a tensiometer. Before the tensiometer is placed in service, it is completely filled with water. As the soil dries, water is pulled out of the porous cup by the adhesive forces acting between the soil particles and water molecules. This creates a measurable partial vacuum inside the instrument indicated on the vacuum gauge. As the soil is wetted by irrigation, water moves into the instrument and relieves the partial vacuum.

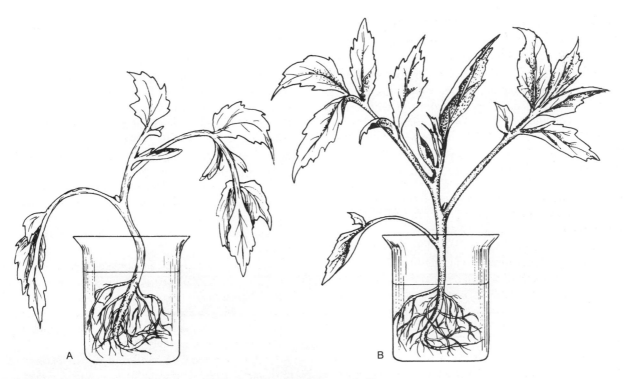

Figure 8–23 Effect of osmotic energy. Even though the plants in both beakers started with the same quantity of water, it was not equally available. In beaker (A) salt was added, creating an osmotic pressure too great for the plant to overcome. Since water was limited in its availability, the plant wilted. Beaker (B) contained pure water, which was readily available to the plant.

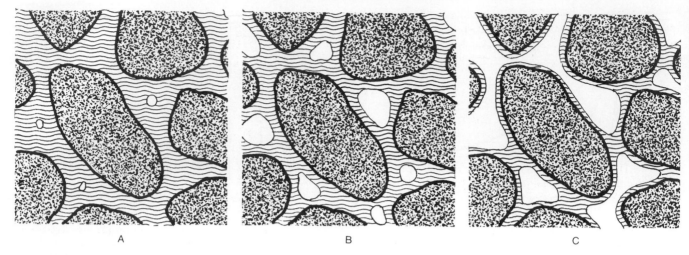

A   B   C

Figure 8–25   (A) In a saturated soil, all the void spaces are filled with water, indicated by the horizontal lines, and the tension (negative pressure) needed to remove it from the soil particles is zero. (B) As the soil drains, the air:water ratio becomes larger and air occupies some of the void spaces. The tension at field capacity (i.e., when drainage ceases) is about 0.33 atm. (C) If the soil continues to dry, then the water film becomes so thin that extracting the water requires a tension of 15 atm (15.2 bars). This is more than most plants can exert. Therefore, the plants wilt permanently (PWP).

plied at a rate equal to the rate of water loss by drainage, by plant use, and evaporation. When no more water is added, losses continue, first from the larger macropores and then from the smaller micropores. Loss of water continues until the adhesive and cohesive forces equal gravity. This usually takes two to three days in a loam soil. At this moisture content, the soil is said to be at **field capacity.** The water has drained from the macropores but the micropores still contain water, and the soil moisture tension varies from 0.2 to 0.35 atm depending on the soil type. (For most loam soils field capacity is defined as the moisture content at 0.33 atm.) If plants are growing in the soil, water loss continues, and if no water is added, eventually the soil will reach a moisture content that does not sustain plant life and the plants will permanently wilt. The soil moisture content at which a plant wilts (the sunflower is often used as a reference) and cannot recover when placed in an environment of

100 percent relative humidity is termed the **permanent wilting point** (PWP).

Note that the permanent wilting point of approximately 15 atmospheres of tension will occur at different soil moisture contents, depending on the soil texture (Fig. 8–22). For an average loam soil at the PWP, all macropores and all but the smallest micropores are emptied of water. If the soil moisture depletes further, the soil becomes air dry. In this condition, all liquid water is gone except that held as a thin, tightly bound layer around the soil particles. This adsorbed water is called **hygroscopic water** and the soil moisture tension is far beyond the availability range for plant use (Fig. 8–26). Another form of water, even less available to plants and called the **crystal lattice water,** is held in the soil's crystalline structure. This water can be removed only by applying sufficient heat to destroy the crystalline structure.

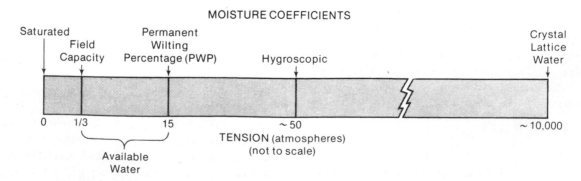

Figure 8–26   Relationship between soil moisture tension and the moisture coefficients. Water available to plants is defined as the water between field capacity and PWP.

Farmers are interested mainly in the water available for plant growth. **Available water** (AW) is defined as the soil moisture between field capacity (FC) and the permanent wilting point (PWP):

$$AW = FC = PWP$$

Water between FC and saturation is not considered available because it is lost through drainage. Water at tensions greater than PWP (15 atm) is held too tightly by the soil for plants to remove. The definition of AW could imply that the water between FC and PWP is equally available to all plants, but this is not necessarily true. Water near PWP is not as available to most plants as that near field capacity. Also, the PWP is not the same for all plant species. For many irrigated crop plants, a soil moisture tension of between 0.2 and 0.8 atm in the root zone is best. Deep-rooted crops may have an advantage when soil closer to the surface is dry because soil moisture in the area of the deeper roots may be adequate for growth.

WATER QUALITY

In evaluating the suitability of land for irrigated agriculture, both the availability and quality of water must be considered. Water used for irrigation almost always contains measurable quantities of dissolved substances which, in general, are soluble salts. These include small but important amounts of dissolved salts originating from the weathering of parent rocks. The value of water for irrigating crops is determined by the amount and kind of salts present.

Saline Water and Salinity

If irrigation water contains soluble salts in sufficient quantities to accumulate within the root zone and interfere with crop yields, a salinity problem arises (Fig. 8–27). The concentration of salts in the water can increase soil salinity to the point where the osmotic pressure leaves plants unable to extract sufficient water for growth (Fig. 8–23). The plants show much the same symptoms they would in a drought—wilting, reduced growth, and in some plants a color change from bright green to bluish green. The wide range of salt tolerance among agricultural crops permits the use of some saline water for irrigation: water too saline for one crop can be used for a more salt-tolerant crop. Table 8–5 gives the salinity tolerances for several crop plants.

Permeability

Poor soil permeability adds to cropping difficulties by increasing soil crust formations, which interfere with seedling emergence. Water-logging of surface soil can also occur, in-

Figure 8–27 Excessive soluble salts in soils and from applied irrigation water may accumulate in lower areas of a field or where drainage is poor. High levels of soluble salts interfere with plant growth. The loss of plants in parts of this field was attributed to excessive salts but was also aggravated by extensive soil crusting. *Source:* William E. Wildman.

creasing diseases, limiting gaseous diffusion, and causing nutritional problems.

The first step in evaluating soil permeability problems is to determine the total amount and kind of soluble salts in the water. Water with low salt content can result in poor soil permeability, just like water with an excessive salt content. Pure water brings into solution precipitated calcium and other soluble salts, resulting in dispersion of soil particles, temporarily blocking and reducing soil pore space. The second step is to consider the ratio of the sodium to calcium plus magnesium content [Na(Ca + Mg)] in the water. When the ratio is higher, dispersion is increased. Third, the carbonates and bicarbonates are evaluated because their presence also contributes to poor permeability.

Toxicity

High concentrations of some salts in water are toxic to some plants. The problem occurs when certain specific ions such as boron, chloride, or sodium are taken up by plants from the soil solution in sufficient quantities to reduce yield or stop growth.

Other Related Problems

Various other situations related to salt concentrations in water can arise. For example, excess nitrogen in irrigation water can cause lodging in grain, delayed maturity of some fruit and vegetable crops, and increased vegetative growth at the expense of root or tuber growth. Sometimes, calcium, magnesium carbonates, or bicarbonates deposit on the leaves and fruits of orchard crops, grapes, and vegetable crops after sprinkle irrigation. These white deposits, while not particularly harmful, do tend to lower quality by creating an unsightly product.

TABLE 8–5   Tolerance of Selected Crops to Soil Salinity.

Values in the table represent the salinity level required to give a 10%, 25%, or 50% reduction in yield. The higher the ECe number, the higher the level of dissolved salts. The higher the salinity, the more salt-tolerant the crop.

	Percent Yield Decrease				Percent Yield Decrease		
	10%	25%	50%		10%	25%	50%
Crop	Soil Salinity (ECe)[a]			Crop	Soil Salinity (ECe)[a]		
VEGETABLE CROPS				**FORAGE CROPS**			
Beets[c] *Beta vulgaris*	5.1	6.8	9.6	Bermuda grass *Cynodon dactylon*	8.5	10.8	14.7
Broccoli *Brassica oleraceae*	3.9	5.5	8.2	Crested wheat grass *Agropyron cristatum*	6.0	9.8	16.0
Tomato *Lycopersicon esculentum*	3.5	5.0	7.6	Trefoil, birdsfoot *Lotus corniculatus*	6.0	7.5	10.0
Cucumber *Cucumis sativus*	3.3	4.4	6.3	Tall fescue *Festuca elatior*	5.8	8.6	13.3
Muskmelon *Cucumis melo*	3.6	5.7	9.1	Sudan grass *Sorghum sudanense*	5.1	8.6	14.4
Spinach *Spinacia oleraceae*	3.3	5.3	8.6	Vetch, spring *Vicia sativa*	3.9	5.3	7.6
Potato, Irish *Solanum tuberosum*	2.5	3.8	5.9	Alfalfa *Medicago sativa*	3.4	5.4	8.8
Sweet corn *Zea mays var. saccharata*	2.5	3.8	5.9	Lovegrass *Eragrostis* spp.	3.2	5.0	8.0
Sweet potato *Ipomoea batatas*	2.4	3.8	6.0	Clover, berseem *Trifolium alexandrinum*	3.2	5.9	10.3
Pepper, bell *Capsicum annuum*	2.2	3.3	5.1	Orchard grass *Dactylis glomerata*	3.1	5.5	9.6
Lettuce *Lactuca sativa*	2.1	3.2	5.2	Clover, alsike, ladino, red, strawberry *Trifolium* spp.	2.3	3.6	5.7
Onion *Allium cepa*	1.8	2.8	4.3				
Carrot *Daucus carota*	1.7	2.8	4.6	**FLOWER CROPS**			
AGRONOMIC CROPS				Rose, hybrid tea *Rosa odorata*	3.5	5.0	7.0
				Poinsettia *Euphorbia pulcherrima*	2.5	6.5	12.0
Barley[b] *Hordeum vulgare*	10.0 mmhos	13.0 mmhos	18.0 mmhos	Chrysanthemum *Chrysanthemum x morifolium*	2.0	2.5	8.0
Cotton *Gossypium hirsutum*	9.6	13.0	17.0	Carnation *Dianthus caryophyllus*	1.5	3.0	10.0
Sugarbeet[c] *Beta vulgaris*	8.7	11.0	15.0	Gladiolus, garden *Gladiolus x hortulanus*	1.5	3.5	7.0
Wheat[bd] *Triticum aestivum*	7.4	9.5	13.0	Lily, Easter *Lilium longiflorum*	1.5	2.5	5.0
Soybean *Glycine max*	5.5	6.2	7.5	Geranium, bedding *Pelargonium x hortorum*	1.5	2.5	5.0
Sorghum *Sorghum bicolor*	5.1	7.2	11.0	Gardenia, common *Gardenia jasminoides*	1.0	1.5	2.5
Groundnut (peanut) *Arachis hypogaea*	3.5	4.1	4.9	Azalea *Rhododendron* spp.	1.0	1.5	2.0
Rice (paddy) *Oryza sativa*	3.8	5.1	7.2				
Corn, dent *Zea mays var. identata*	2.5	3.8	5.9				
Beans *Phaseolus vulgaris*	1.5	2.3	3.6				

(Continued)

| TABLE 8–5 | Continued | |

	Percent Yield Decrease				Percent Yield Decrease		
	10%	25%	50%		10%	25%	50%
Crop	Soil Salinity (ECe)[a]			Crop	Soil Salinity (ECe)[a]		
FRUIT CROPS				TURFGRASSES			
Date palm *Phoenix dactylifera*	6.8	10.9	17.9	Sunturf bermuda *Cynodon mogennisii*	7.5	12.5	17.5
Fig *Ficus carica*	3.8	5.5	8.4	Ormond bermuda *Cynodon dactylon*	5.0	7.0	13.0
Olive *Olea europaea*	3.8	5.5	8.4	Fescue, meadow *Festuca elatior*	4.0	5.5	7.5
Grape *Vitis spp.*	2.5	4.1	6.7	Tifgreen bermuda *Cynodon hybrid*	2.5	4.5	17.0
Orange *Citrus sinensis*	2.3	3.2	4.8	Kentucky bluegrass *Poa pratensis*	1.5	4.0	7.5
Apple *Malus pumila*	2.3	3.3	4.8	Bentgrass, colonial *Agrostis tenuis*	1.5	2.5	5.0
Walnut, Persian *Juglans regia*	2.3	3.3	4.8				
Peach *Prunas persica*	2.2	2.9	4.1				
Apricot *Prunus armeniaca*	2.0	2.6	3.7				
Almond *Prunus dulcis [P. amygdalus]*	2.0	2.8	4.1				
Plum, common *Prunus domestica*	2.1	2.9	4.3				
Blackberry *Rubus spp.*	2.0	2.6	3.8				
Avocado *Persea americana*	1.8	2.5	3.7				
Strawberry *Fragaria spp.*	1.3	1.8	2.5				

[a]Units of electrical conductivity (ECe) are mmhos/cm at 25°C.
[b]Barley and wheat are less tolerant during the germination and seedling stage. ECe should not exceed 4 or 5 mmhos/cm.
[c]Sensitive during germination. ECe should not exceed 3 mmhos/cm for garden beets and sugar beets.
[d]Tolerance data may not apply to new semidwarf cultivars of wheat.
Source: Courtesy of the Food and Agriculture Organization of United Nations, Rome, Italy.

Soils by definition are composed of materials generated from parent rock. However, in many areas of agriculture, especially in the greenhouse and nursery, plants are grown in potting mixes that contain little or no soil. These "artificial" or soilless growing media exhibit the same chemical and physical characteristics as soil. Texture, structure, CEC, and pH are as important with soilless media as they are with soils. Because soilless media are manufactured, they can be custom-blended to provide pore space, CEC, pH, and so on to provide ideal or nearly ideal characteristics for specific crops or stages of crop growth. For example, a germination medium has a very fine texture with moderate water-holding capacity to provide the proper environment for the germination of seedlings. The same medium is too fine and may hold too much water for the needs of the same plant as it matures. Transplanting into a coarser medium allows the plant to continue to grow in a medium that more adequately meets the needs of the larger plant.

Soil compaction in pots is a problem just as it is in the field. Too often pots are filled by packing the media. Packing compresses the media and reduces the available pore space. Thus, the medium no longer exhibits the characteristics it had when originally blended.

SUMMARY AND REVIEW

Soil is that portion of the earth's crust that has formed from the decomposition of rocks and minerals. Soil consists of 4 fractions: solid (rock fragments and minerals), organic (from dead and decaying organisms), liquid (water and dissolved minerals), and atmospheric (O_2, CO_2, and other gases).

Soils are formed from parent material that becomes chemically and physically weathered over time. Chemical weathering includes carbonation, hydration, hydrolysis, and oxidation. Physical weather comes with changes in temperature, mechanical action such as glacier movement, moving water, and other forces. The mineral content of the soil depends on the composition of the parent rock and the type and duration of weathering subjected to the parent material. Parent rock can be igneous, sedimentary, or metamorphic. Climate and topography also influence soil characteristics.

Physical properties of soil include its texture and structure. Texture is defined by the percentage of sand, silt, and clay in a soil. Soil structure is the arrangement of the primary soil particles in the soil mass. Arrangements can be prismatic, subangular blocky, angular blocky, columnar, platy, or granular. A soil with good structure for plant growth has pore spaces (atmospheric fraction) that are large enough to transmit water and air yet hold some water against gravity.

Soil compaction occurs when the soil is compressed. In field situations this most often happens with the passage of heavy equipment over a field. Soil compaction reduces the size and percentage of pores. Root growth is impeded in compacted soils.

The chemical properties of soil include its cation exchange capacity (CEC) and its acidity and alkalinity (pH). Parent material predominately influences the chemical properties of soil. However, climate and weathering affect the parent material characteristics that are retained or expressed. CEC is the ability of the solid fraction of the soil to attract and hold cations, an important consideration in plant nutrient management. The acidity or alkalinity of a soil is defined by its pH or the concentration of H^+ in the soil but not bound to soil particles. pH is a changing characteristic in soils and can vary over time. The buffering characteristic of a soil determines how fast pH changes can occur. pH influences the availability of many plant nutrients and is an important factor in determining fertilization programs for all crops.

Water can be present in soil in different ways. When the pore spaces are filled with water, the soil is saturated. Saturated soils are not healthy for plant roots because air (O_2) is not available for root respiration. Likewise, pore spaces that are mostly filled with air usually do not provide enough water to supply the transpirational requirements of plants. The ability of soil to retain water is its water-holding capacity. Gravity, adhesion, and cohesion are the forces that move water through the soil. Gravity moves water downward. Adhesion and cohesion move water upward or laterally through the soil by capillary action.

EXERCISES FOR UNDERSTANDING

8–1. Visit an area of unplanted fields, a home or golf green construction site, or a greenhouse. Examine the soil or potting mixes. Predict if the soil or mix will hold water or drain rapidly, hold nutrients, and firmly or weakly support root growth.

8–2. If possible, fill some pots with the aforementioned material then plant seeds or transplants in the pots. As the plants grow and you are caring for them (use a household fertilizer available at any garden center or supermarket and apply at the recommended rate), observe and record how often you have to irrigate, if the water flows quickly or slowly through the material, and any other characteristics that you think are affecting the plant growth. If you have taken material from a field, consider how being in the pot changes the characteristics of the material compared with being in the field.

8–3. Mix your own blend of potting mix using whatever components you think will make a good growing medium. Record the material and the percent used. Fill some pots and grow a plant species of interest to you. Observe and record as in Exercise 8–2.

8–4. From you local library or extension office, find the geological and climate history of your area. From that information predict what parent material generated your area's soil and what forces have weathered it.

REFERENCES AND SUPPLEMENTARY READING

BRADY, N. C., and R. R. WEIL. 1999. *The nature and properties of soils.* Twelfth Edition. Upper Saddle River, N.J.: Prentice Hall.

HARPSTEAD, M. I. 1997. *Soil science simplified.* Third Edition. Ames, Iowa: Iowa State University Press.

MCLAREN, R. G., and K. C. CAMERON. 1996. *Soil science. Sustainable production and environmental protection.* Auckland, New Zealand: Oxford University Press.

CHAPTER 9

Soil and Water Management and Mineral Nutrition

KEY LEARNING CONCEPTS

After reading the chapter you should be able to:

♦ Understand the principles and methods of land preparation.
♦ Understand the principles and methods of irrigating field crops.
♦ Understand the mineral nutrition requirements of plants.
♦ Understand the principles and methods of fertilization of crops.
♦ Understand the principles and practices of land conservation.

To a large extent, crop productivity is determined by the way the soil is managed. Good soil management involves a combination of tillage, cropping systems, and soil treatments that complement each other. Desirable combinations minimize the objectionable effects of crop production, while poor management leads to low productivity and degrades the soil.

LAND PREPARATION

Major purposes of land preparation are to: (1) level the land where needed; (2) incorporate crop residues, green manure, and cover crops; (3) prepare and maintain a seedbed in good **tilth;**[1] (4) help control weeds, diseases, and insects; (5) improve the physical condition of the soil; and (6) help control erosion where needed.

In general, **tillage** is defined as the mechanical manipulation of soil to provide a favorable environment for crop growth. Soil moisture condition is a factor in the effectiveness of tillage. Unfavorable soil conditions—too dry, too wet—will result in ineffective tillage and will damage soil structure. Tillage is done with a wide variety of equipment and for various purposes. The seedbed should provide an environment conducive to rapid germination of seeds and growth. For most crop plants, such a seedbed is one in which the surface soil is loose and free of clods and trashy crop residues (Fig. 9–1). The subsoil is permeable to air and water and has adequate drainage and aeration. It should not be water-logged or anaerobic (without oxygen).

Plowing

Generally, the first step in seedbed preparation is to plow the land. When large amounts of crop residues are left on the field from a preceding crop, they are often chopped with a disk or a rotary stalk cutter before plowing. Plows invert the soil and cover the trash, but they often leave the soil in large linear lumps that must be reduced in size.

A farmer has the choice between two plow types, the moldboard or the disk plow, each adapted to certain soil characteristics. Moldboard plows, of which there are many variations, range in size from a single moldboard, or bottom, to a gang of plows that turns 12 furrows (12 bottoms) simultaneously. Each moldboard shears and inverts a slice of soil commonly 15–30 cm (6 to 12 in.) deep and 30 to 60 cm (12 to 24 in.) wide as it moves along, leaving the top of the slice at the bottom of the furrow (Fig. 9–2). Moldboard plows are used when the soil is sufficiently moist to allow the plow to pass through easily but not so wet as to cause the furrow slice to stick to the face of the moldboard. If the soil is either too dry or too wet, excessive power is required and poor plowing results. The ideal moisture content for plowing loam soils is slightly less than field capacity.

Two-way reversible plows (flip-over plows) are used to eliminate "dead" furrows (unfilled furrows). They are also used in hilly areas for contour plowing and in irrigated areas where dead furrows hinder irrigation. These plows have right- and left-hand moldboards mounted so that one series plows in one direction, then at the end of the row, the mold-

[1]Tilth is a subjective term for the physical condition of the soil with respect to its capability to provide a good environment (aeration and porosity) for optimizing crop production.

Figure 9–1 A well-prepared raised seedbed that might be used for small seeded vegetable crops such as lettuce or carrots (*above*). It should be free from previous crop residues or excessive cloddiness (*below*).

Figure 9–2 A four bottom moldboard plow turning under a green manure legume crop. Note how well the green manure crop is being buried. *Source:* Allis-Chalmers.

Figure 9–3 Five moldboard plows, traveling from left to right, are in the soil turning over furrows. At the end of the field on the return trip from right to left, the plow is rotated so that the moldboards now in the air turn the furrows. This plow turns all furrows in the same direction, thereby eliminating "dead" furrows in the middle of the field. This type of moldboard plow is used in irrigated areas to help maintain leveled fields. *Source:* Allis-Chalmers.

Figure 9–4 The disk plow accomplishes essentially the same objective as the moldboard plow. However, the disk plow can operate when soil conditions are either too wet or too dry for the moldboard plow. The disks rotate as the plow moves forward, rolling the furrow slice over.

boards are mechanically rotated into position so that the second series can plow the return trip (Fig. 9–3). These two-way plows always throw the furrow slice in the same direction irrespective of the direction the plow travels.

Moldboard plows are used on bare fields, small grain stubble, corn stubble, sod pastures, and hay crop fields. The operation can be done in spring, summer, or fall if soil moisture content is satisfactory.

Many farmers prefer to plow in late summer or early fall, especially on small grain stubble fields. Fall plowing allows the clods to "slake" (crumble) because of alternate freezing and thawing in the winter. It also distributes the labor load by moving the plowing from the busier spring season to the fall.

The disk plow (do not confuse with the disk harrow) consists of a series of large concave disks 60 to 75 cm (24 to 30 in.) in diameter that are set at an angle to the forward movement and cut into the soil while rotating as the plow moves forward. There can be 3 to 10 or more disks on a plow. Soil moisture conditions are less critical for disk plow operation. These plows are better adapted to dry, hard soils or ones too sticky for a moldboard plow (Fig. 9–4). In other respects, the two types of plow serve the same purpose and accomplish the same results, except that the disk plow generally does not invert the soil or cover crop residues as completely.

Disking

Disk harrows are used to reduce the size of larger soil clods by fracturing them with cleavage and pressure. Disking generally follows plowing, but under some conditions disking can substitute reasonably well for plowing. If the soil is in good tilth a satisfactory seedbed can be prepared by disking alone.

Disk harrows are general-purpose tillage implements consisting of gangs of concave disks. Most disks have two gangs, one behind the other. In operation, the front gang breaks in the middle into a V-shaped configuration so that the sideway forces will balance each other when half of the gang throws the soil to the right and the other half throws it to the left. The rear gang breaks into a Λ-shaped configuration and throws the soil in the opposite direction (Fig. 9–5). Many larger disks break into operating position by allowing one side of the front gang to break forward while the same side of the rear gang breaks back. This forms a <-shaped configuration (Fig. 9–6).

The depth of penetration is regulated by adjusting the angle of the gangs. The size of the implement varies considerably. Some small tractor-mounted disks cut swaths of 180 cm (6 ft), whereas other units cut swaths up to 12 m (40 ft) wide. A special purpose disk, called a stubble disk, has semi-

Figure 9–5 A disk harrow cutting and covering crop residue near prior seedbed preparation. *Source:* Allis-Chalmers.

Figure 9–6 A stubble disk is similar to a disk harrow except that the stubble disk has half-circle notches around the periphery of the disks to help cut through dry stubble residues.

Figure 9–7 A spring-tooth harrow is sometimes used in seedbed preparation because it is effective in breaking up soil crusts, reducing clod sizes, and destroying small weeds. *Source:* Allis-Chalmers.

circular notches cut around the periphery of the disk blade (Fig. 9–6). The notches help cut crop residues more effectively, and in some situations stubble discs are used in lieu of plowing.

Harrowing

The function of the harrow is to further reduce the size of soil clods left after disking, to smooth the soil surface, and to do small-scale leveling. Harrowing also destroys small weeds. This operation generally follows disking. Frequently, farmers attach a harrow behind the disk and do both operations simultaneously. This is the final touch to seedbed preparation unless beds are to be formed for irrigated row crops.

A wide variety of harrows are used. The principal types are: (1) spike-tooth; (2) spring-tooth (Fig. 9–7); (3) chain or drag; and (4) cultipackers, packers, mulchers, and corrugated rollers. The fourth group crushes clods by applying pressure and tends to break hard, dry clods better than drag-type harrows. They also pack the soil slightly, reducing large air spaces.

Listing

In some areas, row crops are planted on ridges formed by listers. A lister is a plow equipped with two moldboards that cuts a furrow slice two ways—half to the right and half to the left. This forms a ridge of soil commonly about 20 to 25 cm (8 to 10 in.) high and of variable width at the base. Listers can be equipped with attachments to list, plant, and fertilize in one operation. Some farmers flatten the tops of the ridges with a roller, drag, or bed shaper before planting.

Cultivation

Cultivation is the tillage between seedling emergence and crop harvest. The main reason for cultivating is to control weeds, but other benefits are improved water infiltration, soil

Figure 9–8 The shank spacing depth makes considerable difference when loosening compacted soil. These shanks were spaced about 50 cm (20 in.) apart, and operate about 40 cm (15 in.) deep. In this case, however, the shanks were set too shallow. They did not get under the compacted soil layer to break it up, and may in fact have created more compaction. Generally, the closer the spacing and sufficient depth, the more effective the soil loosening. *Source:* William E. Wildman.

aeration, the conservation of soil moisture, loosening compacted soils (Fig. 9–8), and in some cases to help with insect control. Some farmers claim that cultivation is neither necessary nor beneficial. Certainly, some row crops are cultivated much more frequently and deeper than necessary—a practice wasteful of time and energy.

Cultivating equipment can be divided into four main classes: (1) row-crop cultivators, (2) field cultivators, (3) rotary hoes, and (4) rototillers.

Row-crop cultivators have various shaped steel shovels that manipulate the soil (Fig. 9–9). The shovels on most equipment are short, narrow, curved, or pointed. For shallow

Figure 9–9 A multiple-row crop cultivator equipped with cultivation shovels is loosening soil and destroying weeds in this soybean field. The enclosed driver cabs in modern tractors are often air conditioned. Some have two-way radio-telephones and other types of electronic equipment. *Source:* International Harvester.

Figure 9–10 A field cultivator (ripper) is used to break up plowsoles or hardpans formed by repeatedly plowing at the same depth. This cultivator is particularly useful in semiarid regions where, because of drought and high winds, it is best to leave the soil protected with a covering of plant residue to reduce wind erosion. *Source:* Deere and Company.

cultivation, wide, thin, horizontal, knifelike blades are used. In fields where vine weeds appear, disks replace the shovels.

Field cultivators are not designed for row crops. They penetrate deeper than row-crop cultivators and are used on fallow ground or stubble fields to control weeds. These cultivators have longer and stronger shovels or sweeps. Some have spring teeth (Fig. 9–10).

Rotary hoes can be used on crops that are drilled or broadcast planted as well as for row crops. They are especially useful on young crops that are too small for other types of cultivators. Also, they are more efficient when operated at higher speeds than other types. Rotary hoes are made up of gangs of rimless wheels whose spokes resemble slightly curved fingers mounted on a horizontal axle. Generally the equipment is made and used in tandem gangs.

Rototillers or rotary plows are used on small plots. Such implements, powered by gasoline engines or by tractors, have an assembly of rotating knives or tines mounted on a horizontal axle. These machines cut a swath of variable width, depending on machine size. The knives rotate vertically and cut into the soil on the downward stroke. The depth may be relatively shallow or down to 15 cm (6 in.). Rototillers are used for seedbed preparation as well as for cultivation. They mulch and shape seedbeds after listing. Often herbicides are applied to the soil prior to powered incorporation. The chemicals are mixed into the soil to a depth about half the length of the tiller's teeth.

Deep Tillage

Some farmers, especially those in the western part of the United States where irrigation is prevalent, use deep tillage to improve problem soils. Extra heavy equipment is used for deep tillage when the soil is dry. One type of implement called a slip plow uses a V-shaped blade that slips along horizontally from 120 to 180 cm (4 to 6 ft) below the soil surface and lifts the soil mass about 15 cm (6 in.) as it passes

Figure 9–11 The V-shaped blade of this slip plow is pulled horizontally well below the soil surface to break up hard soil layers. *Source:* USDA Soil Conservation Service.

through. This shatters the soil profile and breaks any deep, hard, cemented layers (Fig. 9–11).

Another deep-tillage tool is the deep moldboard plow used to turn a furrow slice 150 cm (5 ft) or more deep. The tool serves to bury surface salts.

Another deep-tillage tool is the ripper or deep chisel (Fig. 9–12). This implement consists of one or several shanks that penetrate the soil from 60 to 120 cm (2 to 4 ft). It shatters hardpans best when the soil is dry. Unfortunately, some soils become recompacted and the operation needs to be repeated every three to seven years, depending on the nature of the compacted layer. In the alluvial desert of the southwest United States, deep ripping is used instead of deep plowing to keep the accumulated soil salts buried.

Deep tillage is expensive, and sometimes it does not materially increase crop yields. In established orchards, it can damage trees by severely cutting the roots.

Figure 9–12 An exmple of a deep chisel (ripper) used to break deep compacted hardpans. This procedure is used only when justified because the high amount of energy required makes this an expensive operation. *Source:* William E. Wildman.

Figure 9–13 A multi-row planter planting corn directly into an alfalfa field with no previous land preparation (No-Til). *Source:* Allis-Chalmers.

Minimum Tillage

In many ways, modern tillage methods are similar to those used when horses were the source of power. In those days seedbeds were prepared by plowing, then by disking and harrowing until the clods were broken up, and finally the crop was planted. This procedure required many trips over the land, often causing undesirable soil compaction. Now, the farmer uses large machines powered by heavy tractors that compact the soil at an even faster rate. Much experimentation to use less tillage has been done in the Great Plains and in the Corn Belt states, as well as other areas. Several different methods of minimum tillage have been used, for example, planting seeds directly in plowed ground after a single pass with a rotary hoe. In other trials, no tillage was performed and herbicides were used to kill sod and other undesirable vegetation. In recent years no tillage (Fig. 9–13) has been tried with several test crops and the results have been compared with crops tilled by usual methods.

A minimum tillage technique has been successfully used on severely compacted soils in some Australian orchards. The soil was mechanically tilled once, then covered with a thick layer of legume mulch. Angleworms were introduced and their growth encouraged by the decomposition of the mulch. No further traffic was permitted within 10 m (33 ft) of the tree trunks. After two years the angleworms numbered as many as 2,000 per m^3 (1,530 per yd^3) of soil. The soil had a network of worm paths, water penetration was high, and soil compaction was much alleviated.

Not all minimum tillage results have been favorable. In some cases, yields have been reduced and in others, little or no difference was observed. Even though the yield differences may be small between normal and minimum tillage, the saving in reduced fuel and machinery costs could be appreciable. An added benefit could also be the improved physical condition of the soil.

Figure 9–14 A rice field ready to be planted. The levees follow contour lines to allow uniform water depth within each paddy. *Source:* USDA Soil Conservation Service.

Land Leveling

Irrigated land generally benefits from being level, especially if flood (Fig. 9–14) or furrow irrigation is used and row crops are grown. Exceptions are many; some vineyards, orchards, and some high-value vegetables are grown very successfully on rolling land in contoured rows[2] with sprinkler or drip irrigation.

Land is leveled to permit water to flow and spread evenly over the soil surface without causing erosion. In considering the land's suitability for leveling, the land's productive capacity and the method of irrigation to be used are evaluated. Features that render a site unsuitable include: (1) excessively permeable soil, (2) soil that is very shallow, (3) rough topography (excessive grading will be needed), and (4) poor drainage.

---

[2]Contoured rows put each plant in the row at the same elevation as other plants in the same row.

Figure 9–15 Land leveling equipment using laser technology has greatly contributed to the practice of land leveling. The procedure can now be performed more rapidly and more accurately. At the top of the raised mast in the photograph is a receiver that intercepts the horizontal laser beam emitted from a stationary source located in or near the field. The interception of the laser beam will influence the position of the land plane blade. The received signal will automatically cause the land plane blade to be lowered (to cut deeper into the soil) or raised as is needed to level the field. *Source:* William E. Wildman.

If leveling the land is feasible, heavy equipment will be needed (Fig. 9–15). Timing is important. Land should not be leveled in the rainy season, because leveling of wet soil subjects it to compaction.

Soil Fumigation

For some crops, the soil has to be fumigated before seedbed preparation. These usually are high-value crops where the potential for pest damage is severe enough to justify treatment. Fumigation is normally too expensive to use extensively on most row or field crops. Certain chemicals are used to fumigate soil and destroy harmful bacteria, fungi, and nematodes as well as many weed seeds. A widely used soil fumigant is methyl bromide (CH_3Br). This toxic gas is colorless, odorless, and is usually mixed with chloropicrin (tear gas). Chloropicrin is used to indicate the presence of the more toxic CH_3Br gas because it is nontoxic in small amounts, but it causes discomfort to the eyes. Chloropicrin also has some effect as a fumigant. Methyl bromide will no longer be allowed for most, if not all, soil fumigation after 2002. At the time this text was revised, no suitable substitute was available.

Nurserymen often prefer to use steam instead of chemicals to "partially sterilize," or pasteurize, their soil. A closed container with some means of admitting live steam is used. The temperature at the center of the soil mass is brought to above 71°C (160°F) for 30 minutes to destroy disease-causing organisms. Some nurseries use aerated steam at about 60°C (140°F) for 30 minutes to do the same job.

IRRIGATION

Farmers have irrigated crops for over 4,000 years. Records indicate that crops were irrigated along the Nile, Ganges, Tigris, and Euphrates rivers as early as 2600 B.C. It has been suggested that crop irrigation contributed to the founding of the great civilizations in these areas. It is interesting to note that early civilizations began alongside the rivers in arid or semiarid regions. Irrigation canals over 1,000 years old have been found along the Gila River in Arizona. But, even more amazing, while irrigation has been practiced so long, modern practices date back less than 200 years, and even today farmers in many parts of the world lift water by treadmills or water screws.

The importance of irrigation is evident when precipitation patterns over the earth's total land area are considered. Seventy-five percent of the total land is semiarid to arid and receives, on the average, less than 50 cm (20 in.) of rainfall annually; 20 percent receives less than 25 cm (10 in.).

By the end of the twentieth century, the total irrigated area in the world will have more than tripled since 1900. Most of the increases have occurred in China, India, Pakistan, United States, and the Soviet Union. The expansion of water management and irrigation is illustrated by the increased number of plans and proposals executed since the 1950s. China has built no fewer than 46 dams, and India has almost doubled its irrigated acreage. Ambitious programs are in progress in many parts of the world, including Africa, Pakistan, Spain, Australia, Italy, and the United States.

With the completion of the Aswan High Dam in Egypt in 1967, the Nile River could no longer flood its banks. While this project permitted reclamation of large areas of desert, the settling of fertile silt in Lake Nasser behind the dam removes plant nutrients that formerly were responsible for the creation of the fertile Nile Delta. Also, with irrigation water being available to Egyptian farmers upon demand instead of annually as before the High Dam, intensive cropping is the common practice. This is causing a multitude of soil problems, such as salinity and high water tables.

In 1957 a California water plan was adopted to better control, conserve, and utilize the state's water supplies. The plan provides for storage and transportation of water from water-rich northern California to water-deficient areas to the south through a system of aqueducts on both sides of the Central Valley (Fig. 9–16).

Methods of Application

Selection of the proper water distribution system can save expensive labor and assure better crop yields as well as saving water.

The method of application is important, especially if the cost of water is high. Some factors that determine the method and type of system used are: (1) climate, (2) type of

Figure 9–17　An orchard irrigated by flooding. The levees are contoured, causing the water between the levees to be the same depth over the entire field. This provides even water distribution and uniform water penetration into the root zone. In an older orchard the root zone could be 2 or more m (6 or more ft) in depth. *Source:* USDA Soil Conservation Service.

Figure 9–16　The Tehama-Colusa canal shown here (*above*) is a part of a water transport system through which water from a northern California reservoir (*below*) flows down the west side of the Central Valley to southern California. This and similar large lined canals effectively transport large volumes of water for agricultural as well as industrial and domestic use.

crop, (3) cost of water, (4) slope of field, (5) physical properties of soil, (6) water quality, (7) water availability, (8) drainage capability, and (9) salinity or other problems.

Border or Flood Method

Flood irrigation is used where the topography is flat and level. This method is often used for drilled or broadcast crops, such as hay, pasture, and small cereal grains. Orchards and vineyards are also sometimes flood irrigated (Fig. 9–17).

The land must be graded and leveled for flood irrigation. The amount of grading needed depends upon the topography, cropping system, and cost of grading. A uniform downslope of 0.1 to 0.4 percent[3] is used for most soils and

crops, with little or no cross-field slope. Permanent or temporary levees are constructed running downslope, with a border disk forming ridges that divide the field into strips or checks, preferably not over 10 to 20 meters (33 to 66 ft) wide and 90 to 300 meters (300 to 990 ft) long.

Water from an irrigation pump or canal is turned into the supply, or head, ditch at the higher end of the field. It is released or siphoned into one or more checks and allowed to flow slowly downslope, spreading evenly and uniformly over each entire check as it advances toward the lower end. Ponding, excessive percolation, and inadequate wetting of the soil in different areas in the field should be avoided. Designing and operating such a system efficiently requires considerable experience, skill, and knowledge.

Furrow Irrigation

Furrow irrigation is a modification of flooding—water is confined to furrows rather than wide checks. Water is used more efficiently with furrows than with flooding because the entire surface is not wetted, thus reducing evaporation losses.

Furrow irrigation is frequently used for row crops, orchards, and vineyards. The length of furrow varies from 30 m (100 ft) for small gardens to 450 m (1,500 ft) for field crops, but lengths of 90 to 180 m (300 to 600 ft) are more common. Long furrows cause greater loss of water because of deep percolation and excessive soil erosion at the head of the field.

Furrow spacing is determined by the plant row spacing. One irrigation furrow is generally provided for each crop row (Figs. 9–18 and 9–19). The furrow spacing can be 60 to 180 cm (2 to 6 ft), depending upon the type of crop and wetting characteristics of the soil.

The depth of the furrow should be such that the water can be controlled. Water should flow in the furrow for sufficient time to allow it to percolate into and across the bed,

---

[3]A 0.1 percent downslope drops 0.1 meter in elevation for each 100 m of field length (0.1 ft/100 ft).

Figure 9–18 Preplanting furrow application of water in process. The raised beds will be reshaped just before or during planting to provide a smooth uniform surface. Note the penetration of water from the furrows into the beds.

Figure 9–20 It would be impractical to use methods other than sprinkling for irrigating this permanent pasture because of the uneven topography. This is a permanent set type of sprinkler system similar to those used for golf courses and athletic fields. *Source:* USDA Soil Conservation Service.

Figure 9–19 A tomato field being furrow irrigated with water brought to the field in an open ditch, then siphoned over the ditch bank into the field. The water level in the ditch must be higher than that in the field. *Source: The Daily Democrat.* Woodland-Davis, California.

although not directly wetting the surface of the bed. For most row crops and orchards, furrows from 20 to 30 cm (8 to 12 in.) deep provide the necessary control. Furrows from 10 to 15 cm (4 to 6 in.) in depth are better for small-seed crops.

Uniform crop maturity, necessary for mechanical harvesting, is easier to achieve with furrow than with flood irrigation. Furrow irrigation applies water more uniformly than flood irrigation and improves uniformity of crop development.

Sprinkler Irrigation

Sprinklers are often used when flood or furrow irrigation is impractical. Sprinklers can have some advantages over other irrigation methods; for example, through nozzle size selection, sprinklers are adaptable to high or low soil permeabilities. Through uniform wetting of the surface area, seed germination is more uniform; less total water may be used, and sprinklers are effective in washing salts away from salt-sensitive crops. Additionally, the surface need not be level; a properly de-

signed system offers a method of adequately and accurately applying water, even on sloping land with grades up to 3 percent. Sprinklers are sometimes also used for frost control.

In some situations, sprinkler systems may require less labor, whereas with systems that are highly portable, more labor may be used. In general, equipment and energy costs are higher than with flood or furrow methods.

There are different types of sprinklers, each with certain advantages.

The **hand set** was the first type of sprinkler system developed. The main lines are either buried or portable. This system is used on many crops, and is particularly useful for germinating small-seed crops, especially if the seedbed is rough. A disadvantage of the hand-set system is the added labor required to move the lateral lines as each set is completed. This can be overcome by having a solid-set system, one that is not moved. However, this requires more investment in equipment. The main line is located at the edge of the field and the laterals are usually set perpendicular to the main line. With row crops the laterals may be placed either parallel or perpendicular to the rows.

The **permanent set** type has all lines buried below the surface. In some systems the lines are suspended above the surface. Because of the high investment cost, this type of system is usually restricted to orchards, vineyards, high-value crops, or recreation facilities (Fig. 9–20).

The **wheel line** system was designed to reduce labor by moving pipe across the field with a small gasoline engine instead of by hand. The lateral that carries the water is mounted on wheels spaced about 6 m (20 ft) apart. The lateral pipe, which also acts as the axle, is long, up to 300 meters (985 ft) in length. The sprinklers are mounted in the lateral pipe and another set is often dragged behind the line. To move to the next position, the gasoline engine turns the axle and propels the entire line across the field, saving labor and time (Fig. 9–21). Under windy conditions, application uniformity is poor.

Figure 9–21   A wheel line system of sprinkler irrigation (*above*) in use. The linear line system (*below*) is a more efficient variation of the wheel line system. The larger equipment delivers water more uniformly and in greater quantities. It does require considerably more energy to operate.

The **center pivot** system is another labor-saving variation of sprinkler irrigation. The system is used more often in areas where land values are low or the availability of labor low. The line, mounted on wheels driven by water pressure or electric motors, follows a circular path, pivoting around a fixed central point (Fig. 9–22). One disadvantage of the system is that the circular irrigation pattern leaves the corners in rectangular fields unirrigated (Fig. 9–23). In some areas where land and crop values justify the added costs, the system is modified by adding sprinklers that fold out as the corners of the field are approached, then close back as the line

Figure 9–22   A field sprinkle-irrigated with the center pivot system. The sprinklers rotate in a circular path around a central water supply focal point. *Source:* Valmont Industries.

Figure 9–23   Aerial view of grain fields in Nebraska being sprinkle-irrigated by central pivot systems. The white radius lines are the operating sprinklers. Modification of this equipment will enable the corners of these fields to be irrigated. *Source:* Valmont Industries.

moves past the corner. Sometimes the corners are irrigated with a permanently buried auxiliary line of sprinklers, again increasing equipment costs.

The **hose drag** system is another variation for moving the sprinkler line mechanically. A tractor attached to one end of the flexible line drags it from place to place. This system is mainly used in orchards (Fig. 9–24).

Drip Irrigation

Drip or trickle irrigation is the latest development in irrigation systems. Small amounts of water are allowed to trickle slowly into the soil through mechanical devices called emitters, wetting the soil without runoff (Fig. 9–25). The emission rate of water ranges from about 2 to 8 liter per hr (0.5 to 2 gal/hr).

Figure 9–24   A citrus orchard sprinkle-irrigated with a drag line system. The line is mounted on small wheels so that a tractor can be attached at one end to pull the entire line to a new location. This is a labor-saving arrangement. *Source:* University of California Cooperative Extension.

Figure 9–26 In the western United States, the use of drip irrigation on multirow raised beds for strawberry production has increased and has largely replaced furrow irrigation practices. Strawberries such as these are often mulched with plastic film. The film is placed over the beds and openings in the film are made where the plants are located. Thereafter, further foliage growth is above the plastic mulch. Clear film is used to warm the soil and the film prevents fruit from contacting the soil. Soil fumigation is used to control soil diseases and weeds.

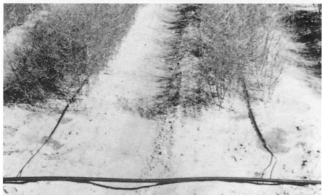

Figure 9–25 *Above:* One type of water emitter used in drip irrigation. The tiny plastic tubing is attached to a larger plastic tube from which the emitter receives the water. The number of emitters depends on the type and size of crop irrigated. *Below:* A less permanent but popular method of trickle irrigation utilizes thin wall plastic tubing having small perforations in the wall that allow water to trickle out. The advantages of this system are its low cost and ease of installation. It is operated at low pressures and easily repaired. Some disadvantages are its short life, a tendency to tear, and frequent plugging of the small perforations. It is well adapted for some annual crops, particularly those of high value, like the asparagus, shown here in fern.

Emitters are connected to a small plastic lateral tube, laid either on the soil surface or buried just beneath it for protection. Some systems have the emitters built into the lateral line or tube. The lateral lines are connected to a buried main line that receives water from a head source. The head source is the control station for the system. Here the water is filtered, may be treated with fertilizers, and is regulated for pressure and timing of application. Some advantages of drip irrigation are (1) the system need not be moved; (2) there is little interference with orchard cultural operations because much of the soil surface is not wetted; (3) there is less fluctuation of soil moisture in the root zone area because of the constant and slow drip application of water, and (4) less water is needed to grow a crop. The area of wetted soil can be as little as 10 percent of the total area of newly planted tree crops or up to 60 percent of the area of a mature crop. The amount of soil wetted depends on the soil's physical properties, the time of application, and the number of emitters used. Some objections are (1) expensive filtration equipment is needed to avoid frequently clogged emitters; (2) water distribution may be un-

Figure 9–27 A five-year-old peach orchard is drip-irrigated with emitters placed around each tree. Only the soil around the root area receives water. As the trees grow and enlarge their root systems, more emitters are added to wet a larger volume of soil.

even on hilly land; (3) salts tend to concentrate on the soil surface and near the wetted area boundary, and because leaching with excess water does not occur; and (4) the distribution of roots may be restricted to the small volume of wetted soil.

Drip irrigation may not fit the needs of every crop or situation, but its use by orchardists, strawberry growers, ornamental nurseries, and for some high-yield field crops is rapidly increasing (Figs. 9–26 and 9–27).

MINERAL NUTRITION

Use of fertilizers has probably increased crop yields and reduced hunger more than any other single agricultural practice. In the United States from 1970 to 1985, the annual consumption of chemical fertilizers has increased about 5 percent per year.

In addition to supplying nutrients to crops to increase yields, fertilizers can also cause marked changes in soil characteristics, some beneficial, some not. These secondary influences play an important role in the choice of fertilizer.

Sixteen chemical elements are known to be essential for the growth of most plants, and a few others are used by some plants under certain conditions. The essential elements are carbon (C), hydrogen (H), oxygen (O), nitrogen (N), phosphorus (P), potassium (K), calcium (Ca), magnesium (Mg), sulfur (S), iron (Fe), manganese (Mn), molybdenum (Mo), copper (Cu), boron (B), zinc (Zn), and chlorine (Cl). Some halophytes (plants that require salts) have been shown to need sodium, and some microorganisms that fix nitrogen symbiotically or nonsymbiotically require cobalt.

Mineral nutrients are divided into groups according to the quantity plants use. The primary **macronutrients**—mineral nutrients used in largest amounts—are nitrogen, phosphorus, and potassium; the **secondary** mineral nutrients— used in lesser amounts than primary—are calcium, magnesium, and sulfur; and the remaining used in the smallest amounts are the **micronutrients.** Micronutrients are sometimes referred to as trace elements.

Mineral nutrients are supplied to the soil by applying crop residues, animal manures, chemical fertilizers, or naturally occurring minerals. Other sources are the atmosphere, irrigation water, rainfall, and the elemental nutrients of the soil itself that can enter the water solution. The actual source of the nutrient (organic or inorganic) is unimportant to the plant so long as the mineral nutrients are available in sufficient quantity and can be easily assimilated. Nutrients are lost from the soil when any part of a crop is removed. Returning the crop residue to the soil does not replace all of the removed nutrients. To maintain or improve the soil's fertility, nutrients must be added in one form or another in amounts equal to or greater than those removed by crop harvest (Figs. 9–28 and 9–29). Generally, commercial fertilizers are easier to apply and manage than manures and crop residues, but the latter two should not be disregarded. They are especially beneficial in adding organic matter to help improve soil structure.

A complete fertilizer contains the three primary nutrients: nitrogen, phosphorus, and potassium. It may also contain some secondary or micronutrients. Each bag of commercial fertilizer carries a label stating the analysis of its contents. This analysis is represented by three figures; for example, 5–10–5. The first figure is the percentage nitrogen by weight; in this case, 5 kg of nitrogen per 100 kg (5 lbs/100 lbs) of fertilizer. The second figure represents phosphorus, specifically 10 percent phosphoric acid (P_2O_5); the third fig-

Figure 9–28 Significant amounts of fertilizer are applied by air. Large areas of range land not readily accessible by other equipment, water-covered rice paddies, and large fields seeded to cereal grains are particularly suitable for aerial application of fertilizers and other chemicals.

Figure 9–29 Broadcasting of a dry fertilizer to a field before plowing. This procedure is sometimes known as a "plow-down" application. The application may provide all of the fertilizer or a portion of it, the remainder to be applied as a side-dressing as the crop develops. The material is visible as white pellets on the soil surface. It will dissolve and enter the soil with the first rain or irrigation. *Source:* Fred Meyer, National Fertilizer Development Center, Tennessee Valley Authority.

ure is potassium, specifically 5 percent potash (K_2O).[4] Thus, one metric ton (MT = 1,000 kg) of a 5–10–5 fertilizer contains 50 kg (110 lb) of nitrogen, 100 kg (220 lb) of phosphoric acid (P_2O_5), and 50 kg (110 lb) of potash (K_2O), a total of 200 kg (440 lb) of nutrients. The remaining 800 kg (1,764 lb) consists of other chemicals in the formulation or of filler.

[4] % P = % P_2O_5 × 0.43
% K = % K_2O × 0.83
% P_2O_5 = % P × 2.29
% K_2O = % K × 1.21

While the label on the bag states the percentage of each primary nutrient, it may not indicate the compounds used to make up the fertilizer. For example, the nitrogen in the fertilizer might be supplied as urea, ammonium nitrate, ammonium sulfate, or sodium nitrate, and so on. The formulation is important because it informs the user of what compounds are used and their chemical form. It indicates the fertilizer's nutrient availability, effect on soil pH, ease of incorporating into the soil, and freedom from caking.

Assume that nitrogen in the form of ammonium sulfate $(NH_4)_2SO_4$, phosphoric acid as monocalcium diphosphate $[Ca(H_2PO_4)_2]$, and potash as potassium chloride (KCl) are used to formulate a complete fertilizer. Their atomic weights are used to calculate the weight of each ingredient needed to supply the proper amount of nutrients. We find that it requires 238, 165, and 79 kg/MT (476, 329, 158 lb/t) of ammonium sulfate, calcium dihydrogen phosphate, and potassium chloride, respectively, to produce a fertilizer with 5–10–5 percent of N, P_2O_5, K_2O. The three constituents total 482 kg/MT (963 lb/t). The balance needed to make up the weight is "filler." The filler most often used is dolomitic limestone or gypsum, but other materials, even sand, can be used.

Fertilizer recommendations are often given as a ratio. A ratio differs from an analysis in that it expresses the amount of one nutrient in relation to the other. For example, the ratio of nutrients in 5–10–5 fertilizer is 1.2.1. Thus, if the recommendation was for a 1:2:1 fertilizer, a product labeled 5–10–5 or 10–20–10, or 15–30–15 would be equally acceptable as long as the rate of application was adjusted accordingly.

The fertilizer's physical properties are worthy of consideration because a lumpy or caked fertilizer is difficult to apply evenly. Some constituents, such as ammonium nitrate (NH_4NO_3), calcium nitrate $[Ca(NO_3)_2]$, sodium nitrate $(NaNO_3)$, and urea $[CO(NH_2)_2]$ absorb water from the air (i.e., they are hygroscopic) and thus must be protected from moisture. They are packaged in moisture-proof bags, and often a conditioning material is added to decrease moisture absorption and caking.

Any fertilization program requires regular and consistent **soil testing.** Soil testing allows a grower to monitor the fertility of the soil or growing media and adjust fertilizer formulations or application frequency if necessary. Tests should be made at a minimum at the beginning of each crop or growing season. More frequent testing is recommended if fertilization is done frequently, as with golf course and greenhouse or nursery management, or if problems develop. Fertility maps of fields can be created and incorporated into a global positioning system (GPS), enabling precision applications of appropriate fertilizers.

Primary Nutrients

Nitrogen

In the past, farmers "grew" much of their nitrogen fertilizer; that is, they plowed under legume crops that had been inoculated with bacteria (*Rhizobium*) which fix atmospheric nitrogen biologically. This remains a practice in many areas. Animal manures are also a nitrogen source, returning to the soil that amount taken out by plants used to feed the animals minus that used by the animal for its own growth. Electrolysis of atmospheric nitrogen by lightning during thunderstorms fixes nitrogen gas as oxides. However, in modern agriculture, the most important source of nitrogen is the synthetic fixation of atmospheric nitrogen gas. This is the process that occurs in fertilizer plants, where natural gas is combined with nitrogen gas under tremendous pressure and at high temperatures to form ammonia. The synthetic ammonia is then further processed into various forms in the manufacture of the commercial fertilizer.

Plants absorb nitrogen as inorganic nitrate ions (NO_3^-) and, in a few cases, as ammonium (NH_4^+) or amino (NH_2^+) ions. Most natural soil nitrogen is in the organic form—that is, combined in some manner with carbon. Organic nitrogen occurs in manures, decomposing organic matter, and urea $[CO(NH_2)_2]$ and must be oxidized before most plants can use it.

The transformation of organic matter to the mineral or inorganic form—for example, organic nitrogen to NH_4^+, NO_2^-, NO_3^-—by microorganisms is called **mineralization.** The conversion of the mineral form to the organic form is called **immobilization** (Fig. 9–30). When organic material with a carbon:nitrogen (C:N) ratio greater than 30 is

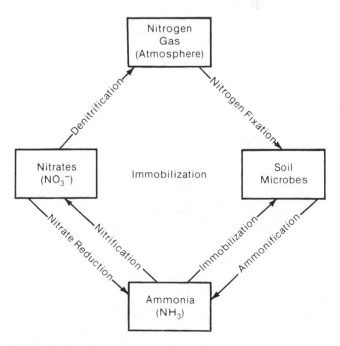

Figure 9–30 The nitrogen cycle in soil.

TABLE 9–1	Nitrogen Fertilizers, Their Composition, and Residual Effect on Soil pH					
	Organic				**Inorganic**	
	%N	Residual Effect			%N	Residual Effect
Dried blood	12	acid		Anhydrous ammonia	82	acid
Guano, Peruvian	12	acid		Urea	46	acid
Fish meal, dried	10	acid		Ammonium nitrate	33	acid
Tankage	8	acid		Aqua ammonia	30	acid
Soybean meal	7	acid		Ammonium chloride	28	neutral
Peanut meal	7	acid		Calcium cyanamide	22	basic
Cottonseed meal	7	acid		Ammonium sulfate	20	acid
Activated sludge	6	acid		Diammonium phosphate	20	acid
Bone meal, raw	4	acid		Sodium nitrate	16	basic
Garbage tankage	3	basic		Calcium nitrate	16	basic
				Potassium nitrate	14	basic

applied to a soil, the nitrogen is immobilized during initial decomposition, then it is mineralized as decomposition proceeds.

The major role of plants in the processes illustrated in Figure 9–30 is the reduction of nitrate to ammonium; this is an extremely important process for all life on the earth and is a process that cannot be carried out by animals. The ammonium is then rapidly converted into a wide variety of important nitrogen-containing metabolites including amino acids, storage proteins, catalytic proteins (enzymes), and nucleic

acids. Also, it should be noted that nitrogen is a key component of chlorophylls, the molecules responsible for the collection of sunlight energy in photosynthesis.

Continued use of nitrogen fertilizers can affect the pH of the soil (Table 9–1). Some fertilizers are residually acid forming, others residually basic, and some have little or no effect (see Fig. 9–31). Circumstances permitting, avoid acid-forming fertilizers on acid soils and basic fertilizers on alkaline soils. This suggestion is not always followed in practice. Fertilizers containing ammonium ions as the

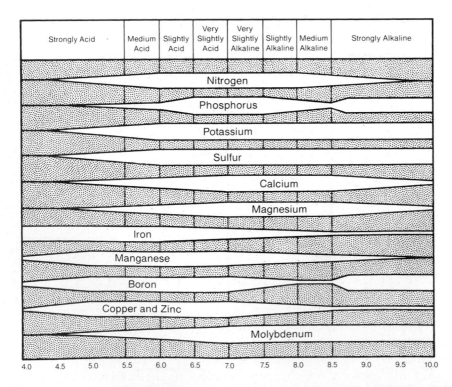

Figure 9–31   Availability of some essential nutrient elements as influenced by soil acidity or alkalinity.

source of nitrogen are residually acid and may be used on alkaline soils. Fertilizers containing basic cations (Ca^{2+}, Na^+, or K^+) with nitrate (NO_3^-) anions as the source of nitrogen are residually basic and may be used on acid soils. In either case, the residual effect moves toward soil neutrality. Potassium sulfate (K_2SO_4) and potassium chloride (KCl) are residually neutral.

The kind of ion determines the mobility, solubility, and availability of nitrogen to the plant. For example, by ionic exchange, positively charged ammonium ions are held by negatively charged soil particles and are prevented from leaching out of the root zone. Under these conditions, the ammonium ion is considered to be relatively immobile. The nitrate ions are negatively charged and are not attracted to the soil particles; therefore, they move freely in the soil solution. More of the ammonium ions are retained in fine-textured soils than in coarse-textured soils because the fine-textured soils have more exchange sites. Conversion of ammonium to nitrate ions (nitrification) speeds up possible nutrient losses by leaching.

Nitrates, while essential for plant health, are detrimental to humans. Nitrate leaching from fields and other areas of crop production has resulted in groundwater contamination in several areas of the United States. As a result, in many areas runoff from farms, greenhouses, and nurseries is clearly monitored and the amount of nitrate runoff is strictly limited. This has forced many crop producers to invest in water collection or filtration systems.

Nitrogen fertilizers are applied as solids, liquids, or gases. They can be applied as preplant fertilizers, top-dressed after the crop has emerged (Fig. 9–32), broadcast evenly over the field and mixed with the soil by disking, drilled or injected into the soil at desired depths, dripped into irrigation water (Fig. 9–33), or injected as a gas or liquid into the soil (Fig. 9–34). These methods can be combined, depending upon the crop, or fertilizing can be combined with planting. Nitrogen fertilizers must not be placed in direct contact with the foliage because they burn the leaves. Ammonia is toxic to living tissue.

Many plants deficient in nitrogen show pale green to yellow leaves, but each crop has its own characteristic symptoms. Nitrogen deficiency generally causes the plant to grow slowly and restricts crop yields (Table 9–2).

Phosphorous

Phosphorous is needed by plants in smaller amounts than nitrogen or potassium, but this does not reflect its true importance in plant nutrition. Phosphorus is a key element in the formation of AMP, ADP, and ATP (adenosine mono-, di-, and triphosphate), which play essential roles in photosynthesis and respiration (Ch. 7). Phosphorus is a constituent of nucleic acid and phospholipids. Also, many intermediates in plant metabolism are phosphorylated compounds, as we have seen in Chapter 7. Phosphorus is often associated with early crop maturity, increased root

Figure 9–32 Application of fertilizers as a liquid is a convenient and widely used method for providing supplemental plant nutrients to crop lands. In some situations fertilizers are applied directly onto the leaves of some crop plants. These foliar applications usually contain secondary or micronutrients.

Figure 9–33 Ammonia gas introduced directly from the storage container through the hose into the flowing canal water. The gas will dissolve and be distributed with the water to the cropland.

Figure 9–34 Anhydrous ammonia fertilizer injected into the soil through the tubing attached to the rear of the shanks being drawn through the soil. The fertilizer is injected either in gaseous form, directly from the pressurized tank, or as liquid (aqua-ammonia), where the gas was previously dissolved in water. The placement can be made at the desired depth and location within the row in relation to the plants, making for more efficient use of fertilizer. *Source:* Fred Meyer, National Fertilizer Development Center, Tennessee Valley Authority.

TABLE 9-2	Summary of Roles of Mineral Elements in Plant Nutrition			
Nutrient Element	**Function in Plants**	**Deficiency Symptoms**	**Losses from Soil**	**Fertilizer**
PRIMARY NUTRIENTS Nitrogen (N)	Synthesis of amino acids, proteins, chlorophyll, nucleic acids, and coenzymes.	Stunted growth, delayed maturity, light green leaves; lower leaves turn yellow and die.	Erosion, leaching, crop removal.	Inorganic salts of ammonia, calcium, sodium, potassium, urea; organic fertilizers; legume crops; animal manures, crop residues, animal waste.
Phosphorus (P)	Used in proteins, nucleo-proteins, metabolic transfer processes, ATP, ADP, photosynthesis, and respiration. Component of phosphyolipids.	Purplish leaves, stems, and branches; reduced yields of seeds and fruits, stunted growth.	Crop removal, fixation in soil. Reversion to unavailable form in soil.	Superphosphate, treble superphosphate, ammonium phosphate, animal manures.
Potassium (K)	Sugar and starch formation, synthesis of proteins. Catalyst for enzyme reactions, neutralizes organic acids, growth of meristematic tissue.	Reduced yields; mottled, spotted or curled older leaves; marginal burning of leaves; weak root system, weak stalks.	Crop removal. Soil fixation leaching.	Potassium sulfate, potassium chloride.
SECONDARY NUTRIENTS Calcium (Ca)	Cell walls, cell growth and division; nitrogen assimilation. Cofactor for some enzymes.	Deformed terminal leaves, reduced root growth. Some plants turn black, dead spots in midrib in some plants. Failure of terminal bud to grow.	Leaching, crop removal.	Calcium sulfate, calcium nitrate, calcium carbonate, dolomitic limestone.
Magnesium (Mg)	Essential in chlorophyll, formation of amino acids and vitamins. Neutralizes organic acids. Essential in formation of fats and sugars. Aids in seed germination.	Plants usually chlorotic (interveinal yellowing of older leaves); leaves may droop.	Leaching, plant removal, and erosion. Some losses by fixation to unavailable form in acid peaty soils.	Foliar sprays with magnesium sulfate, dolomitic limestone.
Sulfur (S)	Essential ingredient in amino acids and vitamins. Flavors in cruciferous plants and onions.	Light green leaves, reduced growth, yellowing of leaves. Weak stems. Similar to N deficiency.	Erosion, leaching, crop removal.	Ammonium sulfate, calcium sulfate, super phosphate, sulfuric acid, elemental sulfur.
MICRONUTRIENTS Boron (B)	Affects flowering, pollen germination, fruiting, cell division, nitrogen metabolism, water relations, hormone movement.	Terminal buds die, lateral branches begin to grow, then lateral buds die, branches form rosettes. Leaves thicken, curl, and become brittle.	Crop removal, leaching.	Sodium or calcium borate, animal manure, superphosphate.
Copper (Cu)	Constituent in enzymes, chlorophyll synthesis, catalyst for respiration, carbohydrate and protein metabolism.	Terminal leaf buds die. Chlorotic leaves. Stunted growth. Terminal leaves die.	Tied-up by highly organic soils and acid soils. Leaching.	Copper sulfate or other copper salts.
Chlorine (Cl)	Not too much known except that it aids in root and shoot growth. Required for growth and development.	Plants wilt. Chlorotic leaves. Some leaf necrosis. Bronzing in leaves.	Almost never deficient under field conditions.	Chloride salts.

TABLE 9–2	Continued			
Nutrient Element	**Function in Plants**	**Deficiency Symptoms**	**Losses from Soil**	**Fertilizer**
Iron (Fe)	Catalyst in synthesis of chlorophyll. Involved in formation of many compounds. Components in many enzymes.	Paling or yellowing of leaves, chlorosis between veins at first. Grasses develop alternate rows of yellowing and green stripes in leaves.	Crop removal, leaching, and erosion. Unavailable in alkaline soils.	Foliar applications of iron chelates, ferrous sulfate, or ferrous ammonium sulfate.
Manganese (Mn)	Chlorophyll synthesis, acts as coenzyme.	Network of green veins on light green background of intervenous tissue. Leaves later become white and abscise.	May be unavailable in alkaline soil. Toxic in acid soils.	Manganese sulfate as a foliar spray or as a soil application.
Molybdenum (Mo)	Essential in some enzyme systems that reduce nitrogen. Protein synthesis.	Plants may become nitrogen deficient. Pale green, rolled or cupped leaves, with yellow spots. Leaves of crucifers become narrow, cereal glumes do not fill out.	May have been lacking when soil was formed or become unavailable.	Solution of sodium molybdate sprayed on plants or soil. Also dusted on seeds before planting.
Zinc (Zn)	Used in formation of auxins, chloroplasts, and starch. Legumes need Zn for seed production.	Abnormal roots; mottled bronzed, or rosetted leaves. Intervenous chlorosis.	May not be available in alkaline soils; and toxic in acid soils. Crop removal.	Zinc sulfate as a foliar spray or added with other fertilizers.

(MICRONUTRIENTS (Continued))

proliferation, and seed formation. Deficiencies result in stunted growth, accumulation of anthocyanin pigment (purpling the leaves of some plants), and reduced yields of seeds and fruits.

Phosphorus is absorbed mainly as orthophosphate ions ($H_2PO_4^-$) and to a lesser extent as monohydrogen phosphate (HPO_4^{2-}). The quantity of either of these formed in the soil at any time is small, but the supply is constantly being renewed. The rate of renewal depends upon soil pH. Phosphorus is most available to the majority of crops in the pH range of 5.5 to 7.0. (Fig. 9–31). Iron and aluminum phosphates precipitate in alkaline soils.

The primary source of phosphorus for fertilizers is mined apatite (rock phosphate), of which much of the world's reserve is located in the United States. Apatite exists in several forms, all of which are more or less insoluble in water. To increase phosphorus availability to plants, apatite is treated with acids or heat when made into fertilizer. Rock phosphate treated with sulfuric acid yields superphosphate, the most abundant phosphorus source, containing 16 to 20 percent phosphoric acid (P_2O_5) (Table 9–3). Liquid phosphoric acid (H_3PO_4) is made by treating apatite with sulfuric acid (a wet process that gives calcium sulfate as a precipitate). Ammonium phosphate results from treating phosphoric acid with ammonia. Superphosphate fertilizers do not appreciably affect soil pH, but phosphoric acid (H_3PO_4) is residually acidic. Another source includes basic slag, a byproduct of the steel industry and an important phosphorus source in many European countries.

TABLE 9–3	Approximate Composition of Some Common Phosphate Fertilizers	
	Phosphorus (%)	
Source	**P**	**P₂O₅**
Ammonium phosphate	7.0	16
Superphosphate	7.0	16–20
Triple superphosphate	18–23	42–50
Dicalcium phosphate	23	52
Phosphoric acid	24	54
Potassium phosphate	18–22	42–50
Raw rock phosphate	10–17	25–30
Calcium metaphosphate	27	62

Potassium

Potassium is absorbed by plants in its ionic form (K^+). It plays roles in regulating the opening and closing of stomata and in water retention. It promotes the growth of meristematic tissue, activates some enzymatic reactions, aids in nitrogen metabolism and the synthesis of proteins, and aids in carbohydrate metabolism and translocation (Table 9–2). Potassium does not appear to be an integral part of plant constituents—protoplasm, fats, or carbohydrates—as do other nutrient elements. Plants absorb macroquantities of potassium.

The first visible deficiency symptom appears in the leaves. Weak stems (lodging in grains), decreased yield, lack of disease resistance, and crop quality have been associated with potassium deficiency.

Quantities of potassium salts are found in several areas of the world. Some lie below the earth's surface and some in dead lakes or seas. Potassium is found naturally in most soils. It comes from the decomposition of rocks and minerals, such as feldspars, muscovite, and biotite. West Germany, the Soviet Union, and Canada have the world's largest reserves, but the United States also has large deposits in New Mexico, Oklahoma, and Texas.

The most widely used potassium fertilizers are potassium chloride (KCl), potassium sulfate (K_2SO_4), and potassium nitrate (KNO_3), commonly known as muriate of potash, sulfate of potash, and saltpeter, respectively. The nitrate form is the most expensive and is mainly used on orchards, vegetables, or other high-value crops.

In the soil, potassium is more mobile than phosphorus but less mobile than nitrates. It is readily leached from light sandy soils. In soils, potassium is bound by the negative charges on the clay colloids. In soils that are very high in clay and low in potassium, this binding of K^+ to the ion exchange sites may make potassium unavailable to plants, but this situation is rare in most agricultural soils. Potassium can be made more available to plants when other cations are added to the soil, for example, Ca^{++} added when the soil is limed.

Secondary Nutrients

Calcium

Calcium is absorbed by the plant in the ionic form (Ca^{2+}). It is essential to all higher plants. A deficiency kills terminal buds in shoots and apical tips in roots, reducing plant growth.

Calcium comes from dolomite, calcite, apatite, and some feldspars. Besides being an essential plant nutrient, calcium carbonate ($CaCO_3$), also called limestone, is used to correct soil acidity (Figs. 9–35 and 9–36). Calcium sulfate (gypsum) is used to help reclaim sodic soils and also to im-

Figure 9–36　Liquid lime (a suspension of $CaCO_3$) being applied to an acid soil. Wide tires distribute weight and lessen soil compaction. *Source:* Fred Meyer, National Fertilizer Development Center, Tennessee Valley Authority.

prove soil structure and aggregation in saline soils (Ch. 8). The best way to correct calcium deficiency in the soil is to add either limestone ($CaCO_3$) or gypsum ($CaSO_4$). The specific form depends upon the soil's pH, and other soil characteristics.

Magnesium

Like calcium, magnesium is absorbed by the plant as an ion (Mg^{2+}). It is an essential nutrient, the central atom in the structure of the chlorophyll molecule. The magnesium salt of ATP is the chemical form of ATP that actually participates in hundreds of biochemical reactions in plant cells. Magnesium is mobile and can translocate from older to younger leaves. Its deficiency causes an intervenous chlorosis (yellowing) in older leaves.

Magnesium is an exchangeable cation in the soil resulting from the decomposition of such minerals as biotite, dolomite, olivine, and serpentine. It also appears in the soil solution. In some arid areas of the world, magnesium occurs in such large quantities that it precipitates in the soil profile and, in rare instances, is so abundant that it is toxic.

A good source of magnesium (and calcium) for fertilizer is dolomitic limestone ($CaCO_3 \cdot MgCO_3$). Magnesium sulfate ($MgSO_4$) and potassium magnesium sulfate ($K_2SO_4 \cdot 2MgSO_4$) are also sources. Magnesium fertilizers are applied much the same as calcium.

Sulfur

Sulfur is absorbed by the plant as the sulfate ion (SO_4^{2-}). It is reduced to the disulfide (S—S) or sulf-hydryl (SH) group before being utilized in the plant. Sulfur is a component of several essential amino acids and several coenzymes, and is a key component of most proteins where it plays a role in maintaining the protein structure. Sulfur also contributes to the characteristic flavor compounds of some vegetables, such as cabbage and onions.

The deficiency symptoms of sulfur can be confused with those of nitrogen, except that sulfur is not as readily

Figure 9–35　The truck is applying agricultural lime ($CaCO_3$) as a finely ground limestone for soil acidity management. *Source:* Fred Meyer, National Fertilizer Development Center, Tennessee Valley Authority.

translocated from older to younger leaves. Therefore, the younger leaves of a plant suffering from sulfur deficiency will appear yellow. Reduced plant growth, uniform yellowing of younger leaves, and weak stems are characteristic symptoms of sulfur-deficient plants.

The primary source of sulfur is the decomposition of metal sulfides in igneous rocks. Appreciable quantities also come from the atmosphere and some from irrigation waters. Sulfur is present in the soil as sulfates and sulfides and in organic combinations in the soil humus.

Sources of sulfur for fertilizers are sulfate salts of aluminum, ammonium, calcium, iron (ferrous), magnesium, manganese, potassium, sodium, and zinc. Other sources are lime sulfur, sulfuric acid, and elemental sulfur.

Micronutrients

Micronutrients are needed by plants in minute quantities, but this fact does not detract from their importance. Deficiency of a necessary micronutrient is as devastating to a plant as a deficient macronutrient. One characteristic common to all micronutrients is that while they are essential in small quantities, they are toxic in large quantities (Table 9–4).

Boron

Boron occurs in most soils normally in quantities of 20 to 200 ppm. In some arid areas it occurs in toxic amounts. In most humid regions, boron occurs as a borosilicate in the form of tourmaline. This material is quite insoluble. Boron-deficient soils occur along the Atlantic coast from Maine to Florida, in the Gulf coast states, the North Central states, and along the Pacific coast from Washington to California. Borax ($Na_2B_4O_7 \cdot 10\ H_2O$) is water soluble, easily leached from sandy soils, and is a good source of boron.

Copper

Many areas throughout the world have copper-deficient soils. In the United States, deficient soils are found in the Great Lakes area, the West Coast, and Florida. Copper deficiencies appear more frequently in highly organic soils but have been found in mineral soils in some countries. The amount of organic matter, soil pH, and the presence of other metallic ions influence the availability of copper. Large quantities of copper in the soil can cause iron deficiency in some plants, and there have been reports of copper toxicity. Copper deficiencies are corrected by applying copper salts to the soil by foliar spraying with a solution of soluble copper salts. Copper sulfate ($CuSO_4 \cdot 5\ H_2O$) or copper ammonium phosphate ($CuNH_4PO_4$) are often used.

Chlorine

Chlorine evaded detection as an essential nutrient for many years because it is so universally abundant in nature. Spray from ocean waves carrying huge amounts of sodium chloride and potassium chloride are blown many miles inland by the wind. Large quantities are also deposited in the soil by precipitation. Larger amounts of this nutrient are used by crop plants than any other micronutrient except iron.

Most chlorine exists in the soil as simple chloride salts and is absorbed by the plant as chloride ions (Cl^-). Because of their similarity, bromine can substitute for some chlorine in some plants. Chlorine deficiency is seldom seen in the field, but symptoms have been observed in tobacco, tomatoes, buckwheat, peas, cabbage, sugar beets, barley, corn, cotton, and potatoes. Deficiency symptoms of chlorine appear to be stunted root growth, leaf bronzing, and chlorotic leaves with some necrosis; often the plant wilts. Excess chlorine is toxic, especially to tobacco and potatoes.

Iron

Iron is more abundant in most soils than any other micronutrients, but often it is deficient because it is unavailable to plants because of its extremely low solubility. Iron deficiency has been noted in crops grown on alkaline or calcareous soils (Fig. 9–31) and on acid soils with high phosphate levels. Citrus, deciduous fruits, soybeans, strawberries, vegetable crops, and many ornamentals have shown chlorosis caused by iron deficiency. The deficiency appears first in the younger leaves as an intervenous yellowing that later progresses over the entire leaf; in severe cases the leaves become almost white. Iron is essential in photosynthetic processes and also functions in several enzymatic reactions. The plant can absorb iron through its roots or leaves as either an ion (Fe^{2+}) or as a complex with organic salts (chelated).

Manganese

Manganese is similar to iron in many ways. It is also a heavy metal and rather immobile in the plant. Manganese exists in the soil in several forms, depending upon the soil environment. It is most available to plants if in the manganous state (Mn^{2+}, MnO) as an exchangeable cation in the soil. In this state, however, it is more subject to oxidation by microorganisms to its trivalent state (Mn^{3+}, Mn_2O_3); in well-aerated soils it can be further oxidized to its least soluble, four-valent state (Mn^{4+}, MnO_2). High soil pH and good oxidizing conditions encourage MnO_2 formation. Poorly aerated or

TABLE 9–4	Approximate Range of Micronutrient Deficiency and Toxicity in Stems and Leaves (Dry wt.)		
Microelement	*Deficiency (ppm)*	*Normal (ppm)*	*Toxicity (ppm)*
Boron	5–30	30–75	75
Copper	4	4–15	20
Manganese	15	15–100	depends on Fe: Mn ratio
Molybdenum	0.1	1–10	low toxicity
Zinc	8	15–50	200

waterlogged soils favor the reduced manganous (Mn^{2+}) state. Manganese is absorbed by the plant in the manganous ionic form (Mn^{2+}), and is often applied to plants as manganese sulfate ($MnSO_4$) or a chelated (complexed with some organic molecule) compound. Manganese may be applied as a foliar spray that is absorbed through the leaves.

Like iron, the first deficiency symptoms show intervenous chlorosis in the younger leaves. This nutrient participates in photosynthesis, activation of enzymes, carbohydrate metabolism, and phosphorylation. Manganese toxicity from large amounts of the micronutrient has been observed in cotton (crinkle leaf) and tobacco on highly acid soils. Liming the soil corrects the malady.

Molybdenum

Molybdenum is an essential nutrient, very noticeable when deficient in crops such as clovers, alfalfa, cereals, vegetables, soybeans, and forage grasses. A condition known as whiptail in cauliflower is caused by molybdenum deficiency. Soil environment affects the availability of this element to a large extent. It is unavailable to plants in strongly acid soil, where it reacts with iron and aluminum silicates to form insoluble compounds. Liming the soil usually increases the availability of molybdenum.

Phosphates seem to aid plants in the absorption of molybdenum, but sulfates tend to hinder its uptake. Symptoms of deficiency vary among crops but intervenous chlorosis is often the first observable symptom. Legumes are stunted and the leaves turn yellow, as with nitrogen deficiency.

Zinc

Soil characteristics influence zinc availability. In alkaline soils deficiencies are expected, and in strongly acid soils toxicity is possible. Deficiencies occur in a wide range of soils but are most frequent in calcareous soils high in phosphorus. Zinc deficiencies have been observed in deciduous and citrus fruits, vegetables, and field crops such as corn, cotton, sorghum, and legumes. Zinc was one of the first micronutrients to be recognized as essential. Zinc attracted scientific interest early because of its importance in human nutrition, and considerable research was conducted in an endeavor to increase the concentration in plants. Zinc primarily acts as an enzyme activator in both plants and animals.

Zinc can be absorbed by the roots from the soil as the exchange cation (Zn^{2+}), or through the leaves when it is sprayed on the foliage as a $ZnSO_4$ solution or chelated compound. Deficiency symptoms first appear in the younger leaves as intervenous yellowing. Later, reduced shoot growth becomes evident. The midrib and margins of corn leaves remain green while a broad band of bleached tissue appears from the base to the tip.

Chelating Agents

The word **chelate** derives from a Greek word meaning claw. A chelate is a large organic molecule that attracts and tightly holds specific cations like a chemical claw, preventing them from taking part in inorganic reactions but at the same time allowing them to be absorbed and used by plants. Chelates combine with metallic cations—iron, manganese, zinc, and copper—to prevent the cations from reacting with inorganic anions that would render them insoluble and unavailable to plants. For example, chelated iron cannot react with hydroxyl anions (OH^-) to form insoluble ferric hydroxide [$Fe_2(OH)_3$]. Chelated cations are more soluble at higher pH than are inorganic ions. Chelates are also known as **sequestering agents.**

Several important agricultural chelates or sequestering agents are commercially available: (1) **e**thylene-**d**iamine**te**tra**a**cetate, or EDTA, sequesters copper, iron, manganese, and zinc; (2) **e**thylene**d**iamine**d**i-o-**h**y-droxyphenyl**a**cetic acid, EDDHA, sequesters iron; (3) **d**iethylene**t**riamine**pen**ta**a**cetic acid, or DTPA, sequesters iron; **n**itrito**tri**a**c**etic acid, or NTA, sequesters zinc; and (4) **h**ydroxy**e**thyl**e**thylene**d**iamine**te**tra**a**cetic acid, or HEDTA, sequesters iron and zinc.

SOIL CONSERVATION

Soil degradation began long before people started farming, but the process was accelerated by permanent agriculture and land tillage. In America, erosion and soil depletion became problems as soon as settlers migrated from Europe, mainly because of the clearing of land for crop production. As population increased, more land was cleared, cropped, and depleted of nutrients. Much of the westward movement in the United States was a search for new, more fertile lands; Eastern soils had lost their productivity by continuous cropping. Under the Homestead Act, the United States government encouraged the movement by offering free land to those who would move west and settle.

George Washington and Patrick Henry were among the earliest American land conservationists. In his final message to Congress in 1796, Washington urged the creation of a board of agriculture. A half-century later Lincoln established the Department of Agriculture in 1862. Little interest in soil conservation was felt for the next 75 years because of the availability of new, western lands. During the 1920s a soil surveyor, H. H. Bennett,[5] called attention to the waste and depletion of America's greatest natural resource—land. Finally, in 1929, Congress established ten soil conservation experiment stations, and assigned personnel to study and gather information on erosion control measures.

The Soil Conservation Service (SCS) was established in 1933 in the Department of Interior as one means of helping the United States recover from the Great Depression of 1929 to 1935. The need for immediate soil conservation became evident on Black Sunday, April 14, 1935 because that

---

[5]H. H. Bennett was appointed chief of the first erosion control agency, which later became the Soil Conservation Service.

Figure 9–37   Two consecutive years of drought, followed by high winds, blew immeasurable amounts of fertile topsoil from areas of Texas and Oklahoma into the Atlantic Ocean. This catastrophe created what came to be known as the "Dust Bowl" on Black Sunday, April 14, 1935. The late afternoon sun, a circular ball above the auto, is barely visible through the dust cloud. *Source:* Library of Congress.

Figure 9–39   The beginning of a deep gully. Special care and treatment will be required to conserve this land. If it is left in grassland, however, further erosion can be prevented.

day the most severe dust storm in U.S. history completely blotted out the noonday sun (Fig. 9–37). During the summer and fall of 1935, the skies over Washington, D.C. and New York were darkened with topsoil blown from Texas, Oklahoma, and other prairie states. That same year, the U.S. government created the first erosion control agency ever established by any nation. The first soil conservation act soon followed, charging the agency with the responsibility of cooperating with farmers to demonstrate good land management and erosion control practices (Fig. 9–38).

Soil conservation is the preservation and extension of the life of soil by using land wisely, keeping it in its most productive state for the present and future generations. Lands best suited for grazing of animals are planted to sod crops. Hilly or mountainous land is kept in trees, which are harvested as timber. Plowing up grassland prairie soils and planting them to row crops in semiarid regions without irrigation has proven to be disastrous. They should be left as grasslands. It does not take long for wind or water to erode and remove the fertile topsoil and form gullies (Figs. 9–39 and 9–40). One of the complicating factors in soil conservation is our dependence upon the soil for food and fiber. The land must be used, but at the same time saved for future use. To do both takes wise land management.

Extent of Erosion

Recent soil surveys in several countries show that vast areas of productive land have been damaged beyond recovery. Erosion continues to be a critical concern in almost every agricultural region of the world. Erosion is particularly severe where intense torrential rains are frequent. Unfortunately, the need for food in some nations has overshadowed the danger of uncontrolled erosion. This is particularly true in East Africa, the Yellow River basin in China, Eastern Europe, Latin America, and parts of Australia, India, and the United States—all plagued with serious and widespread erosion. In the United States erosion has damaged nearly 110 million hectares (272 million acres). The soil erosion problem in these nations is not over the entire nation, but it is usually severe in hilly high rainfall areas and/or in the semiarid regions and especially in those areas where irrigation is not

Figure 9–38   Catastrophic soil erosion caused by drought and high winds occurred in the Texas panhandle during the early 1930s. Vast areas of previously productive farmland were devastated. *Source:* Library of Congress.

Figure 9–40   Rill erosion has begun in this orchard, a forerunner of severe gully formation if not checked soon. *Source:* USDA Soil Conservation Service.

used. At times when the demand for food crops (at home and abroad) is high, economic pressure is put on the farmer to seed land to grain or other row crops when the soil should remain as grassland. This contributes to erosion.

Factors Affecting Erosion

An important factor in erosion control is the amount of plant cover. Land covered with sod or trees loses little, if any, soil, while barren land can quickly lose considerable topsoil. The intensity, duration, and distribution of rainfall are also factors. A torrential rain of short duration on land with little plant cover causes severe soil losses while a gentle, evenly distributed rain causes less. Topography of the land is also a factor. Level land is less likely to erode than sloping land (Fig. 9–41). The soil's physical properties affect erosion. Deep permeable soils that absorb water are less likely to erode than shallow slowly permeable soils.

Gently sloping land can be cropped if proper erosion controls are used, but row crops that require tillage for weed control should never be planted in rows that run up and down steep hills. Row crops on gentle slopes require contoured rows. Sod crops or crops planted by broadcast methods should be used to reduce hillside erosion losses.

Methods of Conservation

The appropriate method of soil conservation depends upon the topography, soil type, cropping and livestock system, and climate. To help with these decisions, the U.S. farmer can call on the Soil Conservation Service (SCS). A professional conservationist will survey and classify the soil into one of eight broad land-capability classes according to its best use with least erosion (Table 9–5). The important consideration is that each parcel of land is managed according to its needs. This means that land not suitable for any type of agriculture, even

though unaffected by erosion, should be left for wildlife and recreation; forest land should be used to produce trees, range and grassland should be used to produce forage, and land suitable for cultivation should be reserved for crop production.

Each kind of farming needs its own special conservation practices, but even with careful land management, additional measures are often necessary to improve land use.

Grass Waterways

These are strips of land of varying width permanently seeded to a grass sod. They conduct water to drainage outlets and control runoff from sloping land with cultivated crops. Waterways are used with contours or terraces that drain into them.

Contour Tillage

One easy cropping practice that reduces losses of topsoil is to till the land on the contour (level elevation) instead of up and down the hill. The land is plowed and the crop rows planted and cultivated around the slope, always at the same elevation from end to end. The rows are curved and sometimes come together in points. The ridges left by the tillage tools form small dikes to catch water, allowing more time for it to percolate into the soil instead of running down the hill.

Contour Strip-Cropping

This effective practice is used to conserve both soil and water. Soil conservation is enhanced by alternating strips of solid-planted crops with row crops; for instance, strips of grain or hay crops can alternate with corn or sugar beets (Fig. 9–42). The strips always run on the contour. In some cases

Figure 9–41 Water does not erode level land. Leveling, as shown here, is one way of preventing erosion by water where irrigation is necessary. *Source:* USDA Soil Conservation Service.

Figure 9–42 Contour strip cropping is useful for conserving water as well as reducing soil losses by wind erosion. In this case a cereal grain crop has been alternately planted with a hay crop. Sometimes if the area is subject to sudden but infrequent torrential rains, permanent flat, broad terraces are formed on the contour. *Source:* R. L. Haaland.

TABLE 9–5	Land Capability Classes	
Class	**Use**	**Conservation Practices Needed**
I	Few limitations. Suitable for wide range of plants. Can be used for row crops, pasture, range, woodland, and wildlife. Not subject to overflow.	Needs ordinary management practices—fertilizer, lime, cover, or green-manure crops, conservation of crop residue, animal manures, and crop rotations.
II	Some limitations. Choice of crop plants reduced. With proper land management, land can be used for cultivated crops, pasture, range, woodland, or wildlife.	Limitations few and easy to apply. Problems may include gentle slopes, moderate susceptibility to wind or water erosion, less than ideal soil depth, slight salinity. May require special conservation practices, water-control devices, or tillage methods, terraces, strip cropping, contour tillage, special crop rotations, and cover crops.
III	Severe limitations reduce choice of plants and/or require special conservation practices. May be used for cultivated crops, pasture, range, woodland, or wildlife.	May require drainage and cropping systems that improve soil structure. Organic matter additions might be needed. In irrigated areas, soils may have high water table, high salinity, or sodic accumulations. Soils may be slowly permeable.
IV	Severe limitations that reduce choice of plants. Requires very special management. Limited use for cultivated crop but can be used for pasture, range, woodland, or wildlife.	Limited cultivated crops because of steep slopes, susceptibility to wind or water erosion, effects of past erosion, shallow soils, overflows, poor drainage, salinity, adverse climate. May be suited for orchards and ornamental trees and shrubs. Special practices need to prevent soil blowing and to conserve moisture.
V	Land limited in use—generally not suitable for cultivation. Little or no erosion hazard but has other limitations. Use limited to pasture, range, woodland, or wildlife.	May be nearly level but has excessive wetness, frequent overflow, rocks, or climate variations. Cultivation of common crop not feasible but pastures can be improved and benefits from proper management can be expected.
VI	Severe limitations make the land unsuitable for cultivation. Restricted to pasture, range, woodland, or wildlife.	Pastures can be improved by seeding, liming, fertilizing, water control with contour furrows, drainage ditches, etc. Have severe limitations that cannot be corrected, thus not suited for cultivated crops. Some soils can be used for such crops as sodded orchards, berries, etc.
VII	Severe limitations make land unsuited for cultivation. Use limited to grazing, woodland, or wildlife.	Physical condition of soils prevents range or pasture improvement practices. Restrictions are more severe than those of Class VI. Can be used for grazing. May be possible to seed some areas.
VIII	Limitations preclude use for commercial plant production. Use restricted to recreation, wildlife, or water supply.	Cannot be expected to yield any significant return from crops, grasses, trees, but benefits from wildlife use and watershed protection or recreation are possible. Class VIII includes badlands, rock outcrops, sand beaches, river wash, mine tailings, etc.

Source: USDA Handbook 210, 1973.

this practice reduces erosion by more than half of what it would be if either crop were planted alone.

Terraces

Terraces are used on long gentle slopes to decrease runoff and to increase water infiltration. On gently rolling land, terraces are low broad mounds that follow the contour and retain water that would otherwise run down the slope. The terraces are constructed with a slight grade so excess water will flow slowly to an outlet, often a grass waterway. Ter-

races are also used in some places on steep slopes (Fig. 9–43). They have been used for centuries in the Andes, China, and many other parts of the world where insufficient flat arable land is available. In Thailand, rice is grown on steep, terraced hillsides.

Wind Erosion

Wind erodes land by removing topsoil just as water does. As with water erosion, the best protection against wind erosion

Figure 9–43 Terrace cultivation of avocado trees in southern California. This method enables the use of land areas not normally suitable for cultivation. Usually, high value crops and favorable climatic environments are important factors in the decision to use the terrace method.

is to provide vegetative cover for the land during periods of high winds. Tillage methods such as stubble mulching have been helpful. Leaving the soil surface rough or cloddy reduces wind velocity at the soil surface, and windbreaks are helpful. These vary in size from tall trees to hedges planted close together perpendicular to the prevailing wind.

SUMMARY AND REVIEW

Crop productivity is closely tied to land preparation. One of the first steps in land preparation has traditionally been tilling the soil. Plowing to turn over the old plant material was followed by disking to break up the clods created by the plow. Harrowing even further reduced clod size and smoothed the soil surface. Cultivators are used between crop rows to loosen the soil surface and destroy weeds. Each step has special equipment designed for that step. There are variations of each equipment type that can be used for specific field conditions.

Tillage equipment is heavy. After several years, heavily tilled soils become compacted and crop productivity decreases. As a result, methods of minimum-till and no-till soil preparation have been developed and have become commonly used in many parts of the United States. With minimum- and no-till, herbicides are often used to kill weeds, then the seeds are sown in shallow furrows or directly on the surface of the soils. The soil remains relatively undisturbed, which may allow the condition of the soil to improve with time.

Many areas that have soils suitable for growing crops lack the necessary or timely rainfall required. Irrigation, although an added cost, is necessary to utilize these production areas. There are several methods of application. The border or flood method requires grading the land flat but with a slight slope from one end to the other. Water is pumped or diverted into the field where it flows along the slope and is kept in bounds by small levees constructed in and around the field. Furrow irrigation is similar but the water is channeled down furrows between crop rows. Sprinkler irrigation is used when the flood or furrow method is impractical. Water is supplied through sprinkler heads located above the crop. Sprinkler systems can be portable (hand-set), stationary (per-

manent set) or mobile (wheel line, center pivot). Drip or trickle irrigation supplies water to precise areas close to each plant or small group of plants through emitters.

Sixteen minerals are required (essential) for the growth of most plants (C, H, O, P, K, N, S, Ca, Fe, Mg, B, Mn, Cu, Zn, Mo, Cl). C, H, and O are supplied by atmosphere and water; the others are minerals in the soil. The minerals are divided into categories based on the amount required by the plant (all are essential). The macronutrients (N, P, K) are required in the greatest amount.

Secondary minerals (Ca, Mg, S) are needed in somewhat smaller quantities than the macronutrients. Micronutrients (the remainder) are required in small or trace amounts.

If the soil does not contain these minerals in sufficient quantities to grow crops productively, then the minerals must be added in a fertilization program. Minerals can be added by applying crop residues, animal manures, chemical fertilizers, or naturally occurring minerals. Chemical fertilizers are the most precise form of adding necessary minerals, but the other forms can be used successfully. To use any method of applying minerals, the mineral content of the fertilizer and the form of the mineral must be known. Commercially prepared fertilizers give the percentage of macronutrients in the mix and may give the percentage of secondary or minor elements. Regular and consistent soil or growing media testing is required in all fertilization programs.

Nitrogen must be mineralized by microorganisms to be available to plants. Mineralized forms of nitrogen include NO_3^- and NH_4^+. NO_3^- is an anion so it is not readily held by soil. It can be easily leached away from the root zone. Nitrogen can be supplied as a solid, liquid, or gas. Phosphorus is supplied by fertilizers containing superphosphate (P_2O_5)

or by injecting phosphoric acid (H_3PO_4) into the irrigation water. It is taken up by the plant as $H_2PO_4^+$. Potassium is found in many different types of rocks and there are several chemical formulations of the element that can be used for fertilizers. It is taken up by the plant as K^+. Calcium, magnesium, and sulfur can be supplied by adding $CaCO_3$, $CaSO_4$, $CaCO_3 \cdot MgCO_3$, and/or $MgSO_4$. Other essential elements can be added by using the appropriate chemical.

Soil conservation is the preservation and extension of the productive life of the soil for present and future generations. Most soil conservation efforts are directed at reducing soil erosion. Water erosion is controlled by the use of grass waterways, contour tillage, contour strip-cropping, and terracing. Wind erosion is controlled by the use of windbreaks. Both water and wind erosion are reduced by the use of minimum-tillage and no-tillage cultivation.

EXERCISES FOR UNDERSTANDING

9–1. Talk with a local farmer, golf course superintendent, or greenhouse or nursery manager. Find out how they prepare their fields or courses, or how they prepare the growing media they use. What factors do they consider as they prepare the soil or media?

9–2. In the same conversation as Exercise 9–1 find out what fertilization program is used and what criteria are used when setting up a fertilization program.

9–3. Using the label from a commercial brand of fertilizer available in feed and seed stores, garden centers, or home and garden sections of many stores, make a list of all the nutritional components. Determine if all the essential minerals not supplied by water and atmosphere (13 minerals) are supplied in the fertilizer. If any are missing, what nutritional deficiency symptom(s) would you expect to see in plants receiving this formulation?

9–4. When driving past cultivated fields, road cuts, bridge embankments, or anywhere the land has been disturbed by humans, identify the methods (if any) used to reduce or prevent erosion. Note if the methods have successfully controlled erosion. If not, what would you suggest?

REFERENCES AND SUPPLEMENTARY READING

BROOKS, K. N., P. F. Ffolliott, H. M. Gregersen, and L. F. DeBano. 1997. *Hydrology and the management of watersheds.* Second edition. Ames, Iowa: Iowa State University Press.

HUDSON, N. 1995. *Soil conservation.* Third Edition. Ames, Iowa: Iowa State University Press.

MARSCHNER, H. 1995. *Mineral nutrition of higher plants.* Second Edition. New York: Academic Press.

REED, D. W. (ed.). 1996. *Water, media, and nutrition for greenhouse crops.* Batavia, Ill.: Ball Publishing.

CHAPTER

Climatic Influences on Crop Production

KEY LEARNING CONCEPTS

After reading the chapter you should be able to:

◆ Know the difference between weather and climate.
◆ Know what determines the climate of an area.
◆ Know the major climatic factors and how those factors affect plant development and health.
◆ Know how those factors may be controlled or manipulated to affect crop productivity.

History shows that early people continually moved about, exploring and settling new areas, often great distances apart. As they moved they took with them the seeds, cuttings, and nursery plants of familiar crops for planting in their new homeland. Many introduced plant species succeeded in their new environment; some, for various reasons, did not. Sometimes these introductions formed the basis of new plant-growing enterprises. For example, coffee, bananas, macadamia nuts, papayas, and pineapples were introduced to the Hawaiian Islands from other tropical regions by voyagers during the 1800s.

The multitude of plant species and cultivars growing throughout the world differ genetically, each reacting to various environmental situations. To grow crops successfully the farmer must (1) grow crop plants already known to be adapted to the climate, (2) grow plants altered genetically (new cultivars) for adaptation to a different climate, or (3) change the climatic environment by artificial means (greenhouses, hothouses, and other devices).

It has been a worldwide challenge for growers to correlate climatic environments with crops that grow best in specific areas (Fig. 10–1). Farmers, agribusiness enterprises, and governments are continually looking for new crops that can be adapted profitably to their particular environment. For example, annual world citrus production has increased more than thirtyfold during the past 100 years. This increase was due primarily to the development by governments and private enterprise of vast new citrus plantings in many parts of the world, especially in the Americas, Africa, and Australia, where favorable climates are present. Safflower (*Carthamus tinctorius*) and sunflower (*Helianthus annuus*) are also examples of crops successfully introduced into new areas of production. Triticale, a cross between wheat and rye, is an example of an entirely new, genetically altered plant being introduced into certain arid and cooler regions primarily because of its wider climatic adaptability.

Figure 10–1 Meteorological data collection station receiving maintenance. Stations such as this one provide valuable timely weather as well as historical climatic information representative of the area where the station is located. Daily temperature, solar radiation, precipitation, wind velocity and direction, and evapotranspiration measurement are some of the data obtained. In conjunction with data from other stations, predictive information can be provided. For example, if a frost is predicted, fruit producers may want to begin frost protection measures. *Source:* University of California Cooperative Extension.

Matching crop to climate has been by trial and error in the past. Better knowledge of the climatic and growth requirements of the various plant species and cultivars permits better prediction of the feasible production of certain crops in different areas.

Plant breeders develop new cultivars that are productive in regions previously less suited to the crop. For example, peaches are not generally suited to subtropical areas because they require considerable winter chilling to overcome the "rest" influence in the buds. However, plant breeders have developed peach cultivars that do not require as much cold, thus extending peach culture to warmer-winter regions.

WEATHER AND CLIMATE

Weather is the immediate day-to-day, local (inplace) combination of such natural phenomena as temperature, precipitation, light intensity and duration, wind direction and velocity, and relative humidity. In any given location these weather factors assume a certain pattern, changing day by day, week by week, month by month, and season to season, and the same pattern repeats year by year. This pattern is the location's **climate.** If the area is small or near the ground surface, the pattern is called the **microclimate.** Thus, climate represents the summation of many decades of weather in a given area. Past climatic records allow one to predict with some accuracy the weather of a given area for a certain time of year.

An area's climate is determined by various factors. For example, climate will change with changes in altitude. This is demonstrated by the delayed flowering of about four days for some plant species for each 130 m (400 ft) increase in altitude. For each degree of latitude, moving from the equator, flowering of the same plant species is delayed about four days (Fig. 10–2). There is also a reduction of about 10°C (18°F) in the average dry-air temperature for each 1,000 m (3,280 ft) rise in elevation. The temperature reduction in moist air is less, about 6°C (11°F) for each 1,000 m.

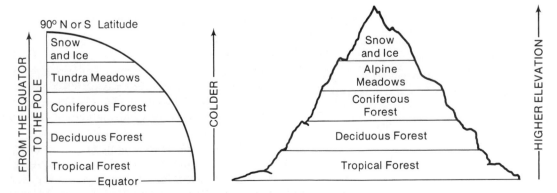

Figure 10–2 Vegetation types from equator to North or South Pole (*left*) are similar to those from the base to the top of a high mountain (*right*) in tropical regions. The principal factor involved in this comparison is temperature.

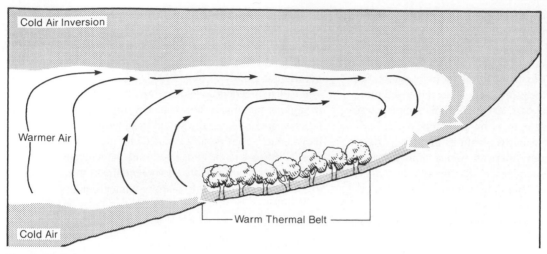

Figure 10–3   Cold air, being more dense, drains down the slope and sinks beneath the warmer air, pushing it upward. As the warmer air rises, it reaches a colder layer. At the higher altitude, the cold layer flows downward, continuing the convectional flow of air near the slope. As a result of air convection, warm thermal belts can occur on slopes, permitting the culture of frost-susceptible crops in places otherwise too cold.

Large bodies of water tend to moderate temperature extremes of the nearby land mass during both winter and summer. Hills or mountains modify precipitation in some areas by increasing rainfall on the windward side and reducing it on the leeward side. Occasionally, the climate of long, narrow bands of 1 to 20 km (0.62 to 12 mi) wide across gently sloping hillsides is quite different from that in the bottom of the valley. Because cold air is more dense than warm air, the cold air drains to the valley floor, leaving the higher band of land warmer. Sometimes orchards are quite productive in these higher thermal belts, whereas fruit trees on the valley floor would be subjected to accumulations of cold air and possible blossom-killing frosts on spring mornings (Fig. 10–3).

Each plant species has certain climatic requirements for optimum growth, and the success or failure of the crop is based on how well the climatic requirements of that species or cultivar are met. A successful farmer makes a thorough study of the climate and weather for the area before planting any crop. This reduces the risk of crop damage.

Although the climate for a given region is established and well-known, extreme variations in weather patterns can eliminate a crop even though, on the average, such problems are unlikely. For example, a spring freeze during blossoming could eliminate the crop of an entire vineyard or orchard. This might not occur again for years but if it happened too frequently, such areas would become too risky for fruit growing, and a better adapted crop should be substituted.

CLIMATIC FACTORS AFFECTING PLANT GROWTH

The world's distribution of crop plants is determined largely by climatic factors such as temperature, rainfall, light, and air movement.

Temperature

Each chemical, physiological, and biological process in plants is influenced by temperature. Plants of different species vary quite widely in their tolerance of temperature extremes, although within a species, they are restricted to rather narrow limits. Some algae survive temperatures of 90°C (194°F) in hot springs and certain arctic plants survive −65°C (−85°F), but such examples are rare. Most plants live and grow within a temperature range of 0°C to 50°C (32°F to 122°F). Biological activities, in general, are limited at the lower temperature by the freezing point of water and at the upper range by protein denaturation.[1]

In some instances, small temperature differences can change the quality of the harvested product. For example, the sugar content of sugar beets is reduced if temperatures during the growing season are too low or the length of the growing season is shortened to the point where there are too few days of a daily mean temperature high enough for sugar production. On the other hand, high night temperatures interfere with sugar accumulation and storage in the beets.

Minimum Temperatures

For each plant species there is a minimum temperature that kills plants outright. From experience and experimental freezing trials, this temperature is known for most crops. Temperature sensitivity prevents the growing of certain species in many regions (Fig. 10–4). For example, oranges cannot be grown commercially in Canada because the trees would be killed by winter temperatures below −7°C

---

[1]The modification or change of a natural protein by heat, alkali, acid, or radiation so that it no longer performs its original function.

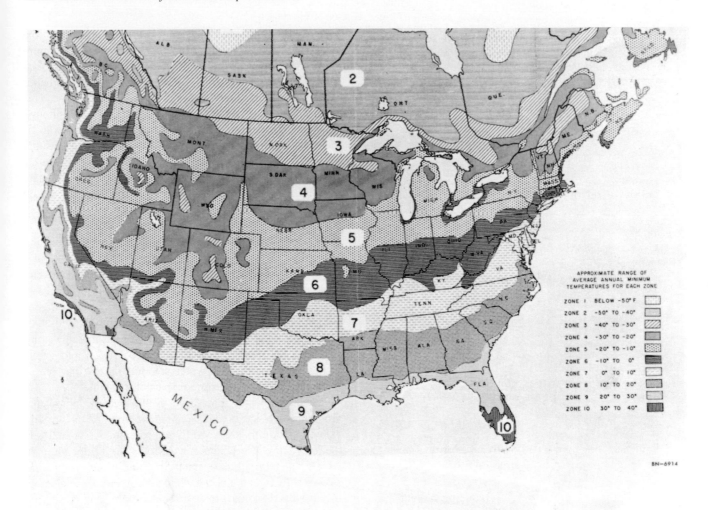

Figure 10–4 Plant hardiness zones of the United States and southern Canada, showing average annual minimum temperatures. *Source:* USDA, ARE. 1960. Plant hardiness zone map. Misc. Publ. 814.

(19°F). In other regions, winters with such temperatures are rare and worth the risk. Unfortunately, such conditions do occur, as in southern Texas in 1951, when thousands of orange trees were killed, and similarly in Florida in 1977 and 1985. With other fruit crops, low temperatures at flowering interfere with reliable production. Growers often take precautions against such occurrences and use frost protection measures.

Low temperatures cause plant damage varying from death of the entire plant to partial tissue damage. The physiological status of tissue at the time of exposure to low temperature is also important. At the same temperature, actively growing succulent tissues, such as flower buds, are usually damaged more severely than other tissue that may be hardened or dormant.

The plant's minimum lethal temperature can be modified by the length of exposure time. For example, a plant with a lethal minimum temperature of −2°C (28°F) can be exposed to that temperature for a few minutes without damage. However, several hours at the same temperature could cause severe damage.

Sometimes, plant tissues are exposed to temperatures several degrees below their freezing points very rapidly—a condition called supercooling. However, if warmed gradually without disturbance, injury is avoided since no ice crystals form. However, if the plant is moved by wind or shaken while supercooled, ice crystals quickly form and in doing so damage the tissue.

Early Frosts

Each year early frosts or freezes cause concern to farmers in the world's temperate zones. One clear, cold night can nullify the effort and expense of producing a crop if freezing occurs at a critical time in the plant's development. Flower buds, beginning from the time they start to open, are generally vulnerable to cold (Table 10–1).

TABLE 10–1 Relative Resistance of Fruit Buds to Cold. Temperatures Endured for 30 Minutes or Less Without Injury

Fruit	Buds Closed But Showing Some Color		Full Bloom		Small Green Fruits	
	°C	°F	°C	°F	°C	°F
Apples (*Malus pumila*)	−4	25	−2	28	−2	28
Peaches (*Prunus persica*)	−4	25	−3	27	−1	30
Cherries (*Prunus avium*)	−4	25	−2	28	−1	30
Pears (*Pyrus communis*)	−4	25	−2	28	−1	30
Plums (*Prunus americana*)	−4	25	−2	28	−1	30
Apricots (*Prunus armeniaca*)	−4	25	−2	28	−1	31
English walnuts (*Juglans regia*)	−1	27	−3	27	−1	30
Oranges (*Citrus sinensis*)	−3	27	−3	27	−1	30

Source: Taken in part from N. W. Ross. 1974. Stanislaus orchard handbook. University of California Cooperative Extension.

Some seedlings are killed by frosts after they emerge from the soil; this is especially true of warm-season vegetable crops. A gentle wind or cloud or fog cover reduces the danger of dropping temperatures by reducing radiant heat loss.

Low Temperature Requirements

Low temperatures are sometimes required to initiate flower bud as well as vegetative growth. For example, deciduous fruit trees develop a physiological condition ("rest period") toward the end of the summer, after which their obvious growth is arrested until they are subjected to a certain amount of chilling through the winter. This phenomenon limits the culture of some crops to the temperate zones, where adequate cold temperatures occur. The cold temperature requirement is also the reason why some ornamental bulbs must be placed in cold storage during the winter before planting the following spring. Rhubarb and strawberry transplants require a low temperature treatment for subsequent vigorous growth and/or fruiting. Some winter wheat cultivars require a cold treatment (vernalization) during seed germination and early seedling development in order to flower and produce grain during the summer. Many biennial plants require a certain amount of winter chilling to flower.

Freezing Damage

Many plants can tolerate freezing temperatures. Metabolic responses to low temperatures change the physiology of the tolerant plants, a process known as **hardening.** These phys-

iological responses to cold temperature are changes in the cell solute concentration, cell membrane permeability, or enzyme activity. Plants are hardened by exposure to increasingly severe environments, following which they can better survive conditions that would otherwise cause damage. This practice has been known and used for many years, although a simple explanation of hardening physiology is rather difficult. Pine trees undergo this process during normal seasonal change in order to survive cold and freezing temperatures. Examples of easily hardened plants are tomatoes, cabbage, peppers, wheat, and many deciduous and evergreen trees and shrubs.

How Cells Freeze   At temperatures below the freezing point of water, ice crystals form in the cells and intercellular spaces. The water, in turning into ice, expands and thereby causes rupturing of the cells. Actually, this causes relatively less damage in comparison to physical changes in cell contents caused by water movement and loss from the cells.

At standard atmospheric pressure, the freezing point of pure water or melting point of ice is 0°C (32°F). Ice melts at this temperature, but water seldom freezes. For ice to form there must be microscopic particles of either ice or materials in the water that serve as the nucleus or "seed," around which ice crystals can grow. Ice crystal growth, called freezing, is preceded by microscopic ice crystal formation, a process called nucleation. The temperature at which water (a solvent) exists in equilibrium with ice (a solute) is its freezing point. The addition of a substance being dissolved (solute) to a substance doing the dissolving (solvent) always causes a lowering of the freezing point. The higher the concentration of solutes in a solvent, the more depressed the freezing point becomes. In nature, absolutely pure water does not exist, so varying degrees of supercooling occur in all biological systems before ice crystals form. The degree of supercooling depends upon the concentration of solutes, colloids, and other material in the cell sap. Once ice crystal formation begins in the plant's cells and water changes from liquid to solid, the temperature in the plant part increases because of the release of the latent heat of fusion[2] as water changes to ice (Fig. 10–5).

After ice has formed in the intercellular spaces, water molecules move from the cell toward the regions of ice crystal formation because of a reduced water potential in intercellular regions. The removal of liquid water by freezing increases the solute concentration within the cells and lowers the freezing point even further. Under conditions of severe and quick freezing of unhardened plants, tissue is usually killed by intracellular ice formation within the cell protoplasts. Rapid thawing is also harmful to plant tissue because sudden changes in cell turgor redistribute the water.

In addition to below-freezing damage, some plants are injured by exposure to short periods of cold at above-

---

[2]The latent heat of fusion for water is about 80 cal/g.

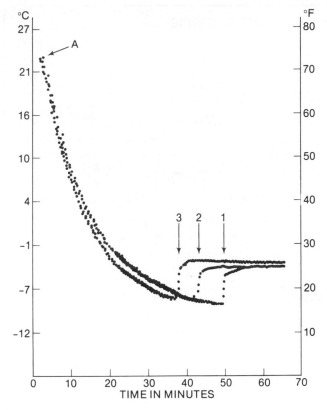

Figure 10–5 Thermograph readings showing supercooling temperature patterns of fruits. Thermocouples were inserted into three separate raw olive (*Olea europaea*) fruits (1, 2, and 3) which, at point A, were placed in a cold box at a temperature of −12°C (10°F). Fruit temperatures dropped steadily for 35 to 50 minutes, supercooling to about −8.5°C (17°F). At this point the fruits started to freeze. As ice crystals formed, fruit temperatures rose abruptly to the freezing point for olives, as shown to be about −4°C (25°F).

freezing temperatures. The fruit tissue of tropical and subtropical fruits, such as pineapples, avocados, bananas, papayas, and tomatoes, is injured during storage at temperatures of 0°C to 10°C (32°F to 50°F). Injury of this type is called **chilling injury**.

Avoiding Crop Losses Due to Frost

Each year frost causes considerable crop losses. Damage due to frost is most likely from unexpected cold periods in the spring, after young crop seedings have emerged or flower buds opened. The risk is greatest during the hours just before sunrise on clear, still nights. The duration and magnitude of temperature drop is most critical at this time. If cloud cover is absent, heat is radiated from the surface into the upper atmosphere during the night. Radiation losses are greater and more rapid from clean-tilled fields than from those covered with crops. Crop losses due to frost can be avoided by planting vegetable crops when the danger of frost has passed. However, this is not always possible nor even desirable. For example, extra-early harvest can bring such a price advantage that it is worth risking damage by frost or bearing the added cost of frost protection. Many fruit, grape, and nut

growers are familiar with frost protection practices. These farmers are often forced to minimize their risks by protecting their crops against losses from freezing. Producers of ornamental potted plants prevent losses by using heated greenhouses, and some grain farmers avoid damage by planting frost-tolerant crops such as winter wheat.

Various techniques and equipment are used to minimize damage to crops when temperatures fall below freezing, as discussed below.

Planting Dates Working with worldwide long-range weather patterns, meteorologists have determined the average last day for early and late frosts in many areas (Fig. 10–6). Planting crops within these dates decreases the likelihood of frost damage.

Hot Caps Vegetable growers sometimes grow crops at temperatures less than optimum in order to harvest earlier or off-season crops. Hot caps are simple but effective devices to capture sunlight, provide heat for accelerated plant growth, and provide some light frost protection, down to −3°C (27°F). These domelike structures (Fig. 10–7) are made of moisture-resistant paper supported with wire strands to retain their shape. They are usually held in place by soil covering their base. As the seedlings grow, the tops are torn open and are removed after the danger of frost has passed. Their use can provide an advantage of two to three weeks earlier production. Hot caps have become expensive and do require considerable labor; therefore, they generally are used with high-value crops such as early cucumber, pepper, summer squash, and tomatoes.

Brushing Brushing is sometimes used as a means to increase temperatures near the plant and thereby to increase plant growth. The term brushing was derived from the use of various kinds of riverbank brush in the construction of the fencelike barrier. As presently done, heavy kraft paper substitutes for the brush material. Crop rows are oriented east-west. The barrier is placed on the north side of the row slanting about 45 deg. toward the south so that sunlight can heat the plant area. The plants are grown at the base and under the extended barrier. The barrier's height, about 50 cm (20 in.), helps in reflecting sunlight toward the soil, and to limit radiation of the captured heat to the atmosphere during the night. Protection against wind is another benefit. Air temperature under the barrier during the day can be as much as 3°C to 6°C (6°F to 11°F) warmer than outside. Brushing, like hot caps, is usually limited for use with high-value crops (Fig. 10–8).

Polyethylene Tunnels, Row Covers, and Mulches An innovative method of increasing the temperature of the plants' microclimate and offering some frost protection is to grow plants in tunnels and row covers of transparent polyethylene sheeting. The tunnels are constructed from thin (1.5 mil)[3]

---

[3]A unit of length. 1 mil = 0.0254 mm (1/1000 in.).

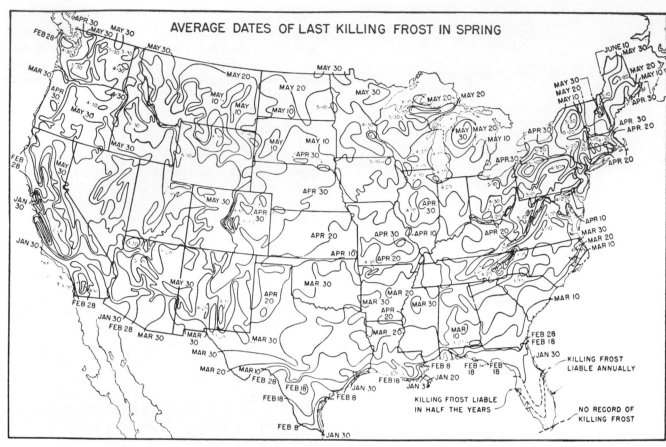

Figure 10–6 Average dates over a 40-year period of last killing frost in spring. The risk of frost damage is minimized if susceptible crops are planted after the average date of the last killing frost. *Source:* USDA Yearbook. 1994. Climate and Man.

Figure 10–7 A field of young pole tomatoes being grown for early fresh market production using waxed paper hot caps. The hot caps serve as a heat trap, capturing the sun's heat within the hot cap environment to accelerate early plant development as well as providing some frost protection. Plant growth has advanced to where it was necessary to ventilate the caps by tearing open the tops. After a little more growth, their benefit is minimal and the caps will be removed.

Figure 10–8 Young summer squash plants growing near a barrier made from heavy brown kraft paper that protects the plants from prevailing cold winds. However, the major benefit is the temperature increase of the air and soil near the plants because of the heat trapping function of the barrier. The barrier is erected to be inclined toward the south and the sun's radiation. The technique is called *brushing*.

continuous polyethylene sheeting of various widths usually supported over the plant rows (Figs. 10–9 and 10–10). Materials other than polyethylene are also used (Fig. 10–11). To prevent removal by wind, the covering is held down with soil

placed along the edges. Such tunnels resemble miniature greenhouses. Temperatures under the covering can be very high, and ventilation may be necessary. Various means of venting are used. This can be done by having perforations or

Figure 10–9 Clear polyethylene tunnels used for fresh-market tomato plants to enhance growing temperatures and for protection from low temperatures. The tops of the tunnels are closed during the night and, if necessary, are opened during the daytime.

other openings in the covering, or the cover may be partially opened or removed.

Various materials are used as a soil mulch to capture heat and assist plant growth. These are applied directly to the soil; plastic film is commonly used. Transparent films permit greater heat accumulation but also present problems, since weed growth as well as crop growth is enhanced. An opaque film reduces this problem (Fig. 10–12). Being labor-intensive and costly, these practices often are restricted to high-value crops. Where there is interference with rainfall, it may be necessary to use furrow or drip irrigation to provide adequate and uniform soil moisture. The removal and disposal of these products is a major concern, which could be reduced by the use of biodegradable materials.

Wind Machines and Heaters On clear, calm nights the soil radiates heat into the atmosphere, leaving cold air to accumulate over the surface. This condition, with the cold air underlying warmer air, is called an air inversion. A strong inversion would be one where the temperature differences between the cold and warm air layers were large. With a weak inversion, the temperature difference is less. Frosts developing under these conditions are called radiation frosts. Normal daytime heating will break up a weak inversion as convection movement of the heated air results in a mixing of the warm and cold layers. With a strong inversion, it may be necessary to use other means to achieve a mixing of the air masses. Wind machines are used to bring about rapid and thorough mixing of the air layers, thereby changing the average temperature of the air near the surface. Heaters of various types are also used to increase the temperature of the air near the surface. When used in conjunction with various types of heating, the wind machines are even more effective in reducing frost danger. The number of wind machines and/or heaters to be used is determined by the amount of temperature change desired.

Wind machines are essentially huge fans mounted on platforms or poles above the trees or vines and powered by

Figure 10–10 Cucumbers growing under clear polyethylene tunnels benefit from increased temperatures resulting within this environment. This cultural practice can accelerate the development of crop plants for earlier marketing. When temperatures are too high, these tunnels can be ventilated by opening the tops or sides or through perforations made in the polyethylene film.

Figure 10–11 Various nonspun fabric materials are used for early crop growth because of the increased temperatures and increased growth occurring beneath the covers. The row covers, although supported with wire hoops in this illustration, can be laid directly onto some crop plants and will stretch somewhat with continuing plant growth. These materials are also useful in excluding some insect pests.

Figure 10–12   Opaque film mulch laid directly onto bed surfaces. The two dark bands on the aluminum-colored film are the plant rows for the squash or pepper transplants to be planted soon. The aluminum color is effective in repelling virus vectoring aphids.

Figure 10–13   This orchard is equipped with large wind machines. The machines mix the cold air near the surface with warmer air above to increase air temperatures and give some frost protection. More efficient and economical machines (*left*) are replacing less efficient machines (*right*) that use surplus aircraft engines and propellers. If the temperature decreases below −1°C (30°F), oil burning radiant heaters can be used along with the wind machines to provide warmer air temperatures. *Source:* James F. Thompson.

Figure 10–14   Oil-burning heaters are sometimes used in orchards or vineyards to aid in frost protection. These heaters with a side return arm recycle the smoke. The result is an intense, practically smokeless heat. *Source:* James F. Thompson.

engines or electric motors of various sizes (Fig. 10–13). In earlier periods, surplus airplane engines with propellers were used as wind machines. Also, all manner of combustible materials (old rubber tires) were used for heating. Present concerns about air pollution have greatly restricted such burning.

When freezing temperatures are predicted, farmers, particularly in fruit production areas, are alerted by weather and local radio stations if it will be necessary to operate frost protection equipment. As the temperature approaches 0°C (32°F), growers ignite heaters and/or activate wind machines, leaving them on until the temperature rises about freezing (Fig. 10–14).

Sprinklers   Sprinkler-applied water can reduce frost damage to some crops because water releases its latent heat of fusion upon freezing (changing from liquid to ice). Additional heat is available for frost prevention if the applied water is warmer than 0°C (32°F) (1 cal/degree/g water). Heat from these two sources (latent heat of fusion and heat lost upon cooling to freezing temperature) warms the developing buds and other tissues on grapevines, fruit trees, or strawberry plants as ice forms. Moreover, the ice that forms insulates the buds from air temperatures below freezing. This practice often provides sufficient heat and insulation to reduce frost damage, but continuous sprinkling is needed during the danger period. If water is discontinued, the temperature of the tissue can drop to the danger point. On the other hand, excess water should be avoided to prevent a heavy ice buildup that can damage trees or vines by breaking the branches. Excess water can also cause flooding or saturated soils.

Furrow Irrigation Occasionally a crop can be protected from frost damage by furrow irrigating the field. Applying warm water warms both the soil and the plants growing in it. The latent heat of fusion released by water changing from the liquid to the solid state yields further calories. Furrow irrigation can be used on row crops, but only once or twice before the field becomes excessively wet. Also, the freezing weather must be anticipated well in advance to allow time to apply water into the furrows.

Rainfall

Topography greatly influences the amount and distribution of rainfall. Air circulation patterns also affect the seasonal distribution of precipitation. It is common for certain valley areas to receive 100 cm (40 in.) of rainfall while 50 km (31 mi) away only 15 cm (6 in.) of precipitation falls. Mountain ranges present barriers to clouds, causing them to rise to higher elevations and generally colder temperatures. This causes the vapor to condense and water to fall on the windward sides as the clouds pass over, leaving the leeward side relatively dry. An area receiving adequate rainfall during two-thirds of the growing season generally would have enough water for many crops. Otherwise, some irrigation will be needed to supplement the rainfall. Areas receiving less than two-thirds of the required water supply during the growing season usually require irrigation. In such areas, dams, aqueducts, canals, or groundwater wells are used to supply supplemental water.

Moisture influences the growth and distribution of plants because it is essential in every biological reaction within the plant. Water is the most abundant constituent in plants, ranging from about 75 to 95 percent of a plant's mass by weight. The role of water in plants includes: (1) involvement in all biological reactions, (2) a structural component in the proteins and nucleic acids in the plant cells, and (3) a regulator of plant temperature. In addition to its in-plant role, water also acts as an environmental regulator of the climate around the plant. Moisture is a major factor in determining climate.

Plants are divided into three categories based upon their need for moisture.

Desert plants, or xerophytes, have remarkable adaptation mechanisms enabling them to function with relatively less water. These plants, such as the cacti, often have very shallow and fibrous root systems that can act as sponges and readily absorb even slight amounts of rainfall falling on the soil surface. Their leaves have been modified to reduce water loss by transpiration. Their stems are often covered with a thick, waxy, resinous material, or they are pubescent (stems and leaves covered with fine hairs), which reduces transpiration. The stems of some xerophytes can store large quantities of water for long periods of time.

At the opposite end of the scale are the hydrophytes, the "fish" of the plant world. These plants thrive in or close to water. Water lilies, swamp and marsh plants are examples of hydrophytes.

Between these two extremes are the mesophytes, the most populous of the three groups. These include most of the economically important agricultural plants; some can adapt to quite diverse environments.

Humidity is a major force in regulating the earth's temperature. Moisture in the atmosphere intercepts and filters some of the solar radiation before it reaches the earth's surface. Where and when it blankets the earth in the form of fog or clouds, it prevents heat losses by reradiation. The physical properties of water, such as the latent heat of vaporization, make it an excellent climatic insulator, thereby reducing wide and extreme temperature variations. Another property, the latent heat of fusion, provides for the release of energy, which can be useful in reducing frost damage.

Light

Quality, intensity, and duration are important properties of light as they influence plants. Quality refers to the spectrum of light, or the wavelength of its photons. Specific wavelengths interact with various plant pigments to elicit certain plant responses. For example, the orange-red and the blue portions of the spectrum are most active in photosynthesis; red and far-red influence photomorphogenic responses such as stem elongation, flowering, and seed germination. Light quality in field crops is influenced by spacing and row orientation.

Intensity refers to the amount or brightness of light. Plants are generally spaced or oriented so that maximum leaf area is exposed to light. Leaf angle to the incoming radiation is important in light absorption. Some plants, such as sugarcane, benefit from high light intensity; others, such as coffee, grow better with lower intensities. Many ornamental house plants require subdued light to survive. Intensity during the daily cycle will vary, and is affected by many factors. However, it can be modified; that is, reduced by shading or increased by supplemental lighting. Duration refers to the periodicity of light, its absence or presence during a 24-hour period, and is often referred to as daylength. Daylength determines plant flowering and other processes in many species. Different cultivars of the same species of soybeans or onions can differ as to daylength requirements. The length of day changes slightly each day throughout the year and varies according to location. Although usually not directly controlled in field plantings, effect of daylength can be modified by: (1) altering crop distribution by planting crops in areas having desirable daylengths; (2) planting during seasons of favorable daylengths; (3) planting cultivars having little or no daylength response; and (4) artificially controlling daylength with either blackout shading and/or supplemental lighting periods.

Length of Growing Season

Plants require specific growing season lengths to complete their life cycles. These requirements vary widely with different species. For example, red raspberries (*Rubus idaeus*) are satisfactorily produced in Scotland's short, cool growing season while, at the other extreme, sugarcane (*Saccharum officinarum*) requires the long, hot, humid growing seasons of the tropical and subtropical regions. Other plant species can grow and produce well over a wide range of temperatures and length of season. Barley (*Hordeum vulgare*), for example, can be grown satisfactorily in temperature zones ranging from the subarctic to the subtropic.

Air Movement

Basically the air circulation patterns in the atmosphere result from the sun's radiation impinging more directly on some portions of the earth's surface (tropical regions) and much less on others (polar regions). The warmer air at the equatorial regions rises and flows toward the poles, cools, sinks as cold polar air, and then returns toward the equator as ground flow. The direction of the ground airflow is affected by several factors: (1) the earth's rotational spin from west to east, (2) the effects of seasons caused by the earth's inclination on its axis, (3) differences in heating and cooling between land and water masses, (4) differences in elevation, (5) effects of mountain ranges and valleys, and (6) local storms resulting from interactions between warm and cold air masses. The final result of these interactions is the establishment of regions, some large, some small, each with a different climate. Sometimes areas with greatly different climatic patterns lie only short distances apart.

Crop yields can be affected by air movement. Gentle winds may benefit crops by replenishing CO_2 to leaves in densely planted crops. On the other hand, crop yield can be impaired by strong winds at critical times in the crop's production cycle, such as the blooming period when bee activity is essential for pollination. Plant leaves or fruits are tender during the early stages of growth, and whipping winds can cause severe injury or desiccation. Continuous winds accompanied by high temperatures during the growing season markedly increase water losses by transpiration. Tall growing trees are often planted in rows perpendicular to the prevailing wind direction to reduce wind velocity and to make an otherwise satisfactory area suitable for growing crops. For example, in Hawaii tall Norfolk Island pines (*Araucaria heterophylla*) are planted in rows to protect macadamia nut plantings and, in California, tall eucalyptus trees (*Eucalyptus* spp.) serve as windbreaks in the central coastal valleys. Man-made barriers are sometimes erected to protect plants from strong and/or prevailing winds and rainfall. In greenhouses, where air movement is limited, circulating fans are installed to increase airflow. The increased airflow reduces the incidence of disease and improves photosynthetic efficiency by refreshing the CO_2 at the surface of the leaves.

CLIMATIC REQUIREMENTS OF SOME CROP PLANTS

Fruit and Nut Crops

Fruit and nut species are climatically adapted to the tropical, subtropical, or temperate zones.

Tropical

Plants in the tropical group do not withstand freezing temperatures, and many do not grow well if temperatures drop below 10°C (50°F). This confines their growth to the equatorial belt, between about 20° north and 20° south latitudes, and at lower elevations where frosts are unlikely to occur. These plants do not require cold temperature exposure for either vegetative growth or flower initiation.

Subtropical

Subtropical fruit plants tolerate some subfreezing temperatures; in fact, they generally require some cool weather for best growth and fruit development. Many subtropical fruit plants are likely to be killed by temperatures below about −7°C (19°F). In completely tropical climates, subtropical fruits do not grow or produce well. Plants in the tropical and subtropical groups are mostly evergreen, although some subtropicals, such as figs, almonds, and pistachios, are not.

Subtropical fruits are grown at the lower elevations north and south of the tropics. The northern and southern latitudinal limits of these regions are ill defined, being greatly modified by other factors that influence climate such as ocean currents, large inland bodies of water, altitude, and mountain ranges. The subtropical belt includes both humid and semiarid regions.

Temperate Zone

Fruit plants of temperate zone have different levels of winter hardiness, some tolerating temperatures as low as −34°C (−29°F) if properly hardened in the fall. Vegetative and flower buds of many temperate-zone plants enter a "resting" state in late summer or fall and require a substantial amount of winter cold before they will resume growth the following spring. This requirement rules out commercial growing of such fruits in tropical and many subtropical regions. The production areas for this group of plants lie in two belts around the world, at approximately 30 deg to 50 deg north and south latitudes, where the local climate is otherwise favorable.

Climatic environments resembling temperate zone conditions can occur in the tropics at sufficiently high elevations. For example, grapevines normally grown in the temperate zones can also grow and fruit well in Bolivia (16 deg south of the equator at an altitude of 2,800 m (8,500 ft), where typical temperate zone climates prevail). Likewise, in Kenya, 4 deg to 10 deg north of the equator, some temperate-zone fruits produce good crops at high altitudes.

Grain Crops

Most of the grain-producing areas lie in the temperate zones of the world where the annual precipitation is neither less than 38 cm (18 in.) nor more than 100 cm (40 in.), excluding those areas where irrigation is used. In some areas wheat and barley are grown with as little as 25 cm (10 in.) of rainfall, but irrigation or other cultural practices such as strip cropping are used to supplement the water supply. In regions where rainfall exceeds 100 cm (40 in.), grain crops do not thrive because of the prevalence of disease, leached soils, and lodging (falling over).

The growing season for grain crops should have at least 100 to 110 days of frost-free weather, but exceptions do exist. Some cultivars that mature in 90 days succeed in short growing seasons. Also, spring-seeded cereals are grown in short-season areas. Moving toward the north and south poles, summer days become longer. The long days favor earlier maturity in some cereals, helping to overcome the effects of the short season. Winter temperature is a factor in areas where winter grains are grown. Rye (*Secale cereale*) is grown farther north than other cereals because it is relatively cold-tolerant. Some cereal grains are killed by cold temperatures that freeze the crowns of the small plants. Alternate freezing and thawing of the soil during the winter causes the plants to be pushed upward out of the soil, a process called "heaving." This also causes winter killing. Sometimes cereal plants suffer from lack of water during the winter if the ground remains frozen for long periods of time.

High temperature increases transpiration, which depletes the available soil moisture supply. High temperatures are especially harmful during seed grain formation and between grain formation and maturity of the grains. The fall weather pattern is important for fall-planted cereals. It is essential that these cereals grow as much as possible and are well established before cold weather sets in. Good growth before winter helps avoid damage from soil heaving later on. Thus, cereal grain farmers hope for warm temperatures, ample rainfall, and good growing conditions during the fall.

Sugar Crops

Climate determines the choice between sugar beets (*Beta vulgaris*) and sugarcane (*Saccharum officinarum*).

Sugar beets are cool-season crops for the north and south temperate zones. In the northern hemisphere, sugar beets are grown from about 35 deg to 60 deg north latitude. For good seed germination, soil temperature at planting should be about 7°C (45°F). Sugar beet seedlings are most sensitive to cold at emergence, but once out of the ground and hardened, the plants become quite cold resistant. During the summer months, mean daily temperatures of 21°C (70°F) or slightly higher promote rapid growth, but extremely high mean temperatures tend to retard growth. Temperatures in some sugar beet areas often reach 40°C (104°F). Under those conditions crop progress is nil.

Soil moisture affects the production of good-quality beets more than temperature. With warm temperatures and ample soil moisture, the vegetative growth of sugar beets during the summer and early fall months is rapid and luxurious. In irrigated areas, water is withheld at the approach of maturity, and farmers plan their fertilization program so that the plants exhaust the available nitrogen by this time. This practice, coupled with cool nights, retards the rate of vegetative growth and favors sugar accumulation. Warm days with bright sunlight augment photosynthesis; thus, by late fall the sugar content in the roots should increase. In some parts of California, sugar beets are cultured differently—planted in the late fall or early winter and harvested in July, when temperatures are highest. This practice is an apparent contradiction, but in reality it is not. Instead of allowing cool weather to retard plant growth, California growers halt vegetative growth of mature sugar beets by withholding irrigation water during the dry summer. The high light intensity of summer causes photosynthesis to proceed at a rapid rate, thus producing a high percentage of sugar.

Sugarcane requires the kind of climate found in the tropical and subtropical belts around the world. Much of this area has a tropical rainforest climate. However, sugarcane is also produced commercially in arid tropics under irrigation (Peru) or even in temperate zones (Louisiana, Florida, and parts of India). Sugarcane requires a long continuous frost-free growing season, an ample water supply, and full sunlight. Low temperatures are not conducive to good cane growth.

The wide climatic variation of the areas where sugarcane is produced must be reconciled with the climatic requirements of the plant. A part of the explanation is that no sexual cycles are involved; that is, the marketed product is sap from the vegetative cells. Thus, neither fruit nor seed need to be produced. As a result, daylength imposes no limit on sugarcane production. As long as the total hours of daylight are available for maximum growth, sugarcane can succeed.

Forage Crops

In almost all climates, irrespective of latitude or altitude, various forage crops are grown for livestock feed. Many species are used, each chosen for its adaption to a particular environment.

Legumes

Alfalfa (*Medicago sativa*), widely distributed in the temperate zone, has remarkable adaptation to climate, provided its soil and water requirements are met. The crop requires considerable amounts of water and produces best in relatively warm, arid climates with supplemental irrigation. Alfalfa is an important forage crop in the humid Northeast and North Central states. However, it grows better in the arid West and Southwest. Diseases are more prevalent in humid areas, and soils in high-rainfall areas are less favorable to alfalfa because most are acidic. The combination of high humidity, high soil moisture, and high temperature is particularly undesirable. During periods of drought, the deep alfalfa tap root—often 4.5 to 6 m (15 to 20 ft) long—can extract sufficient water from the subsoil to survive, if not enough to produce a profitable crop. However, alfalfa plants can be winter-killed, because lack of water the previous summer so weakens the plant that it cannot survive the severe winter.

Alfalfa is more tolerant to high and low temperature extremes than many perennials. In the United States, alfalfa cultivars are classified on the basis of temperature tolerance as hardy, medium hardy, or non-hardy.

Another legume, commonly called sweet clover or melilot (*Melilotus*) even though it is not a true clover, includes about 20 species adapted to a wide range of climates. Sweet clover can be grown without irrigation in most of the United States where the annual rainfall is well distributed and at least 45 to 50 cm (18 to 20 in.), and the soil reaction is neutral or slightly alkaline. Sour or bitter clovers (*M. indica*)—also not true clovers— are adapted to the climates of the Gulf coast, southern California, New Mexico, and Arizona. If rainfall is lacking, irrigation is used. Cultivars differ widely in their adaptation to temperature, moisture, light, and daylength.

Sometimes stands of red clover (*Trifolium pratense*) are winter-killed by low temperature, low soil moisture, alternating cold and warm temperatures, or heaving. Biennial sweet clover is more tolerant to freezing than red clover, but it is also more susceptible to heaving. Low winter temperatures definitely limit adaptation of the winter annual clover species. Sour or bitter clover (*M. indica*) and berseem clover (*T. alexandrinum*) are among the least tolerant to low temperatures of all the clover species. The plant can better stand low temperature if it enters the cold season with good vegetative growth. Clover plants tolerate high summer temperatures if adequate soil moisture is available, but summer temperatures inhibit germination of winter annual clover seeds.

Soil moisture is the most critical single climatic factor for the clovers. They all require a moist soil from time of seed germination to the end of their life cycle. If rainfall is not sufficient, supplemental irrigation must be used. Deep-rooted sweet clovers, if well established, are more drought-resistant than the others. Clovers are long-day plants, but the different species and cultivars vary in their response to daylengths. Clover is often grown with a grain crop for weed control, and the growth of the clover plants is sometimes stunted when they are shaded too much by the taller grain plants. Even when the clover stand is well established, growth might be slow. Then, when the grain is cut for harvest, sudden exposure of the clover plants to the hot summer sun could be fatal.

Grasses

Over 1,000 species of grasses grow in the United States, and about 100 are economically important forage crops. Temperature and soil moisture are the dominant factors determining their distribution. For example, the cold northern humid regions are more suited for the growth of bluegrasses (*Poa* spp.), orchard grass (*Dactylis glomerata*), redtop (*Agrostis gigantea*), and timothy (*Phleum alpinum,* syn. *P. pratense*). In the warm southern humid regions we find bermuda (*Cynodon dactylon*), carpet (*Axonopus affinis*), and Dallisgrass (*Paspalum dilatatum*). In the wet South, there is carpet (*A. affinis*), St. Augustine (*Stenotaphrum secundatum*), and certain reed grasses (*Phalaris* spp.). In the drier northern areas, there is Canada bluegrass (*Poa compressa*), and bromegrass (*Browmus* spp.), while in the drier southern regions, bermudagrass (*C. dactylon*) and Bahiagrass (*Paspalum notatum*) grow well.

In the dry western regions of the United States, the different grass species are spread over a wide temperature range. Soil moisture is more important than temperature. The native bunch grasses, such as blue-bunch wheatgrass (*Agropyron spicatum*) and crested wheatgrass (*A. desertorum*), are found in the northern half of the semiarid region, while blue grama (*Bouteloua gracills*) and buffalo grass (*Buchloe dactyloides*) do well in the drier southern regions. In general, it appears that grasses adapted to the humid regions are influenced primarily by temperature, while those growing in the dry areas are influenced principally by soil moisture levels.

Fiber Crops

Cotton (*Gossypium hirsutum*) and flax (*Linum usitatissimum*) are very important fiber crops. Cotton is grown around the world because of its useful fiber and adaptability. For optimum yields of fiber crops, climatic requirements are rather exacting. Freedom from frost, ample soil moisture, and an abundance of sunlight are three climatic requirements.

Production areas extend from about 35 deg south to about 45 deg north latitudes, although about 90 percent of the total cotton crop is grown north of the equator. Mean annual temperatures in these areas are about 16°C (61°F). For best cotton production, the mean maximum summer temperature should be about 32°C (90°F) and the mean minimum about 25°C (77°F). The growing season should have at least 180 to 200 frost-free days. Some new cultivars mature in fewer days and are adapted to areas with shorter growing seasons. The annual rainfall in the United States cotton belt ranges from about 50 cm (20 in.) in western Oklahoma to 140 cm (55 in.) in parts of Texas. Cotton grown in the arid Southwest must be irrigated.

Ideal weather for cotton production is a warm spring with frequent light rains after planting, followed by a moderately moist summer with both warm days and nights. Cold wet weather in the spring can cause seed decay before germination. Wet summers encourage excessive vegetative growth and increase disease and insect problems. High temperatures cause considerable moisture loss through transpiration and drought. A large number of warm, cloud-free days during the growing season are beneficial. Heavy rains interfere with fruiting, fruit set, and cotton quality.

Vegetable Crops

The climatic range in which different vegetable crops grow is quite narrow. Farmers widen the range for some crops by establishing suitable microclimates. Plastic or glass greenhouses, hot beds, cold frames, hot caps, brushing, plastic mulches, and wind machines used with smudge pots (oil-burning heaters) are examples of equipment and materials used to modify the climate. Certain cultural techniques, such as changing the shape and direction of beds, are sometimes used to modify the climate sufficiently to permit successful culture of some vegetable crops. Irrigation can be used to modify soil temperature. Furrow irrigation with warm water could provide enough heat to ward off frost damage, whereas evaporation of water from the soil surface can provide cooling if that situation is needed.

Sometimes a certain climate is not suitable for a given physiological plant function. For example, perhaps a crop could not flower and produce seed under certain daylength conditions but could grow well and produce a good crop of leaf or stem parts. Seed for this crop could be produced in areas where climatic conditions permitted seed production while the vegetative crop is produced in other areas. Crops are sometimes grown to follow the seasons. For example, in the northern hemisphere short-season crops move northward in the spring and southward in the fall, and vice versa in the southern hemisphere. Also, plant breeders continually introduce new cultivars with broader climatic adaptability. More risks may be taken by vegetable growers with the weather than by other growers, because vegetable crops generally represent a short-term investment.

The important vegetable-growing regions of the world are located in those parts of the temperate zones where sudden and extreme temperature changes do not occur. In the United States these areas are (1) the coastal valleys of the Pacific coast states, (2) along the Atlantic and Gulf coasts, (3) around large inland bodies of water such as the Great Lakes, and (4) in the Rio Grande Valley of Texas.

Vegetable crops are divided into two broad groups by their temperature tolerance—cool season and warm season.

Cool-season crops grow best at temperatures from 15°C to 18°C (59°F to 64°F). Certain crops, such as Brussels sprouts, turnips, rutabagas, spinach, and beets, can tolerate some freezing, but others, including cauliflower, lettuce, carrots, potatoes, peas, and onions, are injured.

Warm-season crops thrive best at temperatures between 18°C to 27°C (64°F to 81°F). Plants in this group are intolerant of frost or prolonged exposure to near freezing. In this group are melons, cucumbers, squash, pumpkins, tomatoes, sweet corn, peppers, sweet potatoes, eggplant, and okra. However, even warm-season vegetables do not grow well if temperatures are extremely high. For example, several consecutive days of temperatures over 40°C (104°F) will destroy pollen in sweet corn and cause blossoms to drop from tomatoes. Hot weather unduly hastens fruit maturity, shortens the harvest period, and causes rapid deterioration of the product.

CLIMATIC INFLUENCES ON PLANT DISEASES

Plants throughout the ages have always been threatened by plant diseases. At times, plant diseases have profoundly affected history. Wars have been fought, countries destroyed, and people have migrated en masse to new lands because of famines caused by plant disease. A noted example was the Irish potato famine of 1846 and 1847, due to the fungus, *Phytophthora infestans,* which caused widespread famine and led to about 1 million deaths due to starvation and a huge emigration of much of Ireland's population. More recently, disease-induced famine occurred in India in 1943 as a result of *Helminthosporium* brown spot of rice. There are many examples of plant disease epidemics influencing human food supply in the past, and even in modern times.

Modern knowledge has demonstrated a definite relationship between climate and plant diseases. Weather influences the incidence of disease by: (1) favoring the growth of the pathogen itself (e.g., high humidity favors fungus development), (2) affecting the incidence of an insect carrier (aphids, for example, transmit viral disease), and (3) increasing the plant's susceptibility to disease (unfavorable weather weakens the plant and lowers its resistance).

Factors necessary for an epidemic to start include a favorable environment (temperature, moisture, nutrition, and the like), large numbers of the pathogen, a susceptible host, and the interaction of each factor. The environment is subject to modification by cultural decisions and procedures. Cultural practices (sanitation) and/or pesticides can reduce pathogen number, and resistance or tolerance may be achieved by plant breeding. Breeding for disease resistance often receives more attention than any other breeding objective.

Rapid advances in plant molecular biology and other biotechnological aspects suggest enormous possibilities for new, rapid, and effective plant breeding procedures to achieve disease resistance and other useful characteristics.

CLIMATIC INFLUENCES ON INSECT PESTS

Insects living in a given region have adapted to the average climatic conditions of that region. When severe departures from these climatic norms occur, that insect population may change materially. Climatic factors such as temperature, moisture, and air movement affect insects in varying degrees and at different times. Many insects have short life spans, some as short as a few days, others up to several weeks.

Those with short life spans usually produce many generations in a single year. Each of the many generations has the potential to evolve slightly, which permits adaptation to climatic and other changes. The ability of some insects to adapt to an environment can be illustrated by the increased resistance of the common housefly (*Musca domestica*) to the insecticide DDT (**d**ichloro-**d**iphenly-**t**richloro ethane). Introduced in the 1940s, DDT was an effective insecticide against the common housefly, but after 20 years its effectiveness was reduced because resistance to its toxicity developed during many generations of flies.

Insects, at each stage of development (egg, larva, pupa, and adult), have a definite temperature tolerance. An insect dies at temperatures below a certain minimum. Depending on its lifestage, most insects can withstand the low temperatures that normally occur in the inhabited parts of the world. Severe cold is usually endured during periods of hibernation. The insect remains inactive or in hibernation between the minimum temperature at which its life can exist and the temperature at which development occurs. The optimum temperature is the range in which most activity and maximum development occur most rapidly. This temperature range can vary, depending on the insect species. Temperatures above the optimum retard growth and development. From the upper limit of the optimum temperature to the temperature that causes death, the insect lives in a state of suspended activity called aestivation. Some water beetles, water bugs, and mosquitoes live normally in hot springs with temperatures ranging from 38°C to 50°C (100°F to 122°F). The longer the optimum temperature is present, the more generations of insects are produced.

The effect of moisture on insects is similar to those of temperature—there is a point of excessive dryness at which the insect dies, a moisture condition for optimum development, and an excessive moisture condition that causes death.

There are specific examples of the manner in which insects respond to climate. For example, San Jose scale (*Aspidiotus perniciosus*) is restricted to the fruit trees growing in the warmer climates. It will not extend into the colder climates unless, through many generations of evolution, the insect develops a tolerance to lower temperatures.

Temperatures of −18°C (0°F) are fatal to the coon boll weevil (*Anthonomus grandis*). This factor of climate restricts damage to cotton by this insect to the warmer regions of the Cotton Belt, where it is a major pest problem. Dry conditions during the early spring are favorable for the development of the chinch bug (*Blissus leucopterus*). All of the recorded outbreaks of this insect have occurred during periods of less than normal rainfall.

Very often the effect of climate on insects is a complicated one—not directly affecting a certain insect, but affecting others that prey on it. For example, one insect population increases drastically in a cool wet spring when temperatures are decidedly below those for optimum development of a parasite of this insect. Under normal temperatures the parasite develops in numbers sufficient to keep the host under control.

SUMMARY AND REVIEW

Weather is the immediate and local occurrence of such natural phenomena as temperature, light, rainfall, wind, and relative humidity. Climate is the pattern these phenomena assume day by day, week by week, month by month, and season to season, which is repeated annually in an area. If the area is small or near the ground surface it is called a microclimate.

Climate is determined by a location's latitude, altitude, and relation to large bodies of water and land masses.

Climatic factors are temperature, rainfall, light, and air movement. Each plant species is adapted to the climate in which it evolved. Some plants have a wide tolerance to climate conditions, other are very intolerant of change.

Temperature. Most plants live and grow when the temperatures fall between 0°C and 50°C. However, the type of growth desired for the production of most crops falls in a much narrower range. That range can vary by crop. Some single-season crops do best in areas where there is a prolonged period of warm days and nights while others require cooler temperatures. Many perennial crops tolerate freezing or require a chilling period during the winter while others do not tolerate freezing or chilling.

Rain/water. Plants are divided into three categories based on their water needs or tolerances. Xerophytes are those plants that have adaptation mechanisms enabling them to function with relatively less water than most other plants. Hydrophytes are those plants that thrive in or close to water. Mesophytes are the most common type and have water requirements and tolerances between the two extremes.

Light. Light quality, intensity, and duration all influence plant growth and crop productivity. Light quality affects photosynthesis, with red and blue being the most effective region. Light quality also affects plants photomorphogenically mainly through the relationship of red and far-red light. Some plants including most field crops require relatively high light intensity. Others, such as coffee and many ornamental foliage plants, tolerate or require relatively low light intensity.

Air movement. Air movement in the field can be beneficial when it replenishes CO_2 for rapidly photosynthesizing leaves. Likewise, it can be detrimental if it is too strong and at the wrong time of the production cycle.

Climatic factors influence the development and spread of disease and insects. Climatic factors that favor the development and spread of diseases and detrimental insects can negatively impact a crop, even if those factors favor crop growth.

In field crops temperature can be controlled or manipulated under some circumstances. To prevent or reduce injury from freezing, wind machines, heaters, mulches, row covers, or water sprinklers are used to slightly raise the temperature in or near the plant. Wind damage can be lessened by the use of windbreaks. Irrigation is used to provide water when rainfall is insufficient. Row orientation and plant spacing can increase light penetration into a crop. Choosing species or varieties/cultivars that are adapted to an area improves crop productivity.

EXERCISES FOR UNDERSTANDING

10–1. Check out the USDA's (cold) hardiness zone map and the American Horticulture Society's heat zone map from the following Web sites. Determine the hardiness and heat zones for your hometown. What effect would local geographic features such as a large lake, hills, or valleys have on the microclimate in the area? Would the zones be higher or lower in number? How would these changes affect plant growth and development?

(American Horticulture Society Web site) www.ahs.org/publications/heat_zone_map.htm (USDA Web site)www.ars-grin.gov/ars/ Beltsville/na/hardzone/ushzmap.htm

10–2. List the ways in which encroachment of cities and suburbs on crop production areas, including farms, nurseries, and greenhouses, may change the microclimates in these areas. What effects could these changes have on crop development?

REFERENCES AND SUPPLEMENTARY READING

KAISER, H. M., and T. E. DRENNON, eds. 1993. *Agricultural dimensions of global climate change.* Del Ray Beach, Fla.: St. Lucie Press.

REDDY, K. R., and H. F. HODGES eds. 2000. *Climate change and global crop productivity.* New York: CABI Publ.

CHAPTER

Biological Competitors of Useful Plants

KEY LEARNING CONCEPTS

After reading the chapter you should be able to:

♦ Understand what the biological competitors of crops are.
♦ Understand how each competitor affects crop productivity.
♦ Understand the principles and primary methods of controlling each type of competitor.

Since the beginnings of agriculture, farmers have had to wage war continuously against weeds, plant diseases, and insects. Even today, with all the scientific research on crop protection, it is estimated that insects, diseases, weeds, and animal pests eliminate half the food produced in the world during the growing, transporting, and storing of crops. In the tropics, where heat and high humidity favor many pathogens, two-thirds of some crops are lost. Losses to plant pests greatly aggravate the world hunger problem.

All of the thousands of kinds of cultivated plants—the ornamentals, the vegetables, the fruits, the fiber, the forage and grain crops, and the forest species—often have difficulty in growing and producing their products. In addition to contending with the un-certainties of weather and soil conditions, plants have many biological competitors—weeds, insects, mites, and disease-producing fungi and bacteria—that contend with them or attack both their tops and roots. In addition, other soil-borne organisms, such as nematodes, as well as systemic pathogens such as viruses and mycoplasma-like organisms attack cultivated plants.

Various plant parasites, such as mistletoe and dodder, attack some plants and often eventually kill them. Birds are generally considered to be among the world's desirable animals, but certain species can become major pests to agriculture by feeding on and destroying certain crops. Deer and rabbits sometimes severely damage plants such as young fruit trees, especially during the winter, unless protective measures are taken. Rats and other rodents destroy a major part of growing and harvested grain crops in some countries.

Many plants carry genetic resistance to many of the pathogenic or parasitic organisms. As a result, no plant species is attacked by all possible pests. The decision to grow a certain crop or a particular cultivar in a certain area is often based upon the genetic resistance of that plant to a common pathogenic organism or insect pest. Alternatively, the reason that certain crops are not grown there is that they are damaged by pests to an intolerable extent. The development, by plant breeding, of new plant types with genetic resistance to pathogens and predators is one of the most effective means of combating these problems.

In the developed countries of the world over the last 100 years, we have learned to control the ravages of insects, diseases, and weeds fairly well by applying various agricultural chemicals: insecticides, fungicides, herbicides, and nematacides. However, the ideal, long-term solution for saving food, fiber, and ornamental plants from attack by pests is control based upon biological methods. These can be used in a number of ways, for example:

1. New plant types can be bred or genetically engineered that are resistant to attacks of insects, diseases, nematodes, and other parasites and predators.
2. Insect populations can be controlled by releasing into crop fields various natural enemies of plant-feeding pests, such as insect parasites, predators, or disease-causing bacteria or viruses.
3. Some insect populations can be almost eliminated by releasing massive numbers of radiation-sterilized male insects. After mating, the females fail to produce offspring.
4. Insects are lured by natural or synthesized attractants into bait traps, where they are killed. Other synthesized attractants, some of which are called pheromones, may be released into the environment to prevent successful mating of male and female.
5. Some insects are prevented from reproducing by juvenile hormones, which keep them from reaching the adult stage, and thus preclude their reproduction. This procedure is in the experimental stage.
6. Some weed species are controlled by releasing into the fields certain insects that feed on the weed but not on the crop plant. (Pulling weeds by hand from our gardens is also, of course, a form of biological control.)
7. A few plant diseases caused by fungi or bacteria have also been controlled with antagonistic or competing fungi or bacteria that are not pathogenic to the crop plants.

While these types of control are helpful and offer considerable promise for the future, they cannot be relied upon entirely at present. For now and into the foreseeable future, pesticides must be used to enable the crop producers and distributors to furnish the kind and amount of food products consumers demand. No food products that show the slightest evidence of insects are tolerated in most markets—no wormy apples, no insect parts in canned or dried fruits or vegetables, and no burrowing passages in fresh vegetables. To maintain such quality standards, for the present at least, we must continue to use pesticides despite the problems some cause.

The modern concept of **integrated pest management (IPM)** employs many approaches to control a pest, rather than relying solely on a single procedure. Integrated pest management develops information on such facets of the problem as:

1. dynamics of the pest population, predictions of the occurrence, population levels, and potential economic damage of pests
2. biology of the pest organism, of its natural enemies, and of the host plant, and their interrelationships in a given environment
3. effects of weather patterns on pest activities
4. effects of cultural practices such as crop rotation, irrigation, cover and companion crops, and harvesting methods on pest activities
5. effects of various control tactics on each other and on the environment.

Integrated pest management uses a range of cultural, biological, mechanical, and chemical measures to hold the pests below economically damaging levels, while at the same time avoiding disruption of the agroecosystem. In the United States, integrated pest management systems are being developed for tree fruits and nuts, cotton, rice, tomatoes, onions, alfalfa, corn, potatoes, and wheat. So far, the concept has been applied primarily to insect pests and only in a limited way to pathogens and weeds.

WEEDS

Ever since people began cultivating plants, they have had to fight weeds competing with crops for space, water, mineral nutrients, and sunlight. A weed has been described as any plant out of place. Wheat plants in a field of oats would be considered weeds, as would oat plants in a field of wheat.

FIBROUS ROOTS ∂∂ ABSORB H2O QUICKER

Some of the biggest problems with weeds in fields of cultivated crops occur in areas where substantial rain falls all through the growing season. The high moisture level keeps the weed seeds germinating, and unless the weeds are controlled, they either choke out and eliminate the desired plants or reduce yields. In areas of low rainfall any competition by weeds significantly reduces yields or does not allow a crop harvest.

Until recently, weed control was entirely physical. The weeds were simply cut off or dug out after they started to grow. Most kinds of weeds are annuals, developing from germinating seeds, but others—more difficult to control—are perennials developing from shoots that continually sprout from underground roots or stems (rhizomes).

The primitive tools used for mechanical weed control were sharpened sticks or metal objects, culminating with the common hoe. Even today in small gardens and flower beds and in fields of such crops as sugar beets and lettuce, the hoe is often the principal means of eliminating weeds as they appear, although in modern agriculture an integration of physical, mechanical, and chemical control is practiced. Commercial crops were planted in rows to allow horse-drawn cultivators to dig out the weeds. Cultivation continued through the summer as the weed seeds kept germinating. The final stage in mechanical weed control in commercial row crops was the multirow, tractor-drawn cultivator. Mechanical weed control in large expanses of soil entails tractor-drawn disks, harrows, and other weeders. With large fields and precision planting, cultivating can remove most of the weeds with only a small band of herbicides applied over the crop row. This reduces the amount of pesticide used in a crop. Tall weeds, particularly in open spaces such as orchards, can be effectively controlled by mowing. This keeps them from producing seeds and reduces their photosynthetic capacity, decreasing the accumulation of stored foods in the roots. Mowing is not effective, however, in controlling low-growing or perennial weeds. Pasturing animals such as sheep or geese along ditch banks or in orchards or cotton can sometimes be used as an alternative to mowing, provided the animals do not disturb the crop plants.

Mulching is another effective way to physically control weeds in small areas. The spaces in and around cultivated plants are covered with wood chips, sawdust, gravel, straw, rice hulls, or similarly inactive materials. If they are thick enough, such mulches shut out light from young weed seedlings, thus preventing their growth. Mulching is ineffective against perennial weeds with heavy root systems that send strong-growing shoots right up through the mulch. Black plastic film is often used for this purpose in nurseries. A recent technique using sheets of clear plastic film "solarizes" (heats) the soil in the summer before planting to control weed seed and many soil-borne pathogens.

Fire has been used to control weeds along roadsides, ditch banks, and other work areas, but high fuel costs and air pollution have curtailed this practice. Special, highly maneuverable flamethrowers were built for this purpose. Such burning has limited value in controlling weeds in crop lands, however, although special burners were developed to burn off very small annual weeds in row crops such as cotton. Controlled burning has been of considerable value in removing unwanted brush from forest areas and range lands.

The accumulated costs in equipment, labor, and energy to control weeds in the world's crops by physical methods are billions of dollars annually in the United States. These costs are, of course, added to the selling price of crops and have tremendously increased the prices people have to pay for their food. Removing weeds from crops is still one of the world's greatest users of energy.

Modern Weed Control Methods

In the early 1940s, agriculturalists in many countries realized the staggering costs of mechanical weed control. Intensive research efforts to develop cheaper and more effective methods of controlling weeds began. These efforts led to a new body of scientific study involving large groups of research and extension workers in various agricultural colleges plus many other scientifically trained persons employed by agricultural chemical companies. Four basic methods of weed control were developed: (1) preventive measures, (2) crop competition, (3) biological control, and (4) chemical control.

Preventive measures entail attempts to reduce all sources of weed seeds such as roadside or ditch stands of weeds, seeds in irrigation canals, seeds blowing in from nearby weedy fields, and weed seed mixed in with crop seeds. Also, efforts are made to keep any weed plants from reseeding themselves in the crop fields or landscape so that new seeds are not present to reestablish a weed population.

In crop competition the cultivated crop plants are simply induced to grow so fast and so vigorously that they shade out and overcome the weeds. This method is most useful in agronomic crops like the cereals, forage crops, and turf-grasses that completely cover the ground. It is of less value in row crops, vineyards, or orchards, where much of the soil surface is bare—although once such crop plants become large enough, they can shade so much of the soil surface that young weed seedlings have difficulty growing.

Biological control of weeds succeeds best where a fungus or an insect predator, a "natural enemy," is introduced into an area where the population of the weed it attacks is large. A natural biological balance is set up between the weed (the host) and its pathogen, the weed population being reduced in numbers but not completely eradicated. There is always the hazard that the introduced controlling organism can develop into as serious a pest as the weed being controlled. Therefore, considerable advance study precedes any such introductions. Government quarantine officials are extremely reluctant to allow any fungi or insects into areas where they do not already exist.

An interesting phenomenon shown by some plants is their built-in natural herbicide properties. That is, some plants excrete chemicals that are toxic to other specific plants. This is known as **allelopathy.** It was observed as long ago as 1832 that thistles growing in a field of oats caused oat plants around them to die. In recent years scientists have become interested in allelopathy as a weed-control method and are trying to determine the nature of the plant-specific toxins.

Chemical Weed Control

Sporadic research with chemicals as a way to control weeds started about 1910 in both Europe and America, but the real basis for chemical weed control was laid with the initial studies of auxins and plant hormone physiology in the 1930s. The introduction of the auxin-type synthetic plant growth regulator 2,4-dichlorophenoxyacetic acid (2,4-D) as an herbicide in 1944 gave the initial impetus to chemical weed control. Production of pesticides is now a multibillion-dollar industry, with sales of herbicides far surpassing both insecticides and fungicides.

Weed-controlling chemicals can be classified as either selective or nonselective. For example, 2,4-D is a selective herbicide for broad-leaved plants. Sprayed on a lawn or a wheat field, for example, it will kill broad-leaved weeds, such as dandelions or wild mustard, but will not harm the grass or wheat. Nonselective herbicides, such as the high aromatic weed oils, kill all vegetation they are applied to.

Herbicides can be further classified according to the timing of their application in relation to the growth cycle of weeds or crops.

Preplanting Treatments The herbicide is incorporated into the soil before the crop is planted. The crop seeds or plants must be highly tolerant of the herbicide. The herbicidal action on the weeds can be due to a direct contact killing, or it can be absorbed by the weed's roots or shoots, then translocated throughout the plant, interfering with various plant processes, including suppressing of cell division.

Preemergence Treatments Herbicides are applied to the soil surface after the crop is planted but before the emergence of the weed seedlings, the crop seedlings, or both. For clarity in any particular case, it should be stated whether *preemergence* refers to the weeds, the crop, or both. Since the seed germination and seedling stage is the weakest link in the annual plant's life cycle, preplant and preemergence treatments are frequently used since they are safer on the crop and generally require less herbicide per given area.

Postemergence Treatments Herbicide treatment follows emergence of the seedlings of the crop plants, the weeds, or both. Application could be postemergence for the crop plants but preemergence for the weeds. In an orchard or vineyard, herbicide applications would always be postemergence for the crop plants but could be either preemergence or postemergence for the weeds.

Another basis for classification of herbicides is the method of application:

1. *Broadcast.* This method covers an entire area uniformly, either by spraying a liquid or disseminating a granular form of the herbicide.
2. *Band treatment.* A relatively narrow band just covering the crop row (or orchard or vineyard row) is treated with herbicide. Weeds between rows are controlled some other way, perhaps by tillage equipment.
3. *Spot treatment.* Herbicide sprays are directed to the foliage of a clump of weeds arising in a relatively clean area, such as an orchard, that has previously been cleaned of weeds by other control measures; or the soil may be spot treated in a small area where a particularly difficult clump of perennial weeds has established itself.

Herbicide Application Equipment   Since herbicidal chemicals are often toxic to plant tissue, it is inadvisable to use the application equipment for other purposes. Enough of the herbicide may remain in the equipment to damage crop plants in other spray applications.

1. *Sprayers.* Power or hand-operated sprayers (Fig. 11–1) are most commonly used to apply herbicides as liquid formulations (wettable powders in water or water-dispersible granules in water) either directly on the weeds or onto the soil. The sprayer is either low volume, applying a concentrated form of the herbicide in a small amount per unit area, or high volume, applying less concentrated solutions but in larger amounts per unit area. Most herbicides are applied with sprayers.

   The amount of the herbicide applied is generally calculated in kilograms per hectare (pounds per acre) of a given concentration as directed on the label.
2. *Granular applications.* Equipment is available to apply granular forms of herbicides as a preemergence application to the soil over the crop rows when the seeds are being drilled into the soil. At high application rates, the granules of the herbicide itself are applied; at lower rates, the granules are first mixed with a carrier such as sand or vermiculite to increase the bulk, thus permitting better distribution of the active herbicide particles. Granular application eliminates the need for hauling quantities of water, and the application equipment is much simpler and less expensive than spray machinery. Granular application, however, is not as uniform as water sprays. Winds can blow the granules and carrier from where it should be applied; in addition, heavy rains or irrigation can wash the granules away. Granular forms of herbicides are generally more costly than other types because of higher shipping costs.

Figure 11–1   Preemergence herbicides are often applied as a spray. This destroys the earliest weeds and gives the crop an advantage throughout the entire season. *Source:* Fred Meyer, Tennessee Valley Authority.

3. *Mixing with soil.* Preemergence herbicides, applied either as a spray or granules, are worked by disking into the upper few inches of the soil to bring them into contract with weed seeds. This application method is effective only if the seed bed is well prepared, lacks large soil clods, and has the proper soil moisture—neither dry nor excessively wet.

4. *Aircraft applications.* Applications of herbicides by airplane or helicopter are particularly useful for large areas or in situations where ground equipment cannot be readily used. Weeds and brush along utility lines or in firebreaks in mountainous areas are easily controlled by air. Large areas of grain or forage crops are covered economically in this manner, such as flooded rice fields that cannot be easily treated with herbicides in any other manner (Fig.11–2). However, air application of herbicides is not used in populated areas owing to drift hazards.

5. *Water application.* Herbicides are applied in sprinkler or furrow irrigation water. This method can be very effective on some crops when water placement is accurately and uniformly controlled.

Types of Weeds

Weeds, like all plants, are classified into annuals, biennials, or perennials. Control measures differ for the three types.

Annuals

These can be summer or winter annuals, completing their life cycle in one year. Annual weeds generally are easy to control, but they produce many seeds and are persistent.

Figure 11–2   Application by aircraft of a postemergence herbicide to crop plants. *Source:* Cessna Aircraft Company.

Seeds of summer annuals germinate in the spring. The plants grow all summer and produce seeds, then die in the fall or winter. The seeds lie dormant all winter, then germinate the following spring. Typical summer annual weeds are crabgrass, foxtail, lambsquarter, ragweed, and cocklebur. These are a particular problem in summer crops such as corn, soybeans, cotton, vegetables, as well as in lawns.

Seeds of the winter annuals germinate in the fall. The plants grow through the winter (in mild climates), and mature their seed in the spring. The seeds lie dormant in the soil through the hot summer, then germinate in the fall. Examples of such weeds are shepherd's purse, cornflower, hairy chess, and chickweed. These are often a problem in such winter crops as winter barley, oats, wheat, winter vegetable crops, and in fall-sown nursery plantings.

Biennials

The seeds of biennial plants germinate in spring. The plants grow vegetatively through the first summer. The following spring, after a winter chilling period, the plants flower. Seeds develop and mature by the end of the second summer. This category includes a few troublesome weeds such as mullein, burdock, wild carrot, and bull thistle, which are a problem in both summer and winter crops.

Perennials

Perennial plants can live indefinitely, although the tops may die down in winter. Once started, perennial weeds remain until they are killed. They usually start from germinating seeds, but many types spread naturally by vegetative means such as root pieces, rhizomes, stolons, or tubers. Some perennial weeds—such as dandelion, plantain, and dock—live for many years as single, individual plants. Other perennial weeds, in addition to producing seed, also propagate by vegetative means, making them difficult to control. Examples are bermuda grass, bindweed, Johnson grass, quackgrass, nutgrass, and red sorrel. Cultivating creeping perennial weeds often just breaks them up and spreads them. However, repeated cultivation for several years prevents development of much leaf area and, together with herbicide applications, brings them under control.

Table 11–1 lists some of the world's most difficult-to-control weeds.

Herbicidal Control of Different Weed Types

The development and use of herbicides is a changing situation. Older chemicals are being discontinued and new ones are replacing them. Herbicides in current use number in the hundreds. When herbicides are being considered for controlling weeds in any area, from the backyard garden to large acreages of crops, expert advice should be obtained, preferably from the local agricultural extension agent, an agricultural chemical supply dealer, or garden center operator. There is usually a choice of herbicides for the task, and it is essential to read the label on the package carefully and follow the directions explicitly. For large agricultural operations, it is often advisable to employ specialized spray operators or pest control advisors, whose business it is to know the available herbicides for each crop and to be familiar with all the regulations governing their use.

PLANT DISEASES AND INSECT PESTS

Estimates of the annual losses in crop plants including ornamentals and turf grass to diseases and pests are at hundreds of billions of dollars. The world's governments and enterprises dependent on the growing of plants try to reduce these losses by every means possible.

The methods for combating plant diseases and insects fall into four groups: biological control, control by cultural practices, government agricultural quarantine and pest eradication programs, and application of pesticides.

Biological Control

Resistant Plant Species and Cultivars

Genetically resistant types of plants are selected naturally. Plants susceptible to a prevalent disease or insect pest tend to disappear, while resistant types remain. The work of geneticists and plant breeders in purposely developing resistant cultivars of agricultural species has been one of their most useful pursuits, and the world's population owes much to them. In many instances cultivars have been developed that grow and produce heavy crops in the presence of certain diseases or insects that eliminated previous, susceptible cultivars. The development of rust-resistant wheat cultivars, for example, has added food for untold millions of people. The development of resistant plant types also reduces the need for costly and sometimes dangerous pesticides. Ideally, all pests and diseases would be overcome in this manner.

The development of plant types resistant to disease and insect pests is a very active field of research in various agricultural experiment stations, using the combined talents of plant breeders, entomologists, and plant pathologists. For example, geneticists have developed wheat strains with hairy leaves and stems that resist attacks of the cereal leaf beetle by inhibiting the egg-laying activities of the females, making the plants practically insect-free. Several approaches are open to plant breeders in developing plants that are resistant to pests. They change the plant's color, surface texture, taste, or odor so that it is no longer attractive to an insect pest. They even introduce characteristics into the plant that repel or poison insects; for example, Bt corn produces a toxin that affects lepidoptera.

TABLE 11–1	Ten Hard-to-Control Weeds

Purple nutsedge (*Cyperus rotundus* L.)
Bermuda grass (*Cynodon dactylon* (L.) Pers.)
Barnyard grass (*Echinochloa crusgalli* (L.) Beauv.)
Jungle rice (*Echinochloa colonum* (L.) Link)
Goose grass (*Eleusine indica* (L.) Gaertn.)
Johnson grass (*Sorghum halepense* (L.) Pers.)
Guinea grass (*Panicum maximum Jacq.*)
Water hyacinth (*Eichhornia crassipes* Mart) Solms
Cogon grass (*Imperata cylindrical* (L.) Beauv.)
Lantana (*Lantana camara* L.)

Source: A. S. Crafts. 1975. Modern weed control. Berkeley, Calif.: University of California Press.

Antagonistic Organisms

A balance usually develops in nature among organisms, both plant and animal. Certain organisms are antagonistic to others and retard their expansion. Environmental or human-induced changes that upset this balance by eliminating one of the organisms can lead to explosive proliferation of the others and to subsequent attacks on vulnerable crops. In a similar manner, if an organism is introduced into an environment but its antagonizing organism is left behind, the results can be devastating to a vulnerable crop plant. Biological control in such a situation would consist of introducing the antagonizing organism of the pest into the area, thus bringing it under control again. Olive parlatoria scale in California olive orchards threatened the existence of the industry, but two parasitic wasps introduced from Asia became well established and, for practical purposes, eliminated the scale. In many greenhouses the introduction of predatory insects has become a regular part of insect and disease control.

Control by Cultural Practices

Crop rotation is a simple but often effective means of controlling certain insect and disease pests. If the same annual crop is grown year after year on the same plot of land, a particularly serious pest may keep increasing year after year, overwintering in crop residues, until it reaches such overwhelming populations that the crop cannot be produced on that piece of land. But by rotating the crop in a one to several year rotation with other nonsusceptible crops, the insect or disease pest, lacking a crop host for a long period, practically dies out. Crop rotation alone may give adequate control, but no stray susceptible host plants must be allowed to grow on the land during the interim period, since they would maintain a population of the pest.

Government Agricultural Quarantine and Pest Eradication Programs

Many serious plant disease and insect pests were not known to occur in certain countries until they were brought in by travelers or shipped in, usually unknowingly, on contaminated plant material. The early colonists from Europe brought into North America certain insects and diseases on seeds and plants carried from their homelands. German soldiers hired by the British during the Revolutionary War brought the Hessian fly to North America with them in their straw bedding. This pest later moved westward through the United States, devastating wheat fields and ultimately causing far more problems for the new country than the soldiers did.

   Strict government inspection[1] and fumigation procedures, plus quarantine of imported plants, plant products, and soil, have kept such pests out of many countries and many areas. However, modern jet travel, taking people and their belongings swiftly from country to country, greatly increases the possibility of introducing pests dangerous to both crops and farm animals into new areas. In some cases, certain plants or plant products are forbidden entry into the country or areas of the country, and agricultural inspectors check luggage for such outlawed products. For example, mangoes, guavas, and passion fruit from Hawaii are not permitted entry into the U.S. mainland, since they may be carrying the Mediterranean fruit fly, which is widespread in Hawaii but absent on the mainland. If introduced into California, for example, the fly could devastate the state's huge fruit industry. However, such plant products can be imported commercially if they are properly fumigated to kill the pests before shipping. Certain kinds of living plant material can be imported into the United States under permit, or if quarantined after entry for two growing seasons to reveal any diseases. Travelers coming into the United States should not attempt to bring or send in such agricultural materials as fruits, vegetables, plants, bulbs, seeds, or cuttings unless advance arrangement have been made and a permit obtained.[2] Any such plant materials being carried or imported through commercial channels must be reported to agricultural quarantine or customs officials upon arrival.

   Government eradication programs are conducted when a serious insect or disease pest breaks out. Often the trouble is eliminated before it has a chance to spread. Such programs require highly trained personnel who know the potential insect and disease problems and are able to recognize the pathogens and the symptoms of their activities. The dangerous Mediterranean fruit fly has been accidentally introduced into Florida on several occasions. Each time it has been eradicated, but always at considerable expense. In California, too, discoveries of the Mediterranean fruit fly, probably brought in on illegal tourist importations of fruit from Hawaii, have caused costly eradication procedures on several occasions.

Application of Pesticides

The potential dangers of pesticide chemicals to humans, food products, farm and domestic animals, wildlife, beneficial insects, and the atmosphere makes them the least desirable method of controlling harmful insects and plant diseases (Fig. 11–3). However, pesticides are often the only method of control and without them the world would not be able to feed its ever-expanding population. Some of our most valuable crops are so susceptible to devastating diseases and insects that without chemical control measures such crops would simply disappear (Figs. 11–4 and 11–5).

---

[1] In the United States inspection is done by the USDA's Animal and Plant Health Inspection Service (APHIS), Plant Protection Quarantine Programs.

[2] Arrange for a permit in advance of the trip from Plant Protection Quarantine Programs, APHIS, USDA, Hyattsville, MD 20782.

Figure 11–4 Certain insect and disease pests are controlled in commercial orchards by insecticides and fungicides applied by power air blast sprayers. *Source:* USDA.

Figure 11–3 When spraying pesticides to control insects and diseases in enclosed areas like greenhouses, workers should wear protective masks and clothing.

Figure 11–5 Dusting grapes with sulfur dust to control powdery mildew caused by the fungus *Uncinula necator. Source:* Blue Anchor, Inc.

PLANT DISEASES

Cultivated plants are subject to a wide array of plant diseases induced by such infectious parasitic pathogens as bacteria, fungi, viruses, and mycoplasma-like organisms. Most plants are immune to most pathogens, but the majority of cultivated plants are susceptible to attacks by at least one pathogen in each of these groups and some crops are susceptible to many. The potato, bean, and cotton, for example, are each a host for at least 30 fungus species. The world's forest tree species, too, are vulnerable to many diseases that can cause immense timber losses.

For an infectious disease to develop there must be a susceptible host, a causal agent, and a favorable environment for the pathogen. If any one of these factors is not present, disease will not develop. Infectious plant diseases are caused by pathogenic agents that are transmitted in some manner from a diseased plant—or in some instances, from plant debris or soil—to a healthy plant.

A plant disease is a harmful alteration of the normal physiological and biochemical development of a plant that is caused by a pathogen. The harmful change is exerted over time, rather than suddenly as in the case of a wound. Plant cells, tissues, and/or organs are destroyed or damaged by infection of the pathogen.

The appearance of an infectious disease requires a source of primary inoculum, which is a portion of the pathogen capable of being disseminated and causing infection. This inoculum may be bacteria, fungal spores or mycelia, or virus particles. To cause disease symptoms, the inoculum must penetrate the host plant and become established, causing infection and setting up a life cycle that includes the host plant. Sometimes the inoculum penetrates the tissues of the host plant but fails to become established because of high resistance of the host tissues or unfavorable

environmental conditions. Host resistance may be a genetic characteristic or a reflection of the particular growth condition of the host plant—succulent or mature, vigorous or weak.

Disease Symptoms

Symptoms of disease often change as the disease progresses. The initial symptoms may be quite different from those in the final stages. These symptoms can generally be placed in the following categories:

1. *Abnormal tissue coloration.* Leaf appearance commonly changes. Leaves may become chlorotic (yellowish); mosaic or mottling patterns may appear, especially with virus diseases.
2. *Wilting.* If the infectious agent interferes with the necessary uptake of water by the host plant, a part or the whole of the plant may die. Verticillium wilt, which blocks the water-conducting tissues (xylem), is one disease that causes this symptom.
3. *Tissue death.* Necrotic (dead) tissue can appear in leaves, stems, or root, either as spots or as entire organs. Decay of soft succulent tissue, as in damping off in

young seedlings, is common. Caukers caused by death of the underlying tissue sometimes appear as sunken, dead tissue on the trunks or limbs of woody plants.

4. *Defoliation.* As the infectious disease progresses, the plant may lose all its leaves and sometimes drop its fruit.

5. *Abnormal increase in tissue size.* Some diseases increase cell numbers or cell size in the plant tissues, twisting and curling the leaves or forming galls on stems or roots (Figs. 11–6 and 11–7).

6. *Dwarfing.* In some cases the pathogenic organism will reduce cell number or size, stunting parts or the whole of the host plant.

7. *Replacement of host plant tissue by tissue of the infectious organism.* This occurs commonly where floral parts or fruits are involved; an example is corn smut, where the ears and tassels become infected and proliferation of the infectious pathogen takes place.

Classification of Infectious Plant Diseases

The pathogens responsible for causing most plant diseases are bacteria, fungi, viruses, mycoplasma-like organisms, parasitic seed plants, and nematodes.

Bacteria

Bacteria are typically one-celled organisms with a cell wall. They occur singly or in colonies of cells—in pairs, chains, or clusters. Bacteria are found almost everywhere on earth in vast numbers. There are both beneficial and pathogenic bacteria. The beneficial bacteria are involved in such diverse processes as digestion in animals, nitrogen fixation in the roots of certain plants, breaking down animal and plant remains, and sewage disposal systems. Pathogenic bacteria, on the other hand, cause severe, often fatal, diseases in both animals and plants. Plant pathogenic bacteria belong to six genera: *Agrobacterium, Corynebacterium, Erwinia, Pseudomonas, Streptomyces,* and *Xanthomonas.* Table 11–2 gives some examples of bacterial diseases.

Bacteria are classified into four groups: the spherical cocci, the rod-shaped bacilli, the spiral-shaped spirilli, and a small group of filamentous forms. Only the bacilli and the filamentous types are known to cause diseases in plants. Some types of bacilli and spirilli are motile—they have whiplike flagella that propel them through films of water. Bacteria multiply at alarming rates under suitable conditions by binary fission—splitting into two parts.

Bacterial diseases in plants are difficult to control. Measures include using resistant species or cultivars and bacteria-free seed, eliminating sources of bacterial contamination, preventing surface wounds that permit the entrance of bacteria into the inner tissues, and propagating only bacteria-free nursery stock. Prolonged exposure to dry air, heat, and sunlight will sometimes kill bacteria in plant material. They are also killed by antibiotic treatment.

Figure 11–6 Peach leaves distorted by peach leaf curl, a disease caused by the fungus *Taphrina deformans.* The disease can be controlled by fungicidal sprays applied before the buds open in the spring.

Figure 11–7 Galls on a young fruit tree infected by crown gall bacteria (*Agrobacterium tumefasciens*).

TABLE 11–2	Some Important Plant Diseases Caused by Bacteria		
Common Name of Disease	**Pathogen**	**Hosts Attacked**	**Control Measures**
Crown gall	*Agrobacterium tumefasciens*	Woody ornamentals and tree fruits	Produce nursery stock free of bacteria; chemotherapy of galls; biocontrol with antagonistic bacteria
Bacterial wilt of cucumbers	*Erwinia tracheiphila*	Cucumbers, muskmelons, squash, and pumpkins	Control insects (spotted and striped cucumber beetles) that carry the bacteria
Bacterial wilt of corn	*Erwinia stewartii*	Corn (sweet and dent corn hybrids)	Resistant cultivars
Common blight of beans	*Xanthomonas phaseoli*	Beans—field (dry), garden (snap), and lima	Disease-free seed; sanitations; disposal of crop residues; three-year crop rotations
Fireblight (*PLANT LOOKS BURNT*)	*Erwinia amylovora*	Pome fruits—apple, pear, quince—some ornamentals, such as pyracantha	Resistant cultivars; cut out diseased tissues; streptomycin sprays at bloom
Bacterial canker	*Pseudomonas syringae*	Almonds, apricots, avocado, cherry, peach, plum	Resistant cultivars; cut out infected areas; spray with Bordeaux mixture
Bacterial soft rot of vegetables	*Erwinia carotovora*	Vegetables with fleshy storage organs	Avoid bruising and mechanical damage during harvest. Use good ventilation during storage at 2°C to 4°C (36°F to 40°F).

Bacteria that cause plant diseases are spread in many ways—they can be splashed about by rains or moved on wind-blown dust, the feet of birds, or on insects. People can unwittingly spread bacterial diseases by, for instance, pruning infected orchard trees during the rainy season. Water facilitates the entrance of bacteria carried on pruning tools into the pruning cuts. Propagation with bacteria-infected plant material is a major way pathogenic bacteria are moved over great distances.

Most plant tissue under the protective bark layers is sterile. Bacteria may coat the bark surface, but they do not cause infection unless, with moisture present, they gain entry to inner tissues through stomata, or leaf, flower, or fruit scars, or wounds. Spraying with antibiotics is sometimes helpful; streptomycin is used, for example, to reduce fireblight (*Erwinia amylovora*) infection in pears. But few chemicals are really effective in controlling bacterial plant diseases.

Fungi

Fungi are classified into more than 50,000 species. Unlike the green plants, they do not contain chlorophyll, do not photosynthesize their own food, and must obtain their nutrients from some other source—a substrate of living or dead plant or animal tissue.

Fungi can be grouped into:

1. Obligate saprophytes, which live only on dead organic matter and inorganic materials.
2. Obligate parasites, which live and develop only on living tissues.
3. Facultative saprophytes, which are normally parasitic on living tissues but can live as saprophytes on dead tissue under certain conditions.
4. Facultative parasites, which are normally saprophytic but can live parasitically under certain conditions.

Some fungi are able to live on only one host species, while others develop on many different kinds.

Fungi, like bacteria, can be beneficial as well as pathogenic. Beneficial fungi participate in biological cycles, decaying dead animal and plant materials and thus converting them into plant nutrients that are absorbed by living plants. Some beneficial fungi grow in a symbiotic relationship with the root cells of higher green plants; this combination is termed a **mycorrhiza.** Roots of many cultivated plants—corn, soybeans, cotton, tobacco, peas, red clover, apples, citrus, pines, aspens, birches, and others—have mycorrhizal relationships with soil fungi. The mycorrhizae appear to be highly beneficial, often necessary, for optimum growth of many plants. Interest and research in such fungi-root symbiotic relations is considerable because establishing proper mycorrhizal fungi with cultivated plants offers a great potential for improved plant growth.

Certain fungi produce useful antibiotics and enzymes. *Pencillium* fungi produce the famous penicillin G, which has prevented countless deaths from bacterial infection, acting by inhibiting formation of the bacteria's cell wall. Many food-producing processes, such as the making of bread, wine, beer, and cheese, are based upon the activities of fungi. Mushrooms, which are fungi, are important as food.

Most plant diseases are caused by fungi, and the food loss to fungal diseases is staggering (Fig. 11–8). Some of the world's great famines can be blamed on pathogenic fungi. Wheat crops of the Middle Ages were ruined when the grains became infected with a dark, dusty powder now known to be

Figure 11–8 Fungal diseases can cause considerable damage to grain crops. Shown here is barley scald caused by the fungus *Rhynochosporium secalis*. The head on the left is only slightly affected while the one on the right is so severely damaged that the lack of photosynthesis has prevented the grains from filling. *Source:* University of California Cooperative Extension.

the spores of the fungus called bunt or stinking smut (*Tilletia* spp.). The potato blight in Ireland and northern Europe, rampant during two successive seasons (1845–1846 and 1846–1847), was caused by the fungus *Phytophthora infestans.* It resulted in the death of more than 1 million people by starvation and caused mass migration from the country. In the 1870s, an epidemic of downy mildew, caused by the fungus *Plasmopara viticola,* struck the grape vineyards of central Europe, causing great losses to the grape growers and wine makers. In the United States alone, hundreds of millions of bushels of wheat have been lost in epidemic years to stem rust (*Puccinia graminis tritici*).

Fungi are simple organisms. The vegetative body, or mycelium (pl., mycelia) of a typical fungus is made up of very small filaments or threads called hyphae that branch in all directions throughout the substrate. The hyphae are filled with protoplasm containing nuclei. The mycelia of pathogenic fungi absorb food from the cells of the host plant. Some fungi produce structures called sclerotia, which are dense, compact masses of hyphae. These species readily resist unfavorable environments, remaining alive until conditions are more favorable and then resume growth. Fungi reproduce by both sexual and asexual spores, which are similar to the seeds of higher plants except that they lack an embryo. The spores germinate under favorable conditions producing hyphae and mycelia.

Fungal diseases of plants are generally easier to control than bacterial or viral diseases. One of the most satisfactory methods of dealing with fungus diseases is strict sanitation to eliminate the pathogenic organisms, starting

with the initial stages of propagation and growth of the potential host plants.

Control measures include: OF FUNGI

1. Soil pasteurization (moist heat at 82°C[180°F] for 30 minutes)
2. Soil fumigants and drenches containing fungicides
3. Planting only disease-free, certified seed
4. Seed treatments with fungicides
5. Foliage sprays with fungicides (protectants or eradicants)
6. Planting only resistant species and cultivars
7. Maintaining good soil drainage
8. Removing crop residues by burning or burying
9. Crop rotation
10. Growing crops in climates unsuitable for pathogenic fungi
11. Careful handling of the crop (vegetables and fruits) to prevent cuts and bruises during harvest and transit
12. Storage of crop products at the proper low temperatures
13. Postharvest treatment of fruits and vegetables with fungicides
14. Biological control by means of an organism, usually another fungus, that is antagonistic to the fungal pathogen

In many of the major crops, cultivars resistant to prevailing diseases are available, and more are continually being developed by plant breeders. Obviously, growers should use such cultivars if they are available.

Several examples of cultivars genetically resistant to fungus disease are notable. Certain hybrid potato cultivars are resistant to late blight *(Phytophthora infestans).* Soybean cultivars resistant to downy mildew *(Peronospora manshurica)* have been developed. In the United States, apple cultivars have been developed at the Indiana and the New York agricultural experiment stations that show high resistance or immunity to apple scab *(Venturia inaequalis),* a devastating disease of apples grown in cool humid climates with summer rainfall. In the cereal crops (oats, wheat, rye, barley), powdery mildew *(Erysiphe graminis)* can be controlled only by the use of resistant cultivars developed by plant breeders. Tomatoes can be grown in *Fusarium*-infested soils only if *Fusarium*-resistant cultivars are planted. Plant breeders are continuously breeding wheat cultivars resistant to stem rust *(Puccinia graminis tritici),* but the fungus continuously mutates, attacking the formerly resistant cultivars. Still newer types then have to be developed. The importance of resistant cultivars in controlling fungus diseases is shown in Table 11–3.

Although complete eradication of the pathogen and the use of resistant cultivars are the most satisfactory ways of dealing with fungus disease, in many instances these measures are not possible. Often the disease appears and its

TABLE 11-3 Some Important Plant Diseases Caused by Fungi and Some Measures for Controlling Them

Common Name of Disease	Pathogen	Hosts Attacked	Control Measures[a]
Stem rust of wheat	Puccinia graminis tritici	Wheat	Eradicate barberry plants (an alternate host), and use resistant cultivars.
Corn smut	Ustilago maydis	Corn	Use resistant cultivars.
Fusarium wilt	Fusarium oxysporum	Tomato, pea, celery, banana, cotton, watermelon	Use resistant cultivars.
Powdery mildews	Erysiphe polygoni	Many hosts	Foliar fungicides except on cereals, where resistant cultivars should be used.
	Podosphaera leucotricha	Apples	
	Uncinula necator	Grape	
	Sphaerotheca pannosa	Rose	
	Erysiphe cicoracearum	Cucurbits	
	Erysiphe graminis	Cereals	
Rust	Puccinia striiformis and P. graminis	Turfgrasses, especially bluegrass and ryegrass	Keep lawn growing rapidly by fertilization; use fungicides; avoid frequent irrigations.
Brown patch	Rhizoctonia solani	Turfgrasses, especially bent grass, bluegrass, bermuda grass, fescues	Avoid excess nitrogen fertilization. Improve drainage. Use certain fungicides.
Brown rot	Monilinia fructicola and M. laxa	Stone fruits: peaches, apricots, plums, cherries, almonds, nectarines	Use resistant cultivars where available. Sanitation. Fungicide sprays. Prevent fruit injury during harvest. Refrigerate harvested fruit.
Apple scab	Venturia inaequalis	Apples	Protectant and eradicant sprays with fungicides. Use resistant cultivars where available.
Peach leaf curl	Taphrina deformans	Peaches and nectarines	Dormant spray of fungicide after leaf fall and again before bud break.
Verticillium wilt	Verticillium albo-atrum and V. dahliae	A wide range of woody fruit and ornamental species; many kinds of herbaceous plants	Plant only resistant species and cultivars. Remove infected limbs or plants and burn. Soil fumigation.
Downy mildew of grape	Plasmopara viticola	Grape	Foliar sprays with a fungicide.
Late blight of Potato	Phytophthora infestans	Potato and tomato	Use resistant cultivars. Destroy all cull potatoes in field. Spray with fungicides during growing season. Store potatoes at 2°C to 4°C (36°F to 40°F).
Damping off	Pythium and Rhizoctonia	Seedings of various species	Fungicidal seed treatments. Soil fumigation or pasteurization. Improve soil drainage. Germinate seedlings at temperatures unfavorable for pathogen growth.
Dutch elm disease	Ceratocystis ulmi	Elm trees	Remove sources of infection and control European elm bark beetle, which spreads the fungus.

POWDERY MILDEW — SULFUR DUST

[a]Detailed directions on control of plant diseases can be obtained from local offices of the agriculture extension service or from commercial garden centers.

Agricultural fungicides are strictly controlled. In the United States, regulations of the federal Environmental Protection Agency (EPA) and of the States' agriculture departments specify on what crops and at what times and concentrations these chemicals can be applied. The regulations protect the applicator and the consuming public. Read package labels carefully and strictly adhere to their directions.

development must be slowed or stopped by whatever means are available. Chemical sprays (fungicides) are an old and well-proven procedure for controlling many plant diseases, and they are relied upon heavily for certain fungus problems. It was only about a century ago that chemical weapons were first devised against plant diseases that had been destroying crops for countless centuries. There are now more than 100 useful fungicides, and new ones are continually being added and older ones dropped. It is best to seek advice from the local agricultural extension service, pest control consultant, or garden shop about which fungicides to use.

Viruses

Viruses are pathogenic particles that infect most higher plants and animals. Virus particles are extremely small (about 20 to 250 millimicrons) and can be seen only with an electron microscope. They consist of an outer sheath of protein and an inner core of nucleic acid that is usually, but not always, ribonucleic acid (RNA). Viruses do not carry on the usual functions of living organisms such as respiration and digestion. They are obligate parasites; that is, they cannot multiply and grow except within the host of insect vector cell. They can be removed from the host, however, and still cause infections. They force the host cell to transform its own components into virus substances that are translocated throughout the plant, often injuring or killing it. The virus carries the genetic information for its replication by the host. Many kinds of plants carry virus particles and show no symptoms. Some viruses, such as the cowpea mosaic virus, occur as a complex of two component particles, each containing different nucleic acid cores. Both components have to be present in a plant for infection and replication to take place.

Whether viruses are "living" entities or not depends upon the definition of *living*. Viruses are certainly not independent living systems. They are unable to develop their own energy for multiplication, having to depend upon the enzyme systems of recognized living organisms. Yet they do exhibit the three interrelated characteristics of living things: reproduction, variation, and selective survival.

To move from one plant to another, virus particles must have some transmitting carrier (vector). The vectors can be insects—most commonly aphids, leafhoppers, thrips—or mites, as well as certain primitive soil-borne fungi, nematodes, pollen grains, or infected seeds (carrying the virus from one generation to another). The activities of humans in propagating plants by budding and grafting or by

cuttings is one of the chief ways viral diseases spread. In fact, virus investigators use grafting and budding procedures to transmit and detect viruses in their studies. The seedling offspring of a virus-infected plant is usually, but not always, free of the virus, depending upon the plant species and the kind of virus.

Viruses are difficult to classify and, for want of anything better, they are given descriptive names based upon a disease they cause—for example, tobacco ringspot, tobacco mosaic, citrus psorosis, citrus tristeza, sugar beet curly top, lettuce mosaic, maize dwarf mosaic, peach yellow bud mosaic, *Prunus* necrotic ringspot, carnation streak, and tomato spotted wilt. Many of these viruses also infect plants of other species. For example, tobacco ringspot virus causes a bud blight in soybeans; maize dwarf mosaic infects sorghum, Sudan grass, sugar cane, and Johnson grass in addition to corn, but it still retains its original name.

Once whole plants are infected, little can be done to free them from the virus. No chemical sprays eradicate viruses, although insecticides can be used to control insect vectors. Since different cultivars and species show different degrees of resistance to some viruses, resistant types should be planted whenever they are available. For orchard species, the best control measure is the planting of nursery trees that have been propagated from known virus-clean sources. The citrus industries in both Florida and California, for example, have set up certification and registration programs to assure that citrus nursery stock is propagated from the most pathogen-free propagative materials available.

All plants of many vegetatively propagated cultivars are infected with viruses, either latent or active. Virus particles, occurring systematically throughout the plants, are readily moved along during the vegetative propagation procedures from the mother plants to all daughter plants.

Fortunately, some viruses are permanently inactivated by prolonged exposure of infected tissue to relatively high temperatures—for example, 20 to 30 days at 38°C (100°F). This procedure, called heat therapy, frees individual plants or cuttings of the viruses. The clean tissue is then used as a propagative source, allowing large-scale production of new plants of the clone without the virus infection. This has been done with many cultivars of fruit and ornamental species. If insect vectors are kept out of the new virus-clean plantings, subsequent reinfection is unlikely, particularly if the planting is at a distance from virus-infected plantings.

Another successful way to eliminate viruses, particularly from herbaceous plants, is to excise the minute shoot tip of vigorously growing plants under aseptic conditions, then allow the tip to develop into a new plant on a nutrient medium. The new plant will usually be free of the virus and will provide a starting point for a clone minus the virus. This procedure is based on the fact, still not well understood by plant virologists, that the virus is usually not present in the actively growing shoot tip of an infected plant. This procedure has been used to clear many herbaceous cultivars of viruses.

Viruses in an infected mother plant usually (but not always) fail to appear in the daughter seedlings. However, if the clone is heterozygous, the new seedlings lack the characteristics of the clone. If, however, the parent plant is a type, such as citrus, that produces apomictic seedlings (i.e., the embryo in the seed arises from the nucellus rather than from a zygote, so that the seedling has the same genetic makeup as the mother plant) then the clonal characteristics are retained in the seedling plant. This procedure is the basis for the development of many vigorous nucellar cultivars of citrus species.

Mycoplasma-like Organisms

The mycoplasmas in animals and the mycoplasma-like organisms in plants are small parasitic organisms intermediate in size between viruses and bacteria.

The cells of mycoplasmas have a three-layered membrane enclosing the living protoplasm; they lack constant shape. In plants, mycoplasmas occur in the phloem and multiply rapidly, disrupting translocation of food materials and usually causing symptoms of yellowing, wilting, distortion, reduced leaf size, and stunting.

Mycoplasmas have long been known to infect animals, including humans, but it has only been since about 1967 that organisms of this type have been recognized as a cause of plant diseases. Many plant diseases formerly attributed to viruses are now known to be caused by mycoplasma-like organisms. Some of the diseases they cause are aster yellows, western-X of peaches, cherry buckskin, pear decline, mulberry dwarf disease, corn stunt, and stubborn disease of citrus. As with viruses, a disease caused by mycoplasma-like organisms is named after the plant on which it was first studied, but it can also occur on many other plants. For example, aster yellows also affects other ornamentals—gladiolus and phlox, for example—and tomato, spinach, onion, lettuce, celery, carrots, strawberry, and many weeds.

Mycoplasmas, in contrast to viruses, have some characteristics of living matter. They can reproduce themselves, they have energy and enzyme systems, and their genetic information, like viruses, is stored as DNA and RNA. Their lack of a cell wall and their smaller size differentiate them from bacteria.

Species and cultivars vary distinctly in their resistance to mycoplasmas. For example, the common pear *(Pyrus communis)* is highly resistant to the pear decline organism, whereas the oriental pear *(Pyrus pyrifolia)* is very susceptible.

Like viral disease, the infective bodies in diseases caused by mycoplasma-like organisms are moved about by sucking insects such as leafhoppers, aphids, and psylla. Studies of corn stunt provide evidence that once the insect vectors establish the infective particles in their bodies, the insects retain the ability to transmit them the rest of their lives.

One obvious method of controlling the spread of these diseases is an effective spray program that eliminates the insect vectors. It has been established, too, that mycoplasma-like organisms are susceptible to certain antibiotics, particularly tetracycline, which has been used to treat pear trees with the pear decline disease.

Parasitic Seed Plants

Some higher plant forms live on the surface of or parasitize other plants and often cause harmful reactions in their hosts. These plants can be placed in three groups: the epiphytes, the hemiparasites, and the true parasites.

The **epiphytes** do little or no harm to their host plants, using them merely for physical support and protection. Examples are Spanish moss and epiphytic orchids, which in their native habitat commonly grow on tree limbs (Fig. 11–9).

The hemiparasites, sometimes called water parasites, do injure their host plants, absorbing water and mineral nutrients from them. However, they possess chlorophyll and can manufacture their own carbohydrates by photosynthesis.

Figure 11–9 The epiphytic plant, Spanish moss *(Tillandsia usneoides)*, a member of the pineapple family, is commonly found attached to trees throughout the southern United States. The tree shown here was near Baton Rouge, Louisiana. Spanish moss absorbs most of its nutrients and water directly from the atmosphere and only uses its host plant as a place to obtain support and light for photosynthesis. It gets no food or water from the host plant as do the parasitic plants. Similar appearing plants, which are called lichens (an algae and a fungus living in an intimate symbiotic relationship as a composite plant), are members of the *Usnea* genus and are sometimes called "old man's beard."

Witchweed *(Striga asiatica)* is a hemiparasitic seed plant that severely damages sugarcane, corn, sorghum, many other grasses, and some broad-leaved plants. It attaches itself to the host's roots and utilizes most of the host's water and mineral nutrients, causing it to wilt, yellow, stunt, and die. The best control is to plant a crop, such as Sudan grass, that stimulates the witchweed seed to germinate, then plow under the entire field. Crops should be rotated and susceptible crops should not be planted.

Mistletoe *(Phoradendron* spp.), another member of the hemiparasitic group, attacks many broad-leaved trees such as Modesto ash, silver maple, honeylocust, hackberry (see Fig. 11–10), cottonwood, walnut, oak, birch, and some conifers. The seeds germinate on limbs of susceptible hosts, forming an attachment disk on the bark. The sticky berries are disseminated throughout the tree and from tree to tree by birds and wind. The usual control, although not very effective, is to cut out the mistletoe branches deep into the tree under the point of attachment. No good herbicidal control has been developed, although a dormant-season application of ethephon, an ethylene-releasing material, is a possible control measure on some plants. The best control is to plant only tree species resistant to mistletoe attacks.

True parasites lack chlorophyll and depend upon their hosts for all nourishment—carbohydrates, minerals, and water. Examples of this group are the dwarf mistle-toe *(Arceuthobium* spp.) and dodder *(Cuscuta* spp.). Broomrape *(Orobanche* spp.) is a serious parasitic pest in Europe and has caused extensive damage to tomatoes in California.

Dwarf mistletoe attacks many coniferous species in the western United States, reducing tree vigor and lowering lumber quality. The sticky seeds (not the fruits) are forcibly ejected and can travel up to about 19 m (60 ft). This is the principal means of dissemination. Birds are known to carry the sticky seeds on their feathers, but wind plays a very minor role in dissemination. The best control is removal of infected trees.

Dodder has many species, but about six cause the major damage, attacking such crops as alfalfa, lespedeza, clover, flax, sugar beets, and some vegetable crops as well as some ornamentals (see Fig. 11–11). Dodder seriously reduces yields and quality of crops. Strict regulations prohibit the sale of crop seed contaminated by dodder seed. Great effort should be taken to avoid planting seed that has dodder seed mixed with it. Patches of dodder in field crops or along fences or ditch banks should be eradicated by burning or by herbicides.

Nematodes

Plant parasitic nematodes are microscopic, eel-like worms that attack the roots, stems, foliage, and inflorescenses of

Figure 11–10 Branch of a hackberry *(Celtis sinensis)* tree being killed by growth of mistletoe.

Figure 11–11 Olive branches attacked by dodder.

plants. They range in length from 0.5 mm to 3 mm (0.02 to 0.125 in.). Nematodes are not closely related to earthworms, wireworms, or flatworms.

A number of genera and species of nematodes are highly damaging to a great range of hosts, including foliage plants, vegetable crops, fruit and nut trees, and forest trees. Some of the most damaging nematode species are:

Root knot *(Meloidog-yne* spp.)

Cyst *(Heterodera* spp.)

Root lesion *(Pratylenchus* spp.)

Spiral *(Helicotylenchus* spp.)

Burrowing *(Radopholus similis)*

Bulb and stem *(Ditylenchus dipsaci)*

Reniform *(Rotylenchulus reni-formis)*

Dagger *(Xiphinema* spp.)

Bud and leaf *(Aphelenchoides* spp.)

Typical root symptoms indicating nematode attack are root knots or galls (Fig. 11–12), root lesions (Fig. 11–13), excessive root branching, injured root tips, and stunted root systems. Symptoms on the above-ground plant parts indicating root infection are a slow decline of the entire plant, wilting even with ample soil moisture, foliage yellowing, and fewer and smaller leaves. These are, in fact, the symptoms that would appear in plants deprived of a properly functioning root system. Bulb and stem nematodes produce stem swellings and shortened internodes. Bud and leaf nematodes distort and kill bud and leaf tissue.

Parasitic nematodes are readily spread by any physical means that can move soil particles about—equipment, tools, shoes, birds, insects, dust, wind, and water. In addition, the movement of nematode-infected plants or plant parts will spread the parasites.

Various methods are available to reduce crop losses from nematodes:

1. **Plant only resistant species and cultivars.** For example, in an area with soil heavily infested with the root-knot nematode, plant apricots, cherries, apples, pears, or plums, which are resistant, rather than peaches or nectarines, which are highly susceptible. (A root-knot nematode-resistant peach rootstock called 'Nemaguard' developed by USDA plant breeders is available, thus permitting peach production even on infested soils.) Certain vegetable crops—sweet corn, asparagus, and cabbage—are resistant to root-knot nematodes whereas radishes are susceptible. Resistant ornamentals include the African marigold, azalea, camellia, and oleander. In Long Island, New York, where the golden nematode is a serious problem for potato production, resistant cultivars are available.

2. **Use only nematode-free nursery stock for planting.** In most countries government nursery inspectors will condemn and destroy any nursery stock showing evidence of nematode infestation.

Figure 11–12 Galls on tomato roots caused by root-knot nematodes.

Figure 11–13 Nematodes can severely damage white potato tubers. These microscopic roundworms drill into the tubers, damaging the skin, disfiguring the potatoes, and making them unmarketable. *Source:* University of California Cooperative Extension.

3. **Avoid importing soil (or plants with soil on their roots) from areas that could be loaded with a dangerous nematode species new to the area.** United States plant importation regulations forbid the

introduction of plants with soil on their roots from other countries.

4. **Treat the soil area with fumigant before planting.** V. TOXIC — Methyl bromide is often used to reduce the nematode population to levels not harmful to plants. Soil mixes for container-grown plants can either be treated with a fumigant or steam-pasteurized at 82°C (180°F) for about 30 minutes. This method is too expensive for field crops other than commercial strawberry fields. The impending loss of methyl bromide may seriously affect the crops where it is used.

5. **In nursery operations, use benches raised off the ground and pot plants only into pasteurized soil mixes.** Keep containers, bins, benches, and flats clean. Fumigate outdoor growing fields where nursery stock will be grown.

6. **Use nematicides in certain cases.** All nematicides are poisonous and must be used carefully, following the directions on the containers exactly. Most such materials will injure or kill plants if applied too close to their root zone.

7. **Rotate crops to control certain nematodes.** Rotation is useful for types that have a narrow host range, such as sugar beets attacked by the cyst nematode. Where the crop value is too low to justify large-scale soil fumigation, crop rotation is the only practical method of nematode control.

Figure 11–14   Insect swarms can devastate crops, as in this stand of corn ruined by grasshoppers. Grasshoppers can be controlled biologically by dropping into infested areas wheat bran sprayed with spores of *Nosema locustae,* which causes a deadly disease of grasshoppers and crickets. This disease does not harm people, plants, or other animals, affecting only grasshoppers and Mormon crickets. *Source:* USDA.

INSECTS AND MITES

Plants and plant products are the chief sources of food for insects. The result is a fierce competition between humans and insects for plants. A never-ending battle has gone on since early humans started cultivating agricultural crops (Fig. 11–14). The battle is likely to intensify in the future as the world's population increases and increased food supplies become more important. Insects also pose a secondary hazard in that they spread diseases among plants and animals.

Some insects, however, favor agriculture. Certain harmless insect species prey on other insect species that destroy plants or crops. The praying mantis and ladybird beetle, both larvae and adults, feed on aphids. The larvae of the green lacewing feeds on mealybugs, scale, and aphids. Predatory *Aphytus* wasps attack red and black scale. Starts of these beneficial insects can be purchased through mail order catalogs, or at nurseries or garden centers, packaged and ready for release in the garden. Once a biological balance is established between the host (harmful insects) and the predatory or parasitic insects, then no further chemical spraying is necessary. In fact, if it is done, the beneficial insects are likely to be killed along with the harmful ones.

The honeybee and certain other insects do a tremendous service in pollinating fruit trees and other crops, such as alfalfa. Some insects produce useful products, like honey

and beeswax from the honeybee and silk from the silkworm. Certain insects are relished as food in some parts of the world. The *Drosophila* fly, because of its short life cycle, ease of handling, and the giant chromosomes of its salivary gland, has been of inestimable value to geneticists in their studies.

Nevertheless, there are innumerable kinds of insects and mites that, if left uncontrolled, would soon reduce the world's food supply to a shambles. Control methods for harmful insects are of four types:

1. **Biological control** (Figs. 11–15 and 11–16). This entails the use of insect parasites, predators, and disease-causing pathogens. These are produced in culture for release into, or application to, infested fields. In the case of parasites and predators, another form of biological control involves the preservation of those that are naturally occurring through careful use of chemical or other forms of control. Most insect pathogens in use today are not capable of perpetuating themselves in the field, and must be applied—usually as a spray—each time they are needed.

2. **Cultural control.** Many cultural practices, such as crop rotation, destruction of crop debris after harvest, careful attention to planting or harvesting date, proper pruning, and the use of pest-resistant cultivars, are useful in preventing or mitigating serious infestations

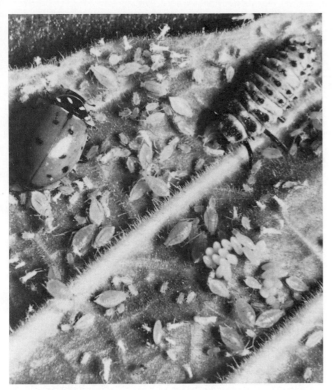

Figure 11–15 Biological control of destructive insects: the lady-bird beetle adult *(upper left)* and larva *(upper right)* feeding on aphids in various growth stages. *Source:* USDA.

To tell the difference between insects, mites, spiders, and ticks:

INSECTS
The adults are characterized by:
1. three definite body regions: head, thorax, and abdomen
2. three pairs of jointed legs
3. one pair of feelers or antennae
4. eyes that are usually compound
5. one or two pairs of wings

MITES, SPIDERS, AND TICKS
These are characterized by:
1. two main body regions—the cephalothorax (head and thorax fused together) and the abdomen
2. four pairs of jointed legs
3. a lack of antennae or wings
4. simple eyes

3. **Physical or mechanical control.** Leaving fields fallow during critical periods in an insect's life cycle, screening of seed beds or ventilators on greenhouses, disking soil beneath tree and vine crops, and burying or chipping prunings are examples of physical practices which control specific insects. Hand-picking insects, syringing plants with water to dislodge insects, and picking up and destroying fallen fruit are examples of physical controls useful in small-scale plantings.

4. **Chemical control.** When confronted by an imminent pest infestation, the farmer often has little recourse but to rely on the application of a chemical insecticide or miticide. For many pests there are no good alternatives, because biological, cultural, or physical controls are known not to be sufficiently effective against certain insects.

Classification of Insects

Just to classify insects is a monumental task. They dominate the land fauna with 850,000 species representing about 80 percent of the world's known animal life. If biological success is judged by numbers of species, numbers of individuals, wide distribution and adaptability, and persistence, the insects are perhaps the most successful animals.

The insect orders with the most species are:

Coleoptera (beetles)

Lepidoptera (butterflies and moths)

Hymenoptera (ants, bees, wasps)

Diptera (true flies)

Insects can be further classified by the type of metamorphosis they exhibit. The four types and some examples are shown in Fig. 11–17. This classification is of considerable value in identification.

Another distinction, important in planning control measures for insects attacking plants, is made between the

Figure 11–16 Natural biological control of a destructive insect: a hy-menopterous parasite injects eggs into the body of a much larger larval insect (the large, hairy form with its head on the right). *Source:* USDA.

of certain insects. Proper fertilization and irrigation schedules are useful in maintaining tree crops in a state of vigor sufficient to prevent certain insects from exploiting them.

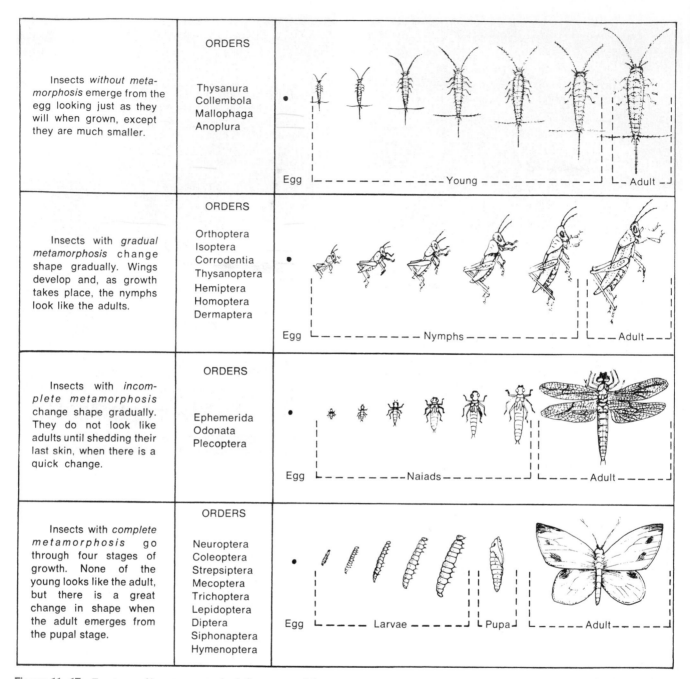

	ORDERS	
Insects *without metamorphosis* emerge from the egg looking just as they will when grown, except they are much smaller.	Thysanura Collembola Mallophaga Anoplura	Egg └ ─ ─ ─ Young ─ ─ ─ ┘ └ Adult ┘
Insects with *gradual metamorphosis* change shape gradually. Wings develop and, as growth takes place, the nymphs look like the adults.	Orthoptera Isoptera Corrodentia Thysanoptera Hemiptera Homoptera Dermaptera	Egg └ ─ ─ Nymphs ─ ─ ┘ └ Adult ┘
Insects with *incomplete metamorphosis* change shape gradually. They do not look like adults until shedding their last skin, when there is a quick change.	Ephemerida Odonata Plecoptera	Egg └ ─ ─ Naiads ─ ─ ┘ └ Adult ┘
Insects with *complete metamorphosis* go through four stages of growth. None of the young looks like the adult, but there is a great change in shape when the adult emerges from the pupal stage.	Neuroptera Coleoptera Strepsiptera Mecoptera Trichoptera Lepidoptera Diptera Siphonaptera Hymenoptera	Egg └ ─ Larvae ─ ┘ └ Pupa ┘ └ ─ Adult ─ ┘

Figure 11–17   Four types of insect metamorphosis from egg to adult.

chewing insects and sucking insects. The chewing insects, such as caterpillars and larvae, and adults of certain other orders feed on foliage, shoots, flowers, and fruit (Fig 11–18). Poisonous sprays or dusts applied to the entire plant are consumed by the insect, causing its death. Sucking insects, such as aphids, leafhoppers, and scales, consume little, if any, of the foliage or surface plant parts. Their mouth parts probe into the interior of the plant and suck out plant fluids (Figs. 11–19, 11–20, and 11–21). Some inject toxic saliva into plants as they feed. They are killed by direct contact with an insecticide as it is applied, or by moving across a plant surface to which the insecticide has recently been applied.

Action of Pesticides Used to Control Insects and Mites

Stomach Poison Action

These materials enter the insect by mouth and kill by absorption into the body through the digestive tract. Formerly, such poisons were applied only to plant surfaces, but in recent years **systemic insecticides** have also become available. These penetrate plant parts and will kill chewing or sucking insects. Systemic materials can be applied as foliar sprays or to the soil where they are absorbed by the roots and translocated throughout the plant.

Figure 11–18 The cabbage looper, an insect with chewing mouth parts, feeds on soybean leaves. Stomach poison insecticides sprayed on the leaves control such pests. *Source:* USDA.

Figure 11–19 A sucking insect, the green peach aphid. *Source:* Robert O. Schuster.

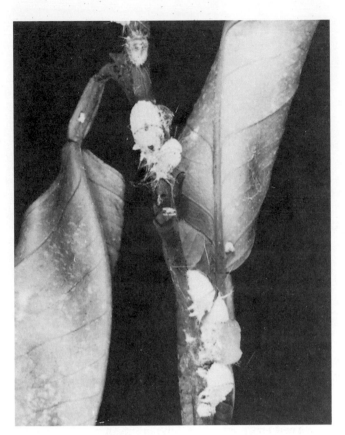

Figure 11–20 Cottony cushion scale *(Icerya purchasi)* feeding by sucking on a branch of an orange tree. *Source:* University of California Cooperative Extension.

Figure 11–21 Olive fruits damaged by parlatoria scale.

Contact Action

These materials are applied so as to contact the body of the insect and act by affecting its nervous or respiratory centers.

Fumigation

Certain toxic volatile chemicals enter the insect's body in a gaseous form through the respiratory system. Fumigants are generally used in enclosed spaces but can also be used in the soil if the surface is sealed with water or covered tightly.

Suffocation

Insects require oxygen to sustain life. Any material that coats their bodies and seals out air will cause death. Spray oils are commonly used on fruits and ornamentals, either as dormant or summer foliage sprays, to control certain scale insects and mites. Oil coatings plug the spiracles (breathing holes) on the sides of their bodies, thus killing by suffocation.

Desiccation

The outer body wall of insects and mites is covered with an oily or waxy protective layer, which prevents loss of body moisture. Any material that will absorb this oily coating can cause excessive water loss and lead to desiccation and death. Such materials have not been used much on agricultural crops although they are effective in controlling household pests such as termites, cockroaches, ants, and crickets. The most effective materials are silica aerogels prepared as extremely lightweight powders that are blown into areas infested with such pests.

Repellent Action

Some materials applied to plants repel certain insects and prevent them from feeding or laying eggs. For example, Bordeaux mixture, a common fungicide, can act as a repellent against the potato flea beetle, psyllid, and leafhopper. Certain insecticides also have some repellent properties.

Attractant Action

Studies of insect physiology have shown that the female of many species—the codling moth, pink bollworm moth, and the cabbage looper, for example—secrete a material that even in extremely minute quantities strongly attracts the male and induces mating. Some of these materials, called **pheromones,** have been analyzed chemically and synthesized. They are now widely used in traps to detect the pres-ence or abundance of certain insects in the vicinity so that spray chemicals can be applied at the proper times. The use of pheromone attractants will, no doubt, be extended in the future as a control measure for many insect pests.

Hormone Action

Hormones control growth and other activities in insects, just as in other animals. Studies have discovered hormones involved in minute amounts during the normal maturation of the insect. For example, as long as the juvenile hormone is present, the insect remains in an immature stage and cannot advance to the adult, reproductive stage. Some insect hormones have been identified chemically and subsequently synthesized. Several are now in the marketplace for control of agricultural pests and pests of public health importance.

Insecticides

A great many chemicals have been used over the years in controlling insects. Some have been found to be so potentially harmful to humans and other animal life, including beneficial insects, and on the environment that their use is no longer permitted. Use of all insecticides is tightly controlled in most countries. In the United States, insecticides must be approved and registered by the Environmental Protection Agency (EPA), and by state agencies also, before they can be used. Even then there usually are heavy restrictions to confine use to certain plants at certain times and at certain concentrations. Residues on food or feed crops exceeding fixed tolerances subject the product to seizure and destruction. Although no pesticide tolerances have been established for ornamental or forest crops, use of these chemicals on such commodities is nevertheless tightly regulated.

Insecticides can be classified as follows:
1. *Inorganic compounds.* These include arsenic, fluorine, phosphorus, and sulfur compounds. These insecticides have been used little in recent times.
2. *Organic compounds.*
   a. *Plant derivatives.* These include such materials as pyrethrum (from the dried and powdered flowers of Chrysanthemum cinerariaefolium), which is a safe and effective insecticide; rotenone (from the roots of several plants in the pea family); and nicotine from Nicotiana spp. Many others could possibly be developed for commercial use if there were sufficient demand.
   b. *Synthetic organic chemicals.* After World War II, an entirely new concept in insect control emerged with the development of synthetic organic insecticides.
      i. *Chlorinated hydrocarbons.* DDT, first developed in Germany in 1874, was widely and successfully used during and after World War II to control mosquitos, flies, fleas, and many agricultural insects. However, the build-up of DDT in the world's ecosystems to levels many scientists considered dangerous has led to banning its use in a number of counties, including the United States. Other chlorinated hydrocarbons, closely related to DDT, have also been very effective in insect control. They are not easily biodegraded, however, tending to build up in plants and in the soil and to be transmitted into fish, fish-eating birds, meat, and milk products.
      ii. *Organic phosphates.* Some of the compounds in this group were developed in Germany near the end of World War II and were found to have good insecticidal properties. They decompose more rapidly than the chlorinated hydrocarbons, but some of the materials are very toxic to mammals and must be used with great care.
      iii. *Carbamates.* Carbaryl was the first chemical in this group to be widely used. It has low mammalian toxicity and its residual action is short-lived. It is effective against a wide range of both sucking and chewing insects, acting as a contact insecticide.
      iv. *Pyrethroids.* This is a relatively new class of insecticides which mimics properties of pyrethrins such as low mammalian toxicity. They will kill a broad spectrum of insects and have a much longer residual life than pyrethrins.
   c. *Spray oils.* Long used to control scale insects and mites on fruit trees and ornamentals, spray oils are

prepared by the distillation and chemical refining of crude oils. A distillation range is chosen to give an oil fraction that is relatively nontoxic to plant tissue, yet lethal to insects. Spray oils are treated with hot sulfuric acid to remove many of the unsaturated molecules in the oil that cause plant injury.

d. *Fumigants.* Fumigant materials vaporize readily into toxic gases that kill insects by contact or fumigant action. They are used in enclosed spaces or injected into the soil. Some fumigants are extremely toxic to humans and must be applied only by trained and licensed operators using the proper equipment. Cancellation of uses of several fumigants has caused farmers to reevaluate their grain storage procedures and to shift from fumigants to protectants. The insecticide malathion is the best known and cheapest of the grain protectants.

e. *Microbial insecticides.* Some kinds of insects are susceptible to certain bacteria, fungi, and viruses. Often these pathogens cause the spectacular disappearance of insects. Such natural biological control of insects has been encouraged and used commercially. For example, *Heliothis* virus is used commercially to kill the tobacco budworm and the cotton bollworm. It infects only these two insect species. Several species of caterpillars *(Lepidoptera)* can be controlled by the toxins produced by the bacterium *Bacillus thuringiensis.* The toxins are applied in dust suspensions or sprays to foliage, which is then ingested by the caterpillars. These bacteria do not sustain themselves naturally and must be reapplied each time caterpillar control is needed. BT corn and other crops have been genetically engineered to produce the microbial toxin that affects pests.

Important Insect Pests of Agricultural Crops and Plant Products

There are innumerable insect pests of food crops, but certain ones are outstanding for the havoc they have wreaked over the years. A number of these species are listed below. When infestation of these or other pests occur, it is best to consult a local agricultural extension agent, pest control consultant, or garden supply center. Insecticide recommendations are continually changing.

Corn Ear Worm (Heliothis zea)

This insect, with three to five generations per year, occurs all over the world whereever corn is grown. Caterpillars feed on the corn silks and kernels, making the ears wormy. The worm also feeds on the other crops—beans, cotton, lettuce, tomato, alfalfa, clover, peanuts, and tobacco. It is best controlled by insecticides.

Codling Moth (Laspeyresia pomonella)

This insect, with two to three generations per year, attacks apples and pears wherever they are grown throughout the world. It is also a pest on walnuts. The larvae tunnel into the fruits, making them wormy, Unless insecticides are used, up to 90 percent of the fruits can become affected.

Peach Twig Borer (Anarsia lineatella)

This species occurs all through the peach-producing areas of the United States. It also attacks most other stone fruits. Overwintering larvae bore into buds and shoots as they start to grow in the spring. There are two to three broods a year, the later ones feeding directly on the fruit.

Lygus Bugs (Lygus Hesperus and L. elisus)

These insects principally attack alfalfa, Ladino clover, sugar beets, safflower, beans, cotton, and carrots. The sucking mouth parts are inserted into buds, flowers, and young fruits, so damaging the crops that they may be unmarketable. In some seed crops, such as alfalfa, no seeds may develop owing to "blasting" of the flower buds by the lygus bugs.

San Jose Scale (Quadraspidiotus perniciosus)

This scale insect is well established throughout North America, Europe, and Asia, attacking most fruit crops and many ornamental trees and shrubs. Heavy infestations of scale can reduce tree vigor and cause death unless they are controlled. Scale spots on fruits reduce their market quality.

Green Peach Aphid (Myzus persicae)

This insect is found throughout the world. It is a sucking pest on many vegetable crops, all stone fruits, and many ornamentals. While its feeding reduces plant vigor, the chief source of damage is its transmission of viral diseases. Viral particles in its salivary fluid are injected into the host plant. Viruses known to be transmitted by the green peach aphid are sugar beet yellows, potato leaf roll, bean mosaic, lettuce mosaic, and cucumber mosaic.

Egyptian Alfalfa Weevil (Hypera brunneipennis)

This is a very destructive pest on alfalfa grown as a hay crop. Principal damage is done by the larvae, which feed on shoot tips, buds, and leaves.

Spider Mites (Bryobia praetiosa, Panonychus ulmi, and Tetranychus urticae)

These pests (which are not true insects) are widely distributed and feed on many species of host plants, including a wide array of vegetable and field crops, greenhouse and nursery plants, fruits, nuts, and ornamentals (see Fig. 11–22). A typical webbing sometimes appears as the mite population increases, with a yellow stippling on leaf surfaces as defoliation begins, resulting from removal of plant fluids by the mites' piercing and sucking mouth parts. Six to ten generations per year can occur. Spider mites have developed resistance to many miticides.

Figure 11–22 Spider mites can be very damaging to many plants. The lily plant on the left has a heavy infestation of mites. The plant on the right is free of mites. *Source:* University of California Cooperative Extension.

Cereal Wireworms (Agriotes spp.)

These are the main insect pests of such cereals as wheat, barley, oats, and rye, especially in the northern growing regions. They are the larval stage of so-called click beetles, which themselves do no harm. Their life cycle spans five years, most of which is spent in the larval wireworm stage in the soil.

Colorado Potato Beetle (Leptinotarsa decemlineata)

This well-known insect, with chewing mouth parts, occurs from Colorado to the eastern United States in all potato-growing areas. It is established on the European continent but has been eradicated from the British Isles. Both adult and larval stages feed on the foliage of the potato plant, completely denuding it. It seldom feeds on other plants. In both stages the insect is large and easily seen.

Cotton Boll Weevil (Anthonomus grandis)

This insect attacks cotton plants in the United States, Mexico (where it originated), and Central America. Probably no insect, other than perhaps the codling moth, has had such an impact on agriculture and probably no other insect has received more study. The larvae develop inside the flower or boll, arising from eggs deposited by the female. They feed on the developing floral parts and fibers. Many generations occur in a single season. Losses are heavy unless insecticides are used.

Sugarcane Shoot Borer (Diatraea saccharalis)

This is the most important insect pest of sugarcane in the Caribbean. Caterpillars feed on the leaves, then enter the stalk and bore into the center, killing the central growing shoot. Later generations bore into side shoots. Such mechanical injury causes the stalks to break and fall over. Yield and quality of extracted juice drops. Control attempts include resistant cultivars, insecticides, and the release of parasites of the borer; none is very effective.

Mediterranean Fruit Fly (Ceratitis capitata)

This notorious insect is of the greatest importance on all fruit species in the Mediterranean countries, south and central Africa (where it originated), western Australia, the west coast of South America, and throughout Central America. It has been kept out of the United States (except for Hawaii) by strict government inspection, quarantine, and eradication measures. The fly attacks peaches, apricots, apples, citrus, bananas, and many other fruits and vegetables. The fly is slightly smaller than the common housefly and is mostly yellow and brown in color, with black markings. It is controlled mainly by poisonous sprays containing attractants and by the release of sterile male flies. Both the Oriental (Dacus dorsalis) and Mexican (Anastrepha ludens) fruit flies (see Fig. 11–23) can also be devastating to fruit crops.

Figure 11–23   Fruit flies, such as the Oriental, Mediterranean, and Mexican, attack many fruit crops and must be kept out of major fruit growing regions if at all possible. The insects in this photograph are adult Mexican fruit flies on an orange fruit. *Source:* USDA.

Grape Berry Moth (Endopiza viteana)

This is the principal insect pest on grapes in Europe, eastern North America, North Africa, and Japan. The larvae feed on developing fruit. Two or three generations can occur during the season. Insecticide spray is the only control measure.

Insects Attacking Dried Fruits

A number of fruits are preserved by drying, either outdoors on trays in the sun or in forced hot-air dehydrators. Important dried fruits are raisins, prunes, dates, figs, apples, peaches, apricots, and pears. All these dried fruits are food for various insects, which are best controlled by fumigation treatments. Several types of insects feed on dried fruits:

> *Beetles:* dried fruit beetle, saw-toothed grain beetle, small darkling beetle, hairy fungus beetle, corn sap beetle, pineapple beetle, and date stone beetle.
> *Moths:* raisin moth, Indian meal moth, almond moth, dried fruit moth, navel orange worm, dried prune moth, and dusky raisin moth
> *Flies:* vinegar fly *(Drosophila)*, soldier fly, blowfly, housefly.

Insects Attacking Stored Grains

It is conservatively estimated that insects destroy at least 5 percent of the world's production of cereal grains, amounting to about 15 million tons annually. Important insects that feed on stored grains are the sawtooth grain beetle, lesser grain beetle, flat grain beetle, red flour beetle, foreign grain beetle, larger black flour beetle, Angoumois grain moth, hairy fungus beetle, granary weevil, and the rice weevil.

Control measures include prompt harvesting, drying the grain with heated air to a low moisture content (11 to 13 percent), and storage in tight, insect-free bins raised above ground. After two to six weeks the grain is fumigated.

RODENTS AND VERTEBRATE WILDLIFE

Rodents, particularly Norway rats, roof rats, and house mice cause great losses to food crops. Such losses occur mainly in stored grains and other food products in open storage, although rats also feed on unharvested fruits and vegetables. Sugarcane fields in Hawaii, for example, are often invaded by rats. Several million MT of grain are lost annually to rats.

The strategy in rodent control is, first, to remove all food and water available to them from the areas they inhabit, and second, to place bait traps containing an anticoagulant rodenticide such as Warfarin in their runways. Control should be done around granaries just before harvest begins. Some rat species, however, show increasing resistance to rodenticides, and in the future the chemicals may not be effective. Chemical sterilants are under development to reduce rodent populations by acting as oral contraceptives.

Young fruit trees and fall-planted seeds in nurseries are often damaged by deer, mice, gophers, squirrels, and rabbits. Effective repellents of rodents and birds have been developed as coatings of forest seeds sown in logged-and-burned-over areas.

Certain birds—particularly crows, ducks, geese, starlings, blackbirds, ravens, magpies, and scrub jays—are a major menace to grain crops and many fruit and nut crops, such as cherries, grapes, prunes, plums, strawberries, almonds, pecans, walnuts, and pistachios. Blackbirds, ducks, and starlings, in particular, can decimate grain crops such as wheat and field and sweet corn just ready for harvest. They can also cause considerable losses in peanut crops. Canadian geese cause extensive damage to golf course greens and fairways. The use of nonlethal chemical repellents is one type of bird control. Scare devices—carbide explosives, shell crackers, and amplified recordings of bird distress calls (Fig. 11–24) are also used with varying degrees of success. Research on chemical reproduction inhibitors may eventually provide the best method of controlling depredating bird populations.

THE SAFE USE OF AGRICULTURAL CHEMICALS—HERBICIDES, INSECTICIDES, FUNGICIDES, MITICIDES, AND NEMATICIDES

The application of agricultural chemicals to food-producing plants must not create a health hazard. Many countries have elaborate procedures for determining whether agricultural chemicals are reasonably safe before they can be registered

Figure 11–24 A nondestructive method of keeping birds from eating the grapes in a vineyard. Amplified bird distress calls are played at intervals during the day at harvest time. *Source:* Blue Anchor, Inc.

for sale to growers of agricultural crops. In the United States, the Environmental Protection Agency (EPA)[3] is responsible for determining the safety of agricultural chemicals; state and local government agencies can also add their own safety requirements.

Before the EPA grants approval for the sale of agricultural chemicals, exhaustive tests are conducted to show:

1. That the product, at the recommended application rate, has low toxicity levels (both acute and chronic) as determined by experiments with test animals

2. An absence of residues in food or feed crops—or, if there is a detectable residue, that is no more than the tolerance level established as safe

3. The fate of residues and breakdown products in the environment—in the soil, runoff water, ground water, or wildlife

4. Whether the product affects the environment by inducing changes in the natural populations of higher plants and animals or of microorganisms

[3]"The Environmental Protection Agency is charged by the United States Congress to protect the nation's land, air, and water systems. Under a mandate of national environmental laws focused on air and water quality, solid waste management, and the control of toxic substances, pesticides, noise and radiation, the Agency strives to formulate and implement actions which lead to a compatible balance between human activities and the ability of natural systems to support and nurture life." From the *EPA Journal*.

The term LD_{50} may be seen on labels of agricultural chemicals. LD means "lethal dose" and refers to the chemical's toxicity. Oral LD_{50} is the dose that will kill 50 percent of test animals ingesting the chemical by mouth. LD_{50} is expressed in milligrams of the chemical per kilogram of body weight of the test animal. The higher the LD_{50} value, the safer the chemical. According to EPA toxicology guidelines, a chemical with an oral LD_{50} of 50 or less must be labeled "DANGER-POISON (FATAL)"; one with an LD_{50} between 50 and 500 is labeled "WARNING (MAY BE FATAL)"; an LD_{50} from 500 to 5,000, "CAUTION"; and one with an LD_{50} over 5,000 is also labeled "CAUTION." All labels must also state "KEEP OUT OF REACH OF CHILDREN."

It is estimated that an agricultural chemical company developing a new pesticide spends $50 million and that six to ten years of research are required to develop information sufficient to satisfy EPA standards.

When the EPA registers a pesticide for use, the label lists very specific restrictions on the product. It is registered for use on a certain crop or crops, to be applied at specific times and at specific concentrations. While a certain herbicide, for example, may be known to control a given weed species, it may not be permissible to use the herbicide to control such weeds if they are growing in a crop for which the herbicide is not registered.

In using pesticides, including herbicides, the warnings given on the following page should be carefully read and followed.

PESTICIDE IMPACTS ON THE ENVIRONMENT

Insecticides, miticides, fungicides, nematicides, and herbicides can be thought of as necessary evils. Without these chemicals, large-scale agriculture and our standard of living as we know it today would not exist. Too little food would be produced to feed the world's 5 billion people, and mass starvation would result.

Unfortunately, however, chemical applications in agriculture do not always do just what they are supposed to do and nothing else. This problem is recognized more and more now, and greater and greater precautions are being taken to avoid unwanted side effects from these chemicals. Government regulations on pesticides have become steadily tighter. Pesticides that are chemically stable and persist in ecosystems are the ones largely responsible for environmental contamination. The law has ordered the replacement of such persistent chemicals as DDT, DDD, dieldrin, aldrin, chlordane, BHC, and heptaclor with the low-persistence organophosphates (malathion, diazinon) and the carbamates (carbaryl and methomyl), which break down rapidly and are not taken up in food chains. There is the risk, however, that large-scale applications of pesticides, particularly herbicides, to agricultural crops, followed by irrigation or heavy rains, can result in leaching of the chemicals into the underground water supply and eventually into drinking water. In fact, traces of several pesticides

---

WARNINGS ON THE USE OF PESTICIDE CHEMICALS AND SUGGESTIONS FOR THEIR PROPER USE

Pesticides are poisonous and should always be used with caution. The following suggestions for using and handling pesticides will help minimize the likelihood of injury from exposure to such chemicals to humans, animals, and crops other than the pest species to be destroyed.

1. Always read and exactly follow all precautionary directions on container labels before using sprays or dusts. Read all warnings and cautions before opening the container. Repeat this process every time you use the pesticide regardless of how often you use it or how familiar you think you are with the directions. Apply materials only in amounts and at times specified.

2. Keep sprays and dusts out of reach of children, unauthorized persons, pets, and livestock. Store all pesticides outside the house in a locked cabinet or shed and away from food and feed.

3. Always store sprays and dusts in their original labeled containers and keep them tightly closed. Never store them in anything but the original container.

4. Never smoke, eat, or chew anything while spraying or dusting.

5. Avoid inhaling sprays or dusts. When directed on label, wear protective clothing and a proper mask.

6. Remove contaminated clothing immediately and wash the contaminated skin thoroughly if liquid concentrates are accidentally spilled on the skin or clothing.

7. Always bathe and change to clean clothing after spraying or dusting. If this is not possible, wash hands and face thoroughly and change clothes. Wash clothing after applying pesticides; never reuse before laundering. Launder this clothing separately from the family wash.

8. Cover food and water sources when treating around livestock or pet areas. Do not contaminate fishponds, streams, or lakes.

9. Always dispose of empty containers so that they pose no hazard to humans, animals, valuable plants or wildlife. Never burn pesticide containers, especially aerosol cans.

10. Read label directions and follow recommendations to keep residues on edible portions of plants within the limits permitted by law.

11. Call a physician or get the patient to a hospital immediately if symptoms of illness occur during or shortly after dusting or spraying. Be sure to take the container or the label of the pesticide used to the physician.

12. Do not use the mouth to siphon liquids from containers or to blow out clogged lines, nozzles, etc.

13. Do not spray with leaking hoses or connections.

14. Do not work in the drift of a spray or dust.

15. Confine chemicals to the property being treated and avoid drift by stopping treatment if the weather conditions are not favorable.

16. Protect nearby evergreen trees and shrubs from the dormant sprays used on fruit trees.

17. Do not use household preparations of pesticides on plants because they contain solvents that can injure plants.

Source: Division of Agricultural Sciences, University of California.

---

have been found in a number of underground water supplies in many states in the United States.

It was pointed out earlier in this chapter that many insect pests are held in check very well by their own natural enemies—often other insect predators. Reducing populations of one serious primary insect pest with insecticides may, at the same time, so reduce the numbers of insect predators feeding on a secondary insect that the secondary pest increases explosively. For example, insecticidal control of codling moth, pear thrips, and pear psylla in pear orchards may be followed by large increases in spider mite populations, because the spray applications have reduced other predaceous mites that feed on the spider mite.

Production of a number of major food crops relies on pollination of the flowers by honeybees and other bees. Many fruit tree species—almonds, apples, plums, and sweet cherries—as well as certain vegetables, such as muskmelons and honeydew melons, and such forage crops as seed alfalfa and seed clover require thorough working of the flowers by bees during bloom. Elimination of bees by haphazard insecticidal applications and drift cannot be tolerated. Some states impose strict legal requirements wherever honeybees could be involved.

Insecticides should not be applied in areas where bees are working, particularly with chemicals highly toxic to bees such as diazinon, Guthion, malathion, parathion, and car-baryl. Some insecticides are relatively nontoxic to honeybees: Aramite, ethion, methoxychlor, Omite, pyrethrins, rotenone, and Tedion.

The state of California requires that beekeepers post their names and telephone numbers on all hives. Anyone planning to apply pesticides in the vicinity must inform the beekeepers of upcoming spray applications. If potentially hazardous insecticides are to be used, spray applicators must notify all beekeepers within a one-mile radius and allow them 48 hours to move their hives.

Effects on Wildlife

There is no doubt that pesticides have harmed wildlife, even though such effects may be difficult to document. Some reported losses of fish and fish-eating birds have resulted from the improper or illegal use of pesticides and sometimes from legal use. Certain of the pesticides most lethal to wildlife, such as DDT, have been withdrawn from general use in the United States.

Some of the adverse effects of pesticides on wildlife have been indirect. For example, in some areas the pheasant population has declined when the weed cover, which had offered protection and nesting places, was cleared out along ditch banks, fence rows, and fallow lands by herbicides.

SUMMARY AND REVIEW

The biological competitors of plants include weeds, disease pathogens (fungal, bacterial, and viral), insects, mites, nematodes, parasitic plants, birds, and animals.

Weeds compete with crops for light, water, and nutrients. Disease pathogens weaken a plant by destroying or damaging plant cells, tissues, or organs. Insects, mites, nematodes, parasitic plants, birds, and animals damage or destroy the plant by feeding on plant parts. In some cases, disease pathogens are spread through the feeding.

Controlling each type of competitor and the crop requires an understanding of the biology of each. Life cycles, biochemistry, and other factors provide vulnerable points to target in a control program. The control of weeds requires knowing when the weed seed germinates what natural controls may exist in the cropping system, and if no natural controls exist, what imposed controls are best. If herbicides are used, you must know what herbicides control the weed and at what life stage the herbicide is effective. You also have to know how the herbicide affects the crop.

For a plant to become infected with a disease pathogen three things must be present: a favorable environment, the pathogen, and a vulnerable host. Removing or reducing any of the three will stop or help control the disease. In most field crop situations it is difficult to control the environment, but in greenhouses and some field situations the environment can be manipulated to reduce favorable conditions for the pathogen. Chemical or biological control of the pathogen may be necessary if the environment is favorable. It is imperative to apply the proper chemical at the proper time for efficacy. Pesticides have little or no effect on bacteria and viruses. Controlling the vector is the best way to control those pathogens. Planting resistant or tolerant species or cultivars can reduce the incidence of disease as can using seeds and vegetatively propagated material that is certified to be free of the pathogen. Crop rotation to interrupt the pathogen's life cycle can be an effective control.

Some insects and mites may be effectively controlled by the removal of crop debris or host weeds. Soil nematodes are most effectively controlled by soil pasteurization or fumigation. Crop rotation will often interrupt the life cycle of a pest. As with disease pathogens, the effective use of a pesticide means applying a chemical with known efficacy against the pest and applying the chemical correctly and at the appropriate time. When using chemicals, be aware of what effect the chemicals may have on natural pathogen predators. If using biological controls as a part of an integrated pest management program, then it is even more critical to be aware of how pesticides affect the biologicals.

EXERCISE FOR UNDERSTANDING

11–1. Using an Internet fact sheet database (such as http://plantfacts.ohio-state.edu/) search for fact sheets on any weed, insect, or disease problem that affects a crop (including home gardens) that interests you. From the fact sheet(s) develop a short strategy for controlling the problem using chemicals, then develop a strategy without using chemicals other than biologicals. Include in each strategy a description of the life cycle of the problem pest.

REFERENCES AND SUPPLEMENTARY READING

FLINT, M.L., and S.H. DREISTADT. 1998. *Natural enemies handbook: the illustrated guide to biological pest control.* Berkeley, Calif.: University of California Press.

JARVIS, W.R. 1990. *Managing diseases in greenhouse crops.* St. Paul, Minn.: Amer. Phytopathological Soc.

NYVALL, R.F. 1999. *Field crop diseases.* Ames, Ia.: Iowa State University Press.

CHAPTER

Harvest, Preservation, Transportation, Storage, and Marketing

KEY LEARNING CONCEPTS

After reading the chapter you should be able to:

♦ Understand the principles of harvesting crops.
♦ Understand the principles of crop preservation and storage.
♦ Understand the principles of transporting crops.
♦ Understand how crops are marketed.

People and their domestic animals use plants to obtain energy to live and enjoy life. Every part of the plant serves these needs—roots, stems, leaves, flowers, fruits, and seeds. We have developed efficient methods and machines for harvesting, storing, and preserving the various plant parts.

HARVESTING

The growth and development of agriculture has been greatly stimulated by the mechanization of many operations, from land preparation to processing various commodities. Agricultural practices constantly change to accommodate machines. Plant breeders develop plants that are better adapted to machine harvest operations. Since the beginning of organized agriculture, crops have been selected for better quality, as well as for easier production. A maximum effort

231

Figure 12–1 Many grain crops are harvested with self-propelled wide swath combine machines. In some large fields, several will operate simultaneously in cutting the plants and separating the grain from the straw. *Source:* James F. Thompson.

Figure 12–2 This tomato harvester is equipped with an electronic device that automatically sorts the tomato fruit by color, discarding the immature and green fruit while retaining those of acceptable red color and maturity.

in recent years has been directed to the development of labor-saving harvesting machines.

Mechanization of crop harvesting began with the cereal grains. The forerunner of the grain combine was the hand sickle and flail, used as far back as 5,000 years ago. The reaper, threshing machine, and combine followed. Other examples of significant machinery development were the cotton picker and cotton gin. Harvest mechanization, whether for grain (Fig. 12–1), sugar beets, cotton, peaches, or tomatoes (Fig. 12–2) has greatly reduced the level of human labor previously needed in the harvest of these crops.

Generally, harvest mechanization was developed for crops intended for processing long before those grown for fresh market purposes. For example, essentially all sour cherries and tomatoes for processing are machine-harvested. Fresh market tomatoes and other fresh market products are still harvested by hand to limit physical injury. With regard to processing tomatoes, plant breeders and engineers cooperatively identified and developed plant and machinery characteristics that would make a mechanized system of harvest feasible. This systems approach was applied to other commodities. On the other hand, it is difficult to adapt some crops, such as table grapes, strawberries, or oranges to mechanization, although breeders and engineers are collectively working toward this goal because continuing economic pressures will encourage mechanization.

Functions of Harvesting Machinery and Equipment

The functions involved in the harvesting and handling operations of plant products are varied and highly dependent upon the particular commodity. These functions include digging, cutting, lifting, separation, grading or sorting, cleaning, conveyance, loading, etc. Such functions are dependent on various factors related to the crop which could include field conditions at the time of harvest, crop maturity, crop use (fresh or as processed product), plant part or portions involved, and crop worth.

Mechanically Harvested Crops

Cotton

Machine harvesting of cotton became common by 1940. Presently, all commercial cotton grown in the United States is harvested by machine, and except in some underdeveloped nations, machines are used extensively. The harvester operates by guiding the plant into vertical rollers equipped with rotating spindles (tines) which, by a combining action, remove the fibers from the plant. Special counter rotating drums remove the fibers from the spindles which are then blown or conveyed into large basket-like trailers for transport to the gin or to storage sites.

A relatively recent practice is to provide for short-term inexpensive storage of harvested cotton until it can be ginned. This is done by placing the harvest cotton into a compression apparatus, near the harvested field site. The compressor will accommodate several trailer loads and produces a compact stack of cotton about 3 meters wide and high, and about 10–12 meters long. The stack is covered with canvas or plastic for moisture protection. When gin capacity permits, this is then taken to the gin for cleaning.

To facilitate harvest, growers use **growth retardants** to arrest vegetative growth and will chemically defoliate cotton plants prior to mechanical harvest to reduce the content of leaf trash that otherwise contaminates and makes cleaning the fibers more difficult. Materials such as ammonia (NH_3), sodium chlorate, and various herbicides are used for defoliation.

In some situations when uniform boll opening is not obtained, a second harvest can be made if the yield and quality justify the harvesting costs. Second-pick cotton is of lower grade and value.

Grain Crops

Cereal grains—barley, corn, oats, rice, rye, sorghum, milo, and wheat—are mechanically harvested. The principle of harvest machine operation is essentially similar for most grain crops, although slightly different for corn harvested for grain.

Interest in reducing the enormous labor requirement for harvesting cereal grains led to the very early (about 1800) development of the grain combine. The tremendous volume of the cereal grains was a considerable incentive for this development. Its significance was obviously illustrated by the huge expansion of cereal production to the central and western United States, and to other major grain-belt production areas in the world.

The modern self-propelled combine, cutting a wide swath, enables one operator to rapidly harvest large areas and volumes of grain (Fig. 12–1). The harvester cuts the grain heads from the plant. The heads pass through rotating cylinders that beat and shake the grain kernels from the head. The kernels are separated from other plant parts (hulls, chaff, straw) by sifting through sieves; the straw is conveyed through and out of the machine, and a blower removes the smaller lighter-than-the-grain material. The grain is conveyed to a holding bin mounted on the machine for later off-loading, or is conveyed directly into trailers or trucks that move parallel with the harvester. The grain is then transported to its intended market or to storage. Grains are harvested as mature seed and when possible are allowed to dry on the plant to a low moisture content before harvest. A high moisture content can result in poor storage and spoilage. Therefore, the grain must be dried, sometimes artificially, after harvest and before storage.

Hay and Forage Crops

The harvest of hay and forage crops is highly mechanized. For most of the various grasses and legume crops grown for hay, timing of the harvest is made to optimize both yield and quality. Either a tractor-mounted or pulled mower cuts the hay, depositing it back onto the field or into windrows in one operation (Figure 12–3). This is done to allow for drying and

Figure 12–4 Baled hay is often picked up in the field with automatic loaders or stackers and transported to the storage area for stacking. This machine eliminates the hand labor of lifting and loading bales, then unloading and stacking them. *Source:* John Dobie.

to facilitate later pickup. Another machine gathers up and compresses the hay into compact bales or rolls and binds them. They may later be collected from the field or equipment may do this operation at the time they are formed. Equipment that requires only a single operator is available to lift, stack onto trailers, and transport the hay out of the field (Fig. 12–4). The bales or rolls, if not used immediately, are stored in large stacks, often in plastic wrap or indoors. In the limited rainfall areas of the Southwest, the stacks frequently may not be covered, since the highly compressed bales are sufficiently dense to resist moisture penetration, so that the interior of the stack remains dry and is not damaged.

Silage is a fermented product of the hay and foliage crops. The crop plant harvested for silage is cut but not permitted to dry. Instead, it is placed into structures such as silos, pits, or bunkers, where it is allowed to ferment by **anaerobic** microbial activity. The product, once fermented, is preserved for subsequent animal feeding. It is important that the crop material be well packed to exclude air and to favor microbial activity. In the United States, corn is by far the preferred crop plant for this purpose.

Root and Tuber Crops

To facilitate mechanical harvesting of many root and tuber crops, the above-ground plant parts are often removed prior to harvest of the below-ground portion. The foliage of some root crops (sugar beets) is used for animal feeding, and some (table beets and turnips) are used for human food. Hand harvest of root crops is hard physical work, and reason enough for developing machinery to assist human labor. It is common for large plantings of potatoes, sugar beets, and carrots to be harvested and handled mechanically (Fig. 12–5). Several root crops, such as radish, table beets, and carrots when marketed with attached tops, are not mechanically harvested. Sweet potatoes are susceptible to physical damage and were traditionally hand-harvested. Because improved equipment has reduced damage, harvest machines are used much more with this crop. Modern root crop harvesters are multiple row machines, either tractor-drawn or self-propelled. The princi-

Figure 12–3 This mower-swather cuts the hay and then places it in windrows for curing, thus combining two operations into one. The time required for curing depends on the temperature and the relative humidity of the air. After the hay is cured, it is baled. *Source:* Deere and Company.

Figure 12–5 The harvest of white potatoes is completely mechanized. Tubers are removed from the ground, separated from vines and soil, then conveyed to trucks alongside the harvester. *Source:* University of California Cooperative Extension.

pal function of the harvester is to pass a broad horizontal blade below the roots or tubers. That blade is angled so that in passing beneath the plant it lifts and/or allows easy pulling of the crop from the soil and onto conveyors, sorting belts, etc. Upon removal from the soil, the product experiences many of the operations common to most root crops. These include cleaning, trimming, washing, and being graded, conveyed, cooled, packed, transported, stored, and/or processed (Fig. 12–6).

With white potatoes, the tubers are harvested when mature and when plant tops begin to die and dry. Vine killing is hastened by the application of a foliage dessicant material, or by mechanically destroying the tops. This reduces the possible interference by the tops with the digging and separation of tubers from the soil. The tubers are lifted along with a layer of soil and placed on a rod-chain shaker endless belt, through which the loose soil can fall but the potatoes cannot.

Machinery used to harvest carrots uses the plant tops to lift the roots from the soil after they have been loosened by an undercutting blade. Following lifting, the tops are then

Figure 12–6 Sorting potato tubers according to size and examination for defects. The removal of defective tubers ensures good quality for packaging and marketing or for storage.

removed. In situations where carrots are intended for processing, the carrot tops may be mowed and the roots harvested in an operation similar to that for potatoes.

Twenty to twenty-five years ago, vegetable root crops were seldom washed. Presently in the United States, unwashed root crops are not competitive with those that are washed. Essentially all potatoes and carrots are washed before packing, and this is generally true for root vegetables.

Fruit and Nut Crops

The variable nature of fruit crops presents a challenge to mechanical harvesting. Fruits vary in their characteristics and rate of maturity. Some fruit will abscise when mature; others do not. Lack of uniform ripening makes a one-time harvest difficult. The delicate nature of many fruits requires careful handling, and most fruit for fresh market is hand-harvested. Frequently, mechanical aids to harvesting are used. Hydraulic platforms or booms replace ladders and enable workers to be lifted and easily moved about while harvesting. Conveyors are used to transfer the harvested product.

Fruits produced for processing are more likely candidates for mechanical harvesting, since the product is processed soon after harvest, and if damage occurred, it would not be as evident in the processed product whereas it might make the fresh product unacceptable. The harvest of the tree nut crops—walnuts, almonds, pecans, filberts, and pistachios—is fully mechanized. The nut crop can sustain considerable impact damage of the shell without significant loss of quality or value of the nut meat. These crops are harvested by being shaken from the tree by a machine that attaches to the tree trunk and vigorously shakes the tree. Canvas catching frames are used to catch and collect the nuts, or they may fall directly on the ground. The nuts are collected by equipment that sweeps and/or vacuums the nuts from the ground into bulk containers. Similar types of shaker machines are used for the harvest of fruit crops such as cling peaches, prunes, and canning cherries (Fig. 12–7). The catching frames are abundantly padded to minimize physical damage. Harvest machines are used for bushberry, raisins, and much of the wine grape crop. Mechanical aids are widely used in most harvest operations. The design and development of harvest machinery is a continuing process and it is likely that fruit crops such as citrus and other difficult-to-harvest crops, presently not mechanically harvested, will be harvested with machinery in the future.

Vegetable Crops

The mechanical harvest of tomatoes grown for processing received worldwide attention because it accomplished a seemingly impossible objective. This was admittedly a remarkable achievement, but one that was preceded by the mechanical harvest of other crops, such as spinach, peas, snap and lima beans, sweet corn, celery, cabbage, and onions for processing. For most vegetables the principal functions of

Figure 12–7 Harvesting sour cherries in Michigan for canning. Fruits are shaken from the tree onto catching frames, then conveyed into a bin. *Source:* USDA.

the harvester are to cut, collect the intended crop portion, and to sort, convey, and transport it. Like fruit crops, most of the vegetables for fresh market continue to be hand-harvested. This is continually changing, and harvest machinery for some difficult-to-harvest crops is under development.

The tomato harvester is an excellent example of the mutual adaptation of the crop to the equipment. The processing tomato was bred to provide both a determinate and prolific fruiting habit, and thereby maximize yield and uniformity of maturity. The fruit size is small and thick-walled, with a thick skin; thus more tolerant to physical handling. The harvester, whether self-propelled or drawn, passes a horizontal blade beneath the plant. The entire above-ground plant is conveyed onto chain belts that shake the tomatoes from the vines. The vines are discharged back onto the ground, and the tomatoes are conveyed past human or electronic sorters before transfer into bulk trailers alongside the harvester (Fig. 12–2). Workers examine the tomatoes, removing those that are immature or otherwise unsuitable. Equipment using photocells rapidly distinguishs color differences and electronically removes immature (green) fruit from the mature (red) fruit.

POSTHARVEST PRESERVATION

The objective of crop preservation is to retain crop quality and retard senescence of plant tissue while providing an attractive and useful product.[1] The postharvest storage or handling of fresh fruit and vegetables commodities must slow the reactions that tend to break down the product.

[1]Your state Cooperative Extension Service is a good resource for information about home food preservation.

These degradation reactions are essentially those resulting from respiration, whereby the carbohydrates synthesized during photosynthesis are broken down to carbon dioxide and water. Procedures that slow or stop the rate of respiration also frequently slow or stop the rate of deterioration. Preventing water loss is important in maintaining product qualities, except when removal of water is a procedure in the preservation method.

The preservation method used depends upon the nature of the product. The cereal grains, nuts, and spices need no preservation if stored dry. Some products such as fruits are soft and fleshy; others are hard and require different methods of preservation. Fleshy dessert fruits (apples, pears, peaches) have much in common with some vegetable fruits (tomatoes, green beans, and peas). Others (asparagus, table beets, spinach, cabbage, broccoli) are less similar. Whereas some products are preserved by canning or freezing, others are less adapted to either canning or freezing. For example, tomatoes do not produce a good frozen product, while broccoli is a poor candidate for canning. Other methods of preservation include drying or dehydration, and fermentation or pickling in brine or another preservative solution.

Preservation by Cooling

Temperature is the important factor in reducing the rate of respiration. As a general rule, lowering temperature reduces enzymatic activity and consequently respiration rate. Some fruits differ in the lowest temperature they can tolerate without damage. Tropical fruits, such as bananas and avocados, and others, such as tomato, eggplant, okra, etc., develop chilling injury when stored even for short periods at temperatures below 10°C (50°F), while other fruits, such as pears or apples, tolerate 0°C (32°F) or even lower. For most other products, storage at the lowest temperature, usually near 0°C (32°F) or slightly above generally provides for the best maintainance of that product for the longest period. Many ornamental crops can have their storage life prolonged by cooling. However, as with other crops, some, especially those that originated in the Tropics and subtropics, may be susceptible to chilling injury.

The rate of respiration increases with temperature over a fairly broad range. At excessively high temperature (usually above 40° to 50°C (104° to 122°F), the rate of respiration declines. However, at these temperatures, moisture loss is the dominant factor in deterioration. Generally, for each 10°C increase in temperature the respiration rate doubles, and therefore is important in the storage of fresh-market fruits and vegetables. Obviously, chilling injury temperatures should be avoided with sensitive warm-season crops. For best results, cooling should follow harvest as soon as possible. For some fruits, harvesting induces a response called the climacteric, when a rapid rise in the rate of respiration characteristically occurs.

Figure 12–8 An example of a large capacity cold storage room widely used to store many fruits and vegetables. The cones protruding from the ceiling allow cold air to be introduced. The relative humidity of the air can be regulated. Large fans, such as the one visible on the side wall, circulate the cold air throughout the room, thereby maintaining a low temperature for the stored product. These rooms are not as effective as other methods in the initial cooling of the product, but are useful for maintaining the temperature and storage of products already cooled.

Refrigeration

The most common method of cooling is the mechanical refrigerator, which can vary in size from a large warehouse to a small home model (Fig. 12–8). The mechanical refrigerator allows control of relative humidity as well as of temperature. Some crops require cool storage at high relative humidity while others store better in a cool, dry atmosphere. Practically all long-distance shipments of fresh fruits and vegetables are made in mechanically refrigerated trucks or railcars.

Hydrocooling

In this process, produce in bulk or packaged in moisture-resistant containers (waxed fiberboard, slatted wood, plastic) is slowly passed through a tank holding a large volume of cold water, or a tunnel-like structure in which cold water is showered onto the product (Fig. 12–9). As this is done, heat is removed from the product. The water, usually cooled by mixing with crushed ice, is recirculated and ice is added on a continuing basis. As needed, this is replenished with clean water. The level of cooling is regulated by adjusting the exposure time. The cooling characteristics of the product, its initial temperature, the temperature of the cooling water, the

Figure 12–9 The hydrocooler achieves cooling by allowing the transfer of heat from the product to cold water (contact by emersion or shower of cold water). The water temperature and exposure time influence the level of cooling obtained. The blocky structure (*below*) encloses the water shower. Two bins containing produce are emerging from the structure after cooling. *Source:* Robert F. Kasmire.

quantity of the product, and the desired final temperature of the product influence the rate of cooling and determine the period of exposure to the cooling water. Product fully immersed in the cold water cools more rapidly than product which is showered. The cooled product can then be loaded in refrigerated trucks or railcars, or stored in a refrigerated cold room.

Forced Air (Pressure) Cooling

Some products, such as strawberries or cauliflower, do not store well if the surfaces are wet, and therefore are often cooled by forced cold air. The product, usually packaged in corrugated fiberboard crates having ventilation holes, is

Figure 12–10   Air which has been cooled is drawn into this forced-air cooler through circular vents (upper right of photo) in the wall. After the cold air flows through the produce, it exhausts through holes in the opposite wall. This type of cooler is used extensively for strawberries, but in this case, cauliflower is being cooled. Cut flowers are also cooled by this method. *Source:* Western Grower and Shipper.

Figure 12–11   A railroad car, after loading, is having crushed ice applied over and around the packages in order to maintain low transit temperatures. This technique would apply several tons of ice into and around the load. *Source:* Robert F. Kasmire.

placed into the cold room as soon as possible. The containers, closely stacked together, are placed against one wall of the cold room which has vents through which cold refrigerated air is forced or drawn by fans through the stack. The stack of containers is placed so that the air must pass from the wall vents through the container ventilation holes and through rather than around the stacked containers. After cooling the product, the air is exhausted from the opposite side of the cold room (Fig. 12–10).

Package Icing

Another cooling method, now largely discontinued, is the placement of crushed ice onto and around the product packaged in moisture-resistant containers. In addition, after loading in the transport vehicle, special equipment would pump or blow crushed ice onto and around the load (Fig. 12–11). Several tons of ice would commonly be used to cool a typical truck or railcar load. A disadvantage is the addition of considerable weight to the load. The ice provides additional cooling and moisture for the product until it melts and drains away. Other disadvantages, in addition to being a messy and wet operation, are its slow and nonuniform cooling, the need to renew the ice or otherwise provide further cooling, and the expense of the ice. This method was extensively used on difficult-to-cool crops such as broccoli, leafy vegetables, and occasionally melons.

A recent improvement of package icing is the use of a slush mixture of water and shaved ice that is applied to the packaged product. The mixture, being rather fluid, fully permeates the packaged contents, and as the water drains, the ice remains. In contrast to top icing, this ice is in more direct contact with the product. Equipment in use can rapidly pressure-inject the slush mixture into a large number of stacked containers in one operation, making package icing a more rapid and efficient procedure.

Vacuum Cooling

Many fresh-market leafy vegetable crops are cooled by vacuum, because it is a rapid and effective method. The procedure allows for direct field packing of many leafy crops which formerly were cooled with ice and therefore required shed packing. Vacuum cooler equipment consists of a large strong round or rectangular steel tube containing ammonia coils, which has an airtight door at either end and is connected to a large vacuum pump (Fig. 12–12). The tube must be strong enough to withstand the air pressure of the external atmosphere when its interior air is evacuated.

After the product is put into the tube and the air-tight doors are closed, a vacuum pump evacuates the air until the internal atmospheric pressure is reduced to about 4 to 6 mm (0.16 to 0.23 in.) of mercury from the normal atmospheric pressure of about 760 mm (30 in.) of mercury. At low atmospheric pressure water changes from liquid to vapor. The phase change absorbs heat from the only available source—the crop being cooled. This heat is called the latent heat of vaporization and amounts to about 580 calories per gram of water at 20°C. After removal (evaporation) of the moisture from the product surfaces, the water vapor,

Figure 12–12   A vacuum cooler can cool lettuce from 30°C to 1°C (86°F to 34°F) in about 30 minutes. While the cooler was originally designed for lettuce, other crops, such as the cauliflower shown here, are now cooled in this manner. *Source:* Western Grower and Shipper.

Figure 12–13   The typically brilliant cloudless summer days in the San Joaquin Valley in California are ideal for drying grapes to make raisins. The grape berries are hand harvested and laid on heavy brown paper to dry. The use of a continuous roll of paper permits some mechanization of the removal of the raisins after drying and before they are taken to the packing house for processing. A rain during this period can be a catastrophe to raisin producers.

contacting the ammonia coils, condenses to ice. A significant amount of water, as much as 5 percent, is lost by the crop on cooling. This is dependent on the initial temperature of the product; the higher its temperature, the greater the water loss and time needed to cool. Some vacuum coolers supply water to replace that lost from the product. For crops such as celery, only a minimum amount of water loss can be tolerated before the product would become limp and unacceptable. Leafy crops such as lettuce are better candidates for vacuum cooling because they provide a large surface area for evaporation of water. The greater the evaporation, the greater the rate of cooling. Other crops, such as cauliflower or cabbage, are dense products with considerably less surface area, and require more time under vacuum to be cooled.

Preservation by Drying

The drying of vegetables, fruits, and berries by solar radiation was probably the first attempt at food preservation, and its usage and popularity remains high. Dates and figs were grown and sun-dried by early civilizations in the eastern Mediterranean area. Preservation by drying succeeds because decay-causing organisms usually do not grow at moisture contents below 10 to 15 percent.

Some advantages of dehydration over other methods are a reduction in weight and volume, thus reducing handling and storage costs. Storage generally requires minimal or no refrigeration. Additionally, the solar drying is less expensive than other methods of dehydration.

A climate having high temperature, clear days, and low humidity is ideal for solar dehydration of various commodities. Crops preserved by solar dehydration include apples, apricots, currants, grapes, peaches, figs, dates, pears, and plums (Fig. 12–13).

Forced Hot Air Dehydration

The popularity of dehydrated foods, particularly with various fruits, has stimulated the development of efficient methods for rapid dehydration with forced hot air. This is more appropriate where natural solar drying is not sufficient or in areas of high humidity.

In this process, the product is spread thinly on trays and passed through a tunnel dryer. Sometimes a short-term blanching with steam is used to stop enzyme activity that causes tissue browning. Air heated to 60°C to 70°C (140°F to 158°F) is forced through the tunnel. The time for dehydration varies from a few hours to as much as 36 hours, depending on the kind and size of the product.

Until recently, vegetables were not dehydrated in any volume, but this has changed, and large volumes of onions, potatoes, mushrooms, and other commodities are dehydrated. The use of ready-mixed vegetables for soups has increased the use of dehydrated celery, carrots, peppers, tomatoes, peas, parsley, chives, and other vegetables.

Dehydration by Freezing

Freeze-drying is a process that is increasing in use despite being an expensive method. This process involves removing water by sublimation[2] of ice at temperatures below the freezing point. Upon rehydration, the quality of the product is equivalent to that of food preserved by freezing. While freeze-dried products are expensive, they are valuable where weight reduction is sought. These products are used in special markets where convenience of preparation is desired, such as in backpacking.

---

[2]Sublimation is the process by which a solid (ice) passes directly to a vapor without going through the liquid phase. Heat is absorbed in this process.

Preservation by Modified and Controlled Atmospheres

Modified atmosphere[3] storage has been known for over 100 years. The first study of this method was made by Jacques Berard in France in 1820, but his findings were not exploited until 1920. Since then research on modified or controlled atmospheric storage has been extensive.

Although temperature management is of principal importance in the storage of fresh products, the oxygen concentration is also important. Since oxygen is essential for aerobic respiration of both the food product and microorganisms that can cause decay, it is important that oxygen concentration as well as temperature be lowered to prolong storage life. This method is extensively used in the major apple production areas, and were it not for CA storage, apples would not be as readily available to consumers throughout the year.

The oxygen concentration is reduced by increasing the carbon dioxide concentration, introduction of nitrogen into the storage chamber, and evacuating air from the chamber. The chambers are sealed and not opened until the product is marketed. Workers entering such storage chambers must wear respirator equipment. The concentration needed in the storage chamber varies considerably with each crop and may be different for a particular apple variety. The CO_2 concentration used to retard respiration and decay for one crop may be injurious to another crop.

Hypobaric or Low-Pressure Storage

This method has limited usage because of the high cost of the equipment required relative to the benefits achieved. After produce is placed in this type of storage, the temperature is lowered to the desired level, air is pumped out, and a partial vacuum is maintained. As the air is evacuated from the chamber, some water-saturated air is allowed to enter. Low-pressure storage reduces further water loss from the product, reduces the risk of CO_2 injury, lowers the oxygen concentration, and sweeps ethylene[4] from the chamber. Removal of this natural plant hormone delays ripening.

Importance of Relative Humidity

Controlling relative humidity enhances prolonged storage. High humidity lowers water loss from plant tissue, thereby reducing desiccation and wilting. High relative humidity allows growth of pathogenic microorganisms when temperatures are higher than desirable for a particular crop. At desirable temperatures, high relative humidity reduces the

danger of decay because the product remains healthier and therefore can better resist infection. Most fresh fruits store best at relative humidities of about 90 to 95 percent. These include apples, avocados, bananas, figs, mangoes, papaya, pears, and pineapples. Easily wilted leafy vegetables should be stored at relative humidities of 95 to 100 percent, but absent of free water. These include artichokes, broccoli, Brussels sprouts, carrots, celery, lettuce, radishes, spinach, turnips, and the like. Some vegetables store better at a lower relative humidity of 75 to 85 percent. Examples are garlic, dry onions, pumpkins, some squashes, and sweet potatoes. Ventilation is needed in all storage facilities to avoid condensation and to keep undesirable gases from accumulating.

Fumigation

Situations occur where it is useful to treat crop products with fumigants for the control of rodents, insects, or decay-causing microorganisms. A number of products are used for these purposes. For example, methyl bromide (CH_3Br) is effective in the control of many insects; sulfur dioxide (SO_2) reduces the risk of decay in grapes, and diphenyl ($C_{12}H_{10}$) in citrus. Elevated concentrations of carbon monoxide or carbon dioxide are also useful in controlling pests and disease development, with certain crops able to tolerate these gases.

Preservation by Processing

The preservation of foods by processing has an interesting early history. In France, about 1795, Napoleon was concerned with the availability and continuing supply of food for his armies. He offered a reward to find a preservation method that would be safe. The prize was won by a Parisian named Appert in 1809 who preserved several products in sterilized glass bottles. Although at that time the process was not fully understood, we now know that its success was due to the use of heat to inactivate the products' enzymes and the destruction of microorganisms, followed by providing and maintaining a sanitary or sterile condition. This method of preservation was rapidly adopted and was widely used in many homes; it is now a large and important industry.

Canning

The purpose of canning is to destroy spoilage organisms by heat, although complete sterilization is not mandatory. The product is placed into gastight containers that are sealed. Hermetically sealed containers are in wide usage. The major containers are cans made from tin-plated sheet steel or glass bottles. Plastic- or foil-coated paper containers are also used, especially to save weight. The interior surface of the cans are coated with acid-resistant lacquers to reduce corrosion and discoloration when high-acid products are canned, or where the products have sulfur compounds that will react with the tin surfaces, turning them black if not coated. After sterilization, usually performed in a pressure cooker, the cans are quickly cooled, labeled, boxed, and stored in warehouses until needed.

---

[3]Modified atmosphere is the situation where the concentrations of N_2, O_2, or CO_2 are varied but not precisely maintained; temperature and relative humidity may be regulated. In controlled atmosphere (CA), temperature and concentrations of N_2, O_2, and CO_2 are precisely regulated within established limits.

[4]Ehtylene (C_2H_4), a natural plant hormone produced by some plants, promotes ripening.

Quick Freezing

Many foods are successfully preserved by quick freezing. Quick freezing is an effective method because it arrests deterioration of product quality as well as the growth of microorganisms that cause spoilage. With some products, hot water or steam blanching is used to stop enzymatic activity, which if continued would reduce product quality.

Freezing is not a new method of food preservation. Eskimos have used it for centuries to preserve their fish and game. However, freezing was not used extensively in a commercial sense until mechanical refrigeration was developed. It was not until the early part of the twentieth century that fruits and vegetables were frozen. Upon thawing, the tissue structure of the frozen product, depending on the specific product, breaks down. Some products are not frozen because their after-thawing characteristics are undesirable; tomatoes are one example. To maintain optimum quality, frozen fruits and vegetables should be kept frozen until prepared for consumption. Properly frozen fruits and vegetables can be stored for long periods of time, and retain much of their fresh-like characteristics.

Processing with Sugar

Some fruit products are processed with high concentrations of sugar for the production of jams, jellies, marmalades, crystallized fruits, and the like. Sugar increases the osmotic pressure to levels where water is unavailable for microbial activity, thereby reducing spoilage possibilities.

Processing with Salt

The processing of vegetables with brine has changed little over the years. This process preserves cucumbers and gherkins as pickles, cabbage as sauerkraut, and onions, beets, beans, peppers, and olives. Fresh cucumbers or cabbage are allowed to ferment anaerobically in a salt solution concentrated enough to prevent the activity of spoilage organisms, but not high enough to destroy the bacteria producing lactic acid. Lactic acid lowers the pH, thus helping to prevent the growth of organisms. Olives in California are usually pretreated before salt storage with a 1 to 2 percent solution of sodium hydroxide (NaOH) to neutralize the bitter glucoside found in the fresh fruit.

Processing by Milling

Milling, especially that of cereal grains, is one of the oldest processing techniques. The process largely consists of using differences in the physical properties of the seed grain, such as density, size, and shape of the grain seeds, for cleaning and separating it from contaminants. The process also involves the removal of the hulls or bran layers of the grain kernel until only the endosperm remains. The cereal may be further ground to a flour. Flours used mainly for baking are made from the endosperm, and are white. The use of whole grains, or grains partially milled, has become very popular because of the interest in having more of the nutrients retained rather than being removed by the milling. Numerous products are made from the various milled cereals.

Rice, an important human food grain crop, is processed by milling away successive layers of the grain coatings until only the endosperm remains. This is known as polished rice. Rice that is not as fully processed (milled) is known as brown rice. In many areas it is preferred because it contains more thiamine, protein, and other nutrients since the bran layers of the grain are not removed.

STORAGE OF HARVESTED PRODUCTS

Proper storage prolongs the life of harvested unprocessed plant products. Storage also offers a possibility of moderating price by management of the supply available to the market, and is often used for this purpose. Storage time varies widely, depending upon the kind of crop stored, its condition, its quality characteristics, and conditions of storage.

Fruits and Vegetables

Fruits and vegetables are stored for relatively short periods, compared with dried forage or cereal crops. Products of low quality probably should not be stored, because they are seldom worth the added storage costs, and storage would not improve the initial quality.

Factors to consider when storing fresh products are (1) length of storage time, (2) temperature, (3) relative humidity, and (4) light.

In general, cool-season crops are stored at cool temperatures, 0°C to 1°C (32°F to 34°F), while warm-season crops are stored at relatively warm temperatures of 10°C to 12°C (50°F to 54°F) to avoid chilling injury. There will be exceptions. For example, sweet corn (warm season) should be stored at cool temperatures immediately after harvest to retard the conversion of sugar to starch, which lowers quality and taste. Potatoes to be made into potato chips or french fries should be stored at temperatures of 10°C to 13°C (50°F to 56°F) and lower if long-term storage is desired. Before being processed, they should be removed from that storage temperature and held for about a week at a higher temperature of about 20°C (68°F), to permit the conversion of sugar to starch. This avoids darkened chips or french fries, which would result from the caramelization of sugars during frying.

Commodities with high water content should be stored in rooms cooled to 0°C to 1°C, and having a relative humidity of 90 to 95 percent. Products such as apples, lettuce, radishes, spinach, and celery store best under these conditions. Beans, melons, cabbage, cauliflower, and potatoes will store well at relative humidities of 85 to 90 percent. Other crops, such as garlic, dry onions, winter squash, or pumpkins, are stored at relative humidities of 70 to 75 percent.

Figure 12–14 Extended storage of dried seed grains is commonly made in large "grain elevators," as pictured. These facilities often have the capability to further dry seed if it is necessary to lower the moisture content. *Source:* James F. Thompson.

Figure 12–15 Baled alfalfa hay is often stored unprotected in the field in arid to semiarid regions to reduce the cost of storage. *Source:* John Dobie.

Light should be either subdued or absent during storage. Light should be kept away from stored potatoes, because light exposure favors the development of chlorophyll or greening of the potato, but more importantly light promotes the formation of the toxic alkaloid solanine.

Seeds

Seed crops—barley, beans, shelled corn, oats, rice, rye, sorghum, soybeans, and wheat—are relatively easy to store for long periods of time, if they are suitably dry. For safe prolonged storage of most seed crops the moisture content should not be more than 12 to 14 percent. Seed crops may be sufficiently dry to store at the time of harvest; if not, artificial drying is necessary to reduce the moisture content (Fig. 12–14).

Hay

Forage crops harvested to feed livestock may be stored in outdoor stacks on the farm, with or without protection against the elements of weather, or in various structures (Fig. 12–15). Loose hay is stored unchopped and air-dried, although hay is more frequently chopped and baled to reduce storage space and handling. Stored hay needs to be dry and have adequate ventilation. Fires can result from the spontaneous combustion of hay stored too wet or inadequately ventilated.

Silage

Silage (ensilage) is the anaerobically fermented livestock feed made from any green forage. Grasses, grain, or legume crops are commonly used, although corn is perhaps the dominant crop for silage production. A silo is a structure used for production and storage of silage. Often the silo is a tall, cylindrical building into which the harvested chopped green forage is blown and tightly packed. The forage is then sealed off to minimize contact of the outer surfaces with the atmosphere, and to permit its fermentation into silage. A silo can also be a horizontal trench, wide and long enough to store the required amount. Almost any type of containment that can be sealed or covered to permit fermentation will function as a silo.

MARKETING OF AGRICULTURAL PRODUCTS

Primitive societies had little marketing. Individuals or family groups gathered or produced their own food and other necessities. As a barter system and, later, money exchange developed, the variety of products available for exchange widened. Hunters could barter their game with growers of grain, vegetables, and fruits, enabling each to obtain and exchange products. Individuals not directly producing food could buy or trade for food, thus allowing them to participate in their societies in numerous and varied work activities other than agriculture and food production.

In modern agriculture, specialization has progressed beyond the self-reliant community, and greater interaction exists between the farmer and the city dweller. A continuing trend is for fewer farmers to supply more as urban populations increase. More and more farmers tend to concentrate on producing the products best adapted to their particular areas, soils, climate, and markets rather than attempting to supply themselves with the many products they directly consume.

Early in the development of marketing, the producer provided an array of products (eggs, poultry, fruit, vegetables, milk, honey, etc.) directly to the final consumer. To make the exchange of commodities easier, the producers brought their products to a central place (village square or farmers' market) where they and buyers negotiated prices. In much of the world this has changed very little. Farmers' markets are a small factor compared to their previous importance, in terms of the total volume of product. For the past 50 years this type of market has been on a decline. Recently, a growing interest in fresh products has created a resurgence in farmers' markets, roadside outlets, and self-harvest operations. Indications are that the general public will continue to support this form of marketing even though it will account for only a relatively small portion of the total distribution. For some agricultural commodities, the Internet has become an important means of marketing.

Today most marketing involves a series of intermediates. Intermediates are prevalent in countries with highly developed systems of agricultural marketing that includes imports and exports in addition to the domestic markets. Some intermediates buy the product and further process it, some transport or store it, and others distribute it as wholesalers or retailers. There are a multitude of intermediate functions in this process. Whereas intermediates are often looked upon as unnecessary, they are essential in the modern orderly distribution of most agricultural commodities.

The Process of Marketing

Marketing plays a key role in the distribution of a product. Neither the modern producer nor the urban consumer can do without it; both depend upon the market for their food and other necessities. The process is better understood by discussion of its major components, which are assembly, distribution, transportation, storage, exchange, and financing.

Assembly

Assembly is the concentration of smaller quantities of a commodity into a central place for the necessary handling, processing, or other procedures, before transfer to market. Concentration at convenient locations gives prospective buyers the opportunity to examine products and arrange their purchase. Not all agricultural products need to be assembled, and in fact, a considerable volume is sold directly to processors, wholesalers, or retailers.

Distribution

Systems have been developed to distribute products from places of assembly to places of consumption. The extensiveness of the distribution system is the foundation of marketing in the industrialized countries. The system must be able to adjust the supply of commodities to market demands quickly and easily as either supply or demand changes.

Food sale chains and institutional and fast-food suppliers have captured a large share of food distribution in developed nations. In North America and Western Europe, management of the retail distribution of food has concentrated in relatively few hands. In the United States, large food chains retail food through thousands of stores (Fig. 12–16). Fast-food restaurant corporations prepare a significant proportion of the total number of meals consumed nationally each day in the United States. The same sort of food retailing has occurred to a large degree in Canada, England, and other European countries, and the trend is certain to expand to other areas.

Figure 12–16 Supermarkets have revolutionized the methods of marketing. In the United States a customer can buy practically any type of fresh fruit or vegetable at any time of the year and at comparatively low prices.

Transportation

An essential part of marketing is rapid, dependable transportation. The products are generally transported from their origin or assembly areas to various markets or processing facilities by the producer, the buyer, or by transport firms contracted to provide such services. Who provides and how the transportation is performed will vary considerably.

Storage

Storage must be considered a part of the marketing system, and certainly is necessary at various stages in the marketing sequence. Producers often find it necessary to store their products, and buyers, whether processors, wholesalers, or retailers, operate storage facilities in order to control the supply and time of resale of the commodity. Processing plants operate storage facilities to hold raw product stocks in order to operate the plant when incoming supplies are low as well as to hold the processed product until distributed.

Exchange

Before any buying or selling can occur, the buyer and seller must meet or communicate. The exchange process involves two phases: contacting possible buyers and sellers, and negotiating an agreeable price. In many countries, the town square acts as the marketplace, where buyers and sellers agree on prices and conditions of the transaction. If agreement is reached, a sale is consummated. In some larger markets, sales are negotiated for a fee by intermediates called brokers. They do not take possession of the goods but facilitate the transaction. This enables buyers and sellers, who may be separated by great distances and unable to physically meet, to arrive at an exchange.

Grade standards developed by the United States Department of Agriculture or individual state Departments of Agriculture are helpful in that they allow the buyer to know about the quality of the product without actually seeing it. Specialized market reporting and news carried by the press, radio, and television provide helpful information.

Financing

Financing is essential in all marketing processes. Producers, wholesalers, processors, retailers, and other intermediaries must either have the capital or credit to produce, hold, and handle a commodity until it is sold and payment is received.

Market financing entails a certain amount of risk. A major risk is a falling market; another is rapid price fluctuation. Prices for agricultural commodities generally are less stable than those for manufactured goods, because supply and demand are subject to greater and less controllable fluctuations. Deterioration of quality entails another financial risk. A buyer must be aware that the quality is the most important attribute of most commodities and that the quality of many commodities depreciates rapidly.

Commodity Markets and Exchanges

Commodity markets transfer ownership of products from producers to processors, manufacturers, or consumers, and determine the price for the exchange. Commodities often handled by this type of market are foodstuffs and nonfood raw materials. Some examples are cereal grains, oil seeds, tea, coffee, rubber, and tobacco, as well as some metals. The commodity market is concerned with trading a given amount of a certain grade of commodity for present delivery. The goods need not be on hand at the time but can be in transit or stored. Terms are arranged, payment is made, and title transferred.

A commodity exchange, or futures market, differs from a commodity market in that the exchange involves the purchase or sale of a contract to deliver a certain amount of a given grade of commodity on an agreed date in the future. This allows for each party to hedge against a future market supply and demand situation. Hedging means that one party covers his or her position in the cash market by taking an opposite position in the futures market. In actual practice, the seller of the contract seldom actually delivers the commodity. Usually before the actual transfer date the contract will be dissolved. The futures market can provide contact with a larger number of prospective buyers or sellers than would another type of market, thus giving the parties to the respective contract greater opportunities to either buy or sell the contract before the transfer date. This may or may not result in a profit, but more importantly, it provides a means to avoid excessive losses. The Chicago Board of Trade is an example of a commodity exchange handling mainly agricultural commodities.

Governmental Marketing Services

Most governments participate in the regulation of marketing to some degree. Some go to the extreme of setting prices, and regulating production and distribution of products. For example, the price of bread is regulated in Egypt; that of milk in Denmark. In the United States, California and some other states regulate milk prices, and the production of tobacco and other crops is regulated through acreage allotments. There are many examples of governmental regulations, which often arise during wars or other emergency periods.

Services provided by the federal government are intended for the protection of public health, establishment of standards and grades, and enforcement of weights, measurements, and other regulatory needs. Federal and state governments cooperate to supply production statistics and estimates for various crops, their condition, and other important market characteristics (Fig. 12–17). For some commodities, information on shipments and market prices is provided on a daily basis.

Grade standards are degrees of quality, each established by definition. The standards define the color, size, and freedom from defects or any other attributes that pertain

Figure 12–17 A U.S. Department of Agriculture inspector examining potatoes for quality and condition. *Source:* Adel A. Kader.

to quality. These standards are published and maintained by the USDA, although some states also establish grade standards. Regulatory departments of the USDA or state may provide inspection and grade determination services. These official standards permit marketing, even over long distances, without the buyer's actual inspection of the commodity. For example, a buyer in Japan buying rice from a broker in Arkansas specifies the quality grade (USDA No. 1) of rice desired, knowing that the grade ordered will have those quality standards, supported by a certificate of inspection issued at the seller's location. Lettuce from Arizona or Florida oranges are sold to New York buyers who order these commodities by well-defined grade standards, knowing before arrival that the desired quality has been guaranteed for delivery.

Agricultural Cooperatives

Farmers may join together to form businesses known as co-operatives (co-ops), which differ from other businesses in that the farmers are owners as well as customers. Co-op members provide the capital, elect officers, and may employ a manager. Farmers use the co-ops to market their products, to procure supplies (fuel, fertilizers, containers, etc.), and to obtain services (insurance, credit, etc.). Co-ops generally do not operate as a profit-making business but function to lower member costs through large volume buying discounts. Through collective marketing and having a large presence in the marketplace, co-ops can often strengthen their selling abilities.

Marketing Boards

Marketing boards were first developed in England and some of the Commonwealth nations. Their primary objective was stabilizing producer prices. Examples of marketing boards established in some countries are the National Coffee Board of Ethiopia, the Sri Lanka Tea Propaganda Board, and the Australian Federal Marketing Board. There are many other examples throughout the world.

In the United States, a similar program, called a marketing order, exists. Used originally for fruits and vegetables in California and for milk in many milk-producing states, marketing orders are agreements among producers to solve a specific problem or to achieve a goal for the common good of the members. Under the authority of a marketing order, either state or federal, an elected administrative board can levy assessments to collect funds to support advisory services, promotional activities, research, or control of supply or sales. Supply control is now largely disallowed for most marketing orders.

TRANSPORTING COMMODITIES

In primitive times the primary movers of goods were humans. As civilizations developed, movement of goods and people progressed to the use of animals and then to machines. Throughout this progression various needs were identified and met, among which were those of efficient roadways, rail lines, equipment, and the like.

Transportation has not been limited to land surfaces. Rivers, canals, sealanes, and airways are also significant avenues for transportation.

Rail Transport

In the United States, during the period from the Civil War to World War I, railroads were the principal transporter of most of the heavy goods. Prior to this period, animal-drawn vehicles were the principal land transportation method. Almost all long-distance shipment of nonperishable agricultural products were made by rail. Early refrigerated railcars permitted shipment of some perishable commodities. These early railcars were boxcars with ice bunkers at each end. Cooling of the product was achieved with the use of fans, powered by the railcar wheels, blowing air over the ice and through the product. Stops were scheduled to resupply ice to the railcar. Often these cars, when loaded with perishables, would be attached to a passenger train for fast hauling to distant markets.

After World War II, most U.S. railroad companies replaced ice bunker cars with mechanically operated refrigerator cars. Because the trains were not required to stop for icing, the conversion reduced the time required for transport.

The cooling system in the mechanically refrigerated cars is operated by a diesel engine located in one end of the car which operates during the trip. Temperature is regulated by thermostats with sensors generally located in the return air ducts. Air is distributed throughout the load by fans. If needed, heating can be provided.

Another adaptation is a system of truck-train transportation called the piggyback method, in which a refrigerated truck-trailer is carried on a railroad flatcar. In many aspects the trailer functions like a refrigerated boxcar. At destination the trailer is removed from the flatcar and attached to a tractor for further transportation to final destination. These piggyback units combine the truck's flexibility of delivery with the train's reduced cost of transport.

Truck Transport

Transport by truck and truck-trailer combinations was minimal before World War II, but expanded rapidly soon after to become the dominant method for transport of goods in the United States. Many shippers, experiencing problems with rail shipment, and preferring greater flexibility and direct control, turned to new, large trucks for shipping. Improved highway systems gave the trucks some advantages over the railroads. The truck is as fast as, and in some cases faster than, the railroad, offering dock-to-dock service, and often at less cost. Trucks now haul the major portion (about 90 percent) of all perishable goods in the United States. Modern trucks for transporting perishables have mechanically operated refrigeration systems similar to those already discussed (Fig. 12–18).

Sea Transport

The history of sea transport and trade is both interesting and colorful. It is impossible to state when early humans first discovered the use of waterways to carry them and their possessions. The Egyptians in the period about 2000 B.C. were among the earliest to use sea transportation. Mastery of the seas was significant in the development of nations throughout written history.

After the dark ages and before the Industrial Revolution, England, Spain, and other European nations entered into competition for world trade, much of which was concerned with the transportation of agricultural products. Shipping by sea rapidly grew to include much of the world. In modern times, sea transportation remains a vital and considerable factor in the transport of food and other agricultural products.

Since the late 1940s there has been a rapid shift toward containerized transport, a trend stimulated by the desire to reduce labor and marketing costs. At first, the containers were designed for dry, unrefrigerated cargo, but the apparent advantages of containerization led to the development of refrigerated containers for all types of fresh products. Com-

Figure 12–18   Trucks account for a substantial amount of transportation of all forms of goods, and particularly for perishables in the United States. Trucks are equipped to provide refrigeration or heat as needed. *Source:* Yoder Bros., Inc., Pendleton, S.C.

plementing this development, container ships were designed and constructed specifically to efficiently handle and transport such containerized cargos (Fig. 12–19). Containers for these container-carrying ships are similar to refrigerated trailer vans for trucks. They are transported via flatbed trucks and railcars designed for this purpose to dockside for shipboard loading. The containers, depending on the carrier, are refrigerated by shipboard units or have individual refrigeration units. Some ships with refrigerated holds still transport certain commodities, but their limitations are numerous. A serious disadvantage of the refrigerated-hold ship is that it usually cannot be mechanically loaded. However, grains, wood chips, cubed alfalfa hay, and other dry commodities are transported in bulk in unrefrigerated ship holds.

Air Transport

Transportation by air, except for small packages and mail, was not important until after World War II. From about 1950 to the mid-1960s, some agricultural commodities were shipped by air freight and scheduled passenger aircraft (Fig. 12–20). The cost was high and usually justified only for high-value commodities.

The development in the late 1960s of the large-capacity, jet-engine-powered aircraft made such transport somewhat more feasible by reducing the operating cost per unit weight of cargo. Containers especially fitted to the large carriers also contributed to handling efficiencies. Perishable cut flowers, strawberries, pineapples, and various fruits and off-season vegetables benefit from, and can afford, this transportation method.

Air transport has opened up previously inaccessible markets. In utilizing this method, other components are required. These include precooling, short-term refrigerated storage at departure and arrival destinations, and effective

Figure 12–19 Regularly scheduled container ships transport more than one thousand sealed containers great distances. The containers can be individually refrigerated if the cargo requires temperature control. *Source:* Matson Navigation Company.

Figure 12–20 Large quantities of cut flowers and some fresh fruits and vegetables are carried throughout the world in specialized containers by air cargo and regularly scheduled passenger aircraft. *Source:* Robert F. Kasmire.

scheduling with unloading and distribution to final destinations. Any delays will negate the benefits of the rapid transport provided by aircraft.

In the modern world, transportation is a necessary and important activity compared to earlier times when distances between producers and consumers were small. Modern methods of transportation allow for a tremendous exchange of products throughout the world and provide access to many new and distant markets.

SUMMARY AND REVIEW

Depending on the nature of the crop and its use, harvesting can be done mechanically or by hand. Mechanical harvesting can include digging, cutting, lifting, sorting, cleaning, conveying, and so on. The degree of mechanization can vary from complete, where everything is done by one machine, such as a combine harvester, to where only one step is done, such as a mechanical shaker that removes fruit from trees, but removal of the fruit from the area requires other machinery or hand labor. More and more, mechani-

cal harvesting is replacing hand labor. Even the most tender or fragile crops usually have some mechanical aid for harvesting.

Preservation and storage prolong the usefulness of the crop. There are many techniques used to preserve and store crops. These include cooling, drying, canning, freezing, and processing with sugar or salt. The method used is determined by the type of crop being preserved and stored and the intended use of that crop.

Transporting crops that are either fresh or in a preserved state is a crucial part of crop production. Transportation systems include trucks, trains, boats, and planes. Proper handling during transit is very important, especially if the movement of the crop takes more than a few hours or if there is exposure to environmental extremes such as high or low temperatures. Maintaining proper temperatures, humidity, protec- tion against mechanical injury such as bruising, and other factors that can damage the crop is critical during transit.

Marketing crops involves several steps. These include assembly of the product, transportation and distribution, storage, exchange, and financing. For many commodities, either formal or informal, standards are set. Many commodities also have marketing boards that can levy assessments or collect funds to provide advisory services, develop promotional activities, and fund research.

EXERCISES FOR UNDERSTANDING

12–1. Look at the contents of your refrigerator and freezer. If you had to ship all the fruits and vegetables, (fresh, in containers, and frozen) to a new location across the country, what would you do to make sure it arrived at your new home in a condition where you would still be able to eat them?

12–2. Choose a crop that interests you. It can be food, forage, fiber, or ornamental. Assume you can grow or produce it in the area where you are living. Develop a protocol for shipping it from your location to another location at least 2,000 miles away. Consider such things as temperature requirements of the crop, environmental temperatures, methods of transportation available in your area and the area you choose to ship to, and the perishability of your crop.

REFERENCES AND SUPPLEMENTARY READING

KAYS, S.J. 1991. *Postharvest physiology of perishable plant products*. New York: Van Nostrand Reinhold.

WILLS, R., and R.B.H. WILLS. 1998. *Postharvest: An introduction to the physiology and handling of fruit, vegetables, and ornamentals*. New York: Cab International.

UNIT

II

AN OVERVIEW OF THE FRUIT CROPS AND ORNAMENTAL PLANTS

CHAPTER

Cultural Practices in Orchards and Vineyards

KEY LEARNING CONCEPTS

After reading the chapter you should be able to:

♦ List the factors used in selecting an orchard or vineyard site.
♦ Understand how to select fruiting cultivars and rootstocks.
♦ Understand how to plant and maintain orchards and vineyards.

The first essential step in the successful establishment of a fruit planting is to be certain that the crop to be planted is adapted to the climate of the region and has an available market. After this has been determined, a number of major critical decisions remain to be made, and there is only one chance to make many of them. All available pertinent information should be sought out before final commitments are made.

Since the production of most fruit crops is a long-term undertaking, poor initial decisions can be costly and impossible to correct later. Factors that should be carefully considered before any planting is done are:

1. Site selection
2. Selection of fruiting cultivars and rootstocks
3. Allowance for pollination requirements
4. Planting distances and tree arrangements
5. Market and market demands

SITE SELECTION

A thorough study should be made of the following aspects of the proposed site. Any one of them could affect the success or failure of the enterprise.

Climate

It is extremely important to learn as much as possible about the weather patterns of the proposed planting site. Such information can often be obtained from neighbors who have lived and grown various fruit crops in the area for many years or, if recording instruments are located in the vicinity, from government weather services. Local cooperative extension or agriculture department agents usually can furnish much valuable information concerning weather patterns in their localities. Several weather conditions are of particular importance in fruit growing.

Temperature

The first determination needed in evaluating a site is the minimum temperature that frequently occurs and if the fruit crop will survive that temperature. For example, if a site frequently experiences a midwinter temperature of −15°F to −20°F, peaches or vinifera grapes should not be grown; however, some cultivars of apples would survive these temperatures.

The second important determination is the length of the growing season, which is determined by the last killing frost in the spring and the first in the fall. These dates can be greatly influenced by the features of a particular site. Avoid sites in low-lying areas, such as river bottoms or low spots in rolling hills, where cold air settles during frosty nights. This can be particularly dangerous when frosts occur during blooming periods in the spring or for cultivars with late-maturing fruits, which could be damaged by early fall freezes. It is much safer to select an orchard site on the upper portions or slopes of rolling terrain. Orchard heating, wind machines, or irrigation systems can often moderate or prevent low-temperature injury problems in frosty sites. While controlling frost is an additional expense, a more important factor is the cost of crop loss or injury.

In regions with hot summers, site location can influence the temperature; a northern or eastern slope may be a few degrees cooler than southern and western slopes (in the northern hemisphere). For growing crops that cannot tolerate high summer heat, such as sweet cherries and apples, the site location and orientation of tree rows is of considerable importance. (The south side of east-west rows can be excessively hot.)

The location of the proposed planting site in relation to large bodies of water should be considered. A planting site on the leeward side of a large body of water is likely to have a microclimate modified considerably both in summer and winter, with the temperature lower in summer and higher in winter than similar sites at a distance from such bodies of water.

In areas subject to severe winter cold, only the most hardy kinds of fruit crops should be attempted.

Wind

Avoid sites that have a history of strong winds. Wind can be detrimental from many aspects. Reduced bee activity during windy days in the pollination season can seriously reduce fruit set and yields. Wind can damage young, tender shoots in the spring and can scar and bruise young fruits. It can limit application of pesticides with speeds over 10 miles per hour. Wind can interfere with the uniform application of sprinkler water. Young trees planted in windy areas must be staked and tied. Windbreaks can help reduce this problem, but they can be a costly added expense and can complicate air drainage problems.

Precipitation and Available Water

Fruit plants require adequate soil moisture throughout the growing season to maintain vegetative growth and to produce a full crop of quality fruits. If no supplemental irrigation is possible, attention must be paid to the rainfall history of the proposed site to determine whether the total and summer rainfall is likely to be adequate and consistent. A better situation exists where water supplies for irrigation are available during times of drought. Some of the best fruit-growing areas in the world are located in areas where no rainfall normally occurs during the growing season. These regions depend entirely on irrigation.

A pattern of continual rains during the pollination period in the spring could result in poor crops by interfering with bee activity. Continual rains during the fruit harvesting period lead to problems, not only in getting the fruit picked but also in promoting various fungal diseases on the maturing fruits. In cold winter areas, accumulated snowfall during the dormant season provides protection against low soil temperatures and potential root injury. Snow can also delay bloom in the spring and reduce the possibility of frost damage to the buds and flowers.

The proposed site for a fruit planting should not be subject to periodic flooding from nearby rivers or streams. Fruit plants will not tolerate water around their roots for any length of time, as the water stops air penetration to the roots. Areas with a high water table are usually unsuitable for fruit growing as only the soil mass above the water table is available for root development, and this is usually quite limited.

Hail

Hail can cause serious damage to orchards and other fruit crops from both mechanical injury and disease infections entering open wounds. Small hailstones early in the growing season can reduce crop value, whereas large hailstones can destroy fruit, shred foliage, break spurs and new shoots, and remove bark from branches. It is thus very important that the frequency and severity of hailstorms at the proposed site be determined in advance.

Location and Soil Characteristics

Slope of the land is important. Ideally it should be over 2 to 3 percent for good air and water drainage but no greater than

10 to 15 percent for successful operation of equipment without contouring.

Topography should be relatively even, without nondraining depressions and no more than 10 percent grades. A north slope reduces potential for frost losses and sunburn but delays bloom and opportunities for early harvest. West slopes should be avoided except in cold climates or at high elevations where heat units are limiting.

Location can be a compromise between extending urbanization and distance for supplies, services, skilled labor, and markets (or warehousing). The ideal orchard soil should be a deep (1.8 m, 6 ft) well-drained, nonsaline, fertile silty loam to a fine sandy loam (see Ch. 8). However, a "rooting depth" of between two and three feet is usually quite adequate for most fruit crops. The surface should slope gently, allowing good runoff from heavy rains and permitting good infiltration of irrigation water. There should be no impervious hardpans or claypans under the surface. Fine-textured clay soils generally make poor orchard soils and should be avoided.

Planting orchard trees on unsuitable soil handicaps the orchard's productivity for the life of the planting and can make the enterprise marginal or unprofitable. If a less than ideal soil type is used, such as a clay loam, it is best to consider planting a fruit species that does relatively well on fine-textured soils, such as plums, pears, or apples, and avoid planting peaches or almonds, which will not tolerate such soils. If the available soil is a loamy sand or sandy loam, plant peaches or almonds (if the climate is suitable). Some species, however, such as grapes, oranges, and olives, do well on a wide range of soil types.

It is essential that all available information be obtained about the soil in the proposed site before planting. A soil map of the area should be consulted at the library of the nearest agricultural college or the office of the cooperative extension or the Soil Conservation Service. The soil of the area to be planted should be systematically sampled with a backhoe and enough pits opened to inspect the soil profile to a depth of about 1.8 m (6 ft). The sample will divulge any hardpan or rock layers or sandy pockets and will show whether the soil is deep enough to support fruit trees. A shallow soil may support smaller fruit plants such as strawberries or bushberries.

A study of the history of the site, including information of other crops previously grown, can be useful in analyzing soil problems. If cotton, potatoes, or tomatoes have been grown there, expect trouble from verticillium wilt. If an old orchard has been pulled out, the soil could be compacted, infected with *Phytophthora* fungus, or high in pesticide residues. Often it is necessary to fumigate the soil to eliminate fungus and nematode problems before planting a new orchard in order to obtain good, vigorous tree growth.

Resistant rootstocks can often reduce problems with soil pests. For example, many sandy loam soils, well suited for growing peaches, are infested with root-knot nematodes and large-scale fumigation may not be practical. But in some instances, by planting peach trees propagated on a nematode-resistant rootstock, such as 'Nemaguard,' good tree growth and productivity can be obtained in spite of the nematodes in the soil. Often, however, such resistant rootstocks are not winter-hardy in cold climates.

Irrigation Water—Quality and Availability

In areas with low rainfall during the growing season, assurance should be obtained that there is a potential source of ample high-quality irrigation water. The water should not contain high soluble salts, including sodium, chlorides, boron, calcium, and magnesium bicarbonates. Water samples can be analyzed by commercial laboratories and unsuitable water sources detected before the planting is made. It is difficult, if not impossible, to correct poor-quality water without high costs.

The quantity of irrigation water to be applied to a fruit planting will vary according to the climatic zone, the crop, and the irrigation system. The frequency of irrigation will depend upon the moisture-holding capacity of the soil. Young trees, coarse-textured soils, and shallow soils require more frequent irrigations but not greater amounts. Most critical are the last 30 days before fruit harvest, when water shortages can significantly reduce fruit size and quality, while increasing various fruit disorders.

Availability of Markets for the Crop

Utilization of the crop is generally assured for plantings in the home garden—by the family, friends, and neighbors, plus canning, drying, or freezing and, perhaps, sales of surplus fruits and nuts. In a commercial planting, however, it is essential to know that a market will be available for the crop and that the desired grade, quality, and volume of product can be produced by the time production begins. The choices for marketing the crop should be thoroughly considered in advance. Are there marketing cooperatives or private packers and shippers available who will take the crop, or will the grower need to grade, pack, store, and then transport the fruit to city fruit markets? Or, perhaps, in locations with considerable highway traffic, particularly close to large cities, on-the-farm roadside or "pick-your-own" sales may utilize most or part of the crop. Sometimes mail-order or Internet enterprises can be developed with proper advertising.

The marketing situation for most fruit crops is often quite fluid and should be thoroughly studied before heavy planting commitments are made. New plantings of a particular fruit crop in a certain area would be questionable when experienced growers in that area are either not planting or are pulling out trees. Perhaps the influx of an insect or disease problem has added control costs that eliminated the profit margin for that crop. Heavy plantings of a crop in a given area because of enticingly high returns at the moment may lead to market gluts when the plants come into production.

State agricultural experiment stations and state departments of agriculture and soil conservation offices are located in all states in the United States. Equivalent institutions are found in many other countries as well. Generally these institutions have one or more experts for each major crop grown in the state who are available for consultation free of charge on all phases of crop production and utilization. There are agricultural economists, too, who can give valuable advice on the economic outlook for particular crops. All states in the United States, as well as in most other countries, have local county agents or farm advisors who are employed by the state-supported cooperative extension service. Their services are generally available at no charge for consultations in planning and operating agricultural enterprises. Most of these people have expert knowledge of the local soils and climate and the requirements of the various crops and, in addition, are well informed of the economic picture. They usually have publications available dealing with the production and utilization of all fruit crops grown in the locality.

In addition to these publicly supported sources of information, private agricultural management services in many areas are prepared to give assistance, for a fee, at different levels—from completely operating the enterprise to consulting. They generally specialize in certain crops or types of crops.

Availability of Labor

In establishing a large, commercial fruit-growing enterprise, the availability of reliable workers to do the extra labor required should be assessed. Extra labor is usually needed to harvest the crop, to prune the trees, and for fruit thinning. Other operations, such as weed control, fertilization, and irrigation, can usually be handled with a minimum crew and are often done by the owner and manager.

In a fruit-growing enterprise, harvest labor is usually the most costly production expense and may determine the profit (or loss) margin. Harvest labor costs and even the availability of harvest labor can fluctuate widely and may remain unknown until harvest is actually under way.

The harvest of some fruit and nut crops such as sour cherries, prunes, almonds, walnuts, pecans, macadamia nuts, canning peaches, and pineapples has been wholly or partly mechanized, and efforts are being made to mechanize the harvest of other fruit and nut crops. Mechanization will greatly stabilize fruit-growing production costs.

Costs in Establishing a Fruit Planting

There are certain costs to be considered in establishing an orchard, vineyard, or berry planting. The actual amounts vary considerably with the country, locality, the crop, and the year. These costs are largely the result of:

1. **capital investment**
   —land, including water and power
   —facilities, including roads, buildings, housing, and irrigation
   —equipment, including shop and farm tools
   —trees or vines, stakes or trellis
2. **overhead costs** (during establishment period)
   —interest, taxes, depreciation, insurance
   —utilities, irrigation water
   —management
3. **production or direct operating costs**
   —labor, supplies, equipment and repairs from preplanting until commercial production
   —land preparation
   —tree or vine planting
   —staking or trellis construction
   —pruning and training plants
   —pest and weed control
   —irrigation, fertilization, and soil management

The time required to bring various fruit crops into production is from two to three years for peaches, apricots, and grapes to five or six years for apples and pears and even longer for sweet cherries and some pear cultivars.

SELECTING FRUITING CULTIVARS AND ROOTSTOCKS

There comes a time in the establishment of a large, commercial fruit-growing enterprise—or in the planting of a single, backyard fruit tree—or a vineyard, or a bushberry, or strawberry planting—when one must decide which cultivar (and rootstock, if needed) to order from the nursery. Once the kind of crop to be grown has been decided, the cultivar will have to be selected, a decision that can determine the success or failure of the enterprise. This decision should not be taken lightly. It should not be left to the local nurseryman, and it should not be based upon what surplus nursery stock happens to be on hand at a bargain at the time. Sometimes this decision must be made and an order placed a year or two in advance so that the propagating wholesale nursery will have time to propagate the desired cultivar on the desired rootstock and to permit obtaining disease-free plant material.

Information concerning the various fruiting cultivars can be obtained from the sources listed on this page, directly from other fruit growers in the vicinity, fruit buyers or packers, and from neighbors. It is risky to accept new, untried cultivars, or to base the decision solely on nursery advertisements.

Prospective fruit growers should realize that market demands for their produce can shift over the years, so they should study and determine what are the present, and possibly the future, demands for fruit products—by commodity and by cultivar. Figure 13–1 shows that in the United States from 1979 to 1997 fruits marketed as fresh and as juice increased slightly, while canned products declined. Frozen and

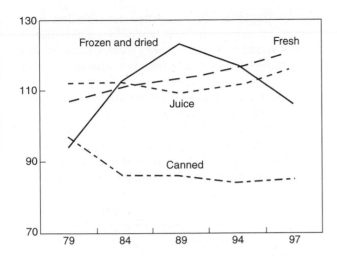

Figure 13–1 Changes in the demand for fruit products for the period from 1979 to 1997. Sales of canned fruits steadily declined while demand for fruits as the fresh product and as juice increased. *Source:* USDA.

dried fruit consumption increased in the early 1980s and then declined in the 1990s. Recent consolidation of large wholesale grocers has dramatically reduced the number of buyers and thus, competition. Overall, this has depressed the prices received by fruit producers. Also in the late 1980s produce sections of supermarkets in the United States and Europe started displaying fresh fruit and vegetables that were unknown a few years earlier, such as kiwifruit, lychees, Asian pears, carambolas, and mameys. This diversity of products coupled with increasing world production and concentration of buyers has created pressure on fruit growers to change cultivars to achieve higher prices. Standard cultivars such as 'Delicious' apples or the 'Hayward' kiwifruit are treated as commodities and receive much lower prices compared with newer cultivars such as Gala or Braeburn apples or yellow-fleshed kiwifruit.

Most fruit and nut trees and some grapevines consist of two parts—the top (fruiting) part and a lower part, which becomes the root system. These two parts are joined together by budding or grafting when the plants are propagated in the nursery. Other fruit plants, however, are propagated "on their own roots," with no graft union. The plants are started as rooted cuttings, suckers, layers, or by runners. Some examples of these are fig, mulberry, olive, quince, pomegranate, currant, gooseberry, and some grape cultivars, as well as the brambles and strawberries.

In fruit plants with two parts, it is very important that the prospective fruit grower decide not only what the top fruiting cultivar is to be but also what rootstock the trees will be on. Some fruit species offer a wide range of choices for the rootstock; with others, there is little choice. For example, there are about a dozen possible rootstock choices each for apples and oranges, but sweet cherries have only two or three rootstock possibilities. Commercial growers of oranges or apples are as concerned about the selection of

rootstock for a new planting as they are about the fruiting cultivar.

Many nurseries attach a label to each nursery fruit tree that identifies the fruiting cultivar and the rootstock. Ideally the prospective fruit grower would be knowledgeable enough to specify to the nursery the desired fruiting cultivar and the desired rootstock. For almost every tree fruit or nut species, certain rootstocks could be used but may result in poor tree performance as the trees get older. A well-informed fruit grower will know of these situations and avoid them.

Rootstocks are generally selected on the basis of several factors:

1. *Tree size.* The ultimate tree size of some fruit species can be limited through the selection of specified rootstocks or interstocks. These are often referred to as "dwarfing." Many are short-lived or have only a limited effect on tree size. The most satisfactory series of dwarfing rootstocks is for apple. There are a limited number of stocks for citrus. Pears can be dwarfed on quince rootstocks, but such trees lack hardiness, are very susceptible to drought, and are often weakly anchored.

2. *Resistance to soil-borne organisms.* Certain rootstocks permit plants to be grown in soil that otherwise would be unsuitable for them. For example, the European-type grapes (*Vitis vinifera*) on their own roots are killed by phylloxera, the grape root louse (*Dactylosphaera vitifoliae*), if planted in soil infested with this pest. But if the plants are grafted onto native American grape rootstocks resistant to this pest—such as *Vitus riparia, Vitis rupestris,* or hybrids between them—the plants thrive and produce well in infested soil. Some rootstocks are less susceptible than others to various pests such as nematodes, verticillium wilt, and armellaria root rot. They should be used where these problems occur.

3. *Resistance to unfavorable soil conditions.* Rootstocks that are more tolerant to poorly drained, heavy, or saline soils are available for some fruit species. Such rootstocks should be specified for plantings where these problems could exist.

4. *Resistance to low winter temperatures.* Certain tree fruit species, particularly apples and citrus, survive cold winters better on some rootstocks than on others. There are stocks that are very resistant to cold themselves, but in some cases have little if any influence on increasing the cold resistance of the scion cultivar.

PLANTING AND CULTURE

Major decisions must be made when the time comes to plant a new orchard, vineyard, or berry planting. Planting distances and patterns must be determined. The distance between plants depends on several factors.

The ultimate tree size of the species and cultivar at maturity is an important consideration. It is obvious, for example, that mature walnut or pecan trees growing to a height of 12 to 15 m (40 to 50 ft) and a breadth of almost the same amount need to be planted much farther apart than plum trees growing only 4.5 m (15 ft) tall. Even within the same species, cultivars differ in size. For example, apple cultivars, such as 'Jonathan,' are smaller and less upright than 'Northern Spy,' which is more vegetative and slower to come into bearing. Even within the same cultivar, apple trees, such as 'Delicious' with a spur-type growth habit, are smaller than nonspur selections.

The rootstock is a second factor determining planting distance. Are dwarfing or invigorating rootstocks going to be used? In apples, for example, the trees to be planted could have been propagated on the very dwarfing 'M. 9' rootstock, or they could have been propagated on an invigorating apple seedling rootstock or on a clonal rootstock such as 'M.M. 111,' giving intermediate vigor and tree size. Apple trees on the most invigorating rootstocks would need to be planted about six times farther apart than those on the most dwarfing rootstocks.

Environment is a third factor determining planting distance. Is the planting site a sandy, shallow, infertile soil—where the trees would be slow-growing and never get very large—or is it a deep, highly fertile clay loam, where the trees are likely to reach their maximum size? Sites for fruit plantings on north slopes, on the leeward side of large bodies of water, and at higher elevations are characteristically cooler, resulting in less tree growth and smaller tree size. Greater latitudes (north or south) are not necessarily cooler but have shorter growing seasons.

The planned tree density is a fourth factor, one under the grower's control. In recent years, the so-called high-density orchard plantings, particularly with apples and to a lesser extent with citrus and pears, have become popular. Trees are planted close together, as hedgerows, or tree walls. Dwarfing rootstocks are used to keep the trees small. The land can be utilized to the maximum by high-density plantings, especially when the trees are young. High-density plantings of the stone fruits—peaches, plums, apricots— have not been as successful mainly because no completely satisfactory dwarfing rootstock is available for these species.

The different categories of planting densities and management systems in use today, particularly for apples, are:

1. *Low density.* Trees are widely spaced (fewer than 250 trees/ha; 100 trees/ac) so that after maturity each tree has ample space and light contact around it. Pruning is kept to a minimum to allow rapid development of maximum tree size. Dwarfing rootstocks are not used. Maintenance labor is minimal, but the yields and gross returns per unit area are also likely to be minimal, particularly for the first 15 to 20 years of orchard life, compared to higher-density plantings (of apples and pears). Fifteen to 20 years may be needed to reach full production.

For the stone fruits, tree nuts, and citrus, low-density plantings may be the most profitable although there has been considerable interest and experimentation in developing high-density management systems with these crops.

2. *Medium density.* Tree spacing (250 to 500 trees/ha; 100 to 200 trees/ac) is at least 1.2 m (4 ft) closer than for low-density plantings, and pruning and training are more intensive. Dwarfing rootstocks, such as 'M.M. 106' or 'M. 7A,' may be used with apple, or the trees may be spur-type cultivars budded on seedling rootstocks. More labor is required in pruning and training the trees, particularly during the early developing years. More care and supervision are required, and the investment per hectare—in nursery stock and, perhaps, irrigation equipment—is greater than for low-density plantings. However, commercial yields are achieved sooner after planting, which reduces the cost of establishing the planting.

3. *High density.* Trees are planted very close together (500 to 1,235 trees/ha; 200 to 500/ac) and specific training systems are used, such as slender trees, multiple rows, or trellis-trained plantings. Training and pruning are very important. Reduced tree size can be aided by the use of the more dwarfing rootstocks, such as Angers quince with pear and 'M. 9' or 'M. 26' with apple.

The grower must be committed to the system and determined to make it work. Neglect of the planting can result in considerable financial loss. The initial investment per hectare is higher, but the total investment can be lower. The yield per hectare during the first 15 years of orchard life can be much greater than for the low-density planting.

Generally, orchard efficiency—judged by increased fruit produced for each man-hour spent—is increased and if properly managed, fruit quality and packout are increased. A trend exists for nearly all fruit and nut crops to achieve higher-density plantings.

Table 13–1 integrates the various factors that must be considered in determining the proper tree spacing for apples. Recommended planting distances for other fruit species are given in Table 13–2. It must be emphasized for these crops, too, that several factors can modify these recommended distances. Greater spacing would be used with conditions of high soil fertility, long growing seasons, vigorous, large-size cultivars, invigorating rootstocks, ample rainfall or irrigation, and heavy use of fertilizers; spacing would be closer in the opposite situations.

The **contour planting system** in sometimes used on rolling slopes or hillsides where some terracing may be needed. This planting arrangement, while subject to problems, permits production from land that otherwise could not be utilized. Considerable care must be taken to stop erosion by heavy rains or by irrigation by diverting the water to run along the tree rows rather than straight downslope.

Certain fruit crops, other than tree fruit, are planted close together in long rows—for example, grapes, raspber-

TABLE 13–1 The Integration of Cultivar, Rootstock, Soil Type, and Management System in Determining the Tree Spacing for Apple Trees

Cultivar	A VF	B Rootstock						C Soil Type			D Management System					E Tree Spacing
	Cultivar Factory	EM 9	M. 26	EM 7	M.M. 106	M.M. 111 & EM II	M.M. 104 & Seedling	Low Productivity	Med. Productivity	High Productivity	Low Density	Med. Density	Med. High Density	High Density	Tree Walls	
Sundale SturdeeSpur	2	2	4	6	8	10	12	2	4	6	0	−4	−6	−8	−10	
Red Chief Gallia Beauty Idared Miller SturdeeSpur SpureeRome MacSpur	4	2	4	6	8	10	12	2	4	6	0	−4	−6	−8	−10	
Golden Del. C449 Jonnee Dbl. Red Jonathan Quinte Macoun Imperial Red Delicious Paulared	6	2	4	6	8	10	12	2	4	6	0	−4	−6	−8	−10	
Beacon Cortland Red Prince Lodi Red Queen Spartan Turley Tydeman's Red Red Winesap	8	2	4	6	8	10	12	2	4	6	0	−4	−6	−8	−10	
Empire R.I. Greening Mutsu Northern Spy Spigold Red Stayman Red York	10	2	4	6	8	10	12	2	4	6	0	−4	−6	−8	−10	

Directions: Add the cultivar factor (column A), rootstock factor (one of the columns under B), and the soil type factor (one of the columns under C), then subtract the management system number (found in one of the D columns). This will give a suggested distance to plant trees in the row. Next, add 8 ft to this figure for the distance needed between the rows. On steep slopes it is better to add 10 ft to get the distance between rows. For tree wall plantings, use 14 ft between rows. If E, and the total of A + B + C − D, = 0 or less, the combination is unprofitable.

Formula: A + B + C − D = E (planting distance between trees in the row). This number plus 8 ft equals planting distance between rows.

Example:

A. Cultivar   Red Prince   Factor of 8
B. Rootstock   M.M. 106   Factor of 8
C. Soil Type   Medium Productiveness   Factor of 4
D. Type of Management   Medium High Density   Factor of −6

8 + 8 + 4 − 6 = 14 ft between trees in row   14 ft + 8 = 22 ft between rows.

Source: Hilltop Orchards and Nurseries, Inc., Hartford, Michigan 49057.

TABLE 13–2	Planting Distances for the Common Fruit and Nut Crops (for Apples, see Table 13–1)
Species	**Planting Distances**
Almond	7.5 × 7.5 to 9 × 9 m (25 × 25 to 30 × 30 ft)
Apricot	6.6 × 6.6 m (22 × 22 ft) (on plum roots) 7.5 × 7.5 m (25 × 25 ft) (on apricot roots)
Avocado	12 × 12 m (40 × 40 ft)
Blueberry	1.2 m (4 ft) apart in rows 3 m (10 ft) apart
Cherry, sour	6 × 6 m (20 × 20 ft)
Cherry, sweet	7.5 × 7.5 to 9 × 9 m (25 × 25 to 30 × 30 ft)
Date palm	9 × 9 m (30 × 30 ft)
Filbert (hazelnut)	4.5 × 4.5 m (15 × 15 ft)
Grape	1.2 to 2.4 m (4 to 8 ft) apart in rows 2.4 to 3 m (8 to 10 ft) apart
Kiwifruit (vines on trellis)	5.4 to 6 m (18 to 20 ft) apart in rows 4.5 m (15 ft) apart
Lemon	6.6 × 6.6 to 9 × 9 m (22 × 22 to 30 × 30 ft)
Olive	9 × 9 m (30 × 30 ft)
Orange	6 × 6 to 7.5 × 7.5 m (20 × 20 to 25 × 25 ft)
Papaya	2.4 × 2.4 to 3 × 3 m (8 × 8 to 10 × 10 ft)
Peach	3 × 6 to 5.4 × 7.2 m (10 × 20 to 18 × 24 ft)
Pear	6.6 × 6.6 m (22 × 22 ft)
Pecan	9 × 9 to 15 × 15 m (30 × 30 to 50 × 50 ft)
Pineapple	30 × 81 cm (12 × 32 in)
Pistachio	7.2 × 7.2 m (24 × 24 ft)
Prune	6 × 6 m (20 × 20 ft)
Raspberry, black	0.6 to 1.2 m (2 to 4 ft) apart in rows 2.1 to 3 m (7 to 10 ft) apart
Raspberry, red	0.75 m (2.5 ft) apart in rows 1.8 m (6 ft) apart
Strawberry (matted-row system)	61 to 71 cm (24 to 28 in) apart, permitting a matted row, 38 to 61 cm (15 to 24 in) wide to develop from runners
Strawberry (double-row bed system)	Beds 96 to 112 cm (38 to 44 in) apart, center to center; two rows in each bed 20 to 30 cm (8 to 12 in) apart. Plants in each row 23 to 36 cm (9 to 14 in) apart
Strawberry (single-row bed system)	Beds 100 to 107 cm (39 to 42 in) apart, center to center; plants 20 to 25 cm (8 to 10 in) apart in rows
Walnut (Persian)	6 × 6 m (20 × 20 ft) to 10.5 × 10.5 m (35 × 35 ft) (Paynet type) 10.5 × 10.5 m (35 × 35 ft) to 12 × 12 m (40 × 40 ft) (Hartley type)

ries, blueberries, blackberries, strawberries, passion fruit, and kiwifruit—at the planting distances given in Table 13–2. Some, such as grapes, passion fruit, and kiwifruit, are trellised because their long fruiting canes must be supported; some grapes, though, such as the 'Tokay,' can be free-standing, without a trellis. Other fruits, such as the pineapple, whose culture is highly mechanized, are grown like an agronomic field crop, the plants being set close together and completely covering the area.

In planting fruit trees of a species requiring cross pollination to set good commercial crops, it is of the utmost importance that trees of the pollinizing cultivar be appropriately spaced among trees of the principal fruiting cultivar.

Laying Out an Orchard Planting

When the young trees are due to arrive from the nursery, the spot where each tree is to be planted throughout the block of land must be marked in readiness for digging the holes for the actual planting. It is important that the trees in the rows be lined up properly. This will facilitate many future orchard operations. A tree out of line can be a target for cultivating disks and other equipment moving down the rows.

> When the time comes to plant the trees, all preparatory steps must have been completed. If the orchard is to be irrigated, all land leveling must be finished and irrigation pipelines installed. If a claypan is present, it should have been broken by ripping. If pathogenic soil organisms, such as nematodes, *Phytophthora,* or *Verticillium* were found, soil fumigation must have been completed with ample time allowed for the dissipation of fumes. Weed control of the area must be completed. If perennial weeds like Johnson grass, poison ivy, or brambles were present, the proper herbicides should have been used to bring them under control. Once the trees are planted, many options to perform these operations are closed.

Planting the Trees

The best time to plant the nursery trees—or grapevines, or bushberries—is any time during the dormant season, when the ground is not frozen and air temperatures are above freezing. It is important to plant as early as possible so the roots will be well established by the time hot weather arrives and the plants will have a full growing season before they are faced by cold weather.

Deciduous plants should have dormant buds and no leaves when they arrive from the nursery. The roots must be protected from drying out by some moist packing material such as wood shavings. Roots must be continually protected up to the time the trees are planted. Broad-leaved evergreen fruit nursery trees, such as citrus or avocados, should arrive with their roots undisturbed in a soil ball covered with burlap or heavy plastic or should be containerized.

High-quality nursery stock has a strong, vigorous straight trunk with an abundance of roots well distributed around the trunk. No part of the trunk or root system should show evidence of damage from careless handling. A small shallow slice into the trunk should show bright green tissue below the bark with no evidence of brown areas from winter damage or sunburn. A slice into the roots should reveal a moist whitish color. No root tissue should be shriveled or look brown, gray, or black below the bark. In budded or grafted plants the union should be well-healed and strong, with no more than a slight bend at the union. The graft union should be at least 10 cm (4 in.) above the previous soil level so that scion rooting[1] is unlikely after planting. Scion rooting can

---

[1]Roots developing above the graft union.

lead to a loss of any desired effects of specific rootstocks such as dwarfing, resistance to nematodes, diseases, etc.

In deciduous nursery stock the dormant buds should be plump and well-developed and should look bright green when cut into. Dead buds may indicate low temperature or herbicide injury or lack of water during the growing season. Broad-leaved evergreen nursery stock should have normal-size leaves of a healthy deep green color and should show no nutrient deficiency symptoms or herbicide damage. Nursery stock should not be infected with any pathogens, insects, or mites. Each nursery plant or lot should be clearly labeled with the species and cultivar name and the rootstock species (and cultivar, if applicable). The size grade should also be stated on the label.

In planting the tree or vine it is important that the planting hole be dug to the proper depth. The base of the main supporting roots, which usually have been trimmed back, or the soil ball, in the case of evergreen plants, should rest on solid, undisturbed soil. If the hole has been dug too deep, necessitating some back-filling before planting, then the plant is apt to sink after watering and settling, putting the graft union below the soil level and leading to attacks of crown rot fungi, principally of the *Phytophthora* species.

The planting hole should be wide enough to easily accommodate the roots without bending and twisting. Tractor-operated soil augers are often used for digging holes. These have the advantage of working fast and saving labor, but unless operated properly the holes can be dug too deep, leading to settling. Also if the soil is too wet when the auger is used, the sides of the hole become severely glazed, making air and water permeability and root penetration difficult. A shovel should be used to break up the compacted sides of the holes when the trees are planted.

For large-scale plantings mechanical tree planters are used. The tractor driver must exercise care to ensure that the rows are as straight as possible. In-row spacing is never as accurate with mechanical planting and it is critical that personnel follow the transplanter to make slight adjustments in graft union height and to firm the soil around the plant.

When evergreen nursery trees like citrus or avocados are planted, the bud union should be several inches above the soil level. Any burlap or other material around the soil ball should be loosened and split open around the top and sides. Metal containers should be removed before planting.

Filling loose soil around the roots once the plant has been set in the hole is an important operation. The soil should be worked around the roots as the hole is filled to ensure good soil-root contact, with the soil pressed firmly about the roots. If there are persistent summer winds from one direction, the tree can be leaned slightly into the wind at planting.

A shallow basin or furrow should be left at the top for filling with water. For those plants with a soil ball around their roots, the top of the soil ball should be slightly exposed at the surface of the basin so that added water can enter di-

rectly into and through the root ball. Often the nursery soil found in such root balls is light and porous and if, after planting, they are covered by the heavier clays of native soils, irrigation water will remain in the clays because of their smaller pores and high capillarity, leaving the lighter soil with the roots completely dry. A few days of hot desiccating winds can increase transpiration and severely injure trees planted too deeply even though the basins have been filled with water.

The soil basins should be filled with water within a few hours of planting. This prevents dehydration of the roots and also settles the soil around the roots. Watering by filling the basins should continue until vigorous shoot growth is well underway, when the regular irrigation system or rainfall can be used. In heavy soils with drainage problems care must be taken to avoid overwatering, which can impede good root aeration and lead to attacks by *Phytophthora* and other soil pathogens.

The trunks of the young trees must be protected from sunburn by wrapping them with paper, cardboard, or styrofoam coverings during the summer or painting with whitewash or water base paint (one part interior white latex paint and one part water). Inspect for bark damage at intervals during the winter by rabbits, mice, or gophers at the soil level and take the necessary preventive measures.

It is best not to fertilize or apply herbicides to the trees at planting time. The developing roots could be injured by excessive salts or by herbicides. Generally, the young trees obtain enough mineral nutrients from the soil for their first year's growth.

PRUNING AND TRAINING

TERMINOLOGY

Pruning and training has its own vocabulary. Here are some important terms and their meanings.

Heading back.   Shoots or limbs are not removed entirely, but the terminal portions are cut off at varying distances from the end. This procedure forces out new shoots from buds below the cut and retards terminal growth of the branch (see Fig. 13–2).

Thinning out.   Shoots or limbs are completely removed at the point where they attach to the next larger (and older) limb. Thinning out corrects an overly dense area or removes interfering or unneeded branches (see Fig. 13–2).

Trunk.   That portion of the tree up to the first main scaffold branch.

Primary scaffold branches.   The main branches arising from the trunk of the tree.

Secondary scaffold branches.   Supporting branches arising from the primary scaffold branches.

(continued)

Central leader training system. A training system for fruit trees where one main vertical trunk remains dominant with all scaffold branches arising from it in an almost horizontal direction (see Fig. 13–3).

Modified leader training system. A central leader continues upward from the trunk, but its identity is lost as it becomes one of the primary or secondary scaffold branches.

Open center or vase training system. A trunk is developed, then at the top of the trunk (0.6 to 0.9 m; 2 to 3 ft from ground level) several primary scaffolds develop outward and upward at a 30° to 45° angle to form a vase configuration (see Fig. 13–4).

Weak crotch. A situation that can develop when two equal-size branches with a narrow angle between them grow for some years at the same rate. Eventually a considerable amount of bark inclusion develops between the two branches at their junction, with no connecting wood. With heavy loads of fruit—and strong winds—the limbs are almost certain to split apart. This problem can be prevented by completely removing one of the branches when they are young or heading one of them back heavily so the other becomes the main limb and the headed-back branch a lateral.

Fruiting wood. One-year-old shoots of certain species, such as peaches, arising from larger branches of the tree, on which flowers and fruits are produced. They often hang downward and are termed hangers (see Fig. 13–10).

Fruiting spurs. Trees of such species as apples, pears, plums, cherries, apricots, and almonds bear some or most of their crop on these short, thick shoots (see Fig. 13–5), which usually live many years.

Stub. Pruning cuts of the thinning-out type must be properly done. Cutting off the branch leaves a wound, which must heal over by callus growth proliferating from the surrounding live tissue or else decay organisms can become established in the wound. If the cut is not made close to the adjacent limb and a stub is left, the wound cannot heal properly.

Watersprouts. Vigorous shoots usually arising from latent buds in the trunk or older limbs in the lower parts of the tree, growing upright through the center of the tree (see Fig. 13–6).

Suckers. Shoots arising from the underground parts of the plant, usually coming from adventitious buds on roots. In grafted trees they generally arise from the rootstock below the graft union. They should be removed as soon as they are noticed (see Fig. 13–6).

THE DEVELOPING FRUIT PLANTING

Once the young trees or vines have become established and started to grow, they will need watchful care to avoid nutritional, disease, or insect problems. Weeds must be controlled by cultivation or herbicides. Irrigation and fertilization must be routinely and properly done.

Every effort should be made to bring the fruit planting into bearing in the fewest possible years. Delayed production is often a result of heavy pruning and overfertilization with nitrogen, both of which keep the plants excessively vegetative.

Pruning and training of fruit plants is an integral part of the procedures used for high production of quality fruits. These procedures vary with the kind of plant being pruned. Grapes are pruned in a certain manner, evergreen fruit species (citrus, avocados, olives, etc.) in another, and deciduous fruit trees in still another. Different species in these groups are pruned differently; for example, apple trees and peach trees are pruned in quite a different manner because of basic differences in their fruit-bearing habits (see Table 13–4).

Fruit plants, as they mature from planting through the early years of rapid growth to the stable growth of full maturity, require different pruning and training procedures. In successful fruit production it is essential that the fruit grower know the correct pruning and training procedures as well as to understand the reasons why they are done as they are. *Pruning* refers to the actual cutting; while *training* refers to the positioning of limbs, branches, or canes.

Reasons for Pruning

Fruit and Nut Trees

1. To develop a strong trunk and scaffold system of branches, well distributed around the tree, which are able to support heavy loads of fruit without limb breakage.
2. To help regulate fruit production. Proper pruning encourages development of the type of shoot system that produces the fruit. In older trees with little vegetative growth rejuvenation, pruning can force the development of productive fruiting shoots. Pruning can also be used to limit excess numbers of fruit by removing some fruit-bearing branches, giving a thinning effect that can improve fruit size and quality. Pruning also provides an important means of balancing vegetative growth to fruiting, which aids in the development of high quality, high annual production, and increased resistance to disease, heat, and cold.
3. To remove dead, broken, or interfering branches.
4. To improve light penetration to the inside and lower parts of the tree.
5. To facilitate insect and disease control by opening the tree, thus increasing penetration of spray materials to the interior branches.
6. To limit tree size and shape to the space allocated to it and to limit tree height to that from which fruit can be conveniently harvested.

Grape Vines

1. To aid in the establishment and maintenance of the vines in a form that facilitates vineyard cultural operations.
2. To distribute the fruiting wood to obtain maximum production of high-quality fruit.

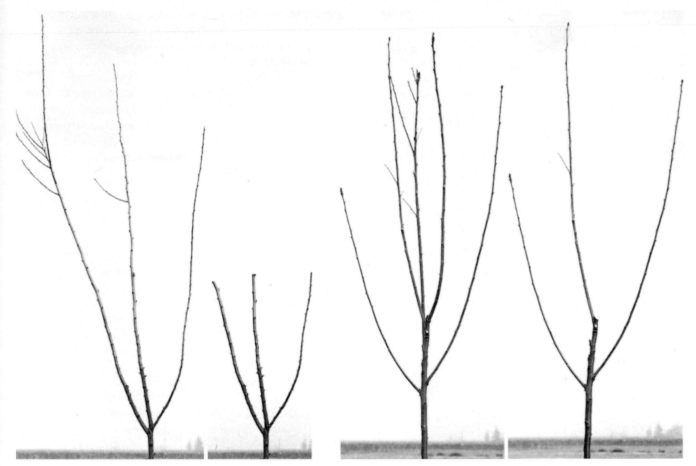

Figure 13–2 In pruning by heading back (*Left*), scaffold branches are cut off to one-half or more of their length. In pruning by thinning out (*Right*), certain scaffold branches are completely removed. Sometimes a combination of the two is used, thinning out, followed by heading back. *Source:* Gerdts, M., A. Hewitt, J. Beutel, J. Clark, and F. Cress. 1977. Pruning home deciduous fruit and nut trees. Univ. of Calif. Div. Agr. Sci. Leaflet 21003.

3. To maintain vigor and production of fruiting canes and to balance vegetative growth with fruiting for high-quality fruit and to impart greater resistance to heat and winter cold.
4. To aid in control of crop size and increase berry size by reducing the number of fruiting clusters.
5. To remove old, nonproductive canes.

Bushberries (Raspberries, Blackberries, Blueberries, Currrants, and Gooseberries)

1. To remove dead, weak, or diseased canes and old shoots that die following fruiting.
2. To thin out weak canes to give adequate light and space to the remaining canes, resulting in larger, better-quality fruit.
3. To develop lateral fruiting branches by summer topping (of black and purple raspberries and upright blackberries, but not red raspberries).

Physiological Responses to Pruning

All plants, if not pruned, tend to develop a balance between growth of the shoot and the root systems. Cutting away part of the top, including the plant's photosynthetic apparatus and food storage tissues, together with reducing the number of vegetative growing points and flower buds while leaving the root system intact leads to some interesting physiological reactions. The fruit grower should be aware of these reactions in order to understand how to prune the plants properly.

1. *There is a reduction in total vegetative growth.* Removing a portion of the top reduces the total amount of subsequent growth, compared to an unpruned plant. The total number of growing points is reduced, resulting in fewer developing shoots, fewer leaves, reduced photosynthesis, reduced amounts of carbohydrates translocated to the roots, reduced root growth, followed by a reduction in mineral and water absorption, which, in turn, further decreases shoot growth. These effects dwarf the entire plant. Generally, the more severe the pruning, the greater the dwarfing.

   The remaining growing points following pruning, which utilize all the stored foods in the plant, usually show strong shoot growth. The increased vigor of these shoots may lead one to believe that the pruning has caused increased total growth, but numerous

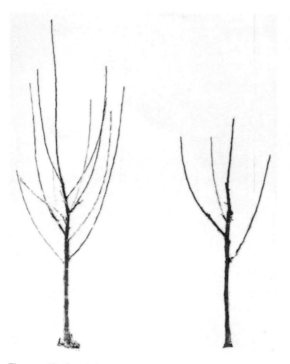

Figure 13–3 *Left:* Young, unpruned apple tree. *Right:* After dormant pruning. Tree is being trained to a central leader system, retaining a dominant central trunk with several primary scaffold branches, well spaced around the tree and along the trunk. *Source:* Gerdts, M., A. Hewitt, J. Beutel, J. Clark, and F. Cress. 1977. Pruning home deciduous fruit and nut trees. Univ. of Calif. Div. Agr. Sci. Leaflet 21003.

experiments have shown that this is not so. (If invigoration of the total plant is necessary, the judicious use of added fertilizers with ample soil moisture, together with only moderate pruning, should be practiced.)

2. *Continual heavy pruning each year of young fruit trees can delay the onset of bearing.* This is particularly true for the broad-leaved evergreens (which store their foods mainly in the leaves, twigs, and branches rather than in the larger branches, trunk, and roots, as do the deciduous fruit trees).

3. *Pruning effects from removal of smaller shoots and branches tend to be localized.* Heading back of shoots during the dormant season, reducing the length of a shoot by 50 percent, for example, results in the development of one or more new shoots from buds just below the pruning cut. The total growth by the end of the summer will not be nearly equal to the total growth of an adjacent shoot that was not dormant pruned. This response is often used in pruning to retard the growth of one of two equal-growing shoots and thus prevent the development of a weak crotch.

4. *Severe pruning in early summer is more likely to weaken the tree and reduce total growth than dormant (winter) pruning, especially with young trees.* Much of the energy for new spring shoot growth comes from stored foods in the roots, trunk, and branches of the tree. These stored foods are not replaced until an

Figure 13–4 *Left:* Young, unpruned peach tree. *Right:* After dormant (winter) pruning to the open vase system. Three primary scaffold limbs were retained, spaced about 120° apart around the trunk and 5 to 7.5 cm (2 to 3 in.) apart vertically. Heading back the scaffold limbs slightly promotes secondary branching. *Source:* Gerdts, M., A. Hewitt, J. Beutel, J. Clark, and F. Cress. 1977. Pruning home deciduous fruit and nut trees. Univ. of Calif. Div. Agr. Sci. Leaflet 21003.

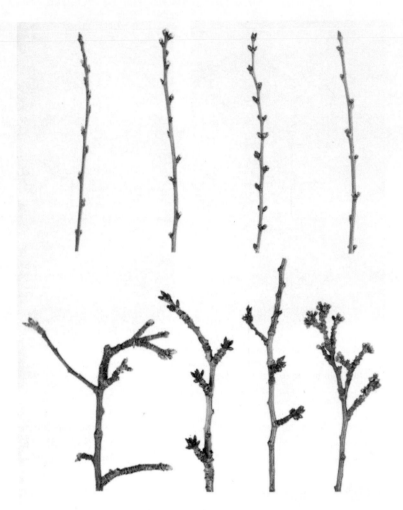

Figure 13–5 Development of fruiting buds laterally on one-year-old shoots (*top*) and laterally and terminally on fruiting spurs (*bottom*). *Top* (left to right): peach, almond, apricot, plum. *Bottom* (left to right): apple, sweet cherry, apricot, plum.

appreciable number of new leaves have formed and photosynthesized enough to reverse carbohydrate movement into the larger branches, trunk, and roots. If summer pruning is done, partially removing these new shoots and reducing photosynthesis and the replenishment of foods already utilized, the tree can be intensely debilitated. However, a moderate pinching back of only the tips of the shoots in the spring to direct growth is not likely to be harmful. Tree response to pruning in the late summer (August) is similar to the response from dormant pruning.

5. *Dormant pruning of bearing deciduous fruit trees can be invigorating.* Although vegetative growing points are removed, flower buds are also removed, thus reducing the crop, consequently reducing the demand on the plant's stored foods. The surplus is then utilized for new vegetative growth.

6. *Moderate dormant pruning of bearing deciduous fruit plants can increase production.* Flowers are produced either laterally or terminally on one-year-old shoots and/or laterally or terminally on fruiting spurs (see Fig. 13–5 and Table 13–3). Such fruit-producing growth (shoots and spurs) must be stimulated to obtain continual crops. Moderate pruning, plus fertilizers and ample soil moisture, result in such growth stimulation. Mature broadleaved evergreen fruit species, such as citrus, olives, and avocados, should be pruned lighter than deciduous species. Experiments have shown that the yield of the evergreen species generally decreases in proportion to the severity of pruning.

7. *Removing shoot terminals causes lateral branching.* Newly planted, one-year-old trees of some fruit species—for example, apples, cherries, pecans, pears, walnuts, quince, European plums, and figs—show strong apical dominance. If the terminal bud is not pruned, the tree will grow mostly as a straight, sparsely-branched trunk. If the terminal bud is removed (plus several inches of shoot below it for convenience) at planting time, however, the lateral buds remaining below the cut will

Figure 13–6 The difference between suckers and watersprouts as shown here with pear trees. Suckers (*Left*) arise from the roots around the base of the tree. Watersprouts (*Right*) are vigorous shoots growing from the trunk and older branches up through the center of the tree.

develop, producing lateral scaffold branches. To cause further branching of these scaffolds the next year, the terminals of the scaffold shoots must again be headed back by several inches.

Other tree species, such as apricots, peaches, almonds, and nectarines, do not show such strong apical dominance. Lateral buds develop into scaffold branches without heading back of the main shoot terminal growth.

Pruning and Training Young Deciduous Fruit Trees

After a young nursery tree has been planted in an orchard, the immediate goal is to develop a tree that has a strong straight trunk with three or four primary scaffold branches well spaced around the tree and vertically on the trunk. Scaffold branches having wide angles with the trunk should be retained so that the branch attachment will be strong. Those with narrow angles should be removed. Remember that the height of attachment of the scaffold branches from the ground will always remain the same, but the trunk and each scaffold branch will increase in diameter each year throughout the life of the tree because of cell division in the lateral cambium layers. There must be space for each of the scaffolds as they grow. All branches should not originate at the same height on the trunk.

If the species is one that is adaptable to mechanical harvesting by trunk shakers (almonds, prunes, cling peaches, olives, sour cherries, citrus), the main trunk should be high enough (before the attachment of the lowest primary scaffold branch) for the shaker clamp to attach.

At planting nursery trees are usually straight trunks 0.9 to 1.5 m (3 to 5 ft) tall. Some nurseries do not cut back the tops when they are dug; however, the top must be reduced in size to compensate for the root system lost when the tree was dug (see Fig. 13–7). The terminal end of the trunk should be headed back to 76 to 90 cm (30 to 36 in.) from the ground to force shoots out from the lateral buds. Some of these shoots will later be selected for the primary scaffold branches. If the nursery trees already have short lateral shoots, they can be retained and used as starting points for the primary scaffold branches.

If the nursery trees already have short lateral shoots, they can be retained and used as starting points for the primary scaffold branches.

Recent work with apple trees shows that special treatment in the nursery to develop "feathered trees" (one-year-old well-branched, featured tree) improved early yield and orchard efficiency. A well-feathered tree should have three to four strong branches with wide crotch angles at the height desired for the lower scaffold branches. After planting, the central leaders are pruned 20 in. above the highest branch left as a scaffold. For terminal-bearing cultivars, such as 'Rome Beauty,' the feathers are shortened 30 percent, and for standard-bearing cultivars, such as 'Golden Delicious,' the feathers are left unpruned. Many growers demand feathered trees for very intensive plantings because of the earlier yield and greater productivity.

As the young trees start to grow in the spring, most of the lateral buds along the trunk are likely to develop. If none of these are removed, the tree may look like a bush by the end

Figure 13–7   Pruning of unbranched nursery fruit trees (*far left*) after planting consists only of cutting back a portion of the top to 50 to 60 cm (20 to 24 in.) above the ground to force out lateral branches, as shown in the second photograph. If the nursery tree is well branched (*right*), pruning after planting consists of heading back the top, removing most laterals, and cutting the remainder back to stubs, from which new scaffold branches can develop (*far right*). *Source:* Gerdts, M., A. Hewitt, J. Beutel, J. Clark, and F. Cress. 1977. Pruning home deciduous fruit and nut trees. Univ. of Calif. Div. Agr. Sci. Leaflet 21003.

of the growing season, with shoots developing all along the trunk, and most of them will have to be removed during the following winter dormant pruning. Much can be done to direct this growth into permanent scaffold branches by judiciously removing or tipping back certain of these succulent lateral shoots as they start to develop in the spring. As soon as the shoots are 5 or 7.5 cm (2 or 3 in.) long, each tree should be carefully inspected. Shoots arising around the base and lower half of the trunk should be removed. Toward the upper half of the trunk three or four of the new shoots that are well placed up and down and around the trunk can be retained to develop into the primary scaffold branches. There should be 15 to 30 cm (6 to 12 in.) vertically between any two scaffolds. One should never be allowed to develop directly above another. All other shoots should have their tips pinched out to retard growth or, if there are many shoots, some could be removed. In apples trained to the central leader or central axis system, removal of the three to four small shoots just below the central leader establishes the dominance of the leader and decreases problems of competition. It is not wise to remove too much leaf area since this can seriously retard growth of the young tree. Close inspection of each tree during the first summer can be most useful—to remove any suckers or strong-growing watersprouts arising from the base of the tree, and to remove or head back shoots not destined to be the primary scaffolds.

If the orchard site is windy, the trunks may need to be staked and tied to keep the trees from being whipped about. A strong stake should be driven in the ground on the windward side of the tree about 0.3 m (1 ft) from it. Then loose, soft cords should be used for tying the trunk at about three places. The trunk should be permitted to move about somewhat, since a swaying trunk becomes much stronger than does one immobilized by tying too tightly.

Pruning and Training During the First Dormant Season

Generally, pruning the first dormant season after planting should be delayed until the coldest part of the winter is over.

Pruned trees are more susceptible to cold damage than unpruned trees. In addition, any killing of tree parts by winter cold will obviously be a factor in deciding what to retain and what to remove during pruning. Early winter and midwinter pruning can aggravate bacterial canker problems in young trees, as well as peach silver leaf and apricot limb die-back, which are associated with rainy weather. The peach tree short-life problem in North and South Carolina is worse with early winter pruning than with late winter or early spring pruning.

If the trees have been lightly shaped by pinching back or removing unwanted shoots the previous summer, the first dormant pruning is much simplified. Basically, the primary scaffolds already selected should be retained and all other competing shoots thinned out. Pruning should be light, however; heavy pruning on young trees has a dwarfing influence and delays cropping.

If no summer training has been done, the first dormant pruning is more involved. All branches arising from around the base of the tree should be removed. Then the strongest three or four shoots at the level desired for scaffolds, well spaced around and up and down—15 to 30 cm (6 to 12 in.) apart vertically—should be retained. Other competing shoots should be thinned out, being careful to avoid excess pruning.

In species with strong apical dominance, such as apples, pears, cherries, figs, European plums, and prunes, it is important to bend limbs to encourage growth at the base of the limbs. In the past these laterals were headed, but research has shown that this is uneconomical because of the delay in cropping. If a lateral is too vigorous, it should be bent by tying, spreading, or attaching weights to decrease growth and encourage flowering. In training systems with a central leader, it is important to avoid strong competing limbs for a distance of 3 ft above the lower tier of scaffolds, and strong laterals in this area should be bent or thinned out entirely.

Other species, like peaches, almonds, apricots, and Japanese plums, do not show strong apical dominance and develop lateral branches readily. In these species the lower scaffolds tend to grow more vigorously than the upper ones and may dominate the tree. To overcome this, head back the lower scaffolds lightly and the upper ones not at all.

Peaches should be headed back to force out lateral branches to form the primary scaffold system. Side branches developing on current season's growth do not make satisfactory scaffold branches.

In all cases try to keep the pruning as light as possible while still accomplishing the desired effects. Lightly pruned trees are apt to come into bearing one to three years earlier than those heavily pruned.

Pruning and Training During the Second Season

During the second growing season the secondary scaffold branches develop from the primary scaffolds. Some judicious pinching out of unwanted shoots arising from the primary scaffolds can be done during early summer, permitting two well-placed shoots on each primary scaffold to develop into the permanent secondary scaffolds. Through the summer, too, any vigorous watersprouts or suckers growing up through the center of the tree should be removed.

In the second dormant season, training to develop the secondary scaffold system can be continued. Two strong, well-placed branches arising from each primary scaffold, which will constitute the secondary scaffold system, can go unpruned.

Peach and apricot trees growing vigorously need to be headed back (60 to 75 cm; 24 to 30 in.) at this time; otherwise, the weight of the fruit may bend them out of shape even with rope or stake bracing.

In cool-weather areas with short growing seasons, or under poor soil and water conditions, three, or perhaps four, growing seasons may be required to complete this basic framework of the tree.

Training on into the third and fourth years is essentially the production of future fruiting wood from branching and rebranching arising from the secondary scaffolds. Pruning consists of thinning out where the branching becomes too heavy, removing limbs that cross over and rub against each other, and removing strong-growing watersprouts and suckers. As the early crop changes the primary scaffold orientation, fruiting wood that is below the scaffolds should be removed since this wood produces small, poorly colored fruit. Laterals off the central leader that have fruited heavily and are procumbent should be cut back to an outward-growing lateral. Heading back is usually not advisable as the trees get into production. An exception is sweet cherries, which tend to continue to produce long, unbranched polelike branches unless they are headed back (see Figs. 13–8, 13–9, and 13–10).

Young trees of some large-growing species, such as the walnut, require pruning for training even after the work can no longer be done from the ground, requiring much ladder work or pruning aids (Fig. 13–11).

Pruning Bearing Deciduous Fruit Trees

As properly trained trees reach an age where they begin to form flowers and produce fruit, they have a strong trunk,

Figure 13–8 Central-leader apple tree after several years' growth. *Left:* Before dormant pruning. *Right:* After pruning. Some lateral branches were removed. In those retained, forked tips were cut to one outward growing branch. Spreader boards have been used with some laterals to develop a wider growth habit. *Source:* Gerdts, M., A. Hewitt, J. Beutel, J. Clark, and F. Cress. 1977. Pruning home deciduous fruit and nut trees. Univ. of Calif. Div. Agr. Sci. Leaflet 21003.

Figure 13–9 Some fruit trees, such as pears, tend to develop an upright, narrow shape, as shown in the unpruned tree at left. Often, this cannot be corrected by pruning alone. With notched, spreader boards, as shown in the pruned tree at right, the primary scaffold branches can be trained to give a wider tree shape. Inner limbs should be removed and those retained pruned to outward facing branches or buds. *Source:* Gerdts, M., A. Hewitt, J. Beutel, J. Clark, and F. Cress. 1977. Pruning home deciduous fruit and nut trees. Univ. of Calif. Div. Agr. Sci. Leaflet 21003.

Figure 13–10 Young, vigorous peach tree before (*Left*) and after (*Right*) pruning. Peaches require rather severe pruning. Many scaffold limbs have been removed, particularly those that grow low and horizontal. The stronger, moderately upright limbs were selected for the permanent primary scaffolds and were headed back to force out additional branching. *Source:* Gerdts, M., A. Hewitt, J. Beutel, J. Clark, and F. Cress. 1977. Pruning home deciduous fruit and nut trees. Univ. of Calif. Div. Agr. Sci. Leaflet 21003.

267

Figure 13–11 Motorized, self-propelled unit used as a pruning aid for the taller young trees. Such devices can only be used on well-prepared level terrain. Used here for pruning young Persian (English) walnut trees in California.

three or four well-placed primary scaffold limbs, and nine to twelve secondary scaffolds, distributing the fruiting wood uniformly around the tree. Removal of any large limbs at this stage should not be necessary (see Figs. 13–12 and 13–13).

Bearing deciduous fruit trees require moderate annual pruning to:

1. Stimulate production of new fruiting wood
2. Allow light penetration into the tree and prevent the development of dense pockets of vegetative growth
3. Reduce excessive amounts of fruiting wood and lessen the need for expensive hand-thinning of fruits
4. Remove broken, diseased, dead, or interfering branches
5. Keep the trees from growing so tall that harvesting becomes difficult
6. Confine the trees to the space available to them

A light annual pruning of bearing trees is better than a heavy pruning every three or four years, giving more regular cropping.

Fruiting Habits

To prune bearing fruit plants intelligently, one must know, for the various fruit species, the locations of the fruit buds on the plant and be able to identify them. Figure 13–14 shows the difference in appearance of fruit buds and vegetative buds, and Figure 13–5 shows the two basic positions of fruiting buds. Apples, pears, apricots, cherries, and plums produce the fruits on spurs. Others produce the fruits laterally (or more rarely terminally) on shoots that grew the previous

Figure 13–12 A mature peach tree before (*Left*) and after (*Right*) dormant pruning. Many of the weaker branches were removed, retaining young fruiting shoots selectively spaced along the stronger scaffold limbs. Relatively severe pruning of peach trees forces growth of new shoots, which produce fruits the following year. *Source:* Gerdts, M., A. Hewitt, J. Bertel, J. Clark, and F. Cress. 197 Pruning home deciduous fruit and nut trees. Univ. of Calif. Div. Agr. Sci. Leaflet 21003.

Figure 13–13 Old fruit trees (*Left*), if not pruned for several years, become dense and brushy, producing excessive numbers of small fruits, whose weight often breaks limbs. To correct such a situation, many of the limbs should be removed (*Right*) during the dormant season to open the center of the tree, retaining strong, well-spaced limbs. If many limbs need to be removed, it may be best to spread the pruning over two years. *Source:* Gerdts, M., A. Hewitt, J. Beutel, J. Clark, and F. Cress. 1977. Pruning home deciduous fruit and nut trees. Univ. of Calif. Div. Agr. Sci. Leaflet 21003.

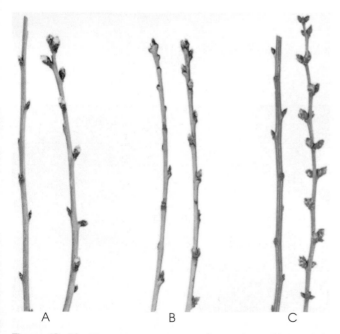

Figure 13–14 The different appearance of vegetative and fruit buds in the almond (A), peach (B), and apricot (C). In each pair, the shoot with vegetative buds is on the left and that with fruit buds is on the right. Vegetative buds are small and pointed, whereas fruit buds are large and plump.

season. Peaches, nectarines, grapes, and olives produce most of their fruits laterally on shoots, while citrus, young pecan trees, some young walnut trees, and loquats bear fruits terminally on shoots (see Table 13–3).

Care must be taken not to prune or break off fruiting spurs, although in trees that consistently overbear to the detriment of fruit size and quality there may be some benefit in partial spur removal. Spurs do not live indefinitely and some renewal is constantly taking place. Annual pruning tends to stimulate new spur formation.

Fruit species producing fruiting buds laterally or terminally on shoots that developed the previous growing season should be pruned to encourage the growth of such shoots. This entails a moderate annual dormant pruning, together with moderate nitrogen fertilization and ample soil moisture. Excessive pruning plus heavy applications of nitrogen can cause excessive rank vegetative growth, which will not form flowers.

The severity of pruning needed by bearing trees to keep them in a state of optimum fruitfulness can be judged to some extent by the length of annual shoot growth. Table 13–4 gives the amount of growth associated with good fruiting. If the annual growth of the average shoot exceeds these amounts, the severity of annual pruning should be reduced and nitrogen fertilizer withheld. If shoot growth is much below these amounts, severity of pruning should be increased as well as, perhaps, the amount of fertilizer applied annually.

Deciduous fruit trees are sometimes topped mechanically (mowing), giving a flat-top appearance to the orchard. There is evidence that summer topping of peaches and plums increases the development of fruit buds throughout the tree, especially in vigorous, young bearing trees. Such top mowing also reduces the amount of dormant pruning labor required

TABLE 13–3   Location of Fruit-Producing Buds on Various Kinds of Fruit Plants. Major Indicates Dominant Location. Minor Indicates Secondary Location

Species	Lateral on Shoots	Terminal on Shoots	Lateral on Spurs	Terminal on Spurs
Almond	Minor	—	Major	—
Apple	Minor	Very minor	—	Major
Apricot	Minor	—	Major	—
Black raspberry	Major	—	—	—
Blueberry	Major	—	—	—
Cherry, sour	Major or minor	—	Minor or major	—
Cherry, sweet	Minor	—	Major	—
Citrus	—	Major	—	—
Cranberries	Major	—	—	—
Fig	Major	—	—	—
Grapes	Major	—	—	—
Kiwifruit	Major	—	—	—
Loquat	—	Major	—	—
Olive	Major	Very minor	—	—
Peach and nectarine	Major	—	Minor	—
Pear	Minor	Very minor	—	Major
Pecan	Minor on young trees	Major on young trees	Minor on mature trees	Major on mature trees
Persimmon	Major	Minor	—	—
Plum, European	Very minor	—	Major	—
Plum, Japanese	Minor	—	Major	—
Quince	Major	Minor	—	—
Walnut				
(Payne type, many lateral buds fruitful)	Major on young trees	Minor on young trees	Minor or equal on mature trees	Equal or major on mature trees
(Franquette type, few lateral buds fruitful)	Minor on young trees	Major on young trees	Minor on mature trees	Major on mature trees

Source: W. P. Tufts and R. W. Harris. 1955. Pruning deciduous fruit trees. Calif. State Agr. Exp. Sta. Ext. Ser. Cir. 444.

TABLE 13–4   Desirable Amounts of Average Shoot Growth for Bearing Trees to Give Maximum Fruit Production

Species	Young Trees—Under 10 Years of Age		Older Trees—Over 10 Years of Age	
	Centimeters	Inches	Centimeters	Inches
Freestone peaches and nectarines	50 to 100	20 to 40	30 to 76	12 to 30
Clingstone peaches	76 to 100	30 to 40	30 to 76	12 to 30
Apricots	30 to 76	12 to 30	25 to 60	10 to 24
Plums (except prunes) and quinces	25 to 60	10 to 24	23 to 46	9 to 18
Almonds, prunes, and cherries	23 to 46	9 to 18	15 to 25	6 to 10
Olives	30 to 60	12 to 24	25 to 38	10 to 15
Figs	20 to 30	8 to 12	15 to 20	6 to 8

and holds the tree heights to levels where pickers can reach the fruit. Dormant hedging should never be done unless hand-pruning follows to thin out and reduce the regrowth. If hand-pruning is not done, the tree periphery becomes very dense and flowering and fruit quality are reduced.

Pruning and Training Young Broad-Leaved Evergreen Fruit Trees

Trees of this type (citrus, avocados, olives, etc.) usually come from the nursery in full leaf with their roots in soil in a container of some sort. When the trees are planted, the container must be removed or opened so the roots can spread out. If the roots have spiraled and thickened in the container, the trees should be rejected and returned to the nursery. Such roots will not grow out sufficiently to prevent the trees from toppling over in strong winds.

Evergreen fruit trees must be pruned considerably lighter than the deciduous fruit trees. Heavy pruning of young evergreen trees causes dwarfing and delays the onset of bearing for years. Many avocado and citrus growers do little or no pruning with quite successful results. Olive trees during the early years require somewhat more pruning and training, but the pruning still should be very light.

The purpose of pruning young trees of this type is essentially the same as for the deciduous trees; that is, to develop a strong trunk and to form primary and secondary scaffolds to support the fruit-bearing surface. Removing broken or interfering branches is important. If rank growth in some branches in the young trees produces an unbalanced tree, the branches should be cut back to better shape the tree. Lemons, in particular, tend to develop a scraggly growth habit and must be pruned more heavily than trees of other citrus. Watersprouts or suckers growing up through the center of the tree should be removed as early as possible.

Pruning Bearing Broad-Leaved Evergreen Fruit Trees

Bearing trees of these species are pruned much lighter than deciduous trees of comparable age.

Avocados need very little pruning, although height suppression of trees of particularly tall-growing cultivars by an annual light topping of upright growing limbs facilitates harvesting. All dead, broken, or low-hanging branches that would interfere with cultivation or irrigation should be taken out.

Bearing orange and grapefruit trees growing on fertile, well-drained soils and under good fertilization, irrigation, and pest control programs generally maintain vigorous growth and good yields with little or no pruning. Watersprouts and suckers growing up through the center of the tree should be removed, as well as any dead or broken limbs.

Lemons may develop masses of excessively dense vegetative growth, which need some thinning out to improve light penetration. A small amount of pruning of lemons is advisable. Lemon trees are vigorous growers and, as they grow larger, pruning back is necessary to keep them in the space available and permit passage of equipment through the orchard.

Mechanical pruning (hedging) and topping (mowing) of citrus orchards is an accepted practice in some localities. It seems to be satisfactory and saves considerable labor costs. The sides of the tree rows are cut back vertically, sometimes on a slant inward toward the top, increasing sunlight on the fruit-bearing surfaces. Small amounts of hedging and topping do not appreciably reduce yields and may increase yields if light penetrates better.

Bearing olive trees, if grown under good irrigation and fertilization practices, also yield best if pruned only moderately each year and the fruiting wood is allowed to reach almost to the ground, making hand picking of the fruit easier. Heavy pruning of mature olive trees reduces yields in proportion to the severity of pruning. It induces strong vegetative growth, which tends to keep the trees unfruitful. As the trees grow larger and taller, more pruning is needed to keep them within space allocations. Some moderate annual pruning of olives is required to force the development of new shoot growth on which the fruits are borne laterally the following year.

Mechanically harvested olive trees are pruned so as to expose the trunk and primary scaffold limbs for attachment of the shaker clamp. In addition, enough of the lower branches are removed to permit operation of the shaker arm and catching frame. Such mechanically harvested trees are allowed to grow much taller than would be advisable for hand harvest.

Mango trees are pruned very little except for training when the trees are young to form a strong trunk and primary scaffold system. Tall, upright-growing young trees may need to be cut back to force lower branching, and any dead or interfering branches are removed. Mango wood is very strong. Thus limbs in older trees rarely break.

Macadamia trees are trained to a central leader formation with four or five whorls of lateral branches 0.3 to 0.6 m (1 to 2 ft) apart up the trunk for 2.4 to 3 m (8 to 10 ft). Once this initial branching is established, little further pruning is needed except to remove dead or obviously interfering lateral branches.

Papayas are fast-growing, short-lived, almost herbaceous plants that grow and fruit as single stems. They require no pruning, although overly tall old trees can be cut back to the ground, forcing out new shoots, the strongest of which is retained to form a new trunk.

Coffee trees are pruned lightly except for removing dead wood and reducing the trees' height if they become too tall. Some shoot removal annually can give more uniform crops of larger berries. Different training systems are practiced. One is to develop a central leader with a series of lateral scaffold branches. Another method is to allow multiple primary scaffolds to develop in low positions from the main trunk.

Pruning and Training Grapevines

Grapes are pruned more heavily than fruit trees. The three main objectives in grape pruning and training are:

1. Establishing and maintaining the vines in a form that facilitates the various vineyard operations, such as cultivation, irrigation, spraying, dusting, and harvesting
2. Distributing the fruiting wood throughout the vine and between vines so as to give high production of good quality fruit over the years
3. Regulating the size of the crop, by reducing or eliminating the need for fruit or cluster thinning.

Grapevines produce their flower and fruit clusters mostly on new shoots arising from buds on canes that grew the previous summer. Thus in dormant pruning of grapevines it is essential that enough buds, but not too many, be retained to obtain a satisfactory crop the following summer.

Grapes adapt to a wide array of training systems. Since they are vines, they assume the shape of the supports used to hold them in place.

In California the principal training systems for the *Vitis vinifera* cultivars are:

1. The head-trained, spur-pruned system
2. The cordon-trained, spur-pruned system—also used for muscadine grapes in the southeastern United States
3. The head-trained, cane-pruned system

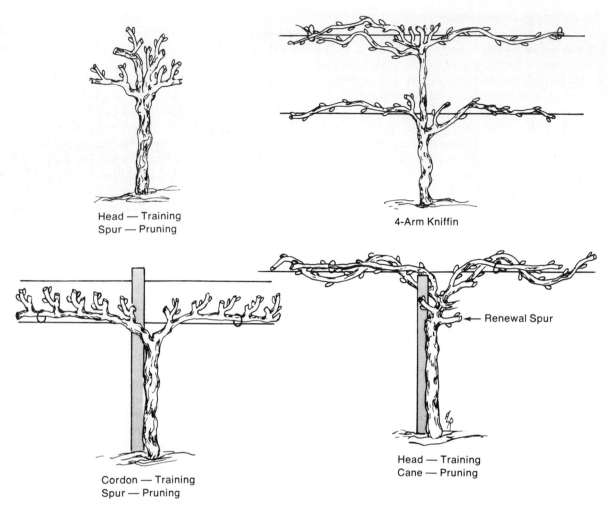

Head — Training
Spur — Pruning

4-Arm Kniffin

Cordon — Training
Spur — Pruning

Renewal Spur

Head — Training
Cane — Pruning

Figure 13–15   Four commonly used methods of training grapevines. The four-arm Kniffin system is used primarily with the labrusca-type grapes grown in the eastern United States. The others are used in training the vinifera-type grapes grown in California. *Source:* Adapted from Weaver, R. J. 1976. *Grape growing.* New York: John Wiley.

These are shown in Figure 13–15. In the eastern United States, several training systems are used for the labrusca-type grapes grown there, but the most popular is the four-arm Kniffin system, as illustrated in Figure 13–15.

In the head-trained, spur-pruned system, a vertical trunk 0.3 to 0.9 m (1 to 3 ft) is developed with support arms uniformly positioned in a circular head. On these arms, short spurs of last year's growth with two buds are left during the winter pruning to produce the shoots that will bear the next year's crop and to furnish canes for the next year's spurs. This system is used in California for the Tokay, Ribier, and Emperor table grape and for most large-clustered wine grape cultivars. The trunk has to be supported by a short stake until it becomes strong enough to stand alone, usually in 10 years. No wire trellis is needed.

In the cordon-trained, spur-pruned system, a vertical trunk is developed to a height of 90 to 120 cm (36 to 48 in.). From the top of the trunk the bilateral arms (cordons) extend in opposite directions for several feet along the lower wire of the trellis. From the upper side of these horizontal arms short

spurs of the previous summer's growth are positioned during winter pruning to provide the shoots that will bear the grape clusters and that, in turn, can be used for the following year's spurs. All other canes are removed. This system is often used for large-clustered cultivars of the French-American wine grapes in the eastern United States.

In the head-trained, cane-pruned system a vertical trunk is developed (Fig. 13–16) reaching to the top wire of the trellis. During the dormant season pruning, two to six canes of the previous summer's growth having 8 to 15 buds each are retained to produce shoots that will bear the fruit clusters the following summer. Often three or four renewal spurs with two buds each are left in the head trunk. From these will grow the canes to be retained for the following year's fruiting. All other canes are removed. This system is especially useful for 'Thompson Seedless,' which does not produce fruitful buds near the base of the canes, and for small-clustered wine cultivars.

The four-arm Kniffin system used in the eastern United States for the native American grapes and American-

Figure 13–16 A vineyard using the head-training, cane-pruning system as it appears in winter after leaf fall but before pruning. *Source:* Blue Anchor, Inc.

French hybrids is very similar to the head-trained, cane-pruned system. A two-wire trellis is required, with the lower wire about 0.9 m (3 ft) and the upper wire 1.8 m (6 ft) from the ground. The four canes retained during the winter pruning—two in each direction extending horizontally on each wire—may have from 8 to 12 buds each, depending upon the cultivar and the vigor of the vine, and vine spacing. Two-budded renewal spurs are retained near the base of each fruiting cane to provide canes for the following year.

When soil conditions and cultivar growth potential result in excessive vigor and subsequent shading and decreased quality, split canopy systems are used to provide improved exposure to light which improved quality and productivity. The Geneva Double Curtain System is widely used in northeastern United States for vigorous Concord and other cultivars. The Lyre system is a system used for excessively vigorous vinifera wine grapes. The Scott-Henry system is a vertical, split plot system that has been used widely in the pacific northwest for vigorous vinifera grapes. The trellis for these systems is more expensive to establish, but the increased yield and fruit quality generally justify the cost.

Grape Arbors

In pruning grapevines to cover an arbor with horizontal supports as high as 2.1 or 2.4 m (7 or 8 ft), the main trunk of the vines should be trained up to the top of the supporting posts, then bent horizontally to go along the top of the supporting wires or slats. Then either spur pruning or cane pruning can be done from this horizontal main stem, depending upon the species and cultivar being grown.

Fruit Thinning

To obtain better yields of high-quality grape berries, one extra bud on half or more of the fruiting spurs (spur pruning) and

an extra cane or two per vine (cane pruning) can be retained. Clusters or fruits should be thinned[2] to reduce the crop to the amount it would have been had the extra fruiting canes not been left. The increased leaf area produced by retaining the extra spurs or canes to support the same number of fruits per vine raises the yield and the quality of the berries.

Because of the production of fruit bearing shoots from latent or non-count buds some cultivars of the French-American hybrids, (e.g., Seyval blanc, Vidal blanc and Chambourcin) need annual cluster thinning. This cluster thinning is necessary to maintain sufficient vine growth for subsequent production and to maintain grape quality.

The following chapters detail pruning procedures for several fruit crops.

WEED CONTROL IN FRUIT PLANTINGS

A heavy growth of weeds utilizes soil moisture and mineral elements, particularly nitrogen, better used for the fruit plants. In addition, weeds interfere with orchard or vineyard operations—pruning, fertilization, thinning, irrigation, and harvesting. In winter as the weeds die they constitute a serious fire hazard and can harbor rodents that will damage tree trunks when food is scarce. In well-managed fruit plantings where irrigation water is scarce or expensive, weeds are meticulously controlled.

Some orchards are managed as a sod culture system in which low growing grasses cover the orchard floor. Sod culture is useful mainly for plantings on sloping hillsides where soil erosion is a problem and for areas where the water supply is plentiful and cheap enough to support both grasses and trees. If rainfall is abundant, water use by the grasses is not a problem. Extra nitrogen and, perhaps, potassium must be added, however, to compensate for what is used by the grasses. Thus sod culture entails increased fertilizer costs, the added cost of mowing the grasses, and possibly added irrigation costs. Sod orchards, due to shading of the soil surface, are particularly cold during bloom and, therefore, subject to crop losses from frost damage.

FERTILIZATION

It should be emphasized that for fruit plants—orchard trees, grapes, bushberries, and strawberries—nitrogen is the mineral element most likely to be deficient and to need replenishment. In some areas, such as the western United States, zinc and perhaps potassium deficiency are likely to occur and to need correcting. Deficiencies of phosphorus, boron, iron, calcium, magnesium, manganese, copper, or sulfur can also affect fruit plantings. Boron, sulfur, and zinc deficiencies are common problems in the U.S. Pacific Northwest fruit plantings. In general, the heavier rainfall areas have greater problems with such major elements as phosphorus,

---

[2]Thinning of grapes can be done as flower thinning—removal of part of the fruiting clusters before bloom—or cluster thinning—removal of clusters after the berries have set.

potassium, calcium, and magnesium, while the arid regions have greater problems with the minor elements.

Unless leaf analyses and field trials show that the fruit plants will respond to applications of these elements, it is generally best not to apply them because of the added cost.

In some instances potassium is deficient in orchard soils and must be added. For example, in some Michigan peach and apricot orchards, New York apple orchards, and California olive orchards potassium is deficient and the trees respond to added fertilizers.

Iron deficiency symptoms can sometimes be found in orchard trees, particularly in water-logged soils, soil with high levels of copper, or soils low in organic matter. Iron deficiency symptoms also occur in alkaline, high-lime soils where the problem is termed **lime-induced chlorosis.** Iron deficiency has been difficult to correct in citrus.

Some Florida soils on which citrus is grown are very low in magnesium, and the addition of this element in fertilizers has been essential for good yields. Florida citrus also requires the addition of copper to eliminate a shoot die-back problem, although corrective measures with added copper can cause an iron deficiency problem.

Zinc deficiency, widespread in many orchard soils, causes shoot die-back and reduces internode length and leaf size ("little-leaf"), consequently decreasing photosynthetic capacity. Low levels of zinc are a particular problem with spur-type Red Delicious in the Pacific Northwest, resulting in small fruit size and limited shoot growth. Annual sprays of zinc sulfate applied in the fall at about normal leaf drop or in the early spring before bud break can reduce the severity of this problem. Sometimes both applications are necessary. Ground applications are ineffective.

SUMMARY AND REVIEW

Selecting the location of a vineyard or orchard depends on finding a climate that is compatible with the requirements of the crop. The winters must provide adequate chilling if it is required and be mild enough if plants are intolerant of cold. Summers must be of adequate length for fruit development with temperatures that the plant can withstand. If rainfall is inadequate, a source of irrigation water must be nearby. The soil type must be of a kind that supports the plants' needs. Markets must be nearby or accessible. A labor supply must be readily available.

Crop and cultivar selection depend on many things. One is the use intended for the crop; for example, fresh market or processing. Another is suitability for the location; not all crops or cultivars do equally well in an area. If the plant is grafted, consideration must be made for both the scion and the rootstock. Consumer preference is also a factor to consider and may present the greatest challenge to a grower because it can be very difficult to predict.

Planting a crop requires good preparation of the land. The proper spacing and arrangement of plants are very important. Growers have to balance the density of plants in an area with consideration of how the distribution affects quality. Generally, higher density plantings produce more fruit, but it is of a lower quality.

Maintaining a high quality crop requires proper irrigation, fertilization, and pruning. Over- or under-supplying water can damage the root system. Improper or poorly timed irrigation can inhibit fruit development. Using the wrong fertilizer or applying fertilizer at the wrong time can do more harm than good. Soil and tissue testing help reduce the risk of fertilizer problems. Pruning is done to maintain the correct balance among leaves, shoots, roots, and fruit. There are several styles of pruning that can be used. Each one serves a unique purpose and choosing the right style for the plant or situation is very important.

EXERCISES FOR UNDERSTANDING

13–1. Choose two different fruits in the produce department of the local grocery store. Using any resources of your choosing, find and describe the planting system most likely to have been used to grow the fruit.

13–2. Using the same or two different fruits as in exercise 1, determine the climate needed for the production of each fruit. From the current season in your area and what you know about each fruit's climate requirement, decide if it could have been produced locally and, if not, where could it have been produced.

REFERENCES AND SUPPLEMENTARY READING

FORSHEY, C. G., D. C. ELFVING, and R. L. STEBBINS. 1992. Training and pruning apple and pear trees. Amer. Soc. Hort. Sci. pp. 166.
GALLETTA, G. J., and D. G. HIMEBRICK. 1990. *Small fruit crop management.* Englewood Cliffs, N.J.: Prentice Hall.

SMART, R. 1991. *Sunlight in wine.* Ministry of Agriculture and Fisheries, New Zealand.
WESTWOOD, M. N. 1993. *Temperature-zone pomology physiology and culture.* Third Edition. Portland, Ore.: Timber Press.

CHAPTER

14

Flowering and Fruiting in Fruit Crops

KEY LEARNING CONCEPTS

After reading the chapter you should be able to:

♦ Understand the steps critical to the production of large quantities of high-quality fruits.
♦ Understand how the production steps relate to various types of fruit crops.
♦ Explain what manipulations can be done at each step to promote the production of quality fruit.

In any fruit planting, whether a commercial orchard or a single tree in the garden, the goal is to produce a large crop of high-quality fruits each year. After the trees (or bushes or vines) have reached an age where they are large enough to be productive, the grower expects the plants to bloom and the flowers to set fruits, which grow to a normal size, and develop the flavor and appearance characteristic of that fruit.

There are several steps that are critical to the production of large quantities of high-quality fruits. Briefly, these are:

1. The initiation of flower buds in the summer, followed (in most deciduous fruits) by the development of a physiological "resting" condition. This is overcome by chilling winter temperatures, and the flower buds continue development early the following spring.
2. Flower opening and pollination in the spring.
3. Fertilization of the egg in the flower, fruit setting, and the beginning of fruit development.

4. Removal (thinning)—in some cases—of excessive numbers of fruits that have set.
5. Growth and development of the fruits through the summer.
6. Fruit maturation, ripening, and harvest.

INITIATION OF FLOWER BUDS

Initiation (also called **differentiation**) involves the change of a vegetative growing point deep inside a bud in the axil of a leaf on a shoot (as in peaches) or on a fruiting spur (as in apples) into miniature flower parts. Initiation does not occur in fruit trees until they have reached a certain size or age (3 years or so for some peaches, and up to 10 or 12 years for some apples) and have accumulated a certain amount of stored nutrients—carbohydrates, nitrogen, and so forth. In some fruit plants this change from a vegetative to a reproductive state is triggered by certain environmental cues. Strawberries require short days in the fall, whereas olives require exposure to a series of days with diurnally fluctuating warm and chilling temperatures. With most deciduous and evergreen fruit species, however, no definite environmental factor is known that triggers the change in a bud from a vegetative to a reproductive (flowering) growing point.

Since fruit species differ in their developmental pattern of flower formation, it is best to consider the different types separately.

Temperate Zone Fruit Species

Studies many years ago showed that initiation of flower parts in the buds of apples, pears, peaches, cherries, plums, raspberries, blueberries, and so forth begins between late spring and late summer during the year preceding bloom. Flower initiation starts after the new shoots have attained a certain diameter and length and a portion of their leaves have matured. Figure 14–1 shows various stages in flower bud initiation and development in the sour cherry.

The size of the subsequent fruit crop depends upon the number of buds changing from a vegetative to a reproductive state. This, in turn, depends upon the general health and nutritive condition of the fruit plant.

Trees that were pruned and fertilized heavily the winter before with nitrogen, then copiously irrigated, are likely to become strongly vegetative, producing long, succulent shoots. Buds on such shoots are not likely to form flower buds for the next year. At the other extreme, weakly growing, slender shoots, especially on older trees that are low in vigor, form few flower buds, particularly if the leaf area has been damaged by insect or disease attacks, thereby reducing the tree's photosynthetic capacity.

Severe and prolonged drought during the critical period of flower bud initiation in early summer can create water deficits in the trees that interfere with flower-bud formation. Slight water deficits are not likely to be harmful and can even stimulate flower initiation by reducing shoot elongation and causing carbohydrate accumulation from photosynthesis. It is evident that the care and management that fruit plants receive strongly influences their productivity.

Once the flower parts are fully formed in the flower buds of deciduous fruit plants, the buds enter a physiological "**rest period**" in which they will not open even if the plants are subjected to favorable temperature, moisture, and light conditions. It is important to note that the same is true for vegetative buds (those that did not differentiate into flower buds) on the same tree. However, trees of some species, particularly almonds and plums, will often bloom in late summer or fall if partly defoliated by mites or drought and then wetted by rain or irrigation.

The beginning of this "rest period" depends upon the species and the general vigor of the plant. The "rest" develops slowly, reaches a peak, then diminishes; the onset of the "rest period" can range from midsummer to late fall. In shorter, slow-growing shoots it starts earlier than in the buds of longer, more vigorously growing shoots.

There is some evidence that the physiological resting condition develops because the buds accumulate certain natural growth inhibitors, such as abscisic acid, and lose native growth promoters, such as the gibberellins.

The physiological buildup of such growth blockages in buds of deciduous fruit trees—which also occurs in temperate-zone ornamental deciduous trees and shrubs, as well as in certain bulb species—is an evolutionary development that increases the plants' chances of survival through the winter. Without this self-blocking mechanism for bud growth in the autumn, tender, succulent shoots and tender, newly formed flowers developing from the buds would start to grow during the winter and would be killed by subfreezing temperatures.

STRAWBERRIES

The strawberry is an example of a fruit species whose flowering is triggered by a definite, easily defined, environmental factor. Most of the important strawberry cultivars are short-day plants. That is, with the onset of short days (and long nights) in the fall, vegetative growing points in the crown of the plant begin changing to reproductive growing points—or flowers. Such plants then bear a single crop the following spring. When the daylength increases during the summer, flowering stops and the plants become vegetative and start producing runners. However, such cultivars also respond to temperature. These short-day plants grown under long days still produce flowers if the temperature is reduced from 21°C (70°F) to 15°C (60°F). This is the situation in the cool, coastal strawberry districts of central California, where very high yields are obtained, since plants of short-day cultivars grown there produce fruit all summer long, even with long days. Certain strawberry cultivars, however, are not responsive to changing daylength. They do not produce runners but continuously form flowers through the long days of summer. These cultivars are termed everbearers.

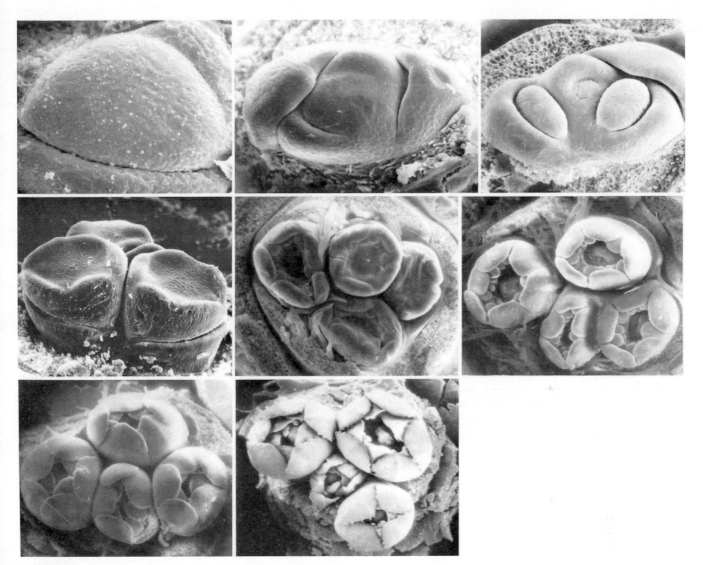

Figure 14–1 Early stages in the development of flower parts in the buds of sour cherry (*Prunus cerasus*), sampled in Hart, Michigan, as shown by scanning electron micrographs. Sampling dates and magnification are:

Top row (left to right): June 10 (x 266); June 16 (x 293); July (x 233).
Center row (left to right): July 30 (x 213); August 15 (x120); September 29 (x110).
Bottom row (left to right): November 4 (x 55); March 27 (x 40).

This shows that the cherry flowers that open in the spring have developed slowly in the buds during the preceding summer and fall. *Source:* D. H. Diaz.

This physiological internal inhibition of growth (rest) occurs only in the buds—not in the roots.

Interestingly enough, chilling of the buds (both vegetative and flower) through the winter is needed to reverse the rest influence. If the chilling is long enough, the influences blocking bud growth disappear. Then, with the beginning of warm spring temperatures, plus adequate soil moisture, both vegetative and flower buds grow rapidly and vigorously. It is believed that the chilling temperatures may act on the buds to lower the levels of growth inhibitors—such as abscisic acid—and increase the amounts of growth promoters, such as the gibberellins.

If the amount of chilling through the winter is marginal, as it can be in regions with normally mild winters, such as the fruit-growing areas of South Africa and the Central Valley of California, bud growth in the spring is erratic. Some flower buds may drop before opening, thus reducing the crop potential, or bud opening in the blooming period may be late and prolonged. Delayed foliation refers to slow development of vegetative buds due to the lack of sufficient winter chilling. A marginal winter chilling may not always be a disadvantage as it can lengthen the blooming period, increasing the chances of good weather during part of the bloom period, giving bees more time to pollinate the flowers. It can also cause the flower buds to open later and avoid being killed by late spring frosts. In areas with long, cold winters, such as the East, Northeast, and Middle West parts of the United States, the buds always receive sufficient chilling and this problem does not exist.

Much study has gone into determining the amount of chilling required by buds of the different fruit species to overcome the rest influence. Table 14–1 summarizes much of this information, which is expressed as the numbers of hours below 7°C (45°F) required by the various species to overcome the rest influence. This temperature is an arbitrary value selected to separate effective and ineffective chilling temperatures. Leaf buds require slightly more chilling than flower buds. Subfreezing temperatures are not necessary. High winter daytime temperatures greater than 16°C (61°F) can be detrimental for deciduous fruits, however, as they tend to counteract chilling temperatures. Fog or overcast weather during the winter days can be beneficial in obtaining good bud chilling by keeping direct sun rays from the buds, which would raise bud temperature above the air temperature.

There is considerable confusion in the use of the word **dormancy** in relation to buds. Dormancy is a general term denoting an inactive state of growth. There are several terms in use describing different kinds of dormancy in buds.

Rest

This occurs when buds fail to grow even under optimal environmental conditions because of internal physiological blocks. Exposure of the buds to a sufficient amount of above-freezing chilling temperatures can terminate rest.

Quiescence

This term describes a condition when nonresting buds fail to grow due to unfavorable environmental conditions such as low temperatures, unavailable water, or an unfavorable photoperiod.

Correlative Inhibition

In this situation, buds do not grow because of an inhibitory influence of another plant part; for example, failure of lateral buds to grow due to inhibition by the terminal part of the shoot.

To summarize, the usual sequence of events for temperate-zone plants is for the buds to enter the rest in late summer or fall; then, after sufficient winter-chilling, the rest influence is terminated. The buds are then said to be quiescent but will resume growth with the onset of favorable growing conditions in the spring.

Subtropical Fruit Species

Subtropical fruit species (e.g., citrus, avocados, olives) differ in regard to flower bud initiation and development.

The olive tree (*Olea europaea*) grows in parts of the world with a long growing season and minimum winter temperatures higher than about −9.4°C (15°F). Lower temperatures kill the trees. To produce flowers and fruits, most olive cultivars require chilling temperatures during the winter—but not to overcome the rest influence of previously formed flower buds. A certain amount of chilling days through the winter directly causes the vegetative growing points in the buds along the shoots to change, in late winter, into flower buds. With no winter chilling, all the buds remain vegetative. Olives, for this reason, are not grown for fruit production in such areas as Florida or Hawaii, where the winter days are continually warm.

Flower buds on citrus trees grown in subtropical climates are initiated in midwinter, with the first microscopic evidence of flower parts appearing about one month later. In areas with very mild winters and cool summers, certain lemon cultivars tend to bloom throughout the year but most heavily in the spring. In hot, tropical regions near the equator, citrus flower buds form and bloom all during the year unless they are influenced by drought.

In citrus—oranges, lemons, grapefruits, and so forth—chilling weather is not involved either in bud initiation or in overcoming flower bud dormancy. Some of the world's

TABLE 14–1	Amount of Winter Chilling Required to Overcome the Rest Period in Buds of Various Deciduous Fruit Species

Species	Approximate Number of Hours Below 7°C (45°F)
American plums (*Prunus americana*)	700 to 1,800 hrs
European plums (*Prunus domestica*)	700 to 1,700
Japanese plums (*Prunus salicina*)	700 to 1,000
Apple (*Malus pumila*)	800 to 1,700
Pear (*Pyrus communis*)	600 to 1,400
Pistachio (*Pistachia vera*)	1,000
Sour cherry (*Prunus cerasus*)	600 to 1,400
Sweet cherry (*Prunus avium*)	500 to 1,300
Peaches and nectarines (*Prunus persica*)	400 to 1,200
Persian walnuts (*Juglans regia*)	700 to 1,500
Apricots (*Prunus armeniaca*)	400 to 1,000
Blueberries (*Vaccinium corymbosum*)	650 to 1,200
Almonds (*Prunus amygdalus*)	200 to 500
Oriental persimmons (*Diospyros kaki*)	less than 100
Figs (*Ficus carica*)	none

Note: Grapes, both *Vitis vinifera* (European) and *Vitis labrusca* (American), require some chilling to overcome the rest influence in the buds, but it is rather light, probably an amount similar to the almond.

largest citrus plantings are in regions such as central Florida, which have no prolonged periods of cold weather.

Tropical Fruit Species

In coffee trees (*Coffee arabica*), flower induction apparently occurs a short time before flower initials appear microscopically in the bud. The flowers open about one month later. The time of flowering depends upon the climate. Following a prolonged dry period, flowers tend to appear about one month after the rains begin. With continuous adequate soil moisture and good growing temperatures, flowering and fruiting occur irregularly throughout the year.

Flower induction in pineapple plants (*Ananas comosus*) can be accomplished by applications of ethylene (or ethephon) to the growing point. Auxin and acetylene are also effective, due to their stimulation or production of ethylene. This is a case where a hormone is effective in inducing flowering.

Flower initiation in the mango (*Mangifera indica*) seems to follow an environmental event that slows growth, such as a prolonged cool or dry period. There is some evidence that a hormone originating in the leaves is translocated to the shoot to induce flowering at the site of actively dividing cells in the buds. This occurs, in Florida, in the fall, winter, or early spring. It has been shown experimentally that spraying mango trees with a solution of potassium nitrate will cause profuse flowering one to two weeks later, although the mechanisms involved are unknown.

POLLINATION

When the previously formed flower parts start to develop and the flowers open, a new critical stage begins in the production of the crop. For most fruit species the flowers must be adequately pollinated before fruit can set and develop. The pollination requirements of the various major fruit species are given in Table 14–2.

Pollination occurs when pollen produced by the anthers, the male parts of the flower, is transferred to the stigma, a female part of the flower. In fruit species, development of flower parts into a fruit requires a complex series of events. Growth of the seed generally stimulates adjacent parts of the flower to develop into a fruit.

Some fruit species have definite pollination requirements. Successful fruit growers know the needs of their fruit crops and provide for adequate pollination. For example, in both almonds and sweet cherries, the flowers of any given cultivar must be pollinated by those of a particular different cultivar. This means that pollen must originate in the anther of the flower of the pollenizer cultivar and be carried by bees (or other insects or wind) and deposited on the stigma of the flower of the cultivar that is to produce the main crop. In this example, all almond and sweet cherry cultivars would be self-unfruitful and certain cultivars would be inter-unfruitful.

TABLE 14–2	Pollination Requirements of Various Fruit Species

Group 1. Species Usually Self-Fruitful. Cross-Pollination Not Required.

Apricot (except for a few cultivars)
Cherry, sour
Citrus—oranges, lemons, mandarins
Coffee
Currant (most cultivars)
Fig, except for the cultivar Calimyrna; other fig cultivars parthenocarpic
Gooseberry (most cultivars)
Grapes, *Vitis vinifera* (European) and *Vitis labrusca* (American) cultivars
Macadamia
Peach (except a few cultivars)
Pecan
Persimmon (fruits of Oriental persimmons are parthenocarpic)
Strawberry
Walnut (but requires pollen shedding when stigmas are receptive)

Group 2. Species Usually Self-Unfruitful. Cross-Pollination Required.

Apple
Almond (certain cultivars are cross-incompatible)
Blueberry
Cherry, sweet (certain cultivars are cross-incompatible, but 'Stella' is self-fruitful)
Feijoa (except for the cultivar Coolidge)
Filbert
Grape, Muscadine (*Vitis rotundifolia*)
Kiwifruit (requires both male and female plants in a planting)
Pistachio (requires both male and female trees in a planting)

Group 3. Species Whose Cultivars Have Varying Pollination Requirements.

a. Cultivars that set a commercial crop without cross-pollination (self-fruitful).
b. Cultivars that set part of a crop without cross-pollination but whose crop is increased by cross-pollination (partially self-fruitful).
c. Cultivars that must be cross-pollinated to set a commercial crop (self-unfruitful).

Fruit species in this group are pear, European plum, Japanese plum, olive, and chestnut

Some fruit species, such as most apple cultivars, must be cross-pollinated, but the pollen can originate in flowers from any other cultivar of that species. But cultivars selected for cross-pollination must have overlapping blooming periods. In contrast, other fruit species, such as sour cherries, apricots, and oranges, are self-fruitful. The pollen can originate in any flower of the same cultivar, either on another tree, another flower on the same tree, or even from the anther of the same flower. These species are normally self-pollinated, but pollen coming from flowers of different cultivars in the same species may also cause fruit to set.

Generally, stigmas of fruit tree flowers are receptive to pollination for only about five days after the flowers open. For fruit to set after pollination takes place, the pollen tube must grow through the style into the egg in the embryo sac.

Temperature is an important factor during all the stages of pollination, pollen tube growth, fertilization, and fruit setting. A temperature range of 15.5°C to 26.5°C (60°F to 80°F) is considered optimum for deciduous fruits. Temperatures much above or below this range impair good fruit setting. Temperatures much above 26.5°C (80°F) inhibit pollen germination. Pollen grains themselves are quite stable at low temperatures—when dry, they can be kept viable for years at −18°C (0°F)—but temperatures dropping to −3°C or −2°C (27°F or 28°F) can kill the ovules in the open flowers of most fruit species.

Before the nursery trees are purchased for a planting of the fruit species listed in Groups 2 or 3, the grower must determine the pollination requirements of his proposed cultivars. Where cross-pollination is required, the planting should consist of the main fruiting cultivar, interspersed with trees of the pollinating cultivar. Quite often the latter trees are set in some arrangement such as every third tree in every third row, or if the main cultivar is one that tends to set fruit heavily, every fourth tree in every fourth row. Where two fruiting cultivars of equal value can be used, harvesting is easier if, for example, blocks of four rows of each of the two cultivars alternate.

Flower Types

Most fruit species produce **perfect** flowers; that is, each flower has both the male and female flower parts.

Some fruit species, however, are unisexual—male plants have staminate flowers that produce only pollen; other (female) plants have pistillate flowers that develop into the fruit. These species are termed **dioecious** (but pollen must be shed when the stigmas are receptive). Examples of dioecious fruit plants are the kiwifruit (*Actinidia deliciosa*), the date palm (*Phoenix dactylifera*), and the pistachio (*Pistacia vera*). (Dates are artificially pollinated, with one male tree planted for 40 or 50 female trees. For the kiwifruit one male vine is planted for every 9 or 10 females, and in the pistachio, one male is required for every 6 female trees).

Another unisexual group produces male flowers and female flowers separately on the same tree. These are termed **monoecious.** Examples are the walnut (*Juglans* spp.), filbert (*Corylus avellana*), and chestnut (*Castanea* spp.). As with the dioecious group, a problem with fruit set is that the male flowers may not be shedding pollen at a time when the female flowers are receptive, and vice versa. This is a factor to consider when selecting cultivars because a second cultivar may be needed to provide pollen when the pistils are receptive.

Certain fruit species, such as the papaya (*Carica papaya*) are bisexual. That is, there is a dioecious type (separate male and female trees) as well as hermaphroditic trees, which have both male and female flowers. In papaya, hermaphroditic flowers have one pistil and about 10 stamens. Yields from these trees can average nearly as much as those from female plants pollinated by nearby male plants. For papaya, one male tree for every 10 to 15 female trees is advisable.

Yet another type of flower development occurs in palms. In the coconut (*Cocos nucifera*), as an example, the inflorescences produce a globular female flower near the base of the inflorescence branch, with the small pollen-producing male flowers above it. There is no problem with timing, because inflorescences are produced continuously in the tropics.

POLLINATION TERMS

Pollination   Transfer of pollen from an anther to a stigma of a flower.

Self-pollination   Transfer of pollen from anther to stigma of the same flower or to the stigma of another flower of the same clone.

Cross-pollination   Transfer of pollen from an anther in the flower of one cultivar to the stigma of a flower in a different cultivar.

Self-fruitful cultivar   One that sets and matures a commercial crop of fruit with its own pollen or as a result of parthenocarpy.

Self-unfruitful cultivar   One that is unable to set and mature a commercial crop of fruit with its own pollen or as a result of parthenocarpy.

Cross-compatible cultivar   One cultivar can pollinate another cultivar so that a commercial crop of fruit is set and matured. The reverse may not be true, however.

Cross-incompatible cultivar   One cultivar is unable to pollinate another particular cultivar so that a commercial crop of fruit is set and matured. (It may be able to pollinate other cultivars, however.)

Intercompatible cultivars   Two cultivars that can pollinate each other so that commercial crops of fruit are set and matured.

Interincompatible cultivars   Two cultivars are unfruitful when pollinated by each other, although either one may effectively pollinate other cultivars.

Parthenocarpic fruits are those that develop without fertilization of the egg. In such fruits the fleshy parts are stimulated to grow but they contain no seeds. Examples of parthenocarpic fruits are the navel orange, most fig cultivars, the Oriental persimmon, pineapple, some bananas, and some seedless grapes.

Insects and Pollination

Fruit species with large, showy flowers generally depend upon insects to transfer pollen. Bees, particularly honeybees, are the most important type of insect involved in the pollination of commercial crops. Some fruit plants, especially those with nonshowy flowers, such as the walnut, olive, pecan, and filbert, are wind-pollinated. They generally produce large amounts of very light pollen that is carried considerable distances in the wind onto the stigmas of other flowers.

In fruit orchards that require cross-pollination, even when the proper mixture of pollenizing cultivars is present, bees must be in the orchard during the blooming period to carry pollen from the flowers of one cultivar to the flowers of the other (Fig. 14–2). For one or two trees in a home garden enough bees are generally present for successful pollination, but an orchardist with a large number of trees that absolutely requires cross-pollination (self-unfruitful) will need to have bees brought in. Beekeepers, for a rental fee, will provide the necessary bees during the blooming season.

Honeybees work the flowers to collect pollen and nectar, which they use as food. Since the bees generally stay within about a 100-yard radius of their hives, hives are placed in an orchard no more than 150 to 180 m (500 to 600 ft) apart. Weather conditions affect bee activity. Below about 13°C (55°F) they are inactive; the optimum, temperatures are about 18° to 27°C (65°F to 80°F). Winds much above 15 mi/hr (24 km/hr) keep bees from flying and continuous rains interfere with their activity, but intermittent showers do not. If bees or other insects are important pollinators, pesticides must be used with extreme caution to protect the pollinators.

Pollination of certain fig cultivars is done by a particular wasp species in a complicated but interesting procedure.

Figure 14–2 To obtain good crops of many fruit species (e.g., sweet cherries, almonds, plums, and apples) cross-pollination between cultivars is necessary. Fruit growers often set hives of bees in their orchards during bloom, as shown here. The bees, working the flowers for nectar and pollen, which they use as food, also transport pollen from tree to tree. *Source:* USDA.

FERTILIZATION AND FRUIT SETTING

Following pollination, the pollen tube grows downward through the style and discharges two sperm cells into the embryo sac. One sperm cell unites with the egg (in the embryo sac) to form the zygote and eventually the new embryo in the seed. The other sperm cell unites with the two polar nuclei to form the endosperm, a nutritive tissue in the seed. The two integuments and all tissues inside become the seed. The ovary and any adjacent accessory structures outside the seed become the fruit.

When fertilization (the union of egg and sperm) has taken place, the seed and fruit start to grow. Hormones produced by the developing seed stimulates growth of the surrounding tissues to become a fruit. However, many of the young fruits, even those containing a fertilized egg, soon drop. This abscission period may last as long as a month and is sometimes called the June drop. If too many fruits drop, the yield may be small; if too few fruits drop, the tree may be so overloaded that fruit thinning procedures are required.

FRUIT THINNING

Very often peach, nectarine, apricot, plum, pear, apple, mandarin, and olive trees set such heavy crops that the fruits do not grow to a satisfactory size (Fig. 14–3). Limbs may break because of an overload of fruit, and the drain on the nutrients in the tree as excessive numbers of fruits and seeds mature may be so great that no flower buds develop for the following year, thus inducing an alternate-bearing pattern.

Alternate bearing occurs in many fruit tree species and is a major problem in obtaining consistent, high-quality fruits. The best control methods are those which remove some of the fruits early in the "on" year.

Even if alternate bearing does not occur, several consecutive years of overcropping can so greatly weaken the trees that they may die. Vines of certain wine grape cultivars can die from overcropping. Fruit quality of grapes can often be markedly improved by thinning flower or immature fruit clusters, or removing parts of clusters. Fruit thinning to remove excessive numbers of fruits is one of the most important cultural procedures for an orchardist or a home gardener in overcoming alternate bearing.

Sweet and sour cherries, prunes, almonds, walnuts, pistachio, filberts, oranges, avocados, strawberries, and the bushberries ordinarily do not require fruit thinning.

Actually thinning begins at pruning time when excess fruiting wood is removed. The effect of fruit thinning on size declines rapidly with time, with little or no effect being observed with thinning done near harvest.

Removal of excess fruits permits the tree to utilize available nutrients to develop the remaining fruits to larger size, as well as to increase root and shoot growth. These, in turn, can absorb more nutrients from the soil and manufacture

Figure 14–3 *Left:* Plum fruits on a branch that was properly thinned. Proper spacing has allowed them to develop properly for fruit size, quality, and early maturity. *Right:* Unthinned fruits are small, late maturing, and of poor quality. In addition, such limbs often break

REFERENCE DATE

In the stone fruits, fruits that are small early in the season are also small at harvest time. Commercial growers of clingstone peaches and apricots have made use of this relationship to predict much earlier in the season average fruit size at harvest. The young, developing fruits are measured at a specific time, called a reference date. For apricots this time is determined by cutting through the ends of the young fruits at intervals. The reference date is seven days after the pit begins to harden, when representative fruit samples are measured for size. The reference date for clingstone peaches is the same except that 10 days are added to the date that the tip of the pit begins to darken. For clingstone peaches, an average suture diameter of 34 to 35 mm (1.36 to 1.40 in.) at reference date is needed to produce an average diameter at harvest of 67 to 68 mm (2.68 to 2.72 in.). This is sufficient to make 90 percent of the fruits larger than the minimum harvest size. If the young fruits at reference date are too small, the fruits are thinned to bring the remaining fruits to a satisfactory size by harvest time.

more carbohydrates through photosynthesis in current and succeeding years.

Amount of Thinning

It may be obvious that a fruit tree is overloaded with young fruits and needs thinning—but how much? It is difficult to judge since a number of factors are involved, such as the cultivar, time of fruit maturity, availability of water, general vigor and age of the tree, as well as growing conditions. Some general guidelines for fruit thinning can be given, however. Peaches are often thinned to about one fruit every 15 to 20 cm (6 or 8 in.) of shoot. Another rule is 25 to 40 leaves per fruit for apples and about 50 leaves per fruit for peaches.

Final fruit size more or less depends upon the leaf-fruit ratio on a branch. The more leaves per fruit, the larger the final fruit size, but fruit size is increased at the expense of total yield. Thinning can be highly profitable, however, where a premium is paid for larger-sized fruits.

Thinning to control yields and fruit size is so important in crops such as peaches and apricots that thinning tables have been developed to determine the optimum number of

fruits to leave on the tree. These consider the distances between trees, fruit size desired (number per pound), and number of tons of fruit desired per acre. For example, if a 22.4 MT/ha (10 t/ac) harvest is desired for apricots and the trees are set on a 7.2 × 7.2 m (24 × 24 ft) planting distance, no more than about 3,200 fruits should be left per tree to attain a fruit size of at least 200 g (0.4 lb). Fruit counts are made on sample trees throughout the orchard to determine if the fruit set is uniform.

Fruit thinning is the best way to increase fruit size, although flower and bud thinning (by pruning out fruiting shoots) is also effective. Overirrigation or excessive nitrogen fertilization will not increase fruit size; in fact, they can stimulate new shoot growth, which competes with fruits for carbohydrates.

Chemical Thinning

The idea of using chemical sprays applied to trees to remove some of the fruits is definitely appealing, and much research has been given to develop thinning sprays to replace hand thinning. For apples and olives, thinning sprays are available and are used commercially.

For apples, effective thinning chemicals are either naphthaleneacetic acid or carbaryl (1-naphthyl N-methyl carbamate), applied 20 to 30 days after full bloom. For olives, naphthaleneacetic acid applied when the fruits are 4 to 6 mm in diameter effectively thins the fruit. It is often difficult to obtain consistent results with chemical thinning sprays. With any chemical thinning applications, proper timing is critical.

Mechanical Thinning

As an improvement over slow and expensive hand thinning, long poles with sections of rubber hose at the end can be used to hit fruiting branches and knock off some of the fruit. Hand thinning to remove missed fruiting clusters may follow. Mechanical tree shakers, used in harvesting operations are sometimes used to shake the trees to remove a portion of the small, immature fruits. Shaking can be satisfactory, but is not very precise and the desirable larger fruits tend to be removed. A light machine shaking, followed by hand thinning or pole knocking, works well in some cases. Hand thinning gives the best results, however, and should be used whenever practicable.

Maturation of fruits refers to the final stage of development while they are still attached to the plant and includes cell enlargement plus the accumulation of carbohydrates and other flavor constituents and a decrease in acids. The flesh may or may not soften.

Ripening of fruits refers to the changes taking place after full maturation, involving a softening of the flesh, the development of characteristic flavors, and an increase in the juice content. Ripening may occur either before or after the fruit has been picked.

Senescence is the period following ripening, during which growth ceases and aging processes replace ripening processes. Senescence may occur before or after harvest.

FRUIT GROWTH

Fruit growth involves both cell division and enlargement for about the first 30 days. From then until maturity, fruit growth consists of enlargement of the existing cells. In all of the many types of fruits, growth involves the enlargement of fleshy tissues in the flowers plus (sometimes) associated structures. Growth patterns of the fruits of the various fruit species have been given much study, and growth curves through the growing season have been developed. Such growth curves are determined by measuring representative tagged fruits on the plant at intervals through the growing period. Usually the diameter is measured. Sometimes representative fruits are detached, and volume or weight measurements taken. The latter method, however, can have a thinning effect and thus give distorted results.

The pome fruits—apples and pears—and the citrus fruits steadily increase in size throughout the season. The stone fruits—peaches, apricots, plums, and olives—grow rapidly at first. A period of reduced growth, in which the "pit" (or endocarp) hardens occurs in midseason, followed by a final period of rapid growth (the final swell) by cell enlargement. Grapes exhibit a similar growth pattern. As both types of fruits mature, sugars, water, and substances responsible for the characteristic flavors and aroma move into the cells.

The fruits of different cultivars within a species—peaches, for example—mature at different times during the summer, generally because of different lengths of the second (pit-hardening) period. The first and third periods of growth tend to be about the same length in all cultivars.

As the fruits grow and enlarge during the summer season, they depend upon carbohydrates, proteins, and minerals within the tree, plus the absorbing ability of the roots for water and additional mineral nutrients. In addition, the photosynthetic capacity of the tree's leaf area must not only supply carbohydrates for the developing fruits but also replenish the stored food depleted in the very early stages of fruit growth when, in deciduous fruits, photosynthetic capacity is limited. In deciduous fruits, new flower buds are being differentiated on the shoots or fruiting spurs while the fruit is growing during the summer. In the evergreen fruit species, such as citrus, avocados, and olives, where food storage in the tree may be less than in deciduous fruits, carbohydrate supplies for the growing fruit depend more upon current photosynthesis.

FRUIT MATURATION, RIPENING, AND SENESCENCE

As fruits near their maximum size, significant changes lead to the end product—a fruit with an attractive color, soft enough to be palatable, sweet and juicy, with an accumulation of the other components that give it its own distinctive flavor and aroma.

The increase in sugars—or soluble solids—can be measured by a refractometer or hydrometer. Decreasing hardness can be measured by various kinds of pressure testers. Changing color, as the masking chlorophyll disappears, can be measured by color charts or color meters. Changing acids can be determined by chemical means or pH meters. All these procedures are used to determine quantitatively when the fruit is ready for harvest. In some states in the United States, legal maturity standards demand such measurements to determine when fruit can be harvested. For example, the basis for legal maturity of oranges in California is a ratio of 8:1 for total soluble solids to titratable acidity.

There can be a difference between "economic" maturity and botanical maturity. For example, some kinds of cucumbers are harvested for pickles when they are far from mature. Also, olives in California are harvested at a botanically immature stage for processing into either edible green or black olives.

Optimum harvest maturity depends upon the intended use of the fruit. For example, pineapples in Hawaii that are to be crushed immediately for juice are allowed to reach maximum maturity and ripeness, with the greatest juice content before picking, but they would not keep long after harvest. But if the pineapples are to be shipped to the U.S. mainland for sale in supermarkets, they will be harvested at a somewhat earlier, firmer stage and ripened on the way to the market.

The home gardener has an advantage over the commercial grower in that he can allow his fruit to reach optimum maturity right on the tree for eating at the optimum degree of ripeness. For commercial production, however, perhaps involving shipments in containers to great distances, the fruits generally must be picked firm enough to endure shipping stress and arrive at their destination in an acceptable condition.

All fruits on a tree do not mature at the same time. Fruits on the top and outside are usually ready to pick before those on the inside. For the home gardener this is an advantage, as it prolongs the harvest period, but for commercial "once-over" harvest, some of the fruits are harvested at a less-than-ideal time.

The various kinds of fruits have different ripening patterns and require different storage procedures. An understanding of these is essential in order to know how to harvest and handle the fruits properly. Most fruits reach maturation and ripen on the tree, vine, or bush, at which time they should be picked. If not picked, they enter a stage of senescence and deterioration. However, other fruits are best harvested when they reach maturity on the tree and are ripened to eating condition off the tree, such as bananas, avocados, and pears.

It must be emphasized that maturing and ripening fruits are living organisms. Their cells are respiring, and many other complex chemical and physical changes are occurring. Some types of fruits are said to be climacteric and other are nonclimacteric (Fig. 14–4). In both types, the respiration rate of the cells slowly decreases as maturation proceeds. However, in climacteric fruits, the respiration rate rises abruptly as the fruit ripens, reaches a peak (the climacteric point), and then declines. Senescence ensues, leading eventually to the death of all the fruit cells. In nonclimacteric fruits the respiration rate gradually declines during ripening, with no particular peak.

Apples, pears, peaches, apricots, plums, mango, banana, papaya, and avocado are examples of climacteric fruits. Most climacteric fruits show the same respiration pattern whether ripening on or off the tree. In avocado, however, the typical climacteric respiration pattern occurs only after the fruit has been detached from the tree. Cherries, figs, oranges, lemons, strawberries, pineapples, grapes, and olives are examples of nonclimacteric fruits.

Refrigeration prolongs fruit life chiefly because the lowered temperatures reduce respiration rates. Controlled atmosphere storage (CA storage)—lowering the oxygen levels from 21 percent (the amount in air) to 2 to 3 percent, as well as increasing the carbon dioxide to 1 to 5 percent instead of the usual 0.03 percent in air—can further lower respiration rates and greatly extend the storage life of fruits. The levels of O_2 and CO_2 that can be tolerated vary with species and

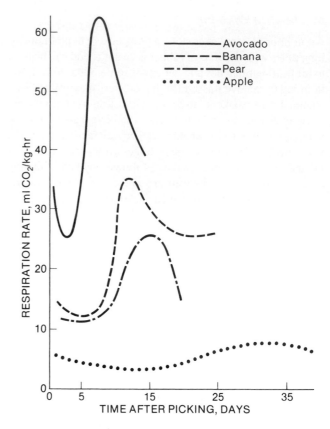

Figure 14–4 Fruits of various species ripen at different rates as shown in the intensity of the respiratory climacteric. The avocado, which ripens rapidly, shows the most intense climacteric peak, followed by the banana, pear, and apple. *Source:* Adapted from Biale, J. B. 1950. Postharvest physiology and biochemistry of fruits. *Ann. Rev. Plant Physiol. 1:* 183–206.

cultivar (Table 14–3). CA storage procedures have come into widespread use in recent years, especially for apples and to a lesser extent for pears. Apples, although harvested mainly in autumn, are now available in the markets the year round because of CA storage facilities.

Ethylene (C_2H_4), which is produced by all plant tissues, including fruits, can stimulate fruit ripening. An ancient Chinese custom was to ripen fruit in rooms where incense was being burned. In the earlier days of this century, kerosene stoves were widely used in California to stimulate color development in lemons. Ethylene, plus other gases, is released by such combustion. Ripe fruits, like bananas and apples, give off ethylene and stimulate the climacteric ripening of other fruits confined in the same containers with them. Experiments in the 1960s provide evidence that ethylene is much involved in fruit ripening and is considered a ripening hormone. Applied ethylene is effective in stimulating the ripening of most climacteric and nonclimacteric fruits, although ripening of the strawberry is not affected by ethylene. Gibberellin, a natural plant hormone, is known to counteract ethylene effects and delay fruit ripening.

The biochemistry and physiology of fruit maturation, ripening, and senescence have received considerable study,

TABLE 14–3   Requirements for Controlled Atmosphere (CA) Storage of Various Apple Cultivars

Cultivar	Carbon Dioxide (percent)	Oxygen (percent)	Temperature °C	Temperature °F
Cortland	5	2–3	2.2	36
	2–3	2–3	0	32
Delicious	1–2	1.5–2	−0.5–0	31–32
Empire	2–3	2–3	−0.5–0	31–32
Golden Delicious	1–3	1.5–2	−0.5–0	31–32
Granny Smith	1–3	2–3	−0.5–0	31–32
Idared	2–3	2–3	−0.5–0	31–32
Jonathan	0.5–5	2.5–3	2.2 one month, then 0	36 one month, then 32
Macoun	5	2–3	2.2	36
McIntosh	2–3 one month, then 5	2.5–3	2.2	36
Northern Spy	2–3	2–3	−0.5–0	31–32
Rome Beauty	1–3	2–3	−0.5–0	31–32
Spartan	2–3	2–3	−0.5–0	31–32
Stayman Winesap	2–5	2–3	−0.5–0	31–32
Winesap	1–2	2–3	−0.5	31
Yellow Newtown (Calif.)	8	3	4.4	40
(Oreg.)	5–6	3	2.2	36
Normal ambient atmosphere levels	0.03	21		

Source: Hardenburg, Watada, and Wang. 1986. The commercial storage of fruits, vegetables, and florist and nursery stock. USDA Agr. Handbook No. 66.

and various theories have been proposed to explain the ripening mechanisms involved. One theory is that, during fruit ripening, new enzymes appear that cause the characteristic changes in the fruit. Energy from respiration provides for the synthesis of the enzyme systems and for their ripening actions. It is believed, too, that in addition to the ripening process itself, hormonal (ethylene, gibberellin, and perhaps others) changes may also make the fruits responsive to ripening signals.

The various kinds of fruits develop differently to the point where they are ready for harvesting and ripening. Some examples will show the variability among fruit species and will point out the necessity for understanding these patterns for proper harvesting and fruit ripening.

Table grapes mature and ripen on the vine; the peak of fruit quality is reached and, ideally, the clusters are picked and consumed at that time. Once the clusters are removed from the vine, deterioration sets in. This can be slowed by placing the clusters in cold storage—0°C (32°F) at 95 percent relative humidity. Such stored grapes can remain edible from one to six months, depending upon the cultivar. In addition, the European (but not American) grapes require sulfur dioxide fumigation treatments to adequately prevent decay.

Apples also develop optimum maturity and ripeness while on the tree, and senescence begins with harvest. This can be delayed by cooling the fruits to 0°C (32°F) within two to three days after harvest and holding them in storage at a relative humidity of 90 to 95 percent. By retarding the respiration rate, CA storage enables fruits of some apple cultivars to remain in good eating condition for as long as nine months.

Pears ripen differently. On the tree the fruits accumulate the maximum of stored foods and the firmness of the flesh decreases, and they should be picked. But for ideal eating characteristics, the pears should be harvested before they ripen and then ripened in a cool place (15.5°C to 22°C; 60°F to 72°F) at 80 to 85 percent relative humidity. Or, they can be placed directly into cold storage (−0.5°C; 31°F) at 90 percent relative humidity, where they will keep for several months. They can be taken from cold storage to higher temperatures at any time for ripening. Some pear cultivars, like d'Anjou, do not ripen properly unless held in cold storage for a time, or treated with ethylene gas.

Oranges, grapefruits, and other sweet citrus fruits are edible and can be picked from the tree weeks, or even months, before full maturity is reached. This slow maturity contrasts markedly with rapid maturity changes noted in most other fruits. Citrus fruits "store" well on the tree and in the home garden can be picked when needed, even after optimum maturity. After a time, however, senescence commences and the fruits deteriorate. Cold storage after harvest can be used for oranges but not at temperatures below 3.5°C to 9°C (38°F to 48°F), depending upon cultivar and growing area.

Avocados, too, can be stored on the tree after full maturity. They are firm when picked but soften and ripen rapidly once removed from the tree. Avocados are not adapted to long-term cold storage.

Plums, when harvested for commercial shipments, are too firm for eating. They continue softening and ripening af-

ter harvest to an optimum point for eating, after which deterioration commences. Plums are often stored under refrigeration at 0°C (32°F) just after picking to prolong the life of the fruits for two to four weeks.

Many kinds of fruit are considered mature enough for picking when they start showing a tinge of color at the apical

TABLE 14–4 Recommended Storage Temperatures and Relative Humidity, Approximate Storage Life, Highest Freezing Point, and Water Content of Fresh Fruits in Commercial Storage

Commodity	Temperature °C	Temperature °F	Relative Humidity (percent)	Approximate Storage Life	Highest Freezing Point °C	Highest Freezing Point °F	Water Content (percent)
Apples	−1–4	30–40	90–95	1–12 months	−1.5	29.3	84.1
Apricots	−0.5–0	31–32	90–95	1–3 weeks	−1.0	30.1	85.4
Avocados	4.4–13	40–55	85–90	2–8 weeks	−.3	31.5	76.0
Bananas, green	13–14	56–58	90–95		−.7	30.6	75.7
Berries							
Blackberries	−0.5–0	31–32	90–95	2–3 days	−.7	30.5	84.8
Blueberries	−0.5–0	31–32	90–95	2 weeks	−1.2	29.7	83.2
Cranberries	2–4	36–40	90–95	2–4 months	−.8	30.4	87.4
Currants	−0.5–0	31–32	90–95	1–4 weeks	−1.0	30.2	84.7
Dewberries	−0.5–0	31–32	90–95	2–3 days	−1.2	29.7	84.5
Elderberries	−0.5–0	31–32	90–95	1–2 weeks	—	—	79.8
Gooseberries	−0.5–0	31–32	90–95	3–4 weeks	−1.0	30.0	88.9
Loganberries	−0.5–0	31–32	90–95	2–3 days	−1.2	29.7	83.0
Raspberries	−0.5–0	31–32	90–95	2–3 days	−1.0	30.0	82.5
Strawberries	0	32	90–95	5–7 days	−.7	30.6	89.9
Carambola	9–10	48–50	85–90	3–4 weeks	—	—	90.4
Cherries, sour	0	32	90–95	3–7 days	−1.7	29.0	83.7
Cherries, sweet	−1 to −0.5	30–31	90–95	2–3 weeks	−1.8	28.8	80.4
Coconuts	0–1.5	32–35	80–85	1–2 months	−.9	30.4	46.9
Dates	−18 or 0	0 or 32	75	6–12 months	−15.7	3.7	22.5
Figs, fresh	−0.5–0	31–32	85–90	7–10 days	−2.4	27.6	78.0
Grapefruit, Calif. & Ariz.	14–15.5	58–60	85–90	6–8 weeks	—	—	87.5
Grapefruit, Fla. & Texas	10–15	50–60	85–90	6–8 weeks	−1.0	30.0	89.1
Grapes, Vinifera	−1 to −0.5	30–31	90–95	1–6 months	−2.1	28.1	81.6
Grapes, American	−0.5–0	31–32	85	2–8 weeks	−1.2	29.7	81.9
Guavas	5–10	41–50	90	2–3 weeks	—	—	83.0
Kiwifruit	−0.5–0	31–32	90–95	3–5 months	−1.6	29.0	82.0
Lemons	13–15	55–59	85–90	1–6 months	−1.4	29.4	87.4
Limes	9–10	48–50	85–90	6–8 weeks	−1.6	29.1	89.3
Loquats	0	32	90	3 weeks	—	—	86.5
Lychees	1.5	35	90–95	3–5 weeks	—	—	81.9
Mangos	13	55	85–90	2–3 weeks	−.9	30.3	81.7
Nectarines	−0.5–0	31–32	90–95	2–4 weeks	−.9	30.4	81.8
Olives, fresh	5–10	41–50	85–90	4–6 weeks	−1.4	29.4	80.0
Oranges, Calif. & Ariz.	3–9	38–48	85–90	3–8 weeks	−1.2	29.7	85.5
Oranges, Fla. & Texas	0–1	32–34	85–90	8–12 weeks	−.7	30.6	86.4
Papayas	7	45	85–90	1–3 weeks	−.9	30.4	88.7
Passion fruit	7–10	45–50	85–90	3–5 weeks	—	—	75.1
Peaches	−0.5–0	31–32	90–95	2–4 weeks	−.9	30.3	89.1
Pears	−1.5 to −0.5	29–31	90–95	2–7 months	−1.5	29.2	83.2
Persimmons, Japanese	−1	30	90	3–4 months	−2.1	28.1	78.2
Pineapples	7–13	45–55	85–90	2–4 weeks	−1.1	30.0	85.3
Plums and prunes	−0.5–0	31–32	90–95	2–5 weeks	−.8	30.5	86.6
Pomegranates	5	41	90–95	2–3 months	−3.0	26.6	82.3
Quinces	−0.5–0	31–32	90	2–3 months	−2.0	28.4	83.8
Tangerines, mandarins, & related citrus fruits	4	40	90–95	2–4 weeks	−1.0	30.1	87.3

Source: Hardenburg, Watada, and Wang. 1986. The commercial storage of fruits, vegetables, and florist and nursery stock. USDA Agr. Handbook No. 66.

end, but they can be left on the tree longer to develop more color and a better flavor. If left on the tree too long—until they are more than one-third colored—they deteriorate rapidly after harvest.

Fruits at maturity have the approximate chemical composition as shown in the table in the appendix. Most fruits contain 80 to 90 percent water, except dates, which are 20 percent. Fruits are low in proteins and fats, but the avocado is 20 percent fat. Fruits are relatively low in carbohydrates, except for dates, and are low in acids, except for lemons, which are 5.5 percent acid. The strawberry and the many subtropical fruits are good sources of calcium. Several of the fruits—strawberry, grapefruit, orange, lemon, mango, and papaya—are particularly good sources of vitamin C. None of the fruits contain appreciable quantities of B_1 or B_2 vitamins.

Fruits of many species, if stored below certain temperatures but above the freezing point, will develop typical damage symptoms, such as internal browning, pitting, scald, dull skin color, a soggy breakdown, susceptibility to decay, and failure to ripen properly. Such fruits become worthless.

Although influenced by many factors, optimum storage conditions are given in Table 14–4 for most fruit species.

SUMMARY AND REVIEW

The six critical steps of fruit production are initiation of flower buds; flower opening and pollination; fertilization followed by fruit set and development in the spring; removal in some cases of excess fruit; growth and development of the fruit through the summer; and finally fruit maturation, ripening, and harvest.

Initiation is the change from vegetative to reproductive growth. For many fruits, an environmental signal such as photoperiod or fluctuating temperatures or development stage of the plant is necessary for initiation to occur. For others, no signal is required. Although a fruit grower usually has little control over these signals, knowing what signal is needed for a crop allows a grower to make wise decisions for crop or cultivar selection.

Flower opening and pollination are critical steps in the production of most fruits. The events surrounding these activities are often complex and require integration of the steps. For instance, when a female flower or the female parts of a flower are mature, the male flower or parts must also be mature and producing pollen. If pollinators are needed, they must be present and active. If the open flower is sensitive to cold, the crop must be protected during this time from frosts.

If natural fruit thinning does not occur shortly after pollination, too many fruits may remain on the tree. When this happens the fruits are too small and of poor quality. Thinning the fruit is necessary. How much fruit to remove depends on the species, cultivar, vigor of the plant, and many other factors, all of which must be considered. Some thinning occurs when the tree or vine is pruned but that may not be enough. Although thinning can be done by hand, that process is slow, laborious, and expensive. Generally thinning is done chemically or mechanically and is done while the fruit is in the early stages of development.

During summer growth and development, fruits often follow specific growth patterns. Supplying the proper nutrients and water during periods of rapid cell division and expansion is critical.

In the final stages of maturation and ripening, growth has nearly stopped but dramatic changes are occurring in the fruit. Some tissues are changing color and others are softening. Sugars levels are increasing, often as starch levels decrease. Respiration rates may change dramatically as in climacteric fruit. Observing or measuring these changes are often used to determine when it is time to harvest the fruit. Many times harvest has to be done just as ripening begins, when sugars are forming but before softening occurs. This is especially the case if transportation and/or storage are required for the fruit. Storage conditions vary among fruit species. Some can be kept cold, others are injured by the cold. Some are sensitive to ethylene, others are insensitive.

EXERCISES FOR UNDERSTANDING

14–1. In the produce department of your grocery store observe the conditions under which fruits are displayed. Are they chilled or at room temperature? Are the conditions proper for each fruit? Ask a stock clerk how the fruit are stored before they are put on display. Are those practices sound?

14–2. Observe your local weather for a week. Write down what would be happening to a fruit tree or vine that could be grown in your area during that week. For example, is it mid-spring and time for flower opening? How would the weather have affected flower opening?

REFERENCES AND SUPPLEMENTARY READING

GALLETTA, G. J., and D. G. HIMELRICK. 1990. *Small fruit crop management.* Englewood Cliffs, N.J.: Prentice Hall.

WESTWOOD, M. N. 1993. *Temperate-zone pomology: Physiology and Culture.* Third Edition. Portland, Ore.: Timber Press.

Nursery Production: Field, Above-Ground Container, and Pot-In-Pot Cultures

KEY LEARNING CONCEPTS

After reading the chapter you should be able to:

♦ Understand nursery field stock production.
♦ Understand nursery above-ground container production.
♦ Understand nursery pot-in-pot (PIP) production.

FIELD PRODUCTION

Once the planting site for a nursery has been selected, one of the first priorities prior to planting is the soil test. Determining the pH and lime test index as well as the fertilizer needs are of prime importance at this time. It is much easier and more desirable to adjust the soil pH and nutrient status prior to planting. Ideally, the initial soil samples and corrective actions are done a year or two prior to planting. This allows time for the needed soil amendments to produce their most desirable effects.

Adjusting pH

In some areas of the United States such as the Pacific Northwest, many soils are acidic and applications of liming materials, limestone (calcium carbonate), burnt lime (calcium oxide), slaked lime (calcium hydroxide), or dolomitic limestone (calcium-magnesium carbonate) may be required. If added, these

TABLE 15–1	Soil Nutrient Values for Most Woody Plants on Most Soils		
Available	P	30–100 lb/A	
Exchangeable	K	200–400 lb/A	
Exchangeable	Ca	800–16,000 lb/A*	
Exchangeable	Mg	150–2,000 lb/A*	
Available	Mn	20–40 lb/A	
Available	B	0.5 lb/A	
Available	Zn	3 lb/A	

*Limits vary depending on CEC, Ca-Mg ratio, and percent base saturation.

Source: E. M. Smith and C. H. Gilliam (1979). Soil pH levels for most woody plants and soils are 5.0 to 7.0 for mineral soils depending on plant species and 5.0 to 6.0 for organic soils.

TABLE 15–2	Corrective Applications of Phosphorus and Potassium to Nonplanted Soils		
Phosphorus			
Soil Test Values lb/A	lb/A 0-20-0	lb/A 0-46-0	lb/A 15-15-15
0–9	1,500	645	2,250
10–19	1,000	430	1,500
20–29	500	215	750
30+	0	0	0
Potassium			
Soil Test Values lb/A	lb/A 0-0-60	lb/A 15-15-15	
0–99	600	2,250	
100–149	500	1,500	
150–199	425	750	
200+	0	0	

Source: E. M. Smith and C. H. Gilliam (1979).

materials will have five effects. First, they supply Ca and increase pH to reduce the effects of Ca leaching. Application of lime is important in preventing and correcting Ca deficiencies. Second, Mg is also supplied with dolomitic limestone and P, Mo, and Mg are increased. Third, harmful concentrations of Al, Mn, and Fe are reduced. Fourth, favorable microbial activity is increased, and elements essential for plant growth are made more available. Fifth, soil structure and tilth are improved. Lime can be applied at any time in the year provided that soil moisture allows the soil to be worked; however, as previously indicated, it is best added before planting. Lime should not be applied with NH_4^+-containing fertilizers, as the pH shift converts NH_4^+ to NH_3, which is partially lost by volatilization. The dissolution of the lime will depend on the soil temperature, the original soil acidity, and the particle size of the lime. Finely ground materials react more rapidly than coarse materials due to larger surface area (Barrows et al., 1968). In most of the United States and Canada, however, the soils are not acidic and applications of lime are not required. On the contrary, in some areas such as parts of Ohio, soils have a high pH, which at times necessitates the use of acidifying fertilizers to lower the pH.

The Preplant Fertilizer Program

It has been repeatedly shown that species differ in their ability to accumulate various elements. It is difficult, therefore, to make a generalization of nutrient requirements. However, following are some probable preplant recommendations.

  If the soil test indicates that additions of mineral elements are required, these are best applied to cover crops 1 or 2 years ahead of planting. This way the fertility levels at planting time will be at their optimum. The combination of fertilizer and cover crops will improve both the nutrient level and the structure of the soil, thus creating a more desirable rooting medium for new plantings.

  Fertilizer applications should be based on soil test results and kind of cover crops grown. Definitely, a high-

phosphorus fertilizer such as 0-46-0, 0-20-0, 4-12-4, and so on should be applied prior to planting nursery stock to provide a source of phosphorus for several years. Nitrogen fertilizers need to be applied annually because of leaching and crop use. To correct low phosphorus and potassium levels prior to planting, see Table 15–2.

  Remember that nutrient availability of plants is affected more by pH than by any other factor. The presence of essential nutrients is not enough; the pH must be in the correct range for the nutrients to be available. The correct moisture levels must also be supplied. Too much or too little water is equally adverse.

Planting Time

Ornamental field production primarily consists of three types: (1) Liner production as bare rootstock or bare root and then containerized, as in Figure 15–1, (2) balled and burlapped (B&B), as in Figure 15–2, balled and then potted or bagged, which are two forms of field production most often done with conifer or broadleaf evergreen culture, and (3) specimen tree or caliper tree production, which usually requires a holding area, as in Figure 15–3. Liner production will also usually require a holding area and/or cold storage.

  Weed control in field culture is critical, especially during the first year, for the production of high-quality trees and shrubs. As in container and container yards, weeds should not be allowed to go to seed and should be controlled on the perimeter of the nursery to prevent spread into growing areas (Neal, 1999). Mowing, cultivating, and applying chemicals can achieve perimeter control. Cultural controls such as crop

Figure 15–1   Bareroot stock may be containerized, as shown here. The container provides protection for the bareroot stock from stress.

Figure 15–3   Tree liners produced on the west coast will be sold to other producers in the United States for caliper tree production.

Figure 15–2   A large specimen tree that has been balled and burlapped for shipping.

rotations and crop establishment can be used in field nursery culture. The principle of crop rotation is that certain weed populations build up with certain crops because they are able to compete with that crop, or more often the herbicides necessary to control them are not registered or are phytotoxic to the crop. Crop establishment works on the general rule that the first plants to establish on a site tend to exclude and/or compete better than all others. This is one reason why, particularly for new growers, we always recommended the planting out of 1 gallon or larger materials of *Thuja occidentalis,* for example, versus 4-in. liners. The 1 gallons were better sized to establish faster and have less difficulty competing with weeds that had escaped preplant chemical and physical controls.

Herbicides are an effective tool for controlling weeds while reducing costs associated with physical weed removal. A chemical weed control program for a commercial field stock nursery consists of four parts. First, a preplant application of a postemergent such as glyphosate. Preplant postemergents are applied to the top growth of weeds prior to working the soil and planting the nursery stock. Second, an application of preplant soil fumigant such as methyl bromide. Under ideal conditions the properly applied fumigant will provide control of soil-borne weeds, insects, disease, and nematodes. Good soil preparation is essential before the use of most soil fumigants. Third, a preplant application of a preemergent. There are few preplant preemergents registered for use in ornamental nurseries. One that is, however, is Treflan. Treflan applications can be made and incorporated from 3 weeks before planting to the time of planting. Treflan controls broadleaf and grass seedlings just after germination. Do not apply to wet soils or soils high in organic matter. The fourth part of the chemical control program consists of post-plant application(s) with selective herbicides. Applications of selective herbicides used either for post- or preemergence weed control can be applied over or between established ornamental crops, depending on what the label reads.

Regardless of whether 1-gal or smaller materials are outplanted into field culture, all materials should be properly hardened off before outplanting. Hardening off is simply exposure of plants to full sun, decreased water and fertilizer, and in some cases low temperatures. Hardening off works on the principle that plants previously exposed to water stress will better survive water stress at a later stage (Levitt, 1980). This is mainly because of the stomata. Transpiration is the main driving force in plant-water relations, but water diffusion from the stomata can sometimes be 50 percent greater than the rate from a free water surface. The stomata of plants previously subjected to water stress generally close sooner than those in plants not previously stressed (Levitt, 1980).

Trees and shrubs transplanted—with bare roots or even with a root ball of soil—into nursery fields for specimen tree production undergo severe physiological shock because their capacity for absorbing water is suddenly greatly reduced, although water loss by transpiration continues. During lifting and handling of planting stock, many of the small absorbing roots are lost, leading to dehydration of transplanted trees (Kozlowski and Davies, 1975a,b).

Repeated root pruning is a practice used in nurseries to develop compact root systems to reduce transplant shock. Root pruning is followed by increased allocation of photosynthates to the roots, resulting in decreased shoot growth, increased root growth, and a progressive increase in the root-shoot ratio. Profusely branched root systems can be lifted with minimum injury and survive drought much better than shallow or sparsely branched root systems. Pruning of the root system of landscape-size Engelmann spruce trees in the nursery (5 years before transplanting) led to a quadrupling of the root surface area and a doubling of the percentage of the whole root system in the root ball (Watson and Syndor, 1987). Responses to root pruning vary with the time of treatment. Root pruning in the spring is much more effective than pruning in the autumn for increasing root growth (Geisler and Ferree, 1984a, b).

Irrigation

One inch of water per acre is usually required to irrigate nursery stock when water is required. One inch per acre equals 27,154 gallons per acre. An irrigation system needs to provide about 70 gallons per minute to irrigate half an acre at a time. Using 11 sprinkler heads, outputting 6 gallons per minute (11/64 in.), on 40- by 50-foot spacing, 70 gallons per minute can be delivered (Table 15–3). The irrigation system should be capable of irrigating up to one-third or one-half of the nursery stock at one time on a hot day.

Work closely with irrigation suppliers to determine suitable irrigation equipment, proper design of the system, and needed water capacity. The water supply and irrigation system should not only meet present needs, but should also provide for possible future expansion.

Container stock usually requires more exacting attention to irrigation than field production, but irrigation is still essential. Timing, frequency, and water quality are important considerations for field stock. With certain kinds of plants including broadleaf evergreens, irrigation is just as critical as with container stock. Regardless of whether you are planning to grow container or field stock, a test of irrigation water quality is essential. Trickle irrigation promotes compact, profusely branched root systems; thus, it is an advantageous cultural practice.

ABOVE-GROUND CONTAINER CULTURE

In the past 10 to 15 years, a change in consumer preference has resulted in a need for more container production. Virtually all varieties and types of herbaceous and woody perennials that can be produced in the United States can be grown in containers. One exception is *Daphne. Daphne* spp. are difficult to produce in containers for reasons not quite understood. Recently, the production of deciduous trees and larger shrub materials in pot-in-pot (PIP) culture has joined the container production rage. In the PIP production system, a planted container is placed in a holder pot that has been permanently placed in the ground (Fig. 15–4). PIP was first started in the southern states to protect roots from extreme summer temperatures, but really caught on in northern states because of the advantages in winter protection.

There has been a dramatic increase in interest in pot-in-pot production and installation in many locations within the United States and Canada in the past 5 years. In 1999 to 2000, Bailey Nurseries in St. Paul, Minnesota, expanded their PIP growing area by putting in 100,000 permanent pots. Bailey's grows predominantly 5-, 7-, 10-, and 15-gal PIP

TABLE 15–3	Effect of Sprinkler Head Size on Water Application and Requirements					
	Head Capacity					
Head Size (in.)	**Spacing of Heads**	**Heads per Half Acre**	**gal/min**	**in./hr**	**Pressure lb/sq in. (psi)**	**Water Needed per Half Acre (gal/min)**
3/32	30 × 40 ft	18	1.6	0.13	40	30
11/64	40 × 50 ft	11	6	0.29	50	70

Figure 15–4 Pot-In-Pot Production showing holder pot and movable pot with emitter tubes from drip irrigation lines.

Figure 15–5 Above-ground container production has many advantages over field production. One gallon containers are usually placed on 1 foot centers.

materials; however, they also produce some 25-gal materials and will be trying some 45-gal materials in the future. This portion of the chapter will be an introduction to the elements of above-ground and PIP production. Container overwintering, PIP in detail, and a review of the aspects of container media monitoring including pH, EC, and aeration will also be discussed.

Potential

Growing trees and shrubs in above-ground containers (Fig. 15–5) and PIP offers a number of production and marketing advantages compared with growing plants in the field (Ruter, 1997). Plants grown in containers can be sold from spring through fall, whereas bare rootstock has a very narrow window of marketability. The prime biological advantage of container stock over bare root is that the root system is packaged and protected from stress. Containerized trees and shrubs, therefore, are more resistant to poor handling practices in the field and suffer less root disturbance and transplant shock (Davidson et al., 1988). Container production

versus bare root also allows the nursery manager to grow three to eight times more plants per unit area, depending on the crop, which reduces the need for expensive and productive field soils. The container producer can produce more plants in a shorter period of time and increase mechanization, resulting in reduced costs and higher returns.

Above-ground container production, however, does have its drawbacks (Ruter, 1997). Root hardiness during overwintering of container-grown nursery crops has become the most important factor limiting container production. Plants overwintered in containers suffer greater winter injury than those in the ground because the roots are surrounded by cold, circulating air rather than the relatively warm, insulating environment of the soil. The shoots are also more susceptible to injury from desiccation because the root zone is frozen in the container. Common winter injury problems found in poorly overwintered container stock include bark splitting, root kill, top kill, collar injury, and desiccation injury. Two other problems with above-ground containerized plants are windthrow and root kill, primarily on the south side of the pot, during hot summer days. PIP production does not suffer from the preceding problems. However, PIP has some disadvantages of its own that will be discussed later.

Growing Media

In both above-ground and PIP production, the choice of container medium is of primary importance (Davidson et al., 1988). Adequate aeration in the container media cannot be overemphasized. Water-holding ability is also important, but it is a secondary factor compared with aeration. Since roots require adequate oxygen to grow properly, a poorly aerated mix will restrict root growth. Total porosity of a mix is the amount of spaces between particles that could potentially be filled with either air or water. A value of 50 percent is adequate for total porosity. Aeration porosity is the amount of space in the mix filled with air after irrigation water has drained out. If aeration porosity is as low as 15 percent, the mix is poorly drained. Values above 30 percent are considered too high. Too much aeration is not bad, but it means that frequent irrigation will be necessary. A mix with a high percentage of very fine particles (i.e., < .5 mm) results in low aeration porosity and poor drainage (Ontario Ministry of Agriculture and Food, 1992).

Peat is a traditional ingredient in container mixes. Combining peat with equal parts of concrete-grade sand is common in the Prairie Provinces of Canada and the midwestern United States. Bailey's uses a mix of 1 part sand, 1 part peat, 6 parts wood chips, and 1 part soil in their PIP operation. This is a heavier mix than the one they use in their above-ground container production. When using sand the alkalinity must be watched, especially when high pH sand is used and aeration is somewhat low. Bark is a common mix ingredient throughout the United States and Canada. It provides good aeration when mixed with peat. Coarse sawdust that is well decomposed is also useful as a mix ingredient. Fresh sawdust should be avoided because of its very high carbon-nitrogen (C/N) ratio. The C/N ratio of sawdust is 1000/1, of conifer bark, 300/1, and for hardwood bark, 150/1. To compensate for the high C/N ratio and to enhance decomposition, add approximately 1 kg of actual N (e.g., ammonium nitrate) per cubic meter of bark. Hardwood bark will decompose more quickly than conifer bark (Ontario Ministry of Agriculture and Food, 1992).

During the 1960s, nurserymen across the United States explored the possibility of using composted tree bark as a peat substitute to reduce potting mix costs (Hoitink et al., 1997). Early on in the utilization of bark composts, improved plant growth and decreased losses caused by *Phytophthora* root rots were observed as side benefits in the nursery industry. More recently, composted yard waste is being used. Composts must be of consistent quality to be used successfully in biological control of diseases of horticultural crops, particularly if used in container media (Inbar et al., 1993). The rate of respiration is one of several procedures that can be used to monitor stability of composts (Iannotti et al., 1994). Variability in compost stability is one principal factor limiting the widespread use of compost products as container media (Hoitink et al., 1997).

A standard above-ground container mix in British Columbia is 3 parts sawdust to 1 part peat. In Ontario, Canada, mixes such as 2 parts bark and 1 part peat or 2 parts bark and 1 part sand provide adequate aeration and water-holding capacity. Bailey Nurseries uses a mix of 5 parts sphagnum peat (pH 4.0), 2 parts topsoil, and 1 part wood chips for most deciduous above-ground container shrub production. In the Prairie Provinces two common mixes are 55 percent sawdust, 30 percent peat, and 15 percent sand; and 34 percent sand, 33 percent soil, and 33 percent wood chips. A standard mix in Oregon is 9 parts uncomposted fir bark and 1 part peat.

Media Testing

There are certain physical and chemical properties of soilless media that are necessary to grow high-quality nursery and greenhouse plants. Next we will discuss the four properties of media that should be monitored in normal greenhouse and nursery production. The four properties are the physical property of aeration and the three chemical properties of electrical conductivity (EC), pH, and nitrate (NO_3^-). Some simple tests to assess the properties of aeration and EC are presented. pH and nitrate monitoring are also discussed. The tests presented are simple, but their impact on container plant management is diverse and crucial.

Aeration Porosity

The composition of the growing medium is extremely important in container production. Not only is the plant root restricted to the limited volume of the container, but a "perched" water table is also created in the bottom of all con-

tainers, which further restricts the total root-growing space (Mathers and Leidenfrost, 1995). The perched water table is an area where all the pore spaces in the media are filled with water. This saturated area occurs no matter how many drainage holes are in the container. The deeper the container, the lower the impact of the saturated area at the bottom of the container. The deeper the container, the greater the overall aeration and the less pores are filled with water. Inadequate or excessive medium aeration can lead to production problems. It is recommended that porosity be measured before any new medium formulation is used, and that aeration porosity be checked at least three times, depending on the components of the mix.

Aeration porosity is a measurement of the amount of air space in the mix after the free irrigation water has drained out. The optimum aeration porosity for woody plants is between 20 to 30 percent (Mathers and Leidenfrost, 1995). Bedding plants need between 8 to 9 percent. Analysis of aeration porosity is very easy. All you need is the container volume and the aeration pore volume.

$$\text{Aeration porosity (\%)} = \frac{\text{Aeration pore volume}}{\text{Container volume}} \times 100\%$$

Procedure

1. To measure the container volume, seal the drainage holes of the container and fill the container with water to the level it would normally be filled to with potting mix. Record this volume.
2. Empty and dry the container and fill it with growing medium. Slowly add water with a graduated cylinder until the mix is completely saturated. A very thin slick of water will appear on the surface when saturation is reached. Record the total volume of water added as "total pore volume." This will take approximately 2 hours to complete when starting with new, dry media.
3. Place the container over a watertight pan and remove the seals from the drain holes. Allow the free water to drain out of the container. This may take several hours. Measure the amount of drained water and record as "aeration pore volume."

The preceding three steps will also allow you to calculate total porosity and water-holding porosity.

$$\text{Total porosity (\%)} = \frac{\text{Total pore volume}}{\text{Container volume}} \times 100\%$$

$$\text{Water-holding porosity (\%)} = \text{Total porosity} - \text{Aeration porosity}$$

The stability of the components in the nursery or greenhouse mix needs to be considered, which is why aeration porosity should be monitored over the life of the mix. Sawdust in a mix is initially high in aeration porosity; however, rapid decomposition can result in a dramatic decrease with time. Bark has a higher lignin content and, therefore, is more resistant to decomposition. Aeration should not vary as greatly with a bark mix as with a sawdust mix. Perlite and vermiculite are not subject to decomposition, but aeration can be lost due to compaction of the mix, especially with high vermiculite percentages. Compost materials are being used extensively in the nursery industry to replace peat. Composts need to be monitored for decomposition. The decomposition rate of compost may be greater than with a coarse-grade peat. Therefore, changes in aeration porosity may be higher with the compost. Aeration of a mix can be improved by using sufficiently coarse media components or deeper containers where possible.

EC Testing

Fertilizers and other dissolved salts change the ability of a solution to conduct electricity. Pure water is a poor conductor, but as salinity of a solution increases, its conductance also increases. Portable salt meters (conductivity meters) can be used by container growers to measure electrical conductivity of solutions. Growing media should be tested at least every 2 weeks or more, preferably every week for salts and pH. Portable EC meters for use in ornamentals range in price from $60 to $1,000. The more expensive meters are durable and should last many years. Keep a chart of the pH and EC values collected. This way you can establish trends; in other words, whether the EC is rising, falling, or staying constant.

Greenhouse and nursery growers need to pay close attention to salt levels. Growers need to realize, however, that not all salts are from fertilizers and it is important to establish the nonnutritional background content of the irrigation water. Occasional complete mineral analyses conducted of the container media by an analytical laboratory are also recommended.

Many nursery growers use controlled-release fertilizers. During and right after periods of temperature fluctuations in the medium, particularly high temperatures, EC monitoring is critical. High-temperature fluctuations may result in heavy salt and nutrient concentrations. During high-temperature periods, irrigating the medium with clear water to leach out excess fertilizer salts and nutrients may be needed. Frequent systematic monitoring of the medium's soluble salt levels—with an electrical conductivity meter—is the simplest way to keep track of nutrient release. Periods of heavy release or when the controlled-release fertilizer approaches exhaustion can be identified. As the fertilizer becomes exhausted, the salinity of the medium approaches that of the irrigation water, indicating the need for supplemental fertilization. A more complete medium test may be needed to determine the proper fertilizer to apply.

Greenhouse growers can use EC monitoring to make informed decisions about fertigation concentrations, water frequencies, and leaching rates. Growers who track their EC values find that they produce superior crops with less fertilizer and lower leaching rates, thus reducing environmental contamination.

Ornamental growers have been slow to conduct regular monitoring of their container-grown plants (Ruter and Garber, 1998). It seems the major reasons for this have been lack of a uniform and simple testing procedure. Many greenhouse growers have used the saturated extract method (SEM) for monitoring, which was developed at Michigan State University. Others are strong proponents of the Virginia Tech Extraction Method (VTEM) or pour-through method for monitoring nursery and some greenhouse crops. There are five main advantages of VTEM over other monitoring procedures: (1) sample extraction time is short, (2) pH and EC tests can be made in the field, (3) no media need to be handled, (4) sample extraction requires no special equipment, and (5) controlled-release fertilizer perils are not ruptured, creating possible false readings (Ruter and Garber, 1998). The procedure for conducting VTEM and a table for interpretation of soluble salt and pH measurements obtained by VTEM as compared with the extraction method are presented later.

pH Testing

pH is a measure of H^+ ions in solution. The pH scale of 0 to 14 is a logarithmic scale. pH 7 is neutral. pH values of less than 7 are considered acidic. pH values greater than 7 are considered basic. Nutrient availability to plants is affected more by pH than by any other factor. In high-pH soils, ions of aluminum (Al), Fe, and Mn precipitate and the availability of these elements decreases. Plants in a high-pH soilless medium may express deficiencies of Fe, B, Zn, Mn, and Mo. Phosphorus may also become deficient in alkaline conditions as it complexes with Ca to form insoluble calcium phosphates. Deficiencies of most of the micronutrients can be corrected by adjusting soil pH. Most plants have a relatively narrow range of preferred pH levels. Figure 15–6 shows the preferred pH scale for most soilless media-grown woody ornamentals (5.4 to 6.0). Acid-tolerant crops, like rhododendrons, will be grown at pH 5.0 to 5.5.

VTEM Pour-Through Procedure

Other articles concerning this method will outline a procedure that involves a PVC ring, a collection vessel, and distilled water. The procedure advocated here, however, requires none of these. Within 2 hours after irrigating containers to saturation a minimum of three containers from each block of plants should be tested. The testing involves pouring through a certain volume of water that the crop is normally irrigated with (Bilderback, 1999 personal communication), and allowing a sufficient amount of time for it to drain to collect an adequate size sample for the EC and pH meters. Typically, 15 oz of water per a 1-gal container is sufficient to collect 5 oz of leachate with most media (Ruter and Garber, 1998).

The values given in Table 15–4 should be used as guidelines only. Several factors will influence the pH and EC values you record. Six such factors are (1) different species have different nutritional requirements, (2) stage of crop growth, (3) time of year, (4) fertilization (liquid feed versus controlled release) and irrigation program, (5) growing medium, and (6) other environmental factors (Ruter and Garber, 1998). The VTEM should be performed every 1 to 2 weeks. It is important to keep good records and chart the pH and EC values observed. Graphically charting the pH and EC

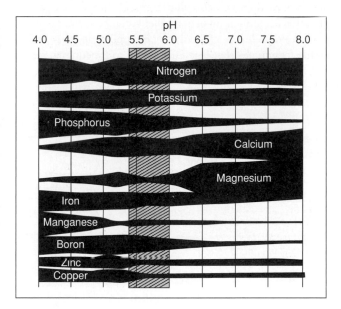

Figure 15–6 pH range of macro- and micronutrients in artificial media.

values will provide a grower with a trend of timely information on whether the pH and EC are rising, falling, or staying steady. The results from a single VTEM procedure should never be used to make a change in culture. The strength of the VTEM procedure is in establishing a picture for the crop to help make informed decisions about watering frequencies, fertilizer needs, and leaching requirements.

Nitrate Testing

Nitrogen is one of the most widely distributed elements in nature. It is present in the atmosphere, the lithosphere, and the hydrosphere. The atmosphere is the main reservoir of N (Mengel and Kirkby, 1987). The soil accounts for only a minute fraction of lithospheric N, and of this soil N, only a small proportion is directly available to plants. Both nitrate (NO_3^-) and ammonium (NH_4^+) are forms of N that can be taken up and metabolized by plants. Nitrate is often a preferential source for plants; however, much depends on the plant species and other environmental factors. A number of reports indicate that the uptake of both N-forms is temperature dependent with rates of uptake being depressed by lower temperatures (Mengel and Kirkby, 1987). The most important difference between NO_3^- and NH_4^+ uptake is in their sensitivity to pH. NH_4^+ uptake takes place best in a neutral medium and it is depressed as the pH falls. The opposite is true for NO_3^- absorption. Uptake of NO_3^- is more rapid at low pH values (Rao and Rains, 1976). Most nitrate ions, like other anions, are easily leached away with heavy rains or excessive irrigation and are quickly removed from the growing media. Because nitrate availability is so important to plant health and yet is so easily leached, a periodic check of the

TABLE 15–4	Interpretation of Soluble Salt and pH Measurements by Extraction Method		
Method	*Soluble Salt*	*pH*	*Electrical Conductivity (dS/M or mMhos/cm)*
VTEM	Sensitive crops (liquid feed) Nursery crops (liquid feed) Nursery crops (controlled-release)	5.2–6.2	0.50–0.75 0.75–1.50 0.20–1.00
Saturated Extract Method (Nursery Crops)	Low Acceptable Optimum High Very High	5.8–6.8	0.00–0.74 0.75–1.49 1.50–2.24 2.25–3.49 3.50+
Saturated Extract Method (Greenhouse Crops)	Low Acceptable Optimum High Very High	5.6–5.8	0.00–0.75 0.75–2.0 2.0–3.5 3.5–5.0 5.0+

Note: The ranges of pH and soluble salt levels should be used as guidelines only. Irrigation water should be < 0.75 dS/M. The soluble salt level of water used in the VTEM procedure should be subtracted from the final leachate value.

Source: The University of Georgia College of Agricultural and Environmental Sciences Cooperative Extension Service.

medium nitrate levels is necessary. EC monitoring alone will give the total salt values. The most important specific salt that needs tracking is nitrate. Perhaps the cheapest and most reliable instruments available for nitrate monitoring are Horiba card meters.

Fertilizers

Plants produced in above-ground containers and PIP require the same essential elements as plants produced in field culture. The total supply of minerals available for plant growth is limited by the size of the container (Johnson, 1979).

Good container media and nutrition management are basic to the production of quality container-grown plants (Hickleton and Cairns, 1992). However, the decisions involved in providing good nutrition to container stock are complicated by a variety of factors: the multitude of fertilizer products available, the variations in container media, the number of species involved, and the various cultural practices used (Swanson et al., 1989). The impact of fertilizers on the environment and groundwater is an important concern. To minimize environmental impact nursery growers are using controlled-release fertilizers (CRFs). Some researchers have indicated that supplemental fertilizing with water-soluble fertilizers, particularly nitrogen, is beneficial for fast-growing crops when using CRFs. Soil tests are an important part of a container production operation. Potting mixes do not retain nutrients and the quantity of water applied causes rapid leaching. Biweekly sampling is suggested. Once the grower becomes familiar with test results, pH and salt readings may be sufficient information at each sample time, and nutrient analysis can be less frequent. This will be discussed in more detail toward the end of the chapter.

As mentioned earlier, root hardiness during overwintering of container-grown nursery crops has become the most important factor limiting container production. There is an array of questions nursery growers ask regarding overwintering procedures that can be broken down into four basic areas: How do I ensure maximum acclimation of my stock? When should I cover the plants? What system of overwintering should I use? When should I uncover my stock in the spring? Providing the answers to these questions should enable most nurserymen to be reasonably successful in overwintering container stock, recognizing that some unknowns still exist (Good et al., 1976b).

How Do I Ensure Maximum Acclimation of My Stock?

To ensure maximum acclimation there are three things a grower should know: (1) the root hardiness values of the stock, (2) the proper cultural practices to follow, and (3) whether suitable conditions for acclimation have been provided before covering begins (i.e., short days).

Certain plants that have hardy shoots may not have sufficient root hardiness to survive in the zonation listed for their

shoots, unless they are well protected. See Table 15–5 for some root hardiness values. Knowledge of root hardiness values allows the grower to know what type of overwintering system to use to provide sufficient protection for the plants.

Plants that are underfertilized or overfertilized could be predisposed to winter injury. Without an adequate supply of nutrients, plants cannot produce metabolites essential for normal physiological activities such as acclimation. If excessive levels of nutrients are provided, promoting fall growth, cold-temperature injury could occur because the late growth would not have ample time to acclimate. The chance of this occurring with container stock is slim. Excessive fertilization accumulation in container soil mixes will not occur if appropriate levels of fertilizer are applied and proper irrigation techniques that leach container soil mediums, reducing fertilizer accumulation, are employed (Good et al., 1976b).

TABLE 15–5 Root Hardiness Values of Specific Ornamental Plants

Species	Killing Point (°C) of Young Roots	Killing Point (°C) of Mature Roots
Cornus florida	−6	−12
Cotoneaster dammeri	−5	−8
Cotoneaster dammeri 'Skogholm'	−7	−11
Euonymus alatus 'Compactus'	−7	−14
Euonymus fortunei 'Vegetus'	−5	−11
Euonymus kiautschovicus	−6	−9
Hypericum sp.	−5	−8
Ilex 'Nellie R. Stevens'	−5	−10
Ilex cornuta 'Dazzler'	−4	−8
Ilex crenata 'Helleri'	−5	−8
Ilex opaca	−5	−13
Ilex × aquipernyi 'San Jose'	−6	−8
Ilex × meserveae 'Blue Boy'	−5	−13
Juniperus conferta	−11	> −23
Juniperus horizontalis 'Plumosa'	−11	−20
Juniperus squamata 'Meyeri'	−11	−18
Koelreuteria paniculata	−9	−20
Magnolia stellata	−6	−13
Mahonia bealei	−4	−11
Pyracantha coccinea 'Lalandei'	−4	−8
Stephanandra incisa 'Crispa'	−8	−18
Taxus × media 'Hicksii'	−8	−20
Viburnum plicatum f. tomentosum	−7	−14

Late-fall fertilizer applications at recommended rates will not force new growth, but actually enhance the nutrient content of plants, which, in turn, can enhance growth the following spring (Pellet, 1981). If containerized plants are fertilized at recommended rates and frequencies through the growing season and into the fall, the plants should acclimate to the maximum capability (Good et al., 1976b).

By September, most woody plants stop visibly growing. This does not necessarily mean the need for water is less. Container soil mixes are very well drained and fall weather can be warm and sunny. Ample fall irrigation is a must because plants will continue to absorb and transpire water. It is becoming apparent that moisture availability is an important factor in the survival of containerized nursery stock during the fall and winter regardless of the method of overwintering.

Plant shoots and mature roots have the capability of acclimating during the short, cool days of autumn. To ensure that the plants have achieved maximum hardiness, covering should be accomplished as late in the fall as possible to take advantage of the natural environmental cues that foster hardier plants (Good et al., 1976a).

When Should I Cover the Plants?

A suggested guideline as to when to cover in the fall is to use the F-date. The F-date is the date of the first frost in the fall, F+30 or F+45. The exact date will vary with latitude of the nursery and with seasonal variation (Davidson et al., 1988). In the Aurora, Oregon, area the 30-year average for the first frost has been October 21. Thus, F+30 or F+45 would be November 21 or December 6, respectively. To find the first frost date for your area of Oregon visit the Web site (www.ocs.orst.edu) and go to climate zone summaries under climate data. For Columbus, Ohio, the first frost date is October 15. Thus, F+30 or F+45 would be November 15 or November 30, respectively. To find the first frost date for your area of Ohio visit the Web site (http://ohioline.ag. ohio-state.edu/b472/climate.html) and go to Figure 3 for the first frost date.

The following checklist will help ensure winter survival of covered stock:

1. All stock should have received a heavy soaking 1 to 2 days before covering.
2. All plants should be dormant at the time of covering.
3. The ground should be well drained and free of organic debris.
4. Wet plants due to rain, snow, dew, or frost should not be covered.
5. A fungal spray over the stock after it is consolidated is suggested.
6. Rodent bait placed every 25 to 50 sq ft under the covering is required.

7. If material is laid flat, all tops should be pointing south so that the tops of plants cover the adjacent plant's roots.
8. Square containers allow for tighter consolidation, which results in less heat loss from the side walls of the containers.

What System of Overwintering Should I Use?

A variety of techniques have been used by nurserymen for overwintering containerized nursery stock: clear or milky plastic poly houses or poly tents; consolidation with or without periphery covering; mass consolidation with a plant foam covering; and consolidation with a top covering of poly-coated plant foam, white plastic, clear plastic/straw/clear plastic ("the sandwich"), clear plastic/grass, geotextiles, white plastic/microfilm, or microfoam.

Temperatures within poly houses are primarily dependent on the amount of sunlight impinging on the houses. More sunlight means higher temperatures within the structures. Unfortunately, polyethylene does a very poor job of holding heat within poly houses and tents. On a clear, cold winter day, heat can radiate out of a poly house in the late afternoon or early evening as fast as it increases in the morning with the rising sun.

Winter night temperatures within poly houses tend to approach outdoor temperatures even though daytime temperatures within the houses are 20° to 30° warmer than outdoor temperatures (Good et al., 1976a). Thus, temperatures low enough to cause root injury can occur in poly houses. The inability of poly houses to protect roots was shown by Gouin (1974). Two plants coming out of overwintering appeared identical but in the subsequent growing season the root loss to the plant overwintered in a poly house was evident due to reduced, poor-quality top growth.

Chong and Desjardins (1981) found little difference between single and double poly houses. Covering the poly house with milky polyethylene is the most common technique used to reduce the frequency of high temperatures. In the case of double-layer, air-inflated houses, only one layer of milky poly is necessary. Because of its opaque nature, less light passes through the milky film and high-temperature buildup in structures is minimized (Good et al., 1976b). Poly houses with microfoam blankets offer better protection, but methods conferring less temperature fluctuations and maximum protection have been found at the University of Minnesota.

Swanson et al. (1988) conducted a study to compare the two best methods for overwintering container stock that they had researched. The two methods were white plastic/ microfilm (WP/MF) or clear plastic/straw/clear plastic ("the sandwich"). The temperatures under both coverings were considerably warmer and less fluctuating than ambient air temperatures. "The sandwich" averaged approximately 8°F to 10°F warmer than the WP/MF covering and was more stable in temperature variation. The clear plastic/straw/clear plastic sandwich provided the highest minimum and the lowest maximum temperatures and the least-fluctuating temperatures under the covering throughout the winter season. They concluded that complete protection of container stock can be obtained with a CP/S/CP sandwich if applied to the laid-down stock prior to severe late fall or early winter freezing temperatures (Swanson et al., 1988).

At Bailey Nurseries they use clear plastic and Sudan grass. They grow 1 acre of Sudan grass per acre of container-growing area. Unlike straw, the Sudan grass has very long sheaths that intertwine and prevent blowing away. Thus, the Sudan grass cover does not require an extra layer of clear plastic on top.

When Should I Uncover My Stock in the Spring?

As with when to cover, a suggested guideline as to when to uncover in the spring is to use the F-date. In this case the F-date is the date of last frost in the spring, F-30 or F-45. The exact date will vary with latitude of the nursery and with seasonal variation (Davidson et al., 1988). In the Aurora, Oregon, area the 30-year average for the last frost has been April 22. Thus, F-30 or F-45 would be March 23 or April 8, respectively. As with the first frost date you can find the last frost date for your area of Oregon by visiting the Web site (http://ohioline.ag.ohio-state.edu/b472/climate.html) and go to Figure 2.

When container stock is uncovered depends on the intentions the grower has for the stock. If plants are left covered in polyethylene overwintering houses, growth will begin earlier in the spring than would normally be the case outdoors. This early initiation of growth is the result of warmer conditions within the structures as winter changes to spring, bringing longer days and more intensive sunshine. If accelerated spring growth were desirable in terms of achieving a larger plant in a shorter period of time, uncovering would be delayed. Plants under these conditions will have to be fertilized, watered, and ventilated because they are very active physiologically (Good et al., 1976b). Growing plants should not be uncovered before the danger of frost is over because the new succulent growth is easily killed by frost. Uncovering should not be done on warm, windy days to avoid excessive drying of succulent growth.

POT-IN-POT CULTURE

The PIP system can eliminate many of the difficulties associated with conventional container growing. In conventional container production winter protection for plant roots is costly and time-consuming. Wind tipping of containerized plants is another time-consuming, laborious drawback of conventional container culture. Wind tipping is also detrimental to quality stock production as top-dressed fertilizers and media are knocked out of the pot and irrigation applications can be missed or delayed, resulting in drought-stressed trees. PIP pro-

duction also eliminates the heat stress and root-killing temperatures experienced in conventional container culture.

The PIP system can also eliminate many of the problems associated with conventional field stock production. In conventional field production, a tremendous amount of soil is lost or "mined" during harvesting of B&B stock (Davidson et al., 1988). Conventional nursery field production also results in more soil compaction than in any other type of farming (Bremer, 1993). PIP production has addressed some of these conventional production difficulties and problems. Some of the problems or the pitfalls of the PIP system are discussed as follows.

Impact on Drought Stress

PIP systems generally use drip irrigation with in-line emitters. Designs vary in delivery rates; however, a rate of 2.3 L/hr has been used with success. Because the root systems of PIP plants were contained, trees grown in the PIP system required less water than plants grown in conventional field culture (Chong and Mathers, 1990). Irrigation was applied every 2 days to PIP plants for 4 hours at a time (Chong and Mathers, 1990). Field-grown trees required watering every 8 hours. John Ruter evaluated cyclic irrigation on water use and found that cyclic irrigation reduced the amount of water leached through the container in the PIP system by approximately 100 percent (Ruter, 1997b). The combination of soil insulation and trickle irrigation ensures that the essential moisture levels are maintained, eliminating the effects of drought stress that may occur in conventional container culture or even field situations. The soil insulation also results in more root mass in PIP-produced plants (Ruter, 1995). Therefore, PIP plants are better adapted to avoid drought stress in the nursery and after outplanting than conventional container-grown plants.

Ruter noted that reducing the amount of water leached through the planted container was also important, since good drainage away from the holder pot was essential to the success of PIP. Sandy soils are well suited for PIP production. Growers need to take precautions if their sites have heavy soils that drain poorly (Ruter, 1997b).

Impact on Heat Stress

The importance of keeping container substrate temperatures below 100°F is well documented; however, substrate temperatures in above-ground containers in Oregon have been measured above 120°F. In Florida and other southern states, temperatures as high as 137°F have been recorded (Martin and Ingram, 1988; Ruter, 1997a). Normal root functioning ceases when root-zone temperatures exceed 96°F for holly (Ruter and Ingram, 1992) and even lower, approximately 90°F, for less-heat-tolerant plants (Levitt, 1979). In above-ground containers, the roots in the western quadrant of the container are often injured or killed by the high temperatures experienced. In the PIP system, roots in the western quadrant were 23° cooler than in above-ground pots (Ruter, 1997b). Recently, Fuchigami and Cheng (1999) emphasized the importance of the plant's ability to photosynthesize and maintain optimum chlorophyll levels to ensure optimum growth and plant health. Plants that experienced high root-zone temperatures suffered loss of chlorophyll and protein production in shoots (Kuroyanagi and Paulsen, 1988). Research indicates this has a significant impact on the overall plant health (Ruter and Ingram, 1992).

Impact on Soil Mining

It is estimated that the harvesting of balled-and-burlapped (B&B) stock can result in the loss of 470 tons of soil per acre during the removal of an acre of 44-in.-diameter B&B trees (Fig. 15–7). This is an average of 94 tons per acre per year for a 5-year rotation (Davidson et al., 1988) or 2.8 in. in 5 years. Soil removal due to "mining" has enormous implications to the economic viability of a field nursery (Murray, 1993). PIP is one management practice that can be used to reduce soil mining. Of course PIP requires relatively permanent modifications to a nursery field that result in soil profile changes. If for some reason you reverted a field from PIP back to conventional culture, however, the soil levels would be virtually unchanged.

Impact on Soil Compaction

Performing intense field activities during late fall and early spring, when rainfall is most frequent and soils are wet, results in soil compaction and reduced soil porosity (Bremer, 1993). The overall effects of conventional nursery field culture result in the reduction of the soil's productive potential and an increase in the cost of production. Cover cropping reduces compaction by reducing the frequency of use of heavy equipment through minimum tillage and by providing support to heavy equipment during wet weather. Compaction of nursery soils is much more harmful than growers realize (Trouse, 1986). When using PIP, compaction is minimized due to the reduced need for heavy equipment to lift the stock. Compaction in the root zone is nil because of the use of artificial media.

Overcoming Rooting Out

One of the problems with PIP has been the rooting out of the plant from the planted container, through the holder pot into the surrounding soil. Rooting out results in plants having to be manually dug and root-pruned before the planted container can be removed from the holder pot (Ruter, 1997a). Various products have been tested for their ability to prevent rooting out. These products include Biobarrier™, a geotextile fabric impregnated with the herbicide trifluralin; Root Control™, a fabric bag material placed between the planted container and the holder pot; and various applications of Spin Out™, a commercial formulation of copper hydroxide that is applied to the side walls of the holder pot, applied to the side walls of the planted container pot, applied to both containers, or applied to the planted container and Root Control™ fabric (Ruter, 1994; 1997b).

Treatments that resulted in water pruning (Pellet, 1983), air pruning (Chong and Mathers, 1990), turning the containers 180° biweekly (Swanson et al., 1992), and other chemical barriers such as oxadiazon and pendimethalin have also been tested. Other fabrics have been tested including Weed-X and Remay Typar 3-oz landscape fabric and Environmentally Friendly Containers, which have raised drainage holes on the sides (Ruter, 1997b). John Ruter has done most of the work involving the prevention of rooting out. He concluded in 1994 that Biobarrier™ was the best treatment for the control of rooting out, but that it also reduced plant growth. He stated that Spin Out™ was useful for reducing rooting-out problems but did not eliminate the problem. The physical controls such as Environmentally Friendly Containers, water pruning, and air pruning have had limited success with vigorous-rooting species (Ruter, 1997b). The 180° biweekly turning of containers has been proven prohibitive due to labor costs (Swanson personal communication, 1993).

Ruter continued his rooting-out studies by looking at the rate of Biobarrier™ necessary to reduce rooting out in vigorous-rooted species but not causing phytotoxicity. He also conducted trials with Spin Out™-treated fabric bags such as Tex-R Agroliners. He concluded that the Tex-R Agroliners did a good job of preventing rooting out of some vigorous-rooting plants and that the bags were easily removed from the root ball, since the roots were not growing through the fabric (Ruter, 1997b, 1998). The Agroliners are used as a bag-in-pot-in-pot production system.

Figure 15–7   Soil loss from machine digging of B&B caliper trees is significant. Large holes left behind by the removal of the trees by the spade are shown in the foreground.

SUMMARY AND REVIEW

As with all other crops grown in the field, site selection for a field-grown nursery is very important. Water availability and quality and soil characteristics, including nutritional status, are critical factors to consider. Adjusting any nutrient imbalance and soil pH are done before planting to allow new plants to begin growth. Additional fertilizer should be applied as needed. Field grown plants are produced as liners or bareroot, balled and burlapped, and specimen or caliper trees. Regardless of type, weed control is important. Irrigation systems can vary, but trickle is often used because it promotes compact, branched root systems.

Above ground container production has several advantages over field production. A broader market window, the root system is packaged and protected during shipping, more plants can be grown in an area, and mechanization can be used to decrease hand labor costs. Drawbacks include the need for more critical attention to irrigation for most species, and root damage during cold winters or hot, sunny summer days. The media in containers is a very important consideration. A proper balance of solids and water- and air-holding pores is essential. To achieve this balance many components such as sand, peat moss, bark, and other materials are blended in proportions that give the media the characteristics desired. There are several tests that a grower can perform to measure the characteristics of the media, not only for solid/pore space but also nutrient status. Using these tests allow a nursery grower to make sound decisions for growing the crop. Many container crops are overwintered under protective coverings such as polyethylene hoop greenhouses. There are several different types of coverings available. A grower must consider the characteristics of the coverings when choosing which one to use.

The PIP system is a variation of the above ground container system. It was designed to overcome the problem of containers freezing in the winter and overheating in the summer. In PIP culture a hole is dug in the field and a pot inserted. Another, slightly smaller container with plant is placed in the first pot. The earth surrounding the outer pot helps to insulate the root system in the inner pot.

EXERCISE FOR UNDERSTANDING

15–1. Visit a local nursery. Observe the method(s) of production. Describe how each is done and what impact the method has on plant growth. Are the methods adjusted for different species or plants of different sizes or are all plants handled the same?

REFERENCES AND SUPPLEMENTARY READING

BREMER, A. H. 1993. A clean choice. *177*(11): 38–41.

BURKE, M. J., L. V. GUSTA, H. A. QUAMME, C. J. WEISER, and P. H. LI. 1976. Freezing and injury in plants. *Ann. Rev. Plant Physiol.* 27:507–28.

BURKE, M. J., and C. STUSHNOFF. 1979. Frost hardiness: A discussion of possible molecular causes of injury with particular reference to deep supercooling of water. In H. Mussell and R. C. Staples (Eds.), *Stress physiology in crop plants* (pp. 197–225). New York: John Wiley.

CHONG, C., and R. L. DESJARDINS. 1981. Comparing methods for overwintering container stock. *Am. Nurseryman* (Jan.): V153(1) p. 8–9, 131–135. ill.

CHONG, C. and H. MATHERS. 1990. Revolutionary Oklahoma technique. *Prairie Landscape Magazine 13*(1): 26, 32–33.

CRANSTON, R. 1995. *Weed control: An introductory manual.* BC Ministry of Agriculture, Fisheries and Food.

DAVIDSON, H., R. MECKLENBURG, and C. PETERSON. 1988. *Nursery management administration and culture.* Engelwood Cliffs, N.J.: Prentice Hall.

DRIX, H., and R. VAN DEN DRIESSCHE. 1974. Mineral nutrition of container-tree seedlings. Great Plains Ag. Co. Pub. No. 68:7784.

FRIES, H. H. 1977. Container size affects growth during a plant's first year. *Am. Nurseryman 146*(2):10, 113–114.

FUCHIGAMI, L. and L. CHENG. 1999. Ornamentals Northwest, Seminars, Portland, Ore. (Handout).

FUCHIGAMI L. H., C. J. WEISER, K. KOBAYASHI, R. TIMMIS, and L. V. GUSTA. 1982. A degree growth state (GS) model and cold acclimation in temperate woody plants. In P. H. Li and A. Sakai (Eds.), *Plant cold hardiness and freezing stress* Vol. 2 (pp. 93–116). New York: Academic Press.

GEISLER, D., and D. C. FERREE. (1984a). Response of plants to root pruning. *Hort. Rev. 6:*155–88.

GEISLER, D., and D. C. FERREE. (1984b). The influence of root pruning on water relations, net photosynthesis and growth of young 'Golden Delicious' apple trees. *J. Am. Soc. Hor. Sci. 109:*827–31.

GOOD, G. L., P. L. STEPONKUS, and S. C. WEIST. 1976a. Using poly houses for protection. *Amer. Nurseryman 144*(7):12, 120–25.

GOOD, G. L., P. L. STEPONKUS, and S. C. WEIST. 1976b. Cultural factors for overwintering. *Amer. Nurseryman 144*(8):14, 117–24.

GOUIN, F. R. 1974. A new concept in overwintering container-grown ornamentals. *Amer. Nurseryman* 140(11): 7–8, 45, 48, 50.

HICKELTON, P. R., and K. G. CAIRNS. 1992. Solubility and application rate of controlled-release fertilizer affects growth and nutrient uptake in containerized woody landscape plants. *J. Am. Soc. Hort. Sci. 117*(4):578–83.

HOITINK, H. A. J., A. G. STONE, and D. Y. HAN. 1997. Suppression of plant diseases by composts. *HortScience 32*:184–187.

IANNOTTI, D. A., M. E., GREBUS, B. L. TOTH, L. V. MADDEN, and H. A. J. HOITINK. 1994. Oxygen respirometry to assess stability and maturity of composted municipal solid waste. *J. Environ. Qual. 23:*1177–1183.

INBAR, Y., Y. HADAR, and Y. CHEN. 1993. Recycling of cattle manure: The composting process and characterization of maturity: *J. Environ. Qual. 22.*:857–863.

JOHNSON, C. R. 1979. How to fertilize plants in containers. *Amer. Nurseryman 150*(4):12–13, 105–108.

KOZLOWSKI, T. T., and W. J. DAVIES. 1975b. Control of water loss in shade trees. *J. Arboric. 1*:81–90.

KOZLOWSKI, T. T., P. J. KRAMER, and S. G. PALLARDY. 1991. *Ecology of woody plants* (pp. 247–302). New York: Academic Press.

KRAMER, P. J., and T. T. KOZLOWSKI. 1979. *Physiology of Woody Plants.* New York: Academic Press. (pp. 334–72).

KUROYANAGI, T., and G. M. PAULSEN. 1988. Mediation of high-temperature injury by roots and shoots during reproductive growth of wheat. *Plant Cell Environ. 11:*517–23.

LEVITT, J. 1979. Responses of plants to environmental stresses. Vol. 1. New York: Academic Press.

LEVITT, J. 1980. *Response of plants to environmental stresses. Vol. 2: Water, radiation, salt and other stresses.* (pp. 93–211). New York: Academic Press.

MARTIN, C. A., and D. L. INGRAM. 1988. Temperature dynamics in black poly containers. *Proc. Southern Nurserymen's Assoc. Res. Conf. 33:*71–74.

MATHERS, H. M. 1999. The shoot-to-root bond. *Amer. Nurseryman 189*(10):66–67, 68, 70.

MATHERS, H. M. and P. LEIDENFROST. 1995. Nursery production guide for commercial growers, 1995/96. British Columbia Ministry of Agriculture and Food, Victoria, BC, Canada.

MENGEL, K. and E. A. KIRKBY. 1987. *Principles of plant nutrition.* (pp. 1–687). International Potash Institute, Switzerland.

MITYGA, H. G., and F. O. LANPHEAR. 1971. Factors influencing the cold hardiness of *Taxus cuspidata* roots. *J. Am. Soc. Hort. Sci. 96*:83–86.

MURRAY, S. M. 1993. Nursery production and soil conservation. *HortWest* (July/Aug):11–12.

NEAL, J. 1999. Weeds and you. *Nursery Management & Production 15*(1):60–62, 64–65.

NOBEL, P. S. 1991. *Physiochemical and environmental plant physiology* (pp. 85–104). New York: Academic Press.

Ontario Ministry of Agriculture and Food. 1992. *Production recommendations for nursery and landscape plants.* Pub. No. 383.

PELLET, H. 1983. An update on Minnesota system of container production. *Am. Nurseryman 157*(1).

PELLET, H. M., and J. V. CARTER. 1981. Nutrition and cold hardiness. *Hort. Rev. 3:*144–71.

RAO, K. P., and RAINS, D. W. 1976. Nitrate absorption by barley. *Plant Physiol. 57:* 55–58.

RUTER, J. M. 1994. Evaluation of control strategies for reducing rooting out problems in pot-in-pot production systems. *J. Environ. Hort. 12*(1):51–54.

RUTER, J. M. 1995. Growth of southern magnolia in pot-in-pot and above-ground production systems. *Proc. Southern Nurserymen's Assoc. Res. Conf. 40:*138–39.

RUTER, J. M. 1997a. Evaluation of Tex-R Agroliners for bag-in-pot production. *Proc. Southern Nurserymen's Assoc. Res. Conf. 42:*162–64.

RUTER, J. M. 1997b. The practicality of pot-in-pot. *Amer. Nurseryman 8*(2):32–37.

RUTER, J. M. 1998. Production of 'Rotundiloba' sweetgum using Tex-R Agroliners. *Proc. Southern Nurserymen's Assoc. Res. Conf. 43:*59–61.

RUTER, J. M., and M. P. GARBER. 1998. *Measuring soluble salts and pH with the pour-through method.* University of Georgia Factsheet.

RUTER, J. M., and D. L. INGRAM. 1992. High root-zone temperatures influence RuBisCo activity and pigment accumulation in leaves of 'Rotundifolia' holly. *J. Am. Soc. Hort. Sci. 117*(1):154–57.

SMITH, E. M., and C. H. GILLIAM. 1979. *Fertilizers, Landscape & Field Grown Nursery Crops.* (pp. 1–12). Cooperative Extension Service, Ohio State University.

SPOMER, L. A. 1980. How container soils influence plant health. *Amer. Nurseryman 151*(12):8–9, 57–6.

SWANSON, B. T., J. B. CALKINS, and D. G. KRUEGER. 1992. *Container growing in the ground.* (pp. 58–59). University of Minnesota, University TRE Nursery Program.

SWANSON, B. T., J. DANIELS, J. Lewis, and J. CALKINS. 1988. *Winter protection of container nursery stock in Minnesota.* (Handout)

SWANSON, B. T., R. J. FULLERTON, and S. RAMER. 1989. Container fertilizers. *Amer. Nurseryman 169*(7):39, 41–49.

TROUSE, A. C. 1986. *Effect of hard–pans and soil compaction on tree longevity and production.* (pp. 87–91). 113th Annual Alabama State Horticulture Society Report.

WATSON, G. W., and T. D. SYNDOR. 1987. The effect of root pruning on root systems of nursery trees. *J. Arboric. 13:*126–130.

WELLS, J. S. 1976. Increasing the potential of container growing. *Amer. Nurseryman 143*(3):12–13, 52, 54, 56, 58–68.

CHAPTER

Landscape Trees: Deciduous, Broad- and Narrow-Leaved Evergreens

KEY LEARNING CONCEPTS

After reading the chapter you should be able to:

♦ Know the names and characteristics of many common deciduous trees.
♦ Understand how to choose and care for quality landscape trees.

TREE SELECTION

Planting a tree in a yard may not be as simple as you think it is. Unless you enjoy buying and replacing frequently, you should choose a species that will grow in the climatic conditions where you live. Undoubtedly you will have the exact site in mind, so the main decisions you have to make are to select the proper species and decide on what function the tree is to fulfill. Trees have many varied shapes and sizes. They have seasonal characteristics such as spring flowers, fall color, persistent leaves (evergreen), or dense foliage that provides shade. Some even have more than one desirable characteristic; for example, providing both shade and edible fruits, as do the English or Persian walnut and the pecan. Unfortunately these two lovely trees cannot be grown in all temperate-zone regions because they are hardy only in mild-winter areas (Table 16–1). If you live in the northernmost regions of the United States, you may have to settle for a hardy tree such as the Siberian crabapple to have a garden tree that also bears fruit. You must first choose a species that will survive the conditions of the region; perhaps the most important climatic condition is the severity of the winters.

A partial tree selection is given in Tables 16–1, 16–2, and 16–3, which list the deciduous, broad-leaved, and narrow-leaved evergreens. It is obvious that the selections based on the hardiness zones are more numerous in the southern states of the United States than in the north. Compare these tables with the hardiness zone map (www.ars-grin.gov/ars/Beltsville/na/hardzone/ushzmap.html) to make your choice.

TABLE 16–1   Some Common Deciduous Trees Grown in the United States and Southern Canada with Temperature Tolerances Indicated by Zones of Plant Hardiness Together with Some Characteristics and Possible Landscape Uses

Common and Latin Names	Plant Hardiness Zones[a]	Approximate Mature Height, Meters (Feet)	Shape of Crown	Growth Rate	Distinct Feature	Possible Landscape Uses
Alder, gray *Alnus incana* (L.) Moench.	3–8	15+ (50+)	Oval	Fast	Hardy	Screen
Alder, white *Alnus rhombifolia* Nutt.	5–9	15+ (50+)	Oval	Fast	Winter fruit	Shade
Ash, Arizona *Fraxinus velutina* Torr.	8–10	12+ (40+)	Pyramidal	Fast	Withstands stress	Shade
Ash, green *Fraxinus pennsylvanica* Marsh.	2–9	18+ (60+)	Oval	Medium	Fall color	Shade
Aspen, quaking *Populus tremuloides* Michx.	3–8	12 (40)	Rounded	Fast	Rustling leaves	Woodlot, shade
Beech, European *Fragus sylvatica* L.	5–10	20+ (60+)	Oval	Slow	Shiny	Shade, woodlot
Birch, river *Betula nigra* L.	4–9	15+ (50+)	Oval	Fast	Curled bark	Shade, specimen
Birch, European white *Betula pendula* Roth. (*B. verrucosa*)	3–9	12 (40)	Pyramidal (see photo)	Medium	Graceful	Groups in lawn
Carolina silver bell *Halesia carolina* L.	5–9	9 (30)	Oval	Medium	Flowers	Garden
Chestnut, Chinese *Castanea mollisima* Blume.	5–9	15+ (50+)	Rounded	Fast	Fruit	Shade, garden
Crabapple, Siberian *Malus baccata* (L.) Borkh.	3–8	12 (40)	Rounded	Medium	Flowers, fruit	Garden
Crape myrtle *Lagerstroemia indica* L.	7–9	9 (30)	Oval (Fig. 15–13)	Slow	Summer flowering	Garden, street tree
Cucumber tree *Magnolia acuminata* L.	5–10	15+ (50+)	Pyramidal	Fast	Flowers	Garden, shade
Dawn redwood *Metasequoia glystostroboides* H.H. Hu and Cheng.	6–10	15+ (50+)	Pyramidal	Fast	Foliage	Summer shade

(Continued)

Dogwood (*Cornus florida*).

Honeylocust (*Gleditisia triacanthos* var. *inermis*).

TABLE 16–1 (continued)

Common and Latin Names	Plant Hardiness Zones[a]	Approximate Mature Height, Meters (Feet)	Shape of Crown	Growth Rate	Distinct Feature	Possible Landscape Uses
Dogwood, flowering *Cornus florida* L.	5–9	12 (40)	Oval (see photo)	Medium	Flowers, fall color	Garden, woodlot
Dogwood, Pacific *Cornus nuttallii* Audub.	7–9	12 (40+)	Oval	Medium	Flowers, fall color	Garden, woodlot
Elm, Chinese *Ulmus parvifolia* Jacq.	6–10	15+ (50+)	Rounded	Fast	Shape	Shade, street tree
Fig, common *Ficus carica* L.	7–10	12 (40)	Rounded	Fast	Fruit	Garden
Hackberry, common *Celtis occidentalis* L.	3–9	15+ (50+)	Rounded (see photo)	Slow	Fall color	Street tree
Hawthorne, downy *Crataegus mollis* (Torr. & A. Gray) Scheele	5–9	9 (30)	Rounded	Medium	Fruit	Garden, woodlot
Hawthorne, lavalle *Crataegus × lavallei* Herincq.	7–10	6 (20)	Oval	Medium	Fruit	Garden
Honeylocust, thornless *Gleditsia triacanthos* var. *inermis* Willd.	5–9	15+ (50+)	Rounded	Fast	Fall color	Street tree, garden (see photo)
Horsechestnut *Aesculus hippocastanum* L.	5–7	15+ (50+)	Rounded	Medium	Foliage	Shade, woodlot
Jacaranda *Jacaranda acutifolia* Humb. and Bonpl.	9–10	12 (40)	Rounded	Medium	Graceful, flowers	Garden, street tree
Jerusalem Thorn *Pardsonia aculeata* L.	8–10	9 (30)	Oval	Fast	Withstands stress	Shade, patio
Larch, European *Larix decidua* Mill.	3–8	15+ (50+)	Pyramidal	Variable	Foliage	Large garden
Larch, Japanese *Larix kaempferi* (Lamb.) Carriere (*L. leptolepis*)	5–8	15+ (50+)	Pyramidal	Fast	Foliage	Large garden
Little-leaved linden *Tilia cordata* Mill.	3–9	15 (50)	Pyramidal	Medium	Shape	Shade, street tree
Magnolia, saucer *Magnolia × soulangiana* Soul-Bod.	5–10	8 (25)	Rounded	Slow	Spring flowers	Garden, shade, street tree
Maidenhair tree *Gingko biloba* L.	5–9	15+ (50+)	Pyramidal	Variable	Fall color	Street tree, garden
Maple, Japanese *Acer palmatum* Thunb.	6–9	6 (20)	Rounded	Slow	Leaf color, size	Garden, patio

Chinese pistache (*Pistacia chineasis*).

Weeping Willow (*Salix alba* var. *trista*).

TABLE 16–1 (continued)

Common and Latin Names	Plant Hardiness Zones[a]	Approximate Mature Height, Meters (Feet)	Shape of Crown	Growth Rate	Distinct Feature	Possible Landscape Uses
Maple, Norway *Acer platinoides* L.	4–9	15+ (50+)	Rounded (see photo)	Fast	Shape	Shade, street type
Maple, red *Acer rubrum* L.	4–9	15+ (50+)	Rounded	Fast	Fall color	Shade; wet, acid soil
Maple, sugar *Acer saccharum* Marsh.	4–9	15+ (50+)	Oval	Medium	Fall color, sugar	Shade, woodlot
Mountain ash *Sorbus americana* Marsh.	3–6	9 (30)	Rounded	Medium	Fall color, fruit	Garden, street tree
Mulberry, fruitless *Morus alba* L.	4–10	11 (35)	Rounded	Fast	Fall color	Shade, street tree
Oak, red *Quercus rubra* L.	4–9	15+ (50+)	Rounded	Fast	Fall color, shape	Shade, street tree
Oak, white *Quercus alba* L.	4–9	15+ (50+)	Rounded	Medium	Shape	Shade, street tree
Pear, Bradford *Pyrus calleryana* Decne. 'Bradford'	5–9	9 (30)	Pyramidal	Slow	Flowers, fruit	Garden, street tree
Pecan *Carya illinoinensis* (Wangenh.) C. Koch.	6–9	12 (40)	Rounded	Slow	Fruit	Shade, orchard
Persimmon *Diospyros virginiana* L.	5–9	12+ (40+)	Pyramidal	Medium	Fruit, fall color	Garden, woodlot
Pistache, Chinese *Pistacia chinensis* Bunge.	6–10	15+ (50+)	Rounded (see photo)	Fast	Fall color	Shade, street tree
Plane tree (sycamore) *Platanus × acerifolia* (Ait.) Willd	5–10	15+ (50+)	Rounded	Fast	Bark, fruit	Shade, street tree
Plum, purple leaf *Prunus cerasifera* J.F. Ehrh. 'Atropurpurea'	4–9	9 (30)	Rounded	Fast	Leaf color, shape	Garden, street tree
Poplar, Lombardy *Populus nigra* 'Italica' Muenchh.	2–10	15+ (50+)	Columnar (see photo)	Fast	Fall color, shape	Windbreak, garden
Redbud, eastern *Cercis canadensis* L.	5–9	12 (40)	Rounded	Medium	Withstands stress	Windbreak
Redbud, western *Cercis occidentalis* Torr.	6–9	6 (20)	Rounded	Medium	Flowers	Garden, woodlot
Russian olive *Elaeagnus angustifolia* L.	3–9	6 (20)	Rounded	Medium	Withstands stress	Windbreak
Service berry, downy *Amelanchier canadensis* (L.) Medic.	4–8	15+ (50+)	Oval	Fast	Flowers, fruit	Garden, woodlot
Silk tree *Albizia julibrissin* Durazz.	7–10	12 (40)	Spreading (see photo)	Fast	Graceful, flowers	Garden
Sumac, staghorn *Rhus typhina* L.	3–9	12 (40)	Rounded	Fast	Fall color	Garden, woodlot
Sweet gum, American *Liquidambar styraciflua* L.	6–10	15+ (50+)	Pyramidal	Medium	Fall color	Garden, street tree
Tulip tree *Liriodendron tulipifera* L.	5–9	15+ (50+)	Pyramidal	Fast	Flowers	Garden, street tree
Walnut, English or Persian *Juglans regia* L.	7–9	15+ (50+)	Rounded	Fast	Fruit	Large garden, woodlot
Willow, weeping (golden) *Salix alba* var. *trista* (Ser.) Gaudin	3–10	15+ (50+)	Weeping	Fast	Leaf color, graceful	Garden, along creek
Willow, weeping (Wisconsin) *Salix × blanda* Anderss.	5–10	12 (40)	Weeping (see photo)	Fast	Shape, graceful	Garden, along creek
Zelkova, Japanese *Zelkova serrata* (Thunb.) Mak.	6–10	15+ (50+)	Rounded	Fast	Shape	Shade, street tree

[a]See hardiness zone map on page 191 or on the Web at www.ars-grin.gov/ars/Beltsville/na/hardzone/ushzmap.html. Plants can be grown in the zones indicated (e.g., 3–8).

Note: For some species superior individuals have been selected and named as cultivars. Often one should select these chosen cultivars of the region for best overall results.

Common and Latin Names	Plant Hardiness Zones[a]	Approximate Mature Height, Meters (Feet)	Shape of Crown	Growth Rate	Distinct Feature	Possible Landscape Uses
Acacia, Bailey *Acacia baileyana* F.J. Muell.	9–10	9 (30)	Rounded	Fast	Flowers	Garden, tolerates poor soils
Ash, evergreen *Fraxinus uhdei* (Wenz) Lingelsh.	9–10	15+ (50+)	Rounded	Fast	Glossy leaves	Garden
Avocado *Persea gratissima* C.F. Gaertn. (*P. americana*)	9–10	9 (30)	Rounded	Medium	Fruit	Garden, grove
Buckthorn, Italian *Rhamnus alatemus* L.	7–10	8 (20)	Rounded	Fast	Shape	Hedge, windbreak
California bay (laurel) *Umbellularia californica* (Hook. and Arn.) Nutt.	7–10	15+ (50+)	Rounded	Slow	Leaves	Garden, woodlot
California pepper tree *Schinus molle* L.	9–10	12 (40)	Weeping	Fast	Shape	Garden, street tree
Camphor tree *Cinnamomum camphora* L.	9–10	15+ (50+)	Rounded (see photo)	Variable	Shiny leaves	Garden, shade
Eucalyptus, silver dollar *Eucalyptus polyanthemos* Schauer.	9–10	15+ (50+)	Columnar	Fast	Gray foliage	Garden, woodlot
Gum, red-flowered *Eucalyptus ficifolia* F. J. Muell.	10	11 (35)	Rounded	Medium	Flowers	Street tree
Holly, American *Ilex opaca* Ait.	6–9	12 (40)	Pyramidal	Slow	Fruit	Garden
Holly, English *Ilex aquifolium* L.	6–9	12 (40)	Oval	Slow	Foliage, fruit	Garden, patio
Iron bark, pink *Eucalyptus sideroxylon* A. Cunn. ex Woolls.	9–10	15+ (50+)	Weeping	Fast	Shape	Woodlot, windbreak
Laurel, English *Prunus laurocerasus* L.	7–10	9 (30)	Rounded	Fast	Shiny leaves	Garden, hedge
Laurel (sweet bay) *Laurus nobilis* L.	8–10	9 (30)	Oval	Slow	Foliage	Garden, street tree, hedge
Madrone *Arbutus menziesii* Pursh.	6–9	15+ (50+)	Oval	Variable	Bark	Woodlot, garden

Norway maple (*Acer platinoides*). *Source:* R.A.H. Legro.

Silk tree (*Albizia julibrissin*).

TABLE 16–2 (continued)

Common and Latin Names	Plant Hardiness Zones[a]	Approximate Mature Height, Meters (Feet)	Shape of Crown	Growth Rate	Distinct Feature	Possible Landscape Uses
Magnolia, southern *Magnolia grandiflora* L.	7–10	15+ (50+)	Oval	Medium	Foliage, flowers	Garden, street tree
Melaleuca, pink *Melaleuca nesophylla* F.J. Muell.	9–10	6 (20)	Irregular	Fast	Bark	Garden
Oak, California live *Quercus agrifolia* Née.	9–10	15+ (50+)	Rounded	Slow	Shape, foliage	Woodlot
Oak, Cork *Quercus suber* L.	8–10	15+ (50+)	Rounded	Medium	Bark	Street tree
Oak, southern live *Quercus virginiana* Mill.	8–10	15+ (50+)	Spreading	Slow	Shape	Woodlot
Olive *Olea europaea* L.	9–10	8 (25)	Rounded	Slow	Gray, foliage	Garden, patio
Orange *Citrus sinensis* (L.) Osbeck.	9–10	8 (25)	Rounded	Medium	Fruit, foliage	Garden, grove
Palm, Canary Island *Phoenix canariensis* Hort. ex. Chabaud.	9–10	15+ (50+)	Rounded	Variable	Leaves	Street tree
Palm, Coconut *Cocos nucifera* L.	10	15+ (50+)	Rounded	Variable	Leaves, fruit	Garden, street tree
Palm, Washington *Washingtonia filifera* (L. Linden) H. Wendl.	9–10	15+ (50+)	Rounded	Variable	Leaves	Street tree
Pear, evergreen *Pyrus kawakamii* Hayata	8–10	9 (30)	Rounded	Medium	Shape, leaves, flowers	Garden
Photinia, Chinese *Photinia serrulata* Lindl.	7–10	11 (35)	Rounded	Medium	Shiny leaves	Garden, windbreak as a large shrub
She-oak, drooping *Casurina stricta* Ait.	9–10	9 (30)	Oval	Fast	Drooping foliage	Woodlot, street tree
Weeping fig *Ficus benjamina* L.	10	6 (20)	Weeping	Medium	Graceful	Garden, patio

[a]See hardiness zone map on page 191 or on the Web at www.ars-grin.gov/ars/Beltsville/na/hardzone/ushzmap.html. Plants can be grown in zones indicated (e.g., 9–10).

Hackberry (*Celtis occidentalis*).

Camphor tree (*Cinnamomum camphora*).

TABLE 16–3	Selected Narrow-Leaved Evergreen Trees of the United States and Southern Canada with Temperature Tolerances Indicated by Plant Hardiness Zones Together with Some Characteristics and Possible Landscape Uses					
Common and Latin Names	Plant Hardiness Zones[a]	Approximate Mature Height, Meters (Feet)	Shape of Crown	Growth Rate	Distinct Feature	Possible Landscape Uses
Arborvitae, oriental *Platycladus orientalis* (L.) Franco (*Thuja orientalis*)	7–10	12 (40)	Pyramidal	Medium	Shape	Garden, hedge
Cedar, Deodar *Cedrus deodara* (D. Don) G. Don.	7–10	15+ (50+)	Pyramidal (see photo)	Fast	Shape	Garden
Cedar, eastern red *Juniperus virginiana* L.	3–9	15 (50)	Pyramidal	Slow	Shape	Garden, woodlot
Cypress, Italian *Cupressus sempervirens* L.	8–10	15+ (50+)	Columnar	Medium	Shape	Garden, windbreak
Cypress, Monterey *Cupressus macrocarpa* Hartweg.	8–10	15+ (50+)	Spreading	Medium	Shape	Garden, windbreak
Cypress, smooth Arizona *Cupressus glabra* Sudw.	7–10	12 (40)	Pyramidal	Fast	Shape	Garden, windbreak
Fir, Douglas *Pseudotsuga menziesii* (Mirb.) Franco	4–9	15+ (50+)	Pyramidal	Fast	Shape	Woodlot
Fir, white *Abies concolor* (Gord.) Lindl.	4–8	15+ (50+)	Pyramidal	Medium	Shape	Garden, screen
Pine, African fern *Podocarpus glacilior* Pilig.	8–10	15 (50)	Irregular	Medium	Graceful, leaves	Garden
Pine, Canary Island *Pinus canariensis* Sweet ex K. Spreng.	9–10	15+ (50+)	Pyramidal	Fast	Shape	Windbreak, shade
Pine, eastern white *Pinus strobus* L.	4–8	15+ (50+)	Irregular	Fast	Shape	Woodlot
Pine, Japanese black *Pinus thunbergiana* Franco	5–9	15+ (50+)	Irregular	Medium	Shape	Garden

Deodar cedar (*Cedrus deodara*).

Umbrella pine (*Pinus pinea*).

TABLE 16-3 (continued)

Common and Latin Names	Plant Hardiness Zones[a]	Approximate Mature Height, Meters (Feet)	Shape of Crown	Growth Rate	Distinct Feature	Possible Landscape Uses
Pine, Norfolk Island *Araucaria heterophylla* ('Salisb) Franco	10	15+ (50+)	Pyramidal	Medium	Shape	Garden
Pine, Scots *Pinus sylvestris* L.	3–8	15+ (50+)	Pyramidal	Medium	Shape	Garden
Pine, Swiss stone *Pinus cembra* L.	4–8	15 (50)	Rounded	Slow	Shape	Garden
Pinc, umbrella *Pinus pinea* L.	8–10	15+ (50+)	Rounded (see photo)	Medium	Shape	Street tree
Redwood, coast *Sequoia sempervirens* (D. Don.) Endl.	8–10	15+ (50+)	Pyramidal	Fast	Shape	Woodlot, large garden, shade
Spruce, Black Hills *Picea glauca densata* (Moench) Voss.	2–7	10 (30)	Pyramidal	Slow	Shape	Garden, screen
Spruce, Colorado *Picea pungens* Engelm.	3–8	15+ (50+)	Pyramidal	Slow	Shape, color	Garden
Yew, Irish *Taxus baccata* L. 'Fastigiata'	6–8	8 (25)	Columnar	Medium	Shape, fruit	Garden, hedge
Yew, Japanese Siebold. and Zucc.	4–7	8 (25)	Columnar	Slow	Shape	Garden, hedge

[a]See hardiness zone map on page 191 or on the Web at www.ars-grin.gov/ars/Beltsville/na/hardzone/ushzmap.html. Plants can grow in zones indicated (e.g., 7–10).

Lombardy poplar (*Populus nigra*).

Birch (*Betula pendula*).

Once you know the possible species, you have to choose the shape and ultimate size of the tree you wish to plant. The ultimate size is an important consideration because of the possible harm a tree can cause as it "outgrows" the space that is allotted to it. It can branch out in such a way that it will grow into utility lines, or interfere with your neighbor's or your own garden space or view, or cut off light penetration to your garden. Maples, which generally grow at a moderate rate, cast a dense shade under which very few plants can grow. The eucalyptus, poplar, mulberry, or plane tree all grow rapidly and may cause damage to sewers, overhead electrical wires, gutters, and roof tops. It is important to choose a species that will stay within the bounds available to the homeowner.

The tree's size and shape characteristics can be maintained or altered by pruning the tops. Pruning requires more effort than allowing the trees to grow naturally. Severe pruning can cause sprouting of "suckers" from the base or watersprouts in the crown. Perhaps the personal satisfaction of training a tree to grow as you desire may be worth the added effort.

Some trees cast a slight shade without shading out certain grass species; red fescue grass will dominate in a lawn under such conditions. You can have both shade and a lawn with the proper selection of tree and grass. Other trees sprawl and grow so large that they can only be used if a large landscape situation exists in which they can develop fully. The best garden trees do not grow too rapidly or too large and fulfill a certain function in the garden over a long period. Some species rarely require pruning. Some cities require a permit to plant a street tree in front of your own home and require that certain kinds be planted. Street trees are chosen primarily for the lowest maintenance cost to the municipality while still fulfilling the function of beauty or shade. Some trees are useful for producing a screen or windbreak; examples are boxelder, Russian olive, *Eucalyptus* spp., and Monterey cypress.

Trees are usually planted for the pleasure they give to the gardener or homeowner. One such pleasure is spring flowering. The experience of seeing a flowering dogwood, flowering crabapple, 'Bradford' pear, magnolia, or redbud at its prime is exciting. Flowering comes early or late in the spring season depending on the location, but one can almost count on it every year at the appearance of the first robin or crocus in the garden. Some of the flowers develop into fruit, which may also attract birds in the winter.

Some trees produce edible fruits and nuts, which may be harvested for home consumption. The pecan, walnut, orange, and crabapple are excellent large garden trees with the bonus of fruit. The possibility of planting shade trees that produce fruits for canning or eating at their peak of ripeness (e.g., peaches, apricots, plums, cherries, apples, etc.) should also be considered. These are candidates for the garden in certain zones. Some trees such as the mulberry, maidenhair (*Gingko*), and the olive, produce undesirable fruits that drop and cover the ground. Male gingko trees that bear no fruit can be purchased at the nursery. A fruitless olive cultivar, Swan Hill, which sheds very little pollen, is available. Fruitless mulberry trees are also widely planted but have strong root systems which can be destructive.

As the summer draws to a close, certain trees produce fall colors beyond imagination. The maples, oaks, ashes, poplars, dogwoods, pistachios, liquidambars, aspens, and the maidenhair tree are splendid examples. People travel long distances to New England and the Rockies in the United States to enjoy fall coloration. Trees with these characteristics are easily incorporated into the home landscape with the proper selection (Table 16–1). Some kinds of trees form excellent silhouettes against the snow in the winter; examples are the alder, beech, birch, plane tree, mountain ash, and many conifers.

ENVIRONMENTAL FACTORS TO BE CONSIDERED IN SELECTING SUITABLE TREE SPECIES

Temperature

The hardiness map and the zones of tree hardiness given in Tables 16–1, 16–2, and 16–3 are guides to selecting a tree for a given region. Small climatic pockets or islands within these broad zones protected by hills or large bodies of water permit the planting of trees less hardy than could be used otherwise. On the other hand, one should consult the closest botanical garden, the local nursery dealer, knowledgeable neighbors, or the local weather bureau to find out the lowest temperatures recorded in that area. This additional information will help determine whether the chosen species can survive an extremely cold winter. Will the species you have chosen survive an unusually severe winter? If the odds are not good that it will, it is better to select a hardier species. Look at what is planted in the neighborhood and see which large trees have survived. However, this is not always a true test of hardiness because a severe winter may only occur once in 20 or 30 years.

Spring frosts are also a problem. Almond, apricot, and some peach cultivars bloom early and the flowers may be injured in spring, eliminating the fruit crop. These late frosts prohibit growing such fruits commercially. Again, the local nursery dealer usually knows which fruit trees are susceptible to early spring frost damage.

Certain locations in the country have winters that are too warm for proper chilling of some fruit species, flowering trees, and shrubs, preventing them from blooming or leafing out properly in the spring.

Light

Most trees require full sunlight to establish themselves and grow well, although the maples and beeches may tolerate semishady locations. The light intensity on a sunny day is much greater than needed for growth, but in a dense crown or tree canopy, the light diminishes quickly toward the center. In general, a sunny location should be chosen for most tree sites. Young trees receiving inadequate light usually become spindly and weak and cannot compete with larger trees.

Some trees are affected by dim light from street lamps. Light shining all night, even though it is of low intensity, stimulates vegetative growth if the temperature is adequate. This stimulation delays the onset of fall dormancy and the branches affected by the light can be subject to winter killing. Some maples, birches, elms, dogwoods, and sycamores are affected by street lighting.

Moisture

The amount of rainfall is an important consideration in tree selection. In locations that lack summer rainfall, one should select a species that can adapt to moisture stress. Where rainfall is not plentiful, one should be prepared to irrigate as needed to supplement rainfall. Native plants from regions of low rainfall are excellent candidates for locations where dry summers occur.

Wind

Wind affects the transpiration of leaves, and therefore more water is required to sustain growth. A gentle breeze up to about 6 km (4 mi) per hour helps replenish CO_2 around the leaves, but a prolonged and higher velocity can deform the tree. Recent experimental work at the University of California has shown that excess movement of tree tops reduces shoot growth up to 25 percent. Some protection from high winds is required to establish the trees. Staking after planting may be necessary.

CATEGORIES OF NURSERY TREES AVAILABLE FOR PLANTING

Nursery trees for landscape plantings are available as bare-root, balled and burlapped (B&B), and as container-grown plants.

Bare-Root Trees

Trees sold bare-root are usually deciduous species with trunks less than 5 cm (2 in.) in diameter and with the soil removed. Conifer seedlings up to three years old are often sold as bare-root trees. Both kinds are grown in the field at a wholesale nursery, and are dug in late fall or early winter, after they become dormant and the leaves fall (for deciduous trees). After digging, the roots are pruned back and the trees may be packed in bundles with their roots in a plastic bag containing moist sawdust or may be individually prepackaged in plastic bags with moist sawdust. Some trees are sold in bulk lots and the roots are kept moist by plunging the roots in sawdust at the local nursery. Good examples of bare-root plants are budded fruit trees, which are sold for winter or early spring transplanting. Many deciduous landscape trees are also sold this way in winter. Be sure that the buds are alive and the root system has not dried out when selecting a tree.

Balled and Burlapped (B&B) Trees

B&B trees are usually evergreen, both narrow- and broad-leaved, grown in the nursery field until they reach a size suitable for sale; however, deciduous trees over a diameter of 5 cm (2 in.) may also be sold in this manner. The soil is cut away from the base of the plant at a prescribed distance from the trunk and a ball is formed (Fig. 16–1). The soil ball is wrapped securely in burlap so that it can be removed from the ground and transplanted in another location. B&B trees are available throughout the year. This common method of digging and transplanting evergreen trees is being replaced by container-grown trees. Care must be taken to see that the root ball is not broken and the top is not dried out or dying.

Container-Grown Trees

Some trees are grown to a suitable size in plastic or metal containers from small cuttings or seedlings. They may remain in the nursery for a year or more, depending on the time to reach a salable size. Some specimen container plants may have large rootballs which are housed in reinforced wooden boxes. This method of growing is best suited to locations with mild winters because the roots in containers must be kept from freezing. In areas with severe winters, the plants are usually protected by placing them inside small quonset-type polyethylene-covered greenhouses, which may or may not be heated.

Container-grown nursery stock offers several advantages:

1. There is a wide selection of plant material.
2. Nursery stock is available throughout the year for planting.
3. The plants are remarkably uniform.
4. The plants are economically grown and can be transported long distances at reasonable prices.

Container-grown plants do have some disadvantages, though. A well-grown plant with a large top may wilt severely in a small container if not irrigated frequently at the retail nursery. If the plant is grown too long in the container, the root system becomes deformed (Figs. 16–2, 16–3) and there may be difficulty in establishing the plant in the soil without proper root pruning. In addition, after planting contact between the porous soil medium in the container and a clay soil may be difficult to establish.

Figure 16–1 Balled and burlapped trees, showing how the root ball is tied to keep it intact. In some areas of the country, nails are woven into the burlap instead of using rope as shown. The ball and the tree should be handled carefully through the final stages of transplanting. If the ball is dropped, the soil will break away from the roots. The tree should be protected from the wind and direct sun to prevent desiccation. The soil ball should be kept moist.

Figure 16–2 Nursery trees held in containers too long will develop an undesirable circling root system. *Source:* Hartmann, H. T., and D. E. Kester. 1983. *Plant propagation: Principles and practices.* Fourth Edition. Englewood Cliffs, N.J.: Prentice-Hall.

Figure 16–3 Roots that have circled in the container continue to enlarge after planting, as has happened with this *Camellia japonica.* The roots are nearly the size of the trunk, which is 4 cm (2 in.) in diameter near the base. This problem is avoided by cutting selected roots to discourage circling at the time the tree is planted.

WHAT TO LOOK FOR WHEN PURCHASING CONTAINER-GROWN TREES

1. *Avoid plants with circling or kinking roots* (Fig. 16–2). These roots are sometimes seen near the top of the soil. Such roots can be troublesome in later years if they are not pruned properly before planting, because they may continue to circle and eventually girdle the trunk (Fig. 16–3) or the tree can be blown over in winds (Fig. 16–4). If such a plant is purchased (which is not recommended), portions of the root system can be cut to discourage the circling habit of the roots. As much as 50 percent of the roots can be removed without affecting tree growth in some species. Root pruning stimulates new roots to form after transplanting. Care should be taken after root pruning and planting to keep the soil moist until the plant is well established with an abundance of new roots (Fig. 16–5).

2. *Look at the size of the top.* The largest tree top is not always the best because of the small and confined root system. A large top may indicate that the tree has been in the nursery too long or that it should have been shifted to a larger container. A large top-to-root ratio is not good because the large leaf area usually transpires water faster than the roots can supply it, making top pruning necessary.

3. *Look at the taper of the trunk.* It should be larger at the soil level than it is a few feet above the base. A good taper indicates a strong trunk whereas a tree with a trunk with little or no taper may not be able to stand upright after the stake is removed (Fig. 16–6). Untapered trunks are usually seen on trees that are tightly staked. This staked trunk is unable to develop reaction wood (xylem), which is formed by continual gentle trunk movements in the breeze.

Figure 16–4   This tree with a weak root system was toppled in a wind storm. It was planted directly from the container without pruning or disturbing the roots. Before a tree is planted from a container, the circling roots should be cut to stimulate branching outward from the root ball.

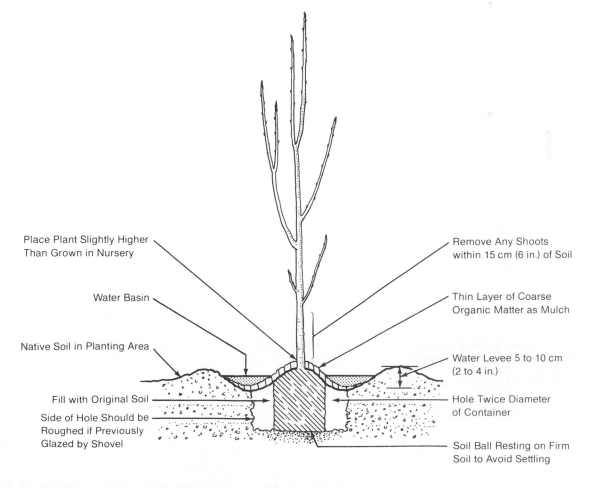

Figure 16–5   Correct way to plant a container-grown tree. *Source:* Modified from Univ. of Calif. Coop. Ext. Ser.

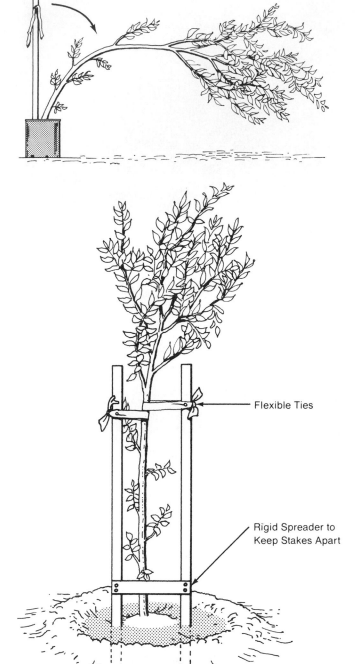

food to the lower trunk area. The crown (top branches) should be open to reduce wind resistance after transplanting. However, a large top may be pruned to a suitable size after transplanting.

5. *Look for a vigorous tree, free of disease, insect damage, or trunk injury.* Vigor is probably the most difficult factor to assess, requiring considerable experience. In general look for signs of good growth. Slow twig growth indicates poor vigor.

PLANTING THE TREE

Trees are best planted in spring when the soil is warming, but the air is still cool and the transpiration rate low. Under these conditions the roots will grow into the soil rapidly and provide ample water to the tops. Conifers and many container-grown deciduous and evergreen broadleafed trees can be successfully transplanted in cold areas (zones 5–6) in periods from mid-August to early October. In warm areas (zones 8 to 10) fall planting of many hardy trees is recommended to allow the root system to grow during the mild winters.

Bare-root trees are usually planted in midwinter in warm areas and in spring in cold areas.

Care of the Newly Purchased Tree

After purchasing the tree one should take good care of it before planting. Bare-root trees should be placed in the shade to keep the tops cool, and the roots should always be kept covered and moist. If the trees are not to be planted at once, the roots should be plunged (heeled-in) in moist sawdust, moist soil, or peat moss. Balled and burlapped trees should not be allowed to dry out. A leafy tree continues to transpire in the balled condition, and the soil can dry out quickly, depending on the **evapotranspiration** rate. Keep the tree in a protected location and tie it to keep from tipping (Fig. 16–1). Extreme care should be taken to avoid breaking the ball when transporting it, because the small roots can be broken and lose contact with the soil.

Container-grown plants are perhaps the easiest to maintain temporarily. Still, the plants should be kept in a shady, wind-protected area to reduce evapotranspiration. The containers should be well watered and should be protected from direct sun by shading them with aluminum foil or by plunging them in sawdust. The temperature of the exposed side of a black container can reach 49°C (120°F) in the direct sun. Either method of covering prevents excessive heat buildup on the periphery of the container—which is apt to kill the roots.

- Flexible Ties

- Rigid Spreader to Keep Stakes Apart

Figure 16–6 A tree that has been tied to a rigid stake (*above*) may be very weak when the tie is removed. After planting, such a tree should be staked to allow the trunk to be somewhat free to move and develop strengthening tissue (*below*). Small branches on the lower portion of the trunk higher than 15 cm (6 in.) should be allowed to remain until the tree is well established. The leaves of such shoots shade the trunk and supply food for good trunk growth.

4. *Look for a good branching habit.* The branches should be well distributed, and some small branches should still be attached low on the trunk (Fig. 16–6). The lower branches shade the trunk after transplanting and also supply

The Native Soil

A family usually purchases a home for its proximity to employment and because of the climate and often because it is what they can afford. Soils can range from fertile farm lands to rocky hillsides with a view. Little consideration is given to the garden soil when a family purchases their dream home. Soil problems of all kinds can exist, and it is best to observe the trees that grow best in adjacent neighborhoods with similar soils.

Sometimes soils are compacted by trucks during the building of the house. Such soil should be tilled with a rototiller or a spade below the zone of compaction. To this loosened soil, incorporate 10 cm (4 in.) of organic matter into the top 20 to 30 cm (8 to 12 in.) of soil or beyond the limits of compaction. It is then best to irrigate the soil and allow it to settle and dry out to a good **tilth** before planting the tree.

The soil should be probed to 1 m (3 ft) deep to find any natural impervious layer (**hardpan**), which may prevent root penetration, which could result in a shallow root system. Similarly, the impervious layer (**subsoil**) may not provide adequate drainage, and the roots will lack adequate oxygen. Layers such as these are often caused by an accumulation of cemented minerals (calcium, iron, or aluminum compounds) and can be penetrated if they are thin. A power soil auger or a posthole digger will penetrate layers up to 10 cm (4 in.) thick. These holes can be filled with amended soil or plant residues so that the roots can enter the open soil layers under the hardpan. Layers thicker than 10 cm (4 in.) are very difficult to penetrate without heavy equipment such as bulldozers. It is possible, however, to haul in a large amount of soil and make a mound at least 1 m deep on which to plant a tree. Often such a scheme can be fit into a landscape plan. In such cases, a **tile drain** should be placed above the impervious layer to drain away the excess water.

Preparing the Hole and Planting

It is to be hoped that the local soil is one that does not pose problems as illustrated above. The hole for B&B (Fig. 16–7) or a container-grown tree should be dug just deep enough to allow the soil ball to set firmly on the native soil and should be about twice the diameter of the ball. Usually when one is digging it is difficult to estimate both dimensions accurately. If the hole is too deep, backfill with soil and tamp it firmly with both feet until the ball can be placed in the hole slightly higher than the original ground line (Fig. 16–5). In heavy clay soils a glaze from the shovel may appear on the walls of the hole; this glaze should be roughened with a stick or shovel before backfilling to allow good root penetration from the soil ball.

Recent research has shown that it is not necessary to mix in any organic matter into the backfill soil before placing it around the soil ball. The native soil should be broken up with the shovel and gradually placed back and tamped in with a stick or gently with a foot. If the ball is sitting on firm soil, the ball will not settle with subsequent irrigations.

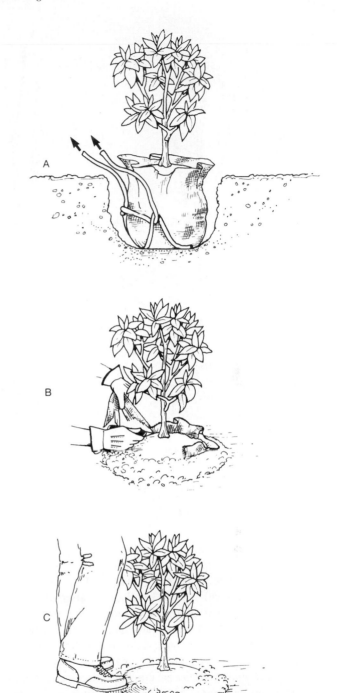

Figure 16–7 The ball of the tree should be placed on firm soil in the hole so that the ball will not settle lower than the intended planting depth. If the crown of the tree is below the final soil level, the trunk is subjected to decay organisms. To plant a balled and burlapped tree, (A) Place ball in hole carefully, taking care that it is oriented properly. Remove ropes and cut the burlap in many places or remove it if it has been treated to prevent decay. (B) If the burlap is decomposable, as most are, cut away the top portion so that no burlap remains above the final soil level. (C) Fill in the hole and tamp the soil firmly until the desired soil level is attained. Water the soil ball and replace soil.

Figure 16–8 Plant a bare-root tree in a hole large enough to accommodate the roots. Prune very long roots for ease in planting. Native soil taken from the hole should be broken up to make backfilling convenient. Firm the soil well around the roots by tamping it with a stick as it is replaced. If the soil is properly firmed, the first irrigation will not settle the soil significantly.

For bare-root trees, make a hole just large enough to accommodate the roots when spread out (Fig. 16–8). Prune broken or very long roots before planting. Gradually fill in the soil and tamp with a stick; finally, only after all the soil is well above the roots, apply a firm foot to press soil around the tree. Irrigate to settle the soil around the roots. Use the excess soil to make a berm (saucer) around the tree to contain water in a basin. Always plant the **crown** of the tree higher than the surrounding soil to ensure that water does not accumulate around the trunk. Avoid overwatering newly planted trees before the roots become established in the undisturbed soil. The reduced oxygen in a saturated soil discourages growth of new roots.

When a bare-root tree is planted in a lawn area, enough **sod** should be removed to make room for a basin that should be constructed around the tree. The turfgrass should be kept away from the crown for at least two years to allow the young tree to become established. The bare soil area should be about 75 cm (30 in.) across. Three upright stakes about 30 cm (12 in.) high, spaced 30 cm (12 in.) from the tree trunk, will reduce the risk of bark damage from lawn mowers. If the tree is staked, then these short stakes are not required.

Mulching the Soil Surface

All newly planted narrow- and broad-leaved evergreens benefit from organic **mulches** of wood chips, fir bark, pine needles, or sphagnum peat moss. A mulch layer 5 to 10 cm (2 to 4 in.) thick may be placed at the base of the tree, extending out to the edges of the longest branches. Bricks or redwood boards, which resist rotting, can be used to keep the mulch in place. The mulch eliminates most weeds, prevents the soil from caking in the hot sun, and moderates soil temperature and water loss near the surface. Water usually penetrates the soil more easily below a mulch.

Tree Staking and Trunk Protection

The newly planted tree may be staked, but it is not always necessary. The purpose of staking is threefold: to protect the trunk from mechanical damage by mowers or other equipment, to anchor the root system, and to support the tree temporarily in an upright position.

Large leafy B&B trees are usually anchored with guy wires. The top of the tree offers so much wind resistance that the small root ball cannot hold the tree from toppling. Excess movement keeps destroying any new roots. Three well-placed guy wires (Fig. 16–9) usually maintain stability and also keep machinery from running into the tree. When the roots have grown into the native soil, the supports should be removed, usually after one year's growth.

Many newly planted trees require trunk support because they have been staked too tightly or have been grown too close together in the nursery. Trees such as these usually have tops that are too heavy for the trunk. The tree supports should be placed as low as possible but still maintain the top in an upright position. This point is found with the method shown in Figure 16–10. The tie is flexible enough to allow some trunk movement. The best material is an elastic webbing that allows movement without abrading the trunk. If the top is leafy and allows much wind resistance, it should be thinned to reduce the wind load.

If the tree lacks small shading branches on the lower trunk, the trunk should be painted with white interior latex paint (diluted with water) or wrapped with strips of burlap. Painting, burlapping, or using paper wrap prevents sun scald of the trunk on bright days. If small branches are still growing along the entire length of the trunk (Fig. 16–6), those up to 15 cm (6 in.) above the soil should be removed, but the ones higher should be left. The small branches shade and supply food to the trunk; however, they should be pinched

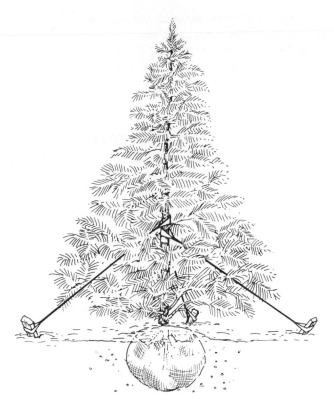

Figure 16–9   Three well-placed wires or plastic ropes will prevent a B&B conifer with dense foliage from toppling in the wind. Wire should be covered to avoid abrasion of the trunk.

Figure 16–10   *Left:* To determine the point at which a tree should be staked, grip the trunk at various heights and bend the top to determine at which point the trunk will spring back. *Right:* Tie the trunk a few cm above the point where the tree trunk returns to an upright position as shown in Figure 16–6.

regularly so they do not develop into large limbs. These lower laterals may be gradually thinned and removed over a two- to three-year period. A wire mesh encircling the tree to about 45 cm (18 in.) above the ground will protect the bark against rabbits and mice if needed.

The support system should be checked occasionally. The ties can be removed after one full growing season, but the trunk should again be checked for strength, as shown in Figure 16–10.

Fertilizing Trees after Planting

There is a chance of applying too much fertilizer at transplanting time, which would burn the roots; therefore, trees are seldom fertilized when planted. However, a *small* handful of nitrate-nitrogen fertilizer, which is readily soluble, can be applied on the soil surface and watered in directly after planting. Fertilizer should never be placed near the crown of the trunk. Trees planted in turf must be fertilized more often than those growing in bare soil (without weeds). Turfgrass strongly competes for mineral nutrients added to the soil, leaving little for the tree roots.

Irrigation after Planting

It is important to irrigate recently planted trees deeply to encourage the roots to grow downward. In dry, hot climates careful irrigation is necessary for the survival of the tree during the first summer. Watering with a soaker hose, which provides low volume over many hours, allows the water to penetrate deeply in many soil types.

Container-grown or B&B trees should be irrigated frequently so that there is always a moisture contact between soil ball and the backfill soil. This is especially true when the mix in the container is light and porous and the native soil tends to be clay. A bare-root tree, on the other hand, must be watered sparingly especially during the first winter (warm zones) or spring (cold zones). The soil can become waterlogged with frequent irrigations, which reduces soil oxygen and discourages new root growth or may encourage root diseases, especially near the crown of the tree.

Tree Care During the First Year—A Summary

The first year is the most difficult period in the life of the tree. New roots must develop to anchor the tree and develop feeder roots to supply the tree with water and mineral nutrients. Staking to keep the plant upright is necessary only for trees with weak trunks and for those planted in windy areas or for trees with large top:root ratios. Deep irrigation encourages deep rooting if the subsoil is similar to the top soil. Feeding the tree during the first year is usually unnecessary, since a small quantity of nutrients can be obtained from the native soil. Overfertilizing young trees the first

growing season can damage and seriously hinder root growth. Some **thinning** of the top may be necessary to reduce the wind load on the top. The trunks of some trees need protection from sun scald or rodents, depending on the environment.

THE CARE OF ESTABLISHED TREES

After the first full year, the strength of the trunk should be checked to determine if the support stakes are still necessary (Fig. 16–6). Usually stakes are not required after one year. With some very weak trees in the beginning, however, the ties should be adjusted and checked to see if any girdling is evident. Those trees may require two years' staking.

After the first year trees usually have to be fertilized to produce both good roots and branches. The best time to fertilize is shortly before the new burst of growth in the spring. As the soil warms, the roots absorb nutrients and make them available for the flush of growth. In all climates the fertilizer can be applied in late winter or early spring during the rainy season. Only nitrogen fertilizer is necessary for most kinds of trees. It can be applied to the soil surface and if it is in a readily soluble form (nitrate) it will leach into the root zone. The amount of nitrogen to apply is about 1.5 to 2.5 kg/100 m^2 (3 to 6 lb/1000 ft^2), and this amount will leach into the root zone. In regions of heavy rainfall, the amounts given above can be divided equally and applied as halves at one-month intervals.

Fertilizer pills of various sizes that can be placed in holes made around the tree have limited solubility and gradually release the nutrients as they slowly dissolve. The pills generally are complete fertilizers (containing nitrogen, phosphorus, and potassium) and since most landscape trees respond to nitrogen only, they are relatively ineffective for their high cost and therefore are not recommended.

Trees growing in lawns are fertilized by the two methods described or by lightly spreading nitrogen fertilizers over the grass surface. Rains and melting snows will leach much of the nitrogen into the root zone. Once the grasses become active and start growing, fertilizers applied will be taken up by grass roots and little, if any, will benefit the tree roots unless greater quantities are added and watered in heavily thereafter. An area at least 75 cm (30 in.) in diameter around the trunk of young trees should be kept free of turfgrass so that the tree can be occasionally fertilized in this bare surface.

PRUNING

Young deciduous or broad-leaved evergreens may have to be pruned after the first year. The purpose of tree pruning is to control the shape or the size, stimulate vigor, or, in the case of fruit trees, to affect the flowering and consequently the fruits which develop later.

The shape of the tree is directed early by selecting or choosing certain branches or buds to grow in a desired direction. The pruner can select one good leader branch and thin out badly placed branches so that a desirable, less-cluttered scaffold system remains (Fig. 16–11). An irregular tree form is sometimes desired, and no one single leader is then chosen. At this time branches with narrow, weak angles (Fig. 16–12) can be eliminated in favor of wide-angle branches. It is possible to prune certain trees so that there are multiple trunks and no single leader (Fig. 16–13). The crape myrtle is an excellent candidate for training in this fashion. A complicated form of pruning that requires much care is the espalier, in which fruit trees are trained against a fence or wall (Fig. 16–14).

The size and foliage density of shade trees may be controlled by pruning at least once a year, but more frequent

Figure 16–13   Crape-myrtle (*Lagerstroemia indica* L.), grown as a single- or multiple-stem tree. If pruned heavily early in the life of this tree, it can be grown as a shrub.

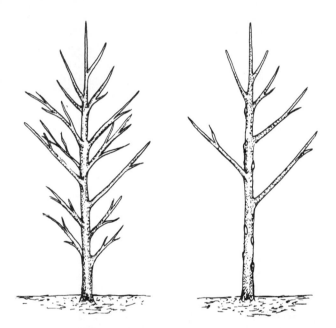

Figure 16–11   *Left:* A young established shade tree before pruning. *Right:* The same tree with those branches retained that will produce the scaffold branches of the tree. It is not a good idea just to prune the leader (topmost) branch to create low branching habits on an ornamental tree.

Figure 16–14   Pruning and maintaining a fruit or an ornamental tree so that it grows in a single plane is called espalier. Shown here are pear trees trained in two different forms. Very often trees are trained on a wall having adequate light. The pruning is done mainly in the winter when the branches are bare, but vigorous troublesome shoots may be removed any time to maintain the shape of the plant. Photographed at Wisley Gardens, Great Britain.

Figure 16–12   *Left:* Branches which have acute angles may split when the branch matures and (*center*): becomes heavy. *Right:* Narrow-angled branches should be removed and wide angled branches should be encouraged.

Figure 16–15 Cutting a tree back severely each year during the dormant season (pollarding) is a way of keeping potentially large trees small. This London plane tree is one of the many on Nob Hill in San Francisco, which are pollarded to keep their size down. The gnarled branches result from this severe pruning.

Figure 16–16 A type of topiary pruning in the Japanese section of Buchart Gardens, Vancouver Island, Canada. This *Chamaecyparis pisifera* Plumosa' often requires pruning during the growing season to remove the small tufts of growth from lateral buds.

Figure 16–17 Normally large trees such as black pine and junipers can be grown in small containers and pruned to reduce their size. These containers average 30 cm (12 in.) long, and the trees are 10 to 20 years old. They are known as bonsai trees.

pruning may be required for vigorous trees. Examples of such pruning control are creating a hedge from closely planted trees to produce a screen or pollarding trees (i.e., cutting them back severely each winter to let in more light in the winter but still to have shade in the summer). Sycamore and fruitless mulberry trees lend themselves to this severe pruning (Fig. 16–15). Because of their vigorous growth, the pruned trees still produce long leafy branches by midsummer. Controlled pruning may be done to shape the plant as animals, clumps, or odd forms as in topiary pruning (Fig. 16–16). The ultimate in controlling tree size is the Japanese treatment of bonsai (Fig. 16–17). These trees are maintained and kept in small containers for many years. Annual pruning of both roots and tops is necessary to maintain the bonsai form.

Even though pruning controls growth, it also invigorates individual shoots by reducing shoot number and opening the center to more light. Thinning operations produce a better top-to-root ratio so that water and nutrients are distributed more efficiently. The remaining shoots have less competition for light, water, and nutrients. In hot dry climates, less leaf surface remains after pruning thus reducing water loss by transpiration. The branch structure is strengthened by eliminating branches which produce narrow, weak angles (Fig. 16–12).

Pruning Methods

The two basic pruning methods of heading back and thinning out are discussed for fruit trees in Chapter 13.

Before pruning any tree, one should ask three simple but basic questions.

1. *What do I want to accomplish by pruning this tree?* Shall I head back the plant to keep it small? Shall I head it back to invigorate and strengthen it? Shall I reduce the flowering or fruiting potential? Just what is my goal?

2. *How will the tree respond to pruning?* Will it greatly alter the growth habit? Will severe pruning markedly reduce or stimulate excessive growth? Will many latent or **adventitious buds** be stimulated to sprout? Based on the normal habit of growth, is the method I am choosing for this particular tree suitable?

3. *How is it actually done?* Some basic pruning tools are necessary to make pruning easy. Bark splitting and ragged cuts should be avoided to allow for rapid healing.

   a. First remove any dead or dying branches or twigs. They usually have a gray lifeless (shriveled) appearance. Scratching through thin bark on small branches will show green tissue underneath on live branches.

   b. Thin out the cross-over branches or those that are growing toward the center of the crown.

   c. Remove narrow angle branches that weaken the tree (Fig. 16–12).

   d. Assess what you have done so far, then thin out the crown to give it the desired size and shape. At all times, keep in mind the natural characteristics and shape of the species.

   e. If not enough material is pruned this year, you can prune correctively next year after observing the resultant growth of this year's pruning operation.

Ornamental trees (and shrubs) should be chosen for their growth habits to fulfill the landscape function, assuming that they are known to be hardy for the region. The trees (and shrubs) chosen to meet the needs of the landscape usually do not require much pruning after the first or second year when the shape is determined. Some pruning may be necessary annually in the case of some flowering shrubs, rose bushes, or conifers.

Pruning can be done almost anytime of year except for such trees as the maples and the elms, which should be pruned in the late fall or early winter when the tree is not actively growing and the sap is not flowing. In areas with severe winters prune after the cold weather has passed. Generally, late summer pruning should be avoided in cold areas, since the new growth will not properly acclimatize and mature before an early frost.

Typical methods of cutting twigs or large branches with shears and saw are illustrated in Figures 16–18 and 16–19.

Narrow-leaved evergreens, or conifers, are shaped or kept within certain size limits by top pruning. Pine tops can be reduced in size (Figs. 16–20, 16–21) by nipping the "candles" of the current year's growth to about half size with the thumb and forefinger. The remaining portion develops a group of short needles. The best time to prune the "candles"

Figure 16–18 When small branches are removed, they should be cut with pruning shears to obtain clean-cut surfaces. The cutting blade is held close to the main branch so that very little stub remains.

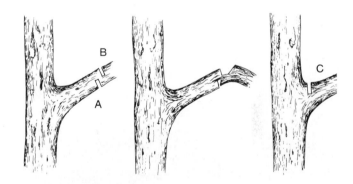

Figure 16–19 The method of cutting large branches to avoid splitting the trunk below the cut. *Left:* Make a cut on the under side at A. *Center:* Remove the branch at B, leaving a stub. *Right:* Remove the stub at C.

is in late spring when they have elongated but the needles are just beginning to grow. A new terminal bud will then form if the pruning is not done too late. Spruce and firs may also be pruned in a similar manner.

Junipers and other conifers without a strong radial symmetry or distant terminal buds may be sheared or irregularly pruned to form any shape or growth habit at most any time of year.

If the leader on a conifer is broken, generally a branch below begins to bend upright. However, one of the adjacent laterals can be encouraged to take over by splinting it upright (Fig. 16–22) until it can remain erect without the aid of the splint. One full growing season is ample for this corrective procedure.

Figure 16–20 Pine trees can be held to a reasonable size by pruning one-half to two-thirds of the new growth (candles) each year in the late spring, thus reducing the height or width of the tree by shortening the distances between whorls.

Figure 16–22 If the leader of a conifer is broken, a new one can be selected and encouraged to become the new leader with the aid of a splint (arrow) for one season. The splint may be removed as soon as it can be determined that the newly selected leader has become the dominant one.

Figure 16–21 Black pine (*Pinus thunbergiana*) candle showing the next year's cone above and last year's below. If cut at the top arrow, the new growth will be reduced by about one-third, and if cut at the lower arrow, growth will be reduced by two-thirds.

Container Growing of Evergreens

Some species of conifers are grown in large wooden tubs as living Christmas trees. Some spruces, firs, and the black pine are suitable for this type of culture. The trees must be irrigated properly to keep them from drying out and to keep salts from accumulating in the container. In irrigating any container-grown plants, it is better to irrigate very heavily at one time so that leaching (dripping) is thorough rather than irrigating frequently with small amounts, which do not allow for leaching. Thoroughly watered plants should be irrigated only when the soil surface becomes slightly dry. Some pruning of the candles of pines may be necessary to keep the top-root ratio reasonable. The root growth will be limited by the container so that the top should also be limited by pruning.

Growing bonsai plants provides an excellent hobby (Fig. 16–17). City dwellers who have balconies or large windows with a south or an east exposure might attempt growing them. During the winter in severe climates they must be taken indoors and protected. Because the roots are grown in shallow and small containers, watering can be a delicate operation. Care must be taken not to allow the soil to become too dry. Frequent checking of soil moisture is necessary for success with these plants.

Many insect and spider mite pests attack both deciduous and evergreen trees. Once the pest is known, the local county agent may be able to prescribe a control. Diseases are not easy to identify unless one can describe the exact symptoms to a plant pathologist knowledgeable in ornamental tree diseases.

SUMMARY AND REVIEW

Landscape trees are chosen for their aesthetic appeal and appropriateness for a site. Appropriateness includes the plant's ability to survive the climate or microclimate where it will be located. There are many genera of landscape trees and within

the genera are species with common characteristics and other characteristics that distinguish the species. All characteristics must be considered when selecting the proper tree for an area. The landscaper must also bear in mind that the tree may change dramatically over the course of its lifetime. These changes must be included in the landscape plan. Pruning and other maintenance are often required to maintain the desired appearance of a tree and prolong its useful life.

Caring for the tree starts with selecting healthy, well-grown and handled trees. Properly preparing the plant site is necessary to maintain the quality. The hole should be dug to the appropriate depth and width, and refilled in such a way that the roots will be encouraged to grow. During establishment proper irrigation is essential. Often the trees have to be staked until new roots can anchor the plant in place. After trees are established, a routine program of irrigation, fertilization, and pruning will keep the tree healthy and in the appropriate shape for its location and purpose.

EXERCISE FOR UNDERSTANDING

16–1. Drive around a local neighborhood or your campus. Observe the trees in the landscape. Do they appear to be in good shape? Are they appropriate for their site? Imagine how they have changed or will change with time. Imagine how their appearances change with the season. What is the most attractive thing about them? What is the least attractive?

REFERENCES AND SUPPLEMENTARY READING

DIRR, M. 1997. *Dirr's hardy trees and shrubs: An illustrated encyclopedia*. Portland, Ore.: Timber Press.

KRUSSMAN, G. 1982. *The manual of cultivated broad-leaved trees and shrubs* (in German). Translated into English by M. Epp. 1986. Vols. I, II, III. Portland, Ore.: Timber Press.

WYMAN, DONALD. 1990. *Trees for American Gardens*. Third Edition. New York: MacMillan.

CHAPTER

Ornamental Shrubs: Deciduous, Broad- and Narrow-Leaved Evergreens

KEY LEARNING CONCEPTS

After reading the chapter you should be able to:

♦ List the names and characteristics of many common landscape shrubs.

♦ Understand how to choose and care for quality landscape shrubs.

SHRUB SELECTION

Shrubs are low-growing woody plants, usually not exceeding 3 m (10 ft), with multiple stems arising from a low crown (Fig. 17–1). Trees, by contrast, usually have a single trunk. Some plants may be grown as either trees or shrubs depending on how they are trained when the plants are young. Crape myrtle, oleander, and Russian olive are examples. Each shrub species has distinct characteristics that may be useful in the garden. Shrubs can be used to accentuate a landscape, to hold soil on a bank from washing away, to act as a screen or windbreak, or to provide colorful blooms or berries. There are many species to choose from.

Because common names of shrubs vary more with location than do the names of trees or other plants, the Latin species names are given preference here. With the Latin name it is possible to communicate the identity of a shrub to the local nursery dealer or horticulturist. Horticultural and nursery catalogs usually carry both the Latin name and the common name used in the local region. Often the common name of a shrub can also be the name of the genus of

Figure 17–1   A large specimen of *Philadelphus coronarius* in bloom. This plant is an excellent candidate for the landscape in the northeast (Table 17–1). This specimen plant is about 1.5 m (5 ft) tall. *Source:* Robert A. H. Legro.

that shrub. For example, one of the forsythias is *Forsythia × intermedia* and the slender deutzia is *Deutzia gracilis*. Many flowering shrubs in Table 17–1 have similar Latin and common names.

When choosing a shrub species, one must first consider whether the plant is hardy in the region. The term *hardiness* refers mainly to the ability of the plant to withstand the lowest temperature of that region. The amount of rainfall in a region may also be important, particularly in drought-ridden areas like the arid southwestern United States. Intolerance to drought, however, can be easily overcome by irrigation. Thus the primary factor in hardiness is the minimum temperatures a shrub can withstand. Recently, the American Horticulture Society produced a heat map (see Appendix) that categorizes areas by summer temperature. This information is also useful for selecting plants.

A partial list of the best flowering shrubs for seven sections in the United States appears in Table 17–1. The climatic sections were determined by their lowest hardiness temperatures. At the borders between regions plants from either section could succeed. Flowering shrubs may be chosen from this table which fulfill certain requirements in the landscape for mature height, flower color, or other traits. In addition, a specific use may be chosen such as a single specimen for accent or for use in a border or as screen and hedge plants.

TABLE 17–1	A Selected Group of Flowering Shrubs for Seven Sections of the United States. These Were Chosen by Experienced Horticulturists for Dependability, Adaptability, and Attractiveness in the Landscape for Their Region*

Latin and Common Names	Approx. Mature Height, Meters (Feet)	Flowers	Other Features	Cultural Requirements	Useful Range (Zones see p. 191)
Northeast States					
Deutzia gracilis Siebold & Zucc. Slender deutzia	1.5 (5)	White, in clusters	Slender branches, pest free	Sunny location	5–8
Forsythia × intermedia Zab. Forsythia	3 (10)	Yellow	Pest free, possible winter kill	Prune for renewal	5–8
Kalmia latifolia L. Mountain laurel	3 (10)	Pink to white	Glossy leaves (Fig. 17–5)	Acid soil, sunny location	5–8
Philadelphus coronarius L. Mock orange	3 (10)	White, fragrant	Vigorous (Fig. 17–1)	Prune after flowering	4–7
Rhododendron catawabiense Michx., Catawaba rhododendron	3 (10)	Lilac-purple	Evergreen, dark green	Acid soil, semi-shade	4–7
Rhododendron pericylymenoides (*R. nudiform*) (Michx.) Shinn. Pinxterbloom	2 (6)	Pink, fragrant	Blue-green leaves	Acid soil, semi-shade	5–8
Spiraea × vanhouttei (C. Briot.) Zab., Bridal-wreath spiraea	2 (6)	White clusters	Arching branches	Sunny location	3–8
Syringa vulgaris L. Common lilac	4 (13)	White to purple	Fragrant	Prune after flowering	3–7
Viburnum carlesii Hemsl. Korean spice viburnum	2 (6)	White, fragrant	Black fruit	Remove suckers	4–8
					(Continued)

*Among the many species mentioned here, there are numerous cultivars in the trade that are far superior for the region. Those should be substituted over the original species whenever possible.

TABLE 17–1 (continued)

Latin and Common Names	Approx. Mature Height, Meters (Feet)	Flowers	Other Features	Cultural Requirements	Useful Range (Zones see p. 191)
		Southeast States			
Abelia × grandiflora (André) Rehd. Glossy abelia	2 (6)	Pale pink, fragrant	Evergreen, glossy leaves	Semi-shade, prune annually	6–7
Camellia japonica L. Common camellia	3 (10)	White to red, large	Glossy leaves	Acid soil, some shade	7–9
Chaenomeles speciosa Nakai (*C. lagenaria*) Flowering quince	2 (6)	Deep pink	Flowers on bare twigs, fruits	Easy culture	4–9
Gardenia jasminoides Ellis. Cape jasmine gardenia	2 (6)	White, fragrant	Glossy leaves	Acid soil, summer shade	8–9
Hydrangea quercifolia Bartr. Oak leaf hydrangea	2 (6)	White to pink	Fall color	Easy culture	6–9
Jasminum floridum Bunge. Flowering jasmine	1 (3)	Yellow, fragrant	Arching branches	Sunny location	7–9
Kalmia latifolia L. Mountain laurel	3 (10)	Pink to white	Glossy leaves (Fig. 17–5)	Acid soil, sunny location	5–9
Lagerstroemia indica L. Crape myrtle	6 (20)	White to pink to lavender	Unique stems and bark	Prune in winter or spring	7–9
Osmanthus fragrans (Thunb.) Lour. Tea olive, Sweet olive	3 (10)	White, fragrant	Evergreen; flowers add scent to tea	Easy culture	8–9
Rhododendron obtusum (Lindl.) Planch., Kurume or Kirishima azalea	1 (3)	White to pink to red	Evergreen, dwarf	Acid soil, semi-shade	7–9
		Midwest States			
Chaemomaeles japonica var. *Alpina* Maxim., Alpine Japanese quince	0.3 (1)	Orange	Dwarf	Easy culture	4–9
Cornus alba 'Sibirica' Loud. Siberian dogwood	3 (10)	White clusters	Blue fruits, red twigs	Easy culture	3–8
Cornus mas L., Cornelian cherry	5 (15)	Yellow clusters	Edible fruit	Easy culture	5–8
Cotoneaster multiflorus Bunge. Cotoneaster	2 (6)	White clusters	Red berries	Easy culture	4–7
Forsythia ovata Nakai Early forsythia	2 (6)	Yellow, small	Compact, hardy	Some thinning necessary	5–7
Potentilla fruticosa L. Shrubby cinquefoil	1 (4)	Yellow, white	Dwarf, fine leaves	Easy culture	2–7
Rosa setigera Michx. Prairie rose	1 (4)	Pink	Arching branches	Prune regularly	5–9
Syringa × persica L. Persian lilac	2 (6)	Lilac, fragrant	Glossy leaves	Thin after flowering	3–7
Viburnum dentatum L. Arrow wood	4 (13)	White clusters	Fall color	Easy culture	4–9
Viburnum lantana L. Wayfaring tree	4 (13)	White clusters	Fruit	Easy culture	4–7
		Rocky Mountain States			
Caragana aurantiaca Koehne Dwarf pea shrub	1 (3)	Orange-yellow	Drought hardy	Sunny location	3–6
Holodiscus dumosus (Nutt.) A. Heller. Bush rock spiraea	2 (6)	White	Drought hardy	Thin after flowering	6–7
Jamesia americana Torr & A. Gray. Cliff jamesia	2 (6)	Waxy, white	Fall color, flaking bark	Sunny location	5–6

TABLE 17–1 (continued)

Latin and Common Names	Approx. Mature Height, Meters (Feet)	Flowers	Other Features	Cultural Requirements	Useful Range (Zones see p. 191)
Kolkwitzia amabilis Graebn. Beautybush	2 (6)	Pink with yellow throat	Arching branches, flaking bark	Thin after flowering	6–8
Potentilla fruiticosa L. Shrubby cinquefoil	1 (4)	Yellow, white	Fine leaves	Easy culture	2–7
Prunus tomentosa Thunb. Nanking cherry	3 (10)	White, pink	Edible fruit	Easy culture	3–6
Rhodotypos scandens (Thunb.) Mak. Jetbead	2 (6)	White	Fruit, handsome foliage	Sheltered location	5–8
Ribes aureum Pursh. Golden currant	2 (6)	Yellow	Edible fruit	Easy culture	3–7
Rubus deliciosus Torr. Boulder raspberry	2 (6)	White	Thornless, arching stems	Sunny location	4–7
Spiraea trichocarpa Nakai. Korean spiraea	2 (6)	White	Hardy	Easy culture	3–8
Southwest States					
Acacia farnesiana, (L) Willd. Scented acacia	6 (20)	Yellow, fragrant	Minute leaves	Sunny location	9–10
Caesalpinia gilliesi (Wallich ex Hook) Benth. Bird of Paradise shrub	2 (6)	Large, yellow	Compound leaves	Sunny location	8–9
Cassia artemisioides, Gaud-Beaup. Silver caccia, wormwood	2 (6)	Yellow	Distinct in all seasons	Thin after flowering	8–9
Justicia ghiesbreghtiana Lemm. Jacobinia	1 (3)	Bright orange	Dwarf plant	Slight shade	7–9
Lagerstroemia indica L. Crape myrtle	6 (20)	White to pink	Unique stems and bark	Prune to train	7–9
Lantana camara 'Nivea' L. Lantana	1 (3)	White to bluish	Long blooming season	Sunny location, prune severely	9–10
Leucophyllum frutescens (Berland) I.M. Johnst., (*L. texanum*), Texas silver sage, Ceniza	1 (4)	Pink to lavender	Drought hardy, compact	Sunny location	6–8
Melaleuca hypericifolia Sm. Bottlebrush	3 (10)	Red, large	Weeping, drought hardy	Sunny location	8–10
Pyracantha coccinea M.J. Roem. Scarlet firethorn	2 (6)	White	Red berries	Easy culture	6–9
Tecoma stans var. *angustata* Redh. Trumpet bush, Yellowbells	5 (18)	Yellow, large	Spreading	Thin after flowering	8–9
Viburnum suspensum Lindl. Sandankwa viburnum	2 (6)	White to pink, fragrant	Evergreen, leathery	Easy culture	8–9
Pacific Coast, South					
Callistemon citrunus (Curtis), Staph., Lemon bottlebrush	5 (16)	Red	Evergreen	Requires some pruning	8–9
Carissa grandiflora (E.H. Mey) A.DC., Natal plum	2 (6)	White	Fruit, glossy leaves	Easy culture	9–10
Cassia artemisioides Gaud-Beaup. Wormwood, silver cassia	2 (6)	Yellow	Distinct in all seasons	Thin after flowering	8–9
Cestrum nocturnum L. Night blooming jasmine	3 (10)	Greenish, fragrant	Climbing	Easy culture	9–10
Choisya ternata HBK. Mexican orange	2 (6)	White, fragrant	Fragrance	Easy culture	9–10

(Continued)

TABLE 17–1   (continued)

Latin and Common Names	Approx. Mature Height, Meters (Feet)	Flowers	Other Features	Cultural Requirements	Useful Range (Zones see p. 191)
Hibiscus rosa-sinensis L. Chinese hibiscus	3 (10)	Rose-red, large	Dark green leaves	Sunny location, good drainage	9–10
Leptospermum scoparium J.R. Frost and G. Frost Tea tree, Manuka	2 (6)	White to red	Evergreen	Sunny location, tolerates dry soil	9–10
Nerium oleander L. Oleander	5 (15)	White, pink	Drought resistant, poisonous	Easy culture	8–10
Osmanthus fragrans (Thunb.) Lour. Sweet olive, Tea olive	3 (10)	Inconspicuous, fragrant	Flowers add scent to tea	Easy culture	8–9
Pittosporum tobira (Thunb.) Act. Japanese pittosporum	3 (10)	White, fragrant	Evergreen, red seeds	Easy culture	7–10
Pacific Coast, North					
Arctostaphylos columbiana Piper Hairy manzanita	3 (10)	Pink, white, urn-shaped	Mahogany stems, arborescent	Acid soil	8–9
Camellia × *williamsii* W.W. Sm. Camellia 'Donation'	3 (10)	Pink to rose	Evergreen	Acid soil, semi-shade	8–9
Ceanothus impressus Trel. California lilac, 'Puget blue'	3 (10)	Gentian blue	Evergreen, glossy leaves	Sunny location, dry conditions	8–9
Cytisus battandieri Maire. Atlas broom	2 (6)	Yellow, fragrant	Silvery gray foliage	Dry conditions	8–9
Eucryphia glutinosa Baill. Eucryphia	6 (20)	White, large	Glossy leaves, evergreen	Acid soil	8–9
Hamamelis mollis D. Oliver. Chinese witch hazel	5 (16)	Fragrant, yellow	Fall color	Sun or some shade	5–9
Mahonia aquifolium (Pursh) Nutt., Oregon grape	2 (6)	Yellow clusters	Fall color, blue berries	Fairly easy culture	5–9
Penstemon fruticosus (Pursh) Greene, Shrubby penstemon	0.3 (1)	Pink to lavender	Evergreen	Required winter rains	5–9
Rhododendron × 'Bow Bells' Rhododendron	1 (4)	Pink, large	Evergreen	Acid soil, some shade	6–7
Rhododendron luteum Sweet, Pontic azalea	2 (7)	Yellow, fragrant	Fall color	Acid soil, some shade	6–8

Source: Harkness, B. ed. 1975. *Handbook of flowering shrubs.* Brooklyn, N.Y.: Brooklyn Botanical Garden.

The partial list of shrubs in Table 17–2 aids in the choice of a shrub based on such characteristics as fragrant flowers or fall color. This list also indicates some shrubs that tolerate particular environmental conditions such as wet or dry soils, acid soils, or heavy shade. The functional use of some shrubs includes some best suited for banks or slopes, for windbreaks or hedges, or simply for attracting birds because they produce edible fruits. A shrub, if it is to be valued in the landscape, should have one or more aesthetic or functional characteristics. These favorable traits become evident by observation in given landscape situations throughout the year. Certain shrubs may be in fashion for several years but, after being used too often, they may become too commonplace and fall from favor.

Certain shrubs can give pleasure throughout most of the year by, for example, flowering in the spring, turning an attractive color in the fall, and displaying brightly colored berries on the bare twigs in the fall and winter. Many of the *Cornus* and *Viburnum* species as well as *Nandina domestica* have these unique and desirable characteristics.

Shrubs may have a functional use as a hedge or screen to eliminate an undesirable view or to reduce the force of the wind. Hedges or screens are usually comprised of a single species. They can be pruned formally to give a boxy appearance or may be allowed to grow freely. In the latter case, the shrubs may have to be pruned occasionally to hold them to a given size and to increase their vigor. In choosing a species for a hedge or screen, the ultimate size should be strongly considered. Indeed, in choosing any plant, the mature size is one of the most important considerations. If growth is slow for any reason, extra plants should be set close together to temporarily fill in the spaces.

Latin Name	U.S. Hardiness Zone (see p. 191)
TABLE 17–2 A Partial List of Ornamental Shrubs Characterized by Some of Their Environmental Tolerances and Landscape Uses	

Latin Name	U.S. Hardiness Zone (see p. 191)
Shrubs for Autumn Color	
Red Leaves	
Acer ginnala	2
Aronia arbutifolia	5
Berberis thunbergii	5
Cornus mas	4
Cotoneaster divaricata	5
Euonymus alata	3
Mahonia aquifolium	5
Nandina domestica	7
Nemopanthus mucronatus	7
Rhododendron calendulaceum	5
Rhus typhina	3
Ribes odoratum	4
Viburnum dentatum	4
Yellow Leaves	
Amelanchier spp.	5
Celastrus scandens	5
Hamamelis mollis	5
Magnolia stellata	5
Shrubs with Fragrant Leaves	
Artemisia spp.	2–5
Juniperus spp.	2–5
Laurus nobilis	6
Lavandula officinalis	5
Myrica spp.	2–7
Myrtus communis	9
Rhus aromatica	3
Rosmarinus officinalis	9
Santolina chamaecyparissus	7
Shrubs with Fragrant Flowers	
Abelia × grandiflora	5
Carpenteria californica	8
Choisya ternata	7
Daphne odora	7
Gardenia jasminoides	9
Jasminum officinale	7
Lonicera spp.	3–7
Osmanthus illicifolius	6
Philadelphus coronarius	4
Philadelphus cymosus 'Conquete'	5
Philadelphus lemoinei 'Avalanche'	5
Rosa, most spp.	4–8
Rubus odoratus	3
Syringa vulgaris	3
Viburnum burkwoodi	5
Viburnum carlesii	4

Latin Name	U.S. Hardiness Zone (see p. 191)
TABLE 17–2 (continued)	
Viburnum fragrans	5
Viburnum odoratissimum	9
Shrubs with Bird-Attracting Fruit	
Amelanchier grandiflora	4
Cornus alba 'Sibirica'	2
Cornus stolonifera	2
Cotoneaster spp.	4–7
Feijoa sellowiana	8
Heteromeles arbutifolia	9
Ilex verticillata	3
Lonicera maackii	2
Lonicera tatarica	3
Mahonia aquifolium	5
Malus sargentii	5
Myrica californica	7
Myrica pennsylvanica	2
Ochna serrulata	8
Pyracantha coccinea	6
Rhus aromatica	3
Rhus glabra	2
Rosa multiflora	5
Sambucus coerulea	5
Sambucus canadensis	3
Symphoricarpus chonaultii	5
Vaccinum corymbosum	3
Viburnum dentatum	2
Viburnum lentago	2
Viburnum prunifolium	3
Viburnum rufidulum	5
Dwarf Shrubs (1 m or Less)	
Andromeda polifolia	2
Arctostaphylos uva-ursi	2
Berberis buxifolia var. nana	5
Buxus microphylla 'Koreana'	4
Buxus sempervirens 'suffruiticosa'	5
Calluna vulgaris	4
Cotoneaster horizontalis	4
Daphne cneorum	4
Daphne giraldii	6
Erica spp.	3–6
Euonymus alata 'Compacta'	3
Forsythia 'Arnold Dwarf'	5
Gaultheria procumbens	5
Hypericum calycinum	5
Ilex crenata 'Helleri'	6
Juniperus chinensis var. sargentii	4
Juniperus horizontalis	2
Kalmia angustifolia	4
Mahonia aquifolium 'Compacta'	5
Picea abies cultivars	2
Pinus mugho	2
Rhododendron obtusum	6
Rhododendron racemosum	5
Rosa wichuraiana	5
(Continued)	

TABLE 17–2   (continued)

Latin Name	U.S. Hardiness Zone (see p. 191)
Ruscus aculeatus	7
Salix tristis	2

Shrubs Tolerant of Some Shade	
Abelia × grandiflora	5
Alnus spp.	2
Amelanchier spp.	4
Berberis thunbergii	5
Camellia japonica	7
Cornus mas	4
Hydrangea arborescens	4
Kalmia latifolia	5
Laurus nobilis	6
Mahonia spp.	5
Nandina domestica	7
Pachysandra terminalis	5
Philadelphus coronarius	4
Pittosporum tobira	8
Rubus odoratus	3
Viburnum dentatum	2
Viburnum tinus	7
Vinca minor	4

Shrubs Tolerant of Moist to Wet Soils	
Alnus spp.	2
Clethra alnifolia	3
Cornus alba	2
Cornus stolonifera	2
Rubus odoratus	3
Salix caprea	4
Salix purpurea	4
Sambucus canadensis	3
Thuja occidentalis cultivars	2
Viburnum dentatum	2

Shrubs Tolerant of Dry Soils	
Arctostaphylos uva-ursi	2
Artemisia spp.	2–5
Callistemon lanceolatus	9
Ceanothus americanus	4
Ceanothus thyrsiflorus	8
Cytisus spp.	5
Elaeagnus angustifolia	2
Euonymus japonica	8
Forsythia ovata	5
Heteromeles arbutifolia	9
Hypericum calycinum	5
Juniperus communis	2
Juniperus horizontalis	2
Kolkwitzia amabilis	4
Nerium oleander	8
Rhus aromatica	3
Rosa rugosa	2
Rosmarinus officinalis	6
Santolina chamaecyparissus	7

TABLE 17–2   (continued)

Latin Name	U.S. Hardiness Zone (see p. 191)
Tamarix pentandra	2
Viburnum lentago	2

Shrubs Tolerating or Requiring Acid Soils	
Amelanchier spp.	4
Calluna spp.	4
Camelia japonica	7
Cytisus spp.	5–6
Erica spp.	5–7
Ilex spp.	3–6
Juniperus communis	2
Kalmia spp.	4
Magnolia virginiana	5
Mahonia aquifolium	5
Myrica pennsylvanica	2
Rhododendron spp.	2–6
Salix uva-ursi	1
Vaccinium spp.	2–3

Shrubs for Banks or Slopes	
Arctostaphylos uva-ursi	2
Berberis thunbergii	5
Ceanothus americanus	4
Cornus sericea (C. stolonifera)	2
Cotoneaster horizontalis	4
Hypericum calycinum	5
Juniperus chinensis var. sargentii	4
Juniperus horizontalis	2
Lantana camara	9
Pachysandra terminalis	5
Rhus aromatica	3
Rosa rugosa	2
Vinca minor	4

Screen and Windbreak Shrubs	
Acer campestre	4
Acer ginnala	2
Cornus mas	4
Elaeagnus angustifolia	2
Kolkwitzia amabilis	4
Laurus nobilis	6
Ligustrum spp.	3–7
Philadelphus coronarius	4
Photinia spp.	7
Prunus laurocerasus	6
Syringa vulgaris	3
Thuja occidentalis	2
Viburnum dentatum	4
Viburnum lentago	2

Hedges can be planted in trenches, which makes working the soil much easier than digging individual holes. The plants can be planted in a single row or in double rows in a zigzag fashion. The spacing between plants depends upon how vigorous the species is and how soon the hedge is required.

A border in the garden can be both beautiful and functional. The shrub border can be used as a screen, and various species within the border can be mixed to create interest. A border can also vary in height, with the tallest plants in the back or along the boundary line. The border can be a foreground for large trees or a background for bedding plants or bulbs. This height is usually ample to provide a background or screen for privacy unless a tall structure on the adjacent property requires additional screening.

The border may follow an irregular curved pattern to create interest. A flowering border can be staged in such a manner to create color much of the spring and summer by choosing species of varying sizes and flowering dates.

THE AVAILABILITY OF SHRUBS

Most nurseries carry the common hardy shrubs of the region, and usually the widest choice is available in the spring of the year. Shrubs are available for purchase as bare-root plants, balled and burlapped plants, and container-grown stock. The planting methods described for trees of each of these types also pertain to shrubs of the same type. However, because the growth habit of a shrub is different from that of a tree (i.e., multiple vs. a single trunk), the method of pruning both at the time of planting and subsequent maintenance is different for these two plant forms.

PRUNING SHRUBS

Since shrubs grow differently from trees, they are pruned differently. Shrubs are best pruned just before planting. In the case of bare-root stock, broken roots should be removed and extra long roots should be shortened to facilitate planting. If the top appears too large for the root system, the branches should be cut back to compensate for the root loss.

Container-grown plants often have roots that are circling in the container. Such plants should be rejected but, if used, the roots should be cut in several places to break the circling pattern and to stimulate new growth. The tops of broad-leaved evergreens often need to be pruned to reduce the wind load and to reduce the transpiration surface. Remove about one-quarter of the leaf area by pruning immediately after planting. The tops of balled and burlapped shrubs should also be pruned for the same reasons, but the roots should not be disturbed other than loosening the burlap and cutting the ropes.

Once the shrubs are established in the landscape, maintenance pruning serves to control their size and shape by removing some stems from the multiple-stemmed crown (Fig. 17–2) to improve the shrubs' health, and to affect flowering. Some shrub species require more pruning than others because of differences in vigor and flowering habit.

Plant size is reduced by thinning out some of the branches. The tips of the branches may be headed back to in-

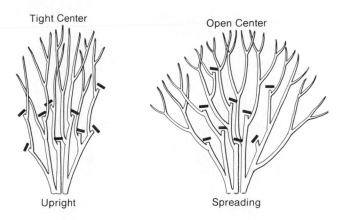

Figure 17–2   Two ways to prune a shrub to create different effects. The cuts are indicated by the black bars. *Source:* Adapted from USDA Home and Garden Bulletin 165.

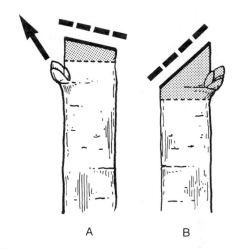

Figure 17–3   Twigs cut at too sharp an angle (B) may die back to the point that will also cause the bud to die. It is better to cut straight across (or nearly so), well above the bud desired for growth (A). That stub will only die back to a point where it will not interfere with bud growth (shown by arrow).

crease peripheral branching if a hedge or formal effect is desired. Thinning will help maintain the natural form and shape of the shrub. In thinning shoots, the wood should be cut above a bud that will grow in the desired direction (Fig. 17–3). Selecting the buds to remain allows partial control of growth even in an informal and natural setting of a rural home.

Shrubs may be thinned heavily by cutting some of the old branches back to the crown near ground level. The taller and older branches (i.e., those of greater diameter) should be cut out first, but not more than one-third of the branches should be cut out in any one year unless complete rejuvenation is desired. Old shrubs may be rejuvenated as shown in Figure 17–4 by cutting back the entire plant. After many new shoots appear, weak ones are pruned out to leave only the strongest and most desirable shoots to form the shrub.

CORRECT

Old Shrubs on Own Roots   Grafted Shrubs

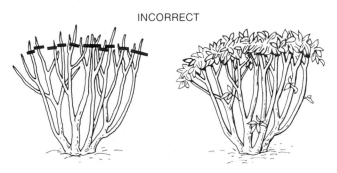

Trim Shoots to
Open Plant

Remove Shoots

INCORRECT

Figure 17–4   Rejuvenating old shrubs. *Upper left:* In some cases it is best to cut back severely (black bars) and develop a new top. *Upper right:* With grafted shrubs, do not cut back below the graft union (knobby growth at bottom). Remove suckers from the rootstock (lines) or thin shoots to open up the plant. *Below:* Cutting or trimming the tops of old shrubs, as indicated by the broken line *(left)*, produces an unsightly leggy plant with a bushy top *(right). Source:* Adapted from USDA Home and Garden Bulletin 165.

Young hedges are pruned gradually by cutting about six to eight inches above the previous cut. Mature hedges that have attained the desired size require frequent trimming to maintain their shape. The shape of the hedge should be slightly tapered, with the base wider than the top. This method of pruning allows light to reach the lower leaves and keep them alive. Keeping a hedge in a formal shape once it has attained the desired size may require as many as four clippings during the growing season. Pruning frequency of a hedge depends on plant vigor, light, temperature, and the water available to sustain vigorous growth.

Diseased wood is pruned out to keep the plant healthy. Diseased branches are cut back to nondiseased wood so that a healthy sprout will grow. This type of pruning should be done when the disease is first noticed, especially in the case of fireblight of pyracantha, cotoneaster, quince, and other members of the rose family.

One of the best reasons for pruning is to affect flowering. Some shrubs initiate their flower buds the summer be-

Figure 17–5   *Kalmia latifolia* flowers are prized spring blooms. *Source:* Robert A. H. Legro.

fore flowering (one-year-old wood) and thus should be pruned at a different season than those that flower on current year's wood.

Flowering on One-Year-Old Wood

Many fruit trees and shrubs initiate their flowers in summer, and the flower buds bloom the following spring when the weather becomes favorable (see Ch. 14). These species are called spring-flowering shrubs. Examples are *Cercis,* dogwood (*Cornus*), *Kalmia* (Fig. 17–5), lilac (*Syringa*), *Pyracantha, Rhododendron* (azalea), *Spiraea,* and *Viburnum.* Severely pruning the branches that have the flower buds in winter or early spring cuts away the potential flowers. One should prune such spring-flowering shrubs *soon after* they have flowered (Fig. 17–6). Removing the old flowers and shoots stimulates the plant to branch and produce additional flowering wood that summer for next year. This flowering behavior is usually identified by observing that the shrub flowers before the leaves appear, such as the flowering quince, flowering peach, forsythia, or *Chimonanthus praecox* (Fig. 17–7). Some species like the lilac produce large flower clusters (inflorescences) instead of single blooms, but these too are borne on the previous year's shoots.

Flowering on Current Year's Wood

The branches of shrubs that flower on the current year's growth (sometimes called summer-flowering shrubs) should be pruned back in late winter or early spring before the buds have sprouted (Fig. 17–8). The resulting shoots have the capability of producing flower buds on the new growth. Some

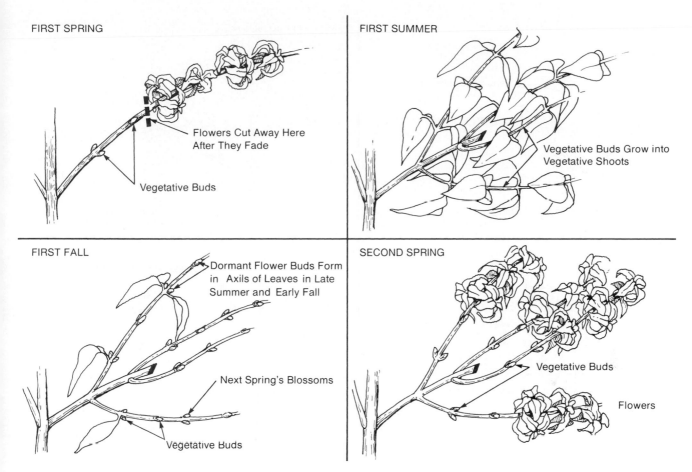

FIRST SPRING

Flowers Cut Away Here
After They Fade

Vegetative Buds

FIRST SUMMER

Vegetative Buds Grow into
Vegetative Shoots

FIRST FALL

Dormant Flower Buds Form
in Axils of Leaves in Late
Summer and Early Fall

Next Spring's Blossoms

Vegetative Buds

SECOND SPRING

Vegetative Buds

Flowers

Figure 17–6 Spring-flowering shrubs and trees should be pruned after they have flowered and as vegetative growth begins. Unfortunately, many gardeners and homeowners prune these plants in the winter when the twigs are bare, and consequently they cut away the flower buds. Before pruning, study the flowering behavior of the shrub or tree.

Figure 17–7 *Chimonanthus praecox* flowers in the spring before the leaves appear. The flowers were initiated the summer before, and they received their necessary chilling during the winter. *Source:* Robert A. H. Legro.

examples of summer-flowering species are *Abelia, Clethra,* crape myrtle, *Hibiscus, Nerium* (oleander), roses (floribunda and hybrid tea), and *Vitex.* This type of flowering is identified in some species by shoots that have light green leaves below the flower stalks. Flowers may also appear in several flushes or cycles during the growing season.

Pruning Nonflowering Shrubs

Shrubs not grown for flowers or fruits are pruned almost anytime of year except late summer. Late pruning often stimulates buds to sprout. Such shoots do not have adequate time to mature or "harden" before the onset of winter, leaving them susceptible to winter injury.

Shrubs, especially those grown as specimen plants, should be given ample space to develop properly. Most shrubs usually spread out as they grow taller. It is tempting to plant small shrubs close together to get the desired effect hurriedly. However, it usually becomes necessary to remove alternate plants when they begin to touch. Crowded plants do not develop the grace and beauty of those grown at

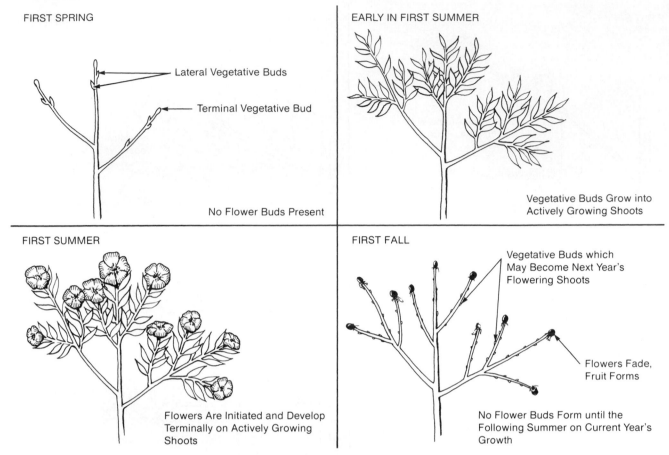

FIRST SPRING

Lateral Vegetative Buds

Terminal Vegetative Bud

No Flower Buds Present

EARLY IN FIRST SUMMER

Vegetative Buds Grow into
Actively Growing Shoots

FIRST SUMMER

Flowers Are Initiated and Develop
Terminally on Actively Growing
Shoots

FIRST FALL

Vegetative Buds which
May Become Next Year's
Flowering Shoots

Flowers Fade,
Fruit Forms

No Flower Buds Form until the
Following Summer on Current Year's
Growth

Figure 17–8   Summer-flowering shrubs and trees can be pruned either in the fall or in the winter when the twigs are bare. The flowers form and develop on current second growth that originates from vegetative buds on the previous season's branches.

recommended distances. Shrubs planted next to the house should be set far enough away to keep them from crowding into the building as they mature.

PRUNING ROSE PLANTS

Hybrid Tea Roses

Most roses are planted bare-root in late winter or early spring depending on the hardiness zone. Generally these bare-root plants need no top pruning at the time of planting, but any broken roots should be pruned back behind the break. Roses are planted "high" like bare-root trees. Low planting results in the development of crown rot diseases.

When the roots begin to grow and the weather warms, new shoots develop from the dormant canes. All leafy shoots are allowed to develop. In the very early stages (when the shoot and leaves are reddish), new growth primarily depends on stored food in the canes for energy. After the leaves turn green, they produce photosynthates that are translocated to the roots.

Hybrid tea roses flower on current year's growth and, therefore, can be pruned anytime. The only pruning before the first flush of flowers should be to remove all lateral flower buds, leaving the terminal bud on each shoot. Remove the lateral buds as soon as possible so that the carbohydrates concentrate in the terminal buds. The remaining buds will produce larger flowers than if the lateral buds were not removed. Allow all the terminal buds to bloom and develop fully in the first spring flush. Do not remove them until they start to drop their petals. This allows the rose bush to translocate most of the food from the mature green leaves to the roots on each flowering shoot. After this, prune away the fading flowers to just above a good bud located about midway from the original source of the cane and the flower. Cut back to a bud that is growing outward from the center of the bush. This allows the rose to spread out more than if inwardly growing buds are chosen. The rule in pruning away old flowers is to leave at least three or more compound leaves (and buds, which are in the axils of these leaves) on the cane. The leaves that remain produce the food for the new flush of shoot growth that originates from at least one of the buds. During the second or third flush of growth (usually by mid-

or late summer) some flower buds can be removed for cut flowers in the home. By then the bush is well established and flowers plus valuable leaves can be removed. Buds that have just begun to unfurl one or two petals should be cut to the desired stemlength, but always leave at least three leaves on the plant below the cut. The cut flower buds have enough sugars in the stems and leaves to permit them to develop into full-sized blooms. A flower preservative solution[1] keeps the flowers fresh longer than plain water alone.

In cold-winter areas where rose bushes must be covered for protection, late pruning reduces the size of the bush, making it easier to cover and protect than a large plant. After the leaves have dropped in the fall and the plant appears dormant, the tops are cut back to two-thirds or one-half of their length. Do not prune back into large-diameter wood (larger than 1.5 cm) at this time unless it is old and requires removal. Directly after pruning, the bushes are covered and wrapped with protective material such as straw, plastic sheeting, or building paper. Remove the protection in the spring and examine the canes to determine if any have been winter-killed. Remove the injured canes and cut back to healthy wood. In addition, remove small, weak, and undesirable canes.

In mild climates, fall pruning to reduce plant size is unnecessary. All pruning is best done in winter after the plants have lost their leaves and are dormant. Remove diseased and weak wood and cut to buds that will direct the plant in the desired shape. Some thinning is required (about one-half of the wood) to maintain an open center and reduce cane competition. Old woody canes should be removed as younger, more vigorous shoots are clearly ready to take their place (Fig. 17–9).

Floribunda and Polyantha Roses

These ever-blooming roses are pruned similarly to hybrid tea roses with two exceptions.

First, floribundas and polyanthas produce multiple flower heads, thus only the central flower bud in each cluster should be removed. This allows the remaining buds to bloom more evenly than if the central bud is left to develop. After flowering, the faded flower cluster is removed to stimulate the potential buds to sprout and produce another flush of flowers.

Second, only one-quarter of last year's growth should be cut out instead of up to one-half as with hybrid teas. If growth in the previous year was vigorous, then up to one-half of the wood is pruned. Canes over two or three years old near the base may not approach the girth of the vigorous hybrid teas. Therefore, it may be necessary to allow canes of 0.5 cm (0.25 in.) or less diameter to remain to produce next year's shoots. Thinning out some canes to remove competition, however, invigorates the bushes.

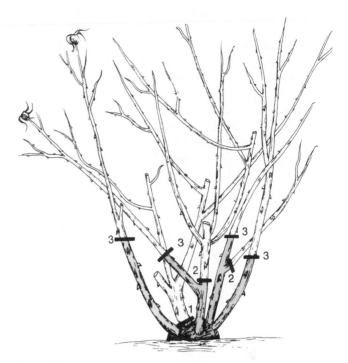

Figure 17–9 Some of the important operations in pruning a hybrid tea rose in the garden. *1:* Cut out all weak and dead wood. *2:* Prune and cut out all but three to six vigorous canes. *3:* Cut back those selected canes to vigorous buds as shown at places marked "3."

Figure 17–10 A climbing or rambling rose with canes that lie in a horizontal position often produce flower stalks at each node. Properly trained canes such as this respond with an abundance of flowers.

Climbing Roses

Climbing or rambling roses flower best on canes that are one to three years old. Flowers originate from laterals on these canes. Care is taken to cut out canes only three or more years old after flowering. During the first few years after planting, the canes are trained on a trellis, fence, or wall with an adequate support system to prevent whipping and wind damage. Canes often attain a length of 2.5 to 3 m (8 to 10 ft) and should be trained horizontally or in an arch. Trained canes produce flowers at each node (Fig. 17–10)

---

[1]Many proprietary materials are available at florists' shops.

because the **apical dominance** of the growing tip is reduced, leaving all nodes in the horizontal plane with the capability of flowering. After the plant has become established, "renewal" pruning should be done to maintain the vigor of the plant by removing large canes over three years old after they have completed flowering. Such pruning will remove about one-third of the canes each year (Fig. 17–11). Retain the newer canes that will flower this or the following year. Ends of long canes are pruned back any time to keep them within desired bounds. Climbing roses are best pruned in early spring before leafing out unless the plants must be taken down and covered for winter protection, in which case some fall pruning is necessary to reduce plant size before covering. In zones 3 and 4 climbing roses are not recommended because of the great chance of winter kill even when covered properly.

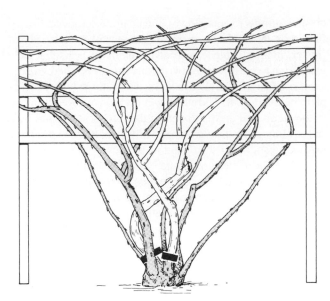

Figure 17–11   The proper method of pruning climbing roses. Canes that are three or more years old should be removed. This thinning allows more space and light for the younger, more vigorous shoots that will flower in later years. If canes become too long, the ends can be removed to maintain the desired size. The darkened canes are retained and those shown in white are to be cut away.

SUMMARY AND REVIEW

Shrubs are low-growing, multiple stem, woody plants that remain short (under 10 ft.). Although multiple stem by nature, some shrubs can be pruned to a tree-like form by removing all but one stem from the crown. The plant is pruned or not pruned depending on its use in the landscape. Shrubs are chosen not only by their form but their color (leaf and flower), fragrance, and hardiness. Hardiness is dependent upon the shrub's ability to survive both winter and summer conditions in an area.

Many of the techniques for successfully establishing and maintaining a shrub in a landscape are the same as for trees. These techniques include proper site location and preparation; plant selection; and cultural practices such as pruning, irrigating, fertilization, and insect and disease control. Because many shrubs are grown for their flowers, it is important to consider when the flowers are initiated and where they initiate on the plant. Pruning at the wrong time or at the wrong locations on the plant can keep the shrub from flowering.

EXERCISES FOR UNDERSTANDING

17–1. Observe some flowering shrubs in your neighborhood. Notice where the flowers initiate. How would you prune the shrubs to make sure that the shrub flowered next year?

17–2. Find your area on both the USDA Hardiness Map and the AHS Heat Zone map. What is your zone on each? What is the general climate for those areas? How do any local topographical features such as a body of water, valley, or hill influence the climate?

REFERENCES AND SUPPLEMENTARY READING

DIRR, M. 1997. *Dirr's hardy trees and shrubs: An illustrated encyclopedia*. Portland, Ore.: Timber Press.

WYMAN, D. 1990. *Trees for American Gardens*. Third Edition. New York: Macmillan.

CHAPTER

18

Floriculture

KEY LEARNING CONCEPTS

After reading the chapter you should be able to:

♦ Understand basic greenhouse structure and components.
♦ Understand how the greenhouse environment is manipulated to regulate plant growth and development.
♦ Understand the principles of growing several greenhouse crops.

THE GREENHOUSE INDUSTRY

Greenhouses (or glasshouses) date from Roman times where wealthy people grew plants in small enclosures, which allowed for spring flowers in the winter and fruit out of season. This desire spread across Europe, helping to create the largest greenhouse industry in the world in the Netherlands. The initial structures were built so that during the day when the sun shone brightly, the glass could be lifted slightly to allow the trapped hot air to escape. Low glass-covered structures that could be opened were called cold frames by the English. To add heat and make it a hot frame, decaying manure was buried about 0.5 m (18 in.) under the structure (Fig. 18–1). Both the cold and hot frames were constructed to allow a person to reach in and tend to the plants (Fig. 18–2). These units developed into pit houses, similar to today's greenhouses,

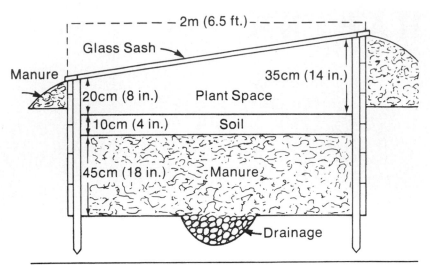

Figure 18–1   A hotbed warmed by decaying manure. Modern hotbeds are warmed by an electrical cable that is thermostatically controlled. The glass sash is lifted during the day to allow excess warm air to escape.

Figure 18–2   Cold frame against the side wall of a greenhouse used to overwinter plants.

Figure 18–3   Detached even-span glass greenhouses constructed in an open area to maximize light intensity and expansion capability.

which were constructed with a peaked roof, doors at each end, and a center walkway cut about 1 m deep to accommodate a standing person.

The modern greenhouse with its high sides and glass roof was developed later (Fig. 18–3). Greenhouses were constructed to allow the rays of the sun to enter the glass roof perpendicularly in the winter but partially reflect them when the sun was high during the summer. Heat was supplied with circulating hot water and, at a much later date, with steam. The sash bars were usually made of thick wood, especially in areas with great snowfalls, to provide structural support (Fig. 18–4). Lumber resistant to rot such as cypress and redwood was found to be ideal, but because these woods are not particularly strong, the girth of the sash bars had to be large, creating many shadows and thereby reducing the light intensity.

The even-span greenhouse was the most popular glass greenhouse design for many years (Fig. 18–5). Even-span greenhouses connected along the eave are termed ridge-and-furrow greenhouses (Fig. 18–6). For decades greenhouses did not vary much in design except that the width of the glass increased as the technology of making better glass developed. With improvements in techniques for extruding aluminum, the wide wooden sash bars were replaced with narrow aluminum bars, which were very strong. Narrow sash bars and wide glass allowed more light to enter during the winter when it was needed most.

As plastics increased in popularity so did their use as greenhouse glazings. Polyvinyl chloride was used briefly, but had an unacceptable life expectancy. Fiberglass became more popular with a life expectancy of 20 years and is still

Figure 18–4 Interior structure of a glass greenhouse consisting of painted cypress sash bars with iron columns providing structural support.

Figure 18–6 A ridge-and-furrow greenhouse range with glass covering and polycarbonate end walls, which provide maximum light intensity and good insulation.

Figure 18–5 Even-span glass greenhouse attached to a head house. The head house is used for storage of planting materials and provides a space for planting and shipping.

Figure 18–7 Quonset-style greenhouse with a polyethylene covering, exhaust fan for cooling, and unit heater for heating.

used if ultraviolet (UV) light-resistant protection is added. Although polyethylene has a short life expectancy (< 4 years) it has become the most popular glazing due to its weight, low heat loss (double layers), and less-expensive price. The quality of polyethylene has increased with added UV inhibitors, infrared inhibitors and absorbers, anticondensate additives, and photoselective inhibitors. Acrylic and polycarbonate single- and double-layer panels have gained in acceptance with many of the same properties as polyethylene. These panels are more rigid with a life expectancy of up to 20 years.

The use of plastics as greenhouse glazings led to the development of quonset-style (Fig. 18–7) and gutter-connected greenhouses (two or more quonset-style greenhouses connected, Fig. 18–8), which are less expensive to build than even-span greenhouses. They remain the most popular greenhouses today with design changes that continually optimize the growing of plants.

Figure 18–8 Gutter-connected greenhouse range with a double polyethylene covering, polycarbonate end walls, and automatic vents providing natural air circulation and cooling.

CONTROLLING THE GREENHOUSE ENVIRONMENT

Temperature

Greenhouses are heat traps that are cooled by applying white shading compound to the glass, by applying shade cloth on the outside of the greenhouse, by using automated shade curtains on the interior, by convection ventilation (allowing the hot air to rise through the overhead ventilator), by forcibly removing air with exhaust fans (Fig. 18–9), or by evaporative cooling using a fan-and-pad system.

The greatest heat load in the greenhouse occurs when the air is heated by the sun's rays and trapped by the enclosure of glass or plastic. However, some form of heat must be added at night during the winter season because too little daytime heat is retained for that purpose. In extremely cold climates, supplementary heat is necessary even in daylight hours. A central heating system derives its heat from a boiler circulating hot water or steam (Fig. 18–10). This type of system is primarily used in large greenhouse ranges. Localized heating can be accomplished through the use of unit heaters, forced-air furnaces (Fig. 18–11), infrared radiant heaters (Fig. 18–12), and root-zone bottom heating.

Environmental controls for operating the cooling and heating of greenhouses range from very simple, such as a basic thermostat, to more complex computerized systems. The more complex environmental control systems can control cooling, heating, shade curtains, fertilization, irrigation, irradiance, and other environmental systems in the greenhouse.

Night temperatures are considered very important for flowering pot plant production. Green foliage plants and some orchids require night temperatures as high as 21°C (70°F) and day temperatures as high as 29°C (85°F). Temperatures not only affect growth of plants but also flowering. For example, *Cattleya* orchids grown as a 29/21°C (85/70°F) day/night temperature will flower much more quickly than

Figure 18–10 Centralized boilers used to heat water that is piped throughout the greenhouse range for heating. Because centralized heating provides a single source of heat, most greenhouse ranges have at least two boiler units so that if one fails the other can be used.

Figure 18–11 Single-unit forced-air furnace used for heating a single greenhouse. Most greenhouses have more than one unit heater to provide even heat distribution.

Figure 18–9 Greenhouse range with exhaust fans in the side walls used to pull air through the greenhouse for greater cooling and air circulation.

Figure 18-12 These heaters emit infrared radiation that upon striking an object is converted to heat. Thus, the plants and containers are heated without having to heat the air—saving energy and money.

those grown at a 21/13°C (70/55°F) temperature. Roses, chrysanthemums, and azaleas do best at night temperatures in the 17°C (63°F) range, whereas carnations, snapdragons, cinerarias, and calceolarias grow well at 10°C (50°F). If height control is not a problem, day temperatures are set 0 to 5 degrees above night temperature on cloudy days and 10 to 15 degrees above night temperature on sunny days to maximize photosynthesis and minimize respiration. Maintaining these temperature differences is easy on overcast days but difficult on bright days because of an extreme heat buildup from the sun's radiation. When bright conditions prevail, plants may have to be (a) shaded during the summer months (e.g., May to September), (b) watered more often to compensate for excessive transpiration, or (c) sprayed with an overhead mist system to increase the humidity. All three of these procedures will compensate for the high light intensity conditions that may cause sunscald damage to leaves or flowers.

DIF (day temperature − night temperature) can be used to help control the plant height of many greenhouse crops. As the day temperature increases above the night temperature, a positive DIF, internode elongation increases. Inversely, a negative DIF will decrease internode elongation.

Light

There are three types of light that affect plant growth and development: light quality, light quantity, and light duration. Light quality, or the light spectrum, drives photosynthesis and can affect plant shape. Plants use red and blue light most efficiently for photosynthesis while far-red light affects internode elongation. Plants under high levels of red light have less internode elongation, are shorter, and usually greener. Plants grown under high levels of far-red light have longer internodes, are taller, and are less branched.

Light quantity or intensity refers to the amount of light that a plant receives and can be measured in foot-candles (fc), photosynthetically active radiation (PAR μmol s^{-1}) or photosynthetic photon flux (PPF μmole s^{-1} m^{-2}). The greater the light intensity the greater the rate of photosynthesis until the light saturation point occurs. Supplemental lighting has been used in greenhouse production for many years. The most common form is incandescent lighting, which is used only for photoperiod control, since it gives off excessive heat and poor light quality. Because fluorescent lights provide only part of the visible spectrum, they are used primarily for germination chambers, growth rooms, and in homes for interior lighting of houseplants. High-intensity discharge (HID) lamps are the most common types of supplemental lighting in the greenhouse, providing maximum light intensity and light quality (Fig. 18–13).

Light duration, termed **photoperiodism,** is the total light energy that may affect growth and development of plants. Short-day (SD) plants flower when the dark period is longer than a specified length. Plants such as azaleas, chrysanthemums, poinsettias, and some orchids require SD

Figure 18-13 Overhead high-intensity discharge lamps used to supplement the lighting in the greenhouse for optimum crop growth.

to initiate flowering. Long-day (LD) plants flower when the light period is longer than a specified length. China asters, Shasta daisies, *Gypsophila,* and *Liatris* require LD. Day-neutral (DN) plants require no specific daylength for flower initiation, for example, *Pelargonium* x *hortorum.*

Greenhouse growers should remember four dates that will help determine the necessity for photoperiod manipulation: June 21, the longest day of the year; December 21, the shortest day of the year; and September 21 and March 21, the autumnal equinox and vernal equinox when the light period is equal to 12 hr. During a long light period, flower induction of SD plants or maintenance of vegetative growth of LD plants can be accomplished by covering the plants with black cloth (excluding all light) from 5 P.M. to 8 P.M. (cooler climates) or 7 P.M. to 7 A.M. (warmer climates). If the light period is short, flower induction of LD plants or vegetative growth of SD plants can be accomplished by night interruption lighting. Incandescent lights can be placed over plants and turned on from 10 P.M. to 2 A.M. (4 hr.) with a light intensity of approximately 2 μ mol ($\approx$10 f.c.).

Control of Plant Height

Control of plant height and growth is important in greenhouse crop production. Plants should be of adequate size for shipping and handling as well as being aesthetically pleasing to the consumer (approximately 1.5 to 2 times the height of the container) (Fig. 18–14). Plant height can be controlled by manipulating the greenhouse environment or production practices. Controlling light quality, quantity, and photoperiod provides a means of controlling plant height and growth. Temperature manipulation using DIF, as discussed earlier, can help control plant growth. Applying a certain amount of water during irrigation and reducing or withhold-

Figure 18–14 Poinsettia (*Euphorbia pulcherrima*) grown for the Christmas holidays. The plant on the left was grown without a plant growth retardant (Control) and the plant on the right was treated with Bonzi. Bonzi is a common plant growth retardant used to reduce plant size for ease of shipping as well as to provide a more aesthetically pleasing appearance.

ing fertilization can also be used to manipulate plant growth. However, there are many situations that prevent the greenhouse grower from using the aforementioned methods.

Plant growth retardants (PGRs) can be applied to control plant growth, most of which blocks the synthesis of gibberellic acid, inhibiting stem elongation. Plant growth retardants can be applied by spraying the plant or drenching the growing medium (uptake by the root system). The concentration and timing of application are extremely important to achieve the desired results. Following is a list of popular plant growth retardants:

A-Rest (ancymidol) is labeled for many flowering pot plants and bedding plants, and it has been a popular growth retardant of bulb crops for many years. A-Rest can be applied as a spray or drench.

B-Nine SP (daminozide) is labeled for many flowering pot plants and bedding plants. It can be applied only as a spray to the surface of leaves.

Bonzi (paclobutrazol) is labeled for many greenhouse crops and can be applied as a spray or a drench. Sprays should be applied to the shoots of the plant as the active ingredient is translocated through the stem and reduces internode elongation.

Cycocel (chlormequat) is labeled for most greenhouse crops and can be applied as a drench or a spray. Combining B-Nine and Cycocel in the same tank mix provides even better growth regulation than using either alone.

Florel (ethephon) is labeled for many floriculture crops and works differently from the previously mentioned PGRs by releasing ethylene to the plant. It has been used for many years to enhance flowering and ripening of fruit. More recently it has been used extensively to reduce plant height and induce lateral branching.

Sumagic (uniconazole) is labeled for some floriculture crops. It can be applied as a spray or drench and works in a similar manner to Bonzi.

Growing Media

Choosing the proper medium to grow quality greenhouse crops is very important. Although the root systems of plants are not visible, the health of the shoot is dependent on the health of the root (Fig. 18–15). The media must have desirable physical characteristics in terms of bulk density or weight so that the container does not easily tip over. However, if the media weight is too great the cost of shipping may be prohibitive. The total pore space, water retention, and air-filled porosity also contribute to the health of the root system by providing appropriate amounts of oxygen and moisture for optimum growth. The pH and electrical conductivity (EC) of media are important chemical properties. For most crops the

Figure 18–15 Soilless growing media moved by a conveyor for filling containers prior to transplanting.

pH range of media is between 5.5 and 6.5, but some crops must have a more specific pH. The EC should be between 0.75 and 2.0 mmho/cm. If the EC of the media becomes too high (> 2.0 mmho/cm), salt accumulation can cause water to be pulled out of the roots resulting in root burn and death.

Most greenhouse crops are grown in soilless media. The benefits of using soilless media over media that contain soil are reduction in bulk density (weight) and increased uniformity of the media. The most common components of soilless media are sphagnum peat moss, pine bark, perlite, and vermiculite. Sphagnum peat moss is low in salts, does not decompose readily, has high water- and nutrient-holding characteristics, is uniform, and relatively disease-free as found in nature. Vermiculite is a sterile expanded mica that holds water and nutrients well and does not decompose. It is uniform and reasonably priced. Vermiculite does compress quickly and, therefore, is used most frequently in seedling mixes because of its water-holding capacity. Perlite is a heat-expanded aluminum silicate rock that is porous and lightweight. It does not hold water but provides excellent pore space. Pine bark and hardwood bark make excellent media components. When composted or aged properly, both barks have high water- and nutrient-holding capacities, and can be graded to specific sizes (1/4 in. and 3/8 in. are the most popular). Most soilless media are amended with dolomitic limestone for increasing the media pH, phosphorus, and trace elements so that the nutrient content is adequate for greenhouse crop production. The type of growing medium that is used for each greenhouse crop can vary significantly from one producer to the next. Therefore, it is extremely important that irrigation and fertilization be closely managed.

Irrigation and Water Quality

Water quality is extremely important to a greenhouse operation due to the use of soilless media, which have much less buffering capacity than that of soil. Irrigation water should

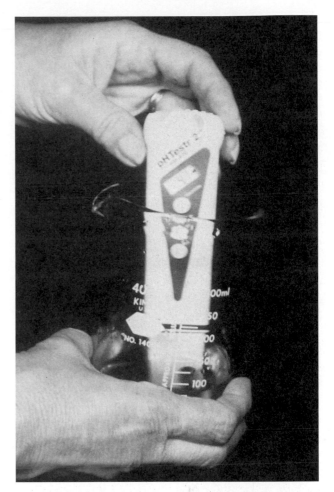

Figure 18–16 Handheld EC meter used to determine the electrical conductivity of irrigation water.

be analyzed prior to beginning a greenhouse operation and at least once a year during regular operation. Samples can be submitted to the local extension agent for testing, and they can also be tested by the grower using a small handheld meter (Fig. 18–16). Salinity, the measure of sodium and chloride in irrigation water, can contribute to increased media EC and root damage if in high concentrations. This has been an increasing problem, especially in areas where saltwater intrusion of aquifers occurs. Alkalinity, the amount of carbonates and bicarbonates in irrigation water, can increase the pH of the growing media. There are methods of controlling salinity and/or alkalinity in irrigation water, but they can be very costly if concentrations are high enough.

Irrigation is an art that is not learned quickly. The ability of a grower to look at the plants across an entire greenhouse range and determine the need for irrigation is of primary importance. Hand-watering of greenhouse crops can be one of the best methods of irrigation, but it is not economical (Fig. 18–17). However, there are many growers that still water some of their crops by hand. The use of overhead sprinklers for the production of some bedding plants is more

Figure 18–17 Hand-watering of greenhouse crops using an extended wand with a water breaker attached to the end. The water breaker helps break up the flow of water to prevent washing the media and/or plants out of containers.

Figure 18–18 The large tray on the top of this ebb-and-flood bench serves as a reservoir for briefly holding water while it is taken up by capillary action through the bottoms of the containers.

economical, but the foliage of most crops must be kept dry to prevent disease. Boom watering can be used effectively on the production of plugs. The use of microtubes has been popular for many years and is an efficient method of irrigating large numbers of plants without getting the foliage wet while minimizing water usage. Mat watering by the use of a capillary mat is another method of irrigating crops when the sizes of containers vary on the same bench. The use of an ebb-and-flood system with flood floors has become popular in recent years, but it can be cost-prohibitive due to the amount of materials used to build the system (Fig. 18–18). Disease can also spread rapidly in this system; thus, it must be monitored carefully.

Most greenhouse operations use some type of liquid fertilizer in the irrigation system, which is called **fertigation.** This adds another factor to irrigation that is discussed in the next section.

Mineral Nutrition

Providing greenhouse crops with all 16 essential elements is very important to optimizing growth due in part to the use of

soilless media and the need to produce a marketable crop as quickly as possible. Most fertilization recommendations are based on the parts per million (ppm) of nitrogen (N) from a complete liquid fertilizer (those fertilizers containing the essential elements) (Fig. 18–19). Therefore, greenhouse fertilizers are formulated to provide the remaining elements in the appropriate amounts.

Some greenhouse crops must have greater amounts of certain elements than other crops and, therefore, fertilizers are specifically formulated for those specific crops. For example, some poinsettia cultivars are inefficient in uptake molybdenum (Mo), therefore, most poinsettia fertilizers have increased concentrations of Mo. A greater amount of nitrate versus ammonium in a fertilizer can produce shorter, stronger plants whereas a greater amount of ammonium versus nitrate can produce taller, weaker plants.

Plants have different nutrient needs during different stages of growth, and fertilization must be adjusted to meet those needs. Because most greenhouse crops are flowering plants, the fertilizer concentrations are low at the planting stage of the crop, increase during rapid crop growth, and decrease at crop maturity or flowering. Application of fertilizers in too high of a concentration without leaching can increase the EC of the media and be deleterious to root growth and thus crop quality. Leaching, however, can contribute to runoff of irrigation water and fertilizers, which has become an increasing problem for greenhouse operations. Therefore, growers are trying to more closely match the fertigation schedule to crop growth, thereby reducing waste or runoff and cost.

Diseases and Insects

Greenhouses are small microclimates that not only provide an optimal environment for producing quality plants, but also an optimal environment for the growth of diseases and

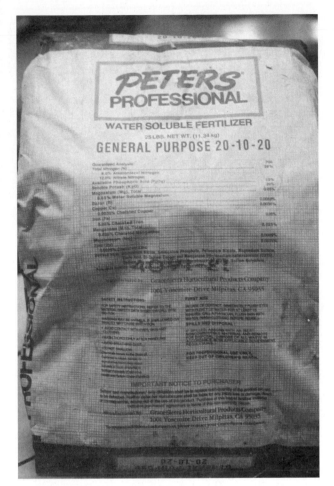

Figure 18–19 Example of the nutrient analysis, 20-10-20 (nitrogen-phosphorus-potassium), on the bag of a general-purpose, liquid-soluble fertilizer.

insects. The most common greenhouse insects are aphids, mealy bugs, spider mites, whiteflies, thrips, fungus gnats, and scales. Bacterial blights, powdery mildew, botrytis, root rot, damping-off, and viruses are just some of the diseases that must be monitored in greenhouse production. The best methods for circumventing these pests are sanitation, integrated pest management (IPM), preventative fungicides, and spray programs.

GREENHOUSE CROPS

The greenhouse industry in the United States began in the Northeast close to large population centers. Cut flowers—primarily roses, carnations, and chrysanthemums—were the major greenhouse crops produced in the early 1900s. Production of fresh-cut flowers spread slowly across the United States to the Midwest, then to Colorado, and finally into California by the 1960s. During this shift of cut flower production from the East Coast to the West Coast, flowering pot plants began to replace cut flowers. Examples of popular

flowering pot plants include poinsettia, chrysanthemums, African violets, azaleas, hydrangeas, Easter lilies, and other flowering bulb crops. Production of flowering pot plants became an even greater part of the greenhouse industry as foreign imports of cut flowers became a major competitor in the late 1960s. During the 1970s and 1980s, green plant or foliage plant production became a considerable part of the greenhouse industry in response to the "green revolution." As technology and transportation improved so did the greenhouse industry. This was most evident in the production of bedding plants with the introduction of plug technology. Bedding plants now make up a majority of total greenhouse crop production.

This brief history of greenhouse production reflects the dynamics of this field in horticulture. The production guidelines of the following crops are examples of some of the more important greenhouse crops grown.

Cut Flowers

A majority of the cut flower production occurs in California, which produces primarily roses and carnations. *Rosa* L. hybrids flower profusely during the long, warm, sunny days of California. Greenhouse nighttime temperatures should be maintained at 15°C to 18°C (60°F to 65°F) for optimum growth and flowering. Flowering can be timed by pinching the terminal buds producing new flowering shoots in 35 to 60 days. The rapidity with which they flower depends on light intensity and temperature. A flower forms on every new shoot after the removal of a flower from the bush (Fig. 18–20). If properly cut and pruned annually, plants may not have to be replanted for 5 to 8 years depending on the health of the plants and the soil structure. Occasionally, disease will hasten the need to replant.

Flowers are cut in the late-bud stage usually when one or two sepals have turned downward or just as one or two petals have unfurled. They are sorted and graded into uniform stem lengths that also have similar flower head (bud) sizes and quality (Fig. 18–21). Flowers are packaged in bunches of 25 and kept in postharvest coolers at 2°C to 4°C (35°F to 40°F).

More recently specialty cut flowers, those cut flowers that are not grown in large quantities by smaller growers, have increasingly become a part of the floral industry. As the demand for these types of crops increases so will the need for production of this type of flowering plant in the greenhouse industry.

Flowering Pot Plants

Poinsettias are the number-one flowering pot plant sold in the United States even though they are considered a holiday or seasonal crop. *Euphorbia pulcherrima* Wild. Ex Klotzch is a tropical plant native to Mexico belonging to the family Euphorbiacea (Fig. 18–22). The poinsettia was introduced

Figure 18–20 Roses grown in a greenhouse for cut flower production. The flower stems are a few days from the cutting stage.

Figure 18–22 Various poinsettia cultivars in shades of red and white grown for the Christmas holidays.

Figure 18–21 Cut roses being graded for packing and shipping to a wholesale florist.

into the United States in 1825 by J. R. Poinsett from whom it gets its common name. Poinsettias have been grown as a Christmas holiday crop mainly due to their photoperiodic re-

sponse for flower initiation. They are classified as short-day (long-night) plants that initiate and develop flowers as the nights become longer than the days. The inflorescence is characterized by a single female flower, without petals and usually without sepals, surrounded by individual male flowers all enclosed in a cup-shaped structure called a cyathium. The showy red, pink, or white portions of the plant, popularly referred to as the flower, are modified leaves called bracts.

Rooted cuttings are planted in well-drained media during August. Plants should be watered immediately after planting and a fungicide should be used shortly thereafter to prevent disease infection. Poinsettia cultivars are classified according to the amount of time required from the start of short days to flowering, which are called response groups. Response groups range from 7 to 10 weeks. Therefore, timing of a finished crop can be manipulated by photoperiod control. Some plants are very sensitive to changes in photoperiod, such as 'Freedom Red,' and may need an extended daylength to prevent premature flower initiation. Daytime temperature should be maintained below 25°C (77°F) to maintain leaf unfolding with nighttime temperature maintained at approximately 20°C (68°F) for flower initiation and development. Reduced temperatures of 16°C to 17°C (60°F to 62°F) are necessary for bract coloration. Light intensities of up to 6,000 fc should be maintained for good bract coloration. After the roots have grown to the sides of the container (approximately 2 weeks), the terminal meristem should be pinched to six internodes to ensure lateral branching. Approximately 2 weeks after pinching (laterals of 2- to 3-in. length), short-day length should be started for inducing flower initiation. According to response group the plants should reach anthesis 7 to 10 weeks later.

Foliage Plants

Foliage plants or houseplants can be grouped by sensitivity to light levels. Light intensity can easily be measured with a foot-candle or lux meter. The high-light group consists of light levels of 4,000 to 8,000 fc (i.e., *Ixora coccinea, Codiaeum variegatum pictum,* Cactaceae, Crassulaceae), the medium-light group of 1,000 to 3,000 fc (i.e., *Begonia* spp., *Saintpaulia ionantha, Draceana* spp., *Ficus* spp.), and the low-light group of 50 to 500 fc (i.e., *Aglaonema commutatum, Philodendron scandens oxycardium, Spathiphyllum* spp., *Dieffenbachia* spp., *Epipremnum aureum*). Those foliage plants that are moved from an environment with a higher light intensity into one of lower light intensity must adjust to this change. Some plants lose chlorophyll and then drop their leaves soon after being moved to a lower light intensity (i.e., *Ficus benjamina* and *Asparagus densiflorus*). These plants develop new leaves that are usually thinner and broader for more efficient photosynthesis. Many plants, which include the aroid, palm, and lily families, do not abscise their leaves. Instead, the leaves become yellow-green and may wither or die but still cling to the stem. Therefore, plants must be acclimatized to the lower light intensities before removing them from environments with higher light intensities. First, reduce the light by 50 percent for several weeks and, finally, reduce the light with shade to 20 percent of the original growing light intensity. The entire acclimatization process should take 4 to 12 weeks, depending on the species in question. Reducing fertilizer and watering during this period also aids the acclimatization process.

Flowering plants have relatively narrow and specific temperature requirements for flowering. Because foliage plants are not grown for their flowers, they can be grown under temperatures that range from 60°F to 90°F (16°C to 32°C). The lower the temperatures the slower the growth and vice versa. The optimum growing temperature for most houseplants is approximately 70–79°F (21°C to 26°C).

Most foliage plants prefer a relative humidity between 70 and 80 percent. This can be easily achieved in a greenhouse full of plants. However, when foliage plants are moved from the greenhouse to an interior environment, low relative humidity can result in slower growth and insect problems. Misting these plants will not provide enough moisture in the air to compensate for this loss. Keeping plants out of the direct flow of air from heating and/or air-conditioning vents and doorways where large amounts of air exchanges occur, and providing adequate water to the container are the best means of compensating for the loss of relative humidity.

Rapidly growing foliage plants in a greenhouse will utilize large amounts of water and fertilizer. However, when moved to an interior environment, growth slows as does nutrient and water uptake. Thus, houseplants will need much lower amounts of fertilizer and water. Overwatering houseplants is one of the most common causes of plant death. Salts may accumulate in the media of plants that are maintained in the same container for long periods of time (greater than one year). Leaching the media once a year or repotting can circumvent this problem. Houseplants can also benefit from rinsing the dust that accumulates on the leaves during repotting.

Bedding Plants

The term *bedding plants* was originally used by gardeners who wanted to use annual or perennial plants in a permanent border or in a specially designed flower bed. This term has been expanded to include flowering ornamental plants, vegetables, and herbs, mostly annuals (Fig. 18–23). Bedding plants are used in large or small containers, on patios, in window boxes, in hanging baskets, in small gardens, and in beds and borders. The bedding plant industry has responded to the needs of a range of "gardeners" from the sophisticated to the novice, by offering a variety of plants in large, medium, or small containers (six-packs are the most common) and a large number of species from which to choose.

Greenhouse production of bedding plants differs from the growing of other greenhouse crops in that a larger number of many different species of plants are grown in the same space in a much shorter time. This means quick rotation of crops and return on investment.

Figure 18–23 A gutter-connected greenhouse range full of bedding plants on benches in flats (containing six sets of six-packs) and hanging baskets overhead grown for spring sales.

Figure 18–24 A 288-plug tray with *Impatiens wallerana* grown for transplanting in larger containers.

Early production of bedding plants was by direct seeding into containers or broadcast seeding or sowing in rows in flats and transplanting into containers. These traditional methods of bedding plant production were improved on and developed into what is now termed the plug production method (Fig. 18–24). This technology along with continuing improvements has meant continued growth and expansion of bedding plant sales in the greenhouse industry.

Plug production entails sowing of seeds into 50 to 800 individual cells that compose a plug tray. This system provides individual root systems reducing transplant shock, maintaining seedlings for longer periods of time before transplant, making scheduling easier, ease of shipping, increased mechanization, and thus reduced labor. Growing plugs is a highly specialized area of greenhouse production that requires specialized equipment and facilities and a very regimented set of cultural practices that vary from one species of bedding plant to the next.

Because of the large number of different species of bedding plants, scheduling, fertilization and irrigation, greenhouse temperatures, types and rates of plant growth regulators, and light levels vary significantly. Although bedding plant production can be highly automated and bedding plants are a quick crop to produce, the amount of knowledge that a grower must have to grow these crops is extraordinary.

Bulb Crops

Many species of plants have fleshy underground storage organs capable of carrying the plants through seasonal cold/warm or dry/wet periods. Such structures are popularly called bulbs, but they are defined more accurately as bulbs, tuberous roots, tubers, corms, or rhizomes. They all produce one or more buds for flower production or renewed vegetative growth.

The *true bulb,* such as the lily, hyacinth, Muscari, Narcissus, tulip, and onion, has numerous fleshy scales or leaf bases attached to a distinct basal plate (stem) that gives rise to roots and shoots. They may or may not have one or more impervious covering layers (tunic).

Tubers are enlarged fleshy stems with adventitious buds (eyes) near the upper surface, as in the tuberous begonia, Eranthis or Caladium, or in a systematic pattern or arrangement, as in the Irish potato. Some organs are enlarged roots and are classified as *tuberous roots,* such as Agapanthus, Dahlia, and Hemerocallis.

Corms have solid shortened stems with buds systematically arranged under a paper-thin protective covering of the leaf base or scale. Crocus, Gladiolus, and Freesia are examples.

The *rhizome* is a fleshy, horizontal underground stem that grows laterally. Examples are the rhizomatous iris, calla lily, and ginger.

The Easter lily or *Lilium longiflorum* Thunb. is one of the most important flowering bulb crops in the United States. These true bulbs are native to Japan and grown in California and Oregon. They are dug in September and October and are either shipped to growers for potting or precooled ("casecooling"). This cooling process or vernalization of the bulbs promotes floral induction and is completed just after Christmas. Bulbs should be planted in well-drained media and drenched with a fungicide. The bulbs are then forced into bloom in the greenhouse. The desired forcing time depends on the date of Easter in that particular year. Growth is measured by counting the number of unfolded leaves during stem elongation. Greenhouse temperatures can be increased or decreased to adjust the rate of leaf unfolding. This allows the grower to adjust the rate of growth so that bloom will occur just prior to Easter. Application of plant growth retardants is necessary to reduce stem height.

SUMMARY AND REVIEW

Greenhouses were initially built to protect plants from cold temperatures while allowing sunlight to reach the plant. However, through the years greenhouse structures became much more sophisticated, enabling growers to manipulate the environment to control plant growth and development. Greenhouse heating and cooling systems are used to regulate temperatures and consequently plant growth and development throughout the year. Shading systems allow the amount of light reaching a plant to be limited when light intensity is too great. High intensity discharge (HID) lamps can be used to provide supplemental light during low light periods. The timing of water and fertilizer applications and formulation of fertilizers provide a means to control plant growth and development. Very few modern greenhouse production systems have plants growing in native soil. The plants are most often in a container which can be as large as a bench or as small as a plug tray. The growing media in the container generally contains very little or no soil, but is a composition of materials such as sand, peat moss, perlite, vermiculite, and wood products. The media components can be blended to give different characteristics suitable for different needs. When manipulating the greenhouse environment is not sufficient to maintain desired plant growth, chemical growth regulators can be used. Manipulating the greenhouse environment is also a means to control insect and disease problems, although at times other methods may also be required.

Greenhouse crops can be grouped into three broad categories. Cut flowers, potted or container crops, and foliage plants. Cut flowers are those plants that are grown for their flowers, leaves, or stems which are harvested and mainly used in flower arrangements. Potted or container crops are grown and sold in the container. This includes potted crops such as poinsettia, chrysanthemum, and annual and perennial bedding plants. Foliage plants are those plants grown for their foliage rather than their flowers. Many of these plants are tropical and adapted to the warm, humid, and low light conditions found under the canopies of large trees in tropical rainforests. Each category and the species within the category have specific cultural requirements that must be met in a greenhouse to produce plants of high quality.

EXERCISES FOR UNDERSTANDING

18–1. If you have a plant in your home or residence hall room, observe its growth and condition and note the environmental conditions of the room. If you do not have plants, visit a local florist or floral department of a large supermarket and observe one of those plants and its environment. Research the plant and find out what cultural practices are recommended to make it grow and flower (if it is a flowering plant). How close does the environment in which the plant is now growing meet its required environment? What can be done to modify the existing environment?

18–2. Visit a local greenhouse. Observe what species are being grown. Through observation and, if possible, talking with the owner or grower, note how the environment is being manipulated to control the growth of each species. How many different environments can you observe?

REFERENCES AND SUPPLEMENTARY READING

Dole, J. M., and H. F. Wilkins. 1999. Floriculture: Principals and species. Upper Saddle River, N.J.: Prentice Hall.

Jarvis, W. R. 1992. *Managing diseases in greenhouse crops*. St. Paul, Minn.: APS Press.

Nelson, P. V. 1998. Greenhouse operation and management. Fifth Edition. Upper Saddle River, N.J.: Prentice Hall.

CHAPTER

Lawns and
Turfgrasses

Written by John H. Madison

KEY LEARNING CONCEPTS

After reading the chapter you should be able to:

♦ Understand the terms commonly used in turfgrass science.
♦ Explain the principles for establishing and maintaining turfgrasses.
♦ List the different types of turfgrass and the environmental and cultural requirements of each type.

Growing turf is a multimillion dollar industry. The replacement value of turf in the United States has been estimated to exceed $12 billion, and annual maintenance costs have been estimated at more than $10 billion. Much of the turf is public grass growing in schoolyards, parks, cemeteries, and golf courses, along highways, on military installations, and on other public lands.

Turf is a unique crop in that the product is not what is harvested but what remains. The crop is grown densely as an entire population instead of individual plants spaced apart so that each grows vigorously. An appreciation of these differences is basic to being a good turf horticulturist.

There are three principal reasons for growing a lawn: (1) as a carpet, to protect the home from mud and dust, and to soften glare and heat; (2) for recreation; and (3) for beauty and pleasure. Studies have shown that lawns so please us that many of our earliest memories are of grass and trees.

John H. Madison is professor emeritus of environmental horticulture, University of California, Davis. He is the author of many publications on turfgrass culture and management.

Turf culture differs from other horticultural pursuits in one important way. In nature, plant growth is often limited by competition with many other plants for water, nutrients, and light. Thousands of seeds germinate and die for every one that lives and grows. Successful horticulture comes from spacing plants and eliminating weeds so that each plant has soil, water, and space allotted to it alone. Grass is the exception. In growing a lawn, even more plants are crowded into a given space than would grow there naturally. We strive to grow dense lawns similar to a fine carpet. The more we succeed in growing a dense lawn, the more stress each individual plant gets through competition from its crowding neighbors.

On an infertile dry soil, lack of nutrients and water limits plant growth, and results in a poor lawn. We can grow a better, denser lawn by fertilizing and irrigating. But as the number of plants (shoots) increase, there is a point where plants shade each other and compete for sunlight. Beyond that point individual plants are smaller, have fewer roots, and use nutrients and water less effectively. Thus, cultural practices that result in a dense, tight lawn can also result in individual plants of reduced quality and vigor.

Turf differs from other horticultural plantings in that, instead of relieving the stress that results from competition, we increase it. The goal in growing lawns is not to grow individual plants of high quality but a plant population of fine appearance. In addition to suffering stress from competition, most lawn plants also suffer climatic stress at some season of the year. In managing a lawn, we can fertilize, mow, and use other practices to either increase or decrease stress. There is a limit, however, to the stress a plant can withstand before weakening to the point of dying.

The stress concept gives us a tool for evaluating the effects of cultural practices on grass. In a figurative way, there is a stress budget for turfgrass. If plant competition is high and midsummer weather imposes a heat stress, little more allowable stress is left in the stress budget. Cultural practices that reduce stress, such as watering, should then be followed. On the other hand, early fall temperatures are often ideal for the growth of turfgrass. Then certain stressful management practices—for example, power raking to thin out plants—that would have damaged a bluegrass lawn in summer's heat can be used. Throughout this chapter different management practices are evaluated in terms of the stress they cause. The practices that increase stress are not necessarily bad, but care should be exercised to select and use these practices in appropriate seasons, and not to use several high-stress operations at inappropriate times.

The word lawn originally referred to a natural area of grass without trees. Today it refers to any expanse of ground on which grass is growing. Turf originally meant a layer of matted earth formed by soil and thickly growing grass plants. This meaning has gradually changed over the years, so that the word *turf* is used by horticulturists to refer to grass that is mowed and cared for. Turfgrass refers to the grass used in growing a horticultural turf.

Turf culture can be divided into two procedures: (1) establishment, and (2) maintenance for decorative or family use with modest traffic, or for heavy traffic by the public, or for athletic competition.

ESTABLISHING A TURF

A first step in growing a turf is to learn something about the species of plants used and their requirements. Turfgrasses are divided into two groups: (1) cool-season or temperate grasses, such as bluegrass and bentgrass adapted to cool climates of the northern United States; and (2) subtropical grasses, such as Bermuda grass and St. Augustine grass adapted to the warmer areas of the southern United States.

Between these regions is a transition zone where neither cool-season nor subtropical grasses perform at their best. Cool-season grasses grow well during cooler parts of the year, but during hot summer weather they become dormant or suffer heat injury and disease. Subtropical grasses in the transition zone do well in the hot summer but become dormant and turn brown in winter (Fig. 19–1).

In the western, arid portion of the U.S. plains states, where water is not available for irrigation, native drought-tolerant grasses such as buffalo grass or grama grass are sometimes grown. While their season of growth is limited, they do cover and protect the soil during the dry period.

Commonly used turfgrasses are described in Table 19–1.

TURFGRASS STRUCTURE

Turfgrass anatomy is illustrated in Figure 19–2. Table 19–2 gives a key for identifying common turfgrasses. The key is limited since it distinguishes only among the common cultivated turfgrasses.

Turfgrasses are well adapted to their role as a carpet. As with most monocots, growth of an individual shoot is **determinate** (ends in a flower), and the plants have fibrous roots without tap roots. Each shoot dies after a time, but is replaced by new shoots growing from axillary buds. These new shoots produce adventitious roots at the nodes. In this way turf continually rejuvenates itself. In many grass species, axillary buds produce horizontal creeping stems. If these grow over the surface of the ground they are called **stolons;** below ground, they are termed **rhizomes.** Both structures enable grass plants to invade open areas and to spread. Other species of turfgrasses grow only new erect shoots or **tillers,** arising from axillary buds. Tillering grasses do spread but slowly.

Turfgrass flowers develop on elongated stems at certain seasons. The rest of the year the tiller shoots are condensed with the vegetative growing point nestling among the leaves. The growing point produces **intercalary meristems;**

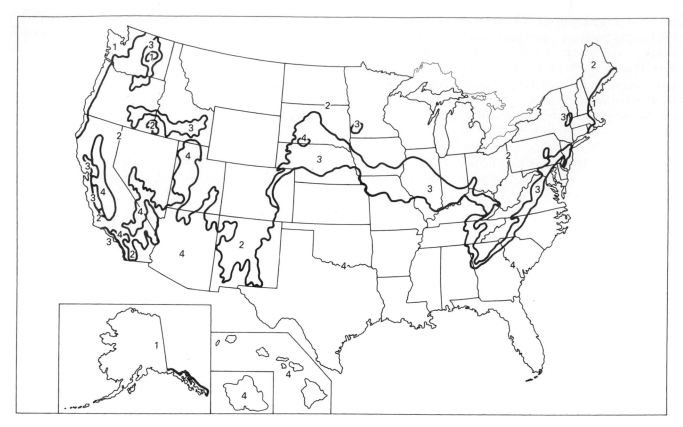

Figure 19-1   Areas of turfgrass adaptation. 1: Areas adapted to temperate grasses, particularly bent-grasses (*Agrostis*). 2:Areas of general adaptation of temperate grasses. 3: Transition zone where subtropical grasses suffer winter cold and temperate grasses are stressed by summer's heat. 4: Area of adaptation for subtropical bermuda and zoysia grasses. The more tender subtropical grasses are limited to hardiness zones 8, 9, and 10 (see Fig. 10–4). This map is based upon mean July temperatures, not the low winter tempera-tures of the hardiness zone map. The boundaries are not sharp and are greatly influenced by local microcli-mates, especially in the mountain states, where elevation and exposure affect adaptation. Alaska is too cold for all but one or two cultivars of turfgrasses, which survive in milder areas of the state. Hawaii is subtrop-ical-tropical.

that is, regions of cell initiation that lie across a stem or leaf and that, by dividing, interpose—or intercalate—new tissue between existing older tissues. The division and elongation of the intercalary meristems "extrudes" leaves and leaf sheaths. Internodes of the stem of grass plants do not elon-gate except to produce stolons, rhizomes, or flower stalks.

CHOOSING TURFGRASSES

Of the turfgrasses, Kentucky bluegrass is unique in produc-ing a large percentage of seedlings by **apomixis.**[1]

Superior plants are tested for turf characteristics such as low growth, disease resistance, cold tolerance, drought

tolerance, and so forth. If they are indeed superior, they can be increased and released as a cultivar. As a result, there are many clonal **cultivars** of bluegrass but few or none of the other turf species.

Named clones of other grasses can be vegetatively propagated with sod or stolons. Named cultivars are sexually propagated as a genetic mixture of seed from parents care-fully selected to a standard of excellence.

Named cultivars of bluegrasses are excellent, but have the disadvantage of complete genetic uniformity so that a turf is totally subject to any weaknesses. For example, a pure stand of 'Merion' bluegrass appears orange at times because of spores of a species of stem rust disease to which it is par-ticularly susceptible. Consequently, bluegrass seed is often prepared as a blend. Seeds of four or five cultivars are mixed to take advantage of the good qualities of each while avoid-ing the extensive damage that could occur when a pure stand of a single cultivar develops a weakness.

When selecting a grass to grow on bare soil—and to grow in intense competition—choose an aggressive grass

---

[1]In this type of apomixis, the embryo resulting from sexual fusion of an egg and pollen nucleus aborts, but an embryo is produced from somatic cells of the embryo sac (female tissue). In this way it is possible to get seedlings which are genetically identical to each other and to the female parent.

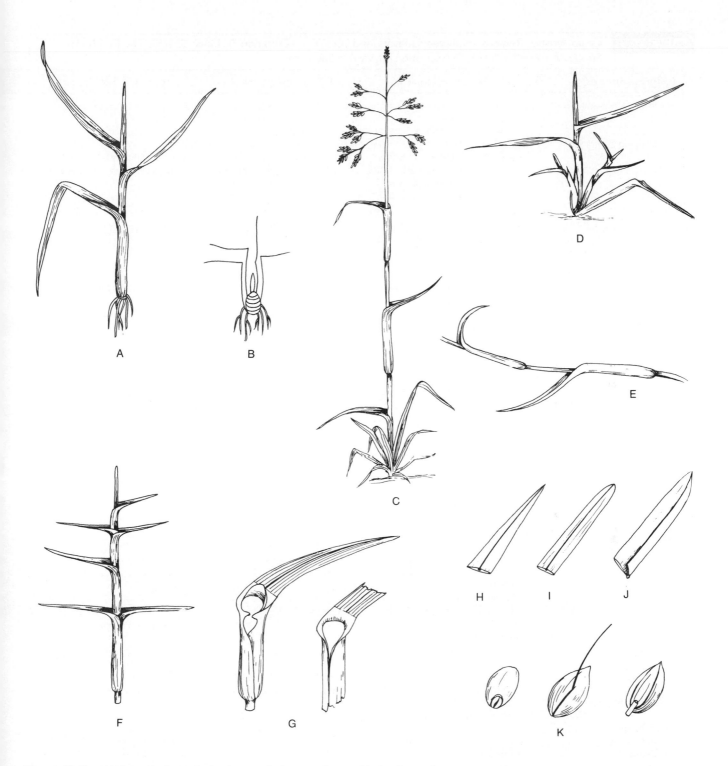

Figure 19–2 (A) The vegetative turfgrass plant usually has a small stem with short internodes and bears one leaf and several adventitious roots at each node. (B) Enlargement of the lower portion of A. (C) The stem elongates prior to flowering. (D) Buds at the nodes produce tillers. (E) Elongated spreading horizontal stem (stolon). (F) Stolons of some grasses produce occasional short internodes alternating with longer ones so that some of the leaves appear opposite. (G) The leaf consists of a blade and a sheath (sheathing the stem). The region where the blade joins the sheath is called the *collar.* An outgrowth of the collar is the *ligule,* which may be a membrane or a fringe of hairs. The collar may develop clasping *auricles.* Leaf blades may be pointed *(H),* rounded *(I),* or boat shaped *(J),* and edges may be parallel *(I, J)* or tapered *(H).* (K) The grass "seed" is properly a fruit—a caryopsis—and may be naked or enclosed in a hull. The hull may have one or more awns and may have a small piece of stem attached.

TABLE 19–1 Characteristics, Regions, Improved Cultivars, and Uses of the Common Turfgrass Species. The First Nine Species Are Northern U.S. Grasses; the Last Six Are Southern

Turfgrass Species	Turf Characteristics	U.S. Regions Where Used	Desirable Characteristics
Agrostis stolonifera L. (Agrostis palustris Huds.) Creeping bentgrass	Able to produce a dense turf of good color and fine texture under extremely close mowing (0.25 in. and less).	North, through the transition zone and with care into parts of the South.	Ability to stand low mowing. Withstands salinity.
Agrostis tenius Sibth. Colonial bentgrass	Variable grass adapted to northern coastal climates. Spreads. Color fair. Coarse-textured mowed high; fine-textured mowed low.	Well adapted to north coast, usable throughout the North.	Adaptable. Withstands acid soils of low fertility and wet soils.
Agrostis spp. 'Highland' bentgrass	In a mixture it forms dense patches of blue-green grass of puffy character unless mowed under one inch. Tolerates more heat and less water than other bents.	Better adapted to transition zone heat and dryness than above.	Performs well in irrigated Mediterranean climate.
Festuca elator L. (F. arundinaceae) Tall fescuegrass	A coarse-textured bunch grass suited only for growing in pure stands.	Adapted to transition zone. Not fully cold hardy.	Deep rooted. Tolerates wear, drought, and neglect.
Festuca pratensis Huds. Meadow fescuegrass	Much like perennial ryegrass when young but becoming coarse with age. Less aggressive than other turfgrasses.	North into the transition zone.	Low cost, fast germination.
Festuca rubra L. var. rubra Creeping red fescue	Fine-leaved, drought-tolerant grass tolerant of some shade. Spreads by rhizomes. Favored by high mowing. Competitive under low fertility.	North. Does poorly in the transition zone.	Fine texture, spreading. Tolerates some drought, shade, infertility.
Festuca rubra L. var. commutata Chewing's red fescue	Fine-leaved, drought-tolerant grass. Tolerates some shade. Spreads slowly. Tolerates lower mowing than above.	North.	Tolerates closer mowing than above.
Lolium perenne L. Improved perennial ryegrass	Turf cultivars have good color and fine texture like bluegrass. A patchy hard-to-mow clump grass unless well fertilized and watered.	Transition zone north. Not fully hardy. Winter grass in South.	Good color and texture. Tolerates traffic.
Poa pratensis L. Kentucky bluegrass	A dense, fine-textured, beautifully colored grass spreading by rhizomes. Seeds produce uniform apomictic seedlings. The aristocrat of lawn grasses.	Best with cool nights and days under 84°F. Fair in transition zone.	Best color. Sod repairs itself from stolons. Heat and dry dormancy.
Cynodon dactylon (L) Pers. Bermudagrass	A dense vigorous grass creeping by stolons and rhizomes. Withstands neglect but a handsome turf when well cared for. Brown in winter.	South and valleys of Southwest.	Vigorous. Withstands drought, wear, and salinity.
Eremochloa ophiouroides (Munro) Hack Centipedegrass	A dense vigorous grass creeping by large stolons. Adapted to low maintenance lawns on acid or infertile soils.	Florida, Gulf states, coastal plains, parts of California.	Withstands low fertility and acidity. Low maintenance.
Lolium multiflorum Lam. Winter grass, annual ryegrass	A coarse short-lived grass used only for overseeding winter dormant grasses for winter color in the South and transition zone.	Winter grass in South.	Grows at low temperatures, near freezing.
Paspalum notatum Fliigge Bahiagrass	Forms an open coarse turf.	Gulf states and coastal plains of South.	Heat tolerant. Low water need.
Stenotaphrum secondatum (Walt.) O. Kuntze St. Augustinegrass	Strong, dense, coarse grass creeping by large stolons. Adapted to shade.	Florida, Gulf states, and parts of California.	Vigorous. Withstands drought, some shade.
Zoysia japonica Steud. Koreangrass, Z. matrella (L.) Merrill	Extremely dense grower of good color. Spreads by rhizomes and stolons.	Z. japonica hardy through zones 6–10, Z. matrella, 8–10.	Very dense—crowds out weeds. Good color and texture.

Undesirable Characteristics	Improved Cultivars	Special Care Requirements	Uses
Subject to disease and insect pests. Invasive.	Penncross, Penneagle, A-4, G-2, Crenshaw, Providence, SR1020, L93	Regular day to day maintenance, mow short, control pests.	Fine turf for putting and bowling greens
Coarse if neglected. Some disease and insect problems.	Astoria, Exeter, Bardot, Egmont, SW7100	Mow 1 in. or less; use low fertility for low maintenance.	Decorative lawns and utility turf
Does not mix well with other grasses.	Highland	Mow 1 in. or less, dethatch in fall.	Decorative lawns and utility turf
Coarse. Weedy in a mixture.	Arid, Bonanza, Falcon, K31, Mustang, Olympic, Rebel	Mow regularly 1.5 in. plus, use infrequent deep irrigation.	Decorative lawns and utility turf; sports turf
Tendency to coarseness and clumpiness.	Arnba, Flyer, Cindy, Sunset	Use in mixes. Give care for other grasses.	To dilute seed mixtures to reduce cost
Intolerant of heat, salinity, close mowing.	Boreal, Dawson, Ensylva, Flyer, Fortress, Merlin, Pennlawn, Ruby	Mow 2 in. plus, light fertilizer. Unmowed for erosion control.	Decorative lawns and utility turf; overseeding subtropical grasses for winter color; unmowed for erosion control; sports turf
Intolerant of heat, salinity.	Banner, Dover, Jamestown, Longfellow, Mary, Koket, Wilma	Mow 2 in. or less, fertilize lightly.	Decorative lawns and utility turf; overseeding subtropical grasses for winter color; sports turf
Limited cold tolerance. Difficult to mow clean. Clumpy if unfertile.	Birdie II, Citation II, Manhattan II, Omega II, Palmer II, Prelude, Yorktown III, and others	Mow 1.5 in. plus with sharp mower. Fertilize and irrigate to keep dense.	Decorative lawns and utility turf; sports turf
Disease susceptible.	Too many to name: Adelphi, Baron, Challenger, Eclipse, Flyking, Merion, Midnight, Pennstar, Touchdown, among many good ones	Mow 1.5 in. plus, keep up lime level, main fertilizing in fall.	Decorative lawns and utility turf; sports turf
Invasive. Severe thatch builder. Disease susceptible in humid climate. Brown in winter.	Common, El Toro, Ormond, Santa Ana, Sunturf, Tifway, and others. Only common bermuda is propagated from seed.	Close frequent mowing, dethatch, overseed for winter color.	Decorative lawns and utility turf; sports turf; fine turf for putting and bowling greens
Coarse. A thatch builder.	Common, Oaklawn	Keep fertilizer low, no lime, dethatch if needed.	Decorative lawns and utility turf
Coarse. Disease susceptible.	Gulf, Tifton 1	Open turf so seed touches soil. Mow as needed.	Overseeding subtropical grasses for winter color
Not cold hardy. Openness favors weed invasion.	Argentine, Pensacola, Tifhi 1, Wilmington	Mow 1.5 in. plus, low fertility. Reduced water need.	Decorative lawns and utility turf; sports turf
Coarse, invasive, disease and insect prone. Vegetative planting.	Bitter blue, Floratam, Floratine	Mow closely and dethatch. Control pests.	Decorative lawns and utility turf
Vegetative planting. Slow recovery of injury. Too dense to overseed.	Meyer, Midwest (Z. japonica); Flawn (Z. matrella); Emerald (hybrid)	Low fertility and water need. Mow closely, or neglect for erosion control.	Decorative lawns and utility turf

TABLE 19–2   A Dichotomous Key to Common Turfgrasses

Both couplets (same numbers) should be read before deciding where to proceed.
Continue to succeeding couplets, eliminating those choices not appropriate until the
species in question is determined based on the information given. Weedy grasses are
not included in this key (see Fig. 21–2).

1—Ligule is a membrane. Leaves are alternate and more or less evenly spaced. Proceed to 2.

1—Ligule is a ciliate membrane or a fringe of hairs. Mixed long and short internodes may result in the appearance of occasional pairs of opposite leaves alternating with single leaves. Proceed to 10.

2—Leaves folded in the bud and opening by unfolding. Proceed to 3.

2—Leaves rolled in the bud and opening by unfurling. Proceed to 5.

3—Leaves of many plants with auricles at the collar. Back side of blades shiny. Red pigment usually present at base of leaf sheath. Perennial ryegrass (Lolium perenne)

3—Distinct auricles not present. Proceed to 4.

4—Tips of leaves boat-shaped. Blades flat with a slight constriction near the tip. Kentucky bluegrass (Poa pratensis).

4—Blades often narrow and needle-like with the margins rolled in. Very fine leaved. Red fescuegrass (Festuca rubra).

5—Auricles present though often little developed. Red pigmentation is usually found at the base of newer leaf sheaths. Blades smooth on the lower surface. Proceed to 6.

5—Not as above. Proceed to 8.

6—Auricles prominent. Margins of the leaf smooth. Annual ryegrass, wintergrass (Lolium multiflorum).

6—Auricles not well developed, margins of leaf rough. Ligule a short truncate membrane. Proceed to 7.

7—A few ciliate hairs are usually found at the margins of the leaf collar. Tall fescuegrass (Festuca arundinacea).

7—No ciliate hairs on the collar. Meadow fescue (Festuca pratensis).

8—Plants spreading by stolons with a well-developed leaf at each node. Ligule elongate to a point. Creeping bentgrass (Agrostis stolonifera).

8—Plant spreading by fine rhizomes or short determinate stolons. Rhizome leaves appear as brown scales. Proceed to 9.

9—Ligule evenly truncate, grass more or less mixing with other grasses. Colonial bentgrass (Agrostis tenuis).

9—Ligule tends to form three peaks of uneven height. Grass forms segregated patches in a mixed turf. Grass has a cast on the bluish side of green. 'Highland' bentgrass (Agrostis spp.)

10—Leaves folded in the bud and opening by unfolding. Proceed to 11.

10—Leaves rolled in the bud and opening by unfurling. Proceed to 14.

11—Leaf sheaths are greatly flattened. Collar is compressed to form a short stalk for the leaf blade. Spreads by stolons. Proceed to 12.

11—Leaf sheaths round or slightly flattened. The collar does not form a petiole. Spreads by both stolons and rhizomes. Proceed to 13.

12—Ligule a fringe of very short hairs. Sheaths have a few hairs at the margins and at the summit of the keeled sheath. The collar is smooth. Saint Augustinegrass (Stenotaphrum secondatum).

12—Ligule is a ciliate membrane. The collar is pubescent and the blades are ciliate. Centipedegrass (Eremochloa ophiuroides).

13—Stolons and rhizomes long and slender (2–4 mm). Bermuda grass, devilgrass (Cynodon dactylon)

13—Stolons and rhizomes short and thick. Bahiagrass (Paspalum notatum)

14—Stolons and rhizomes absent to well developed but slender. Forms a dense turf of fine texture. Manilla grass, mascarenegrass (Zoysia)

14—Stolons and rhizomes short and thick. Forms a somewhat open turf of coarse texture. Bahiagrass (Papalum notatum)

---

that can dominate the area. This is sometimes described as a colonizing species. Many such colonizing species can be considered as weeds. In fact, some of the worst weeds in turf of a desired species are other turfgrass species with a different color, habit, or texture.

Colonial bentgrass (Agrostis tenuis) is a native of northern Europe. Introduced by colonists to the New World, it quickly took over the coastal regions from Rhode Island to Nova Scotia and from Oregon to British Columbia. It also became widespread along the coasts of New Zealand. Kentucky bluegrass (Poa pratensis) is thought to have been introduced at Vincennes, Indiana, by the French about 1720. By the end of the Civil War it was established throughout the Middle West, and the westward settlers thought it was a native grass. Similarly, bermudagrass (Cynodon dactylon) was

introduced into New Mexico in 1750 and within 50 years spread throughout the Southwest.

In any location in the United States, several turfgrasses can be grown and, with good management, all can look good. In a few locations one certain grass is particularly well adapted to the climate. It will gradually invade and dominate lawns in that area. It might be difficult to choose among several grasses when all grow well in an area or to see any reason why one cultivar or another should be preferred. Among good grasses, differences appear only under conditions of stress such as cold, heat, drought, wet soil, shade, low mowing, or disease; each of these take a toll at one time or another. The best species or cultivar for a given location is most likely to be determined, not by appearance or growth habit, but by its ability to survive the few days or weeks in the year when growing conditions are unfavorable. Such characteristics are given in Table 21–1.

Heredity endows certain grasses with the ability to grow and survive stress. Culture determines how well a grass achieves its potential.

SOIL PREPARATION

Soil preparation is a first step in planting a turf, and it includes grading and tilling. Grading should result in a convex swell of the soil surface, free from dips, swales, or pockets where surface runoff water can puddle or pond. Tilling breaks up soil to form a seedbed with good porosity and aeration. In time, tilled soil settles back to its original density, but initial root growth of new turf is aided by the loosened soil.

During tillage, fertilizers and chemical or physical soil amendments should be incorporated in the soil. Chemical amendments include materials such as lime or gypsum used to improve the chemical and physical properties of the soil. Physical amendments are mineral or organic and are generally used to improve the physical properties of a heavy soil or to add organic matter to a biologically impoverished one. Since physical amendments must be used at very high rates to have beneficial effects (often 70 to 90 percent amendment), their use is questionable unless the particular soil problem has been thoroughly analyzed.

Organic amendments serve as food for soil organisms and usually improve soil structure. Peat or manure incorporated into the top 1 to 2 cm (0.4 to 0.8 in.) of the seedbed helps seedlings emerge from crusting soils. Incorporation of other organic wastes improves the root zone environment, but some materials can be toxic (e.g., fresh cedar sawdust). Many organic wastes result in severe temporary nitrogen deficiencies in the soil unless the carbon-to-nitrogen ratio in the waste is less than 20:1. Composting organic wastes for a time before application usually corrects such imbalances and toxicities.

When the seedbed is a sterile subsoil, added organic matter and nitrogen fertilizer enhances biological activity. Where the soil is shallow, a few centimeters of additional soil are often added. A sterile soil may be a fill soil, or it may result from leveling or from spreading out soil from a basement excavation. Sometimes, developers remove topsoil for sale to boost their earnings from a project.

The construction of a field for a sport such as football may require very sophisticated soil preparation. Often synthetic materials are added to the soil components to add stability and improve drainage. Competing weeds are a major problem in raising a new turf. They shade and suppress the grass, and removing them requires time and effort. It is good practice to irrigate the seedbed to germinate weed seeds before sowing grass seed. Weed seedlings are then killed by cultivation or with herbicides. The seedbed should then not be disturbed to the extent that new weed seeds are brought to the surface where they could germinate.

Grass seed is available either in mixtures or pure lots. Most mixtures are designed to be sold to home gardeners rather than to professional turf growers. These seed mixtures represent compromises by the seed companies, and the mixtures will produce an acceptable lawn irrespective of the management given. The label names the species or cultivar of each grass, the percentage of crop and weed seeds, and the percentage of inert matter (Fig. 19–3).

Germination varies from 75 percent for bluegrass seed produced in a poor crop year to over 95 percent for ryegrass seed produced in a good crop year. In general, seed germination should be over 85 percent. Inert matter represents an inevitable amount of chaff or other material. Noxious weed seeds are of small concern in turf as most such weeds are subsequently destroyed by mowing. Turfgrass seed sometimes includes pasture grasses of no concern in a meadow or pasture, but they form coarse, undesirable, persistent weeds in a fine lawn. To avoid contamination by such grasses, seed of sod quality, which is free of such contaminants, is often purchased at a premium price.

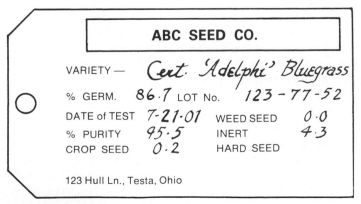

Figure 19–3   Example of a seed tag or label showing the information required by law in the United States.

When named cultivars of grass are used, certified seed carries a certification statement that the plants were inspected while growing in the seed field and found to be pure and true to type.

In the northern two-thirds of the United States, a typical packaged mixture of high quality seed is likely to contain a large percentage of several cultivars of Kentucky bluegrass. This blend forms the basic grass. To this will be added a blend of red fescuegrasses. Red fescue mixes well with bluegrass and grows better in the dry, shady, and less fertile areas. Where winters are not severe, seed of fine-leaved cultivars of perennial ryegrass are also blended with the bluegrass to help the mixture resist seasonal disease problems. A low percentage of a bentgrass is usually added. Bentgrass finally predominates if the resulting lawn is mowed too short for survival of the blue, rye, and fescue grasses. In addition, bentgrass often survives better than the others in areas with wet soils.

Such seed mixtures can be diluted, more or less, with seeds of a filler grass to adjust the price of the mixture. A good filler is meadow fescue (*Festuca pratensis*), which looks like bluegrass during its first few months. Later it becomes coarse, but tends not to persist in a well-tended turf. Red top (*Agrostis alba*) is often used as a filler but is not desirable because it is coarse and persistent. Some mixtures are blended solely for low price and often contain large amounts of pasture grasses or the less desirable turfgrasses.

A high-quality seed mixture selected by a knowledgeable horticulturist might consist of a blend of the first three grasses mentioned, but not include bent or filler grasses. Bentgrass tends to be a weed in bluegrass (and vice versa). Along the coasts of New England, Oregon, Washington, and southern Canada, one might choose a colonial bentgrass alone. Bentgrasses are well adapted in those regions.

In the southern United States a Bermuda grass, centipedegrass, Bahia grass, or carpet grass lawn might be started from pure seed. St. Augustine grass or hybrid Bermuda grass could be started from stolons or sod.

SEEDING

Once the kind of grass is chosen, the rate at which to sow the seed is considered. Seedlings become crowded and are unable to develop properly when seed is sown too heavily; it takes a long time to get a mature usable lawn. Sown too thinly, the plants are far apart, with space left for weeds to start. An initial stand of about one plant per 1 to 5 cm^2 (0.4 to 2 in.^{2}) will develop rapidly into a strong turf. If the soil is free of weeds and if seed is sown in season (best, six to eight weeks before autumn frost; second best, early spring), 0.25 to 0.5 kg per are^2 (0.5 to 1 lb per 1,000 ft^2) is a good rate to sow seed. In weedy soil, double the rate. If one sows off season (e.g., late

fall or early winter) when seed germination is slow, one should increase the rate. For sowing into weedy soil under favorable conditions one should use as much as 1 to 2 kg of seed per are (2 to 4 lb of seed per 1,000 ft^2). The 2 kg rate is high except for tall fescue and perennial ryegrasses. Seed of these grasses is coarse and a still higher rate can be used.

Special machines are available for sowing seed, covering it, firming the soil, and even mulching it. For hand operations, however, either box or cyclone fertilizer spreaders sow seed satisfactorily. The opening is reduced to a size appropriate to the seed. Hand sowing tends to scatter seed unevenly since few persons are skilled seed sowers. When hand sowing is necessary, seed should be vigorously thrown forward in a sweeping arc. The falling seed is more apt to drift into a random pattern than when seeds are dribbled out close to the ground.

When both large and small seeds are used, they are best sown separately in two operations. On small areas, rather than calibrate the spreader, the operator may prefer to reduce the seeder opening to a low rate, then cover the area two or three times in different directions to insure even coverage. Small amounts of seed are often diluted with sand to adapt the volume of seed to the area covered.

Seeding Depth

Another determination is the depth to cover the seed. Turfgrass seeds are small, varying from about 1 million per kilogram for rye and fescue grasses to over 4 million for bluegrasses and over 10 to 17 million for bentgrasses. Relatively few seedlings emerge from depths over 1 cm (0.4 in), and seedlings' emergence for the smaller seeds is best when they are covered to less than 1 to 5 mm. Uniform coverage is not generally possible. A light raking or dragging of the soil surface after seeds are sown covers seeds at depths from 0 to 5 or 10 mm. Some seedlings on the surface die of desiccation, and some deeper ones fail to grow to the surface, but seeding rates allow for such losses.

Time to Sow

At any given geographical location there is a best week in the year in which to sow grass seed. In the southern parts of the United States subtropical grasses are best sown in early summer after annual weeds are removed following their principal flush of seed germination. Seeds of temperate zone grasses are best sown in late summer or early fall so that there is time for the seedlings to become well established and to cover the ground before freezing weather arrives. Fine, vigorous grass stands are easiest to obtain with fall-sown seed. Fall weed problems are reduced, and shorter days and cooler temperatures reduce evaporation and the need for frequent irrigation. The next best time to sow grass seed in the temperate zones is as early in the spring as the soil can be prepared. Grass seed will germinate and the seedlings will grow in cool soil; if a dense stand can be obtained before summer weed seeds germinate, the weed problem is reduced.

---

^{2}1 are = 100 m^2. 100 are = 1 hectare.

After the seed is sown, the seedbed should be rolled, unless one is relying on rainfall to germinate the seed. Rolling firms the soil around the seed and thus encourages capillary movement of available moisture. If there is insufficient moisture for good seed germination, a mulch should be used. A light cover (2 to 5 mm) of clean, weed-free sand or organic waste slows moisture loss from the soil surface. However, light mulching materials such as peat moss tend to wash or blow or to gather in pockets. Such materials are best worked into the top inch (2.5 cm) of soil before seeding.

Seed Germination and Seedling Establishment

Water is the most critical factor during seed germination, whereas nitrogen fertilizer is more critical during seedling establishment. In warm weather most turf seeds begin to germinate five to seven days after sowing and continue for another week. Bluegrass seed, however, is slower and continues to germinate for a month. Germination is slower in cold weather. During germination, soil moisture is necessary in the surface layer at all times. Differences in available water in the root zone soon appear as differences in color and stand of the grass seedlings. At the same time, excess surface water encourages damping-off pathogens, a complex of *Rhizoctonia, Phytophthora, Pythium, Fusarium,* and other species of fungi. Ideally the top soil layer should remain moist, but if the soil surface is allowed to dry at least once a day the mycelia of damping-off organisms shrivel and die.

Once germination has occurred, seedling roots begin to explore the soil. The period between irrigations is gradually extended so there is regular drying of the soil surface between irrigations. Irrigations should wet the soil deeper than the roots extend.

As grass seedlings grow, they may deplete the soil of nitrogen. Growth slows, and the seedlings' color becomes pale green to yellowish. Regular feedings of nitrogen fertilizer to provide about 0.25 kg of N per are (0.5 lb/1000 ft^2) keep the grass growing vigorously and help it suppress weeds. If seedling growth slows and the blade color is dark green, with some red anthocyanin pigment present, the seedbed probably contains insufficient phosphorus for initial seedling establishment.

As the turf grows and becomes thick and tall enough for the first mowing, the most suitable height for mowing must be decided. The mower should be sharp and set for the correct height. A dull mower pulls up seedling plants or tears leaves instead of shearing them. When the new grass is 2.5 to 5 cm (1 to 2 in.) higher than the desired mowing height, the soil should be allowed to dry for a day or two before the grass is mowed. If the clippings are scattered and the weather is dry, clippings will shrivel and fall from sight. But if the clippings form heavy clumps and the weather is moist, they should be removed to prevent smothering and a good environment for diseases.

In the United States it may take 5 months in southern states to 12 months in northern states after seed germination for the turf to form a vigorous mature lawn rugged enough for play.

SODDING

Sodding is a popular and quick alternative to growing turf from seed, especially where only small areas are involved. Sod is pregrown turf cut to include the adhering top 1 to 2 cm (0.4 to 0.8 in.) of soil. A high-quality sod provides a mature weed-free turf of desirable grasses and establishes itself during the first month after planting to give an "instant" lawn. Sod has the same requirements as seed except that the seedbed grade must be lowered next to walks, drives, and buildings to accommodate the thickness of the sod, and some soil must be stockpiled for topdressing the sod to level it and smooth the surface. Thinly cut sod produces roots more rapidly, but dries out faster, than thick sod. Sod cut with 1 to 2 cm (0.4 to 0.8 in.) of soil usually has an adequate reserve of moisture to sustain it during transport and a short period of storage and readily produces new roots. Best results from sod come when it is carefully laid in the fall or spring on moist, well-tilled, weed-free soil to which nitrogen and phosphorus fertilizers have been added. Newly laid sod should be watered within an hour. Sod of subtropical grasses establishes itself most vigorously in the summer.

MAINTENANCE

Maintenance begins once the grass is up. Maintenance consists primarily of mowing, fertilizing, irrigating, and controlling weeds, insects, and diseases. These procedures should be programmed to produce a beautiful dense turf and vigorous healthy plants. As noted earlier, these goals are not completely compatible. There must be some compromise area appropriate to the climate, to the equipment available, and to the level of maintenance one is prepared to pursue.

Not only must the chosen program compromise between beauty and vigor, but there is the added possibility that a goal can be achieved equally well with different management programs. If the desired result is achieved with reasonable economy of effort and resources, no program is more right or more wrong than another. The lack of positive answers makes turf management comparatively difficult or confusing for some and challenging to others. For this reason turf culture is best considered in terms of principles rather than applications.

PRINCIPAL TURF MANAGEMENT PRACTICES

There are three principal turf management practices and many secondary practices. The principal ones are mowing, fertilization, and irrigation. Each of these has a large effect on grass growth, and the manager manipulates them to change grass vigor and appearance.

Mowing

Mowing is a regular chore. There are a number of choices to make in mowing; for example, the kind of equipment to use, the mowing height, frequency of mowing, and whether to remove clippings or leave them.

Mowing Equipment

There are four kinds of modern mowing machines for lawns. Two kinds of power mowers—flail and sickle bar—are used only for large scale and heavy work.

The flail mower is used on roadsides and in rough park areas. It has a rapidly rotating horizontal axle that swings a number of vertical knives inside a housing. The mower requires high energy, mows tall grasses and weeds, and reduces them to a mulch.

The sickle bar mower resembles a hair clipper with a 2m (6 ft) horizontal blade. It uses low energy; mows grass, weeds, and woody seedlings; and lays them in a swath with the stems parallel, convenient for raking. This mower is the same type used to mow forage hay.

The rotary mower rotates a horizontal blade on a vertical axle at high speeds. It uses high energy, cuts by impact, handles tall grass, and may or may not recut leaves to reduce them to a mulch. The rotary mower is available from a 40 cm (16 in.) diameter blade home mower to a large industrial size that cuts a swath 1 m (3.3 ft) or more wide.

The reel mower uses a series of helical blades that gather the grass and shear it off against a stationary blade known as the bed knife. It is a low-energy mower. It does not cut tall grass well, but it produces the highest quality cut of all kinds of mowers.

Reel and sickle bar mowers cut by a clean cut shearing action with low energy input. Rotary and flail mowers cut by impact. Since the blade must move at a high velocity, they require more energy. Impact cuts are often ragged and the high velocity increases the potential danger of accidents.

Mowers should be sharp. Grass recovers from a clean cut from a sharp blade more readily than from a bruising cut made with a dull blade.

Mowing Frequency

Increased mowing frequency increases stress, but results in less stress than mowing grass short. Stress is greatest when a high percentage of the leaf surface is removed in one mowing. The high stress results from the combination of a short cut with a long interval between mowings. In general, it is not desirable to mow more often than every fifth day during the growing season. An interval of more than about 10 days between mowings is undesirable for a neat, formal appearance.

Mowing Height

Decreasing mowing height gives the turf a carpetlike appearance, but greatly increases stress. There are upper limits for mowing height. Bentgrasses mowed higher than 2 to 3

MOWING STRESS

Mowing places a stress on the grass and reduces its vigor. Root growth is slowed for a few hours to several days depending on the severity of mowing. The total leaf production for the season is reduced by each mowing. Mowing opens the grass canopy to allow more light to enter so more plants can grow in the same area, thus increasing competition. The greatest stress is added to the stress budget when grass is mowed very short. High populations of weak plants result. With exceptionally vigorous and invasive grasses, such as bermuda grass or kikuyu grass, frequent short mowing can be used to deliberately weaken them.

cm (0.8 to 1.1 in.) tend to become puffy and somewhat coarse-textured. When mowed above about 8 cm (3 in.), bluegrasses, fescues, and ryegrasses tend to appear open and stemmy. Where an informal or neglected ground cover is wanted on a steep bank or around a vacation cabin, red fescue or zoysia grasses can be left unmowed except for yearly removal of flower stalks. Some zoysia and other subtropical grasses can be mowed less than 2 cm (0.8 in.) to control their natural vigor and encourage a renewed, finer-textured growth.

Clippings

When clippings are left, nutrients are recycled in balanced amounts. Turf appears consistently better fertilized if clippings are allowed to remain and fertilizer added once or twice a year than if clippings are removed and fertilizer is applied every few weeks throughout the growing season. There is no particular virtue in removing clippings. A thatch of dead plant material builds up above the soil layer on many lawns, but thatch results primarily from increased use of fertilizers, water, and pesticides rather than from fallen clippings. If grass is managed so short pieces of mowed leaves fall out of sight between grass blades, turf nutrition benefits from leaving the clippings. But if long clippings fall on a short, tight turf, they appear unsightly and should be removed, but at the cost of greater stress. Grass growth varies throughout the season and mowing should be adjusted to seasonal changes in growth.

Fertilizer

The response of lawns to fertilizer use can be placed in three categories.

Nitrogen

A turf does not become thick and dense without adequate nitrogen from repeated fertilizations. On a new turf, fertilizer applications should add 3 to 5 kg of N per are (6 to 11 lb N/1000 ft^2) before dense turf is achieved.

Once turf is thick and dense, nutrients can be recycled by leaving the clippings. As some nitrogen is lost through leaching, one or two applications of 0.5 kg of N per are (1 lb N/1000 ft^2) should be applied to a turf early in the fall. But if clippings are removed, nitrogen has to be supplied throughout the year. Other nutrients can be added on a yearly basis.

Nitrogen fertilizers produce a deep green grass of high density but with a shortened root system. Fall color holds later and spring greenup comes sooner with higher levels of N. If N fertilization is overdone, however, susceptibility to disease and stress tends to increase.

Phosphorus and Potassium

Phosphoric acid (P_2O_5) and potash (K_2O) are mineral nutrients used in macro amounts by turfgrasses. In many parts of the world, soil P_2O_5 and K_2O are adequate and need not be added, although a light application of P_2O_5 in the soil hastens seedling or sod establishment. The local county agricultural agent knows whether local soils are deficient in P_2O_5 and K_2O. If clippings are removed, added phosphorus and potassium will undoubtedly be needed for good grass growth. The frequency and amount of application varies considerably with local conditions. Some turfgrass fertilizers are especially formulated for gardeners who remove lawn clippings. These contain mineral nutrients in about the proportion in which they are removed in clippings (i.e., 5 N:2 P_2O_5:3 K_2O).

Calcium, Magnesium, and Micronutrients

Soils become acid when calcium or magnesium are deficient. Acid soils are generally low in fertility and, in addition, the availability of some nutrients is reduced. Calcium or magnesium fertilizers are seldom used on turfgrasses unless they are required to correct soil reaction (pH).

In much of the western United States, soils are alkaline (pH over 7.0). This can decrease the availability of phosphorus and certain micronutrients to turfgrasses. Except for using acid residue fertilizers, nothing is usually done to alleviate high pH. As excess sodium adversely affects both soil and plants, calcium as gypsum (calcium sulfate) is applied to counteract the sodium problem.

Micronutrient problems are uncommon in turf if the soil pH is between 5.6 and 6.8. Any micronutrient deficiencies must be diagnosed and dealt with as local problems.

Irrigation

Almost everywhere in the United States rain is deficient at some time during the year; however, in the eastern one-third of the country rain is often excessive. The principal value of irrigation is to prevent severe drought stress, but in arid and semiarid areas irrigation is required to make turf culture possible. The management of water can affect weediness of turf, as either overirrigation or underirrigation favors growth of certain weed species.

Proficiency in turf irrigation is attained when knowledge of the soil, climate, exposure, and grass is integrated into a seasonally changing program. Figure 19–4 shows relative amounts of water needed to wet soils of various texture to various depths. Turfgrasses have most of their roots in the top 5 to 10 cm (2 to 4 in.) of soil. If keeping grasses in active growth under bright summer skies is desired, this soil depth should be wetted with about two irrigations per week (or

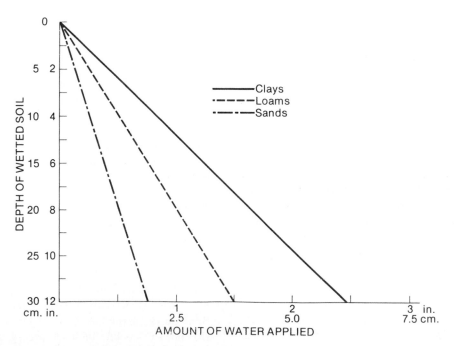

Figure 19–4 Water applied to the soil completely wets it to about the depths shown.

three where the weather is windy or hot). Some subtropical grasses are deep-rooted. If the soil is soaked deeply about once a month, these grasses continue to produce some replacement growth. Both growth and water use by the plants are less than when water is applied more frequently. Rooting depth and water use of turfgrasses are affected by many variables, and exceptions are common and plentiful.

Even when severe drought dries turf, some individual plants of all species recover following rain or irrigation. Bermuda grass recovers well in the South and Southwest, and in the North fescues and 'Highland' bentgrass recover best. Turfgrass breeding could provide more drought-tolerant selections of most turf species.

SECONDARY MANAGEMENT PRACTICES

Pest Control

Problems of pest control are universal. Weeds, diseases, insects, nematodes, and other pests all afflict turfgrasses. Since such organisms tend to occupy the environment to the full extent of its capacity to support them, turf always has insects feeding on grass leaves and roots, fungi consuming dead grass clippings, and weeds filling in any bare spots of soil.

Insects

Insects attacking turf include various caterpillars, beetle grubs, bill bugs, wireworms, flea beetles, chinch bugs, fruit flies, leafhoppers, scales, and aphids. In the spider group, mites attack certain species of grass. Of all these, caterpillars and beetle grubs are likely to affect all lawns.

Caterpillars, the larvae of various species of moths, build silk-lined burrows among the grass crowns and feed nocturnally on the leaves. Caterpillars stay close to their burrows, and each thins out a small area of grass around its burrow, about 2 or 3 cm (0.8 to 1.2 in.) in diameter. Damage is inconsequential until populations build up to several dozens per square meter. At that point, turfgrass is rapidly thinned out as caterpillars mature and their appetites increase. To treat this problem, grass is mowed to reduce leaf area. Leaves are then sprayed with a stomach poison or contact insecticide. Irrigation water is withheld for a day or two so the insecticide is not washed from the leaves.

Grubs that live in the soil and feed on roots are a different problem since it is difficult to get insecticide into the soil at control levels. In their early growth stages grubs are small and seldom a problem. But in the fall or the spring following their emergence, they are large and hungry and they eat so many roots that patches of grass die. The grubs will be found 2.5 to 5 cm (1 to 2 in.) deep in the soil at the edges of the brown dead patches.

Weeds

A weedy turf is symptomatic of poor management and often indicates that not enough attention has been given to practices that produce a vigorous turf. Weeds are frequent when turf is undernourished, overwatered, or mowed too short. Soils compacted by heavy traffic tend to grow poor turf but do support a large population of certain weed species.

Unwanted perennial grasses are the most difficult weeds to control in turf; broad-leaved annuals are the easiest to control.

The first step in weed control is to improve management to encourage the grass to grow vigorously. Then weeds can be dug, pulled, or treated with an herbicide. With the grass growing vigorously, space formerly occupied by a weed fills with grass. Control is then successful. If the space is recolonized by other weeds, management practices should be reexamined as the first step in further control efforts.

Herbicides

Herbicides used to control turf weeds should be used cautiously and with restraint. They are plant poisons that at recommended rates are more toxic to weeds than to the turfgrasses. Herbicides recommended for weed control on turf also injure and cause stress to the turfgrass plants, though the turf outgrows the injury in time. If herbicides are used repeatedly or at times when the grass is under stress or not growing well, the grass might be so retarded that weed problems worsen.

Contact herbicides kill plants they touch and are sometimes used as spot sprays to kill a few individual difficult weeds in a turf. The results, however, are unsightly spots that the grass slowly recolonizes.

Preemergence herbicides control annual weeds by soil application before weed seeds germinate. The herbicide kills seedlings as they push through the treated surface soil layer.

Postemergence selective herbicides are used to remove broad-leaved weeds from grass. Such herbicides are most effective on weeds in the seedling stage. As weeds become older, they become more resistant. Many species of weeds are killed by routine mowing, which continually defoliates them. Weeds that persist tend to be low-growing species that spread out below the mower blades.

If a lawn has only scattered weeds, it is more prudent to dig them out than to mix chemicals and wash spray tanks and thus avoid the risk of spray drift onto garden flowers. The county agricultural agent or local garden center can recommend suitable herbicides to use on lawns.

Diseases

Three important turf diseases are mildew, rust, and smut. Mildew forms a white dust on grass leaves in the shade and can be controlled only by opening up the area to provide the grass with better light and ventilation. Rust is sometimes prominent in the fall when the orange fruiting bodies discolor grass. Fertilizing with nitrogen usually controls rust. Smut occurs in the late spring or summer as a grass disease that produces a line of black greasy spores along the leaf blade.

Soils contain many saprophytic fungi, which live on dead leaves and other soil organic matter. Some of these are **facultative** parasites which cause turfgrass diseases if predisposing factors are present. These diseases are often most common on the best cared for lawns. For example, a *Pythium* water mold destroys turf when fertility, moisture, and temperature are all at high levels. Excess water soaks the soil, causing stress because of poor soil aeration. Add the stress of high temperatures favorable for the growth of *Pythium,* and the pathogen rapidly kills the nitrogen-rich grass.

Among diseases that commonly appear on the ordinary lawn are *Fusarium* and *Rhizoctonia* brown patch, both of which form rings of dying turf in the summer. These diseases are predisposed by heat stress on the grass. When snow covers turf for a long time, the melting snow reveals patches of dead grass killed by winter-active organisms, such as *Fusarium* and *Typhula* species. Cold and darkness encourage these organisms to cause turf disease. Long periods of wet overcast weather in the spring or fall favor growth of the *Helminthosporium* "melting out" pathogen. Growth of *Sclerotina,* or dollar spot, is favored by nitrogen-deficient grass. Organisms causing all of these diseases are generally present but do not infect the grass until a particular factor or combination of factors develops, such as dark overcast days, winter snow cover, high summer temperatures, or excessively wet or compacted soils. The environment must be favorable.

When fungal lawn diseases are a problem, observation shows that the grass areas most affected are those predisposed to fungal attack by such factors as compacted soil, reflected heat onto a lawn by buildings or fences, or excessively short mowing. In the southern states of the United States, hot spots are not a problem for the subtropical grasses; instead, shaded or wet spots lead to diseased turf.

Salt Damage

Salt used to remove snow and ice damages turf bordering sidewalks and roads. The excess salt causes turf injury and often death. The treatment for this problem is to apply lime on acid soils and gypsum on alkaline soils so that calcium replaces the sodium used in the snow-melting mixture.

Scald, a nonpathogenic problem, occurs on turf when puddles of water overheat in the sun, causing heat injury to the grass plants.

Compaction

Soil compaction is a stress factor on lawns. On expensive turf, such as a golf putting green, the compaction problem can be solved by removing the loam soil and substituting a sand, which remains porous and well aerated even when compacted. Most lawn soils cannot be easily changed, though, and are treated by cultivation practices. For small isolated areas, the soil can be loosened by inserting a garden fork in the turf and prying up gently to crack and fracture the soil without disturbing the turf. A hollow tined spading fork,

built specially for hand cultivating, can be used to loosen and aerate the turf.

Turf Cultivation

Cultivation is done mechanically by coring machines or by machines that slice the soil. Coring is desirable for compacted soils. It improves aeration, allows the soil to dry faster, and aids entry of water. The general effect of coring or slicing is to relieve compaction in the surface by moving the compacted layer down to 5 to 7 cm (2 to 3 in.) below the surface.

Overseeding

A common practice in the southern United States is to overseed dormant subtropical grasses whose leaves become brown during the winter. To keep the lawn green, seeds of a cool-season species are sown in the dormant turf in the fall. This provides winter color until the warming soil and vigorous spring growth of the subtropical grasses crowds and suppresses the "winter grass." The traditional grass for such overseeding is annual ryegrass *(Lolium multiflorum)*, but red fescue or perennial ryegrasses are also suitable and are finer-textured. Other cool-season grasses are also used.

Timing is important in overseeding. There should be enough warm weather yet to come for germination and seedling growth, but if overseeding is done too early, hot weather is likely to encourage growth of **damping-off** pathogens.

Renovation

An old, thin, or weedy lawn can be renovated by introducing new seed with improved management practices. To germinate and become established, the new seed must contact soil. A seedbed for either overseeding or renovation is prepared in the existing turf by raking vigorously, by power raking, or by using a coring, thatching, or vertical mowing machine to tear out a dense dead thatch and weak plants and to expose soil between the grass plants. Seed is sown, then kept moist during germination by sprinkling. Sown seed should also be lightly topdressed with compost or sand. As soon as germination begins, a fertilizer should be applied to provide both nitrogen and a small amount of phosphorus. In renovating, an herbicide can be used in advance to kill undesirable grasses in portions of the lawn.

Thatch Control

Stolons of stoloniferous grasses, such as bent and bermuda grasses, are stems that creep along the surface of the ground and, dying, build up a thatch layer of organic matter resistant to decay. Dried thatch often sheds water, creating areas of dry soil. When thatch is wetted by frequent rains, grass plants tend to produce growing roots in the shallow thatch, but few or no roots in the soil. Such plants do not withstand drought. A deep thatch layer is puffy and attracts offensive insects such as earwigs. Thatch is removed by hand or power raking or by a special thatching machine. Thatching and raking machines have a rotating horizontal shaft. Fingers of spring steel wire are attached to the shaft of the rake. A series of coarse-toothed rotary blades on the shaft of the thatching machine cut or tear out the thatch. Thatching may be done annually as preparation for overseeding. If overseeding is not done, thatching is best done at a season when vigorous growth favors grass recovery.

Watering In

When soluble fertilizers or other concentrated chemicals are applied to turf, they can burn the tissues by plasmolyzing the cells. Such chemicals should be thoroughly watered in. Sufficient water is applied to dissolve the chemicals and wash them from the plant and into the soil as a dilute solution.

Miscellaneous Management Practices

When frost heaving has raised grass crowns out of the ground in cold-winter areas, they can be pressed back into the soil by rolling.

Dyes are available for coloring brown or off-colored areas of turf. Such cosmetic practices are considered important when visible turf is a part of television presentations, movies, or public spectacles.

Plugs removed from sod can be planted into another turf to introduce a new grass species. Plugs are also used to repair small damaged areas. Devices are available to cut and remove plugs of various sizes and to cut similar holes in the soil to receive the plugs.

If grass reaches a stage of incipient wilt, it takes on a recognizable gray cast. Spraying the grass with water will relieve such water stress for a few hours. This can preserve grass from desiccation when thorough irrigation must be delayed because of sports or other activities.

Soil, compost, sand, or other such materials can be applied to turf as a topdressing. Topdressing is used to level an uneven surface, to fill a hollow, or is worked into thatch to firm the loose organic surface.

Turf growth is often retarded by competition with tree roots. Judicious pruning of the tree's roots often greatly improves a turf.

Golf, tennis, and bowling greens are precise playing surfaces with grass under severe stress. Special management practices are used to maintain optimum playability of these critical surfaces. These are discussed in detail in some of the cited references.

Seasonal Growth

The seasonality of grass growth is illustrated in Figure 19–5. The dotted curve in Figure 19–5 can be applied either to subtropical grasses growing in the southern United States or to temperate grasses growing in northern states or at high elevations. Low temperatures limit winter growth. As the season warms in the spring, growth increases, peaking in the summer months, provided drought is not limiting. After late

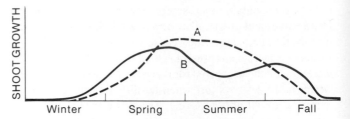

Figure 19–5 Seasonal growth of grasses. The solid line represents a seasonal growth pattern for Kentucky bluegrass, a typical temperate grass. The start of spring growth occurs earlier or later depending on the local climate. Summer dormancy is more or less severe depending on the degree of heat and water stress. The dotted curve represents growth of the subtropical grasses such as Bermuda grass. Most of these make their greatest growth during the hottest weather, provided water is adequate. In the southernmost areas of the United States growth continues through the winter at a low rate.

summer, growth declines until it is again stopped by low winter temperatures. Growth of temperate grasses over most of the urban United States is illustrated by the solid line in Figure 19–5. Depending on winter temperatures, growth can either cease, or slow down in the transition zone. The solid curve is similar to the dotted curve except that during the summer months, temperature optima are exceeded and growth slows. If high temperatures are accompanied by drought, grass can become semidormant, with little growth in summer. As temperatures cool in late summer, active growth resumes. Fall growth of Kentucky bluegrass differs from spring growth. Fall growth spreads and does not produce as many clippings. Growth comprises mostly new tillers and emerging rhizomes. The spreading form of growth decreases with dropping fall temperatures.

An awareness of the annual growth cycle in local regions can help in planning the best time for various operations. Coring, raking, and thatching are operations that should be done when growth is vigorous so recovery will be rapid. Herbicides are best applied during the rising growth curve in spring, when injury to the grass would be low because of cool temperatures, and when rapid growth leads to rapid recovery.

Fertilizing Kentucky bluegrass during the rising growth curve in spring leads to excessive leaf growth. The same fertilizer applied during the rise of the growth curve in late summer results in more tillers and a more spreading growth. New leaves promoted by the fertilizer photosynthesize extra carbohydrate for winter storage.

The falling curve in summer represents a period of stress. This is an appropriate time to apply preventive fungicides—at the time predisposing factors develop. Reduced summer growth results from climatic stress, so the manager is careful not to add extra stress at this time. One would not reduce mowing height nor increase mowing frequency at this time; instead, mow higher and less often.

It is best not to apply nitrogen fertilizer during summer stress. Heat stress results in part from depletion of carbohydrates through higher respiration rates accompanying higher temperatures. Mowing also reduces carbohydrate reserves by reducing the leaf surface available for photosynthesis. Nitrogen fertilizer also depletes carbohydrate reserves, diverting them to protein synthesis. Nitrogen fertilizer at this time puts an extra demand on the plant that can weaken its resistance to disease. Thus, during summer, it is best to withhold

fertilizer or, if necessary, apply it only in small amounts of 0.125 kg/are (0.25 lb/1000 ft^2) or in forms that are slowly available to the plant.

When the growth curve rises in the fall, plants can take extra stress. The low spreading growth pattern and the favorable weather enable the turf to recover from all sorts of abuse.

SUMMARY AND REVIEW

Lawn, turf, and turfgrass in modern usage refers to the ground where grass is growing; grass that is mowed and cared for; and grass in growing in a horticultural turf, respectively. Determinate, rhizome, stolon, and intercalary meristem refer to the type of growth associated with turfgrasses. Subtropical (warm-season) grasses do well in warm areas of the United States but go dormant in the winter in transition zones. Temperate (cool-season) grasses grow well in the cooler parts of the United States, but go dormant in the heat of summer in transition zones. Overseeding refers to the practice of sowing an annual temperate grass on top of a warm-season grass in warm-season transition zones. Sodding is the installation of pre-grown mats of turf to areas for a quick covering of grass.

Establishing turf first requires proper preparation of the land. Selection of the appropriate turfgrass species or cultivar(s) is critical. Correct seeding rates and time of seeding determine how well the turf will establish and cover the area.

Maintenance is a demanding part of turf management because both the condition and appearance of the grass must be excellent at all times during use. Mowing must be done with the proper equipment at the appropriate intervals to the proper height. Irrigation and fertilization programs are determined by the type of grass being grown, the environment, and other cultural practices in use. Weed, insect, and disease control is also a management consideration. Perennial grasses are the most difficult weeds to control in turf. Grubs and caterpillars are the most difficult pests. Fungal diseases are the biggest problem for turf. Wet conditions favor most of the fungal pathogens.

The bentgrasses, fescuegrasses, Kentucky bluegrass, and perennial ryegrass are temperate grasses. Bermuda grass, centipedegrass, Bahiagrass, St. Augustine grass, and zoysia grass are subtropical grasses. Annual ryegrass is a temperate grass used only for overseeding in the winter.

EXERCISES FOR UNDERSTANDING

19–1. Looking at Figure 19–1, determine the major area you live in. Consider the microclimate for your local area. How would your microclimate affect the species of turfgrass you would choose for a golf course or athletic field?

19–2. Visit a local public golf course or high school athletic field. Identify the turfgrass(es) growing. Try to determine what maintenance procedures are used and their effect on the quality of the turf.

REFERENCES AND SUPPLEMENTARY READING

CHRISTIANS, N. E. 1998. *Fundamentals of turfgrass management.* Chelsea, Mich.: Ann Arbor Press.

DANNEBERGER, T. K. 1993. *Turfgrass ecology and management.* Cleveland, Ohio: Franzak and Foster.

EMMONS, R. D. 1995. *Turfgrass science and management.* Second Edition. Albany, N.Y.: Delmar.

TURGEON, A. J. 1996. *Turfgrass management.* Fourth Edition. Upper Saddle River, N.J.: Prentice Hall.

CHAPTER

20

Residential and Public Landscapes

KEY LEARNING CONCEPTS

After studying the chapter you should be able to:

♦ Understand the history and principles of residential and public landscaping.
♦ Understand the many functions plants serve in landscapes.

The gardens and landscape ideals of the Western culture began with the oases of Babylon, Egypt, and the succeeding civilizations of the Mediterranean region. At the beginning of gardening history, there was no distinction between agriculture and horticulture. The early Middle Eastern gardens were filled only with useful plants, which provided fruit, fiber, medicinal products, and forage for domestic animals. Such gardens were often walled and organized around a well, pool, or fountain. By the Roman era, formal gardens of the very wealthy were deliberately ornamental rather than productive. They were organized around a central water feature with topiary shrubs, groves and bosques planted for shade, and container plantings of colorful shrubs and herbaceous plants as well as herbs and fruiting plants. The spread of Islam in the seventh century brought this kind of garden to southwestern Europe, northern Africa, India, and the Far East. Examples of this kind of oasis garden still exist at the Alhambra in Granada, Spain (Fig. 20–1) and at the Taj Mahal in India. The Spanish brought this garden motif to the New World, and many features of the court-

Figure 20–1 A garden in the Alhambra at Granada, Spain created before the time of Columbus and still maintained today as a tourist attraction. The fountains are supplied with water from the surrounding mountains to create a feeling of coolness. High walls surround the small gardens, which contain small fruit trees, closely clipped shrubs, and plants grown in containers. An Islamic garden such as this is always symmetrical; i.e., the left side is a mirror image of the right.

yard and patio gardens of the American Southwest are derived from this tradition.

During the Dark Ages, the Roman gardening style barely survived in the herbal knot gardens and utilitarian vegetable plots of medieval cloisters. Then, at the beginning of the cultural Renaissance in Italy, formal gardening once again blossomed, but plantings were subordinated to terracing, walls, fountains, and other architectural features. Geometric landscape designs defined by sheared shrubs and trees soon became the standard of the noble estates, spreading north to Austria, France, and eventually to England. These gardens of the nobility were ostentatious and demonstrated an intense level of maintenance with vast layouts of clipped hedges, gravel walks edged by carpets of bedding plants, and containers of tender trees, such as citrus, that required overwintering in glasshouses or "orangeries." The gardens at Versailles are a good surviving example of such gardens.

The English nobility soon tired of the rigid, geometric gardens of the "rationalist" Continentals. By the end of the eighteenth century, under the influence of landscape designers Humphrey Repton and Lancelot "Capability" Brown, the formal parterres and hedges of English estates had been largely replaced with "romantic" landscapes. These included sweeping meadows and lawns with ponds, streams and grottoes, and naturalistic copses and groves that framed views of ruins and "follies" built specially for the purpose. Meanwhile, the cottagers of rural England grew vegetables, fruiting trees, shrubs and vines, and annual flowers in great profusion in very small garden areas. It was a hybrid landscape ideal of estate lawns (really pastures for grazing animals) and open woodlands with very dense, colorful, and productive cottage gardens, which the English settlers brought with them to America and tried to re-create while clearing the native forests of the Atlantic seaboard.

East Asian gardening and landscaping, exemplified in China and Japan, began many centuries before European history, and, unlike Western horticulture, was an integral part of the religious and cultural tradition (Fig. 20–2). Completely separate from any production of food or other plant product, Oriental gardens are symbolic miniatures in which forests, rivers, and mountains are represented in formulaic and highly maintained landscapes. In contrast to the public and open nature of Western gardening, the Asian garden tradition has been one of privacy, contemplation, and ceremonial isolation. These elements, too, have been imported to North America, particularly along the West Coast.

Landscaping has many levels of meaning in the United States, from vegetable and flower gardening on the residential scale, to institutional landscapes such as campuses and large public parks in cities, to natural and recreational areas.

Figure 20–2 A garden in Kyoto, Japan, using pines and small shrubs that must be pruned often to maintain the same size. Natural rocks and trees play an important part in Japanese gardens. *Source:* Hazel Hartman.

Figure 20–3 It is dangerous to assume that mature trees can be "saved" when new construction is planned very close to trunks. Any excavation within the drip line, along with compaction and alteration of soil drainage patterns, usually causes the decline or death of existing trees.

Landscape architects, horticulturists and garden designers, and private homeowners all contribute to the human-constructed landscape (Figs. 20–3, 20–4, and 20–5).

THE RESIDENTIAL LANDSCAPE

For better or worse, the lawn is the dominant element of almost every American residential landscape, whether in San Diego, Boston, Grand Forks, or Miami. Although the seeded turf lawn is the least-expensive landscape element to install, it is by far the most expensive and time-consuming to maintain. However, even homeowners who do not set foot outside their house except to mow the lawn generally do not consider any other groundcover suitable. In fact, most municipalities dictate that single-family residences must have lawns maintained to strict specifications of height and "weed-free" condition. With proper plant selection, it is possible to have seasonal color, changing throughout the year, even in the harshest climate zones. However, given the rather low expectations of most American homeowners—which consist of evergreen foundation shrubs, expansive lawn, and an occasional shade tree—the residential landscapes become maintenance problems or do not often reach their full potential (Fig. 20–6). Ironically, a majority of Americans, when polled, list gardening as their major hobby or pastime.

Edible landscape plants (berry bushes, fruit trees, and annual herbs and vegetables) can be planted in most climates, and "organic" landscape practices (such as mulching with nutrient-rich compost) can benefit the soils and plant

Figure 20–4 A naturalistic landscape is best when composed of both evergreen and deciduous trees and shrubs and groundcovers, with central attention to layering of plants in appropriate combinations. *Source:* From Derek Fell, 1998. *Shade gardening*. New York: Friedman Fairfax.

Figure 20–6 *Above:* A cluster of recently planted *Thuja orientalis* in a bank parking lot. These are about 2 ft (60 cm) tall but will grow rapidly if given frequent irrigations. *Below:* This, too, is a *Thuja orientalis,* about 30 years after planting, overwhelming the house. Once it starts growing usually the homeowner does not know how to prune it aesthetically, so it is allowed to take over. Planning is important in knowing what to plant and how to control it.

Figure 20–5 *Top:* A typical, well-maintained highway rest stop providing picnic facilities and pleasant surroundings for automobile travelers. The trees, from left to right, are pines, birches, and purple-leaved plums. *Middle:* One tiny corner of Golden Gate Park in San Francisco, a huge park originally designed in 1877 by John McLaren to keep a three-mile stretch of beach sand from shifting. The park has landscaped facilities for everyone, from playgrounds for children to polo grounds for adults. *Bottom:* The University of California, Davis is a campus of bicycles. Because Davis has a flat terrain, the landscape architect created visual interest by designing two hills and a bikeway with a bridge leading to the library. This effective landscape design was created by first excavating a bike path and then using the soil to make two mounds.

growth in all gardens. Flowers can be grown along with food-producing plants, and herbaceous annuals can be set in mixed plantings with trees and shrubs depending on sun and shade requirements. Too often, Americans' gardening practices reflect their agricultural heritage: single-species planting in straight narrow rows maintained by high chemical and mechanical input.

The residential landscape should be considered an extension of the interior living space—an idea exemplified by the work of California landscape architect Thomas Church and well described in his influential book, *Gardens Are for People.* Privacy and definition of property lines, or different areas of activity, can be defined either architecturally with walls, fences, or trellises, or with plant materials—shrubs, trees, or vines (Fig. 20–7). Evergreens can define views and block sound. Many deciduous trees can provide spring flowers, summer shade, fall color, and winter openness so that sunlight reaches windows and living areas only when desired (see Fig. 20–19).

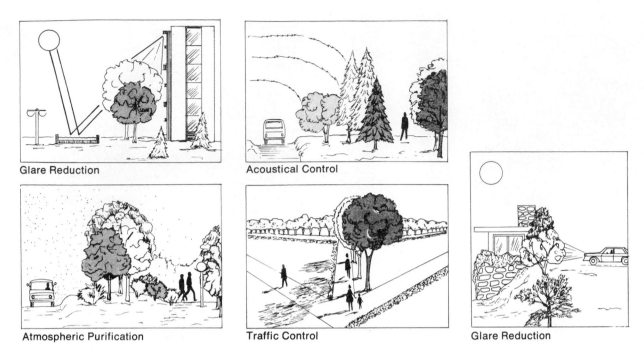

Figure 20–7 Plants can be strategically placed in the landscape to reduce glare, to purify the atmosphere, to control noise, or to direct pedestrian traffic. *Source:* Adapted from Robinette, G. O. 1972. *Plants, people, and environmental quality.* Washington D.C.: U.S. Government Printing Office.

The home landscape should be appropriate to its surroundings. A "prairie" planting may be out of place in a neighborhood of highly manicured lawns, just as a lawn may be out of context in a naturally wooded area. Most of all, maintenance requirements should be considered in planning: the regular use of pesticides, herbicides, and inorganic fertilizers has potentially negative impacts on water quality, wildlife, and human health. The use of "native" plant materials has become very popular, but when choosing such plants one must remember that they are native to a set of ecological conditions—rainfall pattern, soil type, sun or shade tolerance, and temperature regime—and not a particular state (Iowa) or other human-drawn boundary. Thus, for example, the popular service berry (*Amelanchier* sp.) is native to forest understory, not to open lawns or parking lot islands, where the delicate flowers are quickly burned by sun and heat. The residential landscape should also reflect the architecture of the house itself: a formal facade looks best with a formal garden or lawn and symmetrical tree plantings, while a cottage-style house calls for a bright, informal mix of perennials and shrubs. Mountain or woodland houses call for a naturalistic garden in which the boundaries are not obvious. Even in the smallest outdoor space containers can function as a small garden.

Planning the Landscape

In planning the landscape, both the indoors and outdoors should be considered as one unit. This provides for greater usefulness of the total space. The private house and garden has four functional areas: (1) the public area or entry, (2) the public living area, (3) the private living area, and (4) the work or delivery area.

The public entry area includes the front yard or garden area, perhaps a driveway, a walkway to the porch, and the entry. This area may be landscaped to complement the style of the house or to conform to neighborhood landscaping patterns. Shrubs selected for different heights may offer some privacy but are usually used as foundation plantings to allow the house to blend into the landscape.

The living area of a house includes the living, dining, family rooms, and study. One of these rooms may extend out into the garden by way of a patio or deck. The patio may be used for year-round outdoor living in mild climates or only during the summer months in cold-winter areas. Some shade trees or a patio cover can be provided to make the area more comfortable. A surrounding lawn area reduces both heat and glare. Spring bulbs, flowering shrubs, or flowering annuals can provide the garden, patio, and living room with color.

The work area consists of the workshop, laundry, garage, and storage. Shrubs can hide unsightly storage sheds and work areas.

Functions of Plants in the Landscape

Plants are used in the landscape for many functional purposes: to reduce glare, to aid in air purification, to settle dust, to diminish noise by absorbing sound, to direct pedestrian

Screening Objectional Views

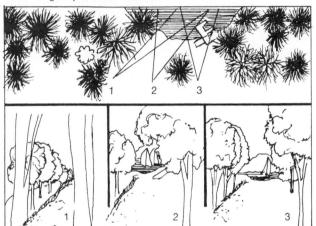

Progressive Realization

Privacy Control

Figure 20–8 Plants can solve certain environmental problems such as screening an undesirable view, directing a person's attention to interesting views, or providing privacy. *Source:* Adapted from Robinette, G. O. 1972. *Plants, people, and environmental quality.* Washington D.C.: U.S. Government Printing Office.

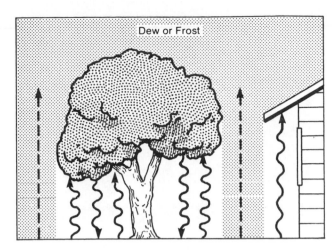

Figure 20–9 Plants grown under trees are protected from radiation frosts. Plants are likewise protected from frosts under eaves or patio covers. *Source:* Adapted from Robinette, G. O. 1972. *Plants, people, and environmental quality.* Washington, D.C.: U.S. Government Printing Office.

Figure 20–10 Evergreen conifers act as snow barriers if planted properly in relation to the prevailing winds. *Source:* Adapted from Robinette, G. O. 1972. *Plants, people, and environmental quality.* Washington, D.C.: U.S. Government Printing Office.

traffic, and to screen certain areas from public view and provide privacy (Figs. 20–7 and 20–8). Plants must reach their mature effective size before they become fully functional.

Plants can partially moderate climate by reducing radiation frosts (Fig. 20–9) or acting as wind or snow barriers (Fig. 20–10). In hot, dry, clear climates, radiation from the sun makes for hot days and radiation to the clear sky during the night results in cold nights. Plants, particularly shade trees, can be used as radiation barriers to modify such patterns, shading the soil during the day and preventing excess radiation at night (Fig. 20–11). Shade trees lead to a more even soil temperature as compared to the bare soil. The soil thermometers in Fig. 20–11 indicate an equable temperature under the trees. Soil under tree cover has a more even temperature from season to season than does bare soil exposed

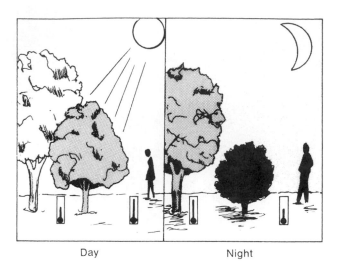

Day   Night

Figure 20–11 Trees and shrubs moderate wide diurnal variation in ground temperatures. They reduce radiation from the sun and prevent excess radiation from the soil to the sky on a clear, cloudless night. *Source:* Adapted from Robinette, G. O. 1972. *Plants, people, and environmental quality.* Washington, D.C.: U.S. Government Printing Office.

Seasonal Variation

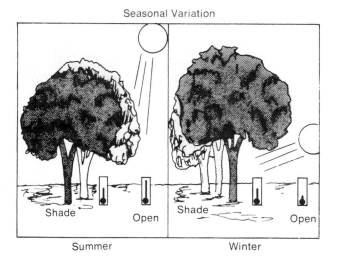

Summer   Winter

Figure 20–12 Broadleaf evergreen trees and shrubs reduce wide seasonal temperature variations by modifying radiation at the soil surface. *Source:* Adapted from Robinette, G. O. 1972. *Plants, people, and environmental quality.* Washington, D.C.: U.S. Government Printing Office.

to the sun (Fig. 20–12). A concrete wall shaded by ivy or other climbing plants has a much lower temperature than a bare wall open to the full sun. A group of closely planted evergreen trees next to a wall can create a dead air space (Fig. 20–13) much like the dead air space in the walls of the house. This tends to maintain an even temperature between the plants and the wall.

Plant materials that can be used for these various functions are available in many forms, shapes, and sizes. A few of these are depicted in Figure 20–14. Landscape architects must be aware of the hardiness of plants that can be

Figure 20–13 An evergreen tree or shrub planted near a wall facing south or west creates a dead air space that becomes "insulation" to maintain even summer temperatures. *Source:* Adapted from Robinette, G. O. 1972. *Plants, people, and environmental quality.* Washington, D.C.: U.S. Government Printing Office.

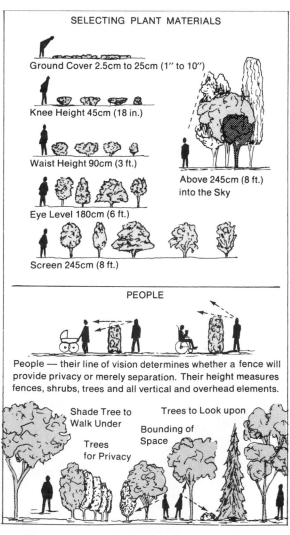

Figure 20–14 Plants of all shapes and sizes can be chosen to create effects or fulfill functions to beautify the environment. *Source:* Adapted from Robinette, G. O. 1972. *Plants, people, and environmental quality.* Washington, D.C.: U.S. Government Printing Office.

Plant Arrangements in Space

The individual plant is a specimen in which, through spacing, it becomes fenestration,* hedges, baffles, tracery, clumps, canopy.

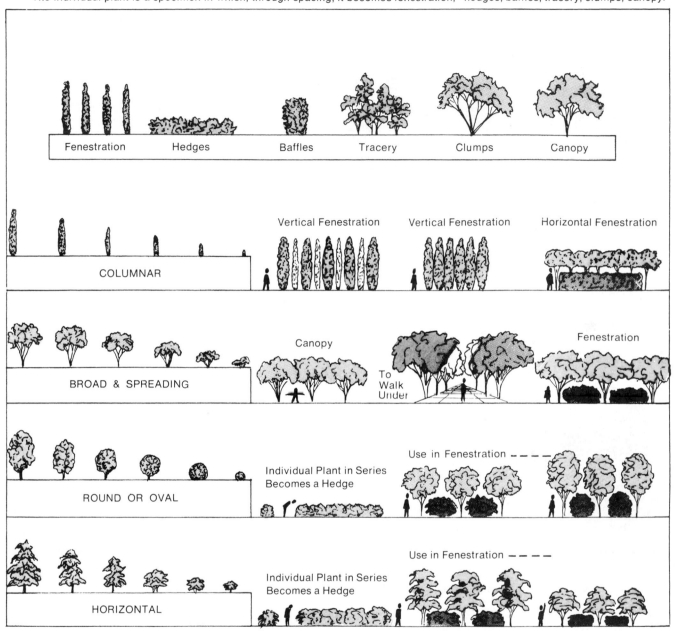

*Arrangement and proportioning of openings.

Figure 20–15 Groups of trees or shrubs can produce desired effects if chosen wisely. These groups are based on trees of various shapes (see Tables 15-1, 15-2, 15-3, and 16-1). *Source:* Adapted from Robinette, G. O. 1972. *Plants, people, and environmental quality.* Washington, D.C.: U.S. Government Printing Office.

functionally used in their plans. In addition, they must supply specifications for soil mixes, drainage requirements, and maintenance care. The ultimate size of these plants must be known if they are to be used to their best advantage. Figure 20–15 shows growth habits of some specimen plants often used in the landscape. These tree and shrub characteristics must be known in order to choose the most desirable plant

for the plan. Some of these characteristics are listed in Tables 16–1, 16–2, 16–3, and 17–1.

Plants are often used to screen undesirable activities or unsightly surroundings (Fig. 20–16). People are becoming more aware of their surroundings, and city, county, and state governments now request certain industries to screen ugly scenes from public view. In the 1960s, a highway beautification

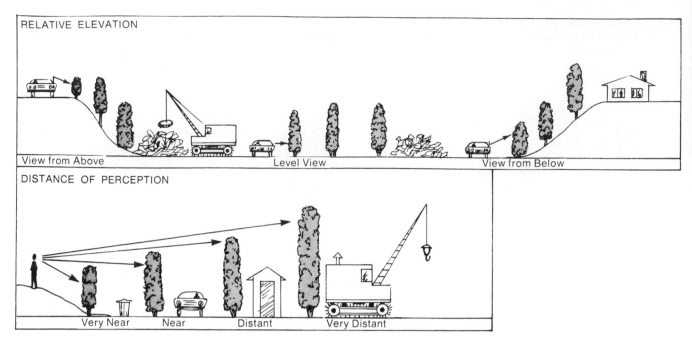

Figure 20–16 Examples of how trees and shrubs can be used to hide undesirable views. *Source:* Adapted from Robinette, G. O. 1972. *Plants, people, and environmental quality.* Washington, D.C.: U.S. Government Printing Office.

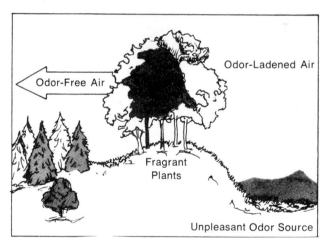

Figure 20–17 Plants can help reduce some unpleasant odors in the atmosphere. *Source:* Adapted from Robinette. G. O. 1972. *Plants, people, and environmental quality.* Washington, D.C.: U.S. Government Printing Office.

Figure 20–18 Plants together with rain can help clean the air. *Source:* Adapted from Robinette, G. O. 1972. *Plants, people, and environmental quality.* Washington D.C.: U.S. Government Printing Office.

act was initiated in the United States to encourage attractive landscaping of federally funded highways.

Areas of high air pollution are usually plagued by unpleasant odors. Plants are useful in ridding the air of such odors (Fig. 20–17) by absorbing them and thus purifying the air. Plants also help remove dust from the air. Dust particles attach to the leaves, and the next rainfall washes the dust away (Fig. 20–18). Carbon dioxide (CO_2) is a waste gas exhaled by humans and animals but absorbed by plants. The plants convert the CO_2 to sugars by photosynthesis and give off pure oxygen in the process. This is a part of the balancing cycle between plant and animal life.

Glare can be controlled by shading created by trees during different hours in the day (Fig. 20–19). The mature plant size necessary to be effective can be calculated from the angle of the sun during the four seasons. Shade is usually desired in summer but not in winter, making deciduous trees and vines most effective (Fig. 20–20). Species of trees vary in the shadow patterns they cast (Fig. 20–21). Density of shade can change with the leaf canopy and species. Plants also vary in the amount of solar radiation they reflect. In general, green plants reflect only about 10 to 25 percent of the radiation (Fig. 20–22), whereas a white building reflects about 90 percent and unpainted concrete about 35 percent.

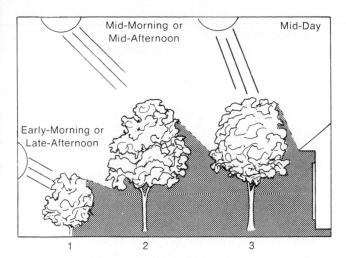

Figure 20–19   Trees and shrubs placed in the proper way can reduce the sun's radiation and glare throughout the day. *Source:* Adapted from Robinette, G. O. 1972. *Plants, people, and environmental quality.* Washington, D.C.: U.S. Government Printing Office.

Glare from street lamps may be blocked by planting trees as illustrated in Figure 20–23. People who live at the end of a dead-end road or a cul-de-sac find hedges or tall shrubs a necessity to reduce glaring lights from oncoming cars.

Windbreaks of various sorts have been used for centuries to slow or divert strong winds (Fig. 20–24). Tall shrubs and hedges used as windbreaks reduce wind speed by increasing the resistance to the wind flow. Coniferous trees, as shown in Figure 20–25, that have thick dense branches clear to the ground are the best all-year plants for windbreaks in close proximity to buildings.

Plant Selection

The landscape is only partially complete until the particular plant species have been chosen to fulfill the functions intended in the plan, but a knowledge of plants and their behavior is essential in making landscaping selections. Certain

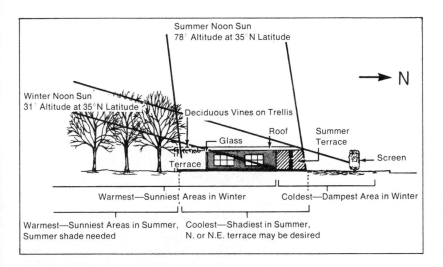

Figure 20–20   For maximum comfort, the design of the house and the placement of trees, shrubs, and vines must be considered. The plan shown here gives good protection from the sun in summer and allows penetration through the deciduous trees of the sun's rays in winter. *Source:* University of California Cooperative Extension.

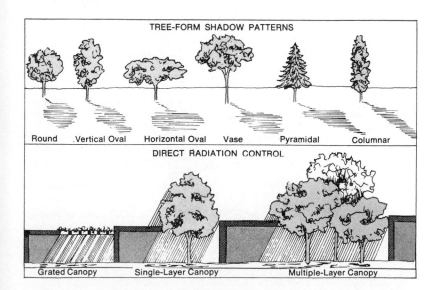

Figure 20–21   Each tree species has a unique shadow pattern. Trees also vary in the density of shade they give. Knowledge of these shade characteristics is valuable in developing a landscape. *Source:* Adapted from Robinette, G. O. 1972. *Plants, people, and environmental quality.* Washington, D.C.: U.S. Government Printing Office.

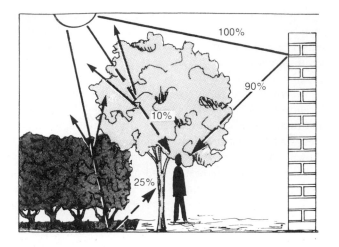

Figure 20–22 Plants can modify solar radiation by diffusion and shading. *Source:* Adapted from Robinette, G. O. 1972. *Plants, people, and environmental quality.* Washington, D.C.: U.S. Government Printing Office.

kinds of shrubs can be trained to become formal hedges because they become dense after pruning. Tree selection is based on their ultimate size, shape, shade, and other desirable characteristics. Dense shade trees, like the maples or beeches, discourage or eliminate most plant species under them. However, many kinds of trees produce filtered shade so that certain shade-requiring shrubs or flowering annuals flourish beneath them. Plant selection, of course, must be based on the hardiness zone (Fig. 10–3), but the particular environmental conditions of the planting site must also be considered. Finally, choices need to be made on the growth habits necessary to enhance the landscape plan. There are, of course, many species and horticultural cultivars that can be used in the various climatic regions. It is very important that a careful, detailed study be made before finally selecting the specific landscape plants to be used for the various aesthetic and functional purposes discussed in this chapter.

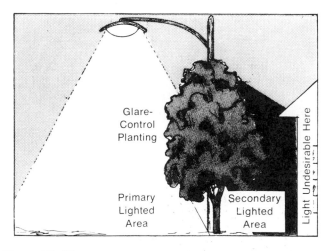

Figure 20–23 The glare from street lights can be modified or completely controlled by planting trees in the proper location.

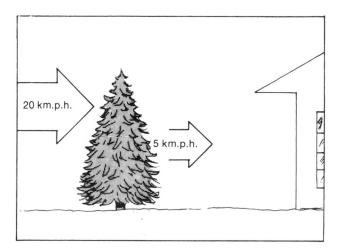

Figure 20–25 Coniferous evergreen trees or shrubs effectively cut wind velocities next to structures when planted a distance from the house which is at least twice the height of the barrier. *Source:* Adapted from Robinette, G. O. 1972. *Plants, people, and environmental quality.* Washington, D.C.: U.S. Government Printing Office.

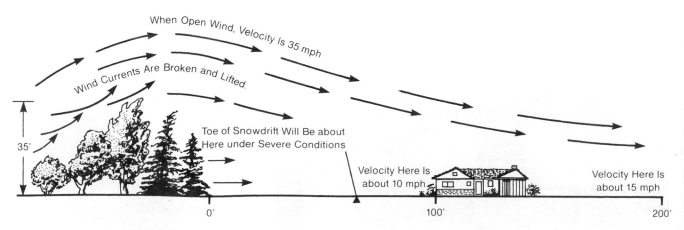

Figure 20–24 Trees and shrubs of varying heights can reduce the velocity of prevailing winds. *Source:* USDA.

SUMMARY AND REVIEW

The history of western landscaping goes back to Biblical times and the Hanging Gardens of Babylon. Early landscaping was gardening and farming with no distinction made between plants grown for food, fiber, and animal feed. Ornamentation was not considered a reason for growing plants. In Western culture, formal ornamental gardens began with the Roman Empire but gardening reverted back to utilitarian purposes during the Dark Ages. With the renaissance, formal gardening with elaborate ornamental features created with plants was revived and became the standard of European nobility. Meanwhile, in East Asian culture, ornamental gardening had been practiced for centuries as an integral part of religious and social culture. The East Asian garden was designed to pro-

mote a peaceful meditative state in those visiting the gardens. In the United States and many other developed countries, many styles of landscape gardens can be found. Landscapers have to understand the history of gardening and landscaping because a client will often request that his or her landscape reflect a historic garden or landscape style.

While aesthetically pleasing, landscape plants can also serve many functions. They can moderate the temperatures of buildings and the ground by providing shade. Annoying street or security light can be blocked by evergreen trees. In windy or noisy areas they can serve as wind and noise barriers. Privacy can be created with hedges. Landscape plants can also reduce odors by filtering air.

EXERCISES FOR UNDERSTANDING

20–1. Make a list of the uses of plants in a nearby residential or public (includes a campus) site. How effectively do you think each plant is serving its function?

20–2. Visit a formal garden and determine what style or period it may be representing.

REFERENCES AND SUPPLEMENTARY READING

AUSTIN, R. L. 1984. *Designing the natural landscape.* New York: Van Nostrand Reinhold.

BECKETT, K. A., D. CARR, and D. STEVENS. 1982. *The contained garden.* London: Francis Lincoln.

CREASEY, R. 1982. *The complete book of edible landscaping.* San Francisco: The Sierra Club.

CROWE, S. 1959. *Garden design.* New York: Hearthside Press.

DANNENMAIER, M. 1998. *A child's garden: Enchanting outdoor spaces for children and parents.* New York: Simon and Schuster.

DICKASON, K. S., and C. C. BURRELL. 1996. *Garden styles.* Lincolnwood, Ill.: Publications International.

ECKBO, G. 1950. *Landscaping for living.* New York: F. W. Dodge.

MILLER, R. W. 1996. *Urban forestry: Planning and managing urban greenspaces.* Upper Saddle River, N.J.: Prentice Hall.

MORSE, H. K. 1962. *Gardening in the shade.* Portland, Ore.: Timber Press.

ROBINETTE, G. O. 1972. *Plants, people, and environmental quality.* Washington, D.C.: U.S. Government Printing Office.

ZION, R. L. 1995. *Trees for architecture and landscape.* New York: Van Nostrand Reinhold.

UNIT
III

MAJOR AGRONOMIC, VEGETABLE, AND FRUIT CROPS

CHAPTER

Agronomic Crops Grown for Food, Feed, or Oil

FOOD AND FEED CROPS

Worldwide food scarcity in the years ahead is a real possibility unless measures are taken to both control population and increase food production. The need for a more efficient agriculture is clear, and a likely area for success with food crops is increased grain production.

Increased food production must come primarily from crop plants with high caloric output per unit area of land. Generally these crops are high carbohydrate producers, and include cereal grains, potatoes, cassava, and sugar crops. Cereal grains are a concentrated source of energy, easily processed, stored, and distributed. They directly supply the world's population with about 80 percent of its total food calories. The word *cereal* comes from the name of the goddess Ceres, "giver of grain," who is said to have given wheat to the early civilizations. It soon becomes evident that the true cereals are grasses, members of the family *Poaceae*. Other families are considered as cereals because their seeds are used in similar ways as those of the grasses. The cereals are man's principal food source because of their adaptability to many climates, soils, and handling methods. They are efficient converters of light energy into food, are hardy, produce many seeds per plant, store well, and are readily processed into many uses, including their conversion into meat through livestock feeding. Some of the important agronomic food crops are discussed in alphabetical order.

Barley (*Hordeum vulgare* L.) POACEAE

Barley, a widely adapted small-grain cereal, is used for human food, livestock feed, and malting.

Barley and wheat are the most ancient of all cereal grains, both dating back 9,000 years. Barley apparently was grown in Mesopotamia in prehistoric times. It is believed to be native to southwestern Asia, but many wild species grow in Ethiopia and southern Tibet. There are two main species: *H. vulgare,* the common six-rowed barley (six rows of kernels per spike) grown mainly for food; and the two-rowed barley *H. distichon,* whose ancestor is probably the wild barley *H. spontaneum,* used in many parts of the world for malting. The crop is classified according to its growth habit as spring (earlier maturing) or winter barley.

Barley is widely adapted because of its drought resistance, tolerance to alkaline and saline soils, and early maturity. Some cultivars grow in subarctic climates, some only in the temperate zones, and others in the subtropics, although barley is not well adapted to hot, humid conditions. The crop may ripen in as few as 60 to 70 days, but normally 90 to 120 days are required. The principal production areas are the former Soviet Union, United States, Europe (West Germany, France, United Kingdom, and Spain), and Canada. In the United States barley is grown mostly in North Dakota, Idaho, Washington, Minnesota, and Montana.

In selecting a cultivar, its yield, disease resistance, winter hardiness, straw stiffness, nonshattering characteristics, and ultimate use are considered.

Barley grows best on a well-drained, fertile loam soil. Its roots penetrate 90 to 180 cm (3 to 6 ft) in a permeable soil. The seedbed should be prepared thoroughly.

In most of the barley-growing areas of the United States and Europe, the crop is seeded in the early or mid-spring months. In the western United States and Canada, barley is seeded in the late fall months.

A complete fertilizer is often applied at seeding. Barley responds well to fertilization. Excessive use of nitrogen is avoided to reduce lodging (Fig. 21–1) or when the grain is used for malting. For malting barleys, cultivars producing more starch and less protein are preferred.

Cultivation for weed control is done before planting, but most weeds are controlled by application of selective herbicides.

Diseases that cause serious losses include powdery mildew, stripe mosaic, yellow dwarf, leaf scald, stem and leaf rust, and Fusarium blight. The parasitic fungus *Helminthosporium* and a smut (*Ustilago*) also attack barley.

Barley has numerous insect pests. The principal ones are wireworms, chinch bugs, and greenbugs. At various times grasshoppers, army worms, aphids, and thrips are also troublesome.

Barley is harvested with a combine in the late spring or early summer. In the northern growing areas this is often in August or September. Some farmers harvest the straw residue, others burn the stubble after harvesting the grain to reduce trash in fields and help control some diseases.

Beans (*Phaseolus* spp. L., *Vigna* spp. Sari, *Vicia* spp. L.) FABACEAE

The seeds of several *Fabaceae* species of field beans and related field peas or cowpeas are grown for food and are a very important caloric and protein source. These include members of several genera. Various types of beans of the genus *Phaseolus* include the lima, scarlet runner, snap, shell, white, black, pea, blackeye, kidney, tepary, black gram, and mung. Other important beans are the broad bean *Vicia faba,* chickpea *Cicer arietinum,* the cowpea also known as the blackeye pea *Vigna unguiculata* or *V. sinensis,* the dried pea *Pisum sativum,* lentil *Lens esculenta,* and pigeon pea *Cajanus indicus.* Many other legumes having local importance are grown for human and animal food. Pulse crops are legumes whose seeds and/or pods are consumed.

The blackeye pea *Vigna unguiculata* is one of the oldest crops. It is native to Central Africa from where it spread to Asia and the Mediterranean basin; it is now grown in many areas. The red kidney bean *Phaseolus vulgaris* originated in Central and South America, and was unknown in Europe before Columbus. The broad bean *Vicia faba,* with its origin probably located in North Africa or the Near East, is an old world species grown by the ancient Egyptians and Greeks. The pea *Pisum sativum* was probably domesticated in Central and Western Asia.

Beans are warm-season crops and are best grown for dry seeds in areas where temperatures are warm but not excessively high, and where the growing season is long enough to avoid frosts. Such areas are found in many temperate re-

Figure 21–1 Lodging, as shown here, is the collapse of stems, causing the plant to fall over. This condition causes serious crop loss of grain in such grain crops as barley, corn, rice, and wheat because it makes harvesting difficult. In wheat, excessive nitrogen fertilization causes the heads to become too heavy with grain. Lodging is aggravated by strong winds. Lodging is also promoted by stalk rots in corn.

gions of North and South America, Europe, Asia, Africa, and Australia. Major production areas are China and India, where dried peas are important. India is also a large producer of lentils. Latin America, the United States, Japan, Italy, and Turkey are also major bean producers. Italy is a leader in the production of broad beans. The major production areas of the United States are Michigan, Nebraska, California, Colorado, Idaho, and North Dakota.

The soil requirements, culture, pests, and harvest of dried beans and peas are similar to those of soybeans and fresh green beans.

Corn (*Zea mays* L.) POACEAE

The botanical origin of corn *Zea mays* family *Poaceae*, better known in the world as maize, is vague and is mostly based on evidence from specimens discovered in caves in the Tehuacan Valley of Mexico. The earliest samples appear to date to 5000 B.C., although maize pollen fossils have been found that are about 80,000 years old. The modern history of corn begins with the first voyage of Columbus, who discovered not only the Americas but also corn. Later explorers found Native Americans growing corn in all parts of the Americas from Canada to Chile. The first Indians to plant and cultivate the crop lived in Mexico and South America.

Corn is a tall annual plant with strong erect stalks, a fibrous root system, long narrow leaves spaced alternately on opposite sides of the stem, and separate male and female flowers on the same plant (monoecious). Now grown worldwide in many environments, corn originated in the Americas as a short-day, long season crop. The increase in diversity of production area environments is the result of the breeding and selecting of cultivars that mature rapidly and are nearly day neutral. Several botanical varieties of *Zea mays* are grown having broad usage and economic importance. The various types of corn can be differentiated on the basis of endosperm composition—usually the relative proportions of soft versus hard endosperm or starch.

—Dent corn (*Z. mays* var. *indentata*) is the principal commercial feed type grown in the United States. The grain is normally yellow, hard, and horny. The kernel is comprised of hard starch with an overlay of softer starch in the crown. The soft endosperm in the crown shrinks on drying more than the hard innermost endosperm, resulting in a dent forming in the top of the kernel. White kernel corn use is increasing in the United States because of increasing Hispanic populations and the popularity of their foods, which use this kind of corn.

—Flint corn (*Z. mays* var. *indurata*) kernels are not indented when dry and are comprised mostly of hard starch.

—Flour corn (*Z. mays* var. *amylacea*) is also known as soft or squaw corn. It has little hard starch. It was preferred because it is easily hand-ground into flour.

—Sweet corn (*Z. mays* var. *saccharata*) kernels have sugary kernels when immature; when dried the seeds are wrinkled and somewhat translucent. This type is a popular and important vegetable crop in some nations.

—Popcorn (*Z. mays* var. *praecox*) is a type of flint corn with a very high proportion of hard starch endosperm. The kernels and ears are smaller than dent corn.

—Pod corn (*Z. mays* var. *tunicata*) kernels are enclosed in husks or pods, and the ears are covered with husks. This curiosity is not grown commercially.

Corn has a multitude of food and industry uses. Corn finds its use in the production of flour, starch, sugar and syrups, corn oil, alcohol production, for uses as adhesives, and coatings for paper and fabrics, etc. Considerable attention is given to improvement of its nutritional value because of its extensive use as food for humans as well as animals. Much of this is aimed at improving its protein value by increasing the content of the amino acids, lysine, and tryptophan, which are normally low in corn grain. Oil content has also been increased using the Top Cross production system, which involves using blends of different types of corn.

Optimal production of corn requires an ample and continuous supply of available soil moisture. The annual precipitation in the U.S. corn belt varies from 60 to 115 cm (24 to 45 in.), with about one-fourth occurring during the summer months; thus, irrigation is usually not provided. Corn is a warm-season crop, requiring relatively high day and night temperatures. The best daytime high temperatures range from 20°C to 27°C (68°F to 81°F), and with night-time lows not less than 14°C (57°F). Seeds do not germinate well at temperatures lower than 10°C (50°F), and temperatures above 40°C (104°F) harm pollination.

The growing season varies according to cultivar; on the average, about 130 to 140 frost-free days are needed, although the range can be from 90 to 170 days. Corn requires abundant sunlight for optimum yields and does not grow well if shaded. Daylength markedly affects the rate of maturity. Cultivars adapted to short days are grown closer to the equator, while cultivars adapted to long days are grown in the more northern (southern in the Southern Hemisphere) areas having relatively long summer days. Modern corn breeding has provided a range of cultivars extremely well adapted to various climates.

In the United States, corn is well adapted to the Midwestern states where the term corn belt appropriately includes Iowa, Illinois, Nebraska, Minnesota, Indiana, Ohio, Missouri, and the Dakotas. Considerable quantities are produced in other parts of the United States.

The development of hybrid corn was a significant advance in efforts to produce more food. Before 1930, hybrid seed was not available, and farmers saved their own seed each year. Now, practically all corn grown is hybrid, especially developed for certain locations and environmental conditions. Before the use of hybrid cultivars, yields of 1,800 to 2,800 kg/ha (1,600 to 2,500 lb/ac) of shelled corn were considered good. Presently, yields of 10 to 11 MT (5.5 to 6.6 tons/ac) or more are common.

Corn grows best on well-drained, friable, fertile loam soils well-supplied with organic matter, but when adequately fertilized will do well on a wide variety of soils. Corn grows best on a slightly acid (pH 6) soil but will perform well in soils with a higher pH. In the corn belt, corn is often followed in a rotation by an inoculated[1] legume crop such as soybean to improve yield potential and reduce buildup of pest problems.

In the past the seedbed was prepared by plowing, followed by sufficient disking and harrowing to provide a firm clodfree seedbed. However, minimum tillage has stimulated many innovations in seedbed preparation to reduce tillage costs and minimize soil compaction.

Commercial corn fields are planted by drilling seed in rows 76 cm (30 in.) apart. In major production areas, populations of 60,000 to 75,000 plants per hectare (24,000 to 30,000 per acre) are commonly grown. In the northern corn belt there is increasing interest in narrowing row spacing to 22 in. or less, especially in conjunction with higher plant populations.

A crop of 2,500 kg (5,600 lb) of corn grain removes 70 to 80 kg (150 to 175 lb) of nitrogen, 25 to 30 kg (55 to 65 lb) of phosphoric acid (P_2O_5), and 32 to 36 kg (70 to 80 lb) of potash (K_2O) plus smaller amounts of secondary and micronutrients from a hectare (2.5 ac) of land. To maintain the original fertility level, this amount and more must be supplied to replace that removed, and to replace losses due to leaching.

Irrigation of corn fields has increased in recent years, and in many areas where irrigation is not provided, the crop would benefit from additional moisture. In areas with insufficient rainfall during the growing season, irrigation is essential. Conversely tile drainage is essential throughout much of the corn belt to allow for timely planting and to minimize saturated soil conditions.

Mechanical cultivation for weed control is done early when weeds are small, since excessive and/or deep cultivation is harmful to corn roots close to the surface. Excellent herbicides are available and are commonly used.

Stalk rots are devastating diseases of corn. Gray leaf spot and northern leaf blight are serious diseases in the corn belt. Ear and common kernel rots are troublesome in many areas. Southern leaf blight reached epidemic proportions throughout the United States in 1970. Fortunately, an alter-

Figure 21–2 A large field of corn, better known around the world as maize, being harvested with multi-row self-propelled combines. The ears are removed from the stalk and the kernels separated from the cob in one operation. *Source:* Allis-Chalmers.

native source of resistance was available and was soon incorporated into new hybrid cultivars. Fusarium stalk and ear rot, charcoal rot, and maize dwarf mosaic virus are other important diseases. More recently grayleaf spot, a disease closely associated with no-till corn, has caused serious economic losses. Destructive insects include corn rootworms, the European corn borer, cutworms, chinch bugs, corn slugs, grasshoppers, and aphids. Insecticides at planting and as seed treatments have been the primary method of controlling insect pests. Recently developed transgenic Bt corns that specifically target lepidopteros insects offer highly effective control of many major corn pests, including European corn borers, and have been widely adopted in the western corn belt.

Corn for grain is harvested when the moisture content is about 20 to 30 percent and is then dried, often artificially, for storage. Most corn is harvested with a combine that separates the ears from the stalk and shells the kernels from the cobs in one operation (Fig. 21–2). Hogs or cattle are sometimes used to salvage corn left by the machinery. In the corn belt areas a large proportion of the crop is fed to animals, notably hogs and cattle, for meat production. Most corn is used as livestock feed, but an increasing percentage is used for industrial purposes including ethanol (as a gasoline substitute) and high fructose corn syrup (sugar for soft drinks). Such on-farm use accounts for about 35 percent of that produced. About 10 percent of the corn grown is converted into silage and/or feed as fresh fodder to animals.

A corn-drying method on the decrease, because it requires extra handling, is the placement of ear corn in covered cribs with slotted sides that permit ventilation. The kernels are shelled at a later time. A moisture level of 20 to 22 percent will permit safe storage in cribs. However, the bulk of the production is stored in enclosed bins or buildings which require that the grain be dried to a moisture content below 14 percent for long-term storage. There is a grade and price penalty for corn with higher moisture. High-moisture corn is subject to heat damage and spoilage.

[1]Legume seed treated with bacteria (*Rhizobia*) that are capable of converting atmospheric nitrogen to nitrates for the plant's use.

Weed control is primarily achieved through the use of herbicides. New transgenic corn with herbicide resistance (such as Round-Up Ready Corn) offer broad spectrum weed control with relatively benign herbicides such as Round-Up® (glyphosate).

Oats (*Avena sativa* L.) POACEAE

The origin of oats is unclear, but they are thought to have originated in Asia-Minor or Southeastern Europe. Oats grew wild in Western Europe during the Bronze and Iron Ages. Following domestication its culture spread to other temperate-zone regions. Oats are principally grown for animal feed, because the crop is easily produced, and the grain and foliage are nutritious. Substantial amounts are also processed for human foods. World production has declined as tractors have replaced animal power.

Oats grow to 60 to 120 cm (2 to 4 ft) tall, and have a fibrous root system. The flowers are borne on panicles, either nearly symmetrical or one-sided. The panicle is made up of numerous (20 to 120) small branches and spikelets composed of two glumes and usually two florets (except hull-less cultivars, which contain more).

Oats, like barley, are well adapted and are produced throughout the temperate zones. Important oat-producing countries are the former Soviet Union, the United States, Germany, Canada, and Poland. Major United States producers are South Dakota, Minnesota, North Dakota, Iowa, and Wisconsin. Northwestern Europe, some northern areas of the United States, and southern Canada produce some late-maturing common white oats.

Oats have less demanding soil requirements than most crops and perform fairly well on acidic and less fertile soils. The majority of oats grown in the United States are sown as early as possible in the spring. As with barley, spring and winter types exist: the latter are used in fall or winter plantings in the more southern latitudes. In the corn belt areas, oats are often a part of the crop rotation following corn. They are often sown with other grasses or companion legumes as a forage crop. Oat culture is relatively simple, and seedbed preparation is often minimal. Oats are planted by broadcasting or seed drill. In many areas oats are fertilized as needed to obtain good yields. Nitrogen is needed if the crop is planted so early that low soil temperatures limit nitrification.

Oats are attacked by a multitude of viral and fungal diseases. Rusts and smuts are the most serious problems. Although smut is controlled by seed treatment, there is little protection against diseases other than growing resistant cultivars. Other serious diseases include yellow dwarf, blast, and septoria leaf spot. Disease problems have stimulated the introduction of many resistant cultivars. Insect pests include grasshoppers, cutworms, aphids, thrips, army worms, and greenbugs.

Practically all oats are harvested by grain combines, or with mowing machines when used for hay or silage. Oat straw makes excellent bedding for livestock, and the straw is often harvested for that purpose.

Rice (*Oryza sativa* L.) POACEAE

Knowledge of the ancient origin of rice is unclear, but most likely it was Southeastern Asia. Some very early Chinese and Indian writings mention rice cultivation for more than 5,000 years in that part of the world. The relative importance of rice, wheat, and corn as human foods cannot be debated. Wheat, rice, and corn are principal food staples for most of the world's population. Each of these major crops prevails in certain areas. Rice is most important in the tropics and semitropic zones and has significant production in some temperate regions. Wheat is generally found in the fertile prairies of the Americas, Europe, Asia, and Australia, while most corn is produced in temperate areas that have warm and moisture-rich growing seasons.

The cultivated plant, customarily grown as an annual, is a semiaquatic annual grass that grows erect. It has narrow parallel-veined leaves. Spikelets are borne on a loose panicle and each contains one flower enclosed by the lemma and palea. The flower has six stamens and one ovary.

Rice is classified several ways. Based on the chemical characteristics of the starch and grain aroma, rice can be classified into three groups: (1) waxy or glutinous types (starchy endosperm contains no amylose[2]); (2) common types, more or less translucent nonglutinous (endosperm contains one-fourth amylose and three-fourths amylopectin[3]); and (3) aromatic or scented types, grown in India and Southeast Asia. The glutinous types, consisting entirely of amylopectin, are grown primarily for a specialty market in the United States but represent about 10 percent of the total production in China and 8.4 percent in Japan. Worldwide, the first two types comprise over 90 percent of the total rice grown.

Rice cultivars are classified as lowland rice (continuously or pond flooded) or upland rice (nonirrigated, or irrigated but not continuously pond flooded) (Fig. 21–3). Upland and lowland do not refer to elevation. More than 80 percent of the world's rice grown is the lowland type. Rice cultivars are also classified on the basis of kernel characteristics into short-, medium-, or long-grain types. The average

---

[2]Amylose is a component of starch characterized by the tendency of its aqueous solution not to gel.
[3]Amylopectin is a component of starch characterized by the tendency of its aqueous solution to set to a stiff gel at room temperature.

Figure 21-3 Young lowland rice plants growing through the water held by earth levees in a flooded rice paddy. *Source:* U.S. Soil Conservation Service.

Figure 21-4 Working the flooded field is the usual method of preparing land for rice in the humid tropics of southeast Asia. Most Asian farmers using simple equipment drawn by water buffaloes plow and harrow the flooded field until the soil is puddled. This helps control weeds, creates a plow pan which reduces water percolation losses, and makes hand transplanting easier. *Source:* R. L. Haaland.

length of unhulled kernels is 7.2, 8.4, and 9.9 mm respectively. Most Northern Asian people who eat rice prefer the short-grain types (also known as pearl rice) because the kernels are sticky when cooked. Most Americans and Europeans prefer the nonsticking, long-grain types.

Another rice classification is based on maturity. The rate of maturity is genetically controlled and is measured by the number of days required for the plants to reach 50 percent heading.

Rice is also classified on the basis of its cultured adaptation as japonica or indica types. The japonica cultivars are usually short-grain and adapted to a temperate climate, while the indica types are long-grain and tropical. Intercrossing of japonica and indica types has led to a breakdown of this traditional classification, and some long-grain varieties are well adapted to temperate areas.

Successful rice production is dependent on ample water availability, and thus important production areas lie along the great rivers and deltas of Asia and other regions. Asia produces over 90 percent of the world's supply. Major producers are China, India, Indonesia, Bangladesh, Thailand, Japan, Burma, and Brazil. Five states in the United States— Arkansas, California, Louisiana, Texas, and Mississippi— produce 95 percent of the U.S. crop, this being only a little more than 1 percent of the world's total.

Rice grows well on moist soils. For best yields, the fields are flooded for most of the growing season; thus clay, clay loam, or silty clay loam soils are most frequently used to conserve water by minimizing seepage losses. Organic and light-textured soils are used provided they have an underlying hardpan or claypan that will prevent water seepage losses. The soil pH ranges between 5.0 and 7.5 for satisfactory plant nutrient availability. Upon flooding, acid or alkaline soils shift slightly toward neutrality. Rice cultivars vary in their tolerance to salinity, but all are adversely affected.

In most Asian countries, the centuries of simple, nonmechanical methods of flooding, terracing, leveling, and other wetland rice cultural operations have produced soils with common characteristics. The immediate soil surface is oxidative and the subsoil strongly reducing (oxygen defi-

Figure 21-5 Most of the world's rice, except in the United States, is transplanted by hand into flooded puddled fields as shown here. The seeds are started in seedling beds and then after 30 to 50 days the young plants are transplanted. *Source:* R. L. Haaland.

cient). These soil conditions create what is called a rice paddy soil.

In the humid tropics the soil is usually worked wet or flooded to prepare land for rice production. The wet fields are plowed with implements drawn by water buffalo (Fig. 21-4). The fields are harrowed crosswise and lengthwise until the soil is well puddled (a soft muddy mass). This helps create a hardpan that limits water percolation and facilitates the hand transplanting operation. Hand transplanting is done in tropical areas and much of Asia in order to establish the crop before seasonal rains (monsoons) occur, or to meet short growing season conditions, and/or to permit double cropping of the soil (Fig. 21-5).

In most developed countries, seedbed preparation is completely mechanized. Direct seeding is practiced either by conventional drilling on dry soil or by seeding, often using

Figure 21–6 Small hand-driven tractors have contributed largely to the reduction of hand labor required to produce rice in southeast Asia. Compare the 2,200 man hours of labor without tractors necessary to produce one hectare of rice (900 hr/ac), with 1,790 hr/ha (725 hr/ac) with small hand-driven tractors, to the mechanized methods used in the United States where 1,000 liters of petroleum fuels plus 12 workhours can produce one hectare (400 gal + 4.75 hr/ac) of rice. *Source:* R. L. Haaland.

Figure 21–7 Under mechanized rice culture, seedbed preparation usually begins shortly after harvest with disposal of the current year's rice straw and stubble. Some states permit the burning of rice straw in the field, as shown here. This procedure is sometimes used for other cereal grains, such as wheat and barley. To minimize air pollution, daily weather data determine the "burn" days, when certain designated areas can be burned. Farmers are assigned "burn days" and are notified a day in advance if they can burn. The straw is also burned to destroy certain diseases (stem rust in rice). *Source:* Marlin Brandon, University of California Cooperative Extension.

aircraft, into flooded fields. The former is practiced predominantly in California. Using this method, seedbeds are finished with large equipment that leaves the surface grooved, or the surface is left rough to prevent seed drift during flooding.

More and more rice farmers worldwide are recognizing and using the improved, high-yielding rice cultivars developed by the International Rice Research Institute (IRRI) in the Philippines or cultivars developed for local adaptation from IRRI germplasm. The choice of seed is an important consideration, and many rice-growing areas have seed certification programs designed to provide high-quality seed. Such seed is varietally pure, produces at least 80 percent germination, and is free of weed seeds and other impurities.

Tropical rice soils in the Far Eastern and Asian rice-growing countries are often deficient in nitrogen, potassium, and phosphorus. The use of fertilizers to improve fertility in these areas is increasing. In the United States yields are increased with the use of fertilizers.

Rice culture has always been and remains a labor-intensive endeavor in many Asian and African countries. Production requires more than 740 man-hours per hectare (300 hr/ac) in most of Asia and Africa (Fig. 21–6). In the United States and Australia, only 18.5 hours/ha (7.5 hr/ac) are needed because of mechanization. This reduction in back-breaking hand labor is possible because of the substitution of machines for human labor. However, it may not be feasible or even desirable for all countries to mechanize rice production to the extent it is in the United States because of their social and economic situations. Many rice-growing countries are overpopulated, the people need jobs, and the farms are too small for efficient use of large machines (Fig. 21–6).

In some areas, rice stubble is disked immediately after the crop is harvested to give the straw more time to decompose, and thus aid in next year's seedbed preparation. Whether disking is done in fall or early spring, a well prepared seedbed is needed. Many rice growers burn the stubble after harvest to help control some rice diseases (Fig. 21–7).

Weeds are one of the major causes of low rice yields. In most countries, weeds are partially controlled by regulating the depth of the water. Hand-weeding is another means of weed control. Chemicals are also available to control problem weeds. Since chemicals move easily with water, especially when paddies are drained, extreme care is needed to avoid environmental pollution.

The principal diseases are seedling blight and seed rot. Brown leaf spot is a serious disease in Texas and Louisiana. Blast is a fungal disease in the southern states that causes long, narrow necrotic spots on rice leaves. Stem rot, caused by a soil-borne fungus, is important in most areas. Several root rot diseases are also troublesome.

Insects and other animal pests are often serious problems. Some of the serious insect pests are the rice leaf miners, rice water weevils, midges, rice leaf folders, armyworms, leafhoppers, thrips, and water scavenger beetles. Animal pests that attack rice include tadpole, shrimp, crayfish, muskrats, and some waterfowl.

It is strongly recommended that the preventive use of insecticides be avoided and that they be used only when the insects have increased to damaging numbers.

In some regions, mosquitoes are a byproduct pest of rice culture since the paddies provide a good breeding ground for them. They are somewhat controlled biologically with a tiny fish called the mosquito fish (*Gambusia affinis*).

Wild Rice (*Zizania palustris* var. *interior* L.) POACEAE

Wild rice, an aquatic member of the grass family, is not closely related to common rice (*Oryza sativa*). Wild rice has been used in the Great Lakes region of the United States and Canada as a food by native Indians for more than 300 years. The seeds sprout under water from grain that fell into the water the previous year, germinating in the spring, and producing a single root and a thin submerged leaf. During the summer a flower stalk elongates and emerges from the water, bearing female flowers at the top of a spike with many drooping male flowers below them. This arrangement almost guarantees cross-pollination between plants. By late summer the plants are 90 to 120 cm (3 to 5 ft) tall and very leafy. The top grains ripen first, and those below ripen subsequently over a period of about 10 days more. In Minnesota, much of the production occurs on wet and marshy public lands, where mechanical harvesting of wild rice is regulated by law. Therefore, much of the harvesting is done from boats by bending the stalks over and into the boat and beating the kernels off with a flail. This process, and the plant's seed-shattering tendencies, results in considerable amounts of seed, more than half, being lost into the water. The portion lost provides seed for the next year's crop. The yield from this kind of culture and harvesting procedure is much smaller than that obtainable by management similar to that for common rice.

In addition to the marshland type of wild rice production (which is rapidly declining), wild rice is also commercially cultivated in California and Minnesota, with cultural techniques similar to those for common rice. New shatter-resistant and semi-dwarf cultivars have been developed.

Rye (*Secale cereale* L.) POACEAE

Southwestern Asia is assumed to be the area of origin for rye. There is evidence of its early cultivation in Western Asia, but domestication was relatively recent, since rye is not mentioned in early Egyptian records. Until the early 1900s rye was principally a European crop grown for making bread. Because of its extreme hardiness it was, and is, widely grown in much of Europe, Asia, and North America. Rye does best in a cool dryish climate and tolerates cold temperatures very well. It also performs well on poor and marginal soils. Therefore, it is often grown where environmental conditions are unfavorable for other cereal crops. Because other cereals are preferred, rye is not an important crop except in certain countries.

Rye is used for human consumption, but most is grown for hay, pasture, and stock feed. It also finds wide use as a cover crop and for the manufacture of beverage alcohol.

There are both spring and winter cultivars, but since rye is very winter hardy, most of the rye grown is fall-seeded. Major world production areas are the former Soviet Union, Poland, West Germany, the United States, and Canada.

Cultural practices for rye are similar to those for winter wheat, but because of its winter hardiness rye is usually seeded several weeks later than winter wheat. Rye is seldom fertilized, except when grown on poor soils, because of its relatively low cash value, although when fertilized it responds well.

Ergot, a fungal disease, produces small black bodies called *sclerotia* in the grain. The sclerotia, if present in quantity, are poisonous to animals and humans, but can be used in pharmaceutical products. Other damaging fungal diseases are now mold, leaf and stem rusts, stalk and head rots, blotch, and some root rots. Many of the same insects that attack barley, oats, and wheat attack rye.

Harvesting entails threshing with combines, as with other small-grain cereals.

Sorghum (*Sorghum bicolor* Moench) [*S. vulgare* Pers] POACEAE

Sorghum is believed to have originated in Africa. Grain sorghums are known by several names, such as durra, Egyptian corn, great millet, or Indian millet. In India sorghum is known as jowar, cholum, or jonna. In the United States, different types of grain sorghum are known as milo, kafir, hegari, feterita, shallu, and kaoliang.

In both Africa and Asia, sorghum is one of the major cereal grains. This important crop was introduced to the United States in the middle 1800s. Major world producers are the United States, India, China, Argentina, Mexico, and Nigeria. Resistant to heat and drought, sorghums having effective and extensive root systems are best adapted to warm and semiarid regions. The principal U.S. production areas are Kansas, Texas, Nebraska, Missouri, and Arkansas, but significant acreages are grown in other states.

Agronomically, sorghum is a common catch-all name applied to all plants of the genus *Sorghum,* which includes numerous grain and foliage types. However, the plants do have markedly different characteristics and uses, and have often been grouped into four types:

1. The grain sorghums (Caffrorum Group) are the nonsaccharine plants, including milo, kafir, feterita, hegari, and hybrid derivatives, among others. These plants are grown for grain used principally for poultry and livestock feed (Fig. 21–8). The grain is similar in composition to corn, but somewhat higher in protein and lower in fat. Grain is ground into meal and made into bread or porridge. Whole grains are sometimes popped or puffed for cereal. Grain sorghums are also used to make dextrose, starch, paste, and alcoholic

Figure 21–8 Because sorghums are grown for grain, silage, pasture, syrup, and straw for brooms, the sorghum breeder has a variety of objectives. These short-stemmed, dwarf grain cultivars have been bred for adaptation to mechanical harvesting, early maturity, and resistance to lodging and shattering. These mature heads of grain are ready for harvest.

Figure 21–9 A young field of sorghum, here planted on beds to facilitate furrow irrigation. *Source:* U.S. Soil Conservation Service.

beverages. The stalks have a dry pith and are not very juicy, except for milo and kafir, which are semijuicy, dual-purpose types for grain and forage.

2. Sweet or forage sorghums or sorgos (Saccharatum Group) are used mainly for forage and silage and to make molasses. The stalks are juicy and sweet. This type is principally grown in the United States and South Africa.

3. Broom corn (Technicum Group). This sorghum is a panicle woody plant, has dry pith, little foliage, and fibrous seed branches 30 to 90 cm (1 to 3 ft) long that are often made into brooms.

4. Grass sorghum (*S. Sudanense*) or Sudangrass is grown for pasture, green chop, silage, or hay. Sudangrass is usually ready to pasture in five to six weeks, but to avoid prussic (hydrocyanic) acid poisoning, it should be at least 45 to 60 cm (18 to 24 in.) tall when grazed. The acid in the form of glucosides occurs in young plants and in new shoots. It is highly toxic because it inhibits cellular oxidative processes.

Plant breeders have developed numerous new grain sorghum hybrids and cultivars with specific adaptation to local regions or conditions. Many cultivars are designated by their rates of maturity. The development of hybrid seed and short stalk cultivars has greatly improved yields and harvesting efficiencies. The most important production factor in grain sorghum is selection of the correct cultivar for a given area.

Grain sorghum grows best on fertile sandy loam soils, but with adequate fertilizers it succeeds on a wide variety of soils. The best soils are friable, well drained, and pH neutral. Sorghum is more tolerant to sodic (alkali) and saline soils than most field crops.

Sorghum seeds need a warm soil to germinate, ideally 18°C (64°F) at planting depth. In the Great Plains, from northern Texas to South Dakota, and in the western corn belt area,

planting begins in mid-May to early June. In the arid Southwest, sorghum is planted from midspring to midsummer.

Factors that result in poor stands are cold soil, poor-quality seed, poorly prepared seedbeds, soil crusting, or improper adjustment of planting machinery. Some farmers try to prevent poor stands by planting more seed than recommended. Assuming a fertile soil, good seedbed, and full irrigation, maximum yields are obtained with rows 25 to 50 cm (10 to 20 in.) apart and seeds 5 to 10 cm (2 to 4 in.) apart within the row. If the grain is to be harvested with a combine and herbicides are used for weed control, the closer row spacing is preferred.

Sorghum seed is best planted about 2.5 cm (1 in.) deep in moist soil, but gives better emergence when planted in dry soils at 5 cm (2 in.). Row crop planters are used if sorghum is planted on beds that will be irrigated or in rows wide enough for tractor cultivation (Fig. 21–9); otherwise, the seed is sown with a grain drill.

Sorghum uses plant nutrients heavily, and the crop must be fertilized for high yields. The rates vary considerably among locations and soil types, especially for nitrogen fertilization. Soil or leaf tissue tests can determine the need for nitrogen, phosphorus, or potassium, as well as other essential plant nutrients.

Grain sorghum must be irrigated in areas with insufficient rainfall; about one-fourth of all sorghum acreage grown in the United States is irrigated. Even though sorghum tolerates drought, it responds well to ample soil moisture.

Sorghum does not compete well with weeds. Thus, weeds must be controlled. A preplanting irrigation germinates a weed crop that is then destroyed by preplant tillage. Mechanical tillage usually keeps weeds under control, but selective herbicides are also effective.

Sorghum is attacked by a variety of fungi, bacteria, and viruses, which cause seed rots, seedling blights, foliar diseases, flower and seedhead diseases, and root or stalk rots.

Resistance to some diseases has been bred into new hybrid cultivars. Some of the more common diseases are loose kernel smut, covered kernel smut, head smut, Pythium root rot, bacterial spot, bacterial streak, crazy top, downy mildew, Fusarium stalk rot, leaf blight, milo disease, and Rhizoctonia stalk rot.

Several insects attack sorghum, often the same ones that attack other cereal grain crops. Chinch bugs cause severe losses if not controlled. Other insects include the corn earworm, leaf aphid, corn borer, armyworm, sorghum webworm, southwestern corn borer, sorghum midge, and, recently, the greenbug.

Most grain sorghum is harvested with combines. The heads are cut from the plant standing in the field. The grain is removed and cleaned as it passes through the machine. Sorghum threshes easily when the moisture content is 20 to 25 percent, but at this moisture content is not safely stored. Safe storage requires further drying to about a 12 percent moisture content.

Sugar Beets (*Beta vulgaris* L., J. Helm) CHENOPODIACEAE

Beets are believed to have their center of origin in the Mediterranean area. Cultivated beets were probably domesticated from the wild beet (*Beta maritima*). Within the one species *B. vulgaris,* there are several useful types. These include the coarse mangel-wurzel (also known as fodder beet—grown for animal feeding), the sugar beet, the table beet, chard, and other foliage-type beets. The last three types are commonly used vegetables (Fig. 21–10).

The sugar beet is a biennial plant grown as an annual plant with simple leaves arranged in a basal rosette alterna-

Figure 21–10 The sugar from sugar beets and sugarcane is chemically the same. This field is almost ready for harvest. Many farmers use the tops, which are almost as nutritious as alfalfa hay, for cattle feed. *Source:* F. J. Hills.

tively on the stem. Leaves are ovate to oblong-ovate. The flowers are borne in clusters in the axils of the leaves. As the ovaries mature the perianths fuse, resulting in a "seed ball" containing several ovaries or seeds.

Sugar beets are by far the most important beet grown commercially. Their average sugar concentration is about 15 percent, with many crops exceeding 20 percent. The roots are typically sharply tapered, white-skinned, and white-fleshed.

Major growing areas are the former Soviet Union, France, West Germany, United States, Poland, and Turkey. Leading United States producers are California, Minnesota, Idaho, North Dakota, and Michigan.

Hybrid cultivars have been developed that are resistant to bolting (development and growth of a seed stalk) and to certain viral diseases. They were also developed for improved adaptation to specific areas and conditions. Most of these new hybrids are monogerm instead of multigerm; that is, the seed ball contains one seed instead of many. When multigerm seed was planted, one seed ball containing several seeds would produce several seedlings, making thinning necessary as well as difficult. The introduction of monogerm seed helped reduce thinning efforts and also allowed the use of more effective "precision" planting equipment.

Sugar beets grow well on a wide range of soils. High organic clays and clay loams produce high-yielding crops if they have good drainage and deep profiles. In the later stages of growth, sugar beets are quite salt-tolerant (up to 8 mmho/cm), but are sensitive in the early stages of seed germination.

Precision planters capable of metering single seed at desired spacings are widely used to achieve uniform stand emergence. The stand of beets is much more even with monogerm seed that has been cleaned, graded into uniform size, and pelleted.[4] Pelleted seeds enable precision planters to meter seed more effectively. Good yields require proper spacing without crowding.

The amount of nitrogen used depends upon the soil type and many other factors. The nitrogen may be applied at planting time, with care taken to avoid salt damage, or at the time of thinning. Farmers plan their fertilizer programs to exhaust the soil nitrogen supply about six to eight weeks before harvest to ensure maximum sugar production and to minimize further vegetative growth.

In the arid Southwest, irrigation is essential for best production. Beets can be sprinkle-irrigated, but more commonly they are furrow-irrigated (Fig. 21–11). It is a good practice to preirrigate the soil before planting, then in the first 10 weeks or so—the period of rapid plant growth when roots are sparse and shallow—several light, fast irrigations are applied. Heavier irrigations are required during midseason.

---

[4]Seed coated with a material primarily to increase its size and uniformity.

Figure 21–11   This young stand of sugar beets planted in a single row per bed is being irrigated by the furrow method.

Weeds compete with beets for light, nutrients, space, and moisture, and provide shelter for insects that in some cases act as hosts for viruses. Herbicides provide full-season weed control when properly applied. Mechanical cultivation, along with hand hoeing, also keeps weeds under control.

Following are several diseases that affect sugar beets. In some western states beet yellow and western yellow viruses have often been severe. Curly top, another prevalent viral disease, is often severe in Colorado, other Rocky Mountain areas, and in the San Joaquin Valley of California. Two major fungal diseases are Cercospora leaf spot and powdery mildew. Downy mildew, rust, and mosaic are minor diseases that sometimes cause significant damage. *Rhizoctonia* and other fungi sometimes attack mature roots. *Rhizomania,* a fungus-vectored virus disease, has recently become a significant production problem.

Some insects, notably aphids and leafhoppers, are virus vectors. Others damage the plant directly. Root-damaging insects are root maggots, wireworms, and white grubs. Insects that attack crowns, stems, and foliage include flea beetles, cutworms, sugar beet crown borers, beet webworms, armyworms, alfalfa loopers, and grasshoppers. Preventive practices include crop rotation and field sanitation.

In addition to insecticides, integrated pest management (IPM) programs take advantage of beneficial insects to help in the control of some insect pests.

Sugar beet harvesting is completely mechanized.

Beets are best harvested when the sugar content is highest, but often the optimum harvest date does not coincide with when the processing mill can or will accept the crop. Since sugar beets are usually grown on contract with a sugar processing plant, the crop, instead of having the harvest delayed, may have to be harvested before optimum maturity to meet contract terms. The price, acreage, cultivar, harvest date, and other management arrangements are included in the contract.

Sugarcane (*Saccharum officinarum* L.) POACEAE

Sugarcane supplies much of the world's sugar. In many areas where production and labor costs are low, sugarcane provides sugar at a lower cost than the sugar provided from sugar beets. Sugarcane is said to be one of the most efficient converters of solar energy and carbon dioxide into chemical energy, possibly producing more calories per unit of land area than any other crop.

Sugarcane probably originated in India, where it has been grown since ancient times. Its earliest mention is found in Indian writings from about 1400 to 1000 B.C. From India, sugarcane spread to China, Java, and other tropical Pacific islands. It also spread westward from India to Iran and Egypt. Columbus introduced the crop to the West Indies. Sugarcane was first planted in the United States in 1751 near New Orleans.

Sugarcane is a tropical plant that matures in 12 to 18 months. It is a tall perennial grass, which often attains a height of 2 to 4 m (Fig. 21–12). Sugarcane has essentially the same structure as other members of the grass family.

The former low-yielding, low-sugar cultivars have been replaced by highly productive, high-sugar-containing cultivars. Biotechnological advances utilizing tissue culture cloning, and employing sources of genetic diversity such as that provided by somaclonal variation, are used by modern breeders to further improve crop performance.

Sugarcane is grown in tropical regions around the world. The main production areas are the warm, humid, tropical lowland regions of North, Central, and South America, the southern United States, the Caribbean, Africa, Asia, and Oceania, including Australia. The leading countries producing sugarcane are India, Brazil, Cuba, China, Mexico, Thailand, Australia, the United States, and South Africa.

Sugarcane grows best on fertile, moist, tropical soils of a wide variety of types, ranging from sandy to heavy clays. Each soil type requires its own particular management and fertilization treatment.

Cultural methods vary widely and are adapted to local conditions. In general, the amount of rainfall determines whether full irrigation, supplemental irrigation, or no irrigation is used.

Figure 21–12   A field of mature sugarcane growing in the delta area of Louisiana. The canes grow to a height of 2 to 4 m (6.5 to 13 ft).

Sugarcane is propagated vegetatively by planting sections of the stems containing three or four nodes into furrows. The stem sections are sometimes planted by hand but are usually planted mechanically. Plantings are made when soil temperature and moisture are suitable. In many areas, a perennial-like cropping is practiced, with the new crop resprouting from previously cut stems left in the soil. However, new plantings of selected and treated stem cuttings are more productive and reduce disease problems.

The timing, method, and rates of application of fertilizers vary widely. Phosphorus fertilizers are generally applied at planting time, but nitrogen fertilizers are delayed since they tend to damage the developing buds and young roots on the stem pieces. Sugarcane uses large amounts of plant nutrients, and heavy applications are needed to maintain high yields.

Maximum sugar yields depend upon an ample supply of warm temperatures, bright sunshine, and soil moisture, provided either from rainfall or irrigation. The crop transpires an enormous amount of water. Growth is optimized when water is frequently and uniformly available.

Weeds are controlled by hand hoeing, by mechanical tillage with various types of cultivators, by flaming, or by chemical herbicides.

Sugarcane is attacked by a wide variety of diseases and insect pests, with each producing country facing its own special problems. Some of the important diseases are mosaic, gumming disease or gummosis, red rot, smut, and ratoon stunting disease. Troublesome insect pests are the corn aphid (a vector for mosaic), cane leafhopper, and moth borer. Rats have been a problem on some plantations in Hawaii and Florida. They are controlled mostly with poison baits.

Before harvest, randomly selected canes are tested for maturity and the juice is assayed for sugar content with a hand refractometer. Testing generally starts four to six weeks before the proposed harvest date. As with sugar beets, the mill needs a steady supply of canes, which farmers provide by coordinating their harvest. Thus, the rate of harvest is governed by the crushing capacity of the mill. Sugarcane is harvested either by hand or mechanically. In the West Indies the canes are generally sent to the mills remarkably free from undesirable leaves and stems because they are removed when canes are harvested by hand. The cane is cut free at the bottom, topped, and stripped, leaving only the main stalk ready for milling. In other areas, especially where labor costs have hastened the adoption of mechanization, delivery of trashy canes to the mill has been a serious problem, although the trash and waste (bagasse) is used as a fuel by the mill. To reduce trashiness and facilitate mechanical harvesting, firing[5] the cane before cutting is a normal practice. Burning

mature cane does not harm it unless the fire is exceptionally hot. The work of cutting and loading is greatly reduced by preharvest burning. The top of the plant and the attached young leaves need to be removed because they contain invert sugars,[6] nitrogen compounds, and starch—all of which interfere with the extraction of sucrose sugar.

Wheat (*Triticum aestivum, Triticum* spp. L.) POACEAE

Wheat and barley are recognized as being among the most ancient crops. The Egyptians and Mesopotamians grew wheat as well as barley and oats.

Two wild species are still found growing in Syria and Turkey, where wheat probably originated and was domesticated. It is known that present-day species originated from the hybridization of several different species.

Wheat leads all of the cereal grains in total volume. Countries that lead in production are China, the former Soviet Union, the United States, India, France, Canada, and Australia. Wheat is grown in almost every state in the United States, with the leading states being Kansas, North Dakota, Oklahoma, Washington, Texas, South Dakota, Minnesota, and Colorado.

Wheat is classified into market classes by the color and composition of the grain and the plant's growing habits; the latter also determines the production area of each class. The classes are (1) hard red spring, (2) durum, (3) hard red winter, (4) soft red winter, and (5) white.

Wheat has a relatively broad adaptation, is very well adapted to harsh climates, and will grow well where rice and corn cannot. Early growth is favored by cool and moist conditions, with warmer and drier weather toward crop maturity.

Generally the winter climate of a particular area determines whether winter or spring types are grown. If winters are severe, spring-type cultivars are planted in the spring to be harvested in the late summer and the fall. If winters are not extreme, winter cultivars are planted in the fall for spring harvest. If winter temperatures are mild, spring-type cultivars are planted in the fall for spring harvest. Another generalization is the different composition of the grain of the different cultivars. The hard wheats contain more protein (13 to 16 percent) and are usually grown in the drier climates. The soft wheats are more starchy, have a lower protein content (8 to 11 percent), and are usually grown in more humid climates. Each of these wheats has different characteristics and different uses.

In the United States the hard red spring cultivars are grown in the northern Great Plains, the Dakotas, Montana, and Minnesota, and produce a high-grade wheat used princi-

---

[5]Several propane burners mounted on wheels are passed quickly over some crops to burn weeds. The practice is timed so that the weeds are young and tender but the crop is more mature and able to withstand the heat.

[6]Glucose and fructose are undesirable invert sugars produced upon the decomposition of sucrose.

pally for bread flour. Durum cultivars, used to make semolina flour for macaroni and other pasta products, are also grown in these areas and some other areas. Hard red winter wheat, which leads in the total volume of production, is used for bread flour and is grown across a broad area ranging from Utah to Illinois. Soft red wheat production is concentrated in Ohio, Indiana, and southern Illinois. This wheat is milled into flour for cakes and pastries. Grain fed to livestock is most often that of the soft wheat types.

Wheat grain is processed by milling, a procedure that removes the outer bran layer. In doing so, much of the grain's protein is lost, since the bran layer contains the highest concentration of protein. This, unfortunately, is the concession made to produce white flour with better baking characteristics. Whole wheat refers to wheat that is partially milled, and it has become a much more popular product because of its greater nutritional contribution.

Because of its importance as a basic food staple to so much of the world's population, extensive breeding programs have been undertaken in practically every wheat-growing country in the world. Breeders continue to introduce new cultivars that are more resistant to diseases, insects, drought, lodging, and shattering. Plant breeders have improved quality (size, texture, weight) and increased yields and winter hardiness.

Wheat is grown on a wide range of soils in temperate climates where annual rainfall ranges between 30 and 90 cm (12 to 36 in.). Such areas constitute most of the grasslands of the world's temperate regions. Many of these soils are deep, well-drained, dark-colored, fertile, and high in organic matter, and they represent some of the world's best soils. The prairie soils of the United States and Canada and the steppes of the Soviet Union are examples of such soils.

Production methods vary throughout the world. In areas with extensive acreages (North America, Australia, Argentina, and the Soviet Union) mechanization is high. In some areas, methods are still primitive, and wheat is harvested by hand and threshed with a flail or by animals walking over the harvested heads to remove the grain kernels.

Seedbed preparation depends upon the method of growing. In areas with less than 38 cm (15 in.) of average annual rainfall, an alternate crop and fallow system is used. Wheat—either the entire field or strips within a field—is planted one year and the ground is left fallow in alternate years (Fig. 21–13). This practice stores and conserves water during the fallow year for the wheat crop the next. Soil is a good reservoir for water provided no vegetation is allowed to grow. The soil is plowed after harvest and left rough, cloddy, and barren for the following year. The rough surface traps and absorbs the scant rainfall and reduces wind erosion. If weeds appear, a light disking or herbicide application kills them.

In semihumid regions (more than 38 cm, or 15 in., of rain) wheat may be grown more frequently, either wheat following wheat or within a rotation of other crops. Disking and

Figure 21–13   Crop fallow strip farming is used in areas where insufficient rainfall occurs in one year to produce a crop. Alternate bands of soil are left fallow in alternate years allowing two years of rainfall to accumulate for a crop every other year. Strip farming also helps prevent wind erosion because all the land is covered either with a growing crop or harvest stubble. *Source:* R. L. Haaland.

Figure 21–14   A well-prepared seedbed ready for fall planting of wheat. Deep, well-drained, dark-colored, heavy soils well supplied with organic matter produce the best wheat crops. The prairie soils of the United States, Canada, and the former Soviet Union are good examples of wheat soils.

harrowing are generally sufficient to provide a good seedbed. In any case, the seedbed at the time of planting needs to be firm and clod-free (Fig. 21–14). It is essential that seeds and seeding equipment of the best quality be used to insure good wheat stands.

The date of planting varies. In areas where the Hessian fly occurs, it is imperative that no wheat be planted before the announced "fly-free" date. In the winter wheat areas, planting begins in early fall in the northern regions and late fall in the southern. It is best to plant when soil moisture is adequate to germinate seed or to sprinkle irrigate after seeding if moisture is not available. In the Pacific Northwest, winter wheat is seeded in early fall. In the spring wheat areas, early seeding gives the best results. In the northern United States, wheat is seeded as soon as seedbeds are prepared in late spring; in the southern states, plantings are made much earlier.

Wheat responds well to fertilizers; the kind and rate of application vary with location and soil. Fertilizer application practices are also variable. In some situations, a complete fertilizer might be supplied in one application, or it may be divided into two or more separate applications as the crop develops. Some situations do not justify the use of a complete fertilizer, and only nitrogen may be provided. Almost without exception, nitrogen is beneficial for increasing yields.

A crop that is drilled with closely spaced rows is seldom cultivated. To prevent weeds from becoming a problem, selective herbicides are used to control them without damaging the crop.

Three rust diseases caused by fungi seriously attack wheat: stem rust, leaf rust, and stripe rust. The smuts are another group of diseases that cause considerable damage to wheat. Bunt (stinking smut) and loose smut are examples. Stinking and loose smuts are controlled by seed treatment and resistant cultivars. Wheat scab is serious in humid areas. Often soil-borne and insect-transmitted viruses give problems and, at times, root rots or crown rots are troublesome.

Several insects damage wheat. In the eastern United States, the Hessian fly is notorious. In the northwestern states and Canada, the wheat stem sawfly is a serious pest. Wheat jointworms, strawworms, chinch bugs, aphids, and grasshoppers are other insects that cause varying degrees of damage in certain areas.

In the large wheat-growing areas of the world, the crop is harvested by combines. Some harvesting is done by contract harvesters who own and operate many combines. In the United States they start early in Texas and continue through fall into the Dakotas. Wheat properly dried to a moisture content of 12 to 14 percent, cleaned, kept cool and free from insects and rodent pests will store very well, and can be held in good condition for long periods. In some parts of the world, there are situations where rodent control is poor, and where large portions of stored grain are destroyed by rats.

OIL CROPS

In addition to the use of animal oils, early man also used various oils from crop plants as a food source and for a primitive fuel. Some of the earliest written records indicated the use of plant oils for illumination, heat for cooking and warmth, and anointing the skin. Olive oil was widely used for these purposes by early Mediterranean civilizations. Linseed oil from flax was recognized for its usefulness in paints. Castor bean oil was used for lubrication and also in paint and varnish products, cosmetics, and as a cathartic. With man's technological development, the use of various oil crops expanded. These oils function in many applications, many that are very specialized and often not very obvious. After the discovery of petroleum and the technology for its use in the internal combustion engine, the primary source of oil for energy purposes became the earth's underground supply. Concern about the future supply of petroleum compels a strong interest in the oil crop plants as a supplemental source. It can readily be seen that these represent renewable sources, whose importance will increase even more in the future. The

better-known uses of plant oils are for cooking, flavoring, margarine, and salad dressings. A multitude of other uses include the manufacture of plastics, paint, varnishes, lacquers, soaps, detergents, inks, cosmetics, lubricants, medicines, fabric, and paper.

Relatively few plant species provide the majority of the vegetable oil production. Oil is found in all living plants, even in bacteria and fungi. In the plant, a major role of the oil is the retardation of tissue water loss. Certain plant tissues, usually the seed, are the most abundant in oil content. Although some seed have very little oil, others have an oil content greater than 50 percent.

The oil is usually removed by crushing and pressing; other extraction methods, often combined with pressing, include extraction with steam and various solvents.

Vegetable oils, readily seen as small droplets when expressed from tissues, are mostly a mixture of triglycerides with small amounts of mono- and diglycerides that are characterized by the content of their various fatty acids. They become degraded (rancid), some more rapidly than others, because of the breakdown of glycerol into other compounds.

A characteristic of oil is its degree of saturation (amount of single bonding of hydrogen) of the fatty acid molecules. Unsaturated fatty acids have one or more double bonds in their structure, and therefore bond less hydrogen. These generally are liquid, or have a melting point about 20°C. The melting point is dependent on degree of saturation and also on the molecule's carbon chain length. An increase in saturation and in the number of carbon atoms increases the melting point.

The addition of hydrogen increases the degree of saturation. The level is influenced by the initial saturation and by the amount of hydrogenation. Hydrogenation is a chemical process that can be selective to increase the degree of saturation. This is usually done to solidify an oil to a fat. Natural fats differ from oils in having fatty acidic constituents that are more or less solid at about 20°C. Plant waxes are fatty acid esters of monohydroxyl alcohols rather than trihydroxyl glycerols. Hydrogenation is performed by introducing hydrogen under pressure into heated oil with the use of a catalyst such as finely powdered nickel or other metallic compounds.

Although their end product uses are not always specific, each oil finds uses as food products and/or for industrial purposes. The various crop plant oils are often roughly grouped by their degree of saturation and also by their ability to absorb oxygen (drying characteristics).

—Oils of highly saturated fatty acids (nondrying) are largely composed of glycerides of mostly saturated fatty acids such as palmitic and oleic that are found in palm, coconut, peanut, olive, and castor oil.

—Oils of highly unsaturated fatty acids (drying) are largely composed of higher amounts of glycerides of unsaturated types such as linoleic and linolenic, as found in linseed, safflower, soybean, and tung oil.

—Oils of less saturated fatty acids are of intermediate composition of the less saturated fatty acids (semidrying) as found in cottonseed, corn, sunflower, rape, and sesame oil.

The less saturated oils are principally monounsaturated, having one reactive double bond in the structure. The unsaturated oils are principally polyunsaturated, having more than one double bond. For human consumption, unsaturated oils are preferred and deemed to be better for health by contributing less to blood cholesterol levels. Margarine is made from the less saturated oil and is preferred by some to butter, which is a highly saturated fat. The less saturated (corn) or polyunsaturated (safflower) oils are made into margarines by hydrogenation.

World production of oil seed crops is dominated by soybeans. Soybean oil accounted for about half of the 195 million MT produced in the 1984/85 growing season. Cottonseed, peanut, sunflower, and rapeseed (canola) are also important sources of oil. Recently, high-oil corn has become an important source of animal feed.

Some of the important oil-producing crops and their production methods are discussed below.

Castor Bean (*Ricinus communis* L.) EUPHORBIACEAE (4, 25, 31, 34, 36, 37)

The castor bean is grown for the oil in its seed. The petalless flowers are produced on racemes, from which brown spine-covered capsular fruits normally containing three large seeds develop (Fig. 21–15). Seed oil content is about 50 percent. Although commercial hybrid seed is available, seed shatter and highly indeterminate maturity limit harvest mechanization and expansion of production. The seeds contain the alkaloid ricin, which is poisonous and makes the expressed cake unsuitable for animal feeding. The oil has many commercial uses in paints, resins, plastics, inks, cosmetics, greases, and hydraulic fluids.

Coconut (*Cocos nucifera* L.) ARECACEAE (6, 10, 26, 31, 33, 28)

The coconut, in addition to being an important nut crop, is an important oil crop.

Corn (*Zea mays* L.) POACEAE

Corn oil content is typically 4 percent. However, high-oil corn contains up to 7 to 8 percent oil. High-oil corn is attractive as livestock feed because it has greater energy than conventional dent corn and can replace more expensive dietary sources of

Figure 21–15 The development of castor bean flowers to seed pods. *Source:* University of California Cooperative Extension.

fat and protein. The most widely used method for producing high-oil corn is the Top Cross system, which involves using a blend of two types of corn. Single-cross high-oil corn hybrids were not widely used because their grain yield potential was lower than conventional corn hybrids. The Top Cross high-oil corn system minimizes this yield disadvantage.

Field corn is primarily grown as a grain crop for livestock feed. Some is cut green and made into silage. Sweet corn is grown mostly for human consumption. Corn grain is also processed into a polyunsaturated oil used for cooking and margarine. For a discussion of the culture of corn, see Chapter 23.

Cotton (*Gossypium* spp.) MALVACEAE

Cotton is principally grown for its lint, used in textiles, although the seed is an important byproduct. The seed, with an oil content of 30 to 40 percent, provides a valuable and widely used cooking oil. The seed residue, called cottonseed meal, is used as a cattle feed.

Flax (*Linum usitatissimum* L.) LINACEAE

In the United States flax is grown for its oil, but in other parts of the world it is also grown as a fiber crop to make linen.

Olive (*Olea europaea* L.) OLEACEAE

Olives have produced fine oil for centuries, but they are also grown for their fruits.

Rape (*Brassica* spp.)

Rapeseed (also called canola) contains better than 40 percent oil, and the pressed meal cake makes an excellent feedstuff, containing about 40 percent protein. This crop is an important and significant oil crop in many areas because of its low temperature and broad soil adaptation. Breeding research has greatly improved the quality of rape crop oil by reducing its glucosinolate and erucic acid content.

Sesame (*Sesamum indicum*) PEDALIACEAE

The seed, containing about 50 percent or more oil of excellent quality and stability, is most often used as a salad or cooking oil, and for flavoring. The seed is also used as a garnish for bakery products. The pressed cake is an excellent protein source for livestock feeding.

Palm Oil (*Elaeis guineensis Jacq.*) ARECACEAE

This tree was called the prince of the plant kingdom by Linnaeus because of its majestic appearance. The oil palm is botanically related to the coconut palm. When mature, the oil palm tree may attain a height of nearly 30 m (100 ft) but generally it grows no taller than 10 m (32 ft). The trees are monoecious (male and female flowers on the same tree). The flowers grow on a short **spadix**[7] that develops into a cluster of more than 1,000 drupes (fruits). The female flower normally has three ovaries, but generally only one is fertilized. The fruit matures about six months after pollination.

The oil palm, native to tropical West Africa, is an important source of vegetable oil. It is grown most abundantly along the west coast of Africa. Oil palms are also cultivated to some extent in the rain forest regions of the Congo, Kenya, Indonesia, and Malaysia. There are plantings in Central and South American countries.

Breeding objectives are to obtain shorter trees and better flowering characteristics. Shorter trees aid in harvesting fruit, and improved flowering would produce a higher per tree yield of oil.

The oil palm grows well on a wide variety of soil types. The trees do best on deep, well-drained soils that are neutral

to slightly alkaline in reaction, but are grown successfully on acid soils.

Trees generally respond to phosphorus applications and sometimes to potash. In South Africa, a bronze spotting of leaves seems to be caused by potassium deficiency. Magnesium has been found to be deficient in West Africa and causes a condition known as the orange frond disease.

Irrigation is not necessary for oil palms because they are grown only in tropical areas where rainfall is abundant and frequent.

Of the diseases reported, a fungus-caused stem rot is the most prevalent. The fungi enter the trunk of the tree when the leaves are cut off, and infected trees often die in two or three years. This fungus (*Marasmium palmivorus*) causes the acid content of the oil to increase in ripening fruit. A red-striped weevil is a serious insect pest.

In many Southeastern Asia areas, the fruit is harvested throughout the year, heaviest in late summer to early fall, lightest in late winter to early spring. Both the shell (pericarp) and the kernel contain oil.

Kernel oil is light yellow in color and used principally in the manufacture of edible products such as margarine, chocolate candies, and pharmaceuticals. Palm oil from the pericarp tissues is deep yellow to red-brown in color, and thick in consistency. It is used for making soap, candles, and lubricating greases. It is also used in processing tin plates and as a coating for iron plates.

Peanut (*Arachis hypogaea* L.) FABACEAE

The peanut is native to South America. It was introduced into Africa, where it contributed a large part to the diet of the peoples of East Central Africa. From Africa the peanut was taken to India, China, and the United States, presently the production leaders. Peanut (groundnut) is a widely grown crop because of its adaptation to tropical, subtropical, and warmer temperate regions.

The peanut is an annual plant with sturdy, hairy branches with growth habits from nearly prostrate to upright. After pollination, the flower withers and drops off, and in a few hours, the base of the flower stalk elongates into a unique structure, called the "peg." The tip of the peg contains the fertilized ovules, which the growing peg carries down and pushes into the soil. After the peg penetrates into the darkness of the soil, it turns horizontally and the ovary begins to swell and grow into the mature underground fruit (Fig. 21–16).

There are three distinct peanut-growing regions in the United States: the Virginia-Carolina area, which primarily grows the larger-seeded Virginia cultivars; the Georgia-Florida-Alabama area, which grows the Southeastern Run-

---

[7]A spadix is a spike with a succulent axis usually enclosed in a spathe. A spathe is the large sheathing bract (or pair of bracts) enclosing an inflorescence, especially a spadix on the same axis, for example, the Jack-in-the-Pulpit flower.

Figure 21–16 Peanut flowers are fertilized above the soil. Fertilization is followed by the development of a subterranean pod after the gynophore (peg) has elongated and buried. Soil penetration provides the necessary moisture and darkness for fruit enlargement, which rarely occurs above ground. *Source:* USDA.

ner and some Virginia and Spanish cultivars; and the Oklahoma-Texas district, which grows small runner and the Spanish cultivars. The three types of cultivars are separated on the basis of clearly identifiable agronomic characteristics. These characteristics include branch form, growth habit, pod size and shape, number of seeds per pod, and color of seeds after storage. Typically the Virginia types have both bunch and runner types and two seeds per pod. The Spanish types have erect bunch growth habits and pods are mostly two-seeded. The runner types have true runners, spreading branch growth habits and two to four seeds per pod.

In the United States, peanuts are produced mainly for grinding into peanut butter, for roasted and salted nuts, and for candy and bakery goods. Some limited amounts are used for livestock feed. In other parts of the world, peanuts are grown mainly for their edible oil and as a valuable protein source. Peanut seed has a content of 40 to 50 percent oil, and from 25 to 30 percent protein. After the nuts are harvested, the stems and leaves are often used for hay.

The soil, especially its texture and structure, is an important factor in peanut culture. The best soils are well-drained sandy loams with deep profiles, which allow the pegs to penetrate readily and the nuts to be cleanly harvested. Clay loam soils produce good crops provided the moisture content is optimal, but if the soil gets dry and hard, yields are reduced because the pods break off and are lost during harvest. In addition, soil particles stick to the pods, dirtying the product and lowering its grade.

Peanuts are often rotated to the benefit of other crops that follow, such as potatoes and corn, or other grass crops; other legumes should be avoided.

Most farmers treat peanut seeds with a recommended seed protectant to kill decay-causing organisms on the seed surface. Care is taken to prevent treated seed from being consumed or used for oil.

Planting dates vary with location. In Texas, peanuts are planted from early spring to midsummer; in the Southeast, most plantings are made during midspring; in Virginia, planting in late spring is most likely.

Peanuts are heavy users of plant nutrients, and they respond well to directly added fertilizers, although it is best to have the preceding crop heavily fertilized. If the soil is more acid than pH 6.0 to 6.5, finely ground limestone is added. The kind and amount of fertilizer vary considerably with soils and location.

In semiarid regions, peanuts must be irrigated, requiring about 60 cm (24 in.) of water. Many farmers irrigate the soil before planting, particularly if winter rains have been scarce and failed to wet the top 90 cm (3 ft) of the soil profile. Irrigation during the growing season is frequent enough to maintain a moist soil during the critical stages of blooming and seed development. Irrigation is discontinued when the plants cease rapid growth so that the ground will dry for harvest.

Some cultivation is necessary and important for weed control in peanuts but excessive cultivation is avoided. Two periods of weed control are critical: when the seedlings are emerging, and when the plants are setting fruit. Cultivations close to the plant can interfere with the developing pegs. Selective herbicides are widely used in all areas of the United States.

The most serious diseases of peanuts are the leaf spots (caused by three fungi), southern blight (also known as white mold or southern stem rot), collar rot, peg rot, and black rot.

The most prevalent insect pests are cutworms, thrips, leafhoppers, corn earworms, armyworms, and lesser cornstalk borers. Several insects infect storage facilities and destroy the stored peanuts. Some of these are cadelle beetle, carpet beetle, confused flour beetle, flour and grain mites, and rice weevil. The control of root knot, lesion, and sting nematodes is very important, although effective nematicides are available for their control.

Peanuts are ready for harvest when the plant no longer grows rapidly, the leaves begin to yellow, and the kernels are fully developed, as indicated by the darkening of the veins inside the hull. Harvesting earlier results in shriveled nuts, but harvesting too late can cause losses because of sprouting. The date of harvest is critical, so the crop is inspected to estimate the optimum date. When 60 to 70 percent of the inside of the hulls are stained tan, peanuts are considered to be mature.

If peanuts are grown for home consumption, they are hand harvested by digging the entire plant, then the vines are allowed to wilt. The pods are plucked from the vines, dried, and cured. For commercial production, peanuts are harvested mechanically. A blade is passed below the soil surface under the peanuts to cut the tap roots and to loosen the soil. The vines are pulled and placed in windows mechanically. After drying a few days, the peanuts are separated from the vines by a combine, collected in trucks, and transported to forced-air dryers, dried to 10 percent moisture, inspected, graded for value and size, taken to storage warehouses, and subsequently sold to manufacturers.

Safflower (*Carthamus tinctorius* L.) ASTERACEAE

Safflower, a relatively new crop to the United States, is actually one of the world's oldest crops. The plant is thought to be native to the Middle East and southwest Asia, where it has been known for centuries. The flowers were first used as a source of red dye for cloth. Safflower is now grown principally for its seed oil, which yields two types of oil: a polyunsaturated (linoleic) type and a monounsaturated (oleic) type, each having varied uses, such as for margarine, salad oil, and mayonnaise.

Safflower is a spiny plant that produces a light-colored oil with a high percentage of polyunsaturated fatty acids. Safflower is an annual plant belonging to the asteraceae family.

Safflower is grown commercially in India, Egypt, Spain, Australia, Israel, Turkey, Mexico, Canada, and the United States. The crop was introduced into the United States experimentally in 1925 but has not achieved important status.

Some cultivars yield a monounsaturated oil similar to olive oil which is used for cooking.

Safflower grows well on a wide range of soils but does best on a neutral soil with a deep, well-drained profile. Safflower tolerates high salinity nearly as well as barley.

Safflower can be seeded by drilling, broadcasting, or row planting. Seed is often drilled in nonirrigated farming areas but planted on raised beds in irrigated regions.

Safflower yields usually increase with additional nitrogen fertilizer. If following an inoculated legume crop, especially one that was plowed under, less nitrogen is needed.

Safflower responds well to ample soil moisture, and irrigation is essential in arid regions.

Safflower does not compete well with weeds, especially when the plants are young and small. The best weed control is prevention, i.e., crop rotation and field sanitation.

The most prevalent diseases are rust, phytophthora root rot, verticillium wilt, fusarium wilt, leaf spot, and bud rot. Rust is more troublesome in irrigated areas or areas of high humidity. Phytophthora root rot is caused by a fungus that occurs widely in the western United States, where the crop is irrigated. Leaf spot is a fungal disease aggravated by rainfall. It causes severe leaf damage and discoloration in areas of high humidity.

Insect damage is not as severe in safflower as in some other crops. The major insect pests are flower thrips, lygus bugs, bean and peach aphids, wireworms, and loopers.

Safflower matures in about 120 days when sown in the late spring and it is harvested when the seeds are white,

the leaves brown, and the stems dry. The plant does not lodge, nor do the seeds shatter unless they get excessively dry, in which case harvesting is often done at night or early morning when humidity is higher. Safflower is harvested with a grain combine of the same sort used for barley or wheat. The moisture content of the seed should be about 8 percent.

Soybean (*Glycine max* [L.] Merrill) FABACEAE

The soybean, also known as the soja or soya bean, is native to eastern Asia. It was cultivated in China and Japan long before written history. Because of its great importance as a high-protein food source, the soybean became one of the five sacred crops of China, joining rice, wheat, common millet, and glutinous millet. Europeans first learned of the soybean about 1700, but not until 1875 was there any great interest in the plant. The soybean in the United States was first mentioned about 1800 in Pennsylvania, where it was reported to grow well.

The soybean is an annual plant of the legume family. It is erect and bushy with many branches, fewer at normal field populations (Fig. 21–17). It varies in height from about 30 to 150 cm (1 to 5 ft) and its root system extends to 150 cm (5 ft) if the soil is permeable. The leaves are alternate and trifoliate except for a pair of opposite simple leaves at the first node above the cotyledons. Most cultivars are pubescent (hairy). Flowers are self-pollinated and are borne on racemes (clusters) of 3 to 15 flowers that are either white or purple, or a blend of these colors.

The annual production of soybeans worldwide has more than tripled since World War II. The United States is the leading producer, growing about 55 percent of the world total of more than 130 million MT. Since the different cultivars

Figure 21–17 Soybeans are erect, bushy, branching plants that in their early growth closely resemble garden beans. Most cultivars have deep root systems, which give them more resistance to drought than other bean cultivars. Nearly all cultivars are pubescent. Here an excellent stand of soybeans about 60 cm (2 ft) high is growing on rich prairie soil in Illinois.

of soybean mature their fruits over a wide range of photoperiods, the soybean is adaptable throughout many temperate areas.

The great increase in production in the United States was due to (1) the development of more productive cultivars that are resistant to lodging, seed shatter, and diseases; (2) the extension of soybean production from the U.S. corn belt to the southeastern states; (3) the utilization research done by the USDA and land grant colleges; and (4) the worldwide increase in demand for both the meal and the oil. Other important growing areas are Brazil, China, and Argentina. In the United States, the primary production states include Illinois, Iowa, Minnesota, Indiana, Ohio, Missouri, and Arkansas. States along the Atlantic coast are also important areas of production.

Soybean cultivars are grouped according to their response to daylength (photoperiod), and it is important to select cultivars adapted to local conditions in order to take full advantage of the available season. Cultivars with maturity times corresponding with the latitude have been developed; those requiring long photoperiods are in the northern latitudes; those requiring short photoperiods in the southern latitudes. Cultivars are also classified as determinate (terminating vegetative development before flowering) or indeterminate (vegetative development continuing for several weeks after the beginning of flowering).

Soybeans do best on a fertile, well-drained, nearly neutral pH soil of light texture but can grow better than many crops on poorly drained soils of low fertility.

Soybeans do not grow well on saline soils with soluble salt concentrations above 5 to 7 mmho/cm electrical conductivity. Boron is toxic to soybeans at concentrations exceeding 0.75 ppm. Soybeans are more tolerant to acid soils than most legumes, but if the pH is much below 6.0, they benefit from an application of limestone. On acid soils, molybdenum becomes limiting and the plant responds to applications of this nutrient. Manganese may become limiting on certain soils when limed above pH 7.0. Iron chlorosis may also be a problem in Iowa and western corn belt soils.

Seedbed preparation is not an important cultural operation. Fields are prepared for soybeans much as for other row crops. The ground is plowed either in the fall or spring. For heavier soils, fall plowing is generally preferred. Disking after plowing breaks up clods and destroys weeds. Harrowing before planting yields a firm, clod-free seedbed. Soybeans are increasingly being produced by using no-till procedures, and grown as a double crop in wheat stubble.

Soybeans benefit from inoculation with nitrogen-fixing bacteria unless it is known that the bacteria are already in the soil from a previous crop. Because *Rhizobium* are almost universally present in the corn belt and most southern environments, seed inoculation is less common. Fungicide treatment for seed is recommended, especially if lower-quality seed is used, even though fungicides generally reduce the effectiveness of inoculation.

Temperatures and daylength determine the time of planting. Farmers in the northern corn belt states plant when the soil temperature is warm enough for fast germination and rapid growth. This generally occurs immediately after corn planting. Mid-season cultivars are used if planting is delayed in the northern corn belt states, but full-season cultivars give higher yields. Soybeans are seldom planted later than June. In the southern states, they are not planted before early May, even if the soil is warm. Because of the short daylength, early planting results in early flowering, which reduces yields because the beans form and mature too soon to take advantage of the full growing season.

In the northern areas, rows are normally 17 to 75 cm (7 to 30 in.) apart on fertile soil; however, full-season cultivars yield almost as much with rows 91 to 100 cm (36 to 40 in.) apart. About 60 percent of the acreage is now solid drilled with 15 to 25 cm rows and is not cultivated. This is less so in southern areas, where plants branch more prolifically and wider 90 to 100 cm (36 to 40 in.) spacings are used. In California, because of irrigation, plantings have single rows 70 to 76 cm (28 to 30 in.) apart or two rows 30 to 40 cm (12 to 16 in.) apart on beds 100 cm (40 in.) from center to center. Because of salt accumulation in the center of the bed in some areas, the rows are placed on the shoulders to avoid salt injury to the plants.

Seeds within the row are spaced between 5 to 15 cm apart depending on the cultivar, seed size, row width, and seed viability.

In some areas a grain drill is used to plant soybeans. The seed is drilled to a depth of 2.5 to 4 cm (1 to 1.5 in.) in moist soils.

If the seed was properly inoculated and the roots show good nodulation, the crop usually needs little or no additional nitrogen. The bacteria in the nodules fix the nitrogen the plant needs from the air. If there are no nodules on the roots and yellowing leaves show a deficiency in nitrogen, an application of 34 to 45 kg/ha (30 to 40 lb/ac) of nitrogen improves yields. Soil or plant tissue tests indicate any need for phosphorus or potassium.

Soybeans are not generally irrigated in the United States except in areas where supplemental irrigation helps the crop through periods of drought. An increasing use of irrigation is now being practiced in many midwest production areas. Need for water is identified in soybeans, as in cotton, by a change in leaf color from a light, brilliant green to a darker, dull green. Soybeans suffer the greatest yield loss from insufficient water during pod formation and pod fill. Sprinklers are the common method of applying irrigation water in the southern and midwestern states where supplemental irrigation is used. In western states both sprinkler and furrow irrigation methods are used.

Weeds drastically reduce soybean yields—sometimes by as much as 50 percent. Early cultivation and careful seedbed preparation help minimize these losses. When planted in rows, soybeans are easy to cultivate after seedling emergence. Cultivation should be no deeper than necessary to kill weeds. Many selective herbicides are available and are effective preplant, preemergence, or postemergence weed control materials.

With the development of disease-resistant cultivars, some diseases are less damaging, but considerable disease damage persists. Phytophthora root rot is a major disease problem. Pythium rot generally occurs early in the season because its development is favored by cold temperature. Rhizoctonia rot results in damping off or death of small seedlings. Fusarium rot is a seed-borne root rot disease most frequent on small seedlings. Anthracnose causes poor seed germination or death of seedlings. Soybean mosaic causes crinkly or ruffled leaves. Bacterial blight appears early as a small leaf spot on young plants. Downy mildew is a mildew growth appearing on the first leaves as they open. Stem canker kills the plant during the last half of the growing season. Pod and stem blight is a major disease affecting seed quality.

Several insects cause losses to soybean growers, but few are considered serious economic pests, although the Mexican bean beetle has adapted to soybeans to become the primary insect pest. Other pests include seed corn maggots, seed corn beetles, wireworms, white grubs, thrips, southern corn rootworms, Japanese beetles, grasshoppers, alfalfa hoppers, blister beetles, cabbage loopers, stink bugs, and velvet bean caterpillars. Spider mites are very damaging in California and semi-arid areas because of the dry summer weather.

The soybean cyst nematode, root-knot nematode, and the sting nematode cause problems for the soybean farmer.

Soybeans are physiologically mature when the seeds are at a moisture content of about 50 percent, and when the leaves begin to yellow and abscise (even without frost). The beans lose moisture quite rapidly, depending upon the weather and relative humidity. Soybeans are harvested with a combine when seed moisture is 10 to 18 percent, although harvest efficiency is better at 17 to 19 percent. For safe storage, seed moisture content should not exceed 13 percent.

Harvesting is more critical for soybeans than most crops because of shattering losses. Breeding has largely solved this problem. Lodging has been greatly reduced. The height of the cutter bar is important; the lower it is set, consistent with field conditions, the better.

Virtually no soybeans are now grown for hay. There are forage-type cultivars with fine stems used for this purpose. The soybean hay would be cut when the bean pods are about half filled. At this time the leaves are about maximum size and the stems not excessively woody. The hay is cured in the swath for two to three days before it is raked into windrows. Good-quality soybean hay is about equal in feed value to other legume hays but is difficult to cure without loss of leaves, reducing quality.

Sunflower (*Helianthus annuus* L.) ASTERACEAE

The sunflower is a native of North America, and may have been domesticated before corn. It is a tall annual plant with rough hirsute stems. The flower heads are a compound inflorescence composed of many individual flowers in a large disc, subtended by large ray flowers (Fig. 21–18). In wild specimens the flower heads are 8 to 10 cm wide, but larger in commercial cultivars. The flowers are cross-pollinated by insects. Wide use of sunflower as an edible oil crop began about 1830 in Russia, and it has become the main source of edible oil in the former Soviet Union, other eastern European countries, and Argentina. Reintroduction to the United States

Figure 21–18 The land area devoted to production of sunflowers is rapidly increasing. The whole seed contains from 25 to 40 percent oil which is high in polyunsaturated fatty acids. The oil is used in margarine, salad oils, and other foods. The oil cake remaining after the oil is extracted contains about 32 percent high-quality protein and is used mostly for livestock feed.

was for its use initially as silage. Its use as an oil crop has greatly increased its production, with more than 1 million hectares now grown. This increase was assisted by the use of Russian germplasm providing oil content above 40 percent and the ability to produce hybrid seed. The introduction and wide usage of hybrid seed improved crop uniformity, yields, and earlier maturity, as well as disease resistance. Breeding for shorter stems (120 to 150 cm) reduced lodging and mechanical harvesting problems.

The cultivated plant has at least three uses: oil, confectionary products, and fodder. The oil form produces a valuable and desirable oil having polyunsaturated and monounsaturated characteristics similar to corn oil and olive oil, respectively. The confection-type cultivars are usually large-seeded and are roasted like nuts, with or without the hull. The smaller whole seed is used as a feed component for pet and wild birds and for small animals. The fodder-type plant is very tall with large stems and leaves, and the entire plant is chopped green for livestock feed. The cake or meal from the oil-pressing process is used for animal feed.

Sunflowers may be grown under rainfed or irrigated conditions. Soil moisture is most critical at flowering; although they have good drought tolerance, irrigation may be necessary for successful production in semiarid regions.

Row plantings are usually made with a width of 70 to 75 cm, to provide field populations between 4 to 7 plants per square meter. Bees may be utilized in the field to aid in pollination.

Major diseases are rust, downy mildew, head mold, and sclerotinia rot. The salt marsh caterpillar, stalkbores, and sunflower moth are serious insect pests. Sunflowers are harvested with combines in a similar manner as the cereal grains, with special types of pick-up reels being used to minimize seed losses.

REFERENCES AND SUPPLEMENTARY READING

DRISCOLL, C. J. 1990. *Plant sciences: production, genetics, and breeding.* New York: E. Horwood.

EVANS, L. T. 1993. *Crop evolution, adaptation, and yield.* New York: Cambridge University Press.

PEARSON, C. J. (ed.). 1992. *Field crop ecosystems.* New York: Elsevier.

WOLF, B. 1996. *Diagnostic techniques for improving crop production.* New York: Food Products Press.

CHAPTER

Forage and Fiber Crops

FORAGE CROPS

With few exceptions, forage crops are grass and legume plants whose stems and leaves are grown to feed herbivorous animals, particularly ruminants.[1] Forage crops provide feed directly, when animals graze on pasture or rangeland; as harvested hay; as chopped green material for direct feeding; or as silage.

In many countries, and particularly in the United States, the production of animals for meat and other products is a major agricultural enterprise. These activities rely heavily upon the production and utilization of forage crops. In the United States, the value of animal products is more than one-fourth of all farm cash receipts, and the value of the forage crops fed to animals is greater than any other single crop. Worldwide, more land is devoted to forage crops than to all other crops combined. The ratio of land for forage production versus other crops varies, depending on geographic features and population pressures. Much of this land is rocky, hilly, or too wet or too dry for growing other crops but is suitable to supply pasture for animals. Additionally, such cropping can be beneficial in sustaining the land against erosion, and for some soil-

---

[1] A ruminant animal has a multichambered stomach containing large populations of microorganisms that break down tough, fibrous materials high in cellulose, hemicellulose, and lignins to simpler, easily absorbed materials. Ruminants also digest concentrated foods, such as cereal grains.

building improvements. In other areas the land may be too valuable to use for a relatively low-value commodity such as forages.

Either grain or forage crops can be used for feeding animals, but usually it is a combination of both. The proportion of each often depends on economic situations. When a strong market demand exists for grain, and the price is high, less grain and more forages are fed to livestock. In the United States, corn, barley, oats, sorghum, and rye are the cereal grains generally fed to livestock, while wheat is used primarily for human foods.

It is often suggested that it would be more efficient for people to consume plants and plant products directly, rather than to feed them to animals and then consume the meat and dairy products. However, the types of forage feeds consumed by animals are high-cellulose vegetative materials that cannot be easily digested by humans. Additionally, many grains fed to livestock, such as sorghums, rye, barley, and oats, are not particularly relished by most people. The natural grasses covering much of the land expanses of the earth can be best converted to human food products only by pasturing animals. Finally, many forage crops grow reasonably well in soils that are often unsuitable for growing other crops, allowing for productivity from land areas that otherwise would be of little agricultural value.

Presently, livestock animals have their forage diets supplemented with concentrated sources of high-energy foods, usually from various cereal grains. Grain-fed animals gain weight more rapidly and provide a higher quality of meat, and dairy animals produce more of a higher-quality milk. However, these animals could survive totally on forage crops alone. Actually, increasing human health concerns regarding blood cholesterol levels have favored greater production of grass–fed animals to produce meats lower in fat.

The maximum productivity from range and pasture land is far from reached. Better planning and management of forage crop, range, and pasture lands has the potential to significantly raise their productivity.

The many forage crop species grown for ruminant animal feed belong to two large plant families—POACEAE, the grasses (Fig. 22–1), and FABACEAE, the legumes (Fig. 22–2). These two families also furnish many important human foods—wheat, corn, rice, sugarcane, oats, and barley in the grasses, and soybeans, peanuts, beans, and peas in the legumes.

Most forage crops are herbaceous perennials. Some are annuals requiring replanting each year, but others reseed themselves.

The production of seed for the various types of forage crops is a large industry involving many seed firms in various parts of the world. They produce the seed in areas with especially favorable climates. Much of this is sold as certified seed. Certified seed is relatively free of other crop and weed seeds and has a high germination percentage. Some American states have seed-certifying agencies that are re-sponsible for arranging sources of reliable forage–crop seeds.

Forage Grasses

Plants in the grass family generally have a well-branched, fibrous root system that grows several centimeters to a meter or more deep, depending upon the species. The initial root system develops from the germinated seed, but the permanent root system develops as adventitious roots from the stems form. Several shoots, or tillers, of varying lengths develop from each plant. Plants of some grass species have horizontal stems—rhizomes, stolons, or both. The leaves are long and narrow with parallel veins, typical of the monocotyledons. The inflorescence of the forage grasses may be either a spike, as in the wheat grasses, or a panicle, as in smooth bromegrass.

Forage grasses can be classified as bunch types, which maintain a distinct clump, or sod formers, which spread vegetatively by rhizomes or stolons to form a solid carpet. The forage grasses may also be classified as **cool-season** or **warm-season** types. Most of the cool-season grasses originated in northeastern Asia or northern Europe, and their growth is favored by a cool climate with ample rainfall (or irrigation) for optimum growth.

Important Cool-Season Perennial Forage Grasses

Smooth Bromegrass (*Bromus inermis* Leyss.)

This grass is widely grown in the United States except along the Pacific coast and in the southern states. It was introduced from Europe about 1890. It has a panicle inflorescence and spreads by rhizomes. It requires a fertile, well-drained soil. Smooth bromegrass is moderately drought tolerant, has some sensitivity to low winter temperatures, gives high yields, and is highly palatable for animals. It is very nutritious and can be used for pasture, hay, or silage. It is often grown in association with a legume.

Orchard Grass (*Dactylis glomerata* L.)

A native of western and central Europe, this forage grass has been grown in the United States for over 200 years, mainly in the northeastern states but now also in the Rocky Mountain area, in most areas along the Pacific coast, and in the transition zone between warm and cool climates. It is a bunchgrass with many broad basal leaves and a tall panicle inflorescence. It does not produce rhizomes or stolons, and therefore is not a sod producer. Orchard grass is high yielding if amply fertilized and can be used for pasture, hay, or

Figure 22–1 Some grass species forage crops. *Above (left to right):* Smooth bromegrass, Orchard grass, Tall fescue. *Below (left to right):* Timothy, Bluegrass, Western wheatgrass.

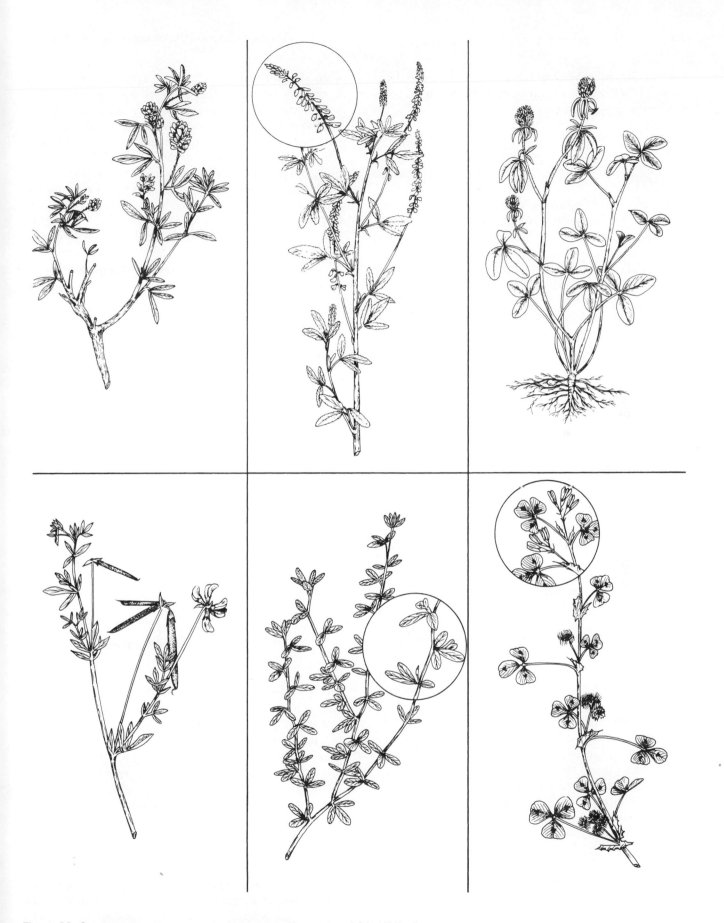

Figure 22–2 Examples of legume species forage crops. *Above (left to right):* Alfalfa, Sweet clover, Red clover. *Below (left to right):* Birdsfoot trefoil, Lespedeza, Bur clover.

silage. It loses palatability rapidly with the onset of seed maturity. It is not as winter hardy or drought resistant as some of the other cool-season grasses.

Tall Fescue (*Festuca arundinacea* L.)

Tall fescue is grown in the transition zone between warm and cool climates. It is popular in the Ohio River Valley and elsewhere in the eastern United States except New England and Florida. It is also grown in scattered parts of the West and the Pacific Northwest. It is grown for pasture and for hay. There are other fescue species, but this one is the most important. The fescues are less palatable to animals than some other grasses because of their alkaloid content. Fescues are deep rooted and tolerate shade, flooding, and salty and alkaline soils.

Reed Canary Grass (*Phalaris arundinacea* L.)

This is grown in the northern part of the United States and in southern Canada. If properly managed, it produces a highly nutritious pasture, or it can be harvested for hay and silage. As with most grasses, it loses palatability upon reaching full maturity, and it also has a problem with high alkaloids. Reed canary grass has a strong root system that tolerates wet, poorly drained land and spreads by rhizomes to form a dense sod. It is useful for stabilizing gully areas in pastures.

Timothy (*Phleum pratense* L.)

Timothy, native to northern Europe/Asia, was first grown about 1750 in the northeastern United States. It is a bunchgrass that attains a height of about 0.8 m (30 in.). It is unusual among forage grasses in that it has an enlarged bulblike thickened stem internode (haplocorm) at the base. Buds on the haplocorm produce new shoots each year, maintaining the plant's perennial growth habit. Individual shoots are biennial, growing vegetatively the first year and flowering the second. Timothy can be used as pasture, hay, or silage. It is highly palatable, especially to horses. It is often grown in association with legumes. Timothy is winter hardy but not drought resistant and does not tolerate close grazing because of its shallow root system.

Bluegrass (*Poa* spp.)

These grasses are widely distributed throughout the world in the cooler parts of the temperate zones. Their origin was southeastern Europe. Kentucky bluegrass (*P. pratensis* L.) is extensively grown over many of the cool parts of the United States as a pasture grass and a lawn grass. Its rhizomatous growth habit produces a thick sod. It is a highly palatable pasture grass and withstands close grazing, but is less productive than some of the other grasses. Kentucky bluegrass does not tolerate heat and drought owing to its shallow root system. It tends to go dormant during the heat of late summer. Bluegrass seed must be planted very shallow, as freshly harvested seeds require light for germination. In addition to the common Kentucky bluegrass, several cultivars have been developed—primarily for lawns and turf rather than for pasture. Bluegrass seedlings are about 90 percent apomictic (forming vegetatively from female parent tissue).

Important Perennial Forage Grasses

Bermudagrass (*Cynodon dactylon* [L.] Pers.)

This grass is grown extensively throughout all the tropical and subtropical areas of the world and the southern areas of the United States. It probably originated in Africa. Bermudagrass is an aggressive plant that can spread rapidly by seeds, rhizomes, and stolons. It is used for both pasture and hay crops, and fine-leaved cultivars for lawns and turf in warmer areas. Bermudagrass grows best on fertile, clay soils in full sun. In winter it turns brown, making it undesirable for lawns. Improved hybrid cultivars are more productive, more frost tolerant, and more drought resistant than common bermudagrass.

Dallisgrass (*Paspalum dilatatum* Poir.)

This is a clump grass native to northern Argentina, Uruguay, and southern Brazil. It grows to a height of 0.6 to 1.2 m (2 to 4 ft) and produces many basal leaves. It was introduced into the United States about 1842. It is best adapted to the warm southern areas of this country. It is a productive pasture grass, starting growth early in the spring and continuing on into late fall. Proper management and cutting is necessary to keep the plants growing vegetatively and to maintain palatability. It does not tolerate close grazing. In California, Dallisgrass often becomes the dominant species in irrigated pastures during the hot summer months when the growth of other, more palatable grasses is markedly slowed. Dallisgrass is often considered a serious weed pest along irrigation canals—the plant grows well in ditch bank habitats. Its abundant seeds are light, flat, and oily and are carried to all areas irrigated with water flowing from the contaminated ditches.

Bahiagrass (*Paspalum notatum* Flügge)

In the United States, this grass is best adapted to the Gulf coast and Florida, where it is used as a pasture grass. Bahiagrass, native to South America, is a strong-growing, deep-rooted rhizomatous plant that forms a thick sod. Its deep root system adapts it well to sandy or dry soils.

Crested Wheatgrass (*Agropyron cristatum* [Willd.] Beauvois)

This is a long-lived, vigorous tetraploid bunchgrass with a deep, spreading root system. It was introduced to the United States from Siberia about 1890 and has become a valuable forage crop. It is drought tolerant (although growth is slowed when the weather is hot and dry), cold hardy, and tolerant of close grazing. It starts to grow early in the spring, producing a palatable and nutritious pasture feed for livestock.

Intermediate Wheatgrass (*Agropyron intermedium* [Host] Beauvois)

This was introduced into the United States from the Caspian region of Russia in the early 1900s. It is well adapted to the Pacific Northwest and the Great Plains range and dryland regions. It produces best when the weather is cool during spring, early summer, and fall. Intermediate wheatgrass is used for pasture and is often planted with alfalfa to form a high-producing hay crop. It spreads by rhizomes, forming a good sod with an extensive root system, but it is only moderately drought resistant.

Other Species A number of other native and introduced wheatgrass species are grown for forage under range and dryland conditions. Some of these are Western wheatgrass (*Agropyron Smithii* Rydb.), bluebunch wheatgrass [*A. spicatum* (Pursh) Scribn. & Smith], slender wheatgrass [*A. trachycaulum* (Link) Malte], pubescent wheatgrass [*A. trichophorum* (Link) Richt], and tall wheatgrass [*A. elongatum* (Host) Beauvois].

In addition to the wheatgrasses, there are several important range or pasture grasses in many areas of the central and western United States. Among these are gramagrasses [*Bouteloua curtipendula* (Michx.) and *B. gracilis* (HBK) Lag. ex Steud.], bluestems (*Andropogon* spp.), buffalograss [*Buchloe dactyloides* (Nutt.) Engelm.], switchgrass (*Panicum virgatum* L.), wild rye (*Elymus* spp.), and tall oatgrass (*Arrhenatherum elatius* L. Presl.).

Additional Annual Grasses Used As Forage Crops

Several annual plants, which must be reseeded each year, make substantial contributions as forage crops.

Corn (*Zea mays* L.)

Among its many uses, corn is made into silage for ruminant livestock. Corn is grown over most of the United States for this purpose. It is harvested by machines that finely chop the entire plant before storage in silos.

Sorghum (*Sorghum bicolor* [L.] Moench)

Sorghum was introduced into the United States from France about 1855. The sorgo type, which has a sweet juice, is grown over much of the United States as a silage crop. Sorghum is a drought-tolerant, warm-season crop that grows well at higher temperatures but does not tolerate low temperatures.

Sudan Grass (*Sorghum sudanese*) [Piper] (Stepf)

This type of sorghum is cultivated throughout the United States as a forage crop for silage and for pasture. It has been replaced to a considerable extent in the northeastern states by more productive sorghum-sudan grass hybrids.

Italian Ryegrass (*Lolium multiflorum* Lam.)

This forage crop is grown for pasture, hay, and silage in the southeastern United States and in the Pacific Northwest.

Forage Legumes

Legumes are valuable forage crops because of their relatively high protein content—often 15 to 30 percent. Legumes have a symbiotic relationship with *Rhizobium* bacteria, which invade the plant through the root hairs and live in gall-like nodules on the roots. These bacteria in the nodules convert atmospheric nitrogen primarily to ammonia. Energy for this reduction of free nitrogen gas (N_2) from the atmosphere to ammonia (NH_3) is supplied by the plant's carbohydrates.

For maximum yields from forage legumes, it is essential to inoculate the seed properly with fresh, pure cultures of the correct strain of *Rhizobium*. Legume seeds may be inoculated by the seed seller or processor, or by the grower just before planting. Soil pH must be above 5.5 for the bacteria to survive; often the soil must be limed.

Stems of the various legume species vary greatly in length, size, branching, and fiberness. The leaves are compound and arranged alternately along the stem. Most legumes have taproots. Flowers are in racemes (pea), in a spikelike raceme (alfalfa), or in heads (clover). Some species, such as white clover, spread by stolons (aboveground horizontal stems). Generally rhizomes (belowground, horizontal stems) do not occur in the legumes except for certain types of alfalfa. Most forage legume species are perennials but a few are annuals.

Some Important Legume Forage Crops

Alfalfa (*Medicago sativa* L.)

Called "lucerne" in many European countries and in Argentina, Australia, and New Zealand, alfalfa is believed to have originated in Iran. It has been a valuable forage crop since earliest recorded history. Alfalfa is sometimes called the queen of the forage crops since it is one of the most widely planted and useful forages in many parts of the world. It is a perennial plant grown in all parts of the United States. Alfalfa has been a mainstay in the development of U.S. agriculture, starting with its introduction into California about 1850 from Chile. Alfalfa has a deep tap root and requires a deep (usually more than 1 m) and well-drained soil for maximum production. It produces more protein per unit area than other forage crops. The variegated type is very winter hardy, and cultivars from this group are recommended for the northern United States and Canada. Other, nonhardy types are widely planted in mild-winter areas since they recover and grow rapidly after cutting, giving high yields.

In the northern states alfalfa is often grown in combination with other legumes such as red clover, or with certain grasses such as smooth bromegrass. In the west, pure stands of alfalfa are grown. Alfalfa is a nutritious feed palatable to livestock, but it must be harvested properly to retain its nutritive properties. An optimum maturity stage has to be selected so that the crop retains the nutritive value and palatability of the early shoot and leaf growth, yet provides the higher yields of the more mature growth. Such a maturity stage is believed to occur when about one-tenth of the flowers have opened. In northern areas, where winter damage can be a problem, the last normal cutting should be at least four weeks before the first killing frost to permit the plants to store carbohydrates and develop cold resistance. With certain cultivars, a final harvest can be made closer to the first killing frost date without damage. In areas where winter "heaving" of the soil occurs, alfalfa may be injured.

One of the disadvantages of alfalfa, as well as of some other legumes, is that it can cause bloat—accumulation of large amounts of gas in the animal's stomach—in pasturing cattle and sheep. Bloat can be minimized, however, by certain practices such as chopping up the hay and mixing it with other rations before feeding.

Alfalfa is susceptible to bacterial wilt, but many resistant cultivars are available. Leaf spot is a problem in the eastern United States, damaging leaves and reducing the feed value. The alfalfa weevil, which does considerable damage, is best controlled chemically. Lygus bug is a serious pest in some areas where seed is produced.

Certified seed of many alfalfa and alfalfa hybrid cultivars is available in the United States for their adaption to many production areas. In California, alfalfa seed production is in itself a major enterprise.

White Clover (*Trifolium repens* L.)

This is a widely grown legume, finding considerable use as a perennial pasture crop for livestock. It is believed to have originated in the eastern Mediterranean region.

There are three types of white clover—the small, intermediate, and large. Small or common white clover is found in lawns over much of the United States and Canada. The intermediate type is grown in the southern states in permanent pastures with bermudagrass, bahiagrass, or dallisgrass, but it requires frequent renovation. The large types—such as the cultivars Merit and Pilgrim—are grown mostly in northern regions and usually in association with the orchard grasses, the ryegrasses, and the fescues. They yield much more than the smaller types, provided they are heavily fertilized.

'Ladino' clover has several characteristics that make it a valuable pasture legume: it is very productive if soil moisture is ample; it is highly palatable and high in protein—20 to 30 percent; it tolerates poor drainage; it is a strong-growing plant, spreading by stolons, that recovers easily after grazing; it competes well in grass mixtures; and it reseeds itself easily.

Like alfalfa, white clover can cause bloat in ruminant animals. It is a desirable feed crop for single-stomach animals such as hogs.

Red Clover (*Trifolium pratense* L.)

This species is native to southeastern Europe and Asia Minor. Although a perennial, red clover is widely grown in the northeastern United States and southern Canada as a biennial pasture, hay, and silage forage crop—disease and insect problems limit its longevity. Red clover has been losing popularity in recent years; other legumes are replacing it. However, red clover is well adapted to a wide range of climatic conditions, although preferably cool and moist. The cultivars are easy to establish, grow rapidly, and give good yields (but less than alfalfa).

Two main types of red clover are grown in the United States and Canada. Medium red clover is largely planted in the Pacific Northwest, along the Atlantic coast, and in the Midwest. It produces two hay crops a year. The planting is generally replaced after the second cutting of the second year. The second type is known as the American mammoth red clover. It grows slower than the medium red clovers, producing only one hay crop. Red clovers alone cause bloat in pasturing ruminant livestock, so grasses are usually also seeded in a pasture mixture. Other important red clovers include strawberry (*T. fragiferom*), subclover (*T. subterraneum*), and rose clover (*T. hirtum*).

Alsike Clover (*Trifolium hybridum* L.)

Alsike clover is grown for pasture, hay, and silage primarily in the New England and north central states as well as in the Pacific Northwest and eastern Canada. It is also an important forage crop in northern Europe, where it has been grown for centuries. Alsike clover is best adapted to cool climates and wet soils, and even tolerates flooding. It grows on a wide range of soil types. Alsike clover is often planted with grass, such as timothy, which aids in holding it erect. There are both diploid and tetraploid strains.

Bird's-foot Trefoil (*Lotus corniculatus* L.)

The broad-leaved bird's-foot trefoil is an excellent long-lived, highly nutritious pasture crop producing most heavily in mid to late summer. It is a native legume in Europe and parts of Asia, being brought into cultivation in Europe some time after 1900. It is a perennial plant grown in the northeastern and north central United States and in southeastern Canada. It is also grown along the Pacific coast under irrigation. In the Pacific coast states, bird's-foot trefoil does not compete with 'Ladino' clover. It survives only where the soil is too dry or too infertile for 'Ladino' growing.

Bird's-foot trefoil is one of the most salt-tolerant forage legumes and, when planted as an irrigated pasture crop, is a most useful plant in alkali reclamation projects. It is used primarily as a pasture forage crop and does well mixed with grasses such as timothy or smooth bromegrass. It is not ordinarily used for hay. Most cultivars are winter hardy in northern United States and southern Canada. Bird's-foot trefoil is generally deep-rooted and grows well on a wide range of soil types but is difficult to establish. The seeds of various cultivars require inoculation with special strains of *Rhizobium* bacteria specific for the species. Unlike most legumes, bird's-foot trefoil does not cause bloat in ruminant animals.

Crimson Clover (*Trifolium incarnatum* L.)

This is a winter annual legume pasture crop grown in the southeast United States and along the Pacific coast. It does well under a wide range of soil and climatic conditions.

It tolerates some soil acidity and grows well on either sandy or clay soils but requires good drainage. It grows more at lower winter temperatures than most other clovers. For fall and winter grazing, seeds are sown from July to November in the northern hemisphere. Crimson clover can be used for winter pasture and then harvested for seed in the summer. It does well in mixtures of such grasses as bermudagrass or dallisgrass. Good yields of high-quality hay may be obtained if it is harvested just before the half-bloom stage. Reseeding cultivars have been developed. The seeds germinate gradually over a long period in autumn.

Lespedeza (*Lespedeza* spp.)

The origin of lespedeza is eastern Asia. Three species of lespedeza are grown in the United States: Korean (*L. stipulacea*), common or striate (*L. striate*), and sericea (*L. cuneata*). The first two are annuals; the third is perennial. Production areas in the United States range from southeastern Nebraska to southeastern Texas and extend eastward to the Atlantic coast. The lespedezas are important warm-season legumes grown for summer pasture and hay in the United States since the early 1930s. Lespedezas are best used in combination with other crops, such as small grains and grass sods.

Other Foliage Legumes   There are many other legume cultivars grown to advantage that meet specific conditions and needs. These include the vetches: common vetch (*Vicia sativa*), hairy vetch (*V. villosa*), and purple (*V. atropurpurea*). Although slow starting and not competitive with weeds, the vetches are hardy, and perform well in acid and low-fertility soils. Winter-growing, reseeding annual legumes such as subterranean clover (*Trifolium subterraneum* L.) and the annual medics such as bur clover (*Medicago polymorpha* L.) are rapidly increasing as important introduced forage plants in the Mediterranean-type range areas of the Pacific coast. Lupines (*Lupinus* spp.), roughpea (*Lathyrus hirsutus*), kudzu (*Pueraria thunbergiana*), and velvetbean (*Stizolobium deeringianum*) are other legumes that are prolific forage producers and nitrogen providers for the associated grasses frequently found in pasture and range production. The invasive nature of kudzu has resulted in it becoming a nuisance species in the South.

Utilization of Forage Crops

Livestock grazing on rangelands and pastures are fed at the lowest cost since the animals themselves do the harvesting of these usually low-production input crop plants. A distinction is made between rangeland and pasture. Rangeland refers to extensive grazing areas consisting largely of native vegetation, whereas most of the pasture lands in the United States are located in the eastern and central sections and are seeded with introduced forage species (Fig. 22–3). Most rangelands are in the western United States. While rangelands may not be fenced, pasture land is almost always fenced.

Rangelands

In the United States most of the rangeland is in the western states. In addition to livestock grazing, rangelands provide wildlife feed, recreation, and timber resources. About half

Figure 22–3 *Above:* Forage from range land usually consists of native vegetation and the area would not be irrigated. *Below:* Forage from the less steep pasture land would often be comprised of introduced species and have a greater probability of supplemental irrigation.

the rangelands in the western United States are owned by the federal government but permits for grazing can be obtained from the Forest Service's Bureau of Land Management, Bureau of Reclamation, and other public and private agencies.

The proper management of rangelands gives maximum livestock production while conserving the land and its vegetative cover. Proper rangeland management can do much to upgrade the land's ability to feed livestock. Undesirable forage species can be eliminated by fire, chemical or mechanical means, or competition from more aggressive plants. Undesirable plant communities can often be replaced by seeding the area to species that are more productive and more nutritious and palatable for livestock. Proper grazing management can also influence the proportion of desirable species.

Pastures

Pastures are land areas seeded to various introduced forage species. In the eastern and midwestern United States, pasture lands are generally supported by rainfall but in arid and semiarid western areas, irrigation of the pastures is often necessary.

Pastures may be classified as:

1. **Permanent pastures,** maintained indefinitely for grazing.
2. **Temporary pastures,** in use as pastures for less than one year.
3. **Cropland pastures,** rotated with other crops on the same land.

Permanent Pastures Permanent pastures develop from a stand of natural grasses, by intentional seeding of one or more perennials or reseeding annuals, or from abandoned cultivated fields invaded spontaneously by one or more aggressive forage plants. Permanent pasture is often the only practical economic use of land that is too hilly or too wet for other agricultural activities.

For a permanent pasture to be productive and support an economical number of animals, it must be given care. In the absence of legumes, fertilizers are likely to be needed, especially nitrogen. Leguminous pasture crops may require phosphorous, potassium, and sulfur. In high-rainfall areas, the soils become acid and require added lime for best growth of the pasture vegetation. More desirable species may be reseeded to upgrade the pasture's productivity or to extend the grazing period. A seedbed should be prepared and the seeding done in early spring or early fall, depending on the species.

The grazing periods and the number of animals on a given land area must be regulated to maintain a pasture in good condition. Poor grazing management can ruin a fine pasture. In regions with severe winters, late fall grazing should be avoided since it lowers nutrient reserves in the plants and decreases their winter survival. Legumes in pastures need at least four weeks without grazing before the average date of the first killing frost.

Rotational grazing—that is, using different sections of the pasture in sequence—permits flexible and efficient use of pasture areas. A series of fenced adjacent pastures allows the moving of animals from one section to the next, thus giving each section a time to recover. Some fields may not be harvested for hay or silage.

Low-quality permanent pastures are often best improved by renovation. This consists of:

1. complete or partial elimination of the existing sod by chemical or mechanical means
2. growing of at least one annual crop, such as a small grain
3. preparation of a smooth, well-pulverized seedbed
4. soil tests to determine the need for fertilizers or lime, which should be applied before seeding
5. reseeding at the proper time of year with improved selected legume and/or grass cultivars

Renovated pastures should not be heavily grazed during the first year or not until the plants become well established and can remain productive with proper management.

Temporary Pastures   Sometimes large amounts of pasture forage are needed on an emergency basis. This need can be met by utilizing such annual crops as Italian ryegrass, Sudan grass, wheat, oats, millet, or crimson clover, depending upon the climate of the area.

Cropland Pastures   Some pastures are included in a rotation scheme with other crops. If the pasture is to serve for a single year, annuals such as Sudan grass, ryegrass, winter cereals, or sweet clover might be used. If the pasture is to occupy several years in the rotational cycle, perennials such as orchard grass, timothy, or bromegrass plus alfalfa or red clover could be used.

When pastures, especially those with legumes, are included in a crop rotation, the land benefits in several ways: the nitrogen level in the soil is increased, soil erosion is reduced, good soil structure is maintained, and the productivity of crops in the cycle is raised.

Hay

Hay is defined as the shoots and leaves—and in some cases, the flowers, fruits, and seeds—of forage plants that are harvested and dried for future feeding to livestock. Hay usually consists of grasses, legumes, or a combination of the two. The harvested material is dried to a moisture level of 15 to 20 percent or less. Hay is the most important type of stored forage and provides considerable flexibility in animal-feeding operations. Properly dried and stored, it can be kept for several years with little loss of nutritive qualities. In severe-winter regions, it permits feeding ruminant animals a nutritious bulk material when outside pastures are covered with snow or otherwise unavailable.

Hay is a cash commodity that can be bought and sold, whereas the value of pasture forage can be converted only by animal feeding.

In general, the best time to harvest hay for maximum yields and still maintain an acceptable degree of palatability is near heading, when the flowers first appear. Some forage crops, such as American mammoth red clover, give only one cutting of hay during the growing season. Others, such as alfalfa, provide several.

Hay quality is determined by plant maturity leafiness, color, and amount of foreign material. During harvest and transportation, every effort should be made to retain as much of the leaf area as possible since most of the protein is in the leaves. High-quality hay has a fresh green color, a good aroma, and a pliable texture. It is nutritious and palatable. Detrimental foreign material includes weeds, poisonous or thorny plants, spiny seeds, and such objects as pieces of wire, nails, rocks, and dirt clods.

Hay may be stored loose, chopped, baled, or as cubes or wafers. Harvesting of hay is now highly mechanized. Many specialized machines are available for handling hay. The hay is cut, put into windrows, and baled by mobile machinery (Fig. 22–4). This saves considerable labor over earlier methods of hauling hay to stationary hay balers. Most hay in the United States is now stored as bales, an efficient use of space. Bales may be small-square, small-round, or large-round types (Fig. 22–5). They are a commodity that can be bought, sold, and transported more easily than hay in loose form. Larger baled units can be stored for later distribution (Fig. 22–6).

Hay can also be chopped, green or dry, in the field into particles small enough to be blown in an airstream into trucks. The hay is then transported to feeding areas to be fed green to livestock or to dehydration equipment for storage and later feeding.

Hay can be cubed, pelleted, or wafered in parts of the United States with rainless summers of low humidity, such as California, Arizona, New Mexico, and eastern Washington. In these areas hay can be field-cured down to a moisture level of about 12 percent. This procedure permits almost complete mechanization of the harvesting, transporting, and feeding of hay.

Figure 22–4   Alfalfa hay curing in windrows after cutting and before baling.

Figure 22–5 Transport of roll-type hay bales to animal feedlots.

Figure 22–6 Hay, after being cut and cured in a windrow, can be collected into stack wagons and compressed into tight, dome-topped, weather-resistant stacks for storage and future feeding. *Above:* A stack wagon collecting hay from the windrow. *Below:* A smaller high stack being unloaded. *Source:* Deere and Company.

Silage

The feeding of silage to livestock is an ancient practice in Europe dating back many hundreds of years, although the first silo was not constructed in the United States until 1876.

Silage is green chopped forage, allowed to ferment under anaerobic conditions. Entire green corn plants are most often used for making silage, but many other green, moist forage crops can also be used, alone or mixed. Some of these are winter cereals, sorghum, Sudan grass, smooth brome-grass, orchard grass, Italian ryegrass, reed canary grass, timothy, alfalfa, alsike clover, and red clover.

In silage fermentation by the direct cut method, the green chopped forage material—at a moisture content of about 70 percent—is blown into the silo. Silos are airtight structures, either vertical or horizontal. They may be made of wood, concrete, or glass-coated steel plates; they may be large plastic-bag-like containers, or they may be mere trenches in the ground. After the oxygen in the mass of tightly packed chopped material is used up by enzymatic reactions and aerobic bacteria, anaerobic bacteria act on the carbohydrates in the plant tissue to form lactic acid, which essentially ferments the plant material. The pH drops to 4.2 or below, inhibiting spoilage bacteria and enzyme action, and

thus prevents deterioration. The fermentation process is completed in two to three weeks. Silage can be kept in good condition for several years if air is kept out, moisture stays high, and the pH remains below 4.2.

Modern silage preparation is highly mechanized. It starts with cutting the corn plants—or other forage material—by special harvesters that blow the ground-up material into trailers for hauling to the silo. Silos are filled at the top and the preserved feed withdrawn from the top or the bottom, depending upon the construction (Fig. 22–7).

FIBER CROPS

Many kinds of plants are cultivated for their fibers, which are used to make yarn, fabrics, rope, paper, insulation, raw cellulose, and hundreds of other products.

Fiber-producing plants can be categorized as:

1. Surface fibers—those in which the fibers are produced on the surface of the plant parts in association with floral structures. The principal examples are cotton and kapok, where the fibers develop as outgrowths of the seed coat epidermal cells.

Figure 22–7 *Above:* Harvesting corn for silage. This machinery cuts the corn stalks, including the ears, then chops and blows the material into wagons for hauling to the silos. *Source:* R. L. Haaland. *Below:* Typical storage silo which is filled with the chopped material.

2. Soft or bast fibers—those in which the fibers are located in the stems or, more precisely, in the outer phloem tissues of the bark. Botanically, these are phloem fibers, which are groups of very long, thick-walled cells just external to the conducting phloem sieve tubes. Plants producing such soft stem-fibers include flax, hemp, jute, kenaf, and ramie.

3. Hard fibers—those which are rougher, more lignified strands of vein-like bundles of phloem and xylem supporting cells primarily from leaves of monocotyledons. These are longitudinal rigid fibers. The agave plant is an important example of this group.

For centuries people obtained many needed materials from these plants for clothing, ropes and sails for ships, and cord for fishing nets. In more recent times, numerous other products have been made from such fibrous materials.

Various types of cotton plants were naturally dispersed throughout many of the warm parts of the world for thousands of years, and their lint was spun and woven into cloth.

The Egyptians made linen cloth from flax fibers, and ancient writings show that the hemp plant was being grown for its fiber in China north of the Himalaya mountains as long ago as 2800 B.C. Ramie was cultivated by early civilizations; the cloth made from it was used to wrap mummies in ancient Egypt. Early Chinese literature also mentions the cultivation of ramie for its fibers.

Plants Producing Surface-Fibers in Association with Floral Parts

Cotton (*Gossypium* spp.) MALVACEAE

Cotton has been one of the world's most important crops since the beginning of civilization. It continues to make immense contributions to people's comfort. In spite of competition from synthetic fibers, cotton is still basic in the world's textile industry. Cotton is grown principally for its lint although the seed produces a valuable and widely used food oil. The seed residue, called cottonseed meal, is used as a livestock feed and processed into a high-protein flour.

The various cotton species originated in several warm regions of the world. Tropical by nature, this heat-loving crop has been extended by man into the warmer temperature regions. There is evidence that cotton was grown and processed into cloth as long ago as 5000 B.C. in Mexico, 3000 B.C. in Pakistan, 2500 B.C. in Peru, and 2000 B.C. in India.

Only 4 of the 20 or so cotton species are cultivated for their spinnable fibers. Two are diploid Old World species—*G. arboreum* L. and *G. herbaceum* L.—which are believed to have originated in southern Africa. These have a 2n chromosome number of 26 and produce a short—1 to 2 cm (0.4 to 0.8 in.)—coarse lint. The other two are tetraploid New World species—*G. hirsutum* L. and *G. barbadense* L. They have a chromosome number of 52. About two-thirds of the cotton grown in the world and almost all that is grown in the United States is known as American Upland cotton and belongs to the species *G. hirsutum*. This species originated in Central America and southern Mexico as a perennial shrub. Improvements through breeding and selection have altered it to an annual plant. American Upland cotton produces fibers varying from short to long—2.2 to 3.1 cm (0.9 to 1.2 in.)—according to the cultivar.[2]

G. barbadense L., originating in the Andean region of Peru, Ecuador, and Colombia, accounts for no more than 1 percent of the cotton grown in the United States, where it is known as American Pima. It is an extralong staple cotton

---

[2]In cotton, a cultivar is not a clone, pure line, or a primary mixture of pure lines. It is usually a progeny row selection, bulked and mass multiplied.

with fibers up to 3.3 cm (1.3 in.) long that are very good for fabrics.

The leading U.S. cotton producing states are Texas, California, Mississippi, Arizona, and Louisiana.

In some countries, such as China, Egypt, and Sudan, cotton is a major crop and a very important revenue source, even though considerable hand harvesting is still done. Hand harvesting of cotton is laborious, but may not be limiting to those countries that have the advantage of an abundant supply of low-cost labor. In the United States, the former Soviet Union, Australia, Israel, and other developed countries, production can also be economical and efficient, because of the advantages of highly mechanized practices.

In planting, cotton acid-delinted certified seed treated with a fungicide is spaced 8 to 20 cm (3 to 8 in.) apart in rows 76 to 107 cm (30 to 42 in.) apart. Planting is done with a multirow mechanical drill planter. Some planters also apply fertilizers and pesticides in the row. Planting is done after all danger of frost has passed and soil temperatures have reached at least 16°C (61°F), preferably 20°C (68°F). In the United States this ranges from early March in southern Texas to early May in the northerly areas of the Cotton Belt.

Maximum productivity is favored with high temperatures during the growing season, high light intensity, ample soil moisture, and good soil fertility. Cotton cannot tolerate frost and requires at least 200 frost-free days from planting to crop maturity. In some areas, as few as 160 to 170 frost-free days will permit production. Generally, the longer growing season provides a greater yield potential.

Optimum day temperatures during the flowering period range from 30°C to 38°C (86°F to 100°F). Because of breeding and selection programs in the temperate regions, cotton is no longer subjected to major photoperiodic control, but high light intensity does promote maximum production of fiber and seeds. Cotton is classified as a C-3 plant, and as such, is not as efficient a user of light energy as C-4 plants.

Depending on the cultivar and environment, flowering branches in cotton originate at about the seventh node on the main stem. Floral buds, called squares, are visible on those branches about 3 to 4 weeks before flower opening, which occurs about two months after the seed is planted. The cotton plant initiates far more flower buds than will develop into flowers, and far more flowers than will mature into profitable bolls—the ovaries or fruits that contain the seeds (ovules) bearing the fibers.

The cotton flower is perfect, containing both male and female organs, and is either self-or cross-pollinated. The majority are self-pollinated (see Fig. 22–8).

In most commercial cotton, flower petal color is cream to pale yellow on the day the flowers open. The notable exception is Pima cotton, whose flowers have deep purple spots at the base of the corolla, and bright yellow petals. Coincidental with flower opening is anther dehiscence and the release of pollen grains, which germinate about one-half hour after they are deposited on the stigma surface.

Figure 22–8 Cotton flowers ready for pollination. *Source:* National Cotton Council of America.

Varying slightly among cultivars and influenced by the environment, Upland cottons initiate development of the important lint fibers as protuberances on epidermal cells of the ovule seed coat. This happens soon after fertilization. Fertilization occurs by the end of the first day after flower opening. According to current thinking, fertilization initiates processes that cause the formation of a plant hormone essential to seed formation—and thus fruit set—and continued fiber elongation (Fig. 22–9).

All commercial cultivars of Upland cotton produce two types of fibers: **lint** and **fuzz** (the latter called **linters**). As described, lint fibers and fuzz fibers originate as epidermal hairs, but while lint fibers initiate very soon, fuzz fibers originate about 6 to 12 days later. Although variable, fuzz fibers are usually less than one-fifth as long as lint fibers.

Lint fibers elongate and reach their final length in about 21 days. Toward the end of this period secondary wall deposition on the inner surfaces of the cell wall continues to increase the primary constituent, which is cellulose, up to 90 percent. At maturity, the lint dries, and the fibers become twisted and thereby are more suitable for spinning. As bolls dry (Fig. 22–9), they split and open, exposing the lint. Before ginning, lint and seeds are called seed cotton.

At maturity, lint fibers (Fig. 22–10) have narrow and delicate shanks at their bases and thus are easily removed during ginning. Fuzz fibers thicken at their bases and are more strongly retained by the seed. After lint removal, seeds for certification and future planting are treated with acid to remove the fuzz fibers, or these fibers are removed by a different set of gin saws. The product is used for mattress stuffing or for cellulose. Removal of fuzz fibers from

Figure 22–9 Open and unopened cotton fruits (bolls). *Source:* National Cotton Council of America.

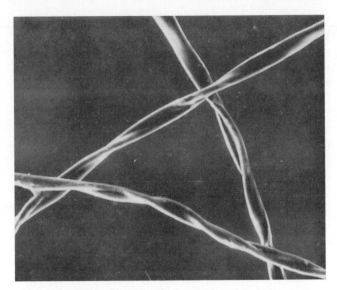

Figure 22–10 Scanning electron micrograph of mature cotton lint fibers. Magnification is 225×. *Source:* C. A. Beasley.

the seed is essential for the operation of precision planting equipment.

Cotton is often grown in rotation with other crops, partly to control diseases and soil pests such as nematodes. It may be a two-year rotation with small-grain cereals or a three-year rotation with small-grain cereals and corn, or a longer-term rotation that might also include alfalfa, sugarbeets, or beans.

Cotton plants are deep-rooted and are moderate to heavy water users, particularly after blooming. Many of the production areas, especially in the western United States, are irrigated, while in the southern and eastern states water is supplied by rainfall, although supplemental irrigation must be used. For maximum yields under irrigation, it is important to time water applications properly so that the plants grow steadily throughout the season. Irrigations applied too early or too late may promote undesired excessive vegetative growth. Late irrigations can interfere with harvest operations.

Cotton grows on a wide range of soil types from moderately acid to moderately alkaline, up to pH 7.8, but a pH from 6.0 to 6.5 is optimum. Very acid soils or those high in alkali or salts should be avoided, although cotton plants have good salt tolerance. Cotton plants have a high nitrogen re-

quirement at the time of flower bloom, and they also need potassium and phosphorus. Fertilizers are best applied in concentrated bands into the root zone during or after planting the seeds. Additional fertilizers are applied as needed. Readily available soil nitrogen should be mostly exhausted by the end of the season so that vegetative growth ceases and fruit maturation is maximized.

Weed control in cotton plantings is necessary. This is best done by applications of preemergence or postemergence herbicides and mechanical cultivation.

Although diseases in cotton are less devastating than insect pests, some can be serious. Damping-off fungi can kill the young seedlings, although the problem can be avoided by using fungicide-treated seed and delaying planting until the soil is warm enough to give rapid seed germination and seedling growth. Cotton plants are susceptible to the soilborne fusarium and verticillium fungi that attack the roots, causing wilt. A fusarium wilt/nematode complex is a serious problem. The best control is to plant tolerant cultivars.

Bacterial blight of cotton is best controlled by seed treatments and crop rotations with nonsusceptible plants. Leaf crumple, a sweetpotato whitefly-transmitted virus, is a serious problem in the low desert of California.

It was estimated that one out of every five to six bales of the potential crop was lost to insects, mostly to the boll weevil. The problem has lessened but is still serious. Various other insect pests—the pink bollworm, cotton bollworm (corn earworm), thrips, cotton leafworm, fleahopper, tobacco budworm, banded wing whitefly, lygus bug, cotton aphid, and spider mite—also contribute to these losses.

The cotton boll weevil (*Anthonomus grandis*), a major pest in Mexico, is believed to have originated in Mexico from which it entered the cotton-growing regions of Texas in 1892, spreading from there to the other cotton-producing states. The larvae feed only on cotton, but the adults have been known to feed on other related plants. The cotton boll weevil is found only on the North American continent. Boll weevil damage is due to the larvae feeding on the floral parts in the unopened buds or on the lint in developing seeds, causing the young bolls to dry up and fall off. The most effective control is application of insecticides to kill the adult females, which puncture the developing bolls and deposit their eggs inside. Eggs and larvae are inside the bolls and therefore cannot be reached by nonsystemic insecticides. Few other insects have had such an impact on agriculture or received more study than the cotton boll weevil. It does not thrive in rainless, low-humidity climates so it has not been the problem in the semiarid western states that it has been in the eastern cotton-producing regions.

The pink bollworm (*Pectinophora gossypiella*) is widely distributed in most cotton-producing countries and has been a serious problem in Egypt and India. It also poses a major threat to the large California cotton plantings. Larvae of this insect enter developing buds, flowers, and bolls and consume both lint and seed.

Various species of nematodes attack cotton roots. Control is either costly soil fumigation or long-term crop rotation with nonsusceptible crops, such as the cereals.

Harvest machinery can be either a "spindle-type" picker (Fig. 22–11) or a "stripper." The picker has vertical drums with many revolving steel spindles that engage and twist the cotton and the seeds from the open bolls as it passes along the rows of cotton plants. Any unopened bolls are left for later pickings.

The stripper harvester has roller or mechanical fingers that pull entire bolls, mature and immature, from the plant. This machine can be used effectively only with cultivars whose bolls mature at about the same time. Cotton harvesting can extend over a long time, from midautumn into midwinter if the weather is dry.

Before machine harvesting starts, chemical defoliants (for the spindle-type harvesters) or chemical desiccants (for the stripper-type harvesters) are sometimes applied. The defoliants accelerate leaf drop so that the spindles can remove the bolls easier and lessen contamination (Fig. 22–12). Desiccants rapidly kill the leaves but they stay attached to the plant. Both types of chemicals are applied seven to fourteen days before harvest.

After the cotton has been picked from the plants by the harvesters, it is emptied into large wire-covered trailers for transport to cotton gins. The pressed form is covered with a tarpaulin to protect it from weather. Gins have special equipment to break up the module, and the seed cotton is handled as loose delivered cotton. The cotton plus seeds is unloaded using vacuum tubes. Some cotton may require passing through driers to lower its moisture content for easier processing. The cotton is transferred to equipment that removes soil, trash, and foreign material, and then on to the gins. Here the lint is removed from the seed.

Figure 22–11 Mechanical cotton harvester unloading its cotton filled collection bin into a trailer that will carry the seed cotton to a gin for separation of lint from seeds. *Source:* James F. Thompson.

The seed-free cotton lint may be further cleaned and is then baled and moved to the textile mills for processing into yarns and fabrics. The seeds that are left behind are collected and moved to seed-processing houses for fuzz removal. The seed coats are cracked and the kernels removed for oil extraction by pressing or chemical procedures.

Kapok (*Ceiba pentandra* [L.] Gaertn.) BOMBACACEAE

Moisture-resistant fibers called kapok or silk cotton are derived from the seed hairs in the pods of the kapok tree. These trees grow wild in tropical American forests and have been introduced to other tropical regions throughout the world. Almost all commercial kapok production—several thousand tons annually—now comes from the Orient, particularly Thailand, Indonesia, and Cambodia. Most kapok is sold to the United States, where it is used as insulation in sleeping bags and as filling in life preservers. Synthetic fibers, foam rubber, and plastics are rapidly replacing kapok for such purposes. The hollow kapok fibers are similar to those produced by the cotton plant. The fibers are not twisted and are too

Figure 22–12 Mature cotton plants ready for harvest. *Left:* Before defoliation. *Source:* USDA. *Right:* After spraying with harvest-aid defoliant chemical. *Source:* National Cotton Council of America.

smooth and brittle to be spun into cloth but have a remarkable buoyancy and resiliency and are very impervious to water. The large kapok trees bear a great many of the football-shaped pods, up to 15 cm (6 in.) long, which are filled with the fiber-producing seeds. The pods either fall to the ground or are cut from the tree. After drying, the fibers are extracted from the seeds by hand. An edible oil, similar to cottonseed oil, can be pressed from the seeds.

Plants Producing Soft Stem-Fibers

Flax (*Linum usitatissimum* L.) LINACEAE

The flax plant is a slender, herbaceous annual grown in many subtropical and temperate zone countries for its oily seed and for its stem fibers. The fibers are used to make linen cloth, book and cigarette papers, and paper currency.

In the United States, culture of flax for its fibers became unprofitable and ceased in the mid-1950s. Currently, all United States flax is produced for seed. Different flax cultivars are used for linseed oil than for fiber production. Linseed oil is pressed from the seed for paint manufacture, and the residue, called linseed meal, is used as a livestock feed.

The former Soviet Union is by far the leading producer of flax fibers. Poland, France, Czechoslovakia, Rumania, and Belgium are also major producers.

Flax is a cool-season plant. In areas with mild winters, seeds are sown in the fall and the crop harvested early the following summer. In cold-winter regions, seeds are sown in early spring and, if used for fiber production, the stems are harvested 80 to 100 days later, when about half the seeds are mature and the leaves have dropped from the lower two-thirds of the stem. At harvest, weather should be dry to "cure" the plants properly.

The fibers are obtained from the stems of the flax plant by a process called retting. Commonly, the retting involves simply leaving the stems in the field, where they are exposed to weathering from the dew and to the action of soil-borne bacteria on the straw. This partial rotting for one to three weeks dissolves gums that hold the fibers to the woody xylem tissues and destroys soft tissues around the fibers. A more intensive rapid method is cold water retting by complete immersion of flax stem bundles in cold water for several days. The stems then pass through machines that break up the woody parts but retain the flexible fibers largely intact. Other machines separate the short, woody sections from the fibers, which are then baled and shipped to spinning mills.

Hemp (*Cannabis sativa* L.) CANNABACEAE

The hemp plant is a seed-propagated herbaceous annual adapted to mild temperate zone climates. The unusually stout stem fibers of certain hemp cultivars are used to prepare tough threads, twines, ropes, and textile products. Cloth from hemp can be used in clothing and other products. Most hemp produced for fiber is grown in such temperate countries as the former Soviet Union, Yugoslavia, China, India, Korea, Poland, and Turkey, but various hemp cultivars can be found cultivated or wild in almost all countries in the temperate and tropical zones.

For fiber production, the plants are seeded thickly and grown to a height of 1.5 to 3 m (5 to 10 ft). Hemp seed is planted early in the growing season, and the plants are uprooted or cut off for fiber harvest during the period from start of bloom until full maturity. The stems are hollow except near the base. The fibers are located in the phloem and pericycle tissues of the bark.

The straw is retted like flax to aid in separating the fibers from the other tissues in the stem. Machines are used to further separate the fibers. Manila hemp (*Musa textilis*), also known as abaca, is used in paper manufacturing.

Because certain low-growing types of *Cannabis sativa* are the source of the drugs marijuana and hashish, cultivation of all hemp is prohibited in many countries, including the United States. However, the demand for hemp fiber has changed the laws in some countries such as Canada.

Jute (*Corchorus capsularis* L. and *C. olitorius* L.) TILIACEAE

Plants of these two jute species are herbaceous, seed-propagated annuals cultivated in hot, moist climates with at least 7.5 to 10 cm (3 to 4 in.) of rain per month. The plants grow from 1.8 to 4.5 m (6 to 15 ft) high. The fiber strands are located in the stem just under the bark, some running the full length of the stem, and are embedded in nonfibrous tissue. The fibers are held together by natural plant gums.

As a fiber-producing crop, jute probably ranks next to cotton in importance. It is primarily grown on small farms in Bangladesh and in several districts in India. Jute is chiefly used in the manufacture of bulky, strong, nonstretching twines, ropes, and burlap (Hessian) fabrics, bags, and sacks for packaging many industrial and agricultural commodities. It has a great many other industrial uses—such as backings for carpets and linoleum coverings, webbings for upholstered furniture, packing in electric cables, and interlinings in tailored clothes.

Jute seeds are broadcast over the soil in the spring and the plants are later thinned to about $33/m^2$ ($3/ft^2$). The crop is harvested by cutting off the stems about the time the flowers start to fade. The bundles of stems are left in the fields for a time to shed their leaves.

To release the fibers from other tissues, the stems must be retted. To do this, bundles of the stems are submerged under water in pools or streams for 10 to 30 days—long enough for bacterial action to break down the tissues surrounding the

fibers, but without damaging the fibers themselves. At the proper time the fibers are loosened from the wet stems by beating the small bundles with paddles, then breaking the stems to expose the fibers, which are pulled from the remainder of the stems. The fibers are then washed in water and hung on poles or lines to dry. The dried fibers are taken to baling centers where they are sorted and graded according to strength, cleanliness, color, softness, luster, and uniformity, then pressed into bales. The bales are shipped to local spinning mills or exported.

Kenaf (*Hibiscus cannabinus* L.) MALVACEAE

Kenaf has been cultivated for centuries in many places throughout the world between about 45° N and 30° S latitudes. Thailand, India, Brazil, China, and the USSR are the major kenaf producers. Kenaf competes with jute as a stem fiber crop but has less exacting soil and climatic requirements. Kenaf produces an excellent fiber, tougher and stronger than jute but somewhat coarser and less supple. Kenaf fibers are used for making twines, ropes, and fishing nets and are also suitable for paper manufacture.

The kenaf plant is an herbaceous annual with a strong taproot and a long, unbranched stem reaching to a height of 1.5 to 4.5 m (5 to 15 ft). The plants require a growing season of 100 to 140 days with considerable moisture from either rainfall or irrigation. Harvesting, retting, and drying are done during drier weather. The stems are harvested just as flowering starts by uprooting the plants and tying them into bundles that are placed horizontally in water for retting, as described for jute. Retting may take 10 to 20 days. The fibers are then stripped off the stalks and dried.

Ramie (*Boehmeria nivea* [L.] Gaud.-Beaup.) URTICACEAE

Ramie fibers are long strands in the inner bark of the ramie or China grass plant, a many-stemmed perennial shrub with slender shoots about 2.5 cm (1 in.) thick and up to 2.4 m (8 ft) long and heart-shaped leaves along the upper third of the stem. New stems arise from the crown after the older ones are harvested. Three harvests can be obtained annually, and the plant can live for several years before it has to be replaced.

Ramie has been cultivated for thousands of years, and is mentioned in Chinese writings as early as 2200 B.C. It apparently is native to the Chinese area of eastern Asia. It was also cultivated by the ancient Egyptian civilizations. Present-day ramie production centers mostly in warm, humid regions of China, the Philippines, Japan, Indonesia, and Malaysia with fertile soils.

The cells making up ramie fibers are among the longest known—up to 0.3 m (1 ft). The fibers are eight times stronger than cotton and have a fine durable texture and a good color. Ramie fibers are superior in many ways to flax, hemp, and jute. However, the chief problem with ramie has been the difficulty in freeing the fiber bundles from the gummy tissues surrounding them, along with problems in mechanizing the processing of the fibers—the extraordinarily smooth surface of the fibers makes spinning difficult with machinery developed for other fibers.

As with other stem-fiber crops, retting is required to free the desirable fibers from the other stem tissues. This is done by bacterial action, which decomposes the thin-walled surrounding cells and leaves the thick-walled fibers intact. Ramie is more difficult to rett than flax and hemp. Just wetting the stems is insufficient to break down the cementing gums. Pounding and scraping is also required, followed by chemical treatments with acid or lye to remove the tenacious gums and resins to produce completely smooth fibers. After degumming, the fibers are washed, then softened with glycerine, soaps, or waxes. Ramie fibers are generally spun on the machinery designed for silk. Ramie fabrics are usually blends with other materials, such as cotton or wool.

Plants Producing Hard Leaf-Fibers

Agave (*Agave sisalana* Perr.) AGAVACEAE

The long hard sisal fibers used in making twines, cords, and ropes are obtained from the 0.6 to 1.2 m (2 to 4 ft) leaves of agave plants. Sisal is mainly produced in East Africa, Brazil, Mexico, and Haiti. A related plant known as henequen (*A. fourcroydes* Lem.) is also grown for its fibers, which are much weaker than sisal. Plants of these two species are similar in appearance to the common century plant (*A. americana* L.).

The agave plant consists of a rosette of stiff, heavy, dark green leaves 10 to 20 cm (4 to 8 in.) wide, 60 to 120 cm (2 to 4 ft) long, and 2.5 to 10 cm (1 to 4 in.) thick, arising from a short trunk. The plant grows very slowly but after about four years, it reaches full size and harvesting of the lower leaves begins. About 200 leaves can be harvested for fiber extraction before leaf growth ceases and a flower stalk grows upward rapidly to a height of 4.5 to 7.5 m (15 to 25 ft) and bears light yellow flowers. During the next six months, flowering and fruiting take place, followed by the death of the entire plant.

Agaves are vegetatively propagated by suckers arising around the trunk of the plant or by bulbils (small bulbs) that develop in the flower clusters after flowering, thus allowing growers to maintain improved forms.

In harvesting for fiber, the lower leaves are cut off and bundled after the spines are removed. Cleaning machines scrape the leaves, removing the pulp and waste material and leaving the fibers exposed. The fibers are then dried in the sun, or artificially, and are brushed, graded, and baled. The final product is yellow to yellow-white in color, flexible, and strong. Sisal is used mostly in making cords, such as binder twine; it has only limited use in fabric manufacture.

REFERENCES AND SUPPLEMENTARY READING

BARNES, R. F., D. A. MILLER, and C. J. NELSON, eds. 1995. *Forages.* Vols. I and II. Ames, IA: Iowa State University Press.

HODGSON, J., and A. W. ILLIUS, eds. 1996. The ecology and management of grazing systems. Wallingford, Oxon, UK: CAB International.

CHAPTER

23

Vegetable Crops Grown for Fruits or Seeds

The importance of vegetable crops is apparent in at least two ways. First, as a food group vegetables significantly enhance the quality of the human diet by contributing variability, bulk, fiber, vitamins, minerals, and other compounds thought to positively impact human health. In the United States, the benefits of a diet rich in fruits and vegetables was officially recognized in 1991 when the Produce for Better Health Organization and the National Cancer Institute (NCI) instituted the "Five-A-Day" campaign, urging consumers to eat at least five servings of fruits and vegetables daily in order to reduce the risk of cancer, heart disease, and hypertension. The "Five-A-Day" campaign also has the support of other agencies, including the American Dietetics Association and the U.S. Department of Agriculture (USDA). In 1996, the USDA included the "Five-A-Day" campaign in their comprehensive school lunch initiatives aimed at improving the diets and health awareness of America's schoolchildren and educators. In addition, nearly 200 years of research was used to develop the USDA's "Food Guide Pyramid," which includes eating fruits and vegetables as essential to a healthy diet. More recently, the terms "chemoprevention," "designer food," "functional food," "nutraceutical," "pharmafood," and "phytochemical" have been used to describe specific health-related properties of foodstuffs, especially vegetables. A growing amount of evidence suggests that vegetables contain compounds with specific beneficial effects on human health in addition to the bulk, fiber, vitamins, and minerals for which they have long been recognized. Add to this the fact that vegetables range in caloric and carbohydrate content and it is easy to understand why vegetables are important to nutrition and health throughout the world.

Vegetable farming, selling, and processing are also important sources of employment and income worldwide. In many areas of the United States and elsewhere, local and regional economies are supported by vegetable-related enterprises ranging from large multinational corporations to individual family farms, markets, or other businesses. The USDA estimates that vegetables were harvested from nearly 54,000 farms in the United States in 1998. Also, vegetable farming was reported to generate more than $15 billion in cash receipts in 1998, a figure that does not account for receipts generated by vegetable processing or the full impact of vegetable enterprises on numerous local economies in the United States. For example, the labor, equipment, machinery, disposable supplies, and other technical assistance required by vegetable farming, selling, and processing creates additional farming-related revenue.

STATISTICAL OVERVIEW OF WORLD PRODUCTION

Worldwide, the total production of vegetables and melons (including potatoes, sweet potatoes, and dry peas) in 1999 amounted to approximately 1 billion MT (1.1 billion tons), an increase of nearly 23 percent from the estimated 827 million MT produced in 1984 (Table 23–1). The area from which vegetables and melons were harvested in 1999 (nearly 74 million hectares) also increased nearly 12 percent from 1984 (Table 23–1). Most vegetable and melon production occurs in Asia—the Food and Agriculture Organization of the United Nations listed Asia as producing 61 percent of the world's vegetables by weight and melons on 57 percent of the land devoted to vegetable and melon production in 1999 (Table 23–1). The next largest producers, Europe and North and Central America, produced approximately 22 and 8 percent by weight of the world's vegetables and melons in 1999, respectively. Approximately 22 and 6 percent of the area harvested to these crops were located in Europe and North and Central America, respectively.

Production of a number of commodities discussed in this chapter (cucumber, eggplant, cantaloupe and melon, green and dry peas, peppers, tomatoes, and watermelon) accounted for approximately 23 percent by weight of the total vegetable and melon production in 1999, based on UN-FAO data. Asia was the leading producer of nearly all these commodities, typically followed by Europe and North and Central America (Table 23–2). Of the commodities listed in Table 23–2, Asia, Europe, and North and Central America averaged 60, 19, and 12 percent by weight, respectively, of the world's production in 1999.

TABLE 23–1	Production Indices for Vegetables and Melons (Including Potatoes, Sweet Potatoes, and Dry Peas) in Seven World Regions in 1984 and 1999

Region	Area Harvested (1,000 ha) 1984	Area Harvested (1,000 ha) 1999	Production (1,000,000 MT) 1984	Production (1,000,000 MT) 1999
Africa	5,320	7,452	37.4	60.3
Asia		42,277		654.7
Caribbean	328	330	2.3	2.2
Europe		16,509		234.3
N & C Amer	3,125	4,255	58.9	81.1
Oceania	461	672	4.2	6.1
S Amer	2,319	2,529	24.6	33.5
World	64,799	73,659	827	1,070

Source: World Agricultural Information Centre, Food and Agricultural Organization of the United Nations, FAOSTAT On-Line (http://apps.fao.org/lim500/nph-wrap.pl?Production.Crops. Primary&Domain=SUA&servlet=1).
Note: Region sums do not equal world values due to rounding and estimates of production in Asia and Europe in 1984.

TABLE 23–2	Production by Region in 1999 for Eight Commodities Grown for Fruits or Seeds. The Three Leading Regions are Shown

Commodity	Region	Production (MT)	% of World Production
Cucumber	Asia	22,380,704	78
	Europe	4,074,912	14
	N & C Amer	1,746,631	6
Eggplant	Asia	19,638,954	92
	Europe	681,644	3
	N & C Amer	106,825	1
Melon	Asia	12,790,422	66
	Europe	2,845,857	15
	N & C Amer	2,296,005	12
Peas, Green	Asia	3,390,342	48
	Europe	1,866,141	26
	N & C Amer	1,199,606	17
Peas, Dry	Europe	6,039,147	52
	N & C Amer	2,515,973	22
	Asia	2,395,964	20
Peppers	Asia	10,086,352	56
	N & C Amer	2,678,861	15
	Europe	2,642,393	15
Tomatoes	Asia	41,488,030	44
	Europe	22,009,364	23
	N & C Amer	13,646,127	14
Watermelon	Asia	40,197,456	78
	Europe	3,709,347	7
	Africa	3,276,634	6

Source: World Agricultural Information Centre, Food and Agricultural Organization of the United Nations, FAOSTAT On-Line (http://apps.fao.org/lim500/nph-wrap.pl?Production.Crops. Primary&Domain=SUA&servlet=1).
Note: Regions included in analysis were Africa, Asia, Caribbean, Europe, North and Central America, Oceania, and South America.

TABLE 23–3	Production by Country in 1999 for Eight Commodities Grown for Fruits or Seeds. The Leading Country in the Three Leading Regions, as Indicated in Table 23–1, is shown

Commodity	Country/ Region	% of Region Production	% of World Production
Cucumber	China/Asia	71	78
	Ukraine/Europe	16	2
	U.S.A./ N & C Amer	62	6
Eggplant	China/Asia	56	52
	Italy/Europe	48	2
	Mexico/ N & C Amer	56	<1
Melon	China/Asia	53	37
	Spain/Europe	34	5
	U.S.A./ N & C Amer	58	7
Peas, Green	India/Asia	59	28
	France/Europe	31	8
	U.S.A./ N & C Amer	91	15
Peas, Dry	France/Europe	43	22
	Canada/ N & C Amer	90	19
	China/Asia	54	11
Peppers	China/Asia	74	42
	Mexico/ N & C Amer	73	11
	Spain/Europe	35	5
Tomatoes	China/Asia	43	19
	Italy/Europe	32	7
	U.S.A./ N & C Amer	73	10
Watermelon	China/Asia	68	53
	Spain/Europe	19	1
	Egypt/Africa	46	3

Source: World Agricultural Information Centre, Food and Agricultural Organization of the United Nations, FAOSTAT On-Line (http://apps.fao.org/lim500/nph-wrap.pl?Production.Crops. Primary&Domain=SUA&servlet=1).
Note: Regions included in analysis were Africa, Asia, Caribbean, Europe, North and Central America, Oceania, and South America.

China was the leading producer of six of the eight commodities shown in Tables 23–2 and 23–3 in 1999. China produced nearly 78 percent by weight of the world's cucumbers, 52 percent of the eggplant, and 53 percent of the watermelon in 1999 (Table 23–3). The United States produced 6 percent by weight of the world's cucumber crop and less than 1 percent of the eggplant and watermelon crops.

The United States Department of Agriculture National Agricultural Statistics Service (NASS) develops annual production and market reports for numerous commodities produced in the United States, including 25 vegetables typically sold in the fresh market and 10 vegetables sold as, or in, processed products. The 1999 NASS report indicated that slightly more than 3 million acres were devoted annually to

vegetable production in the United States from 1989 to 1998. Approximately 53 percent contained crops destined for fresh markets with the remainder intended for use as, or in, processed products. Average annual production of fresh and processing vegetables for 1989 to 1998 was approximately 18 and 16 million tons, respectively. During 1989 to 1998, California, Florida, and Arizona together produced an average of 14 million tons of fresh market vegetables annually with California producing approximately 10 million tons yearly. California, Wisconsin, and Washington together produced an average of 12 million tons of processing vegetables annually with California producing approximately 10 million tons yearly. From 1989 to 1998, the average annual value of fresh market vegetables produced in the United States exceeded $6 billion. The average annual value of the processing crops when delivered to the processing plant was slightly more than $1 billion. It is important to note that the value of the final processed products was much greater.

Beans, Snap or Green (*Phaseolus vulgaris* L.) and Lima Beans (*Phaseolus limensis* Macf.) FABEACEAE

The genus *Phaseolus* includes a number of "vegetable" bean species whose immature fleshy pods and immature seeds, and/or dried seeds are a highly valued food. Dry beans in particular are a nutritionally important protein and carbohydrate food source for the less affluent populations. The common bean and other *Phaseolus* members are native to Central and South America, and were domesticated and widely used long before the arrival of European explorers.

The better-known common bean types are the snap bean and lima bean, and various types of field beans (shell beans, kidney beans, mung, etc.) grown for their mature dry seeds. Snap beans and limas are grown as annuals and have differing vine types, those usually having indeterminate fruiting and grown on supports (pole), and the bush forms, usually determinate and grown unsupported. Since the immature pods and seed of snapbeans are the edible portions, selection was directed toward developing cultivars having thicker, fleshy pods of low fiber and slowly developing seed (Fig. 23–1). Pod fiber development occurs concurrently with seed maturation and is undesirable. For the dry seed bean types, rapid seed development is the more important criterion, and pod fiber is not of interest. Lima beans, whether grown for fresh market or for processing, are harvested when their seeds have enlarged but before they are fully mature. However, fully mature dried lima bean seeds are also an important product. Beans are self-pollinated and propagated from seed.

From 1996 to 1998, annual U.S. commercial fresh market snap and lima bean production averaged nearly 427 million pounds, an amount supplemented by a significant level of home garden production. Georgia, California, Florida, and Tennessee together typically accounted for 72

Figure 23–1 An example of bush snap bean pods exhibiting the important characteristics of uniformity and straight, smooth, thick fleshy pods having minimal seed enlargement. These beans are green and slightly oval or nearly round in pod cross section shape. There are many snap bean cultivars of different shape, size, and color. *Source:* Harris Moran Seed Company.

with flower fertilization and may cause abortion of blossoms or young pods.

Numerous snap bean cultivars are available, and others are frequently introduced which have been developed for optimum adaptation to specific regions, seasonal periods, and market preferences. The criteria for producer's cultivar choice includes the end products' use for fresh or processing purposes. This will determine the desired plant growth habit—determinate types for mechanical harvesting and processing, or indeterminate types for hand harvesting and fresh use. Variables provided by many cultivars are pod color—green and yellow (wax); pod shape—flat, oval, or round in cross section; length; and seed color, which includes a wide range. Processors have a preference for white or light-colored seed coats as these cultivars present a more attractive product. With lima beans, seed size is also an important factor.

Snap beans are commonly planted in rows about 75 cm (30 in.) apart, particularly for hand harvesting. Close in-row spacings are used: 5 cm (2 in.), requiring about 60 to 100 kg of seed per hectare. Seeds are sown about 2.5 cm (1 in.) deep. Since slow germination can contribute to seed decay and poor stands, plantings should be avoided at soil temperatures less than 20°C (68°F). Mechanical harvesters permit harvest of variable row widths, and close, 22.5 cm (9 in.), row spacings are also easily accommodated.

Soil moisture should be near field capacity, particularly during flowering, to optimize production. The crop commonly consumes about 40 to 50 cm of applied or rainfed water. Beans grown for their seed would have the moisture supply intentionally restricted as the seeds approach maturity.

Silty or sandy loams are preferable soil types, although a wide range of soil textures can be used. Beans are intolerant of salinity, and particularly sensitive to boron. Soil pH levels should be above 6.0 and less than 7.5

Beans are not heavy feeders of mineral nutrients, and excess nitrogen increases vine growth at the expense of pods. Shallow mechanical cultivation is usually necessary for weed control, and several selective herbicides are commonly used.

Unfortunately, a number of bacterial diseases such as common blight, halo blight, and bacterial wilt are troublesome. Serious fungal diseases are fusarium yellows, fusarium root rot, white mold, rust, *Anthracnose,* and powdery mildew. Common bean mosaic, cucumber mosaic, and curly top are viruses that also contribute to crop losses. Cultivar resistance is available for some of the important viruses and other diseases. Insect pests include the Mexican bean beetle, cut worm, seed corn maggot, flea beetles, spider mite, leaf miners, leaf hoppers, lygus bugs, and aphids. Chemical pesticides are widely used for control.

Snap beans are harvested before they reach physiological maturity. This stage often occurs in as little as 50 to 65 days from planting; lima beans require about 30 to 40 additional days. Snap bean pods reach harvest stage about 2 weeks after bloom. Further delay results in undesirable seed

percent by weight of the commercial U.S. fresh market snap and lima bean crops during the same time period. Significant production also occurs in Virginia, North Carolina, New York, and New Jersey. Snap and lima beans are also grown for use in processed products, primarily soups and frozen vegetable mixes. Production for processing markets is widespread in the United States but greatest in Wisconsin, Oregon, Illinois, New York, and Michigan. From 1996 to 1998, snap and lima bean production for use in processed products averaged approximately 1.6 billion pounds.

Lima bean production for processing is concentrated in California, with few other areas having the specific marine-moderated climate needed for effective production. Fresh market production of lima beans is relatively rare, due in part to the crop's more exacting climatic requirements and the reluctance of consumers to shell the beans from the pods.

Both snap and lima beans are warm-season crops requiring frost-free growing periods, preferably without large temperature fluctuations. Lima beans are very sensitive to both cool and high temperature fluctuations which interfere size increase and development, and an increase in pod fiber content. For frozen or canned processing beans, the average cross-section diameter of the pods is used as a basis for maturity indication; the smaller the diameter, the better the edible qualities. This is a producer's decision balanced against total yield potentials. Following harvest, rapid cooling to 4°C (39°F) is important for quality maintenance. Even with good postharvest handling and temperature management, 7°C to 10°C (45°F to 50°F) at 95 percent relative humidity, shelf life is relatively short (1 to 2 weeks).

Cucumber (*Cucumis sativus* L.) CUCURBITACEAE

The cucumber is a prostrate, branching vine native to Asia, most probably northern India, and introduced to Africa and Europe before written history. The oblong, cylindrical fruit is primarily used fresh as a salad vegetable but is also processed by pickling. The plant is an annual with hairy leaves and tendrils, commonly with a **monoecious** flowering habit. The first flowers, generally staminate, are followed by pistillate flowers which, when fertilized, develop into fruit. In field production, insect pollinators are necessary. Many newly introduced cultivars have predominantly pistillate flowers, producing many fruit per plant, and therefore are especially desirable for mechanical harvesting of processing-type cucumbers. Cultivars used for glasshouse production are gynoecious as well as parthenocarpic (Fig. 23–2). Besides its scientific curiosity, the sex expression features of the cucumber have been extensively researched for commercial benefits. Silver nitrate and gibberellic acid applications promote staminate flowering, whereas ethephon promotes pistillate flowering. This procedure is used in hybrid seed production.

From 1996 to 1998, annual U.S. production of fresh market and processing cucumber crops averaged approximately 1 and 2 billion pounds, respectively. Florida, Georgia, and California together accounted for an average of 64 percent by weight of the U.S. fresh market cucumber crops. Michigan, North Carolina, and Florida together accounted for an average of 45 percent by weight of the U.S. processing cucumber crops in the same time period. Texas and Wisconsin also produce large amounts of cucumbers sold as processed products.

Cucumbers are a warm-season crop and are intolerant of frost. However, because of their short growing season, they are grown almost everywhere in the United States. Ideal growing temperatures are near 30°C (86°F). Cucumbers are propagated from seed and planted when soils are warm and frosts are not a threat. In some areas, plastic tunnels and/or hot caps are used for very early production to capture better prices. Cucumbers are deep rooted and grown on most good, well-drained soils, but do best on sandy loam soils that have

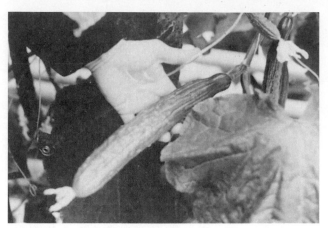

Figure 23–2 Greenhouse-produced parthenocarpic cucumber. These fruit develop without pollen fertilization and therefore are seedless.

a pH of about 5.5 to 7.0. Good yields depend upon adequate soil moisture throughout growth. Because of their vining habit, wide row spacings are used. When grown for processing, higher plant densities are used. Cucumbers are responsive to fertilization, particularly nitrogen. Applications of 75 to 100 kg/ha (67 to 90 lb/ac), with half that amount of phosphorus and potassium, is usually sufficient.

Hybrid cultivars providing greater yields from concentrate fruit set and gynoecious characteristics, and disease resistances are becoming increasingly important in cultivar selection. Most important is the determination of type, since the fresh market and the processed cucumber are quite different (Figs. 23–3 and 23–4). While earliness and disease resistance are important for either purpose, the fruit shape, surface skin features, and spine color differ and must be considered. Cucumbers for processing have lighter green color and thin skins, are blunt and short with a small fruit length to cross-sectional diameter ratio, approximately 2.5 to 3.0, compared to 4.0 or more for the long tapered, thick-skinned, darker green fresh market fruit. Cucumbers for pickling also are characterized by their warty surface.

Diseases are a chronic problem for cucumber producers. Accordingly, considerable emphasis is given to the development of disease-resistant cultivars, of which many are available having multiple resistances. Important diseases include bacterial wilt, angular leaf spot, *Anthracnose,* downy and powdery mildew, scab, fusarium wilt, gummy stem blight, and alternaria leaf spot. Additionally, the virus diseases—cucumber, watermelon, and squash mosaic, as well as curly top and squash leaf curl virus—cause considerable economic losses. Insect pests likewise are a serious concern. Some of those most troublesome are the striped and 12-spotted cucumber beetle, squash bug, leaf miner, leaf hopper, mites, and aphids. The cucumber beetle is responsible for the spread of bacterial wilt; aphids and leaf hoppers are virus vectors.

Figure 23–3   Cucumbers produced for processing, usually by pickling, are smaller, blunt shaped, and have thin skins. A premium price is usually paid for the smallest-size processing cucumbers. *Source:* Ferry-Morse Seed Company.

Figure 23–4   Fresh market cucumbers, sometimes referred to as "slicing" cucumbers, preferably are long, straight, and with uniformly dark green colored thick skin. *Source:* Ferry-Morse Seed Company.

Fresh market cucumbers grown in greenhouses are harvested nearly every day, are field-grown slicing cucumbers are harvested almost as often, each 2 or 3 days. This is necessary in order to minimize the production of overly large fruit, which are strongly correlated with overmaturity, loss of green color, seed enlargement, and consequent lesser quality. For the processing crop, hand harvesting is done frequently; otherwise, fruit continue to grow, and larger fruit are of considerably less value. Mechanical harvesting of pickling cucumbers is well advanced and is rapidly replacing hand harvest practices. However, although premium prices are paid for the smaller size fruit, this benefit must be balanced against higher total yields. Furthermore, a compromise must also be achieved between the high labor cost of multiple hand harvesting and the single harvest accomplished by machine. This apparent shortcoming is met in

large measure by use of improved high-yielding cultivars and higher plant populations. Even so, careful consideration is given to selecting the optimum time to harvest with regard to fruit size and the overall yield. Even with proper postharvest temperature management, between 7°C to 10°C (45°F to 50°F), and high relative humidity, fresh cucumbers have a shelf life of not more than 10 to 14 days. Lower temperatures can cause chilling injury.

Eggplant (*Solanum melongena* L.) SOLANACEAE

The eggplant, also known as aubergine or brinjal, is thought to be native to Southeast Asia and has been grown in China for many centuries. From 1996 to 1998, annual U.S. commercial production of eggplant averaged approximately 76 million pounds. Florida, New Jersey, and, to a lesser extent, Texas together account for the majority of U.S. eggplant production. Eggplant is grown commercially in many states, however, with most sold to nearby markets.

Eggplants are more sensitive to frost than most vegetable crops. Because it is a warm-season crop, production benefits from daytime temperatures of 27°C to 32°C (80°F to 90°F) and warm night temperatures of 21°C to 27°C (70°F to 80°F). Light, well-drained, sandy loam soils with a pH range from 5.5 to 7.0 are preferred.

Eggplant seeds are slow to germinate and seedling plants grow slowly. For this and other reasons, transplants are often used in place of direct seeding. Plant spacings are commonly 60 to 90 cm (2 to 3 ft) between plants in rows 90 to 120 cm (3 to 4 ft) apart. A complete fertilizer applied to provide 50 to 100 kg/ha (45 to 90 lb/ac) each of nitrogen, phosphoric acid, and potash will usually supply the necessary plant nutrition.

Most cultivars grown in the United States are purple-skinned and globe-shaped types (Fig. 23–5). However, a great variation of types, varying in skin color and shape, are produced in other countries. Smaller, dark purple, long, tapered fruit known as the "Asian" types have become very popular in the United States. Hybrid cultivars provide many production benefits and therefore such use has greatly increased.

Important diseases affecting eggplants are phomopsis blight, *Anthracnose,* phytophthora fruit rot, verticillium wilt, fusarium wilt, powdery mildew, cucumber mosaic virus, and other viruses. A physiological disorder, blossom end-rot, often causes substantial fruit damage. Insect pests include flea beetles, leaf hoppers, the Colorado potato beetle, red spider mites, and aphids.

Eggplants are harvested by cutting from the stem, leaving a portion of the stem and the calyx on the fruit. They cannot be readily pulled from the plant without damaging the remaining plant. Additionally the fruit is very susceptible to

Figure 23–5 Eggplant fruit of the globe type. Variations in skin color and fruit shape are provided by the many cultivars available. *Source:* Ferry-Morse Seed Company.

Figure 23–6 In order to obtain an extra early crop of muskmelons, growers will use hot caps or plastic tunnels, mulches, and/or row covers in newly seeded fields to provide higher temperatures for accelerated growth. *Source:* U.S. Conservation Service.

scarring and bruising. Harvesting is done when the fruits achieve full or the desired size, but before seeds mature. Fruit maturity is estimated by gentle pressure; a slight indentation indicates edible maturity, which occurs before the seeds have fully developed and are still soft.

Eggplants are consumed fresh, but are usually prepared by baking, frying, or pickling. The best fruits are dense, firm, and well colored. The skin should be shiny and free of defects. Eggplants can be stored for short periods at 7°C to 10°C (45°F to 50°F) at about 85 percent relative humidity. Eggplant fruit, like many warm-season vegetables, are sensitive to chilling injury.

Muskmelon (*Cucumis melo* L. Reticulatus Group)
CUCURBITACEAE

The domesticated muskmelon. *Cucumis melo,* var. *reticulatus,* also called cantaloupe, is believed to have been developed from its wild-type ancestor, whose origin was in tropical Africa. The ancient Egyptians, Greeks, and Romans grew muskmelons. The muskmelon was introduced to the Americas by Columbus.

From 1996 to 1998, annual U.S. production of muskmelon averaged approximately 2.5 billion pounds. California, Arizona, and Texas may account for nearly 90 percent by weight of the U.S. muskmelon crop, although considerable summer production occurs in many other states.

Being warm-season plants, muskmelons are intolerant of frost and low temperatures. Seed should not be planted until soil temperatures reach 18°C (65°F). Long, hot, sunny days, above 30°C (86°F), with low relative humidity are ideal, provided soil moisture is ample. The high value for

early melons encourages the use of hot caps, plastic mulching, row covers, and other practices to provide a warmer environment for rapid early growth and harvest (Fig. 23–6). Use of these procedures has greatly increased in the northeastern and central areas of the United States that have relatively short growing periods of high temperature. Additionally, transplants are generally used in these areas in order to fit the shorter growing season and to obtain earlier production. In the United States West and Southwest, direct seeding into wide raised beds is a general practice. Seed are drilled 2.5 to 5 cm (1 to 2 in.) deep at row spacings of 150 to 180 cm (5 to 6 ft). In row spacings, after thinning plants are usually 60 to 90 cm (2 to 3 ft) apart.

Muskmelons grow on a wide variety of soils, but do best in well-drained sandy loams that are neutral to slightly alkaline. Saline or sodic soils are avoided if possible, although muskmelons have a fair tolerance to salinity. The crop is a heavy consumer of mineral nutrients and responds readily to applied fertilizers. Approximately 100 kg/ha (90 lb/ac) of nitrogen, and 50 kg/ha (45 lb/ac) each of phosphoric acid and potash satisfy most growing conditions. A portion of the nitrogen fertilizer is usually applied at the time of planting, and the balance in one or two applications during growth and/or just before fruit enlargement. Muskmelons require abundant water during rapid vine growth and fruit setting, but excessive water should be avoided during fruit ripening. If high soil moisture has been maintained during early growth, a sufficient reserve moisture supply usually would be available to sustain the plant during fruit ripening.

Important cultivar characteristics in addition to earliness and disease resistance are fruit size, commonly ranging from 1.0 to 2.0 kg (2.2 to 4.4 lb), flesh thickness, size of seed cavity, and flesh color, usually a salmon to orange color, although some cultivars have green-, lemon-, cream-, or white-colored flesh. Exterior features are also important. Melons that are transported long distances benefit from having a thick rind with coarse surface netting (Fig. 23–7). Cultivars with thin rinds and with smooth surfaces or minimum netting are also produced because they have favorable taste quality features. Cultivars are also characterized by their ribbing, some being deeply ribbed, others smooth. While not an important factor in marketing in many parts of the world, United States markets favor spherical-shaped melons. Mixed

Figure 23–7 Muskmelons selected for long-distance shipment have thick walls, a small seed cavity, and a strong well netted outer rind. These characteristics enable the fruit to better withstand the postharvest handling involved in long-distance shipping. The word *cantaloupe* is a marketing term used for western grown muskmelons. *Source:* Harris Moran Seed Company

melons, such as the persian, *C. melo* var. *reticulatus,* and *C. melo* var. *inodorus* cultivars such as Casaba, Crenshaw, and Honeydew are also muskmelons, although they differ in several attributes. Although important, these types are a smaller factor in the total volume of muskmelons produced. They are called winter melons because of their longer shelf life and ability to be stored for short periods.

The most serious of the cucurbit diseases affecting muskmelons are the wilts, fusarium and verticillium, and the mildews, powdery and downy. The severe damage caused by viruses cannot be discounted. Watermelon virus 1 and 2, zucchini yellows, and squash mosaic virus seriously impair production. Even though resistance has been introduced by plant breeders to combat some of the important diseases, they remain a very important production concern.

Important insect pests include wireworms, cutworms, leaf hoppers, leaf miners, vine borer, squash bugs, cucumber beetles, and aphids. Root knot nematode infestations are also a frequent problem. Since soil fumigation is generally too expensive for melon production, this problem is combatted by the use of nematicides and/or crop rotation practices.

Fruit appearance alone is not a good guide to determine harvest maturity. Although refractometer readings of fruit tissue sugar or soluble solids content, or direct taste sampling are good indicators, the more commonly practiced

criterion of maturity is how easily the melon detaches (slips) from the vine. At the "full-slip" stage, the muskmelon easily detaches from the stem, leaving a clean abscission. Melons harvested at full-slip are at prime maturity, having maximum sugar content with the flesh beginning to soften. These melons, although of high quality, are less suited for lengthy transport to markets, and are often marketed locally. However, with improvements in postharvest cooling, handling, and rapid transportation to markets, this situation is less common. Melons harvested at half-slip are also of good quality and physically better suited for long-distance marketing. The majority of western-grown melons are harvested at this maturity stage. Melons that have not developed the abscission layer are immature and considerably lower in sugar and flavor constituents. However, it should be noted that the persian and winter melons do not form an abscission layer, and the determination of optimum maturity by visual observation is difficult.

Practically all muskmelons are consumed fresh, processed frozen fruit being a small-volume specialty item. In the United States, fresh muskmelons are available in large supply during summer periods. Production from the southwestern states during the early spring and late fall extends the marketing period. Except for the "winter" types, muskmelons cannot be stored beyond two weeks without serious quality deterioration.

Okra (*Abelmoschus esculentus* L. Moench) (*Hibiscus esculentus* L.) MALVACEAE

Okra, also called gumbo and lady's finger, is African in origin with early introduction into Mediterranean areas and India, where it continues to be very popular and appreciated. Precise statistics for the commercial production of okra are difficult to obtain. United Nations FAO data suggest that world production of okra was nearly 9 billion pounds in 1999. Leading world producers of okra in 1999 included Asia and Africa with approximately 73 and 26 percent by weight of total production, respectively. This vegetable is of minor importance in the United States except in the southern and southeastern states.

A warm-season crop with climatic requirements very similar to those for cotton, okra also requires a long growing season with hot days and warm nights. Okra is adapted to a wide variety of soils, but a deep fertile sandy loam with good drainage is optimum. Commonly used row spacings are about 90 cm (3 ft) apart with in-row spacings ranging from 30 to 60 cm (1 to 2 ft), depending on eventual plant size. Slow seed germination is hastened by soaking the hard seeds in water before planting. Fertilizer applications to provide 50 kg/ha (45 lb/ac) of nitrogen, phosphoric acid, and potash usually provide sufficient nutrition to produce good yields. Okra plantings can be kept free of weeds by mechanical cultivation or with herbicide usage. However, okra is produced in greatest volume in the tropical areas of the world in countries where herbicide use is economically prohibitive and hand hoeing predominates.

The cultivars of okra are classified as dwarf and tall types, and by the typical pod length produced. Additionally, pods differ in the amount of ribbing, some being deeply ribbed, others smooth. Pods also differ in color, ranging from dark green to pale white. The natural offensive spininess of the primitive okra has been eliminated through plant selection and breeding.

Okra is extremely sensitive to verticillium wilt. Fusarium wilt and some leaf-spot-causing fungi such as *Alternaria* and *Cercospora* are also problems but less so than *Verticillium*. *Rhizoctonia* and *Phytophthora* are serious root diseases, especially under wet soil conditions. Many of the insect pests of closely related cotton also share their preference for okra and include those such as the cotton bollworm, corn ear worm, stink bugs, and aphids. Nematodes are also a serious concern and are significant contributors to crop loss.

Harvest generally can begin about two months after planting, and may continue for as long as three or four months. Nearly daily harvest is required to avoid rapid overmaturity of the pods, which become very fibrous, with associated rapid hardening of the enclosed seed. The edible pods are at prime market condition when they attain a length of 8 to 15 cm (3 to 6 in.). Pods are harvested by hand snapping or cutting from the plant. Mechanical harvesters are in an advanced developmental stage, although presently of limited commercial use. Hand harvesters wear gloves to avoid direct contact with plant tissues, which contain a skin-irritating oil, and to avoid bruising of the delicate pods. The smaller young pods are the most tender, containing relatively little fiber, and therefore are of greater value. However, unless a premium is paid, the harvest of very small pods results in yield reductions. The crop is valued for fresh market purposes, although considerable quantities are processed as frozen, canned, pickled, and dried products. A significant quantity of okra is used because of its gummy or thickening characteristics in the preparation of soups and stews (Fig. 23–8). In many areas, air-dried pods ground into powder are common inexpensive methods of preservation. Harvested pods deteriorate rapidly, and the best prevention is rapid cooling to 7°C to 10°C (45°F to 50°F) at a relative humidity of about 90 percent. A lower temperature will cause chilling injury.

Peas (*Pisum sativum* L.) FABACEAE

The garden pea, sometimes called the English, green, or common pea, is a viny annual cool-season plant grown for its edible seed, although some cultivars are also grown for their immature edible pods (Figs. 23–9 and 23–10). Neither the

Figure 23–8   Tender okra pods are utilized in many ways: fried, pickled, and as an ingredient in soups and gumbos. Because the fruit contains a mucilage-like substance, okra is commonly used to thicken broths and soups. *Source:* Ferry-Morse Seed Company.

Figure 23–9   Green pea plant exhibiting vining habit and developing pea. *Source:* Ferry-Morse Seed Company.

wild progenitor nor the early history of the pea is well-known. Probable centers of origin are considered to be Ethiopia, the Mediterranean area, and central Asia, with a proposed secondary center of diversity in the Near East. Greek and Roman writers mentioned peas, but not until the seventeenth century were varieties described.

From 1996 to 1998, annual U.S. commercial production of green peas averaged approximately 922 million

Figure 23–10 Edible podded peas, also known as "snow" or "China" peas, are harvested for consumption before the seed enlarges and the pods become fibrous. The desired characteristic of edible podded peas is the slow development of seeds and pod fiber.

pounds. In the United States, processing peas are primarily grown in Minnesota, Washington, and Wisconsin. These states together accounted for an average of 69 percent of the U.S. processing pea crop from 1996 to 1998, with comparable amounts of production in each state. Edible podded cultivars, often referred to as the snow or China pea, are an important crop in China, Japan, and California.

Dried peas, later reconstituted and used in soups and other products, are an important crop worldwide but less so in the United States. In contrast to green peas, which are harvested when immature, dried peas are allowed to mature and dry. This allows for easier storage and transport. From 1996 to 1998, annual U.S. commercial dry pea production averaged 555 million pounds.

Peas are grown on many soil types. Fertile, light-textured, well-drained soils favor early production and are generally preferable for production. Peas are sensitive to lower pH values; a satisfactory range is 5.5 to 7.0.

Peas are commonly planted as early as weather permits, both for early production and to avoid maturing during higher temperatures. Peas grow best at cool temperatures of 13°C to 18°C (55°F to 64°F). High temperatures tend to increase starch accumulation and fiber content of seeds and pods that therefore are of lower quality. The processing crop is planted by using grain drill seeders in narrow rows about 20 cm (8 in.) apart. Hand harvesting of fresh market peas is accommodated by wider row spacings, about 60 to 90 cm (2

to 3 ft). Seed is spaced about 5 cm (2 in.) apart in the row at a depth of 2.5 cm (1 in.). Planting into moist soil is preferred to hasten germination. Accurate determination of optimum harvest maturity is possible by employing heat units or degree day information. In this procedure hourly temperatures above a base level are computed for the particular cultivar, and are used to predict harvest dates as well as to schedule planting dates. Such predictions enable producers and processors to handle the processing crop efficiently, and make it common for processing peas to be processed within hours following harvest.

Peas are less responsive to applied fertilizers than many vegetables, and are sensitive to soil salinity. Most fertile soils will satisfactorily support the crop. On less fertile soils, fertilizer nitrogen applied at 50 kg/ha (45 lb/ac), and 100 kg/ha (90 lb/ac) of both phosphoric acid and potash are advantageous. Excessive or poorly timed nitrogen fertilization may promote vegetative growth to the detriment of pod production.

Pea cultivars are classified by their growth habit—vining or bush. The vining types, except for some edible pea pod cultivars, have limited use. Another category is seed type; both wrinkled and smooth surface seeds are common. Smooth seed types typically have a higher starch-to-sugar content, whereas wrinkled seed types usually have a higher sugar-to-starch content. This is a characteristic of fully matured seed. At the edible (immature seed) stage, pea cultivars, whether smooth- or wrinkle-seeded, have high sugar quality. A belief that small size seed is associated with tenderness and sweetness is incorrect. Large seed can be both sweet and tender. The stage of maturity at harvest rather than size is the determining factor. Flowering normally begins in most presently used cultivars by the seventh to tenth node, although some cultivars flower as soon as the fourth node and some well beyond the twentieth node. Other important cultivar characteristics concern plant height and vegetative (vine) growth. Less vine growth facilitates mechanical harvest: however, the vines have a byproduct value for hay. Earliness, disease resistance, pod shape (blunt or pointed), length, color, the number of seeds per pod, and multiple pod development per node are other factors. Additionally, seed color, whether pale or dark green, is an important feature.

Significant diseases include bacterial blight, ascochyta blight, powdery mildew, downy mildew, septoria blight, fusarium wilt, fusarium and phytium root rots, as well as *Anthracnose* and rust. Bean yellow mosaic, enation mosaic virus, and other viruses are also damaging.

Serious insect pests include pea aphids, pea leaf miners, pea weevils, lygus bugs, spider mites, and seed corn maggots. Nematodes likewise cause severe losses in some areas. Since the crop cannot economically justify fumigation, crop rotation is a practical control for nematode problems.

Mechanical harvesting of peas for processing is a well-established and efficient procedure. Optimum pea quality is determined quantitatively by the resistance to pressure of a sample of peas. Less resistance indicates tenderness: greater resistance indicates advancing maturity and starch accumulation. Fresh podded peas for hand harvesting are easily sampled by taste to determine sweetness and tenderness. Frequent harvesting is scheduled to avoid overmaturity. Determinate cultivars are harvested over a two-to-three week interval, indeterminate types for a longer interval.

Peppers (*Capsicum annuum* L.) SOLANACEAE

Peppers are an important vegetable commodity highly prized for the flavor, color, vitamin C, and pungency they provide to the human diet. Peppers are a valued spice throughout the world and an important contribution from the Americas to the world's spices. The plants are shrubby perennials, although usually grown as herbaceous annuals in tropic, subtropic, and temperate regions. Peppers are native to the Central and South American tropics. Columbus brought the pepper to Europe. From there it was introduced to other areas, and into China in the late 1700s.

Capsicum annuum is the most extensively grown species. *C. Frutescens* is represented by the Tabasco type: *C. chinesse, C. pubescens,* and *C. baccatum* are also cultivated. Common black pepper (*Piper nigrum*), also used for seasoning foods, belongs to a different botanical family and should not be confused with the Solanaceous pepper.

From 1996 to 1998, annual U.S. bell pepper production averaged 1.5 billion pounds. California, Florida, and New Jersey together accounted for an average of nearly 90 percent by weight of the commercial U.S. bell pepper crops during the same time period. The production of "chili" types grown for processing into a dry powder is centered in the southwestern United States, led by Texas, New Mexico, Arizona, and California.

The plant is frost-sensitive and grows best at a warmer temperature, preferably within a range of 25°C to 30°C (77°F to 86°F). Excessively high temperatures, above 38°C (100°F), especially at flowering, can result in poor fruit set.

Fertile, lighter-textured soils within a pH range from 5.5 to 7.0 favor high productivity. Warm soil temperatures are required for rapid germination, and to avoid seedling damping-off. Many crops are directly seeded, but because of slow seedling growth, many plantings are started from transplants. Commonly used row spacings are about 90 cm (3 ft) apart with in-row spacings ranging from 30 to 45 cm (12 to 18 in.). To produce high-yielding crops, application of a complete fertilizer to supply about 100–150 kg/ha (90–135 lb/ac) of nitrogen, and 50 kg/ha (45 lb/ac) of phosphoric acid and potash is commonly supplied. The initial slow growth of pepper seedlings does not compete well with weeds, and the use of herbicides and mechanical cultivation is commonly practiced.

Peppers are interesting in the number of cultivar types grown, each having various preferred uses. These are distinguishable mostly by their characteristic shape and include types commonly known as bell (Fig. 23–11), pimiento, squash or cheese, ancho, anaheim, cayenne, cuban, jalapeno, cherry, wax (Fig. 23–12), and tabasco. Fruit color is also used as a distinguishing characteristic. Peppers are green, yellow, various shades of brown, purple, and red in color. They also differ in pungency; some are extremely pungent, others very mild or sweet. Pungency is contributed by the capsaicinoid content, the principal compound being capsaicin. Other characteristics considered in cultivar choices include fruit wall thickness and fruit-bearing habit, whether upright or pendant. Fruit size, earliness, and disease resistance are other important factors.

Viruses are probably the most destructive and production-limiting diseases. Serious virus diseases are the mechanically transmitted tobacco mosaic, and potato X virus; the aphid vectored potato Y, tobacco etch, and cucumber mosaic virus; and thrip-transmitted spotted wilt virus. Other important diseases include bacterial spot, bacterial soft rot, fusarium wilt, verticillium wilt, phytophthora root rot and blight, *Anthracnose, Cercospora,* and *Alterneria.* Nematodes

Figure 23–11 Peppers are available in many shapes and colors. Pictured are the blocky, thick-walled bell type peppers. These are popular for use in salads and in many cooked dishes. Whether pungent or sweet, peppers provide enjoyable flavor directly and add flavor to many foods. *Source:* Harris Moran Seed Company.

Figure 23–12 Hungarian Yellow Wax peppers typically have a tapered shape, thin fruit walls, strong pungency, and both pendant and upright fruit bearing habits. *Source:* Ferry-Morse Seed Company.

are still another contributor to crop damage. Other fruit defects include blossom end rot, sunscald, and in some situations, cold-temperature-induced injury. Damaging insect pests include aphids, because of their feeding damage and virus transmission; the pepper weevil; pepper maggot; flea beetles; leaf miners; and several lepidopterous insects.

Peppers for fresh usage are hand harvested with the size of the fruit often being the determining factor. Cultivars grown for processing into chili powder are allowed to develop full coloration and to dry on the vine before being harvested. A limited amount of chili peppers are harvested by machine, and this amount will surely increase. Bell types are the most popular types in the United States and are usually harvested when green, although some are permitted to develop red color for certain markets. Due in large measure to an increasing proportion of Hispanics and the increasing popularity of Mexican-style foods, the appearance of peppers of varying types has noticeably increased in U.S. markets.

Peppers are consumed in a variety of ways—as fresh vegetables eaten uncooked in salads, and in many other prepared forms, including canning and pickling. The chili types

are also used fresh and in whole dry forms or after grinding into powder. Popular powdered forms of pepper in addition to chili powder are paprika and cayenne.

Most fresh market peppers can be stored for several weeks if held at a temperature of 5°C to 7°C (41°F to 45°F) at a relative humidity of 90 percent. Dried peppers can be stored for considerable time if maintained under dry conditions.

Pumpkins and Squashes (*Cucurbita spp.* L.)
CUCURBITACEAE

The term pumpkin or squash does not have a precise botanical meaning. These are *Cucurbita* species of the *Cucurbitaceae* family, and the terms squash and pumpkin are used interchangeably. While there is less confusion regarding summer squash (cultivars used for their immature fruit and having soft rinds), there remains confusion between types of winter squash and pumpkins (mature fruit having hard rinds). These are important commodities in the diets of much of the world's population, being grown from the tropics and into much of the temperate zones.

Precise statistics for the commercial production of various types of pumpkins and squashes are difficult to obtain. In addition, a substantial amount of home production occurs. United Nations FAO data suggest that world production of pumpkins, squashes, and gourds was approximately 33 billion pounds in 1999, although the *Cucurbita* spp. included in this figure are not specified. Leading world producers of these commodities in 1999 included Asia and Europe with 64 and 15 percent by weight of total production, respectively. Winter squashes/pumpkins are also widely grown in the United States, with major production occurring during the late summer and the fall. These commodities, in addition to their multiple use as a vegetable, are also used for their seed, as ornamentals, and for livestock feeding.

The center of origin of the genus is primarily Central America, ranging into Mexico and the northern parts of South America. The major volume produced is contributed by four species, with *Cucurbita pepo* as the largest contributor, other species being *C. moshata, C. maxima,* and *C. mixta.* Species are distinguishable by their stem, peduncle, seed, and leaf variations. All are monoecious with generally stable sex expression characteristics and are dependent on insect pollinators for fertilization of the flowers. Hybrid seed production is managed through manual removal of unwanted staminate flowers, or suppression of their development through the use of chemical compounds. Some interspecific hybridization is possible and is used to develop new cultivar types and for the incorporation of disease resistance.

Having a tropical origin, the genus is frost-sensitive: warm temperatures above 25°C (77°F) are more favorable for growth and development.

The summer squashes (mostly *Cucurbita pepo*) are very popular during the warmer periods of the year. Winter or off-season production makes these squash types available almost year-round. In the United States off-season production occurs throughout the southeast and southwestern states during the winter period, because these areas have relatively warmer temperatures. During the summer, production occurs throughout the country, and is easily accomplished because the required growing season is short, less than two months.

Pumpkins and squashes grow best on well-drained, fertile soils of varying textures. Light-textured soils are preferred for summer squash; finer-textured soils are favored for the winter squashes. The suitable soil pH range is rather broad, from 5.5 to 7.5. Direct seeding in the field is the most common method of propagation. The summer squash cultivars are generally bushy in growth habit and commonly are spaced in rows about 90 to 120 cm (3 to 4 ft) apart and at 30 to 40 cm (12 to 15 in.) in-row spacings. The viny winter squashes and pumpkins require wider spacings, these being about 150 to 180 cm (5 to 6 ft) between rows and about 60 cm (2 ft) within the row.

The fertilizer requirements of pumpkins and squashes are similar to other cucurbits (see muskmelon). Adequate soil moisture, especially during flowering and fruit sizing, is most important, and when rainfall is insufficient, irrigation may be needed. Squash root systems are extensive with deep roots, although many are relatively shallow and therefore are easily damaged by deep cultivation. Early cultivation, either mechanical or chemical, is important to avoid weed competition with the crop. Once beyond the seedling stage, squash plants are effective in shading and thereby suppressing weed growth.

Cultivar selection varies with regard to the intended use of the product. For summer squashes, earliness, prolific fruiting, and disease resistance is very important. Other criteria are fruit size, shape, and color. Market preferences generally are for small summer squash fruit, these being viewed as being of superior quality, and are harvested prior to mature seed development. Fruit shapes vary greatly; spherical, flat, scalloped, curved, cylindrical, long, tapered, short, blunt, and combinations thereof are commonly produced. Color variation is also pronounced, with exterior fruit colors that include white, yellow, black, all shades of green, and often multiple colors. Some of the better-known summer squash types are zucchini, straightneck, crookneck, and scallop (Fig. 23–13).

The winter squashes/pumpkins are generally larger-size fruit. A major distinction is that these are harvested after they have developed hard rinds and well-developed seed. Fruit shape and color vary greatly. These variations along with their irregular surfaces (warts) and their durability make some cultivars popular for use as ornamentals. Some cultivars achieve enormous size, some intentionally grown for exhibition purposes. Single-fruit weights as much as 250 kg (550 lb) have been obtained. Earliness is less important for winter squash cultivars in contrast to summer squashes. Win-

Figure 23–13   Summer squash similar to these are a popular vegetable in almost every home garden because they are easily grown and highly productive. Commercial production is also significant. These bush summer squash are a dark green bush "zucchini" type. Other types of summer squash or many different shapes and colors are also popular. *Source:* Ferry-Morse Seed Company.

ter squash normally require at least 75 days, with many well over 100 days to achieve full maturity. Their holding or storage capabilities are highly important. This characteristic is assisted by the development of a hard rind that resists disease and deterioration. Better-known types of winter squash and pumpkins are acorn, butternut, hubbard, banana, marrow, spaghetti, and of course the popular pie and Halloween pumpkins (Fig. 23–14).

Squashes and pumpkins are subject to the same diseases that affect other cucurbits (see cucumbers). Important insect pests are the 12-spotted and the striped cucumber beetle, squash vine borer, squash bug, and aphids. The root-knot nematode is also a production concern.

Summer squashes are harvested at an immature stage when the rind is still soft and the seed are underdeveloped. Delayed harvest results in further development of the fruit with overenlargement, toughness of the rind, and hardening of the seed all contributing to loss of quality and value. The converse is true with the winter squashes and pumpkins, which are harvested when fruit are fully mature, with hard rinds, firm flesh, and fully developed seed. Hand labor to cut the fruit from the stem is the dominant method of harvesting, although a limited amount of winter squash for processing is harvested with labor-aid conveyor equipment. Mechanical handling at the present time is too damaging to both the summer and winter types of squash.

Squashes and pumpkins have broad uses. Both types are processed by canning and freezing methods. Many of the summer squash are consumed fresh and in various methods of cooking. The winter squashes are usually cooked by boiling, baking, and in pies. In previous years, a considerable volume of squashes/pumpkins were produced for livestock feeding; however, such use has declined. A great advantage of the winter squashes is their ability to be stored for a rela-

Figure 23–14 An assortment of winter squashes including banana (A), acorn (B), hubbard, (C), and butternut (D), along with some pumpkins (E). *Source:* USDA.

tively long period when held at temperatures of about 10°C to 13°C (50°F to 55°F) at a relative humidity between 50 to 75 percent. Summer squash have a limited postharvest life, usually not more than two weeks, even when maintained at favorable conditions of 7°C to 10°C (45°F to 50°F) at high (95 percent) relative humidities.

Sweet Corn (*Zea mays* L. var. *rugosa,* Bonaf.) POACEAE

Corn, which originated in the Americas, was domesticated at least 5,000 years ago. It was a staple grain crop for the native Indians and remains a major world grain crop. Only since the middle of the nineteenth century have sweet corn types of corn achieved status as a vegetable commodity. The popularity of sweet corn is at a maximum in the United States, where it is a frequently grown vegetable in many home gardens and where a substantial level of commercial production occurs in many parts of the country. The crop has steadily increased in popularity in other parts of the world, notably in Europe.

From 1996 to 1998, annual U.S. production of fresh market and processing sweet corn crops averaged approxi-

mately 2.4 and 6.6 billion pounds, respectively. Florida, California, and Georgia together accounted for approximately 50 percent by weight of the U.S. fresh market sweet corn crops with significant amounts of production in other states including Ohio, New York, Pennsylvania, and New Jersey. Minnesota, Washington, and Wisconsin together accounted for an average of 68 percent by weight of the U.S. processing sweet corn crops in the same time period with significant amounts of production in other states including Oregon, Illinois, and Idaho.

The cultural requirements for sweet corn production parallel those for field or grain corn. The crop thrives in humid environments having hot 30°C (86°F) days and warm nights. Through plant breeding efforts, the temperature range has been broadened so that cultivars are available that perform well at lower temperatures without the penalty of lower crop performance. Extensive use of hybrid cultivars provides producers with greater adaptation, yield, and kernel quality attributes. Commercial plantings of these high-quality seeds are sown in rows to a depth of 2.5 to 4 cm (1 to 1.5 in.), commonly spaced about 90 cm (3 ft) apart with in-row spacing of about 30 cm (12 in.).

Well-fertilized soils of medium texture are most favorable for high yields of good-quality sweet corn, with nitrogen being the most important nutrient. Nitrogen applications of 125 to 225 kg/ha (110 to 200 lb/ac) are often recommended, with phosphoric acid and potash applications of about 170 kg/ha (150 lb/ac). Timely use of mechanical cultivation or herbicides enables corn seedlings to become sufficiently established to shade later developing weeds. Sweet corn is a heavy consumer of soil moisture, and the provision of an adequate and well-timed supply of water is important whether supplied by rainfall or irrigation. The crop generally will utilize about 60 cm (2 ft) of water.

Proper cultivar selection is key to a grower's success and involves a number of factors. Market (fresh, processing) is perhaps the most important criterion since optimal characteristics may differ between markets. In most cases, though, vigor, maturity (timing, uniformity), disease resistance, endosperm type, kernel color (yellow, white, bicolor), and eating quality are strongly considered when selecting sweet corn varieties. Ear and shank length, row number, tip fill and cover, and flag leaf length and color are important in fresh market production. The height at which the ear develops and the ease with which it may be removed from the plant are also important when harvest is to be done by hand.

The range of characteristics, especially those related to eating quality, available in sweet corn cultivars has steadily increased since the early 1980s. Following the identification of a number of mutations controlling various kernel traits, plant breeders and others have used an array of techniques to create sweet corn kernels with different combinations of sweetness, aroma, texture, and pericarp thickness. The time through which kernels remain sweet after harvest has also been increased, facilitating the development of long-distance shipping to fresh markets. Large improvements in the resistance to certain *Lepidopteran* insect pests have also been achieved, partly through the use of modern plant breeding techniques. Sweet corn cultivars containing genes from the naturally occurring bacterium *Bacillus thuringiensis* (Bt) gene are less damaged by feeding of the larval form of corn earworm, a particularly devastating pest since it feeds directly on developing kernels and ears. Sweet corn cultivars with resistance to certain herbicides have also been developed using comparable techniques.

Crop-damaging diseases include Stewart's bacterial wilt, northern corn leaf blight, southern corn leaf blight, *Anthracnose,* common smut, head smut, rusts, and downy mildew. Virus diseases include maize dwarf mosaic, sugarcane mosaic, cucumber mosaic virus, and corn stunt virus. Insect pests are often a production-limiting problem because of the extreme damage they cause. Corn earworms are the most devastating pest, since they directly attack the ear and kernels. Armyworms and cutworms cause severe plant damage. Other insect pests are spider mites, etc.

The harvest period is critical for sweet corn because of the relatively short time the kernels are high in sugar. Direct examination is the best method to determine optimum maturity (Fig. 23–15). The new types of super-sweet corn allow for some latitude as they tend to retain a high sugar level for a longer period, although other kernel quality characteristics, such as firm pericarp, are a quality detraction for some consumers. Rapid and effective postharvest cooling is necessary to retain sweet corn quality, since a high respiration rate results in rapid conversion of endosperm sugars to starch. Night harvest is often performed to retain quality.

Sweet corn for processing is most frequently harvested by machine. Some fresh market sweet corn is also machine harvested, but hand harvest is still widely practiced to minimize crop damage. Even with good postharvest handling sweet corn has a relatively short shelf life. Optimum holding temperatures are 0°C (32°F) at high, 95 percent, relative humidity.

Tomato (*Lycopersicon esculentum* Mill.) SOLANACEAE

Tomatoes are one of the most popular, versatile, and widely grown vegetables—known and grown throughout the world and in nearly every home garden. Although not a new crop to the native populations of tropical and subtropical America, the tomato is relatively new to the rest of the world. Columbus returned with tomatoes to Europe, where they were first grown for ornamental purposes. Cultivation for a food crop was soon established along with its dispersion throughout Europe and other areas. The crop began to be cultivated in North America in the early 1700s.

Tomatoes are warm-season perennials grown as annuals. They are frost-sensitive and for many cultivars require about 3 to 4 months from seeding to produce mature fruit. Production is generally optimized under conditions that provide warm, above 27°C (86°F), temperatures with clear dry days. Presently, through the efforts of plant breeders, cultivars are available having considerably broader climatic adaptation and earliness characteristics (Fig. 23–16). Much breeding progress was achieved by the use of related species of *Lycopersicon esculentum,* such as *L. pimpinellifolum, L. hirsutum, L. chilense,* and *L. peruvianum.* The latter species are not commercially significant but serve as useful germplasm sources for the incorporation of disease and pest resistances as well as horticultural qualities such as cold and heat tolerance, soluble solids, viscosity, uniform ripening, jointless stem attachment, and salt and drought tolerance.

Most areas of the United States have some level of commercial tomato production, either for fresh or processing use or both. From 1996 to 1998, annual U.S. production of fresh market and processing tomato crops averaged approximately 3 and 20 billion pounds, respectively. Florida, California, and Virginia together accounted for approximately 70 percent by weight of the annual U.S. fresh market tomato crops, with significant amounts of production in other states including North Carolina, New Jersey, and Texas. California accounted for an average of 94 percent by weight of the U.S.

Figure 23–15 Sweet corn ears of good quality display well-developed kernels. The time for harvest for optimum quality is determined primarily by recording the date when at least 50 percent of the crop reaches anthesis. Many varieties will be at peak quality approximately 21 days after anthesis. *Source:* Harris Moran Seed Company.

Figure 23–16 Many tomato cultivars have been developed that are variable in size, shape, color, and internal characteristics to meet many fresh market and processing uses. *Source:* National Garden Bureau.

Figure 23–17 Early fresh market tomatoes growing during the winter season under clear polyethylene tunnels. These fruits will be harvested in late winter and early spring for the off-season market. This cultural method is practiced in many mild-winter tomato production areas in the world to obtain earlier production.

processing tomato crops in the same time period. Significant amounts of production also occur in Ohio and Indiana. From 1989 to 1998, fresh market tomatoes were harvested from approximately 129,000 acres yearly while processing tomatoes were harvested from nearly 322,000 acres yearly. From 1989 to 1998, the average annual value of fresh market tomato crops produced in the United States exceeded $1 billion. The average annual value of the processing tomato crops when delivered to the processing plant was approximately $646 million. It is important to note that the value of the final processed products was much greater.

Highest yields are obtained from the use of well-drained fertile soils of medium texture with soil pH levels between 5.5 to 7.0. Plantings are made by direct seeding and from transplants as conditions dictate. Plant spacings are varied depending upon crop use, whether for fresh market or for processing purposes. A small amount of fresh market tomatoes (mature green) are machine harvested in the United States; otherwise fresh market tomatoes are hand harvested and plant spacings are made to facilitate such harvests. Also, in the United States, a pronounced shift in fresh market production has occurred whereby determinate bush types are used in contrast to the use of indeterminate cultivars supported on trellis or on poles. Row widths for pole-type cultivation vary from 100 to 180 cm (33 to 60 in.) with in-row spacings of 30 to 60 cm (1 to 2 ft). Plantings for processing tomatoes are also widely spaced at 150–80 cm (5–6 ft) but by increasing the number of plants within the plant row, higher plant populations are obtained. Production of processing tomatoes in the United States is highly mechanized. Harvest machines, many with electronic equipment able to separate fruit by color differences, are in common use in much of the United States. Other important tomato-producing countries have also adopted harvest mechanization practices. With mechanical harvesting only one single plant destructive harvest is possible; therefore, it is important to have high plant populations as well

as small-vined cultivars with determinate fruiting and high yield capabilities.

The production of tomatoes in structures such as glass or plastic houses or plastic-covered tunnels is a significant and very important activity (Fig. 23–17). This is an expensive and highly specialized operation and requires a high level of technology and management.

For high yields fertilization is generally necessary. The application of about 100 kg/ha (90 lb/ac) each of nitrogen, phosphoric acid, and potash in most situations will generally provide sufficient plant nutrition. Late or excessive application of nitrogen should be avoided, as it tends to stimulate unneeded vegetative growth and to delay fruit ripening. Frequent shallow mechanical cultivation and/or effective herbicide application help to control weed competition.

Soil moisture should be near field capacity during vegetative growth, but it is often beneficial if moisture levels diminish during fruit ripening. Careful application of supplemental irrigation is helpful in avoiding potential disease and fruit quality problems.

Unfortunately, tomato production is frequently plagued by serious disease problems. Significant losses are caused by bacteria diseases, such as canker, southern bacterial wilt, bacterial spot, and speck. Fungal diseases include fusarium and verticillium wilt, septoria leaf spot, *Anthracnose,* early and late blight, botrytis gray mold, *Stemphylium,* phoma rot, and powdery mildew. Additionally, several serious virus diseases that affect tomatoes include a form of tobacco mosaic virus that affects plant growth and causes internal browning of the fruit, bushy stunt, big bud, cucumber mosaic virus, as well as the thrip-transmitted spotted wilt, and nematode-transmitted tomato ring spot. Nematodes such as root knot and lesion also cause direct crop damage. Introduced resistance for nematodes and many diseases assist in combatting some of these problems. Nevertheless, diseases remain a very significant production problem. Additionally, abiotic disorders such as blossom end-rot, catface, sunscald, and blotchy ripening present other problems. Insect pests likewise present production concerns. Pests such as hornworm, cutworm, flea beetles, and mites are some of many potential pests.

Tomato fruit are harvested at various stages of maturity highly correlated with fruit color development. For certain markets, fruit are harvested when they are fully (vine-ripe) colored, others when partially colored, and others absent of red color but at the mature green stage. Mature green fruit are firm and, given time, have the ability to develop full coloration. Distance and time to market are the primary consideration as to when fruit are harvested. For distance markets, the mature green or partially colored fruit are used with the expectation of color developing during transit and prior to retail and consumption. Such fruit, because of their firmness, permit better market arrival condition than would be achieved with fully matured fruit. However, they do not have the flavor of vine-ripened fruit. Temperature maintenance is an important postharvest procedure; however, tomatoes are chilling-sensitive, and temperatures lower than 13°C (55°F) should be avoided. Higher temperatures, 15°C to 21°C (59°F to 70°F), are used to assist ripening of less mature fruit. In some situations, exposure of the fruit to ethylene gas is a procedure to hasten color development.

Processing tomatoes (Fig. 23–18) are harvested when fully mature, because color and flavor are important factors in product quality in addition to fruit pH, sugar, soluble solids, and viscosity. For mechanical harvesting, the producer determines when the crop has the highest proportion of ripe fruit in order to optimize yield.

Choice of cultivars involves other criteria, aside from the intended use, such as earliness, adaptation, disease or pest resistance, fruit size, shape, and color, in addition to internal quality characteristics. Plant growth and fruiting habit are also considerations. The worldwide importance of this commodity encourages extensive cultivar developments, and many choices are available, with new cultivars being continuously introduced. The use of hybrid cultivars

Figure 23–18 Processing tomatoes. *Source:* Harris Moran Seed Company.

provides many benefits, and the use of hybrids is rapidly increasing.

Watermelon (*Citrullis lanatus* Thunb.) (*Citrullis vulgaris* Schrad.) CUCURBITACEACE

Watermelons, like other cucurbits, are warm-season annuals having long, prostrate vine growth; vines readily attain lengths of more than 6 meters (20 ft). The crop is believed to have originated in the dry areas of southern Africa. Two wild forms occur, one producing bitter fruit, another nonbitter. The nonbitter form was a food and water source for the natives of the area, and was the progenitor for domestication of the present watermelon. In addition to its direct consumption, use is made of watermelon for livestock feeding; the seed have been used as human food, and a beer is made from watermelons in the Soviet Union.

Watermelons require a relatively long growing season, usually 3 to 4 months, to produce mature fruit. Their growth is favored by temperatures between 25°C to 30°C (77°F to 86°F) and higher.

From 1996 to 1998, annual U.S. watermelon production averaged 4 billion pounds. Texas, California, and Florida together accounted for an average of nearly 55 percent by weight of the commercial U.S. watermelon crops during the same time period with significant amounts of production in Georgia and other southeastern states.

Watermelons grow best on fertile, well-drained, lighter-textured soils, at pH levels near neutral. The root systems are generally deep and extensive. Because of the relatively low value of the crop compared to its use of surface area, it is common that the best soils are used for crops other than watermelons. In the same context, fertilizers, although beneficial, may be restricted in usage due to economic conditions. Over 60 percent of world production occurs in the developing nations. Fertilizer applications to provide 100 kg/ha (90 lb/ac) of nitrogen and about 50 kg/ha (45 lb/ac) of phosphoric acid and potash generally support good production in most situations.

Plant growth, with its many long trailing vines, occupies considerable field space. Accordingly row widths are wide, varying from 200–300 cm (6.5–10 ft) with in-row spacings of 60–100 cm (24–40 in.) or more. Weed control is critical during early seedling growth, since wide plant spacings and initially limited plant growth makes the crop noncompetitive with weeds until adequate vine development occurs. Once the vines are fully extended, weeds are usually effectively suppressed. Both mechanical cultivation and selective herbicides are used for weed control.

Watermelons are high consumers of moisture, especially during the periods of flowering and fruit enlargement. Most crops will easily utilize moisture, however applied (irrigation or rainfall), equivalent to 50–100 cm (20–40 in.).

Diseases affecting watermelons are similar to those attacking other cucurbits (see cucumber and muskmelon discussion). Insect pests similarly include wireworms, aphids, mites, leafhoppers, and cutworms. Nematodes, particularly the root knot nematodes, are serious threats.

An important cultivar distinction is vine growth. New types have been introduced with substantially smaller vine growth (bush types), although these have not yet received wide commercial acceptance. Because diseases and pests are serious production problems, resistance is an important criterion in cultivar choice. Other cultivar differences include rind thickness, fruit number, size, shape, and exterior coloration—striped, black, green, or mixtures, and flesh-color (Fig. 23–19). Watermelons, although commonly red, also have yellow cultivars. The seed coat color is a market preference factor in addition to seed size and their abundance. Light seed color and fewer seed are preferred. Triploid watermelons are seedless (Fig. 23–20).

Watermelon production remains dependent on hand harvesting procedures, although harvest-aid equipment is used in some areas. Harvest is timed to obtain mature fruit at optimum sugar content. Once harvested, the fruit do not further mature. Achieving the right stage of maturity is best done by direct sampling, since there are no other dependable indicators, even though some are claimed as reliable, such as ground spot coloration, tendril drying, and thumping sound. The fruit will not readily detach from the stem and are cut free from the vine.

Watermelons do not store well, although they can be held at a temperature between 5°C to 10°C (39°F to 50°F) for 2 to 3 weeks in satisfactory condition.

Figure 23–19 Watermelon fruit provide a sweet, juicy, healthy, and refreshing food treat. Large variation in cultivar sizes, shapes, exterior marking, and coloration exist. Note the striped fruit in the illustration. Yellow-colored flesh watermelons have not achieved wide popularity. A strong thick rind is important in resisting damage. *Source:* Harris Moran Seed Company.

Figure 23–20 New breeding and production techniques allow for the production of seedless watermelon. *Source:* Harris Moran Seed Company.

REFERENCES AND SUPPLEMENTARY READING

DECOTEAU, D. R. 2000. *Vegetable crops*. Upper Saddle River, NJ: Prentice Hall.

MAYNARD, D. N. 1997. *Knott's handbook for vegetable growers*. New York: John Wiley.

CHAPTER

24

Vegetable Crops Grown for Flowers, Leaves, or Stems

A diverse and large number of crops are included in the grouping that includes vegetables grown for the consumption of their floral, leaf, or stem parts. One similarity is that most crops in this grouping are cool-season vegetables. Some, such as cabbage and lettuce, are produced in significant amounts. However, many others—for example, artichokes, asparagus, or rhubarb—are rather specialized, and their production is relatively limited. These vegetative types of vegetables are not considered to be major caloric providers to human diets. Nevertheless, they do have considerable nutritive value, notably for their contribution of vitamin A, vitamin C, other vitamins and minerals, as well as dietary fiber. Their other attributes are the texture, flavor, color, and variety they provide to meals.

This chapter discusses some important and familiar vegetables whose flowers, leaves, or stems are grown for food.

STATISTICAL OVERVIEW OF WORLD PRODUCTION

The production of a number of commodities discussed in this chapter (asparagus, cabbage, lettuce, and spinach) accounted for nearly 7 percent by weight of the total worldwide vegetable and melon production in 1999, based on UN-FAO data. Asia was the leading producer of each of these commodities, typically followed by Europe and North and Central America (Table 24–1). Of the

TABLE 24–1	Production by Region in 1999 for Four Commodities Grown for Flowers, Leaves, or Stems. The Three Leading Regions Are Shown

Commodity	Region	Production (MT)	% of World Production
Asparagus	Asia	3,258,856	84
	Europe	234,118	6
	S Amer	199,214	5
Cabbage	Asia	33,098,417	68
	Europe	11,589,111	24
	N & C Amer	2,715,137	6
Lettuce	Asia	7,794,148	48
	N & C Amer	4,401,243	27
	Europe	3,344,383	21
Spinach	Asia	6,404,240	88
	Europe	522,256	7
	N & C Amer	257,955	4

Source: World Agricultural Information Centre, Food and Agricultural Organization of the United Nations, FAOSTAT On-Line (http://aaas.fao.org/lim500rph-wrap.pl?Production.Crops.Primary& Domain=SUA&servlet=1).
Note: Regions included in analysis were Africa, Asia, Caribbean, Europe, North and Central America, Oceania, and South America.

TABLE 24–2	Production by Country in 1999 for Four Commodities Grown for Flowers, Leaves, or Stems. The Leading Country in the Three Leading Regions, as indicated in Table 24–1, is shown first

Commodity	County/ Region	% of Region Production	% of World Production
Asparagus	China/Asia	97	81
	Spain/Europe	25	2
	Peru/S Amer	87	4
Cabbage	China/Asia	56	38
	Russia/Europe	25	6
	U.S.A./ N & C Amer	79	4
Lettuce	China/Asia	74	36
	U.S.A./ N & C Amer	93	26
	Spain/Europe	30	6
Spinach	China/Asia	87	76
	France/Europe	23	2
	U.S.A./ N & C Amer	92	3

Source: World Agricultural Information Centre, Food and Agricultural Organization of the United Nations, FAOSTAT On-Line (http://apps.fao.org/lim500rph-wrap.pl?Production.Crops.Primary& Domain=SUA&serviet=1).
Note: Regions in analysis were Africa, Asia, Caribbean, Europe, North and Central America, Oceania, and South America.

commodities listed in Table 24–1, Asia, Europe, and North and Central America averaged 72, 15, and 12 percent by weight, respectively, of the world's production in 1999.

China was the leading producer of each of the four commodities shown in Tables 24–1 and 24–2 in 1999. China produced nearly 81 percent by weight of the world's asparagus, 76 percent of the spinach, and 38 percent of the cabbage in 1999 (Table 24–2). The United States produced 26 percent by weight of the world's lettuce crop and 4 and 3 percent of the cabbage and spinach crops, respectively.

Artichoke (*Cynara scolymus* L.) ASTERACEAE

The globe artichoke, cultivated for over 2000 years, is a thistlelike herbaceous perennial plant native to North Africa and other Mediterranean areas. In early times the foliage was eaten. However, the preferred edible portion is the enlarged but immature flower bud, comprised of numerous overlaid bracts, and its fleshy receptacle (Fig. 24–1). The globe artichoke is sometimes confused with the Jerusalem artichoke, *Helianthus tuberosus*. Although both are members of the *Compositae* family, they are different, the edible portion of the Jerusalem artichoke being the underground fleshy tubers. When eaten, these have a taste that resembles that of globe artichoke, perhaps a reason for the confusion.

A frost-free climate with cool, foggy days is a requirement to achieve quality production of artichokes. Hot, dry weather causes rapid growth accompanied by rapid develop-

Figure 24–1 The globe artichoke is an herbaceous thistlelike plant. The edible parts are the fleshy bases of the immature flower heads together with the attached fleshy bases of the large bracts.

ment of the internal floral tissues and fiber, both detrimental to tenderness. Additionally, high temperatures cause the bracts to separate and spread apart, which is detrimental to market appearance. Freezing temperatures blister the epidermal tissues of the bracts and also are a detraction to market appearance. When not seriously frosted, careful and rapid postharvest handling will permit marketing during periods when the crop is in short supply.

Precise statistics for the commercial production of artichokes are difficult to obtain. United Nations FAO data suggest that world production of artichokes was approximately 2.6 billion pounds in 1999. Leading world producers of artichokes in 1999 included Europe, Africa, and South America with 69, 11, and 9 percent by weight of total production, respectively. Nearly all U.S. production occurs along the Pacific Coast of California.

Artichokes are propagated vegetatively from stem pieces or from rooted offshoots from the base of older plants (Fig. 24–2). Plantings last for 5 or more years before reestablishment is necessary. Plants grown from seed are not true to cultivar type and therefore are not used. However, plant breeders have made considerable progress to improve homozygosity of some cultivars, so propagation using seed may replace vegetative propagation practices. Plants are widely spaced. Row widths and in-row spacings of 100 to 120 cm (40 to 48 in.) are commonly used, except in the United States where wider row spacings, 240 to 300 cm (8 to 10 ft) continue to be used, although in-row spacings have presently narrowed to 100 to 120 cm (40 to 48 in.). Wider spacings are used to permit the use of cultivation and harvesting equipment and to enhance production of large-size flower buds preferred in United States markets (Fig. 24–3).

Artichokes grow well on a wide variety of soils but produce best on deep, fertile soils that are well drained. The plant responds well to fertilizers, especially nitrogen. Phosphorus and potassium are usually supplied prior to establishment of the planting, with nitrogen and animal manures applied annually. Since very little plant material is removed in the harvest of the crop and most of the plant residue is returned to the soil, large amounts of plant nutrients are not required.

Farmers generally cut back foliage down to the soil surface following final harvest in late spring or early summer. Regrowth during the summer is delayed by restricting moisture for several months. This is a method that allows for scheduling the resumption of production. Fertilizer is applied prior to the application of moisture or the start of rainfall in the late summer or early fall to stimulate regrowth.

Throughout active growth and the production season, artichokes require ample soil moisture from either rainfall or irrigation. About 90 to 120 cm (3 to 4 ft) of water is needed annually. Water stress results in bud formation that lacks compactness. Excessive moisture is damaging to the roots and favors disease development. Furrow irrigation is most commonly used on level land; sprinklers are used on rolling land, although the use of sprinkler-applied water is increasing in total.

In the United States one cultivar type, 'Green Globe,' dominates. This produces a large-size, nearly spherical-shaped, green-colored bud having small bract spines, although some selections are spineless. In other parts of the world, the choices of cultivars are many. Globe artichokes are grown that are flattened, round, or ovate in shape. Buds are large and small, various shades of green to deep purple in

Figure 24–2 Artichoke fields are rejuvenated or entire new plantings are established by planting divisions of the crown or rooted offshoots, such as this, that arise from the previous stem.

Figure 24–3 A field of artichoke plants growing along the central California coast during late winter. These plants are usually cut back to the soil surface in late spring. Their regrowth is regulated by controlling the moisture supplied to the plant. Wide spacings between plants and rows to facilitate cultivation and harvesting and deep trenching between rows for drainage of winter rains are other common cultural practices.

color, with and without spines, and vary in bract number and bract fleshiness. Additionally, cultivars have early- and late-bearing characteristics, and choices are made according to the cultivars' adaptability to the specific growing environment.

Figure 24–4 Harvested artichokes are transported in bulk bins to a central packing shed for size, quality grading, and cooling before packing.

The best-quality buds are compact, heavy or dense for their size, with tight, fleshy, bright-colored bracts. Most importantly, internal floral development must be minimal.

Several diseases, such as fusarium root rot and botrytis, are chronic production concerns, especially during long periods of warm, wet weather. Curly dwarf virus is also injurious. In the United States, the artichoke plum moth, leafminers, and aphids are principal insect problems. Insects similar to the plum moth are troublesome in European production. Additionally, nematodes, various rodent pests (field mice), and snails cause both plant and crop damage.

Artichokes are harvested over a long period, from the early fall through the winter and into the late spring. A limited amount of summer production occurs. Productivity is less during the colder periods and tends to peak as spring temperatures increase. Buds are harvested by hand and cut from the stem when they achieve the desired size (Fig. 24–4). In the United States the buds are usually between 5 to 10 cm (2 to 4 in.) in diameter and are harvested before the bracts begin to spread apart. Yields of 4 to 5 buds per flowering stalk are obtained. The number of flower stalks varies from 3 to 10 and depends in part on individual plant size and spacings. The buds are graded, precooled in cold water to about 4°C (39°F) and packaged into water-resistant paperboard boxes. In Europe, where they are more popular, the crop is usually marketed within a day of harvest and, unless exported, precooling and packaging is often omitted. Artichokes can be stored in good condition for up to 3 to 4 weeks if maintained at temperatures between 0°C to 4°C (32°F to 39°F) at a relative humidity above 90 percent.

Asparagus (*Asparagus officinalis* L.) LILIACEAE

From 1996 to 1998, annual U.S. asparagus production averaged approximately 202 million pounds. California, Wash-

Figure 24–5 A female asparagus plant. These plants bear only pistillate flowers. The flowers were fertilized by pollen from a nearby male plant. Note the seed berries. Asparagus is a good example of a dioecious plant (male and female flowers on separate plants).

ington, and Michigan together accounted for an average of nearly 90 percent by weight of the commercial U.S. asparagus crops during the same time period, with significant production in other states including Minnesota and New Jersey. Asparagus grown in the United States is sold primarily in the fresh market. From 1996 to 1998, approximately 61 percent by weight of the asparagus grown in the United States was sold in the fresh market with the remainder being used in processed products.

The edible portion of this perennial plant grows from fleshy underground rhizomes called "crowns." The plants are dioecious, having either staminate or pistillate flowers (Fig. 24–5). Asparagus is thought to be native to the eastern Mediterranean region.

Better quality and higher yields are produced when the crop matures during periods when temperatures are about 16°C to 25°C (60°F to 77°F), especially after a long period of fern growth. In colder climates the growing season is shortened because of the limited period for fern production. A reasonable period of active fern growth is necessary in order to supply the "crown" with the products of photosynthesis that the fern produces following the harvest period. The photosynthate produced is translocated to the "crowns" to be utilized for the following year's production of spears. In warmer areas, sufficient fern growth occurs and the crowns can sustain a longer harvest period without exhaustion of their food reserves. During warmer periods, spear growth is very rapid, with greater fiber development within the spears and with a tendency for the spear tips to lose some of their compactness characteristics. Large spear diameter, length, compact tips, and tenderness are determining features for market quality. New cultivars offer the ability to maintain good quality under warmer temperatures. Many new asparagus cultivars are available and the choice of cultivar is strongly influenced by growing location.

Since a planting is maintained for 5 to 10 or more years, well-drained, fertile, light-textured, or loamy soils should be used to assure good health and longevity for the crowns. Peat or muck soils, if well drained, also produce excellent crops. A deep, well-prepared seedbed is essential for the establishment and maintenance of a good stand. Asparagus is propagated by direct seeding into the field, by transplanting nursery-grown seedling plants, or by transplanting one-year-old crowns. Seed, seedlings, or crowns are generally placed in the bottom of a furrow 20 to 30 cm (8 to 12 in.) deep in the center of what will become the row or bed. Seeds are initially covered with 2 to 3 cm (0.8 to 1.2 in.) of soil, crowns with about 10 cm (4 in.) of soil, and as the plants develop the furrows are gradually filled to the surface of the field. The rows are 150 to 180 cm (5 to 6 ft) apart, with plants spaced 30 cm (12 in.) apart in the row. Directly seeded fields usually require 2 to 3 kg/ha (1.75 to 2.5 lb/ac) of seed.

Fertilizer usage varies; in general, 90 to 180 kg/ha (81 to 162 lb/ac) of nitrogen are used annually. Similar amounts of phosphorus and potassium are used at the time of planting, and supplemental amounts are applied as needed during the continuation of the planting. Weed control entails shallow disking early in the spring before the spears appear. Chemical herbicides are widely used and give better results than mechanical cultivation in very weedy situations. Volunteer asparagus seedlings from fern-produced seed often become the major weed competitor.

Asparagus plantings normally receive adequate rainfall during nonharvest periods except in arid or semiarid regions. In the western and southwestern United States about 50 cm (20 in.) of water is often applied during the late summer following harvest and during fern growth to fill the soil profile.

The principal diseases are asparagus rust, phytophthora spear and root rot, fusarium crown rot, *Stemphyllium*, and *Cercospora*. Insect pests include the asparagus beetle, garden symphlid, and recently the asparagus aphid.

In late winter and the cool spring, spears are harvested every two to four days, but as the days warm, daily harvest is often necessary (Fig. 24–6). The spears are cut by hand with a special flat-bladed knife with the end sharpened to a V-shaped notch. The blade is attached to a long shank to permit deep penetration into the soil. The knife is pushed into the soil at an angle of about 60 degrees to cut the spear below the soil surface. Care must be taken to avoid damage to spears not yet ready for harvest. The cut spears are gathered and transported to packing areas, where they are washed, sorted, and graded by diameter and length before being packed into specialized packages and hydrocooled. Spears must be kept at low temperatures, as they will continue to elongate and will do so more rapidly when temperatures are warmer. Mechanical harvesting of fresh market asparagus has not been economically feasible. However, some asparagus intended for processing by freezing is mechanically harvested. The premium-quality processed asparagus is still hand harvested. Good-quality spears are straight, tender, long, large in diameter, and bright green with compact tips. Crooked or misshapen spears are culls and have some usefulness in processed products such as soups. Postharvest handling of fresh asparagus is very critical. Not only must the product be rapidly precooled and kept at a low temperature, 0°C to 2°C (32°F to 36°F), and high relative humidity, 90 percent, but it must be packaged and held in an upright position, since the spears continue to elongate; if they are stored in a horizontal position, the spear tips will bend upward, thereby losing their market appearance.

White asparagus production remains a popular choice for European markets, although the consumption of green asparagus has replaced much of the white asparagus product. Such production depends on blanching the elongating spears, usually with a mounding of soil cover. Harvest is more difficult, since it must be done just before the spear tips emerge from the soil. The high labor requirement for such production is the main discouragement for its continuing use.

After harvest, which should not exceed 6 to 8 weeks for well-established plantings, the spears are allowed to elongate into fernlike plants that achieve a height of 90 to 180 cm (3 to 6 ft) (Fig. 24–7). Young plantings are harvested

Figure 24–6 A field of asparagus with emerged spears, ready for harvest. If weather conditions are ideal (warm), the crop is harvested at least every other day. Fresh market asparagus is usually harvested by hand.

Figure 24–7 After six to eight weeks, it is necessary to cease harvesting asparagus spears and allow the plants to develop fern, as shown here. This allows the foliage to photosynthesize food reserves and to resupply the roots with nutrient materials for another harvest season.

for shorter periods in order not to exhaust the food reserves of the crown. The ferns continue growth until they senesce or die in the late fall or winter. They are then removed by cutting and are burned or disked into the soil. The field should then be clean of the previous plant residues and weeds and ready for the following year's spear production and harvest.

Broccoli (*Brassica oleracea* L., Italica Group)
BRASSICACEAE

From 1996 to 1998, annual U.S. broccoli production averaged nearly 2 billion pounds. California, Arizona, and Texas together accounted for most of the commercial U.S. broccoli crops from 1996 to 1998 but significant production also took place in Maine, Minnesota, and New Jersey. Broccoli grown in the United States is sold primarily in the fresh market. From 1996 to 1998, approximately 94 percent by weight of the broccoli grown in the United States was sold in the fresh market with the remainder being used in processed products.

The flowers of the cole crops—broccoli, Brussels sprouts, cabbage, and cauliflower—have four petals and four petals arranged in the form of a cross. This feature gave the family the name CRUCIFERAE (now BRASSICACEAE), which in Latin means "cross bearers," and from which the English words *crucify* and *crucifix* are derived. The word *broccoli* comes from the Latin word *bracchium,* meaning arm or branch. The tightly grouped fleshy terminal stems, young leaves, and young flower buds and bracts are the plant parts of broccoli that are consumed.

The plant is native to eastern Asia and the Mediterranean areas and was introduced a relatively short time ago into the United States, in about the 1920s. Previously it was not widely cultivated in Europe. However, because of its

high nutritive value and ease of preparation, it has become very popular, and its production has expanded to many other parts of the world.

Commercial plantings are usually seeded directly into well-prepared seedbeds or are transplanted. Transplants are used in areas where shorter periods of favorable climate mandate such use or where labor is plentiful. Row spacings are variable, ranging from 50 to 90 cm (20 to 36 in.) between rows with plants from 20 to 45 cm (8 to 18 in.) apart in the row. Broccoli grows best on fertile, medium-textured soils high in organic matter. Broccoli generally requires fertilization to achieve reasonable yields. Generally, fertilizer applied to provide 60 to 150 kg/ha (54 to 135 lb/ac) of nitrogen and like amounts of P_2O_5 and K_2O produce good yields.

Weeds are controlled by mechanical cultivation or by the use of pre- and postemergence selective herbicides, although in many situations hand cultivation supplements herbicide use.

Cultivar choices are based on adaptation, maturity range, and product characteristics such as the size and compaction of the terminal inflorescence, its shape, color, amount of branching, size of floral buds, and ability to resist flower bud opening when at harvest maturity. Disease resistance is also an important consideration. The introduction of hybrids has provided many production benefits and their use has greatly increased.

Important diseases are seedling damping off caused by *Pythium* and *Rhizoctonia* species, downy mildew, ring spot, and clubroot. Abiotic disorders are hollow stem and brown bead. Cyst and root knot nematodes likewise are production concerns. Insect pests include the cabbage and seed corn maggot, cutworms, flea beetles, aphids, imported cabbageworm, looper, armyworm, corn earworm, and diamondback moth.

The developing inflorescence (head) is hand harvested while it is still immature, compact, and absent of open flowers by cutting from the stem, resulting in a head about 15 to 25 cm (6 to 10 in.) in length, and about 10 to 20 cm (4 to 8 in.) in diameter (Fig. 24–8). The terminal head is first to develop; secondary heads later develop in the axils of lower leaves of the main single stem. These are also harvested when cost effective. Although these are much smaller, they are nevertheless an excellent product. Multiple harvests of both primary and secondary growth are generally made because plant development is not sufficiently uniform for a single harvest. This limitation has been an impediment to mechanical harvesting.

Broccoli is very perishable because of its high rate of respiration and requires immediate cooling, usually by icing. Good quality is indicated by deep green, firm, and compact heads without flower bud opening. In addition to its excellent fresh market uses, broccoli processed by freezing provides a very attractive and valuable product.

Figure 24–8 Broccoli inflorescence "head" at market-mature stage. A delay in harvesting of only a few days will result in further floral growth and a loss of market value because many of the small buds comprising the inflorescence will open.

Brussels Sprouts (*Brassica oleraceae* L., Gemmifera Group) BRASSICACEAE

This cabbage-like tasting vegetable probably evolved from a primitive nonheading Mediterranean cabbage, after its progression to northwestern Europe. The plant received its name from the city of Brussels, where it was first reported to be grown; it continues to be a popular vegetable throughout that area. The plant develops leafy buds resembling miniature cabbagelike heads about 2.5 to 5 cm (1 to 2 in.) in diameter in the leaf axils along the single tall unbranched stem (Fig. 24–9). The plant grows 60 to 90 cm (2 to 3 ft) tall and requires a cool climate to develop compact, quality buds. The crop is relatively tolerant to cold temperatures and can withstand slight freezing. Its growth is extensive in northern Europe, in particular in the United Kingdom and the Netherlands. United States production is concentrated in the coastal area of central California.

For some commercial plantings, direct seeding with later thinning is made, although because of its lengthy growing season, transplants are probably used more often. Common plant spacings are 40 to 90 cm (16 to 36 in.) between rows with in-row spacings of 40 to 60 cm (16 to 24 in.). The culture of Brussels sprouts is similar to that of cabbage. The plants are heavy users of plant nutrients and produce good yields when provided with supplemental fertilization. Fertile, medium-textured soils with good drainage optimize production. Brussels sprouts require a constant supply of water for proper development, and most plantings utilize about 38 to 50 cm (15 to 20 in.) of water.

Brussels sprouts are affected by many of the diseases and insects troublesome to other crucifer crops. The more important diseases are clubroot, black leg, downy mildew,

Figure 24–9 In the early stages of development, Brussels sprouts closely resemble cabbage seedlings but as the plant matures, the main stem elongates 60 to 90 cm (2 to 3 ft) and the axillary buds along the stem develop into small cabbagelike heads about 2.5 to 5 cm (1 to 2 in.) in diameter.

Sclerotinia, ringspot, and verticillium wilt. Tipburn and internal browning are abiotic disorders occasionally causing serious losses. Nematodes are chronic concerns. Insect pests of major importance are the cabbage aphid and the cabbage and seed corn maggot. As with other crucifers, the lepidoptera larvae are constant problems.

Cultivars, which presently are principally hybrids, are chosen based on their maturity, plant height, ability to resist stem lodging, and their uniform or determinate-like production along the stem. Important criteria of quality are buds that are bright dark green and compact.

The crop for fresh market use is normally hand harvested when the basal sprouts are between 2.5 to 5 cm (1 to 2 in.) in diameter, by breaking away the lower leaves and the buds from the stem. These are collected and brought to a central area for further trimming, sorting, packing, and cooling. For fresh market, the crop is generally harvested several times, starting with the lowermost buds and progressing upward as the buds closer to the terminal of the stem enlarge.

The process is similar for the crop intended for processing by freezing, although single harvesting is now more commonly performed. New cultivars more determinate in bud development allow for relatively high yields even when only one harvest is made. Many producers of the crop for processing use mechanical equipment to remove the buds from the stem. However, other aspects of the harvest procedure are not fully mechanized. In the United States, about 80 percent of the crop is processed. In other countries the proportion of the crop for processing is much smaller but rapidly expanding.

Cabbage (*Brassica oleracea* L., Capitata Group)
BRASSICACEAE

Cabbage is well adapted for growth in cool climates. It is a biennial plant and tolerant of slight freezing. On the other hand, cabbage is somewhat more tolerant of warmer temperatures than many other cool-season crops. The edible portion is essentially a large vegetative terminal bud, formed by overlapping of numerous leaves developing over the growing point of its shortened stem, and is called the head (Fig. 24–10). Cabbage is believed to have developed from a wild type of a nonheading plant in the east coast of England and the western European coast. Evidence indicates that the early Egyptians used cabbage for food and medicine, and that the Mediterranean area was its origin.

From 1996 to 1998, annual U.S. fresh market cabbage production averaged nearly 2.5 billion pounds. New York, California, and Texas together typically accounted for one-half of the commercial U.S. fresh market cabbage crops during the same time period. Significant production also occurs

in Ohio, Florida, Georgia, and New Jersey. Fresh market cabbage is sold as whole or half-heads or after minimal processing into slaw. Cabbage is also grown for use in processed products, primarily sauerkraut. Production for processing markets is widespread in the United States but greatest in Wisconsin, Ohio, New York, Michigan, Texas, Florida, California, North Carolina, and New Jersey. From 1996 to 1998, cabbage production for use in processed products averaged approximately 329 million pounds.

To achieve quality and yield, cabbage is grown on highly fertile, well-drained soils. Light-textured soils are preferred for earlier production, whereas late maturing and processing cultivars perform better on finer-textured soils.

Cabbage is grown from four- to six-week-old transplants or from seed sown directly into the field. Transplants are mechanically planted, although in many parts of the world hand transplanting is common. Field spacings are usually about 60 to 90 cm (2 to 3 ft) apart in rows, with in-row spacings about 30 to 60 cm (1 to 2 ft). Eventual size of the cultivar determines the plant spacings used.

Cabbage is a heavy user of plant nutrients, and fertilizers commonly applied provide between 60 to 135 kg/ha (56 to 120 lb/ac) of nitrogen, 100 to 200 kg/ha (90 to 180 lb/ac) of P_2O_5 and K_2O.

Adequate use of selective herbicides and timely mechanical cultivation usually provide satisfactory weed control. Deep tillage is avoided, since a proportion of the root system is relatively shallow and would be damaged. A water supply of 45 to 60 cm (18 to 24 in.) in most situations adequately provides enough water to raise a crop.

Cabbage is susceptible to a number of diseases, insects, nematodes, and physiological disorders. The most serious diseases are clubroot, black rot, black leg, downy mildew, fusarium yellows, and alternaria leaf spot. Tipburn is an important disorder. Insect pests include the many lepidoptera insects as well as aphids, flea beetles, maggots, and cutworms. The cyst and rootknot nematodes are also troublesome. Resistance to some diseases is useful in limiting losses.

The cultivar chosen largely depends on the earliness desired, the size and shape of the head, and the use of the crop. Other variables include head color (pale, dark, blue-green, and red), leaf texture (smooth or savoyed), interior stem height, compactness, storage characteristics, ability to resist bursting when mature, and needed disease resistance. Hybrid cultivars are widely available, and their superior performance has greatly replaced the use of open pollinated cultivars. Head weight of fresh market cabbage varies considerably, ranging from 1 to 5 kg (2.2 to 11 lb) with marketing trends favoring the smaller, 1 to 2 kg sizes. The converse is true for processing cabbage, where larger sizes are preferred.

Cabbage is harvested when the heads have reached full size for the particular cultivar. Fresh market cabbage is hand

Figure 24–10 A field of head cabbage shortly before harvest.

Figure 24–11 A uniform field of fresh-market cabbage being harvested with the use of a labor air conveyor. Hybrid cabbage cultivars such as the one used for this field contribute to a high level of harvest maturity.

Figure 24–12 The cauliflower head, sometimes called curd, develops at the terminal portion of the relatively short stem and is comprised of compact fleshy tissues. Overwrapping of the leaves surrounding the head protects it from discoloration, which would result from direct exposure to sunlight.

harvested by cutting the stem below the head leaves. Usually a few outer "wrapper" leaves are left to provide some protection against bruising injury. Harvesting may be assisted with the use of conveyor equipment (Fig. 24–11). The heads are packed in boxes or bags and are commonly cooled by forced-air methods. Fields may be harvested several times, as the heads mature at different times. The cabbage crop for processing is harvested just once, and in some areas mechanical harvesters are used. Such equipment, while satisfactory for the processing crop, results in too much damage to be practical for the fresh market crop. The storage of cabbage is an important procedure in many parts of the world. Cabbage maintained at low temperatures, 0°C to 3°C (32°F to 38°F), with high relative humidity can be stored for several months. Controlled atmospheric storage of cabbage improves storage capabilities, although the practice is not in wide use.

Cauliflower (*Brassica oleracea* L., Botrytis Group)
BRASSICACEAE

The origin of cauliflower is believed to be the eastern Mediterranean area, with subsequent development perhaps from sprouting broccoli in northwestern Europe, where it has been known for centuries. Both annual and biennial forms are cultivated. From 1996 to 1998, annual U.S. cauliflower production averaged approximately 714 million pounds. California, Arizona, and New York together accounted for most of the commercial U.S. fresh market cauliflower crops during the same time period. Significant production also occurs in Florida, Georgia, and New Jersey. Cauliflower is also grown for use in processed products, including soups and frozen vegetable mixes. Production for processing markets is mainly restricted to California, Arizona, Michigan, New York, and Texas. From 1996 to 1998, cauliflower production

for use in processed products averaged approximately 85 million pounds.

Among the Brassica crops, cauliflower has the most exacting climatic and cultural requirements. Moderately uniformly cool climates are best adapted for cauliflower, and production sites are frequently near large bodies of water to benefit from their climate-moderating influence. Preferred mean temperatures during growth and crop maturity are 15°C to 20°C (59°F to 68°F). Unlike cabbage, cauliflower is much more intolerant of frost or high temperatures. Cultivars known as "tropical types" having better warm-temperature adaptation are grown in large volume in many warmer temperate regions, namely India and China.

The edible product, called the head or curd, is mistakenly considered to be floral tissue, but is instead prefloral fleshy apical meristem tissues. The curd consists of the repeatedly branched terminal portion of the main axis of the plant, comprised of a shoot system with short internodes, branch apices, and bracts (Fig. 24–12). High temperatures result in poor head formation and quality with accompanying market defects such as bract development within the curd, loss of compactness, and ricy curds.

Good productivity demands fertile, medium-textured, well-drained soils. Other cultural procedures, fertilization, and irrigation requirements are similar to those for cabbage. Cauliflower is similarly affected by many of the diseases, insects, and nematodes that attack other crucifers. Phytophthora stem and root rot, and cauliflower and turnip mosaic viruses are particularly serious for cauliflower. Premature development (buttoning) of the head, hollow stem, and curd riciness are abiotic disorders of considerable seriousness.

Direct exposure of the curd to sunlight tends to discolor the otherwise white curd surfaces to a creamy yellow,

Figure 24–13   Harvesting and field packing of cauliflower is performed using labor aid conveyor equipment. The same or similar equipment is used for the harvesting of broccoli and some other vegetable row crops. Field packing has some advantages that packing sheds do not, and vice versa. Packaging costs usually determine whether packing is done in the field or in sheds.

which is a market defect, but does not otherwise affect its edible quality. Leaves cover and protect the curd as they form, but as the curd enlarges, the inner leaves are forced apart and the curd is exposed to sunlight. To protect and blanch the enlarging curd, the long outer leaves are gathered together over the top of the curd and tied together to keep out light. Cauliflower is hand harvested before the curds overmature to ensure high quality. The curds are cut from the stem with a basal whorl of leaves attached, which serve to protect the delicate curd. Often field packing is done (Fig. 24–13) or the harvested crop is taken to central packing sheds for further trimming, grading, packaging, and cooling. Both forced cold air and vacuum cooling methods are used.

Cultivar choice considers the crop's maturity period, head size, shape, compactness, and amount of foliage cover. Late maturing and biannual cultivars have an advantage of abundant foliage growth and leaf cover is less of a problem. However, there are few other advantages in using such cultivars, and they presently are seldom grown. The introduction of hybrid cultivars has provided greater uniformity of maturity and helps to reduce the number of harvests needed to complete a crop. For these reasons hybrids are being used more often, although open-pollinated cultivars still dominate. The majority of the crop is used for fresh market; the crop is also processed by freezing, and some amounts are processed by pickling.

Celery (*Apium graveolens* L. var. *dulce* Mill.) APIACEAE

The former family name, *Umbelliferae*, came from the characteristic umbel form of the inflorescence, and includes celery, carrots, parsnips, parsley. Writings indicate that the first use of celery was as a medicine in the fifth century A.D., and as a cultivated food crop in the early 1600s. The plant is thought to be native to the Mediterranean area, although wild types are found widely distributed in other areas. The crop's world importance is relatively small, both as a vegetable and

for the use of its seeds and foliage in seasoning foods. Celeriac, A. *Graveolens* var. *rapaceum,* also known as knob or root celery, is closely related and finds similar use, particularly among northern Europeans. From 1996 to 1998, annual U.S. celery production averaged nearly 2 billion pounds. California, Michigan, and Texas together accounted for most of the commercial U.S. celery crops from 1996 to 1998, but significant production also takes place in Florida and New York. California produces celery at all times of the year and shares with Texas and Florida most of the winter and early spring crops. Late summer and early fall crops are produced in Michigan and New York.

Celery, a biennial plant, has rather exacting climatic requirements. Optimum daytime growing temperatures range between 16°C to 21°C (60°F to 70°F) with night temperatures slightly lower. Exposure of temperatures less than 10°C (50°F) will initiate vernalization and seed stalk development and should be avoided. Celery produces a large volume of plant material, as much as 135 MT/ha (60 t/ac), and thus is an obvious heavy user of mineral nutrients and soil moisture. Best growth occurs when rain or irrigation is uniformly distributed throughout the growing season. The crop readily consumes 75 to 100 cm (30 to 40 in.) of moisture during the growing season and depending upon soil type and location may require even more. Furrow and subirrigation methods are generally used. Sprinkle irrigation, although often used in the early period of growth, has the potential to spread foliar (blight) disease. Drip irrigation methods, recently introduced, have been effective, and such use will probably increase. To meet the growth needs of the crop, the most productive soils are usually chosen for celery production. Alluvial soils, preferably friable loams or clay loams, high in organic matter, with a high water-holding capacity are best suited, and these are used in California production. In other U.S. celery-producing areas, such as Florida, Michigan, Ohio, and New York, peat soils are widely used, and provide an excellent soil medium. Several decades ago, before the conversion from blanched (white) celery to unblanched (green), peat soils were also used in California production, mainly because the peat soil facilitated blanching. When blanched celery production diminished, producers moved production to other locations and soil types.

In addition to fertile soils, the crop is generally heavily fertilized. A wide range of supplemental fertilizers is applied, depending on soil type and the soil's native fertility. Commonly, fertilizer to provide 200 to 450 kg/ha (180 to 400 lb/ac) of nitrogen, 100 to 200 kg/ha (90 to 180 lb/ac) of P_2O_5, and 50 to 200 kg/ha (45 to 180 lb/ac) of K_2O are supplied to the crop.

Because of its long growth period and slow germination and emergence of the seeded crop, most celery is grown from transplants. Transplant seedlings are produced in greenhouses or other protective structures, or may be grown in a field nursery until about 6 to 8 weeks of age, at which time they are planted into the production field. Celery producers often purchase transplants from nurseries that specialize in transplant production in order to obtain timely delivery of uniform high-quality plants that have not been subjected to vernalization temperatures of less than 10°C (50°F).

The celery plant is a slow grower, and in its early development it does not compete against weeds. Hence weed control is critical. Selective herbicides and careful shallow mechanical cultivation help to keep weeds under control.

Celery is a difficult and expensive crop to grow, and it is subject to many diseases, pests, and disorders that must be combatted. Some of the most serious diseases are fusarium yellows, early blight (*Cercospora*), late blight (*Septoria*), and pink root (*Sclerotinia*). Additionally, virus diseases, such as celery mosaic and aster yellows, can result in severe crop losses. In some production areas, calcium deficiency results in blackheart, and boron deficiencies cause a condition called cracked stem. Another chronic concern is premature bolting, resulting from a period of exposure to temperatures lower than 10°C (50°F). Troublesome insect pests include leaf hoppers, tarnished plant bug, aphids, cabbage loopers, armyworms, and the very damaging leaf miner.

Cultivar choices are not as plentiful as with other commodities and are generally divided into two types—the yellowish green, also known as self-blanching cultivars, or the Tall Utah green cultivars (Fig. 24–14). Blanched celery is produced by covering the developing leaf petioles with soil or other materials. Being extremely labor-intensive, blanched celery is infrequently found in European markets and no longer in U.S. markets. Self-blanching types were developed as a replacement for the blanched product and even these are produced in limited amounts, since the green types are much more popular. Preferred celery cultivars are brightly colored green, with long, thick, wide, fleshy petioles, attached to a short stem, with petioles that overlap to form an upright compact cluster of stemlike petioles, collectively called a stalk or head.

A major expense in celery production is for harvesting and packaging, which uses considerable labor. Mature fields are harvested only once, with workers undercutting the plants with a knife just below the stem (Fig. 24–15). The plants are trimmed to a uniform length after the remaining root, most of the leaf-blade foliage above a specific length, and the unmarketable petioles have been cut away. Size grading is done based on plant diameter. Those of similar size are packaged together; those plants that are very small are marketed as celery "hearts." Packing is done both in the field and in sheds, depending on economic factors and other situations. Each method has advantages. As with many other leafy vegetables, it is important to rapidly cool the crop in order to retain its quality. Hydrocooling, vacuum cooling, and forced-air cooling are methods used to lower product temperatures. Celery has a relatively long shelf life if maintained at temperatures near 0°C to 4°C (32°F to 40°F) and at a high (90 percent) humidity.

Figure 24–14 Celery typical of the "Tall Utah" type has bright green color with long, thick fleshy petioles well compressed in an upright grouping commonly referred to as a stalk or head.

Figure 24–15 There are several different practices used in the hand harvest of celery. In this field the cutters use a special knife. The blade is broad and sharpened on the blunt end so it can cut the plant's roots with a jabbing motion just below the soil surface. The edge of the blade is also sharp so that the plant is trimmed to the appropriate height with a chopping motion. Some growers top the celery while it is standing in the field with a mowing machine adjusted to the desired height. *Source:* University of California Cooperative Extension.

Most of the crop is marketed fresh, although a significant portion is processed as canned, frozen, and dehydrated products, which find wide use in soups, stews, and for flavoring in other food products. The processed crop is handled in bulk, and since it is not necessary that the petioles be connected, the crop for processing is frequently machine harvested.

Chive (*Allium schoenoprasum* L.) AMARYLLIDACEAE

Chives are small bushy onion-like perennials grown for their delicate tubelike leaves, which are used in salads, soups, stews, omelets, various cheese products, and to flavor numerous other food products (Fig. 24–16). It appears that chives are native to the northern hemisphere, and wild forms are found in North America and Eurasia. The crop is of minor economic importance, although it has been cultivated for centuries in Europe and the Orient. In the United States relatively little commercial production occurs except in California, Texas, and Arizona.

Chives are cool-season, cold-tolerant plants with climatic requirements similar to onion and garlic. Chives are usually propagated by seed. The plant, through the process of tillering, develops a clump of small plants. These clumps could be divided and used for further propagation, but seed use is preferred. The plant also does produce small thin bulbs, but these, too, are not useful for propagation. Cultural practices and disease and pest problems are similar to those of onions.

When the leaves have attained marketable size, they are cut from the plant, allowing for the remaining small unharvested leaves to continue growth and permit subsequent harvesting. Field plantings can be maintained for a period of two to three years. The cut leaves are gathered together to form a small bundle for fresh market use. Shelf life is short, and low temperature and high humidity are necessary to reduce the wilting of the delicate leaves. Chives are also a popular processed crop, prepared as frozen, freeze-dried, or dehydrated products.

Collards and Kale (*Brassica oleracea* L., Acephala Group) BRASSICACEAE

Collards and kale are plants that resemble cabbage but are nonheading (Fig. 24–17). They are closely related to wild cabbage and have been cultivated for many centuries; they are considered to be the least-developed members of the *Oleracea* species. Collards and kales are winter hardy biennials and tolerate both somewhat lower and somewhat higher temperatures than cabbage. These plants are grown for their abundant foliage, which is usually consumed as a cooked vegetable or as an ingredient in other foods. Some kale cultivars, in addition to having attractive green coloration, also have foliage with color variations ranging from purple, red, pink, to white in addition to very crinkled leaves and find use as ornamental plants.

In the United States, both collards and kale are very popular in the southeastern and southern states, where the majority of commercial as well as home garden production

Figure 24–16 Chives can be grown in a pot in the kitchen window and harvested as needed throughout the year. The upper portion of tubular leaves is chopped and used for flavoring. *Source:* USDA.

Figure 24–17 Collards are closely related to cabbage but are nonheading. The leaves are broader than those of kale, and the lower ones are generally harvested first, progressing up the stalk to a rosette of leaves at the top. The leaves are popular as winter greens in the southern United States. *Source:* USDA.

occurs. Kale is more popular than collards in most of the world, where it is widely grown although usually on a small scale. It has many nutritional virtues but has not been widely adopted. Cultural management of these vegetables is similar to that for cabbage, and the crops are affected by similar disease and pest problems.

The distinction between collard and kale is not clear. Collards have a relatively smooth leaf blade, whereas kales, with some exceptions (forage kales), generally have crinkled leaves, some cultivars having pronounced curly foliage. Kale cultivars also vary in height, with dwarf, medium-tall, and tall types. Some kale cultivars, which include the thicken stem or "marrow-stem" types, are used for animal forage.

Kale and collards are harvested before the leaves become excessively large, tough, or fibrous. Tender, high-quality leaves develop from well-fertilized, rapidly growing plants. Either the entire plant, if small, is cut off with a sharp knife, or selected large leaves are cut, bundled together, and placed in paper or wood containers for cooling and subsequent marketing. The latter procedure permits additional harvests of leaves following continued leaf growth. Substantial portions of the crop are processed by canning and freezing methods.

Endive (*Cichorium endivia* L.) and Chicory (*Cichorium intybus* L.) ASTERACEAE

Early records indicate that forms of endive and chicory were grown for many centuries and were well-known to the early Egyptians. The probable origin of endive was eastern India, whereas chicory appears to be a native of the Mediterranean region. Both are cool-season crops, somewhat similar in appearance to lettuce. Both endive and chicory initially produce a rosette of leaves that develop from a short stem. When mature, some types form a head, others remain as an open rosette, and some are intermediate between these forms. Additional variation exists with regard to leaf size and shape; some are broad-leaved, others have narrow, divided, and rather crinkled leaves; some have smooth, broad-leaved foliage. Those endives with broad crinkled leaves are commonly known as escarole. Some forms of endive and chicory also develop anthocyanin and have reddish-tinged leaves. Many endive and chicory cultivars are annuals or are grown as annuals, although some important chicories are grown as biennial crops. From 1996 to 1998, annual U.S. fresh market endive and escarole production averaged approximately 58 million pounds. Florida, New Jersey, and Ohio were leading producers, with approximately 50 percent by weight of the production in Florida.

Endive is often confused with chicory, *C. intybus*, since many forms of chicory are similar in growth and appearance to endive. One form of chicory, known as witloof and also as Belgium and/or French endive, is a popular crop in northern Europe (Fig. 24–18). Although associated with

Figure 24–18 Belgium endive, which botanically is a chicory, is a popular northern European vegetable. The compact terminal bud of the chicory plant "chicon" was produced by forcing its growth under controlled temperatures in the absence of light.

the name endive, it is a chicory. It differs from most endive, mainly in how it is produced. Witloof is the product of the second year's growth of the chicory plant. The purpose of the first year's growth is to produce the thick fleshy storage taproot, used to produce the second season's growth. These roots are harvested at the end of the growing season. The plant's foliage, which strongly resembles cos lettuce, is removed, except that the growing point of the stem is retained and remains attached to the root. The roots are subjected during storage to cold-temperature vernalization at 0°C to 5°C (32°F to 41°F) for several weeks or more and then are placed in soil or peat forcing beds or in containers for **hydroponic** culture. The forcing procedure involves placing the roots in an environment that favors growth of the primary bud of the stem, which, were it allowed to continue growth, would become the seed stalk. The growth of the primary bud is kept compact by controlling forcing conditions and procedures. The market-mature compact bud is called a chicon. The temperature in the area of the roots is maintained at a higher level than the air temperature above the roots. This tends to maintain the compaction of the developing bud. Through heating of the soil beds or the hydroponic solutions, along with the use of insulation for regulation of temperature and relative humidity, and, when necessary, the use of cooling equipment, the witloff producer selects and manages the environment needed to initiate bud growth and its development into the compact chicon.

The range of forcing temperatures is 15°C to 20°C (59°F to 68°F) in the root zone, and 12°C to 17°C (54°F to 63°F) in the area of the developing chicons. A compact marketable chicon will develop within 20 to 30 days, depending on cultivar and temperatures. A good-quality typical bud is about 6 to 8 cm in diameter and about twice as long in length, and weighs about 100 to 120 grams.

The production of witloof is a specialty crop that requires considerable expense and labor. Nevertheless, it is popular and preferred over other chicories because of its less intense flavor. For market acceptance the chicon must be white, and therefore to avoid chlorophyll formation, forcing is performed in the absence of light. Belgium, France, and the Netherlands are the major producers of witloof chicory. Other forms of chicory, which are nonforced, are extensively grown in Italy, where the crop is very popular. Endive and chicory have a distinctive bitter taste, which is apparently enjoyed by certain populations and not by others. Persons from the Mediterranean regions have a preference for these salad crops.

Cultural practices for the production of endive and chicory are similar to those used in lettuce production, and these crops face many of the same disease and pest problems. Endive and chicory are more tolerant of cooler temperatures than most lettuce crops, and are often overwintered. However, prolonged exposure to cold temperatures will initiate bolting. To reduce bitterness, the plants may be blanched. This is done by pulling the outer leaves together over the inner leaves or otherwise excluding sunlight for a week or two prior to harvest. In some situations, closer plant spacings can partly contribute to blanching. At harvest the plants are cut below the base of the lowest leaves or at the soil line; outer unmarketable leaves are trimmed, and the plants are packaged. Rapid cooling, by icing, hydrocooling, or vacuum cooling, maintains quality. All forms of endive and chicory are consumed fresh, usually in salads, and also as a cooked potherb.

Kohlrabi (*Brassica oleracea* L., Gonglylodes Group)
BRASSICACEAE

Kohlrabi is a cool-season biennial plant grown for its turnip-like above-ground enlarged stem (Fig. 24–19). Compared with other vegetables, it is a relatively recent food crop. Its delicate flavor compares with that of the turnip, but because it is milder, it should in the future result in increased acceptance. Its cultivation presently is rather limited and has not expanded much beyond the northern European countries. The climatic and cultural requirements of kohlrabi are similar to those of turnips. However, exposure to temperatures of less than 10°C (50°F) are conducive to vernalization and bolting. Optimum growing temperatures are between 15°C to 20°C (59°F to 68°F).

Rapid and continuous growth produces tender and crispy-fleshed kohlrabi. The crop is usually ready for harvest about 60 to 70 days after sowing seed. The plants are harvested when the swollen stems reach a diameter of 5 to 8 cm (2 to 3 in.). Delayed harvest results in further size increase but at the expense of rapid fiber development. The plants are easily pulled from the soil by hand, and the root is trimmed to the base of the swollen stem. The leaves arising from the

Figure 24–19 The enlarged stem of kohlrabi is the part used. Some people prefer kohlrabies to turnips and claim they are milder. The stem on the left is the green type, while the one on the right is the purple type. *Source:* USDA.

kohlrabi are relatively sparse, but are often used to bundle together several kohlrabi that are marketed in bunches, or the leaves are removed and the crop is handled in bulk. Good quality is indicated by kohlrabi that is well colored (both green and purple cultivars are grown) and by its round shape and smooth surface, with limited numbers of attached leaves. Kohlrabi are eaten raw and cooked.

Leek (*Allium ampeloprasum* L., Porrum Group)
AMARYLLIDACEAE

Leek, a biennial plant closely related to the onion, is grown for its sheathlike arrangement of elongated, closely overlapped, blanched leaves (Fig. 24–20). The leaves are fleshy and solid, but generally do not form bulbs as onions do, although under long-day conditions, a slight thickening near the base of the leaves resembling bulbing may occur. Exposure to low temperatures, if of sufficient duration, may result in bolting.

Leek is believed to be a native of the eastern Mediterranean region, where its food use was known as much as 3,000 to 4,000 years ago. While its culture may have spread to the rest of the world, its consumption remains limited. The crop is most popular in the northern European countries, which are the principal producers. In the United States, the crop is largely unrecognized, and its production is very limited. Leek has some cultural differences that are advantageous. The crop neither forms bulbs nor enters a rest period, as does the common onion, but continues growth and can be harvested over a long period of time. This adaptabil-

Figure 24–20 Leek is a relatively large plant that is onion-like in appearance. The lower portions of the plant's leaves are tightly overlapped and resemble, but are not, a thickened oval to round stem.

ity enables it to be grown without the necessity of harvesting at a specific period. Furthermore, it has greater cold-temperature resistance than the onion, but should be harvested before the occurrence of frost, which will cause injury to the surface tissues.

Leeks are grown from seed, and seeds are used to produce transplants. Length of season is a factor in this decision, since leeks require a relatively long growing season to reach marketable size. Leeks grow well on fertile lighter-textured soils, their cultural requirements being similar to those for onions.

The leaves of leeks are blanched to improve market quality, which requires a long straight sheath of leaves that are white for as much of the sheath length as possible. This is achieved by blanching, which is frequently done by progressively drawing soil toward the base of the plant during its growth. This is often further facilitated by starting the transplant or placing the seed in a shallow trench and filling in with soil as the plant grows. A preferred length for the blanched portion is 15 to 20 cm (6 to 8 in.) with the diameter about 2.5 to 5 cm (1 to 2 in.) thick.

Leek is usually planted in the spring; its long growth period precludes harvest until the fall. Harvest can continue from later plantings and in some areas can be extended through the winter and into the early spring, although the possibility of bolting should be considered.

Leek is harvested by pulling the plants from the soil after they have been undercut. The root system of leek, although relatively shallow, is rather extensive. The plants are cleaned by washing. A common problem is the contamination of the crop by soil that falls into the whorl of leaves and between layers of the overlapping leaves, making cleaning difficult. The plants are trimmed of their roots and the top parts of the large coarse outer leaves are cut away to leave the shank or sheath of overlapped leaves. Usually several shanks are tied together in a small bundle for marketing. Considerable amounts of leek are processed. The crop is very versatile and can be used in many foods. It is valued for its flavor contribution to other foods and is highly regarded for the thickening characteristics it provides in the preparation of soups and stews. Leeks have a relatively long storage life. Well-trimmed leeks can be stored for several months under good cold storage conditions.

Lettuce (*Lactuca sativa* L.) ASTERACEAE

Lettuce is a plant with an ancient history. Its origin appears to have been in Asia Minor. There is evidence that forms of lettuce were used in Egypt during the period about 4500 B.C. The Romans grew types of lettuce resembling the present romaine cultivars as early as the beginning of the Christian era, and the crop was well-known in China about the time of the fifth century.

Cultivated lettuce is an annual plant closely related to the common wild or prickly lettuce (*L. serriola*) weed. The many forms of lettuce which can easily intercross are often grouped into several types. These include the usually large and dense heading crisphead cultivars that have brittle-textured foliage which is tightly folded. The outer leaves are dark green and the inner foliage pale and mostly absent of chlorophyll. The butterhead cultivars, whose heads are smaller, are relatively soft and less dense, and comprised of broad, soft, smooth or slightly crumpled oily textured leaves. The romaine or cos cultivars have elongated coarse-textured leaves that form a loaf-shaped, semi-compact, low-density head. Outer leaves are darker green, coarse textured, with heavy ribs. Inner leaves are smaller and lighter in color, with interior leaves nearly absent of chlorophyll. The leaf lettuce cultivars, which are leafy but loose, are nonheading cultivars of varied color, leaf types, and textures (Fig. 24–21). An additional lettuce type, known by several names, such as celtuce, asparagus, and stem lettuce, is not widely produced but is most popular in the Orient. It is grown for its elongated, thick stems that are peeled and used as a cooked vegetable. The leaf blades are of lesser significance compared to the stem.

A

B

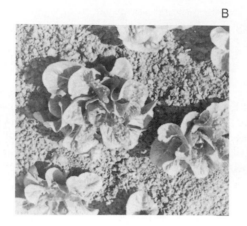

C

D

Figure 24–21   Lettuce is a widely grown salad crop where climatic conditions are favorable. A heading type (A), represented by crispheaded cultivars, is the most popular in the United States. These cultivars are well adapted to outdoor culture, although some are produced in glasshouses. These are also well adapted for long distance shipment. Another type that also forms a head, although considerably less compact, is represented by the butterhead cultivars (B). This type is more popular in Europe, and is less adapted to outdoor culture or to long distance shipments, and is often grown in glasshouses. The romaine or cos types (C) develop an upright, cylindrical form that exhibits an incomplete overlapping of its leaves and partly resembles a head. This type is usually produced outdoors. The nonheading types (D) have an open leafy growth and do not exhibit heading characteristics. These are adapted to indoor and outdoor culture, and they may be shipped longer distances if postharvest handling is well managed. *Source:* Ferry-Morse Seed Company.

Head, leaf, romaine, and other types of lettuce are produced and consumed in great quantities in the United States. From 1996 to 1998, annual U.S. head, leaf, and romaine lettuce production averaged nearly 7, 1, and 1 billion pounds, respectively. California and Arizona together accounted for most of the lettuce crops. However, significant amounts are produced in Colorado (head), Ohio (leaf and romaine), and Florida (leaf and romaine). More recently, romaine lettuce production has increased in South Carolina and Georgia. Much of the lettuce production in Europe occurs in controlled environments (i.e., greenhouses) although outdoor production is increasing.

Lettuce is a cool-season crop and grows best within a temperature range of 12°C to 20°C (55°F to 68°F). Summer production is therefore restricted to areas that have relatively low temperatures, such as those provided by higher elevations (Colorado), or temperatures moderated by large bodies of water (coastal California). Winter production occurs in the desert valleys of California and Arizona, and in southern locations where winter climates are favorable for growth. High temperatures, those above 27°C (80°F), are detrimental, as they interfere with head development and plant edible quality and tend to promote premature seed stalk development. Some slower-bolting cultivars are used if the crop is grown

during warmer temperatures to avoid bolting. Lettuce seedlings and mature plants will tolerate slight freezing, although the epidermal tissues of the outermost leaves may exhibit damage. However, since these leaves are trimmed before marketing, such damage is generally not significant.

Plant breeding programs have been responsible for the introduction of lettuce cultivars having pronounced adaptability to anticipated temperature conditions for specific locations and seasonal periods. It is the vegetative portion of the plant that is of market interest, and vegetative growth is highly responsive to even slight changes in temperature. Hence, the development of cultivars with responsiveness to slight temperature is useful. In addition to this adaptability, other factors that influence cultivar selection consider disease resistance, foliar characteristics of texture, smoothness or frilled, leaf shape, width, length, degree of heading, and compaction, head size, and shape, color, stem size, and bolting resistance, disease resistance being one of the highest priorities.

Lettuce grows well on a wide variety of soils, provided climatic requirements are met. In the production areas of Florida, Michigan, New York, and Wisconsin muck soils are used, whereas mineral soils are used in the southwestern states. Temperatures prevailing during growth are more important than soil types or most other factors. Although fertile soils are desirable, supplemental fertilization allows production on less fertile soils. Soils with high water-holding capacity and good drainage are important for good root growth and plant performance.

Lettuce seeds are small, and to have uniform emergence and growth, a well-prepared smooth seedbed is important. Considerable effort is made in this regard. The majority of plantings are made by direct seeding. Usually an excess of seed is sown, with surplus plants thinned out. To provide greater precision in the planting procedure, it is now common to coat the seed to produce seed pellets of uniform size and shape that will facilitate seed handling by planter equipment. Uniform size and shape of seed pellets allow the use of planters that can accurately handle the coated seed and provide accurate depth of placement and sufficient separation between individual seeds. It is important to have sufficient separation between developing plants, or else heading is hindered and closely spaced plants do not develop marketable heads.

The use of transplants in outdoor lettuce production is often limited because of the high costs for this practice, although for some situations the advantages of transplant use are justified. Lettuce grown in glasshouses or other protective structures is commonly transplanted.

For outdoor production, plant row spacings for the crispheaded cultivars are commonly 50 to 60 cm (20 to 24 in.) apart, with in-row spacings of 25 to 30 cm (10 to 12 in.) (Fig. 24–22). Plant spacings are slightly closer for the production of other lettuce types, such as butterhead, cos, and leaf cultivars, which produce smaller plants. In the southwestern states where irrigation is needed, the crop is often grown on raised beds, which offer other advantages than that of facilitating irrigation. Surface-applied water is the common irrigation method (Fig. 24–23), although sprinkler irrigation is widely used in the germination of the seeded crop and during its early (preheading) growth. Irrigation requirements range between 30 to 60 cm (12 to 24 in.) of water for a crop. In areas where lettuce is grown on muck soils, irrigation may also be supplied by subbing water up through the soil profile.

Lettuce is a moderately heavy consumer of nutrients. Rapid early growth is important to establish a large, vigorous plant prior to heading. Fertilizer recommendations are dependent on the specific location and on seasonal conditions. For most situations, a total of 100 to 200 kg/ha (90 to 180 lb/ac) of nitrogen is sufficient. A portion of the nitrogen fertilizer, usually one-third to one-half, is applied prior to planting and the remainder is applied as side dressings during the crop's development. About 60 to 100 kg/ha (54 to 90 lb/ac) of P_2O_5 and a similar quantity of K_2O supply other major nutrient elements, although potassium is often not required in the southwestern U.S. production areas.

Lettuce is attacked by a wide variety of plant pests. Important diseases include pythium-induced damping-off, scle-

Figure 24–22 In the Salinas and Imperial valleys of California a common cultural practice is the planting of two rows of head lettuce on each raised bed. Normally the beds are 100 cm (40 in.) apart and the plants are thinned to 25 to 30 cm (10 to 12 in.) apart within the row. This field of lettuce is at the rosette growth stage and beginning head formation.

Figure 24–23 A planting of red leaf lettuce being irrigated by the furrow method. Water is siphoned from the head ditch. A new dam has been installed below the one being used. After this section of field has been irrigated, the upper dam will be removed, allowing water to move to the next section. The siphons will be repositioned for the new area to be irrigated.

rotinia drop, corky root, downy mildew, powdery mildew, *Rhizoctonia, Botrytis,* and *Anthracnose.* Serious virus and viruslike diseases are lettuce mosaic virus, lettuce infectious yellows, beet western yellows, spotted wilt, aster yellows, and big vein. Abiotic disorders such as tipburn and russet spotting are large contributors to crop loss. In some situations plant resistance is available; a clean seed program to detect infected seed is effective in the control of lettuce mosaic virus (LMV).

Aphids are one of the most troublesome insect pests because of their direct physical injury and their vector role in virus transmission. Other serious damage is caused by flea beetles, crickets, spring tails, leaf hoppers, several of the lepidopterous larvae, and recently the silver-leaf whitefly.

Figure 24–24 While many different machines and mechanical aids have been tested for the harvest of head lettuce, it remains essentially a hand harvest crew operation. Cardboard boxes are distributed in the field ahead of the cutters. The heads, if determined to be mature, are cut at the soil surface, using a jabbing motion with a sharp broad-tip knife. The cutters place the heads, bottomside up, in the middle of the bed after some trimming. Some wrapper leaves are retained to protect the head from damage. Packer members of the harvest crew pack the designated number of heads, usually 24, into the box. The first layer is placed with the butt (stem) ends down, and the second layer with the butt ends up. Another crew worker sprays water over the packed heads to wash off the latex sap coming from the stem, which if not removed turns an undesirable pink color. Other workers close the cartons and load them onto trucks or trailers that transport the lettuce to a vacuum cooler. *Source:* University of California Cooperative Extension.

Figure 24–25 Individual film wrapping of lettuce reduces the amount and weight of wrapper leaves in contrast to those commonly retained in "naked" or nonwrapped lettuce. To facilitate the wrapping procedure, labor aid conveyors are used. The lettuce is cut, trimmed essentially free of wrapper leaves, and placed onto the conveyor, where each head is separately wrapped in a plastic film. Market receivers often prefer wrapped lettuce because its shelflife is prolonged. *Source:* University of California Cooperative Extension.

Harvest of crisphead lettuce is made when the developed head is relatively firm, but can still be compressed. Plants that develop to a very firm condition will be very dense, bitter tasting, and their marketing appearance will suffer. Leaf or romaine lettuce harvest begins when the leaves approach maximum size but before they become tough or bitter. All types of lettuce are harvested by cutting through the plant stem near the soil surface. Soiled or damaged leaves are removed before packing. Some outer leaves are retained to provide protection to the product. These are removed at a later time. Most lettuce is field packed into cardboard cartons (Fig. 24–24). Shed packing was discontinued many years ago when vacuum cooling was developed as an alternative cooling method to icing. Packed cartons are transported to a central vacuum cooler, where the product temperature is greatly reduced before shipment to markets.

Hand labor is extensively used in the harvesting of lettuce, although various mechanical aids are used in harvesting, particularly for lettuce that is marketed in a film overwrapping (Fig. 24–25), and for lettuce that is bulk handled and intended for processing. Such lettuce will be precut or shredded and packaged for use by restaurants and other institutional food service providers. Although prototypes of several mechanical harvesters were developed and tested, none presently operate. Such equipment used pressure resistance or X-ray emission and/or emissions from a radioactive source to pass through or to be absorbed by the lettuce head in order to determine head density. Head density correlated with the amount of radiation that was absorbed, and the greater the absorbence, the greater the density and the maturity. Presently, maturity is determined by the experienced hand touch of human harvesters. The storage or shelf life of the various types of lettuce is different; the crisphead and cos types have a longer shelf life, up to several weeks if maintained under cold temperatures, 2°C to 4°C (35°F to 40°F), and high humidity. Butterhead and leaf types of lettuce have a considerably shorter shelf life.

Mustard (*Brassica juncea* L., *B. hirta* Moench., *B. Nigra* Koch, *B. carinata*) BRASSICACEAE

Mustards are cool-season annual plants. Their abundant foliage develops in a rosette pattern on a relatively short stem. Leaf blades are usually broad, with some types having smooth-textured leaves and others extremely curled leaves (Figs. 24–26 and 24–27). Characteristically the foliage is strongly flavored, although a wide range of pungency levels occurs. It is this flavoring characteristic that makes the various mustards popular vegetables for use in salads and as cooked greens. Mustard seeds also vary in flavor and pungency and find use in culinary condiments, and also as a vegetable oil. The origin of the mustards *B. hirta* and *B. nigra* is respectively the eastern Mediterranean and Asia Minor regions, with *B. juncea* from the central Asia-Himalaya area, and *B. carinata* from Ethiopia. Mustard has been grown in most parts of the world since Biblical times.

When grown for its foliage, the plant performs best in cooler climates. For seed production long, warmer days are

Figure 24–26 Tenderleaf mustard is a smooth-leaved cultivar. It produces large tender leaves and is high yielding. *Source:* Ferry-Morse Seed Company.

Figure 24–27 Southern Giant mustard produces a large plant with crinkled leaves. To many people mustard leaves add zest to a salad and other foods. *Source:* Ferry-Morse Seed Company.

preferred. Cultural and soil requirements are similar to those for spinach or cabbage. The crop grows well on fertile soils over a wide range of soil textures, and within a pH range from 5.5 to 7.5. Plants are always grown from seed, since the low cash value of the crop seldom can justify the use of transplants. Seed is sown early in the spring for a late spring/early summer crop, or planted in late summer or early fall for a late fall or winter harvest. Common row widths are 30 to 60 cm (12 to 24 in.), with in-row plant spacings of 20 to 25 cm (8 to 10 in.). Fertilizer nitrogen is beneficial, since

vegetative growth is the marketed product, and commonly 50 to 100 kg/ha (45 to 90 lb/ac) of nitrogen is supplied.

Cultivar characteristics are concerned with leaf size, shape, texture, color, growth habit—erect to prostrate—and resistance to premature bolting. Many of the same diseases and pests of other crucifers are problems for mustard producers.

Plants can be harvested as soon as six weeks after sowing seed. In some situations only the well-developed outer leaves are harvested and are tied together into bunches for marketing. In other situations the entire plant is harvested. Considerable volumes of mustard are processed either by canning or freezing methods. The fresh market product must be well cooled in order to retain market quality; otherwise, the foliage will wilt and exhibit a loss of green color. Properly cooled, a shelf life of 10 to 14 days is achievable.

Parsley (*Petroselinum crispum* Mill.) APIACEAE

Parsley, a biennial plant native to the Mediterranean area, has been known as a cultivated crop for more than 2,000 years. The crop, a close relative of celery, is widely used as an herb or garnish and in flavoring various foods. The plant produces a rosette of divided leaves on a short stem. Distinctive foliage types are produced, such as the plain or single-leaf type used mostly for food flavoring, the double-leaf type and the moss or triple-curled leaf type preferred for garnishes (Fig. 24–28). An additional parsley type, known as 'Hamburg Rooted', is grown for its foliage as well as for its slightly thickened edible root. Parsley is a highly nutritious crop and an excellent source of vitamin A and C, but is not consumed in significant quantities to make much impact on human diets.

Parsley seed is small and slow to germinate, occasionally making the crop difficult to establish. The leaf-type cultivars require about 75 days to mature, while the 'Hamburg Rooted' type requires several weeks longer. To hasten germination and uniform emergence, seeds are often presoaked overnight before planting. The crop has good tolerance to cold temperatures and can be sown and/or transplanted relatively early in the spring. Large commercial plantings are usually seeded directly into the field rather than transplanted. Fertile, light-textured soils with a high water-holding capacity are preferred. Crop nutrition requirements are similar to those of carrots. Plant spacings commonly used are 10 to 20 cm (4 to 8 in.) between plants within rows that are 30 to 60 cm (12 to 24 in.) apart. Weed control is particularly important for the crop if it is intended for processing, because weed contamination is difficult to remove from the processed product.

Harvesting begins when the leaves reach the desirable size. The foliage is usually cut either mechanically or by hand above the soil line and sufficiently high to allow for

Figure 24–28   Parsley, in the same botanical family as celery, is a delicious addition to stews and soups. It is often used as a decorative garnish. This is a triple-curled leaf type preferred for garnish use. Other types are widely used in food flavoring. *Source:* Ferry-Morse Seed Company.

continued growth of the plant for subsequent harvesting. The crop can be harvested several times. The cut leaves may be handled in bulk, but for some fresh market use, the leaves are tied into small bunches. Considerable volumes of parsley are processed by dehydration. When produced for this purpose, the crop is usually mechanically harvested and the product is handled in bulk. With good low-temperature maintenance and a high relative humidity, parsley has a reasonable shelf life of up to 3 to 4 weeks.

Rhubarb (*Rheum rhabarbarum* L.) POLYGONACEAE

Rhubarb is a large-leafed plant grown for its thick leaf petioles, which are used as a desert vegetable in pie fillings or sauces. The plant is a cool-season perennial, and one of the first vegetables ready for consumption in the spring. Plant growth is kept inactive by low winter temperatures. By using plant forcing procedures to stimulate rapid growth, a crop can be produced during the late winter–early spring. Temperature control is provided by the use of protective structures or shelters, and if needed, heat can be supplied. In mild climates, warm early spring temperatures will encourage early production in field conditions. The crop can also be managed to produce a crop during the fall and winter, in which case the crop is kept dormant through the restriction

of moisture during the summer. The plant does not grow well at temperatures above 25°C (77°F), and summer-produced rhubarb is usually of poor quality.

Rhubarb is grown on a wide variety of soils, but growth benefits from fertile, well-drained soils. The plant is a heavy consumer of mineral nutrients, and fertilizer applications of 100 to 170 kg/ha (90 to 150 lb/ac) of nitrogen are generally supplied.

Rhubarb is usually propagated vegetatively by dividing portions of the stem, collectively called the crown, that develop from the previous season's growth. This is often preferred to the use of seed, which is not used because the resulting plants are not uniform or true to the parental type. Additionally, the seedlings must be grown for a year before the plant is large enough to begin to be harvested. Carefully selected, vigorous crowns are divided either early in the spring or late winter. The crowns are cut into pieces, each of which should have at least two or three strong buds. These crown pieces are planted about 5 to 10 cm (2 to 4 in.) deep and spaced equidistant about 90 to 150 cm (3 to 5 ft) apart within and between the rows. After several years of growth these crown pieces become larger and, because of self-competition, they become less productive. They in turn are subdivided and used for the propagation of additional plants.

In moderate climates, mature plants can be overwintered in the field if frosts are not severe or frequent or if some protection from freezing is provided. In cold climates, mature plants remaining in the field must be protected from freezing. Occasionally the plants are lifted out of the field and stored. Both outdoor and indoor storages are used. With either kind of storage the plants are maintained at a low temperature but freezing is avoided. These will later be replanted into the field, or may be used to produce a forcing crop.

The attractive deep red coloration of the rhubarb petiole is an important market feature. Rhubarb is forced in the dark to avoid chlorophyll development, which would interfere with the expression of the red pigment.

The most serious disease of rhubarb is brown rot. It causes rotting at the base of the petioles. Other troublesome diseases are fusarium root rot and rhubarb leaf spot.

Rhubarb is harvested by pulling the petioles from the plant and is done by moving the petiole to the side and pulling upward. This provides a fairly clean separation. The best market-quality rhubarb petioles are deep red, long, thick, fleshy, and tender. *The leaf blades are poisonous and must not be eaten.* They are removed and generally are not attached when the petioles are marketed. Rhubarb is usually presented in the market as a group of petioles tied together into a small bundle.

In areas with long growing seasons, light harvesting may be performed during the plant's second year of growth. Otherwise full harvesting is delayed until the third year's growth. The plant is managed in many respects like asparagus in that sufficient foliage growth must be achieved each season in order to resupply the nutritional reserves of the crown to support the growth to follow. Fields are usually replanted after six to eight years because of excessive crowding of the plants.

Rhubarb requires careful postharvest management because the petioles are sensitive to dehydration and easily wilt when exposed to high temperatures and low humidity.

Spinach (*Spinacia oleracea* L.) CHENOPODIACEAE

Spinach is a popular and nutritious vegetable frequently used raw in salads and as a cooked potherb. Spinach is an annual plant, dependent upon wind for pollination of its flowers. The plant is unusual in that it exhibits dioecious characteristics, male or female flowers on separate plants, and also is infrequently **monoecious,** with varying proportions of male and female flowers on the same plant. The variation extends from plants that are considered extreme males, through vegetative males, to female plants. The greater the proportion of male flowers, the greater the tendency is to flower earlier. The male plants have relatively little vegetative growth and die soon after flowering; therefore, they contribute very little to crop production. The preferred types are the female and vegetative male plants, which are much slower to flower. These plant types are able to produce an abundant amount of foliage, which can usually be harvested before flowering occurs.

Spinach is native to central Asia, most likely in the area of Iran. The absence of records suggests that its domestication was fairly recent. Although its growth is widely distributed, spinach is not an intensively grown crop. From 1996 to 1998, annual U.S. spinach production averaged approximately 527 million pounds. California, New Jersey, and Texas together accounted for more than 90 percent by weight of the commercial U.S. fresh market spinach crops from 1996 to 1998, but significant production also takes place in Arkansas, Oklahoma, and Colorado. Spinach is also produced for use in processed products. From 1996 to 1998, annual U.S. processing spinach production averaged approximately 254 million pounds. Texas led U.S. processing spinach production with an average annual production of approximately 105 million pounds.

Spinach is a cool-season, long-day plant that produces its best vegetative growth under cool 15°C to 18°C (59°F to 64°F) temperatures and short daylength. Long days, especially if coupled with higher temperatures, above 25°C (77°F), cause the plants to bolt and flower, which is detrimental to production. Spinach plants have some frost tolerance and in many areas are grown as an overwintering crop.

Climate is important in deciding planting dates to ensure that the crop will not mature during periods when high temperatures or long days occur. Low temperatures can partially offset the bolting influence of long daylength. High temperatures intensify the bolting response. While producers try not to exceed the critical daylength, slightly warmer temperatures do accelerate vegetative growth, which is the primary goal. Spinach grows well on a wide variety of soils. During most winter conditions light-textured soils are preferred as they generally afford better drainage.

Spinach is grown from seed; the high plant populations required preclude the use of transplants. Seeds are sown in narrow rows, ranging from 20 to 50 cm (8 to 20 in.). Plants within the row are relatively close, 5 to 10 cm (2 to 4 in.). The crop grown for processing will be at narrow spacings and mechanically harvested. The fresh market crop is grown at wider spacings in order to accommodate hand harvest operations, although some fresh market spinach is also harvested by machine.

The wide variety of soil types and growing periods makes it difficult to generalize about fertilizer use. A reasonable target might be 75 to 150 kg/ha (67 to 135 lb/ac) of nitrogen and similar amounts of phosphorous and potassium. High nitrogen usage and the form of nitrogen applied should be considered in regard to the accumulation of nitrates in the foliage, which in the human digestive tract are converted into nitrites and present a health concern. The use of ammoniated sources of nitrogen fertilizer are useful in reducing nitrate accumulation. Additionally, spinach cultivars differ in the level of nitrate they accumulate. Spinach also contains relatively high levels of oxalic acid.

Considerable attention is given to having good weed control especially for the processed crop, which is mechanically harvested and for which the removal of weed plants is very difficult. Herbicides are widely used, and in some cases weed-free fields are reserved for the production of spinach for processing.

Spinach is shallow-rooted and requires a constant and uniform water supply for productive growth of high quality. Therefore, when rainfall is inadequate, irrigation may be necessary. The crop's short growth period (50 to 80 days) and growth during cooler temperatures does not make a large demand for moisture, and 15 to 30 cm (6 to 12 in.) of water is usually adequate.

Types of spinach are distinguished by their foliage characteristics. There are cultivars that are smooth-leaved, others that are savoy-leaved, and some having intermediate leaf textures. Other important cultivar characteristics are leaf color, leaf blade size, leaf petiole length, growth habit—prostrate or upright—seasonal adaptation, bolting, and disease resistance.

Spinach is susceptible to many diseases; downy mildew (blue mold), cucumber mosaic virus, curly top virus, beet yellows virus, pythium damping off, fusarium, white rust, and leaf spot are the most serious. For some diseases resistance is available. Aphids, leaf miners, and mites are other important pests. Spinach is also highly sensitive to air pollution damage.

Spinach is grown for its leaves, which are harvested when the majority of the leaves are nearly fully expanded and while the plant is in the rosette stage. Harvest of cultivars exhibiting upright growth is easier than the harvest of cultivars having prostrate growth habits. Close plant spacings are beneficial to upright leaf growth. The crop's usage determines the cultivar and harvest method. The flat or smooth-surfaced leaf types are preferred for processing use. These are easier to wash free of soil grit. Fresh market preference is for the savoy types. The semi-savory leaf types are adaptable for both fresh and processing use.

The mechanical spinach harvester was one of the earliest machinery developments for vegetable harvesting (Fig. 24–29). This equipment is essentially a mowing machine. The spinach is cut above the stem, which can continue to grow, possibly allowing for a second harvest. The cutter blade is raised high enough to minimize harvesting of excessive petiole length, but low enough to harvest the leaf blades.

Spinach for fresh market can also be cut with mechanical equipment, in which case the harvested leaves are handled in bulk. Another method is to undercut the plants,

cutting the taproot just below the surface, and then tying together several plants into a small bunch (Fig. 24–30). For fresh market use, the spinach is usually harvested before the foliage has reached its maximum size. The processed crop is harvested when maximum foliage growth is obtained. Spinach is processed by freezing and canning. Postharvest handling of fresh spinach is highly dependent upon maintaining low temperatures, 0°C to 2°C (32°F to 35°F), and high humidity. Shelf life nevertheless is a relatively short 7 to 10 days.

Swiss Chard (*Beta vulgaris* L., Cicla Group)
CHENOPODIACEAE

Swiss chard or chard is a biennial plant which is very similar to the table beet in developing large, crisp, fleshy leaf stalks and large leaves (Fig. 24–31), but differs in not producing the enlarged tap root.

Swiss chard is a widely adapted, cool-season crop that tolerates hot weather better than spinach. It is a well-known crop and one of the easiest to grow. Most often the crop is started from seed. The most favorable soils are fertile sandy loams or loams with good drainage. Row spacings are wider than those for spinach because the plants at maturity are larger. Row widths commonly are 45 to 60 cm (18 to 24 in.) with in-row spacings of about 10 to 15 cm (4 to 6 in.). Plant nutrition requirements are not excessive, with 60 to 120

Figure 24–29 The mechanical harvesting of spinach for processing was an early and relatively simple development. The crop is quickly and efficiently harvested and handled in bulk. The machine mower bar cuts the spinach, which is conveyed into large trucks or trailers for transport to the processors' facility. Often a second mechanical harvest is possible from the regrowth of the plants after the first harvest.

Figure 24–30 Hand harvesting and bunching of fresh-market spinach.

Figure 24–31 Swiss chard has large, crisp, white (in some varieties red) petioles with large leaves. The crop is grown for its tender leaves and petioles; the leaves are cooked and served like spinach and the petioles can be cooked and served like asparagus. Swiss chard is popular because it can be cultivated easily and is productive and more heat-tolerant than spinach. *Source:* USDA.

kg/ha (54 to 108 lb/ac) of nitrogen and similar levels of P$_2$O$_5$ and K$_2$O generally adequate.

Other cultural procedures are similar to those for beets. Chard is also susceptible to the same diseases and insect pests as beets. Cultivar choices are limited, with the major variation in foliage and leaf rib coloration. The crop is harvested by cutting individual petioles at the base of the plant with a sharp knife, being careful to avoid injury to younger leaves or the growing point. This procedure allows for several harvests as additional leaves continue to develop and en-

large. Harvest of the entire plant is also done when the crop is grown for processing. When grown for processing, the crop is often harvested mechanically with equipment similar to that used to harvest spinach. Like spinach, by cutting above the stem's growing point regrowth is permitted and a second harvest may be possible. For fresh market, six to ten leaves are tied together to form a bunch for marketing. Chard is processed by canning and freezing. Postharvest management of fresh market chard is similar to that for spinach, although shelf life is slightly longer.

REFERENCES AND SUPPLEMENTARY READING

DECOTEAU, D. R. 2000. *Vegetable crops.* Upper Saddle River, NJ: Prentice Hall.

MAYNARD, D. N. 1997. *Knott's handbook for vegetable growers.* New York: John Wiley.

CHAPTER

Vegetable Crops Grown for Underground Parts

Underground plant parts used for food consist of storage roots, bulbs, rhizomes, corms, and tubers. Many plants store part of the excess photosynthate (food) produced during their growth in these plant parts. Rhizomes, corms, and tubers are specialized stems that have buds and the ability to develop roots, leaves, and other stems, and are able to fully regenerate into new plants. Tubers are thickened stems that exhibit compressed or small internodal elongation. The "eyes" of the potato tuber are buds that develop into stems, which can also develop into fully functional plants. Bulbs like the onion are comprised of specialized storage leaves attached to a short compressed stem. The base of such leaves enlarge because of the accumulation of food synthesized by the plant that is stored in these leaves. This enlargement produces the structure known as a bulb. The sweet potato, carrot, beet, and turnip are examples of vegetables where the enlarged storage root serves for the accumulation of plant food reserves.

Underground plant parts have worldwide importance as a food source. These crops are staples, extensively used in the food supply of all nations, whether developed or developing. Characteristically these plants are efficient converters of solar energy into chemical energy, principally as carbohydrates, but also as proteins. For example, the Irish potato plant, using an equivalent land area, produces carbohydrates second to sugarcane and protein second to soybeans. Other plants grown for their edible underground parts are also efficient in their synthesis and storage of food products. This chapter discusses some important crop plants whose underground parts are used for food.

| TABLE 25-1 | Production by Region in 1999 for Five Commodities Grown for Underground Parts. The Three Leading Regions Are Shown |

Commodity	Region	Production (MT)	% of World Production
Carrots	Asia	6,765,210	37
	Europe	6,540,267	35
	N & C Amer	2,885,263	16
Garlic	Asia	7,549,996	81
	Europe	743,047	8
	N & C Amer	388,935	4
Dry Onion	Asia	27,029,022	62
	Europe	6,861,273	16
	N & C Amer	3,389,261	8
Potatoes	Europe	135,542,728	46
	Asia	105,046,171	36
	N & C Amer	28,010,509	10
Sweet Potato	Asia	122,863,617	91
	Africa	9,333,492	7
	S Amer	1,380,394	1

Source: World Agricultural Information Centre, Food and Agricultural Organization of the United Nations, FAOSTAT On-Line (http://apps.fao.org/lim500/nph-wrap.pl?Production.Crops.Primary& Domain=SUA&servlet=1).
Note: Regions included in analysis were Africa, Asia, Caribbean, Europe, North and Central America, Oceania, and South America.

| TABLE 25-2 | Production by Country in 1999 for Five Commodities Grown for Underground Parts. The Leading Country in the Three Leading Regions, as Indicated in Table 25-1, is Shown First |

Commodity	Country/ Region	% of Region Production	% of World Production
Carrots	China/Asia	66	24
	Russia/Europe	18	6
	U.S.A./ N & C Amer	76	12
Garlic	China/Asia	79	64
	Spain/Europe	25	2
	U.S.A./ N & C Amer	80	3
Dry Onion	China/Asia	42	26
	Russia/Europe	15	2
	U.S.A./ N & C Amer	88	7
Potatoes	Russia/Europe	23	11
	China/Asia	53	19
	U.S.A./ N & C Amer	77	7
Sweet Potato	China/Asia	94	85
	Uganda/Africa	27	2
	Brazil/S Amer	47	<1

Source: World Agricultural Information Centre, Food and Agricultural Organization of the United Nations, FAOSTAT On-Line (http://apps.fao.org/lim500/nph-wrap.pl?Production.Corps.Primary& Domain=SUA&servlet=1).
Note: Regions included in analysis were Africa, Asia, Caribbean, Europe, North and Central America, Oceania, and South America.

STATISTICAL OVERVIEW OF WORLD PRODUCTION

The production of a number of commodities discussed in this chapter (carrots, garlic, dry onion, potato, and sweet potato) accounted for approximately 46 percent by weight of the total vegetable and melon production in 1999, based on UN-FAO data. Asia was the leading producer of each of these commodities, except potatoes, typically followed by Europe and North and Central America (Table 25–1). Of the commodities listed in Table 25–1, Asia, Europe, and North and Central America averaged 61, 26, and 10 percent by weight, respectively, of the world's production in 1999.

In 1999 China was the leading producer of each of the commodities shown in Tables 25–1 and 25–2, except potatoes. China produced nearly 85 percent by weight of the world's sweet potato, 64 percent of the garlic, and 26 percent of the dry onion crops in 1999 (Table 25–2). The United States produced 7 percent by weight of the world's dry onion and potato, 3 percent of the garlic, and less than 1 percent of the sweet potato crops in 1999, respectively.

Beets (*Beta vulgaris* L., Crassa Group J. Helm)
CHENOPODIACEAE

Table beets, native to Europe and North Africa, are cool-season biennials grown mainly for their roots. However, the tender young tops are sometimes used as greens. Beets were first used for food about the third century A.D.

From 1996 to 1998, annual U.S. production of beets used primarily in processed products averaged nearly 237 million pounds. Wisconsin and New York together typically accounted for approximately 72 percent by weight of the commercial U.S. processing beet crops during the same time period. Significant production also occurs in California, Minnesota, Oregon, Ohio, Michigan, and New Jersey. Texas leads in fresh market beet production.

Beets are rather hardy and tolerate some freezing, but they also grow well in warm weather. Excessively hot weather, however, causes zoning—the appearance of alternating light and dark red concentric circles in the root. Beet seeds germinate within a broad soil temperature range from 5°C to 25°C (41°F to 77°F).

Beets are unusual in that when seeds are produced, several will be clustered together in what is called a seedball. These seedballs, when planted to grow the crop, produce several seedlings, unfortunately in close proximity, which complicates thinning. The introduction of a monogerm (single seed) characteristic into new cultivars will be a benefit to more accurate stand establishment. Sowings are made into rows about 45 to 60 cm (18 to 24 in.) apart. Thinning is practiced for fresh market production to achieve in-row spacings of about 5 to 8 cm (2 to 3 in.). Beets grown for processing are seldom thinned, and in fact high populations are desirable, because processors often have a preference for small-size beet roots.

During the first growing season the plant normally produces a thickened root and a rosette of leaves. However, temperatures of 5°C to 10°C (41°F to 50°F) for at least 15 days can result in seed stalk formation even with fairly young plants.

Beets are produced on a wide variety of soils, but deep, well-drained loams or sandy soils are best. The best seedbeds are well prepared by plowing 15 to 20 cm (6 to 8 in.) deep followed by sufficient disking and harrowing to break up clods.

Best-quality beets are produced when environmental conditions encourage uninterrupted rapid growth. Beets generally respond to fertilizers. The kind and amount vary with soil type, inherent fertility, and previous fertility program. In general, commercial growers use mixed fertilizers and apply 112 to 225 kg/ha (100 to 200 lb/ac) of nitrogen, phosphoric acid, and potash. Some farmers prefer to sidedress half their nitrogen fertilizer when the crop is about half grown. A physiological symptom known as internal black spot caused by boron deficiency is prevented with applications of borax.

Weeds drastically decrease yields and should be controlled by mechanical cultivation and herbicides.

Important characteristics in cultivar selection include days to maturity, root size and shape, foliage size, exterior surface smoothness, interior color, and degree of zoning.

Cercospora leaf spot is a widespread disease of beets, especially during wet periods. Other important diseases are *Pythium* and *Rhizoctonia.*

The larva of the beet leaf miner is the most injurious insect pest, and sugar beet cyst nematodes cause considerable damage.

Fresh market beets are often marketed with their tops left on, graded according to size, and bunched. Large commercial fresh market plantings marketed with tops removed are dug mechanically. This latter method lengthens shelf life by reducing water loss from the beets. Processed beets are handled in a similar manner. They are topped and dug mechanically with a harvester, transported to a packing shed, where they are washed, and sent to the processor or market. Beets store well at a temperature of 0°C (32°F) and 90 percent relative humidity.

Carrots (*Daucus carota* L.) APIACEAE

The carrot, a cool-season crop, is grown for its fleshy storage root. It is a native of Europe and parts of Asia. While the carrot is mentioned in some of the ancient Greek writings, it is

likely that the crop, as we know it, is relatively new due to improvements by plant breeding. The carrot was introduced in North and South America in the early 1600s. The wild type is an annual, but cultivated types are biennials. The root is slender and varies in length from 5 to 25 cm (2 to 10 in.) when harvested.

Carrots are harvested somewhere in the United States every month of the year. From 1996 to 1998, annual U.S. fresh market carrot production averaged nearly 4 billion pounds. California, Colorado, and Michigan together accounted for nearly 90 percent by weight of the commercial U.S. fresh market carrot crops from 1996 to 1998, but significant production also takes place in Washington, Texas, and Florida. Carrots are also produced for use in processed products, including baby foods, soups, and frozen vegetable mixes. From 1996 to 1998, annual U.S. processing carrot production averaged slightly more than 1 billion pounds. Washington, California, and Wisconsin together accounted for an average of 70 percent by weight of the commercial U.S. processing carrot crops from 1996 to 1998, with significant production in Texas, Michigan, and Oregon.

Carrots grow best at temperatures ranging from 15°C to 20°C (59°F to 68°F). In the seedling stage the plants are sensitive to both high and low temperatures, and mild freezes at harvest cause some leaf damage. Many young plants are injured or killed on hot, sunny days. Long periods of hot weather can cause objectionable strong flavor and coarse root texture. Temperatures below 10°C (50°F) tend to cause longer, more slender, and paler roots. Color develops best at 15°C (59°F).

Carrots are often divided into types according to shape and length of root, with each type represented by several cultivars (Fig. 25–1).

The most important determination in cultivar selection is the intended use of the carrot. For both fresh market and processing purposes, several root types are preferred for different markets and different processed products. Within the selected types, root diameter and length, shape, color—interior and exterior—surface smoothness, taper, and root stubbing are important criteria.

Carrot seeds are small and germination is variable and slow. The tiny seedlings require some time to become established. The seeds are planted in a firm, well-prepared seedbed free of clods. The best soils do not puddle or crust severely when they are irrigated or rained on.

Tillage is done primarily to control weeds, and the number of operations is kept to a minimum.

On mineral soils, 55 to 110 kg/ha (49 to 98 lb/ac) of nitrogen is ample. If additional nitrogen is needed, it is added as a sidedressing at the rate of 35 to 50 kg/ha (31 to 45 lb/ac). On peat or muck soils, less nitrogen is needed. Soils low in phosphorus often require 55 to 110 kg/ha (49 to 98 lb/ac) phosphoric acid. Potash is applied at rates of 55 to 110 kg/ha (49 to 98 lb/ac) on soils where the need for potassium is indicated. In the western and southwestern United States potassium is rarely needed.

Figure 25–1 The Imperator type of carrot is characteristically long and slightly tapered. It is a good multiple-purpose carrot and is widely used for home garden and commercial fresh-market and processing purposes. It has a bright orange color, is high yielding, and is well adapted to long distance shipping and storage. *Source:* Ferry-Morse Seed Company.

Figure 25–2 Hand bunching of untopped fresh market carrots. The worker selects and bunches together a uniform amount of carrots using a rubber band or twist-em. The carrots were undercut to make lifting them from the soil easier. Several individual small bunches will be tied together into a large bunch for easier handling and transport to a shed for cooling and packing.

Figure 25–3 Celeriac is a close relative of celery. In contrast to celery, where the petioles are the preferred edible parts, the edible portion of celeriac is the enlarged rootlike structure comprised of both stem and root tissues. *Source:* Ferry-Morse Seed Company.

A constant water supply, either in the form of rainfall or irrigation, is essential for seed germination and high yields of good-quality carrots. In the semiarid United States about 45 to 60 cm (18 to 24 in.) of irrigation water is needed.

Aster yellows, a mycoplasma disease transmitted by the six-spotted leaf hopper, is sometimes a serious pest. Other serious diseases are bacterial blight, alternaria and cercospora leaf blight, *Sclerotinia,* and several others. Carrot rust fly, leaf hoppers, and aphids are important insect pests. Nematodes are a constant concern and must be controlled to avoid substantial losses.

For harvest, machinery is used that undercuts the roots to loosen them from the soil. At the same time the tops are used to lift the plant and are separated from the roots. The roots are transported in bulk to processing facilities or to packing sheds where they are washed, sorted according to size, hydrocooled, and packaged, usually into polyethelene bags. Some fresh market carrots are sold with their tops left on. These are harvested differently: equipment used undercuts the roots to loosen them, and they are then hand-pulled and bunched (Fig. 25–2). Green tops attached to the roots are principally for a more attractive market appearance.

Celeriac (*Apium graveolens* L., var. *rapaceum* Gaud-Beaup.) APIACEAE

Celeriac and celery are closely related, both belonging to the same species. Celeriac is commonly known as turnip-root (Fig. 25–3). The above-ground parts look similar to celery except that the foliage is less compact, more dwarfed, and darker green. In contrast to celery, the leaf petioles are less fleshy and more fibrous. The edible part is the enlarged stem and root axis. Celeriac is mainly used to flavor other foods, such as soups and stews, but is also used raw or cooked in salads. The crop is not grown extensively in the United States. In Europe the crop is better-known and more widely used.

The climatic requirements and cultural procedures for celeriac are similar to those for celery except that the rows are generally spaced closer—45 to 60 cm (18 to 24 in.) apart. Like celery, celeriac is either directly seeded or transplanted. The plants are harvested by undercutting and lifting when they have reached marketable size, 10 to 15 cm (4 to 6 in.). The many small branching roots and foliage may be trimmed in the field or in packing sheds. The entrapment of soil by these roots presents a problem in cleaning, and washing is often necessary. For some markets, celeriac is sold in bulk; in others, several smaller roots which are 5 to 7 cm (2 to 2.5 in.), are sold in bunches with attached tops. Like many root crops, celeriac stores well in a cool moist environment.

Garlic (*Allium sativum* L.) AMARYLLIDACEAE

Garlic is grown especially for its strong and popular flavoring characteristics, which make it a universally important spice. The edible part is a compound bulb consisting of segments called cloves, each surrounded by a thin white or pink sheath. The cloves are formed in the axil of the inner foliage leaves; the outer leaves form the sheath. The leaves are solid and flattened, rather than hollow and round as in onions.

Garlic has been grown for centuries. Its origin is believed to be central Asia.

Precise statistics for the commercial production of garlic are difficult to obtain. United Nations FAO data suggest that world production of garlic was approximately 20 billion pounds in 1999. Leading world producers of garlic in 1999 included Asia and Europe with 81 and 8 percent by weight of total production, respectively.

Garlic is a cool-season crop usually planted in the fall or early spring. This allows the plant to develop sufficient size and to be exposed to low temperatures, both important, prior to experiencing the progressively longer summer photoperiod which is needed to initiate bulbing. Harvest would occur in the fall or late fall depending on planting dates.

There are several cultivars of garlic, each adapted to certain areas and conditions. The most common are California Late, California Early, and Creole. California Late produces more cloves that are smoother and of better quality than the others, but the total yield is less because the cloves are smaller. California Early does not store well, but has the advantage of high yield with bulbs reaching maturity two or three weeks earlier than California Late. Creole is mostly grown at lower latitudes. It matures about a month earlier than California Early. The cloves are covered with a purple sheath.

Garlic grows well on a wide variety of soils but does best on light sandy loams. Weed control and harvesting are more difficult with heavy clay soils. To avoid some diseases, garlic is best planted on land that has not grown garlic the previous year. The crop is propagated by planting cloves broken apart just before planting. Bulbs, actually cloves, to be used for propagation are exposed to cold temperatures during storage, in case field temperatures are not sufficiently low, to hasten the bulbing response to the longer daylength. Bulbs are stored for 30 to 60 days or more at 5°C to 10°C (41°F to 50°F) prior to being separated into individual cloves. The small center cloves are discarded, because small cloves produce small plants that yield less than plants grown from large cloves. Garlic is usually planted on raised beds 100 cm (40 in.) apart with two rows on a bed. The rows on the bed are about 30 cm (12 in.) apart and the cloves about 5 cm (2 in.) apart within the row. Cloves are planted by hand or machine, with care being taken that they are properly placed correct side (distal) up.

Garlic generally responds to fertilizer but is not so demanding as other vegetable crops. If the previous crop was well fertilized, with some carryover, no additional fertilizer is needed. If not, the best results are obtained by applying up to 90 kg/ha (80 lb/ac) of nitrogen. For the fall-planted crops, half is applied at planting time and the other half in the spring when the bulbs begin to enlarge.

Excessive irrigation discolors the sheath covering the cloves and is best avoided. However, for high yields the upper 60 cm (2 ft) of the soil needs to be kept near field capacity. Generally, winter rains are sufficient to start and grow the crop until spring.

Many of the diseases of onions also attack garlic. Onion thrips are the most destructive insect. Stem and bulb nematodes are troublesome in some areas.

Garlic bulbs sent to dehydrators for processing are mechanically harvested; otherwise, the bulbs are pulled by hand. A tractor-mounted blade, passed under the row when the tops become dry and bent over, permits easier hand-pulling of the bulbs. Several rows are gathered together and placed in a windrow with tops up to protect the bulbs from sun. These plants are allowed to dry and cure in the field for a week or two. The tops and roots are removed by shears, leaving about 1.3 cm (0.5 in.) of the root and 2.5 cm (1 in.) of top attached to each bulb. They are again covered with the tops and left in the field for further drying. To keep well, bulbs must be thoroughly field-dried before storing. At ordinary room temperatures and humidities, garlic will keep for several months. For prolonged storage, temperatures should be at 0°C (32°F) or less than 5°C (41°F) and at a relative humidity of less than 60 percent.

Horseradish (*Armoracia rusticana* P. Gaertn., B. Mey. & Scherb.) BRASSICACEAE

Horseradish is a hardy perennial, native to southeastern Europe. The leaves, rhizomes, and roots have been used as food or condiment since the Middle Ages and before that for medicinal purposes. The plant is grown for its pungent compound, allyl isothiocyanate (C_3H_5CNS).

The area around St. Louis, Missouri, and East St. Louis, Illinois, is considered the horseradish capital of the United States. Horseradish is also grown on a smaller scale in Washington, Wisconsin, Pennsylvania, and northern California.

There are two types of horseradish, with little choice of cultivar. The common type has broad, crinkled leaves and produces high-quality roots. The Bohemian type has narrow, smooth leaves, is more disease-resistant, but produces lower quality roots. In selecting a cultivar, the main concern is to choose a rootstock that is healthy, vigorous, and adapted to the area.

While horseradish grows in almost any soil, it does best in deep, fertile, well-drained loams or sandy loams. It also grows well on peat soils but not so well on shallow, compacted, or heavy clay soils with hard subsoils. Poor soils produce highly branched, smaller, and poor-quality roots.

Horseradish is a perennial but is commercially grown as an annual to prevent it from becoming a weed pest. It is easily propagated from root cuttings and sometimes from crowns, but not from seed. The root pieces are planted as early as possible after frost in the spring and the plants occupy the land until fall. After the larger marketable roots are harvested and trimmed, the smaller slender roots are stored and saved to be used as root cuttings for planting next year's crop. It is necessary to identify the upper and lower end of the pieces, so that they will be correctly replanted right-side (proximal[1]) up. These root pieces are bundled and stored in an outdoor pit or in moist sand in a cold cellar until spring, where they are protected against freezing while a high relative humidity is maintained and any moisture condensation is avoided.

Before planting the root cuttings, the soil is plowed as deeply as possible and a firm clod-free seedbed is prepared. Furrows for plantings are formed about 75 cm (30 in.) apart and 13 to 23 cm (5 to 9 in.) deep. The planting stock is placed, all slanting in one direction, about 30 to 38 cm (12 to 15 in.) apart in the row. Care is taken to place the proximal end up in the furrow. The furrow is then filled in with soil.

Shallow cultivation early in the season prevents weeds. Late cultivation is avoided. During the growing season, the plants are "stripped" to get good-quality roots for market. This involves carefully removing the soil from around the top of the main root and leaving the soil at the lower end undisturbed. The small lateral roots are then rubbed off by hand. This procedure produces marketable

[1]The end nearest the crown of the plant.

roots that are free of side roots. Horseradish grown for processing is not stripped as it is not necessary and because too much hand labor is required.

Horseradish is considered a heavy feeder of plant nutrients, thus some commercial fertilizer is often recommended. For average soils not heavily fertilized from previous crops, an application of 55 kg/ha (49 lb/ac) of nitrogen, 110 kg/ha (98 lb/ac) of phosphoric acid, and 55 kg/ha (49 lb/ac) of potash is sufficient if manure is also used. If no manure is used, an application of 110 kg/ha (98 lb/ac) nitrogen, phosphoric acid, and potash is recommended.

The roots grow best late in the growing season: therefore, harvest is delayed as long as possible in the fall to allow for as much root growth and enlargement as possible before harvesting. For fresh market horseradish, roots are plowed out and trimmed of tops and side roots. Processing horseradish is harvested with a mechanical digger. The entire root system, including side roots, is used for horseradish products.

Onions (*Allium cepa* L.) AMARYLLIDACEAE

Onions are one of the oldest vegetables and are known to have been used before 3000 B.C. Onions are probably a native of the area from southwestern Asia eastward to Pakistan.

From 1996 to 1998, annual U.S. dry onion production averaged nearly 7 million pounds. California, Oregon, and Washington together accounted for nearly 50 percent by weight of the commercial U.S. dry onion crops from 1996 to 1998. Significant production also takes place in Idaho, Colorado, Texas, New Mexico, New York, Michigan, and Georgia. Most dry onions are sold in the fresh market, but a large amount are also sold as, or included in, processed products. The early maturing bulb onions are produced in the southern and southwestern states and are usually marketed directly after harvesting. Little, if any, of the early crop is stored. The later maturing crop is produced in the northern states and is marketed directly, and after storage for later marketing during the winter when supplies are normally short.

Onions are cool-season biennial plants that grow well at temperatures ranging from 13°C to 25°C (55°F to 77°F); they have some frost tolerance. Seeds can germinate from 7°C to 30°C (45°F to 86°F) but do best at about 18°C (64°F). During the early stages of growth (before bulbing) onions grow better at relatively cool temperatures, but during bulbing, harvesting, and curing, higher temperatures and low relative humidity are desirable (Fig. 25–4). Plant growth declines at temperatures greater than 27°C (81°F). Bulbing is initiated primarily by daylength and not by the age of the plant. Once initiated, warmer temperatures hasten bulb growth and enlargement. Bulbing and flowering are not related. Vernalization is required for bolting and flowering (Fig. 25–5). Cultivars have very critical photoperiod requirements

Figure 25–4 An example of a globe shape Sweet Spanish type of onion. Attractive well developed onion bulbs such as those illustrated result when grown under favorable environmental conditions. *Source:* Ferry-Morse Seed Company.

Figure 25–5 Hand harvesting of an onion seed crop is preferrable to machine harvesting because of less damage and higher recovery of seed. The heads (umbels) are cut from the stalks and will be laid on large tarpaulins for further field drying and then trashing. *Source:* Ferry-Morse Seed Company.

for bulb formation and do not bulb if the daylength is too short. The critical photoperiod varies from 12 hours for the short-day cultivars to 15 hours for the long-day types. In northern Europe, some cultivars are grown with photoperiod requirements longer than 15 hours. Thus it is not feasible to obtain bulbs with long-day cultivars in the southern latitudes because the days are too short.

Onions vary considerably in size, shape, color, flavor, keeping quality, and maturity dates. New and better cultivars

are continually appearing from breeding programs of universities, seed companies, and the USDA. Cultivars are often grouped according to the latitudes where they grow best. Cultivars that require short days are well adapted to areas between 24 degrees and 28 degrees latitude. These are best grown in the United States in southern Texas for fall planting and early spring harvest. Cultivars having medium-day requirements are adapted to regions in latitudes between 32 degrees and 40 degrees. They are grown principally in the mild-winter areas of California, Texas, and the eastern coastal states, where they are planted in the fall for late spring or early summer harvest. Cultivars adapted to areas north of 36 degrees latitude are planted in early spring, when daylengths are increasing, for late summer or early fall harvest in northern states.

Cultivar characteristics of importance are daylength adaptation to the growing location, size and shape of the bulb, its exterior color, pungency, and storage qualities. Disease resistance is also an important consideration.

The most desirable soils for onion production are sandy loams, loams, peats, and mucks. Clays or coarse sandy soils are best avoided. Clay soils do not remain friable under cultivation and thus retard bulb enlargement. Coarse sandy soils do not retain sufficient water to keep the rather restricted root system of onions well supplied with moisture.

The seedbed is prepared by deep plowing, followed by sufficient tillage to provide a firm seedbed.

There are three common methods of propagating onions; the method chosen is determined by the location, maturity date, and the crop use. The methods are direct seeding, transplanting, and planting small bulbs (sets). The direct-seeding method is used for most crops grown commercially for market as bulbs or for processing by dehydration into onion powder, flakes, salt, and so forth. Seeding rates vary considerably, depending upon soil fertility, use, and row spacing. Generally the objective is to produce 20 to 30 plants per meter of row. Normally, seeds are planted in rows spaced 30 to 60 cm (12 to 24 in.) apart for a crop grown for dry bulbs.

Dry sets produce higher yields and earlier bulbs than direct seeding. They are also used to produce earlier green onions to be harvested before they bulb. Most home gardeners prefer to plant dry sets in their gardens. Transplants about the size of a pencil are sometimes planted by hand or machine directly in the field, especially for early crops. This method eliminates thinning and, although costly, is justified in some situations to achieve earlier production and marketing advantages and is infrequently done commercially.

Early cultivation for weed control is especially important for onions because they are slow starters and serious losses in crop yields occur if control practices are delayed. Many fields are hand weeded during the season, even though chemical weed control is used extensively in most commercial plantings.

Onions do not grow well on acid soils (pH less than 6.0). Finely ground limestone or hydrated lime is applied to bring the soil toward a neutral reaction. Rates of application and kinds of commercial fertilizers differ widely in different areas. For most situations, fertilization to provide 90 to 135 kg/ha (100 to 150 lb/ac) of nitrogen and 68 to 180 kg/ha (75 to 200 lb/ac) of phosphoric acid and potash is usually adequate. Later in the growing season, additional nitrogen at 35 kg/ha (31 lb/ac) can be supplied by sidedressing. On muck soils potassium is often the limiting nutrient, and applications of potassium sulfate or potassium chloride usually give good response.

In the absence of rainfall, onions require irrigation. The plants should not be stressed by a lack of water. The shallow root system of onions requires frequent applications of water from rain or irrigation. Early onions require 5 to 7 irrigations, while a late crop needs 7 to 10. The amount of water needed varies between 38 and 75 cm (15 to 30 in.) depending upon maturity date, soil type, and location. Irrigation is generally started immediately after seeding and is continued at required intervals until plants begin to mature. Irrigation is then stopped and the soil allowed to dry to prevent secondary root growth. Onions are irrigated by furrow or sprinkler methods or by flooding with raised borders.

Downy mildew, white rot, pink root, smut. *Fusarium,* botrytis leaf blight and neck rot, purple blotch, and yellow dwarf virus are among the serious onion diseases. Cultivars with resistance to some of these diseases are used, but resistance is not available for all. Additionally, nematodes cause serious losses.

Onion thrips, onion maggots, and cutworms are the most troublesome insect pests.

Most onions are harvested as dry bulbs either for marketing as such or for processing into various canned, pickled, frozen, or dehydrated products. Before they bulb, onions are also harvested, bunched, and sold as fresh green onions (Fig. 25–6). Harvesting for mature bulbs generally begins when the top foliage begins to dry and fall over. This varies with the season and environmental conditions. Additionally, market conditions are often a determining factor in earlier harvesting. Ideally, the bulb's market and postharvest storage qualities are optimized when full maturity (all tops down) is obtained. Most bulb onions for processing are usually fairly or fully mature and are harvested mechanically. An onion plow is passed under the bulbs to loosen the soil. They are then pulled and laid in windrows on top of the ground or placed in crates for further curing and drying. The curing time varies from a few days to two or three weeks depending upon temperature and relative humidity. Market onions not fully mature are damaged by mechanical harvesters and are therefore often hand harvested and hand trimmed using labor aid equipment. Care is taken during harvest to avoid damage by bruising. After the tops have been removed, the onions are cleaned and graded according to size, shape, and color. Grade standards vary with each cul-

Figure 25–6 Immature green onions trimmed and tied in a bunch and intended for fresh use. Such onions are grown under conditions that limit bulbing or are harvested before significant bulbing occurs. *Source:* Sakate Seed Company.

tivar. Most bulbs for marketing are packaged in 23 kg (50 lb) open-mesh bags to assist ventilation. Some onion cultivars have characteristics that permit relatively long storage, while others do not. These characteristics are strongly correlated with bulb maturity and pungency levels. Long-day types are usually pungent, have a higher dry matter content, and tend to cure and store better. Short-day types tend to be less pungent (sweet), have a lower dry matter content, and a relatively short storage life. For short-time storage, the onions are maintained at about 0°C (32°F) with a relative humidity of about 60 percent. All onions, whether stored for long or short periods, should be cured as well as possible and not damaged by bruising or disease. Onions grown specifically to be stored are allowed to mature as fully as field conditions permit. After harvest they are cured at a temperature of about 25°C (77°F) and moderate (75 percent) humidity for several days; the temperature is then slowly lowered close to 0°C (32°F) to retard respiration and possible sprouting. Treatment with maleic hydrazide is used to prevent sprouting in storage.

Green onions are young immature plants, grown at very high populations for fresh market use in salads. Long-day cultivars are grown in short-day environments so as to inhibit bulbing. These plants are generally hand harvested, several being bunched together, the tops uniformly trimmed, and marketed directly.

Parsnips (*Pastinaca sativa* L.) APIACEAE

Parsnips are considered a native of Europe and Asia and have been used for food and medicinal purposes since the Greek and Roman eras. They were introduced into the United States in Virginia and Massachusetts in the early 1600s. The plants are cool-season biennials but are grown as annuals. Parsnips are slow growing and require a long season (100 to 120 days) to form their white, fleshy edible roots. This crop occupies expensive vegetable land for longer periods of time than most farmers prefer, which accounts in part for their limited importance.

Parsnips are extensively grown in the temperate zones of Europe and Asia. Most commercial production in the United States occurs in those states where cooler temperatures prevail, such as Pennsylvania, Illinois, parts of California, and New York. Good parsnip stands are difficult to establish in the warmer southern states.

Parsnips grow best when they are planted and grown during a relatively cool summer and harvested in a cool fall. In California they are grown in the coastal valleys where summers are mild.

Cultivars are selected that produce roots with white flesh, small cores, and are relatively free from unsightly lenticel growth, which lowers quality. Roots are tapered and grow 25 to 30 cm (10 to 12 in.) long.

Parsnips do best in a deep, fertile, well-drained, loam or sandy loam soil. Well-shaped, smooth roots are difficult to grow in heavy clay soils. Soil preparation and culture is much the same as for carrots. The seedbed should be deep and carefully prepared by deep plowing. Cloddy or stony soil causes rough, misshapened roots. Parsnip seed loses its vitality quickly, even when stored under good conditions. Fresh seed should be used. Seed is planted into moist soil 1 to 2.5 cm (0.5 to 1 in.) deep as soon as possible after the soil has warmed. Parsnip seed germinates slowly even at the optimum soil temperature for germination of about 20°C (68°F). Rows are placed 38 to 45 cm (15 to 18 in.) apart for hand tillage and 60 to 75 cm (24 to 30 in.) for tractor tillage. After emergence plants are thinned to in-row spacings of 8 to 10 cm (3 to 4 in.).

Parsnips compete poorly with weeds; therefore, weed control is essential. Fertilizer recommendations are about the same as for table beets and perhaps a little less than for carrots.

Water requirements are similar to those for carrots.

Parsnips are affected by the same diseases and insect pests attacking carrots, although they are usually less seriously affected.

Parsnip roots withstand some freezing; thus, the roots are often stored in the soil and harvested from late fall to early spring, although bolting will occur if harvesting is delayed. In areas of severe winters, however, roots are best harvested during the fall because of the difficulty of digging them from frozen ground. The crop is harvested by loosening the roots with a plow and removing them by hand. At harvest time parsnips are high in starch, but after storage either in the soil or in a cool, moist pit or cellar at about 0°C to 1°C (32°F to 34°F), the starch is converted to sugar, markedly improving the flavor.

Potatoes (*Solanum tuberosum* L.) SOLANACEAE

The potato plant is a bushy, herbaceous annual 60 to 90 cm (2 to 3 ft) in height. The edible tuber is a swollen underground stem. The "eyes" on the tuber are buds and are more numerous on the apical (distal) end (opposite the end of attachment) than the basal (proximal) end.

As a source for human food, the potato is a leading vegetable crop. Potatoes are South American in origin and were cultivated in Chile and Peru by the Indians long before the arrival of European explorers. Early Spanish explorers transported them to Europe about 1575, and from there they returned to North America with the colonists settling along the Atlantic coast. In those early days, however, potatoes were not a significant food source in North America until the influx of Irish immigration which began in the early 1700s and reached a maximum later, after the potato famine in Ireland in 1846. Potatoes then became America's favorite carbohydrate food because their cost was low and most people liked them (Fig. 25–7). Today potatoes are grown on every continent and in every state of the United States. And, like carrots, potatoes are harvested somewhere in the United States every month of the year. From 1996 to 1998, annual U.S. potato production averaged more than 48 billion pounds. The fall potato crop of approximately 44 billion pounds annually is significantly greater than the winter (323 million pounds), spring (2 billion pounds), and summer (2 billion pounds) crops. Florida and California account for most of the winter and spring potato crops, although North Carolina produces large amounts of potatoes for the spring market. Colorado, Texas, and California dominate summer production. Idaho, Washington, and Wisconsin together accounted for approximately 55 percent by weight of the U.S. fall potato production. A significant amount of fall potato production also occurs in North Dakota, Colorado, Minnesota, Maine, Oregon, Michigan, and New York.

Potatoes are sold in three primary markets: tablestock/fresh market, processing, and seed. In general, processing, fresh market, and tablestock markets use approximately 58, 28, and 5 percent by weight of the annual U.S. potato crops, respectively. Potatoes are a cool-season crop, slightly tolerant of frost, but easily damaged by freezing weather near maturity. Tubers form on the end of stems called stolons, which botanically are rhizomes. The tuber is a fleshy stem with buds in the axil of leaf scars. They are often referred to as "eyes." Maximum yields of high-quality tu-

Figure 25–7   Each year an increasing percentage of the total potato crop in the United States is processed into many new products. Here, Idaho russet potatoes are prepared for crinkle-cut potato chips. *Source:* University of California Cooperative Extension.

bers are produced when the mean temperature is between 15°C and 18°C (59°F and 64°F) during the growing season. Tuberization (tuber formation) is also favored by long days of high light intensity. An optimum temperature for tuber development is about 18°C (68°F). Tuberization is progressively reduced when night temperatures rise above 20°C (68°F) and totally inhibited at 30°C (86°F).

Flowers are common but fruits rarely set unless precise climate and a long daylength period are met. The fruits are green berries, somewhat similar to small tomatoes, within which seed are produced. True seeds are used in breeding programs but have not, until recently, been planted for commercial production. Recently there has been much interest in using seeds for propagation in developing countries. This interest is with regard to the need and cost of importing disease-free seed potatoes for propagation, where disease-free seed stock cannot be produced.

New potato cultivars appear as a result of widespread breeding programs of national government and experimental station programs, the International Potato Center in Peru, and some private firms. Surprisingly, many older cultivars remain in wide use. Cultivars differ widely in maturity time, yield, appearance, cooking qualities, storage, and resistance to diseases and pests. Cultivars tend to have very refined adaptations to specific climates and locations. Along with the intended usage—fresh for boiling, baking, or processing—the choice of the cultivar also depends on its adaptation to the growing area.

High-yielding tubers of good quality are produced on a wide variety of soils, provided management is proper. The preferred soils are light-textured mineral soils and mucks. A fertile soil with good water drainage and aeration is important. Good-quality potatoes do not come from waterlogged soils because the excess water tends to enlarge lenticels. Heavy clay soils or stony soils produce irregular-shaped and rough-skinned tubers. Compacted soils not only reduce yield, but also often produce unmarketable, knobby potatoes.

The seedbed is prepared by deep plowing 20 to 25 cm (8 to 10 in.) followed by sufficient disking and harrowing to free the soil of clods.

Potatoes are propagated vegetatively by planting either small whole tubers or cut "seed" pieces. It is seldom justified to plant anything but certified disease-free seed pieces. Certified seed tubers are produced in the northern tier of states, as well as California and Colorado (under strict state inspection and regulation) by growers who specialize in this production. Seed tubers are produced in the cool areas of the northern states so that virus-infected plants can be detected and rogued. Under cooler growing conditions virus expression is more evident making it easier to recognize. The certification programs thus insure that tubers to be used for propagation will not be infected with a virus.

Immediately after harvest, the buds ("eyes") on potato tubers enter a "rest period" of five to ten weeks, during which time they will not sprout naturally even if they are placed in a favorable environment. For planting, the tuber is cut into two to six pieces. The size or weight of the seed piece is important; the ideal is about 40 to 55 g (1.3 to 2 oz). Small (55 g) tubers are often planted whole. Each piece should contain at least one healthy bud, preferably two.

The planting date depends upon the climate, location, and the desired harvest date. High soil temperature (30°C; 86°F) prevents tuber formation; thus, planting dates are chosen to keep tubers from forming during hot weather. Potatoes can be planted as soon as soil temperatures reach about 5°C (41°F), but emergence is more rapid at 20°–22°C (68°–71°F). This is the best time for early potatoes in Idaho, Montana, Maine, New York, Minnesota, and Wisconsin. In the southern states and Texas, main planting comes in late fall or early winter. In California, planting dates vary from winter to spring to fall; in the Tule Lake area near the Oregon border, planting is done during late spring to early summer; and in the southern San Joaquin Valley late winter and early spring planting occurs.

Before tubers are cut for seed pieces they should be warmed to room temperature. After cutting, they should be allowed to cure. This is done by holding the cut pieces at a temperature between 18°–21°C (65°–70°F) for 2–3 days at a relative humidity of 90 percent. This allows the cut surfaces to suberize, which is the development of a corky protective surface. The seed pieces are also commonly treated with a fungicide for additional protection against decay organisms. Seed pieces are planted 8 to 10 cm (3 to 4 in.) deep. Commercial producers using mechanical planters plant two to six rows at a time. The rows vary from 75 to 90 cm (30 to 36 in.) apart with in-row spacings between plants of 20 to 30 cm (8 to 12 in.). In many areas, especially where furrow irrigation is needed, potatoes are planted in ridges. In other areas they are planted flat, then ridges are formed during cultivation.

Cultivation is done only deeply enough and often enough to control weeds. Deep cultivation damages surface roots and excessive tillage compacts the soil. During the last cultivation, soil is thrown up close to the plant to form ridges that bury weeds and protect the tubers from light and sunburn. The best weed control is by applications of herbicides.

Potatoes are heavy users of mineral nutrients. Recommendations for fertilizer applications vary widely with location, climate, and soil type. For example, potatoes growing on some mineral soils in the southern San Joaquin Valley of California respond well to applications of 110 kg/ha (98 lb/ac) of nitrogen, but little to phosphoric acid and not at all to potash. Peat soils respond well to all three primary nutrients, and many farmers prefer to use complete commercial mixes. Some mineral soils with little organic matter need applications of 70 to 200 kg/ha (60 to 180 lb/ac) of nitrogen, 225 to 280 kg/ha (200 to 250 lb/ac) of phosphoric acid, and 170 to 280 kg/ha (150 to 250 lb/ac) of potash. Heavily cropped soils in the northeastern United States respond well to somewhat larger applications.

Most potatoes are grown without irrigation and depend entirely upon rainfall. But in some areas farmers have found it profitable to supplement rainfall with sprinkler irrigation. It is futile to attempt to grow potatoes without irrigation in the western and southwestern states. Furrow irrigation is the most popular method. Sprinkler irrigation is used to some extent in many areas where furrow irrigation is not practical. When irrigation is used, the soil is kept close to field capacity. Alternating wet and dry soil conditions during tuber formation produces growth cracks and knobby or multiple tubers. Smooth tubers result from rapid continuous growth. In general, a crop of potatoes requires between 50 and 75 cm (20 to 30 in.) of irrigation water.

Potatoes are subject to a host of disease and insect pests. To reduce losses, the use of disease-free seed pieces, seed piece treatment, crop rotation, and field sanitation are all essential practices. Chemical pest control, if necessary, is best when integrated into the total control program. The important potato diseases include early blight, late blight, scab, verticillium wilt, black leg, and ring rot. Virus diseases are probably the most serious threat for potato producers and account for most of the crop losses due to disease. Several of the more important viruses are leafroll, potato virus X, potato virus Y, tobacco rattle virus, and curly top. The mycoplasm aster yellows is also a serious disease (Fig. 25–8). The most important insect pests are Colorado potato beetles, spider mites, flea beetles, leafhoppers, aphids, white grubs, blister beetles, and wireworms. Nematodes are destructive potato pests.

Almost all commercial potato farmers harvest their crop mechanically, with specialized harvesters. To facilitate harvesting, the potato vines, if not already dead, are destroyed mechanically or by chemical desiccation. This reduces crop trash and helps to firm the tuber skin (periderm)

Figure 25–8   A potato breeding trial in which cultivars are being evaluated for resistance to virus. Such trials are a necessary function in the plant breeder's effort to reduce crop losses from virus and other diseases. *Source:* University of California Cooperative Extension.

so it is less susceptible to damage (Fig. 25–9). The potatoes are dug and lifted from the soil as gently as possible to avoid bruising. Bruising contributes to a physiological malady known as internal black spot, a blackening of the tissue inside the tuber. After the potatoes are removed from the soil, they are elevated into the harvester where shakers remove any adhering soil. They are then conveyed to trucks moving alongside the harvester. The tubers are transported to a central packing shed and emptied into a large vat of chlorinated water. The water reduces bruising to a minimum, washes away any soil on the tubers, and disinfects them. They are removed from the water on belts and scrubbed mechanically with a soft brush, dried, graded, and sometimes waxed. Dumping into water is seldom done for processing potatoes (which account for about 60 percent of the total United States crop) or for fresh market potatoes to be stored for later marketing. It is mainly done for potatoes that will rapidly enter the marketing channels.

A considerable volume of potatoes is stored for later marketing, which accounts for their availability through the year. Storage facilities especially designed to maintain low temperatures and high humidities enable producers to store potatoes for many months without incurring serious loss of product quality. The low temperatures reduce respiration and tuber sprouting, and the high humidity reduces desiccation. Freshly harvested tubers will not sprout even under favorable conditions. This period is called "rest," and lasts for 5 to 15 weeks, depending on the cultivar. Following this period the tubers are "dormant," which means they will sprout if conditions are favorable. To avoid sprouting of potatoes held for long storage periods, sprout inhibitors such as malic hydrazide and Chlorpropham (CIPC) are used.

Figure 25–9 Often potato tubers are harvested before they are fully mature. This generally causes the skins to develop an unsightly condition known as feathering as shown here, although there is no damage to the tuber itself. *Source:* University of California Cooperative Extension.

Tubers entering storage should be held at a curing temperature of 10°C to 13°C (50°F to 55°F) for several days with relative humidity above 95 percent. These conditions allow for suberization of injuries and for the skin of immature tubers to develop. Following this treatment, the temperature is lowered to about 4°C to 7°C (40°F to 45°F). The tubers are kept in the dark to prevent greening due to chlorophyll development, which is accompanied by the formation of the poisonous alkaloid, solanine.

Tubers held at low temperatures will have some of their starch converted into sugars. Prior to removal from storage, whether for fresh market or processing use, the tubers should be reconditioned. This is done by warming the tubers for 5 to 10 days at a higher 10°C (50°F) temperature so that the sugars will reconvert to starch.

Radish (*Raphanus sativus* L.) BRASSICACEAE

Of all vegetable crops, the garden radish is one of the easiest and quickest vegetables to grow. A crop can be produced within 4 to 6 weeks of seeding. The radish, probably native to the eastern Mediterranean region, was cultivated for centuries. Its use can be documented beyond 2000 B.C.

Radishes are widely grown in Europe, Asia, Africa, South America, and the United States. Practi-cally every home gardener plants radishes for an early spring vegetable.

Radishes are a cool-season crop tolerant of some slight frost. Best growth occurs when the monthly temperature averages about 15°C to 18°C (59°F to 64°F). They are somewhat intolerant of temperatures exceeding 25°C to 27°C (77°F to 81°F).

Radish cultivars are often grouped according to their time to mature and the season when grown. Rapid-growing, quick-maturing cultivars are popular with commercial growers and home gardeners for use raw in salads. These are annual plants best known by their small, short, often globe-shaped roots and mild pungency. Although often red-skinned, white and mixed white and red cultivars are available, as are cultivars with larger, longer, tapered roots.

When grown during high temperatures, the interior root tissues tend to be pithy, and the plants may bolt. Over-maturity will also cause a softening or pithiness of the normally crisp tissues.

Radishes are grown on all types of soils from muck to sand. Light-textured, fertile sandy loams produce roots of the highest yield and best quality, especially for cultivars having long roots. Care is taken to obtain a smooth seedbed.

Since radishes are hardy, they are planted early in the spring, and to lengthen the harvest season, several plantings at 10- to 12-day intervals are often made. Most cultivars used in the United States are intolerant of high temperature, so planting in midsummer is generally avoided. Radishes are planted in slightly raised beds where furrow irrigation is used, or on flat ground if sprinkled or not irrigated. Two rows, 25 to 30 cm (10 to 12 in.) apart, are planted on beds that are 100 cm (40 in.) apart. Wider beds, up to 200 cm (80 in.) with multiple rows, are used by large-scale growers. In fields planted without beds, rows are 38 cm (15 in.) apart. The in-row spacing is about 6 to 8 seeds/5 cm (3 to 4 seeds/in.) of row and the depth of seeding is about 1 cm (0.5 in.).

On a worldwide basis the later-maturing radish cultivars are more important. These are commonly called winter radishes. Winter radishes are biennial plants that are usually more pungent, larger, and well adapted for storing. Size, shape, and even flesh and exterior colors vary considerably. The roots of some cultivars are exceptionally large and/or long. Winter radishes are very popular and are consumed in large quantities in raw, cooked, and pickled form throughout the Orient. 'Daikon' is a well-known representative cultivar. Winter radishes are usually planted to mature in the fall or winter. In mild climates, the crop can grow through the winter for spring harvesting. Much of the popularity of the winter radishes is because of their strong flavor characteristic and their good storage qualities.

Cultivar choices are highly dependent on market preferences. However, characteristics such as root size, shape, color, pungency, disease resistance, and storage are important in their selection.

Radishes respond well to fertilizers. To produce tender, crisp roots many farmers broadcast 45 to 55 kg/ha (40 to 50 lb/ac) of nitrogen, 90 to 110 kg/ha (80 to 100 lb/ac) of phosphoric acid, and 90 to 110 kg/ha (80 to 100 lb/ac) of potash in the spring if several plantings (up to six plantings a year in Florida and southern California) are made on the same land during the season. Less fertilizer is used if only one early crop is to be grown and if the previous crop was well fertilized.

Diseases can be a serious problem with radishes. A major disease is fusarium wilt. Others include *Rhizoctonia, Phytophthora,* rust, club root, *Pythium,* powdery and downy mildew, and *Anthracnose.* Aphids, cabbage root maggots, leaf miner, flea beetles, and other pests of cruciferous crops are the most serious insect pests.

For the fast-maturing, early-spring radishes, harvesting begins about 21 to 29 days after planting, but may require 40 to 50 days during cold weather. Winter radish cultivars are usually planted in the early fall. They require 50 to 90 days to mature and are harvested before winter. They become bitter and pithy if harvest is delayed. Commercial growers use multiple-row harvesters that grip the tops to pull the plants from the soil, and cut the roots from the tops, then placing them in bins for transport to a packing shed. Here the roots are washed, graded, and bagged in plastic bags for market. They are cooled immediately. For markets that prefer the radishes with their tops intact, they are generally pulled by hand and tied in bunches in the field (Fig. 25–10). They are then transported to the packing shed, where they are washed and packed in crates and placed in cooling rooms before marketing.

Rutabaga (*Brassica napus* L., Napobrassica Group)
BRASSICACEAE

Rutabagas are a cool-season biennial root crop grown mainly in northern Europe, England, and Canada. Limited amounts are grown in the United States for human consumption and some for stock feeding. They are sometimes called Swedes, Swedish turnips, Russian turnips, Canadian turnips, or yellow turnips. Rutabagas are popular in northern countries because of their good storage qualities, thus providing a vegetable for use during winter periods. The cultural practices for rutabagas are similar to those for turnips. The two differ in appearance; rutabagas have thicker necks, are usually larger, and exhibit more advantageous root growth (Fig. 25–11). However, like turnips, rutabagas have either white or yellow flesh, and the exterior colors of some cultivars can be white, tan, or green, with upper portions purple. The principal production areas in the United States are the north central states and Washington. Seeds are planted in late spring or early summer, and the crop is harvested in midfall or early winter. The roots are pulled by hand and trimmed in the field. They are stored at low temperatures and high humidities.

Figure 25–11 The rutabaga is a close relative of the turnip. In fact the roots sometimes look very similar, except that properly developed rutabaga roots tend to be round while turnip roots tend to be more flattened. Rutabaga leaves are hairless and bluish, while turnip leaves are hairy and greener. *Source:* Ferry-Morse Seed Company.

Figure 25–10 Radishes are almost always included in every home garden and are generally the first crop harvested in the spring. *Source:* Ferry-Morse Seed Company.

Those going to market are washed and waxed to increase shelf life by reducing shriveling due to water loss.

Salsify (*Tragopogon porrifolius* L.) ASTERACEAE

Salsify is a hardy biennial, sometimes called oyster plant or vegetable oyster. It has long, cylindrical roots, used mainly in soups for its delicate oysterlike flavor. The plant grows wild around the Mediterranean Sea, in southern England, and along roadsides in the United States and Canada. During its second season, it bears a purplish flower that looks like an enlarged dandelion head. Salsify is a market vegetable crop of limited use.

Salsify is a cool-season crop tolerant of some freezing weather. Seeds are planted about the time of the average date for the last frost in the spring.

This slow-growing plant requires the entire season to develop, and like parsnips, its quality is improved after a heavy frost or cold storage for at least two weeks. The tapered roots are 20 to 23 cm (8 to 9 in.) long, 2.5 to 4 cm (1 to 1.5 in.) thick, and feature creamy, white flesh.

The crop requires a deep, moist, fertile, loamy soil to produce good yields of high-quality roots. Rocky, shallow, or heavy clay soils tend to produce misshapen roots. Salsify grows best on calcareous soils because of its rather high lime requirement. Acid soils should be limed to increase pH.

The culture of salsify is practically identical to that of parsnips. The home gardener plants the seeds 1 cm (0.4 in.) deep in rows 45 cm (18 in.) apart. The plants are thinned to about 8 to 10 cm (3 to 4 in.) apart. Like parsnips, the viability of salsify seed deteriorates rapidly with age, and seeds over a year old are not recommended for use.

A fertile soil is needed to produce high yields. The crop responds well to high soil organic matter, also to applications of about 110 kg/ha (98 lb/ac) each of nitrogen, phosphoric acid, and potash.

Roots are ready for harvest in late fall. Except where the ground freezes hard, the roots are left in the soil until needed for use or market. They are harvested by digging and marketed topped, washed, and tied in bunches of 10 to 12 roots. Salsify roots generally store very well.

Sweet Potato (*Ipomoea batatas* Lam.) CONVOLVULACEAE

The moist, soft-textured sweet potato cultivars are often erroneously called yams to distinguish them from the dry-textured sweet potato cultivars. This is an unfortunate misnomer since true yams are different plants entirely, not even slightly related to sweet potatoes. True yams belong to the DIOSCOREACEAE family.

Sweet potatoes are native to tropical America, and were transported to the Pacific islands and on to Asia early in history. They were cultivated as food in the southern parts of the United States by the Indians and were later taken to Europe. There is no evidence that the ancient civilizations of Egypt, China, Persia, or Greece knew of sweet potatoes.

Sweet potatoes are perennial vines that grow prostrate on the ground. They belong to the morning-glory family and are grown for their tuberous roots, being unique in producing adventitious shoots that are used to propagate the crop. When grown in temperate and some subtropic regions, the crop is handled as an annual. In the tropics and much of the subtropical regions, the crop is cultured as a perennial.

Sweet potatoes are of particular importance as a food crop throughout subtropical and tropical regions. It is one of the most important carbohydrate sources for many millions of people, particularly those in the developing nations. From 1996 to 1998, annual U.S. sweet potato production averaged slightly more than 1 billion pounds. North Carolina, Louisiana, and California together accounted for an average of nearly 80 percent by weight of the commercial U.S. sweet potato crops during the same time period. Significant production also occurs in Mississippi, Alabama, and Texas.

Sweet potatoes are warm-season plants that do not tolerate frost nor grow well in cool weather. Temperatures below 10°C (50°F) cause chilling injury. They require a long, warm growing season and grow best when mean monthly temperatures are above 20°C (68°F) for at least three months.

The crop is propagated vegetatively. Cultivars are divided into moist-flesh and dry-flesh types. The moist-flesh types (mistakenly called yams) are softer and sweeter with orange to deep-orange flesh. The dry-flesh types have yellow skins with yellow or white flesh and are called sweet potatoes (Fig. 25–12). Exterior skin colors do not distinguish moist-fleshed from dry-fleshed cultivars.

The ideal soil for sweet potatoes is a well-drained sand or sandy loam with a clay subsoil. However, loams and clay loams grow good sweet potatoes if they are well drained.

The field is carefully prepared by plowing 15 to 20 cm (6 to 8 in.) deep followed by disking and then harrowing to eliminate all clods. Any residue of the previous year's crop is completely buried. Most farmers prefer to prepare their land well in advance of planting. In most areas, the crop is planted on broad flat ridges whose height is determined by the soil type, drainage, and rainfall.

Sweet potatoes are grown from plants or sprouts sometimes called "slips" obtained from roots of the previous crop. These plants are grown by placing small- to medium-sized sweet potato tuberous roots in hotbeds filled with sand. After the roots are bedded, they are covered with about 5 cm (2 in.) of sand, and the beds are covered with plastic in the form of a low tunnel. The tunnel increases the bed temperatures and accelerates the growth of the slips. The plastic must be

Figure 25–12 The edible portion of a sweet potato is a tuberous root. The sweet potato makes up a large portion of the human diet in some African countries, India, China, South America, the South Pacific islands, and the southern United States. Sweet potatoes are tasty and nutritious. *Source:* University of California Cooperative Extension.

removed when soil temperatures reach 30°C (86°F) to avoid overheating damage. The beds must be kept moist while the plants are growing. The slips are pulled when they are 20 to 25 cm (8 to 10 in.) long with six to eight well-developed leaves and a good root system. Transplants are harvested from the parent tuberous root two or three times. An adequate number of slips are grown from 450 to 675 kg (400 to 600 lb) of bedded roots to plant one hectare. Unfortunately, this method of vegetative propagation poses some danger of spreading diseases or nematodes that might be on the transplants.

The slips are planted either by hand or with transplanting machines. They are planted in rows 90 to 100 cm (36 to 40 in.) apart and spaced 30 to 45 cm (12 to 18 in.) within the row. Closer spacing tends to increase the yield of marketable sweet potatoes.

Cultivation is necessary for adequate weed control, but excessive cultivation must be avoided because of root injury and the possibility of unnecessary soil compaction. Hand hoeing may be necessary later when the vines interfere with the cultivator blades. Chemical herbicides are helpful.

As a general rule, high yields of good-quality sweet potatoes require fertilizers on all soils. The recommendations are for moderate amounts of nitrogen with higher proportions of phosphorus and potassium. A reasonable application varies from 45 to 110 kg/ha (40 to 100 lb/ac) nitrogen, 45 to 90 kg/ha (40 to 80 lb/ac) phosphoric acid, and 70 to 145 kg/ha (60 to 130 lb/ac) of potash.

Sweet potatoes are mainly grown in warm humid regions where the annual rainfall averages about 100 cm (40 in.) and is fairly evenly distributed throughout the growing season. Farmers in the drier parts of the United States are realizing greater success with sweet potatoes by supplement-

ing the rainfall with sprinkler or furrow irrigation when moisture is deficient. In the arid Southwest, sweet potatoes are irrigated. Frequent irrigations supplying 45 to 60 cm (18 to 24 in.) of water are needed to produce a crop.

Sweet potatoes are attacked by several bacterial, fungal, and virus diseases. Some of the more important diseases are fusarium wilt and stem rot, black rot, rhizopus soft rot, scurf, white rust, cercospora leaf spot, internal cork and russet crack virus, and a number of storage diseases. Root knot nematode is a frequent problem and most fields are fumigated to control this pest. Insect pests include flea beetles, white flies, cutworms, wireworms, and sweet potato weevil.

Sweet potato roots attain highest yield and best quality when harvested mature (fully developed), which usually requires 130 to 150 days. Short days induce root enlargement which, once initiated, will continue under longer daylength. However, some roots are harvested for an extra early higher price market, despite reduced yield and lower quality. Farmers in northern areas often allow sweet potatoes to remain in the field until after the first frost to kill the vines and allow easier digging. The roots are dug as soon as possible to prevent vine decay accompanied by development of organisms in the vines that move down into the roots causing root decay. Also, soil temperatures below 10°C (50°F) cause chilling injury to the roots, which will later lead to rotting. Before digging, the general procedure is to remove the vines, by a vine-cutting machine, or by a sharp colter (a cutting wheel) attached to the plow used for digging. The sweet potatoes are then plowed out with any one of various types of plows. The roots come out clean and easily if they are dug when the soil is dry. The roots are placed in containers as soon as possible to avoid exposure to the bright sunlight. Sweet potatoes are susceptible to damage by bruising and are handled carefully. Boxes or large bins serve as containers for storage and marketing. Successful storage depends upon proper curing at a temperature of about 30°C (86°F) and a relative humidity of 85 to 90 percent for a period of 5 to 7 days. After curing, the storage temperature is maintained at 13°C to 15°C (55°F to 59°F) and the relative humidity at 80 to 85 percent. The temperature is never allowed to drop to 12°C (55°F) because of chilling injury.

Turnip (*Brassica rapa* L., Rapifera Group) BRASSICACEAE

Turnips are one of the easiest vegetables to grow and one of the more widely adapted of the root crops. They are native to northern Asia and extensively grown in Europe, Asia, and almost everywhere in the United States. They are used as food for both animals and humans. The roots may be eaten raw but are usually cooked, and the leaves make delicious greens. The plant is a cool-season biennial. Seeds are sown either early in the spring (for a fall crop) or in the fall (for a spring crop) for the roots to mature during cool weather. Continued temperatures below 10°C (50°F) are likely to cause **bolting.** The plants resist frost and mild freezing. In most areas spring and fall crops are grown since only 60 to 80 days are required for the roots to reach maturity.

As with all vegetable crops, a moderately deep, friable, fertile, well-drained soil is ideal for turnips. Extremely tight clay soils or very sandy soils are avoided if possible. Common cultivars include both white and yellow flesh types. Exterior colors can vary from white to tan and yellow, often with the uppermost portion purple. Size and shapes also vary, some being globe-shaped, others pronouncedly flat. Some cultivars are grown specifically for the abundant and high-quality foliage.

Turnips are generally planted in rows 45 to 60 cm (18 to 24 in.) apart and thinned to an in-row spacing of 5 to 8 cm (2 to 3 in.). The seed is sometimes broadcast, but only if weeds are not a problem.

Heavy application of expensive fertilizer is generally not profitable for turnips. For most soil of average fertility, 17 to 35 kg/ha (15 to 30 lb/ac) nitrogen, 55 to 85 kg/ha (49 to 75 lb/ac) of phosphoric acid, and 17 to 35 kg/ha (15 to 30 lb/ac) of potash usually give satisfactory results.

Diseases are usually not a serious problem. The most troublesome ones include clubroot, leaf spot, white rust, scab, mosaic, and root rot. Flea beetles, wireworms, and thrips are the most troublesome insect pests.

Turnip roots are harvested by pulling them from the soil after undercutting to loosen the plant, especially if the tops are to be left attached. Roots to be sold for early spring market are pulled when they are about 5 cm (2 in.) in diameter, washed free of soil, then tied in bunches of five or six roots. The tops are usually left on but some are topped for bulk marketing. Turnips store well when provided low, 0°C to 5°C (32°F to 37°F) temperatures with high relative humidity.

REFERENCES AND SUPPLEMENTARY READING

DECOTEAU, D. R. 2000. *Vegetable crops.* Upper Saddle River, NJ: Prentice Hall.

MAYNARD, D. N. 1997. *Knott's handbook for vegetable growers.* New York: John Wiley.

CHAPTER

26

Temperate Zone Fruit and Nut Crops

This chapter considers the important fruit and nut crops grown in the temperate-zone regions of the Northern and Southern Hemispheres. Most of these plants withstand very cold winter temperatures and many do, in fact, require winter chilling for good productivity. Most of the species are deciduous, which means they drop their leaves in winter, and all are woody perennials (except for the strawberry). All of these fruits are heterozygous; therefore, they do not reproduce true from seed and must be propagated by asexual methods.

Almond (*Prunus dulcis* [Mill] D. A. Webb) [*Prunus amygdalus* Batsch.] ROSACEAE

Almonds are one of the oldest nut crops, originating in Southeast Asia and later introduced into the Mediterranean region. Almonds grow in regions of the world that are characterized as having a subtropical Mediterranean climate with mild, wet winters and warm, dry summers. They are moderately cold-hardy, but require a small amount of winter chilling to overcome bud dormancy, which causes them to bloom very early in the spring. Almond culture is confined, therefore, to areas without late spring frosts. A temperature of −4°C (25°F) for 30 minutes at full bloom damages 20 to 100 percent of the flowers, depending on the cultivar. The almond is cultivated mainly in southern Europe (Spain and Italy) and California, with plantings in North Africa (Morocco), across the

Middle East (Turkey and Iran), and the equivalent Southern Hemisphere zone (Australia and South Africa). California leads in almond production, followed by Spain, Turkey, Greece, and Italy. The requirement of almond trees for some winter chilling eliminates their culture from the tropical regions of the world. The almond fruit is very similar to that of the peach. Unlike other stone fruit, the flesh (mesocarp) does not grow in stage III, but dries and splits on ripening to expose the endocarp, containing the edible almond "nut" (seed) inside. The endocarp, or shell, may be crumbling and soft, "papershell," moderately soft, "soft shell," or "hard shell." The latter cannot be broken by hand.

Almond trees are relatively small, averaging 4 to 6 m (12 to 19 ft) but may be up to 9 m (30 ft) in height. They are propagated by budding selections or cultivars onto almond or peach seedling rootstocks. Certain plum and peach-almond hybrid rootstocks can also be used. The leading almond cultivar grown in California is "Nonpareil." It has two main advantages: a mild flavor and a papershell, which gives it more than 70 percent crackout (ratio of kernel weight to total weight of kernel plus shell). The other main cultivar in California is 'Mission' with 50 to 60 percent crackout and a stronger flavor. Later cultivars have not made great inroads except as pollenizers and include cultivars such as 'Peerless,' 'Merced,' 'Butte,' 'Carmel,' and 'Ne Plus Ultra.' Europe has numerous regional and local cultivars. However, all have poor crackouts (e.g., 'Marcona,' 27 percent, and 'Tuona,' 35 percent) and tend to be grown in extensive dryland plantings.

Young almond trees are trained to a vase shape, and more recently as central-leader trees. After establishment of the framework, pruning is restricted to maintaining a reasonably open tree, which is accessible to spraying. The tree can be invigorated, and shoots with new fruiting spurs can be produced by removing crowded limbs and upright waterspouts.

Almonds do not set fruit with their own pollen, so at least two cultivars must be planted together for cross-pollination. However, some cultivars, such as 'Nonpareil' and 'IXL,' do not pollinate each other. Therefore, proper cultivars must be selected carefully to ensure that bloom occurs at the same time. In addition, bees are required for adequate pollination—one to three colonies per acre. Almond fruit-set can also be severely reduced by cold and rainy conditions during the blooming period.

Because almonds are grown in regions where summer rainfall is light or nonexistent, irrigation is needed for good tree growth and yields. Almond trees do best on light, well-drained soils, but are tolerant of poor soils, especially if grown on plum rootstock. They do not grow well on heavy, poorly drained soils, and they require annual nitrogen fertilization and, in some areas, potassium and zinc.

Almonds are harvested in late summer after the hulls split open. In earlier days the limbs were beaten with padded mallets to cause the nuts to fall to the ground or onto canvases where they were picked up. In advanced growing areas, the harvest is completely mechanized. Mechanical tree shakers drop the fruit to smoothly prepared ground where mechanized pickup machines gather the almonds into bins. Harvested nuts are run through a huller and then dried. After drying, the nuts are either shelled if they are to be sold as kernels or bleached if they are to be sold in-shell.

Some almond cultivars are subject to a genetic, noninfectious disorder (almond bud failure) intensified by high growing temperatures. Blossom and twig blight (*Monilinia laxa* and *M. cinerea*) are the major limiting fungal diseases worldwide. The most serious foliage disease is shot hole (*Wilsonomyces carpophilus*), which is presently controlled by chemical applications. Several other fungal diseases important in the Mediterranean region are travelure (*Venturia amygdali* E.E. Fosha), polystigma (*P. occhraceum*), fusicocum (*Fusicocum amygdali* Ducomet), and anthracnose (*Gloeosporium amygdalinum* Brizi.).

In California, navel orangeworm (*Amyelois transitella* Walk) and peach tree borer (*Anarsia lineatella*) can be serious insect pests. Mite species, including pacific spider mite (*Tetranychus pacificus* McG.), two-spotted spider mite (*T. urticae* Koch), European red mite (*Pannonycus ulmi* K), and brown almond mite (*Bryobia rubriculus* Scheuten) can adversely affect production.

Apple (*Malus* × *domestica* Borkh.) ROSACEAE

Apple cultivars grown today apparently originated as complex hybrids of several wild species of *Malus* native to western Asia. Apples were grown in Greece about 600 B.C., and the remains of apples have been found in excavated sites of prehistoric lake dwellers in northern Italy and Switzerland. There is evidence that they were cultivated in the Nile Valley during the reign of Ramses III in the twelfth century B.C. The best seedling trees from natural crosses have been selected since early times, and in recent years plant breeders have used controlled hybridization. More than 10,000 apple cultivars have been documented, but in the United States only about 10 major cultivars are grown. Apple cultivars fall into two main groups: those eaten fresh and those cooked. Some cultivars can be used both ways. At present the two most widely grown apple cultivars throughout the world are 'Delicious,' plus its highly colored red sports (mutations), and 'Golden Delicious.' Both originated in the United States as chance seedlings. Other cultivars popular in the United States, are 'Granny Smith,' 'McIntosh,' 'Rome Beauty,' 'Fuji,' 'Jonathan,' 'York,' 'Gala,' and 'Idared.' The development of the attractive red-skinned sports of the popular 'Delicious' apple (e.g., 'Topred,' 'Red Prince,' 'Sharp Red,' 'Starking Full Red') was a major advance in apple production and has greatly stimulated fruit sales. In recent years, almost 100 percent of the 'Delicious' apple plantings have been of these super-red sports. Such highly colored sports have also been found for several other cultivars. Development of the so-called spur-type sports also stimulated apple production, particularly of the 'Red Delicious' and the 'Golden Delicious.' In the spur-type sports, the branches all over the trees, even the main limbs, are heavily covered with fruiting spurs. These trees can be planted close together, particularly if dwarfing rootstocks are used, and give very high yields.

The apple tree is adaptable to the environmental conditions found in the two temperate zones. It is the most important world fruit crop after oranges, bananas, and grapes. Leading countries in apple production are China, the United States, France, Italy, and Turkey. China now produces about one-third of the world's apple production, following large-scale plantings in the 1980s and 1990s. In the United States, Washington is the leading apple-producing state by far and production is steady. Other high production states include New York, California, Michigan, Pennsylvania, and Virginia. Canada also ranks high in apple production; however, the European countries together produce over twice as many apples as do the North American countries.

Apple trees produce the highest yields of high-quality fruits in regions having long daylight hours with high light intensity during the growing season plus relatively warm days with cool nights and low relative humidity. In such areas, irrigation is necessary. Trees of most modern cultivars are likely to be injured by winter temperatures lower than $-26°C$ to $-29°C$ ($-15°F$ to $-20°F$).

Apple trees are generally trained to a central leader or a delayed open center form (see Chapter 13) to develop a strong trunk with several well-spaced lateral primary scaffold branches. As the trees grow older, the main pruning consists of removing crowded or crossing branches. Annual pruning of bearing apple trees is needed to stimulate production of new growth—about 20 to 30 cm (8 to 12 in.) per year per shoot—and a continual renewal of the fruiting spurs. Large amounts of vegetative growth with light crops indicate excessive pruning (and, perhaps, overfertilization). Very little shoot growth and large numbers of small fruits indicate a need for heavier pruning. Moderate annual growth and vigor is best for optimum yields.

Apple trees bloom later than cherries and most peach cultivars, but the flowers can be damaged by late spring frosts. A temperature of $-3.9°C$ ($23°F$) will kill 90 percent of the flowers at full bloom. A few cultivars are partly self-fruitful, but cross-pollination will generally improve set. Flower buds are produced on tips of shoots or on spurs formed on 2-year-old or older wood. Lateral buds on 1-year-old wood can also be fruitful in some cultivars. Fruit growth follows a single sigmoidal growth curve. Shoot growth is either from a terminal or lateral vegetative bud. New growth of the flowering spur is continued by a bourse shoot in the axil of one of the basal spur leaves. Apples tend to set too much fruit in many years, leading to alternate bearing (a heavy crop of small fruits one year and little or

no fruit the next). This problem can be overcome by fruit thinning in the "on" year (removing many of the small fruits shortly after they have set). A few trees can be thinned by hand, but for commercial orchards chemical spray thinning is necessary.

Apples are propagated by budding or grafting onto rootstock plants, either seedlings (from 'Delicious,' 'Golden Delicious,' 'McIntosh,' or 'Rome Beauty' seeds) or plants propagated by such asexual methods as cuttings or layering (see Chapter 5). Great advances have been made with the apple in developing clonal dwarfing and invigorating rootstocks.

Apple trees grow best in deep, well-drained loam soils. Avoid heavy, wet clays unless structure and drainage can be improved. The best pH range is 5.5 to 6.5, but in some circumstances 4.5 to 8.0 is acceptable. Utilize a good nutrient regime but avoid excessive nitrogen fertilizers, which encourage excessive vegetative growth and reduce storage quality and color on red or partly red cultivars. If shoot growth averages less than about 15 cm (6 in.) per year, the trees should be fertilized annually with nitrogen, and in some soils with potassium. In arid or semiarid regions, the trees must receive summer irrigation.

For highest-quality apples it is important that the fruit be picked at the proper stage of maturity. At harvest the best-quality fruits are firm, nonbruised, crisp, juicy, well-colored, and good-flavored. Apples picked too soon are sour, astringent, starchy, and poor-flavored. Those picked too late soon become soft and mealy and do not store well.

To maintain the quality of apples picked at optimum maturity they should be refrigerated immediately at about $-1°C$ to $0°C$ ($30°F$ to $32°F$) and stored at 90 percent relative humidity. Some cultivars, however, such as 'McIntosh,' 'Empire,' 'Jonathan,' and 'Granny Smith' are best stored at higher temperatures about $2.2°C$ ($36°F$). If held at room temperature, apples soon become overripe, soft, and mealy. Apples held under controlled atmospheres (CA)—1 to 5 percent CO_2 and 1 to 3 percent O_2—to lower their respiration rate keep well under refrigeration for many months. This practice has permitted the year-round marketing of apples. Most U.S.-grown apples sold after April come from controlled atmosphere storage. Millions of kilo-

---

Rootstocks for apples (all the Malling and Malling-Merton stocks were developed in England).

Extremely dwarfing: M.27, C.54

Dwarfing: M.9, Mark, M.26, B.9, O.3, P.2

Semi-dwarfing: MM.106, MM.111, M.2, M.4, M.7, P.22

Standard: Northern Spy, M.1, M.793, MM.104, Domestic seedling

Invigorating: MM.115, M.13, M.25, Robusta 5, Alnarp 2, Bud.490, Maruba

grams of 'Delicious,' 'Golden Delicious,' 'McIntosh,' 'Rome Beauty,' 'Yellow Newton,' and 'Jonathan,' in particular, are CA stored each year. A new compound called MCP (methylcyclopropene) may help in maintaining the quality of apples in CA storage. MCP works by inhibiting the ethylene-binding process within fruit. It also eliminates fruit scald, a common disorder in fruit held under long-term storage.

In addition to fresh apples, other important apple products include apple juice concentrate, pasteurized apple juice, cider, canned applesauce and apple slices, jelly, apple butter, and dried and frozen apples.

Apples are subject to many disease and insect pests. Some diseases affecting apples are scab, powdery mildew, and fire blight. These are less problematic in dry climates than in areas with summer rainfall. Important insect pests on apples are the codling moth, plum curculio, and woolly aphid. Mites can also be a problem. To produce high-quality apples, a series of sprays are required, depending on the diseases and insects present and the prevailing climate. The series starts with a dormant spray before the blossom buds start to open and ends with three or more cover sprays applied while the fruit is developing. A total of 10 or more sprays may be required in humid growing regions. Apples are also susceptible to root lesion nematodes, which are best controlled by fumigating the soil before the trees are planted.

Apricot (*Prunus armeniaca* L.) ROSACEAE

The apricot is believed to have originated in western China, where it has been cultivated since about 2000 B.C. It was introduced into the Mediterranean region about 100 B.C. Apricots were being grown in England in the thirteenth century, generally against south-facing walls or in glasshouses. But, even after centuries of selection and improvement, there are still no commercial cultivars well suited to cool maritime climates. Apricots were brought to the United States in the early 1700s. In recent years, Turkey has been the world leader in apricot production, with Iran, Pakistan, Spain, Italy, France, Morocco, and the United States being other major apricot-producing countries. Apricot trees are hardy to winter cold. And although the flower buds have a moderate chilling requirement to overcome their rest influence, they bloom very early in spring, so subsequent frosts may kill the flowers and fruits. Apricots in full bloom will have about 90 percent of the flowers killed if temperatures drop to $-6.4°C$ (20.5°F). This early blooming habit rules out apricot production in regions with late spring frosts. In regions with very mild winters, apricot buds tend to drop without opening due to insufficient chilling.

The apricot is one of the stone, or drupe, fruits. The edible portion of the fruit is the enlarged mesocarp of the ovary wall. The endocarp is the pit or stone, and the exocarp is the skin of the fruit. The true seed is within the endocarp.

Apricot trees are propagated by T-budding named cultivars onto apricot, peach, or plum seedling rootstocks. In California, over half of all apricot trees are propagated on peach seedlings, but some cultivars in Michigan and Canada are incompatible with peach rootstocks.

Commercial apricot cultivars are 'Blenheim' ('Royal'), 'Tilton,' 'Patterson,' 'Perfection,' 'Moorpark,' 'Derby,' and 'Riland.' 'Tilton' is a heavy producer but the fruits have a poorer color and flavor than those of 'Blenheim' ('Royal'). These cultivars are all self-pollinating except for 'Perfection' and 'Riland,' which require another cultivar for cross-pollination. New apricot cultivars with high-quality fruits are needed to fill gaps in the industry.

Apricot trees are generally trained to a vase-shaped form in their early years. As the trees mature, some of the older branches bend downward, and must be cut back to upright growing laterals. Although apricots bear on spurs, some heavy top-pruning may be required as the trees age to stimulate growth and production of new spurs. Growth stimulation is accomplished by completely removing crowded limbs and cutting back those showing excessive horizontal growth. If the trees grow poorly with yellowish foliage and produce small, early-ripening fruits, they will probably respond to nitrogen fertilizers. Add just enough nitrogen to correct the problem, since excessive nitrogen can cause heavy vegetative growth, delayed fruit maturity, and uneven ripening. Apricot trees require ample soil moisture throughout the growing season either from rainfall or irrigation.

Apricots tend to set too much fruit and must be thinned so that the fruits are about 4 to 8 cm (1.5 to 3 in.) apart by the time they are 2.5 cm (1 in.) in diameter. Otherwise, the fruit will be very small with, perhaps, no crop the following year. Proper fruit thinning is very important in apricot production.

Apricots are one of the first fresh fruits on the market in the spring and early shipments often bring handsome prices. Depending on the growing region, they are shipped to markets into late summer. Apricots can be used fresh, canned, or dried. Tree-ripened apricots are highest in quality, but fresh fruits to be shipped to distant markets are picked somewhat earlier than optimum and allowed to ripen en route. Fresh apricots are very perishable, but they can be stored for 1 or 2 weeks at about 0°C (32°F) and 90 percent relative humidity. The best ripening temperature is about 21°C (70°F). For eating, apricots should be plump and a uniform golden color. Avoid dull-looking, soft fruits or those that are hard and greenish yellow in color. Fruit for commercial or home canning or for jams is picked at a riper stage than for fresh eating—fully ripe but firm. Apricot fruits dry well if they are left on the tree as long as possible to attain a higher sugar content, but they should be picked before they become too soft to handle.

Two close relatives of apricot, *P. mume* and *P. ansu,* are also grown as food crops, but are not eaten fresh. Both perform better in humid regions than the apricot and are important crops in China, Japan, Korea, and Taiwan. *P. mume* fruit is too astringent for fresh eating, but "ume-boshi" pickle and "umishu" liquor are considered to be essential contributors to a healthy diet in modern Japan. *P. ansu* has a larger fruit than *P. mume,* but high acidity makes it unsuitable for fresh eating. It is generally used to make jams and syrups.

Some diseases that affect the apricot include blossom blast (*Pseudomonas syringae*), which can be a major problem in wet springs, and this pathogen also causes bacterial canker. Brown rot (*Monilinia* sp.) at harvest (and also at blossom time) can be a major problem. Several sprayings with a fungicide during the blooming period are necessary to control this fungus. Shot hole fungus (*Wilsonomyceds carpophila*), also known as Coryneum blight, is a serious defoliating disease in most production areas. Perennial canker (*Leucostoma cincta*), Eutypa canker (*Eutypa lata*), bacterial spot (*Xanthomonas campestris*), and anthracnose (*Glomeralla cingulata*) are regionally important diseases.

Compared with many other fruit crops, apricots have relatively few major insect and mite pests. Several borers (*Ansaria lineatella, Scolytus rugulosus,* and *Synanthedon exitiosa*), leaf rollers (several species of tortricid moths), spider mites (several species), lecanium scale (*Parthenolecanium corni*), and aphids (several species) can be damaging in some locations and seasons.

Blackberry [*Rubus* (Subgenus *Eubatus*) spp.]
ROSACEAE

The two main centers of origin of the blackberry seem to be in eastern North America and the European continent. It is speculated that blackberries started their development with the retreat of glaciers during the Ice Age. Although blackberry species are native to many parts of the world, little domestication and commercial use has been made of them except in North America and Europe. Evidence suggests that blackberries were domesticated by the seventeenth century in Europe and during the nineteenth century in North America. Oregon is the leading state in the United States in blackberry production, followed by California, Washington, and Texas, in that order.

Blackberries include many species in both erect and trailing types. The erect forms produce self-supporting arched canes and do not need trellising in contrast to the trailing types, which do require support. The erect types are more winter hardy than the trailing, but the fruits are less sweet.

Blackberries are best adapted to temperate zone climates, with the erect types growing best in areas with cool, humid summers. The trailing types can be grown successfully in regions with hot, dry summers provided they are irrigated. Blackberry canes can be injured when winter temperatures fall below $-23°C$ ($-10°F$). One of the goals of blackberry breeders is to develop cultivars with greater cold

hardiness to expand the growing zone. In the United States the most cold-resistant types are not grown north of hardiness Zone 5, with many of the trailing types confined to Zones 7 and 8.

Blackberries grow best on deep, fertile soils on sloping land that provides good air and water drainage. Ample soil moisture and protection from hot, drying winds are also needed. Avoid bottomland with frost pockets where the blossoms could be killed in the spring. Established plantings usually respond to nitrogen fertilizers.

Erect types are grown in hedgerows with the original plants set about 0.6 m (2 ft) apart in rows 2.4 to 2.7 m (8 to 9 ft) apart. The hedges fill in from suckers arising from the roots. Trailing types on trellises have wider spacing: 1.2 to 2.4 m (4 to 8 ft) apart in rows separated by 2.4 to 3.0 m (8 to 10 ft). A good herbicide treatment of the soil before planting to eliminate perennial weeds is advisable.

Some popular erect cultivars are 'Brazos,' 'Darrow,' 'Cheyenne,' and 'Choctaw.' Semierect thornless types are 'Thornfree,' 'Chester Thornless,' 'Smoothstem,' and 'Hull Thornless,' none of which are hardy in northern climates. Suitable trailing types are 'Boysen,' 'Logan,' 'Marion,' 'Oklawaya,' and 'Young.' Blackberries are generally self-fruitful, but some cultivars benefit from cross-pollination. Bees are helpful in pollination, even with self-fruitful types, to obtain good fruit-sets. Blackberry fruits consist of many individual druplet fruits aggregated over an elongated receptacle to which they are attached.

Blackberry plants have a distinctive growth habit, which must be understood to obtain good fruit production. The roots are perennial, with new shoots arising each year from the crown of the plant. These shoots, or primocanes, which produce lateral branches grow vegetatively during the first year and form flower buds. When the first-year shoots reach a height of 1.2 to 1.5 m (4 to 5 ft), they should be tipped back to induce more lateral shoot formation. In their second year, these shoots or floricanes produce fruits and then die and must be cut out. At the beginning of the fruit-producing season, cut the laterals back to about 30 cm (12 in.) in length to obtain the best fruit quality. Excessive root suckers arising in the rows should be thinned out each year; otherwise, a dense thicket eventually develops. Suckers arising between rows must be removed. After fruiting is over in the summer, cut out all fruiting canes and dispose of them. Thin out the new vegetative shoots arising from the crowns. Canes of trailing types are not cut back but are thinned to 10 to 15 per hill, with the long canes tied to a trellis system.

Erect types of blackberries are easily propagated by digging up suckers with a piece of the root attached. Trailing types are propagated by burying the tips of the canes a few centimeters below ground. The tips of the canes root and produce shoots to form a new plant that can be detached from the parent plant. Blackberries can also be propagated by root cuttings, but when this method is used with some types of thornless blackberries, the new plants revert to thorny forms. Blackberries can also be propagated by tissue-culture methods.

Blackberry fruits should be picked when they turn from red to black and are fully ripe and sweet but still firm. Pick early in the morning every other day as the fruits ripen during the season. The fruit can be kept in good condition for up to 10 days under refrigeration at 2°C (36°F). There is much interest in mechanical harvesting of blackberries, and some commercial plantings for fruit processing are harvested by machines that straddle the rows and shake the berries off.

Blackberries have few insect pests, which generally are not a serious problem. Nematodes can be brought in on the roots of nursery plants, so the roots should be carefully checked. Blackberries are susceptible to Verticillum wilt (*Verticillium albo-atrum* Reinke & Berth.), which can be especially severe if plantings are set out on land previously planted to such crops as tomatoes, potatoes, eggplant, peppers, or cotton. Pre-plant soil fumigation with chloropicrin plus methyl bromide kills the fungus. Orange rust [*Gymnoconia peckiana* (Howe) Trott.] is a systemic disease that kills the plants. A yellowish cast with orange pustules appears on the underside of the leaves. Dig and destroy all infested plants. Rosette or double blossom [*Cercosporella rubi* (Wint.) Plakidas] may seriously reduce yields in southern regions of the United States.

Virus and virus like diseases are found on both wild and cultivated blackberries throughout the United States. These viruses reduce growth, yield, and fruit quality and usually have no visual symptoms. In starting a new planting it is best to plant only state-certified, virus-clean nursery stock from nurseries producing such material, however, only a few states in the United States have such programs.

Blueberry (*Vaccinium* spp.) ERICACEAE

The blueberry is the most recent major fruit crop in the United States to be brought under cultivation from native wild species. Selection and hybridization in the United States since the early 1900s has developed vastly improved cultivars that have formed the basis for an entirely new agricultural industry. Considerable interest in the blueberry has also developed in Europe and Canada.

Cultivated blueberries are mostly the highbush (*V. corymbosum* L.) and, to a much lesser extent, the rabbiteye (*V. ashei* Reade) species. There are about 24 cluster-fruited blueberry species native to the eastern part of the United States and Canada. About 12 other species in other *Vaccinium* subgenera exist in the Pacific Coast states and western Canada. Large quantities of blueberry fruits were harvested from natural wild stands in the eastern United States and Canada for about 300 years after the Pilgrims landed and before any attempt was made to develop improved cultivars and grow them commercially. The impor-

tant highbush cultivars now being grown resulted from hybridization among the native wild species. The hybrid cultivars produce berries three to four times larger than the wild types and have an excellent flavor. Crops of the lowbush blueberry (*V. angustifolium* Ait.) are today taken mostly from native wild stands, although breeding efforts have developed improved cultivars. Lowbush plants are generally only 15 to 46 cm (6 to 18 in.) tall.

Blueberries have rather exacting soil and climatic requirements for good growth and productivity. The soil must be acidic (pH 4.3 to 5.5) and must be well drained, porous, and have a high organic matter content. Some southern areas must maintain pH levels above 4.8 to avoid manganese toxicity. Blueberries are rather shallow-rooted and need ample soil moisture from either rainfall or irrigation. Blueberry roots are believed to have developed a symbiotic association with a mycorrhizal fungus that probably transforms organic soil nitrogen into available forms.

The leading countries in blueberry production are the United States, Canada, and Poland. In the United States, highbush production is found along the Atlantic Coast from eastern North Carolina into New England, in southern Michigan and northern Indiana, and in western Oregon and Washington. Michigan and New Jersey are the top producing states, followed by Maine, North Carolina, Oregon, Washington, and Georgia. Blueberries grow well in either warm or cool, moist summer climates with long sunny days and cool nights during the fruit-ripening period. Canes are killed by winter temperatures below −32°C (−25°F). They require about as much winter chilling as the peach to overcome the rest period of the buds and permit normal growth and blossoming in the spring. This rules out their production in areas with mild winters, such as southern California, although low-chilling rabbiteye and hybrid cultivars grow well in mild-winter areas.

The best commercial highbush blueberry cultivars vary with the region but include 'Bluecrop,' 'Jersey,' 'Weymouth,' 'Croatan,' 'Blueray,' 'Elliott,' 'Rubel,' and 'Berkeley.'

Cross-pollination is necessary for good crops of large fruits. Two cultivars should be planted together as two rows of one, two of the second, and so forth. Insect pollinators are required to obtain a high level of set. One to five colonies of bees per acre should be provided to disperse the pollen. The size of the blueberry fruit is determined by seed number, pollen source, and moisture. High seed number results from good bee activity with pollen provided by another cultivar.

Plants are usually set 1.2 to 1.5 m (4 to 5 ft) apart in rows 3.0 to 3.7 m (10 to 12 ft) apart. Blueberries are often planted on the poorer soils, in which case a complete fertilizer is generally required for good growth.

The blueberry produces its fruit on shoots that grew the previous season, with the largest fruits on the most vigorous growth. Blueberries tend to overbear, producing small, late-maturing fruits unless pruned. Pruning consists of thinning out small, weak branches, cutting out some of the older stems in the center of the bush, and removing low-drooping branches close to the ground. Pruning is best done in the winter after the coldest weather has occurred.

Blueberries are propagated by hardwood or softwood cuttings. Hardwood cuttings, 10 to 13 cm (4 to 5 in.) long, made from dormant shoots (without fruit buds) of the previous summer's growth, are usually used and planted in a peat moss–sand mixture under covered frames with automatic watering. Cuttings of some cultivars root much easier than those of others. Blueberries reproduce naturally by shoots arising from spreading belowground stems (rhizomes). Most blueberries root well from softwood stem cuttings under intermittent mist.

Blueberry fruits are mostly hand-picked, but machine harvesting is also used for processing fruits. In one type of harvester, electric-powered vibrators shake the berries off into a catching frame. Large, over-the-row, self-propelled machines are also used to shake the berries off. Lowbush blueberries are harvested by handheld scoops with rake-type teeth passed through the low-growing bushes to separate the berries from the plants.

Blueberries are fully ripe and ready to harvest when they are light to dark blue or blue-black. Overripe fruits become watery and soft. Blueberries are truly a convenience food item with no pitting, peeling, or waste. They are ready to eat with just washing. Blueberries are marketed fresh, canned, or frozen. Fresh berries can be held in good condition for 2 to 4 weeks, depending on the cultivar, at 0°C (32°F) and 90 percent relative humidity, or for several months if frozen. Blueberries are attacked by a number of viral, fungal, and bacterial diseases, as well as by nematodes and certain insect pests. Virus-tested nursery plants are available, and growers should make every effort to obtain them for new plantings. Local agricultural extension service workers should be consulted. Blueberries are attractive to birds, and the plants may have to be protected by covering them with plastic netting suspended on a framework for small plantings.

Cherry (*Prunus avium* L. and *P. cerasus* L.) ROSACEAE

The commercially important cherry cultivars belong to two species: the sweet cherries (*P. avium*) and the sour cherries (*P. cerasus*). The cherry apparently was domesticated from wild forms growing in south-central Europe and Asia Minor, and has been grown throughout Europe since ancient times. Cherries were introduced into North America from Europe by the earliest settlers shortly after Columbus' voyages. Turkey is the leading sweet cherry-producing country, followed by the United States and Iran. The leading countries in sour cherry production are Russia, Poland, and the United States.

In the United States most sour cherries are produced east of the Rocky Mountains, particularly in Michigan and New York, while sweet cherries are grown mainly in Washington, Oregon, Michigan, California, and New York. Sour cherry trees are hardy in winter cold and are adaptable to a range of soil types if they are well drained. The fruits are generally used for processing in pies, jams, and jellies.

Sweet cherry trees are less tolerant of low winter temperatures than the sour cherry. They grow best where they obtain sufficient winter chilling to overcome bud dormancy and where the summers are mild (have some, but not excessive, summer heat) and humidity is low. Most commercial sweet cherry cultivars require cross-pollination to set fruit, along with bees to distribute the pollen. Not all cultivars are interfruitful, so care must be taken to plant sweet cherry trees of at least two cultivars that will bloom together and cross-pollinate each other. A temperature of −3.9°C (25°F) at full bloom will kill 90 percent of sweet cherry flowers.

Some sweet cherry cultivars commonly planted are 'Bing,' 'Lambert,' 'Royal Anne' ('Napoleon'), 'Black Tartarian,' 'Rainier,' 'Stella' (self-fruitful), and 'Van' (use 'Black Tartarian' and 'Van' as pollenizers for the other cultivars. A significant advance in breeding was the development of self-fertile sweet cherry cultivars. Commercial quality self-fertile sweet cherry cultivars include 'Celeste,' 'Isabella,' 'Lapins,' 'New Star,' 'Starkrimson,' 'Stella,' 'Sunburst,' and 'Sweetheart.' The main sour cherry cultivars are 'Montmorency,' 'Schattenmorelle,' and 'Meteor.' They are all self-fruitful and set heavy crops in solid block plantings provided enough bees are present. The Duke cherries, intermediate in types between the sweet and sour, are considered hybrids between these two groups.

Cherry trees are propagated by T-budding on seedling rootstocks. Both *P. avium* (Mazzard) and *P. mahaleb* seedlings are used as rootstocks for sweet cherries, whereas sour cherries are grown mainly on *P. mahaleb* roots. *Mahaleb* stock is hardier and more drought resistant than Mazzard. 'Colt,' a clonal dwarfing cherry rootstock, was developed in England and is being tried to some extent in the United States, but it is not winter-hardy in the northern cherry districts, and it is extremely susceptible to crown gall. A clonal Mazzard rootstock (F 12/1) has been used commercially for many years and seems quite satisfactory. Rootstock breeding programs in Germany (Giessen series, Weiroot series), Belgium (Gembloux series), and the Czech Republic (P-HL series) have introduced rootstocks capable of dwarfing sweet cherry trees and greatly increasing precocious fruiting. However, several of the rootstocks have responded adversely to common pollen-borne ilarviruses (prune dwarf virus and prunus necrotic ring spot virus) found in the Pacific Northwest of the United States. This underscores the importance of including early screening of new rootstocks for viral sensitivities in the process of characterizing rootstocks for their suitability in commercial fruit production.

Young sweet cherry trees tend to grow upright with long, unbranched scaffolds. Dormant pruning for the first several years must include heading back these limbs to outward growing buds to encourage branching and removing any crowded inside limbs. As the trees mature and develop fruiting spurs, little pruning is needed other than the removal of interfering branches, thinning out dense growth, and thinning out vigorous, upright-growing shoots. Some pruning is needed to keep new growth developing, upon which additional fruiting spurs form to replace older unproductive ones. Sour cherry trees are more spreading than sweet cherries and are usually pruned to a vase-shaped form. Mature trees are pruned to thin out dense growth and to invigorate shoots for production of fruiting spurs.

Cherry trees, especially sweet cherries, require a continuous supply of soil moisture during the growing season but do not tolerate wet, poorly drained soils. A deep, well-drained loam soil should be used. The trees respond to nitrogen fertilizer applications and in some areas to zinc. High-nitrogen fertilization can cause excessive tree vigor, poor fruit quality, and delayed fruit maturation.

Cherries are stone fruits, the edible parts of the fruit being the outer layers of the mature ovary wall—the flesh (mesocarp) and the skin (exocarp). The pit (endocarp) encloses the seed.

Cherries are harvested in May and June in California and in mid-June through July in most other cherry-producing states. Cherries for fresh shipment are very perishable and require careful handling. They can be held in cold storage for 2 weeks at 0°C (32°F) at 95 percent relative humidity if placed in storage immediately after harvest. Best-quality fresh fruits are firm, juicy, and well colored. Substantial amounts of cherries are brined and reach the consumer as bottled maraschino cherries, canned cherries, canned fruit cocktails, package glacé cherries, pies, jellies, and ice cream toppings.

Cherries are subject to several fungal diseases, particularly brown rot (*Monilinia fructicola* and *M. laxa*) and cherry leaf spot (*Blumeriella jaapii*). Silverleaf (*Chondrostereum purpureum*) is a serious fungal disease in New Zealand, Chile, and France, and *Verticillium* wilt can be a problem in many locations. Bacterial canker (*Pseudomonas* sp.) is a worldwide problem and *Phytophthora* crown or root rots can be troublesome. Cherries are susceptible to several virus and virus-like diseases spread by pollen and by insect vectors, so it is important that only certified virus-free nursery trees be planted.

The most important pests of the cherry are various species of cherry fruit fly (*Rhagoletis* sp.), leaf rollers (*Archips* sp. and *Pandemis* sp.) and the black cherry aphid (*Myzus cerasi*). Plum curculio can be a problem, and spider mites can build up to damaging levels.

Cranberry (*Vaccinium macrocarpon* Ait.) ERICACEAE

The cranberry (*Vaccinium macrocarpon* Ait.) is a native American fruit species and is closely related to the European cranberry (*V. oxycoccus* L.) and the lingoberry (*V. vitis-idaea*). It was found growing wild and used by the Indians when the Pilgrims landed on the New England coast. Commercial culture was started in 1810 by Henry Hall, a Revolutionary War veteran, in the town of Dennis, Massachusetts, on Cape Cod. Cranberry plantings were established in New Jersey in 1835, Wisconsin in 1853, and on the West Coast in 1883. The United States leads in world cranberry production, followed by Canada (Nova Scotia, Quebec, and British Columbia) and Latvia. Wisconsin leads in U.S. production, followed by Massachusetts, New Jersey, Oregon, and Washington. Cranberries are little grown elsewhere in the world, although there is interest in this crop in some European countries. Cranberry production is confined to temperate zone regions with cool, moist climates and, in nature, to acid soils along the edge of streams and bogs.

The cranberry is a low-growing evergreen vine. In late summer upright shoots form flower buds that develop into flowers by midsummer of the following year. Insect pollination is required for fruit-set. In the last 150 years of commercial cranberry production, 132 wild selections have been made plus seven improved cultivars obtained by controlled hybridization. The four leading cultivars, accounting for the bulk of commercial production are 'Early Black,' 'Howes,' 'Searles,' and 'McFarlin.' These all originated as single vines selected in native bogs in Massachusetts, except 'Searles,' which was selected in Wisconsin. Cranberry breeding programs have been activated in several of the producing states to develop improved cultivars, and a number of new cultivars have been introduced.

Cranberries can be grown successfully in a wide range of soils, from pure sand to peat, provided the substrate is acidic. While most fruit crops prefer a soil pH from 6 to 7, cranberries grow best with a soil pH ranging from 4.0 to 5.5. Cranberries are grown in bogs about 0.8 ha (2 ac) in size that can be flooded and drained at the appropriate times. Levees, pumps, and drainage ditches are required, which increase the cost of production. The bogs are leveled and covered with a layer of sand about 1.3 cm (0.5 in.) deep. Vines are planted by cutting them up into 7.5 or 10 cm (3 or 4 in.) pieces with a device such as a corn chopper. The pieces are broadcast over the bogs at about 1,120 kg/ha (1,000 lb/ac) and are then pressed into the soil with a heavy mechanized roller that has parallel ridges several inches apart. After about 2.7 m^2 (30 ft^2) are planted, the area is sprinkled immediately with water and, after the entire bog is planted, it is thoroughly irrigated. The pieces root quickly and, with subsequent nitrogen fertilization, vigorous vine growth is encouraged to bring the planting into production as soon as possible. Herbicide sprays are used to control weeds. The bog is sanded again every 3 to 4 years.

Cranberry bogs in some areas are flooded in late autumn as soon as the soil freezes so that the plants overwinter in or under a thin layer of ice. This prevents the plants from being killed from desiccation during the winter, and it helps

Figure 26–1 Water wheels driven through the cranberry bogs after flooding shake the ripe berries off the vines and they float to the surface. The berries are then lifted by conveyers from the water into waiting trucks. *Source:* Ocean Spray Cranberries, Inc.

eliminate some insect pests. Water is drained from the bogs in the spring after danger from a severe freeze is over. Sprinkler irrigators in the bog are used for the plants during the summer growing season. Three to five years are required for the bog to develop full production.

To obtain large berries a good seed set is required; thus, adequate pollination is necessary. Two to four hives of bees per hectare are necessary during the blooming period.

The harvest season in Massachusetts extends from mid-September to early November. Harvesting is mechanical with two methods: water picking and dry harvesting. Almost 85 percent of the crop is "wet picked," which entails flooding the bog to between 6 in. and 3 ft deep. A machine called a "water wheel" (Fig. 26–1) is then driven through the water, shaking the ripe berries off the vine in an eggbeater fashion. The berries float to the top of the water, where they are corralled and lifted by conveyor belt from the water to an awaiting truck. The dry harvest method, used primarily for fresh fruit, employs a machine similar in appearance to a lawn mower. The machine is guided over the dry bogs, combing the berries from the vine. The berries are then lifted by a small conveyor to the back of the machine, where they are dumped in a bag or box before being transported to a truck for delivery to the receiving station. Wet-picked cranberries are used exclusively for processing.

Seventy percent of the overall crop is frozen for later use in juice drinks, 20 percent is employed for sauce, and 10 percent used for the fresh-fruit market. A portion of the dry-picked cranberries not used for the fresh market is used in various sauce forms. Cranberries for fresh market can be stored up to 4 months at 2.2°C to 4.4°C (36°F to 40°F) at 90 to 95 percent humidity and with good ventilation.

Cranberries are subject to several fungal vine diseases and storage diseases of the harvested fruits. In setting out new plantings it is important to obtain vine cutting material from a disease-free source. It is estimated that storage rots beginning in the growing fields can cause a 25 percent crop loss.

There are a number of insect pests on cranberries. Flooding the bogs at appropriate times can control some of these. Other methods include application of insecticides by air or rotary sprinkler systems.

Currant (*Ribes* spp.) SAXIFRAGACEAE

There are some 150 species in the genus *Ribes* distributed mostly in the temperate zone regions of the Northern Hemisphere. Currants were domesticated from wild plants within the last 400 years. The leading countries in currant production are Russia, Poland, and Germany. Black currants (*R. nigrum*) were grown in England in the seventeenth century and brought to North America by the Pilgrims in the early 1600s. The red currant (*R. sativum*), derived from *R. petraeum* and *R. rubrum,* is the principal type grown in the United States and is best adapted to climates with cool, humid summers. They do not grow well in hot, arid regions. Red currants are very winter-hardy and are easy to grow. They do well on medium- to heavy-textured soils that range from slightly acidic to slightly alkaline. They are shallow-rooted and need ample soil moisture from rain or irrigation, but the roots do not tolerate standing water very long.

Red currants grow as bushes with upright shoots developing from the crown of the plant. Fruit is produced from buds at the base of one-year-old shoots and from spurs on older wood. Since the older shoots become less fruitful, they should be retained only for about 3 years, and then removed to encourage new shoot growth. This pruning should be done annually in early spring. Currant bushes should be fertilized annually either with a well-rotted manure in the fall or a complete fertilizer in early spring.

Currants are easily propagated by hardwood cuttings planted in late autumn to early spring before the buds develop. Leading red currant cultivars are 'Red Lake,' 'Perfection,' and 'Jonkheer van Tets.' Popular white currant cultivars include 'White Imperial' and 'White Versailles,' while 'Baldwin,' 'Topsy,' and 'Ben Lomond' are leading black currant cultivars.

Comparatively few insect or disease pests affect currants. Currant bushes are an alternate host to the white pine blister rust; thus, the bushes have been eradicated in many parts of the United States.

Gooseberry (*Ribes* spp.) SAXIFRAGACEAE

Gooseberry culture was not mentioned in the writings of the early Greek and Roman botanists. The crop was apparently grown in Britain and Western Europe in the late thirteenth century. The European gooseberry (*R. uva-crispa* L.) did not thrive in America when it was brought over by the early settlers because of American gooseberry mildew and a generally unsuitable climate. Hybridization between the European and the native American gooseberry (*R. hirtellum*) produced 'Houghton' and its seedling 'Downing,' which were long the standard gooseberry cultivars grown in the United States. More recently developed cultivars are 'Poorman,' 'Welcome,' 'Glenndale,' 'Abundance,' 'Chautauqua,' 'Pixwell,' and 'Oregon Champion.'

Climatic requirements and culture for the gooseberry are essentially the same as those for the red currant. The gooseberry is also an alternate host for white pine blister rust. Although gooseberries are popular in Europe, in the United States interest has declined and they are generally grown in home gardens. Commercial production is chiefly in Oregon. Gooseberries are used primarily for jam, jelly, and pies.

Grape (*Vitis* spp.) VITACEAE

Grapes are one of the world's major fruit crops and have been since earliest recorded history. World production exceeds that of any other fruit. The European grape (*Vitis vinifera* L.) is believed to have originated in the region of the Caucasus Mountains between the Black and Caspian Seas, where it still grows wild. From there it was introduced into the Mediterranean region and on throughout Europe and later by explorers to all the continents. There is evidence that the ancient Egyptians were cultivating grapes some 6,000 years ago. Cortez introduced vines in the Americas in 1524. *V. vinifera* is, by far, the most important species, accounting for grapes to make wine, for drying, and for fresh fruit.

About 18 species of grapes important to viticulture as fruiting or rootstock types are native to North America. These include such species as *V. labrusca* L., which has several important fruiting cultivars, and *V. riparia* Michx., *V. rupestris* Scheele, and *V. berlandieri* Planch. and *V. champinii* Planch., from which a number of rootstocks resistant to phylloxera and nematodes have been derived. In addition, the Muscadine grape (*V. rotundifolia* Michx.) has several fruiting cultivars.

Grapes are widely grown on every continent in the two temperate zones, but four countries—Italy, France, the United States, and Spain—dominate grape production, with almost half the total world's supply. Other major grape-producing countries are the former Soviet Union, Turkey, Portugal, Rumania, Algeria, Argentina, Yugoslavia, Hungary, Greece, and Germany. In the United States, California leads, by far, in grape production with about 4.9 million tons produced annually. Washington is second with about 96,000 tons, followed by New York (92,000 t), Pennsylvania (78,000 t), and Michigan (46,000 t).

Although grapes are used mostly for wines and juices, considerable amounts are consumed as raisins, fresh table grapes, jams, and jellies.

The important *V. vinifera* cultivars are generally grown in warm temperate climates, but are not well adapted to tropical environments. The *V. vinifera* cultivars are not grown nearer to the equator than about 20° latitude (unless the elevation is high) because in the Tropics the vines become evergreen and unfruitful. Grapevines require a period of inactive growth to overcome bud dormancy. Heat accumulation (measured as degree days) is also a determining factor as to where grapes can be successfully grown. Most grapes will not satisfactorily ripen when less than 700°C to 900°C days are received. Several areas receiving adequate summer heat cannot grow *V. vinifera* grapes because of winter cold damage. Dormant vines are killed at temperatures ranging from −18°C to −12°C (0°F to 10°F). The native American species are much more winter-hardy than *V. vinifera*, and their cultivars are grown in areas with severe winters. However, dormant vines of *V. labrusca* cultivars can be damaged by temperatures ranging from −22°C to −28°C (−8°F to −18°F).

V. vinifera cultivars have rather exacting climatic requirements during the growing season for heavy production of a high-quality product. A long, warm or hot and dry summer is best. High humidity and rains result in disease problems for the vine and fruit. *V. labrusca* and *V. rotundifolia* cultivars, on the other hand, do well in regions with summer rains and high humidity. For best quality wines from *V. vinifera* cultivars, a long growing season is required to produce enough sugar in the grapes to ferment to alcohol. Relatively cool temperatures produce grapes with high acidity to contribute quality to the wine. Such climatic conditions are found in northern coastal California valleys, the Burgundy and Bordeaux districts of France, northern Spain, central and northern Italy, and Yugoslavia.

The important present-day grape cultivars are listed in Table 26–1. About 8,000 cultivars worldwide have been named and described. In addition to these, some French hybrid cultivars are widely grown in the eastern United States, particularly for making wine. Examples are 'Aurore,' 'Baco Noir,' 'Foch,' 'Seyval Blanc,' 'Vidal Blanc,' 'Dechuanac,' and 'Chancellor.'

The fact that grapes grow well over such a wide area of the world shows that they are adapted to a wide range of soil types. Generally, however, heavy clay, shallow, sodic, or poorly drained soils should be avoided. Traditionally, especially in Europe and the eastern United States, vineyards are heavily fertilized with manures or straw. Grapes respond to nitrogen fertilizers, but apparently have a relatively low nitrogen requirement and they do not show pronounced visual deficiency symptoms. Potassium deficiency (black leaf areas, leaf margin chlorosis, and necrosis) in grapes is common, and the vines respond markedly to potassium fertilizers. Grapes have responded to phosphorus fertilizers in certain acid soils in California. Iron, manganese, and zinc deficiency (little leaf and foliar chlorosis) can also sometimes occur in grapes. Grapes require ample soil moisture during the growing season either from rainfall or irrigation.

TABLE 26–1	Some Important Grape Cultivars of the World

European (*Vitis vinifera*)			American (*Vitis abrusca*)			(*Vitis rotundifolia*)
Wine	Table	Raisins	Wine	Table	Juice	Table and Wine
Red **White**	Almeria	Black Corinth	Catawba	Beta	Beta	Carlos
Aleatico Burger	Black Beauty	(Zante currant)	Concord	Bluebell	Concord	Cowart
Alicante Chardonnay	Calmeria	Black Monukka	Delaware	Concord	Fredonia	Fry
Bouschat Chenin blanc	Cardinal	Muscat of	Diamond	Edelweiss	Niagara	Higgins
Barbera French Colombard	Emperor	Alexandria	Elvira	Kay Gray	Van Buren	Hunt
Cabernet Gewurztraminer	Exotic	Thompson	Horizon	Ontario		Jumbo
Sauvignon Malvasia Bianca	Flame Tokay	Seedless	Ives	Price		Magnolia
Carignane Muscat blanc	Italia	(Sultana,	Niagara	St. Croix		Nesbitt
Carmine Pinot blanc	Lady Fingers	Sultanina,	St. Croix	Swenson Red		Noble
Carnelian Sauvignon blanc	Malaga	Oval Kishmish)		Valiant		Scuppernong
Gamay Semillon	Perlette			Worden		Sterling
Beaujolais Thompson	Queen					Sugargate
Gamey Seedless	Red Flame					Watergate
Grenache White Riesling	Seedless					
Grignola (Johannisberger	Redglobe					
Mataro Riesling)	Ribier (Alphonse					
Merlot	Lavallé)					
Napa Gamay	Rish Baba					
Petit Sirah	Thompson					
Pinot noir	Seedless					
Rubired						
Ruby						
Cabernet						
Zinfandel						

As in deciduous tree fruits, flower parts and clusters in the grape initiate in buds on new shoots shortly after bloom ends the preceding season, continuing on through the summer. Following the winter dormancy period, shoots then develop from buds produced the previous season, and these shoots bear the buds that develop into the flower clusters. *V. vinifera* and *V. labrusca* cultivars are usually self-pollinated but, in some instances, cross-pollination by wind or insects improves fruit-set. *V. rotundifolia* cultivars are dioecious, with male and female flowers on different plants. However, breeding and selection have produced self-fertile varieties with near-perfect flowers, which also serve as pollen sources for pistillate cultivars. It appears that both wind and insects play a role in the pollination of female flowers. Fruits of most grape cultivars have seeds, but seedlessness in the important 'Thompson Seedless' cultivar is due to early embryo abortion.

Grapes, except *V. rotundifolia* cultivars, generally are easily propagated by cuttings, either hardwood or leafy softwood under mist. This is the usual propagation method unless certain resistant rootstocks are required to protect the roots of *V. vinifera* L. cultivars from attacks by the soilborne grape phylloxera (*Dactylasphaera vitifoliae*) or from root-knot nematodes (*Meloidogyne* spp.). In these cases grafting or budding is done using as rootstocks rooted or unrooted cuttings of certain hybrid rootstocks (e.g., 'AxR#I,' 'Har-

mony,' 'Dogridge,' and 'St. George'). These rootstocks have been developed for this purpose using resistant native American species, such as *V. champinii, V. riparia, V. rupestris,* and *V. berlandieri,* as one of the parents.

Grapes must be pruned heavily and in a certain manner, depending on the fruit-bearing habit, in order to obtain maximum production.

Table grapes are harvested when the berries attain their most attractive appearance and eating quality. Grapes do not ripen further after harvest. Quality is highest just as the clusters are cut from the vine, and deterioration starts at that point. Quality loss can be slowed by immediate cooling of the clusters to about −1°C to 0°C (30°F to 32°F) with relative humidity at 85 to 95 percent. Fumigation of *V. vinifera* table grapes with sulfur dioxide (1 percent for 20 minutes) as soon as possible after harvest and before storage or shipment reduces decay and permits fruit storage of some cultivars, such as 'Emperor' and 'Calmeria,' for as long as 6 or 7 months. Grapes held in prolonged cold storage are refumigated at about 10-day intervals. American-type grapes are injured by sulfur dioxide, so they are not fumigated.

Wine production from the fruit of the grape has intrigued people since Neolithic times. Many books have been written about all phases of wine production and utilization. Wine production is based on a rather complex alcoholic

HOW FINE WINES ARE MADE

After harvesting at the desired sugar and acid levels, the grapes (*Vitis vinifera* wine cultivars) are hauled to the winery in large steel tanks or gondola trailers holding from 1 to 5 MT (1.1 to 5.5 t). The grapes must be clean and in good condition. The grapes, either red or white, are dumped onto a conveyor that carries them to the crusher-stemmer. This can be two rollers set close together (crusher only), but usually consists of revolving paddles that slap the berries free of the stems and at the same time crack them open (crusher-stemmer). The crushed berries, free from the stems, collect in the bottom of the crusher-stemmer and are pumped out into the fermented tanks. To counteract oxidation, growth of "wild" yeasts and other undesirable microorganisms, and development of "browning" enzymes, all of which would reduce wine quality, sulfur dioxide is introduced into the crushed grapes. For white wines the juice is pressed from the white grapes immediately after crushing to prevent extraction of bitter tannins and other undesirable materials from the skins. The juice is pumped to tanks for overnight settling and removal of any grape pulp or other semisolid materials. The next morning the clear juice is transferred into closed fermenting tanks where it is inoculated with an exact amount of specially cultured wine yeast to start a controlled fermentation in which the grape sugars are converted to alcohol and CO_2. After fermentation starts, the temperature must be held between 7°C and 10°C (45°F and 50°F) for white wines. The fermentation process gives off heat that must be removed. If the temperature should exceed 32°C (90°F), the wine yeasts are destroyed and fermentation stops (this is called a "stuck wine"). Air is kept out since oxidation darkens the juice and eliminates the fresh, fruity grape flavors. In earlier days of wine making, temperature was controlled by placing fermentation tanks in cool, constant-temperature tunnels and caves dug into rocky hillsides. Modern wineries attain exact temperature control with large refrigeration rooms with jacketed stainless steel tanks individually controlled by refrigeration.

In making red wines the grape skins, pulp, and seeds, along with the juice, are included during the fermentation process to obtain the color, tannins, and flavors characteristic of the red wines. Fermentation is started in the crushed red grapes by inoculation with a pure yeast culture. About 665 L (175 gal) of juice are obtained from each short ton of red grapes—with white grapes 57 to 76 L (15 to 20 gal) less juice per ton is obtained.

Temperatures during red wine fermentation are much different than for white wines. The best temperature for the reds is between 21°C to 26.5°C (70°F to 80°F). These warmer temperatures are necessary to extract the tannins, color, and flavors from the skins.

Rose or pink wines are made from red grapes, but the color—coming from the skins—is controlled by the length of time the juice is in contact with the skins. After fermentation continues for a few hours and the juice becomes pink, it is drained and pressed free from the skins with fermentation continuing in closed tanks at the same temperature, 7°C to 10°C (45°F to 50°F), used for the white wines.

After fermentation has proceeded to a certain point in making all types of wines, the still cloudy juice (wine) is transferred to closed storage tanks. Particles settle out and the clear wine is poured off (racked). This may be repeated several times to eliminate all sediment. Clarifying agents are added to help pull down particles staying in suspension. The wine is finally filtered to remove all suspended material and to give it a brilliant clarity.

High-quality wine must be aged in oak casks or barrels, which are kept completely full to prevent oxidation and spoilage. Some wine evaporates through the pores of the wood, so the casks are continually topped to keep them full. Aging in the wood barrels varies with the type of wine. Reds take 1 or 2 years, most whites 9 months, and the rose wines less. Some wineries then stop the wood-aging process by transferring the wine to large glass-lined tanks for further aging, perhaps up to 12 months, during which time the wine constituents change chemically, allowing flavor, color, and odor characteristics to develop more fully.

Bottling of the wines is the final critical process. Every effort is made to avoid introducing harmful oxygen into the wines. Inert nitrogen under gentle pressure is often used instead of pumps to move the wine from storage tanks to the bottling area. Empty bottles are freed of air by purging each one with nitrogen just before the wine enters. Bottles are stoppered by creating a vacuum just before the corks are inserted. Aging and maturing of the wines still continues after bottling. A peak of quality is reached, after which the quality starts to decrease. The rose wines are the shortest lived. The white wines generally reach a peak of quality after 2 or 3 years. The red wines keep developing quality longer and are slower to decline. Cabernet Sauvignon, for instance, is likely to keep improving up to 20 years in the bottle under proper storage conditions.

In commercial wine-making enterprises all operations along every step of the way are carefully controlled. Laboratory examinations and analytical procedures are used to determine the microbiological processes, chemical composition, and sensory qualities.

Wine making from *V. labrusca* grapes, like 'Concord,' which are low in sugar, differ somewhat from that described for *V. vinifera*. The crushed grapes are immediately transferred into a holding vat where enzymes are added to break down mucilaginous substances in and around the pulp. Heating is done during this process to develop the desired color, after which the juice is pressed out and cooled. It is then transferred to the first fermenting tank where sulfur dioxide, yeast, and extra sugar are added. Following this, the movement of the wine through settling vats and aging tanks is similar to that described previously for the *vinifera* grapes.

Wines can be classed as either the dry table wines, which have no more than 14 percent alcohol, or the sweeter "aperitif' and "dessert" wines (sherry, port, muscatel), which have an alcohol content of about 20 percent because of the addition of brandy distilled from wine. Sparkling wines, like champagne from white wines and sparkling burgundy from reds, are prepared by a secondary fermentation of dry table wines in a closed container. This involves the addition of a certain amount of sugar and a pure yeast culture, and the wine undergoes a second fermentation and produces enough carbon dioxide to develop a pressure of 4 or 5 atm (4 to 5 kg/cm²).

fermentation process, first explained by Pasteur in 1866, whereby the sugars (glucose and fructose) in grapes are converted to alcohol and carbon dioxide by the activity of the enzymes produced by wine yeast (*Saccharomuces cerevisiae* var. *ellipsoideus*).

$$1 \text{ glucose } (C_6H_{12}O_6 \Rightarrow 2 \text{ ethyl alcohol } (CH_3CH_2OH) + 2$$
$$\text{carbon dioxide } (CO_2) + \text{about 56 kilocalorie of energy}$$

Other chemical constituents in the grape produce the characteristic flavor, color, and odor of wine. Wine becomes vinegar if certain bacteria (*Acetobacter*) develop in the wine, changing the ethyl alcohol to acetic acid.

$$1 \text{ ethyl alcohol } (CH_3CH_2OH) + \text{oxygen } (O_2) \Rightarrow 1 \text{ acetic}$$
$$\text{acid } (CH_3COOH) + 1 \text{ water } (H_2O)$$

Keeping equipment and containers clean and preventing the exposure of wine to air minimizes the acetic acid reaction.

Raisins are dried grapes of a cultivar with a high sugar content like 'Thompson Seedless.' The grapes are harvested for raisins when the total soluble solids (mostly sugars) reach about 20 percent. Grapes below 17 percent soluble solids give a poor product and should not be dried for raisins. About 194 kg (432 lb) of raisins can be obtained from a MT (2,205 lb) of harvested grapes. Raisin grapes are usually dried on paper sheets placed in the sun between the vine rows. The grapes are turned once during drying, then they are rolled in the paper for further "curing," finally being emptied into boxes for delivery to the packing plant. At this time the raisins should contain 10 to 15 percent moisture.

Grapes are subject to many diseases and production may be eliminated from areas where they occur. High humidity promotes the development of certain fungal diseases, such as anthracnose, Eutypa (deadarm), botrytis, bunch rot, black rot, and downy mildew, which do not occur in dry, semiarid regions. Powdery mildew, however, develops easily in dry climates and is, by far, the most troublesome fungal disease on grapevines in California. It is controlled by dusting the vines with sulfur throughout the growing season. Except for Pierce's Disease (PD), diseases caused by bacteria generally are not a major problem with grapevines. PD (*Xylella fastidiosa*) is spread by leafhoppers known as sharpshooters. In the southern states, PD is the most formidable obstacle to the growing of *V. vinifera* grapes. There are several viral diseases, such as fan leaf, yellow mosaic, vein banding, yellow vein, arabis mosaic, leaf roll, and corky bark, that cause stunting, deformity, chlorosis, delayed fruit maturity, poor color, and reduced fruitfulness. Best control of viral diseases is not to take propagating material from infected vines but to use certified planting stock. With the modern practice of heat therapy to eliminate viruses from plant material, it is now possible to obtain propagating material from vines free of these known viral diseases.

Insects and other pests harmful to grapevines and their fruit include the grape leafhopper, grape leaf folder, omnivorous leaf roller, spider mites, grape mealy bug, grape bud beetle, grasshoppers, nematodes, and phylloxera.

Phylloxera, a type of aphid, is native to the United States east of the Rocky Mountains where it feeds on roots of the native American species but does not kill them. Phylloxera was identified in France about 1868, having been introduced earlier on the roots of some imported American vines. This pest spread rapidly throughout France and, within 30 years, 75 percent of the *vinifera* grapevines in France were destroyed. Phylloxera was also found in California grape-growing areas in 1873. The only method of growing *vinifera* grape cultivars in areas where phylloxera is present in the soil is by grafting onto roots of special resistant hybrid rootstocks that include some parentage of the native American *Vitis* species. In about 20 percent of the California vineyards the vines are grown in heavy soils where such resistant rootstocks must be used. In the sandy soils, phylloxera is not a problem but often nematodes are. Nematode-resistant rootstocks must be used. Propagation by cuttings is the usual method; however, only where these pests are not present.

Hazelnut (Filbert) (*Corylus avellana* L.)
BETULACEAE

There are about 9 species of *Corylus,* all indigenous to North America, Europe, and Asia. The European hazelnut (*Corylus avellana* L.) was the dominant tree species in much of northern Europe from 8000 to 5500 B.C. Hazelnuts were being cultivated during the ancient Greek civilizations. The term *hazelnut* is used worldwide for nuts of all *Corylus* species. The common name "filbert" originated in England to distinguish long-husked types of *C. avellana* from those with short husks. *C. avellana* is the species usually cultivated for the nuts. Hazelnut production is mainly limited to climates moderated by large bodies of water such as the southern coast of the Black Sea, the northern coast of the Mediterranean Sea, and the Northwest coast of the United States where the climate is influenced by the Pacific Ocean. Turkey accounts for over 75 percent of the world's production, followed by Italy, the United States, and Spain. Oregon produces about 99 percent of the hazelnuts harvested in the United States.

Hazelnut buds require some winter chilling to overcome their dormancy conditions. Hazelnut trees grow in a wide range of soil types. Fertilization is usually unnecessary unless shoot growth is weak, in which case they should receive 0.22 to 0.45 kg (0.5 to 1 lb) of actual nitrogen per tree per year.

In Europe, hazelnuts are trained as a multistemmed bush, but in Oregon and Washington they are trained as trees to facilitate mechanized orchard practices. Regardless of the training system suckers should be removed, as they interfere with nut harvest. When trained as a tree, four to five main scaffold branches are selected. Pruning is done primarily to stimulate new growth and to remove any broken or interfering branches.

Hazelnuts are monoecious plants; that is, the pollen-bearing male flowers (catkins) and the female (pistillate) flowers are produced separately on the same tree. In addition, most hazelnuts are self-unfruitful; that is, two cultivars must be interplanted for cross-pollination. Furthermore, the pollen must be shed on one cultivar at the time when the pistils are receptive on the other. If all these conditions are not met, poor crops of nuts result. Hazelnut pollen is spread by wind rather than by insects.

The shell of the hazelnut is the matured ovary wall of the flower and the nut meat inside is the matured embryo. The husk surrounding the shell is called the involucre.

The most important commercial cultivars of the European hazelnut grown in the United States are 'Barcelona,' 'Ennis,' and 'Daviana,' the latter being the pollenizer. These grow well in western Oregon but not in the eastern United States because of a fungal disease called eastern filbert blight [*Anisogramma anomala* (Peck) E. Müller]. The Oregon Agricultural Experiment Station released 'Lewis' hazelnut in 1997. Nut size is slightly smaller than 'Barcelona,' but it has better resistance to eastern filbert blight. The native American hazelnuts (*C. americana* and *C. cornuta*) resist this disease. Breeding projects crossing these two species have developed some hybrid cultivars, such as 'Reed' and 'Potomac.'

Hazelnuts are harvested by picking the nuts up, either by hand or by mechanical pickup machines, for a smooth well-prepared soil surface after they drop. This should be done promptly to maintain quality. The nuts are washed and dried in forced warm air at 37.5°C (100°F). Dried nuts can be stored in good condition for as long as 15 months at about 21°C (70°F) if their moisture content is kept below 10 percent.

Hazelnut cultivars are usually propagated by simple layering using young mother plants, or by "hot grafting," where heat is applied to the graft union for a time after grafting. Seedling trees are easily started by planting the seeds (nuts) in the spring after a 3-month stratification period at 4.5°C (40°F).

There are relatively few insect and disease pests of hazelnuts, although eastern filbert blight is a major problem for European hazelnut cultivars grown in the eastern United States.

Peach and Nectarine (*Prunus persica* L. Batsch.)
ROSACEAE

Few fruit tree species have spread so rapidly and become adapted to so many climatic situations as have the peach and nectarine. The peach and nectarine differ chiefly in the lack of the typical peach pubescence (fuzz) on nectarine fruits and in fruit flavor, sugar content, and aroma. Both have the same tree characteristics. Mutant limbs producing nectarines have been found on peach trees, but there are no records of nectarine reverting back to peach.

The peach originated as a wild form in China and apparently was cultivated there about 2000 B.C. It was taken westward to Persia and later to Greece between 400 and 300 B.C. The Romans were cultivating the peach about the time of Christ and spread it through their empire in Europe, from which it later was disseminated throughout the world into all countries of the temperate zones. Nectarines have also been grown since ancient times, being mentioned in Pliny's writings about A.D. 50. Peach was introduced to continental America via the Spanish conquest of Mexico and into Florida as early as 1565 with the founding of St. Augustine. In the American colonies, the indigenous seedlings via Mexico and the Southeast met with the cultivars or seed sources brought from England. These two sources of peach material provided further advances through selection. The introduction of 'Chinese Cling' peach to the United States by Charles Downing in 1850 ultimately led to a revolution in peach cultivars grown for commercial production. Among the most successful cultivars grown today, many, if not most, can be traced back to 'Elberta' or 'Belle,' both seedlings of 'Chinese Cling.'

The peach is now one of the most popular commercial and home garden fruits. China leads in the world peach and nectarine production, followed by Italy, the United States, France, and Japan. In the United States, California is the leading producer of freestone peaches, followed by South Carolina, Georgia, Pennsylvania, New Jersey, Washington, and Michigan. Almost the entire commercial clingstone canned peach production in the United States comes from California, with yearly production averaging 541,500 t from 1997 to 1999.

California produces about 95 percent of the nectarines grown in the United States, with most of the crop harvested in July and August. Areas that regularly have late spring frosts are unsuitable for peach or nectarine production. At full bloom 90 percent of the flowers are killed at −4.9°C (23°F). Low-lying frost-pocket planting sites must be avoided. Peaches attain their highest quality in regions with warm to hot summers and in areas with low summer humidity. Fungal disease control is much easier in these areas than where summer rains prevail.

Peach fruits can be canned, consumed fresh, dried, frozen, and preserved and jellied. There are two types of both peaches and nectarines—clingstone (flesh firmly attached to the pit) and freestone (flesh separating easily from the pit). In both types there are yellow-fleshed and white-fleshed cultivars. Peaches are one of the stone or drupe fruits.

Commercial canned peach production in California is based mainly on such clingstone cultivars as 'Halford,' 'Carolyn,' and 'Loadel.' Clingstones generally make a better-looking canned product than freestones because of their firmer-textured flesh, which retains its shape during canning. Also, the flesh is brighter and more uniform and the juice stays clearer. Freestone peaches are used to some extent for canning, but they are mainly eaten fresh.

There are many freestone peach cultivars in production, most of the important ones having originated from controlled crosses made by government-supported or private plant breeders. Fruits of peach cultivars mature at different times, so that fresh peaches are available in the markets from midspring to midautumn. Some of the leading freestone shipping peach cultivars in the U.S. markets are 'Flavorcrest,' 'Summer Lady,' 'O'Henry,' 'Elegant Lady,' 'Rich Lacy,' 'Redglobe,' 'Harvester,' 'Juneprince,' 'Cresthaven,' 'Loring,' and 'Redhaven.'

Prominent nectarine cultivars are 'Red Diamond,' 'Summer Bright,' 'Spring Bright,' 'Summer Fire,' 'August Red,' 'Rose Diamond,' and 'Ruby Diamond.' With these newer nectarine cultivars (far superior to those grown in early days), more plantings are being made of this delectable fruit, which now competes with fresh peaches in the markets. California has a virtual monopoly of nectarines in the United States, with plantings of about 14,366 ha (35,500 ac).

Considerable work in breeding both peaches and nectarines by public and private agencies continues, and further improved cultivars are likely to be introduced.

Peaches and nectarines generally are self-fruitful, and trees can be planted in solid blocks without pollination problems. Exceptions are 'J. H. Hale,' 'Halberta,' 'June Elberta,' 'Marsun,' 'Candoka,' 'Chinese Cling,' 'Giant,' 'Delta,' and 'Alamar,' which have defective pollen and require trees of a different cultivar in the planting to provide pollen, plus a bee population to transfer it.

Peach trees in most years tend to set more fruit than they can properly mature. It is essential, therefore, that excess fruits be removed by hand from the tree by thinning to 13 to 25 cm (5 to 10 in.) apart when the fruits are still small. Although a costly operation, thinning increases fruit size, improves fruit color and quality, and prevents limb breakage.

For best tree growth and production of peaches and nectarines, a deep, reasonably fertile, well-drained sandy loam soil is preferable. Heavy, poorly drained soils should be avoided.

Generally, peaches have a high-nitrogen fertilizer requirement, and in humid regions particularly, they respond to potassium fertilizers but little or not at all to phosphorus. Peaches grown in zinc-deficient soils develop a little-leaf or rosette condition that is corrected by dormant spray applications of zinc sulfate. Iron deficiency can be a problem in sodic soils. Peach trees must have continuous soil moisture during the growing season from either rainfall or irrigation.

Nursery trees of peach cultivars are usually propagated by T-budding or chip-budding onto peach seedling rootstocks, using seeds of such cultivars as 'Lovell' or 'Halford.'

Since peach roots are susceptible to nematode attacks, it is advisable in areas where root-knot nematodes are a problem to use seedlings of a nematode-resistant rootstock, such as Nemaguard, Flordagard, or Guardian. Nemaguard rootstock is not winter-hardy, however, in the colder peach-growing regions, and none of the rootstocks are resistant to the ring nematode. Guardian rootstock is recommended in the southeastern United States where peach tree short-life syndrome is a major problem.

Young peach trees are generally pruned to an open-center or vase-shaped form, keeping the pruning to a minimum to avoid growth retardation and a delay in the onset of bearing. Peaches and nectarines bear fruits laterally on shoots that grew the previous summer so that relatively heavy pruning of mature trees is necessary to ensure a continuing supply of vigorous shoots. If the trees are not pruned, little shoot growth takes place and fruit production is minimal. Other pruning systems used in commercial peach production include central leader, hedgerow, tatura trellis, meadow, central spindle, and Kearney perpendicular V. The V system of training is gaining in popularity, especially among California growers.

Peach fruits are mature and ready to pick when the skin background color becomes creamy yellow. They continue to ripen and soften after picking but the highest-quality fruits are those allowed to become fully ripe on the tree. Peach fruits, although not adapted to long storage, can be held from 2 to 4 weeks depending on the cultivar and harvest maturity, at about 0°C (32°F) at 90 percent relative humidity. Higher temperatures of 2.2°C to 5°C (36°F to 41°F) and above are not suitable for peach storage. Rapid cooling to 0°C (32°F) after harvest is desirable to retard ripening and to prevent development of decay organisms. In commercial operations, a hydrocooling apparatus is installed at the end of the fruit-packing line. The fruits are cooled by a swift-flowing shower of 0°C (32°F) water with a fungistat.

Peach and nectarine trees are subject to a number of disease and insect problems. Peach leaf curl, due to a fungus that distorts the leaves, is easily prevented by dormant fungicide sprays. *Verticillium* wilt and *Armillaria* root rot are caused by fungi attacking the roots. *Cytospora* canker is due to a fungus that invades wounds in limbs and shortens tree life. Bacterial canker is found mostly on young trees, causing lesions in the bark and death of branch and trunk tissues. Brown rot fungus causes fruit decay, especially in nectarines, and particularly when humid, rainy weather occurs; however, this can be controlled by some of the newer organic fungicides. Several viral diseases affect peaches and nectarines, such as peach yellows, red suture, phony disease, peach rosette, peach mosaic, and yellow bud mosaic. Peach X-disease is believed to be caused by mycoplasma-like bodies. Plum pox virus (Sharka), already a major European disease, continues to spread geographically, being reported for the first time in South America in Chile in 1992, and, in North America, in the USA in 1999 and Canada in 2000.

Virus problems are best avoided by planting only nursery-certified trees that are free of such diseases.

Insects affecting peaches are plum curculio, stink bugs, peach tree borer, San Jose scale, aphids, and Oriental fruit moth. Several species of mites attack peaches and nectarines.

Pear (*Pyrus* spp.) ROSACEAE

There are two groups of pears based on their origin: those species originating in western Asia around the Caspian Sea, the best known being *Pyrus communis* L., to which the present-day pear cultivars belong; and the Oriental pears, originating in northern Asia, including *P. pyrifolia* (Burm. f.) Nakai, *P. ussuriensis* Maxim., *P. betulaefolia* Bunge., and *P. calleryana* Decne. There are some hybrid cultivars of *P. pyrifolia* and *P. communis,* such as 'Kieffer,' 'LeConte,' and 'Garber.'

Pears are an ancient fruit, cultivated during the early Egyptian, Greek, and Roman civilizations. Around 1000 B.C. Homer wrote that pears were one of the "gifts of the gods," and about 300 B.C. Theophrastus wrote of the wild and cultivated types, described methods of propagating pears by grafting, and mentioned the need for cross-pollination. Other early descriptions of pears were authored by Cato (235–150 B.C.) and Pliny the Elder (79–23 B.C.). Pears flourished in early France and were later introduced into England and Germany, from which they were dispersed to the other temperate zone countries throughout the world. Pears rank second to apples in world production of deciduous tree fruits. China is the world's leading pear-producing country, followed by Italy, the United States, Japan, Spain, France, West Germany, Turkey, and Australia. In the United States, Washington is the leading producer, followed by California, Oregon, New York, and Michigan.

Like apples, pears must be grown in regions with enough winter cold [about 1,000 h under 7°C (45°F)] to overcome the rest influence in the vegetative and flower buds so they will develop in the spring. This requirement eliminates pear production from tropical and subtropical regions. Pear trees are not hardy below −23.3°C (−10°F), so they are not grown in areas with very severe winters. Currently, the best pear cultivars are very susceptible to the fire blight disease caused by the bacterium *Erwinia amylovora.* This organism is native to the Western Hemisphere and flourishes in regions having warm, humid springs and summers with prolonged rainy spells. It has recently spread to England, northern Europe, and New Zealand. This disease confines large-scale commercial pear production to semiarid regions such as California (where blight attacks can still occur), or to regions where the fire blight bacteria do not exist. Pears, particularly the popular 'Bartlett,' grow well in areas with hot summers.

Pear trees produce best on river bottom silt loam and tolerate wet, clay soils better than most other deciduous fruits. They must have continuous soil moisture through the growing season, from either rainfall or irrigation.

Pear trees require nitrogen fertilizers for adequate vegetative growth, but overfertilization can cause excessively rank, vigorous growth, making the trees more susceptible to attacks by fire blight bacteria. In humid rainy regions, where there is a major problem, it is difficult to maintain sufficient vegetative growth for good productivity without bringing on fire blight attacks. In most soils pears do not require other fertilizers, although there are a few instances where applications of potassium, phosphorus, or magnesium have helped.

The most important pear cultivar is the 'Bartlett,' which is known outside North America by its original name, 'Williams Bon Chrétien,' given when it was bred in England in 1796. 'Bartlett' is considered a summer pear, ripening in California in July and August. Important later-ripening winter pear cultivars are 'Anjou,' 'Bosc,' and 'Comice.' The 'Kieffer' pear is grown to a limited extent in the eastern United States, since it is resistant to fire blight. The 'Conference' pear is the most popular in Europe. There are some solid red pear cultivars available that are proving to be much in demand and may be more widely grown in the future. These include 'Red Clapp,' 'Red Bartlett,' 'Reimer Red,' 'Red Comice,' 'Cascade,' 'Rogue Red,' and 'Red Anjou.'

The popular Asian pear, sometimes called "pear-apple," is a pear species, *Pyrus pyrifolia.* There are several cultivars available: 'Nijiseiki' ('Twentieth Century'), 'Hosui,' 'Shinseiki,' 'Chojuro,' 'Shinko,' and 'Kikusi.'

Most pear cultivars require cross-pollination to set good crops. There are places, however, where certain cultivars set heavy crops parthenocarpically, without cross-pollination, and produce seedless fruits. An example is in the Sacramento Valley of California, where 'Bartlett' is grown with few or no pollenizer cultivars and productivity is maintained by parthenocarpy. As much as 85 to 99 percent of the fruit in these orchards are seedless. However, these seedless fruits lack flavor, soften earlier, and have lower soluble solids accumulation compared with the seeded fruits. But in most regions 'Bartlett' trees must be interplanted with those of other cultivars to obtain good fruit-set. 'Anjou,' 'Bosc,' and 'Seckel' set some fruit when planted alone but produce more if cross-pollinated. Other cultivars, like 'Winter Nelis,' are self-unfruitful and must be cross-pollinated to set any fruit.

Pears are not prone to alternate bearing, as are apples, and fruit thinning is generally unnecessary unless an obviously excessive crop has set. A temperature of −4.9°C (23°F) at full bloom kills approximately 90 percent of 'Bartlett' flowers.

Pear trees are propagated either by T-budding or chip-budding onto seedling rootstocks in the nursery or by winter bench grafting onto short pear root pieces for planting in the nursery in the spring. *P. communis* seedlings are the best rootstock for pear cultivars, but the seed source is important to ensure that only *P. communis* trees are the parents. Where fire blight is a severe problem, rooted cuttings of a blight-resistant type like 'Old Home' can be allowed to grow to form the roots, trunk, and primary scaffold branches of the tree. When the scaffolds are 1 year old, they can be top-budded to the desired fruiting cultivar. If fire blight attacks the top of the tree, it usually stops at the 'Old Home' tissue and does not kill the entire tree. Later, after all blight is cut out, the 'Old Home' branches can be rebudded to the fruiting cultivar.

Rooted quince cuttings can be used as the rootstock to obtain partially dwarfed pear trees, but certain cultivars such as 'Bartlett,' 'Bosc,' 'Winter Nelis,' and 'Seckel' are not directly compatible with quince. These require an interstock like 'Old Home' or 'Hardy' pear and two graft unions between the quince roots and the fruiting top cultivar.

In the United States young pear trees are best trained to a modified leader system with as little pruning as practicable. Young trees tend to grow very upright, so pruning during the first several years should include complete removal of some upright shoots. Once the trees start to bear, the weight of the fruit aids in developing a spreading growth habit.

Pear fruits are produced on fruiting spurs and, to a lesser extent, laterally on shoots that grew the preceding year. Some moderate annual pruning by thinning out shoots stimulates renewal of both spurs and fruiting shoots. Light pruning should be used to obtain maximum yields and to avoid stimulation of vigorous shoot growth, which is very susceptible to fire blight.

Pears are utilized as fresh, canned, or dried fruit. A pear cider, *perry,* is made in some European countries. Pear fruits must be picked from the tree when they are mature but still hard and ripened off the tree at a temperature of 20°C to 21°C (68°F to 70°F) for 10 to 12 days at a relative humidity of about 85 percent. Higher ripening temperatures detract from flavor and appearance. A characteristic yellow color develops during ripening, along with flesh softening and aroma and flavor changes.

Mature but unripened pears can be kept for several months if they are stored immediately after harvest at -1°C (30°F) and 90 to 95 percent relative humidity. Pears are highly susceptible to excessive CO_2. Optimum and safe CA atmosphere for commercial storage is 2 to 2.5 percent O_2 and 0.89 to 1 percent CO_2. 'Bartlett' fruits keep as long as 2 1/2 to 3 months; 'Bosc' and 'Comice,' 3 to 4 months; and 'An-

jou,' 4 to 8 months. Anytime during these periods the fruits can be removed from cold storage and placed at 20°C to 21°C (68°F to 70°F) for ripening, which then occurs rapidly in 4 or 5 days.

"Pear decline" was a severe problem in Italy and in the West Coast pear districts of the United States in the 1960s, killing hundreds of thousands of pear trees grafted on certain Oriental pear rootstocks. Death is due to a mycoplasma-like organism spread from tree to tree by the insect, pear psylla. 'Bartlett' (*P. communis*), for example, is itself resistant, but the organism, moving down the phloem to the Oriental rootstock (e.g., *P. pyrifolia*), kills its phloem tissue and girdles the trunk; thus, the tree's roots die from lack of food materials from the top. Best control is the use of nonsusceptible rootstocks, such as *P. cummunis* or *P. betulaefolia* seedlings.

Codling moth can cause wormy pear fruits, but is easily controlled by properly timed insecticide sprays during the growing season. Several kinds of mites, as well as pear psylla, can become major pests.

Pecan (*Carya illinoensis* Wang.) JUGLANDACEAE

The pecan is a large deciduous tree native to the south-central United States. The early French and Spanish explorers and settlers in this area found the native Indians using the nuts as food. The Algonquin Indians referred to walnuts, hickory nuts, and pecans as "pakans," meaning a hard-shelled nut that must be cracked with an instrument. The name *pacane* was adopted by the early French in Louisiana for one nut in particular, the pecan. In areas of the South outside the French influence, the *e* was dropped and the name became "pecan." In the United States pecans are grown commercially in the southern states. Georgia leads in production, followed by Texas, New Mexico, Arizona, Alabama, Louisiana, Oklahoma, and Mississippi.

Pecans require a long, frost-free growing season with hot days and warm nights to properly mature the nuts. The tree does, however, require some winter chilling to overcome bud dormancy and permit proper vegetative growth in the spring. For this reason it is not adapted to tropical or subtropical regions. Pecan trees tolerate considerable winter cold without damage.

For good tree growth and heavy production, pecans require a deep, well-drained, and well-aerated soil free of hardpan layers. The trees do not do well on saline or highly alkaline soils. Pecans respond to nitrogen fertilizers by increased growth. Deficiency of nitrogen is shown by yellowish foliage. Added nitrogen to maintain vigor and productiveness is particularly useful as the trees mature. Pecans are very sensitive to zinc deficiency in the soil and may require several foliar spray applications of zinc sulfate per year to keep them growing properly. Continuous soil moisture,

Bartlett pears are considered to be mature and ready to pick when the fruits are full-sized [at least 6.0 cm (2 3/8 in.) in diameter] and change color from a deep green to a yellowish green, or when the soluble solids, as measured by a refractometer, are not less than 10 percent, or when a pressure test made by a 0.79 cm (5/16 in.) plunger into the pared flesh is not over 9.9 kg (22 lb). Immature pears do not ripen properly and do not give a marketable product.

from either irrigation or rainfall, is necessary for good tree growth and production.

In the early days pecan groves consisted of native and planted seedling trees. From about 1850 to 1910 selections from the best trees were made and named for further vegetative propagation. From this came the cultivars 'Schley,' 'Success,' 'Stuart,' and 'Desirable.' Four cultivars, 'Stuart,' 'Western Schley,' 'Desirable,' and 'Wichita,' comprise more than 55 percent of the orchards planted with improved varieties. Pecan hybridization began about 1914, culminating with the establishment in 1930 of a large pecan breeding program by the USDA and the Texas Department of Agriculture of Brownwood, Texas. The program was conducted by the late L. D. Romberg (1931 to 1968), G. D. Madden (1968 to 1977), and T. E. Thompson (1979 to present), and has released 23 pecan cultivars. USDA pecan cultivar releases from Brownwood include 'Barton' (1953), 'Comanche' (1955), 'Wichita' and 'Choctaw' (1959), 'Apache' and 'Sioux' (1961), 'Mohawk' (1965), 'Caddo' and 'Shawnee' (1968), 'Cheyenne' (1970), 'Cherokee' (1971), 'Chickasaw' and 'Shoshoni' (1972), 'Tejas' (1973), 'Kiowa' (1976), 'Pawnee' (1984), 'Osage' (1989), 'Oconee' (1989), 'Houma' (1989), 'Navaho' (1994), 'Creek' (1996), 'Kanza' (1996), and 'Hopi' (1999). Many of these have been heavily planted in recent years, particularly 'Wichita' and 'Cheyenne.'

Pecans have both nut-producing pistillate (female) flowers and pollen-producing (male) catkins on the same tree. Staminate and pistillate flowers mature at different times on an individual tree (dichogamy): If staminate flowers dehisce pollen before pistillate flowers are receptive, the tree is protandrous; if pistillate flowers are receptive before pollen is shed from catkins, the tree is protogynous. For a good crop to be set, the pollen must be shed at a time when the female flowers are receptive. Therefore, two cultivars of different dichogamy are used in the planting to provide pollen at the proper time.

Pecan cultivars can be propagated by patch budding or whip grafting onto 2-year-old pecan seedling rootstocks grown in the nursery row. For good germination the seeds are best soaked in water for 24 hours, then stored in damp vermiculite in polyethylene bags for 3 to 4 months at about 1°C to 3.5°C (34°F to 38°F) before planting. Pecan seeds dry out and lose viability, so they should be kept cool and damp from the time they are harvested until stratification begins.

Young pecan trees are best trained by light pruning to a central leader system, which can give strong, wide-angled scaffold branches. Narrow-angled scaffolds are weak and can easily break under heavy crop loads.

Pecans are ready to harvest when the hull loses its green color and starts to split. The trees are harvested commercially by mechanical tree shakers that shake the nuts off the branches onto cloth sheets or the ground, where they are picked up mechanically. The nuts are air-dried after harvest to remove 10 to 20 percent of the moisture. The whole nut keeps in good condition for several months in a cool, dry place. If too warm, the nut meats become rancid. Most of the pecan crop is handled through shelling equipment, with the nut meats used in baked goods, ice cream, and confections.

Pecans grown in the humid areas of the South with frequent rains develop a serious fungal disease called scab [*Cladosporium caryigenum* (Ell. et Lang.) Gottwald] that attacks both the foliage and the nuts, although some cultivars show high scab resistance. Other important diseases include vein spot (*Gnomonia nerviseda* Cole), downy spot (*Mycosphaerella caryigena* Demaree & Cole), and pecan leaf scorch (possibly *Xylella fastidiosa* Wells et al.). Pecans are attacked by various insects including the pecan weevil (*Curculio caryae* Horn), hickory shuckworm (*Cydia caryana* Fitch), pecan nut casebearer (*Acrobasis nuvorella* Nuenzig), and the black, black-margined, and yellow aphids (*Melanocallis caryaefolia* Davis, *Monellia caryella* Fitch, and *Monelliopsis pecans* Bissell, respectively).

Plum and Prune (Prunus spp.) ROSACEAE

Native wild plums encircle the globe in the northern temperate zone region. There are groups indigenous to Europe, America, and Asia. Cultivated plums are grown throughout the two temperate zone regions of the world. Some types of plums require considerable winter cold [700 to 1,100 h below 7.2°C (45°F)] to break bud dormancy and permit growth and flowering in the spring. This requirement prevents their culture in tropical and subtropical regions (except at higher altitudes). Plum trees are generally quite winter-hardy.

The cultivated European firm-textured plums are in the species *Prunus domestica* L. and *P. insititia* L. The Japanese plums belong to *P. salicina* Lindl. The American plums include *P. americana* Marsh., *P. hortulana* Bailey, and *P. munsoniana* Wight and Hedr.

P. insititia is believed to have originated in southeastern Europe and parts of Asia. It is the plum mentioned in the writings of the early Greek poets during the sixth century B.C. The European plums (*P. domestica* and *P. insititia*) have been distributed throughout the world by seeds and scions for grafting and are the source of some of the best commercial cultivars.

Prunes are firm-fleshed fruits of *P. domestica* cultivars that have a high enough sugar content that they can be dried whole, without fermenting around the pit, to produce a firm, tasteful product that can be stored for long periods. All prunes are plums, but not all plums are prunes. *P. cerasifera* Ehrh. cultivars are used as ornamentals and for plum rootstocks.

Certain cultivars of Japanese plums (*P. salicina*) were introduced into the United States about 1870 by Luther Burbank and others. It is likely that this species originated in China because there are Japanese reports that it was introduced there from China about 1720. There are many cultivars of *P. salicina,* some of which are the most popular market types.

Early explorers and settlers found native American plums growing and being used as food by the Indians in the areas east of the Rocky Mountains. Many cultivars of American plums have been developed, but they do not compete well in the markets with fruits of the European and Japanese types.

Among the stone fruits, plums rank next to peaches in total production. China is the world's leading producer of plums and prunes, followed by the United States and Yugoslavia. About 90 percent of Yugoslavia's fresh prune crop is processed into brandy. California is the leading producer of both plums and prunes in the United States, followed by Washington, Oregon, Michigan, and Idaho.

Japanese plums are early blooming and are well adapted to areas with mild winters and hot summers. European plums are late blooming and produce best under colder winters and moderate summer temperatures.

Plums grow well on many soil types but do best on deep, well-drained soils of medium texture. They tolerate heavier soils better than most other stone fruits, unless grown on peach roots. Japanese plums respond to heavy applications of nitrogen fertilizers with increased growth and larger fruits. European plums seem to require less nitrogen but, unlike Japanese plums, respond well to potassium fertilizers in some soils. Some prune orchards in California have suffered heavily from potassium deficiency. Zinc deficiency has appeared in many California prune orchards, but it can be corrected with dormant sprays of zinc sulfate. For good vegetative growth and crop production, plums require continuous soil moisture throughout the growing season from rainfall or irrigation.

Plum cultivars can be grouped into those used primarily for fresh consumption, canning, and jellies, and those used for drying (prunes). Fresh market plums are mainly cultivars of the Japanese types (or hybrids with Japanese parentage). Some popular ones are 'Friar,' 'Angeleno,' 'Black Amber,' 'Fortune,' 'Red Beaut,' 'Santa Rosa,' 'Royal Diamond,' 'Simka,' 'Howard Sun,' and 'Black Beaut.' 'Santa Rosa' is a good pollenizer for all of these. European fresh-eating plums include 'Tragedy,' 'President,' 'Green Gage,' 'Stanley,' and 'Yellow Egg.' Cultivars used for prune production are such *P. domestica* types as 'French' (96 percent of California production), 'Imperial,' 'Sugar,' 'Robe de Sergeant,' and 'Italian.'

Plums are one of the stone or drupe fruits. Most Japanese plum cultivars are self-unfruitful and require pollinator trees of the same species for setting crops. Many of these cultivars are inter-unfruitful, so advice from the local agricultural extension service is necessary so that the proper combinations are used to satisfy the pollination requirements. Some of the European cultivars like 'Agen,' 'French,' 'Damson,' 'Pershore,' 'Stanley,' 'Methley,' 'French' prune, and 'Sugar' prune are self-fruitful and can

be planted in solid blocks. Others, like 'Diamond,' 'Grand Duke,' 'President,' 'Tragedy,' and 'Italian' prune are completely or partially self-unfruitful and require pollenizer trees of another cultivar. It is wise in planting plums to mix two or more cultivars. All European plum cultivars seem to cross-pollinate readily, provided their bloom periods overlap. A good bee population in the orchard at bloom is essential to ensure proper pollination.

There has been considerable interest in the United States, Canada, and in Europe in plum breeding by crossing the various species. For example, Luther Burbank's 'Santa Rosa' plum, the leading shipping cultivar grown in the United States, is a mixture of *P. salicina, P. americana,* and *P. simonii.*

Fruit thinning is often required with plums, especially the Japanese types, spacing the fruits 10 to 15 cm (4 to 6 in.) apart. Unless properly thinned, fruit size is small and tree growth weak, with little or no crop the following year.

Plum trees can be propagated by T-budding or chip-budding onto seedling rootstock trees of myrobalan plum (*P. cerasifera*). Rooted hardwood cuttings of such plum cultivars as Myrobalan 29C, Myrobalan B, Brompton, Pershore, St. Julian A, and Marianna 2624 also make good rootstocks for plums and are widely used.

Plum trees are best trained to an open-center shape with three or four primary scaffolds and seven to nine secondary scaffold branches. Shoots of Japanese plums form lateral branches readily, but those of European plums in the first 2 years should be headed back to force out lateral branching to develop the scaffolds. As plum trees come into bearing, the fruits are borne mostly on fruiting spurs that live for 5 to 8 years, developing laterally on the larger branches. Pruning bearing trees consists mainly of thinning out fruiting wood to reduce the crop load and to stimulate renewal shoots for future crops. Insufficient removal of fruiting wood leads to excessive fruit-set and high thinning costs.

Fresh market plums should be left on the tree until they are well matured, but they will continue to ripen after picking. At optimal harvest time the fruit should soften to yield to gentle pressure, be juicy and aromatic, and have a final skin color typical of the cultivar. Although fresh plums do not adapt to long storage, they can be held for 2 to 4 weeks depending on the cultivar, if they are placed at 0°C (32°F) and 90 percent relative humidity immediately after harvest.

To produce high-quality prunes the fruit must be harvested at the proper maturity stage. The flesh color (of French prunes) should change from green to full yellow or amber and the skin should become red; flesh firmness should sharply decrease and soluble solids (sugar content) of the extracted juice should be at least 22 percent. Prune harvest in large orchards is generally fully mechanized with mechanical tree shakers removing the fruits, which fall onto mechanized canvas catching frames. In smaller operations the fruits are allowed to fall on the ground when ripe or hooked poles are used to shake the branches. The fruits are then picked up by hand. Following harvest, the prunes are partially dehydrated in a special prune dryer. Trays of prunes move through the dehydrator where they are exposed for 2 h to hot air [85°C (185°F) to start and 71°C (160°F) to finish]. Final moisture content of the prunes should be about 16 percent.

Plum and prune trees are susceptible to bacterial canker (*Psuedomonas syringae* pv. *syringae*), crown gall (*Agrobacterium tumefaciens*), oak root fungus (*Armillaria* spp.), crown rot (*Phytophthora* spp.), Leucostoma canker (*Leucostoma* spp.), and such viruses as plum pox (Sharka), ring spot, and prune dwarf. Insect problems include San Jose scale, peach twig borer, and codling moth. Several species of mites also attack plums and prunes.

Quince (*Cydonia oblong* Mill.) ROSACEAE

The quince is an ancient fruit believed to have originated in the region stretching from Iran eastward to northern India and Tibet. There is evidence that it was cultivated about 4000 B.C. by western Asian peoples. The quince was a popular fruit of the ancient Greeks and Romans. It is a small, slow-growing deciduous tree that is well adapted to many areas of the temperate zones. It requires a slight amount of winter chilling for the buds to develop properly in the spring. Unlike most other deciduous fruit trees, quince fruit buds do not form the preceding summer but develop at the terminal ends of new shoots produced in the spring of the current year. The quince grows well and produces crops in either cool or hot summers.

Main production areas at present are Iran, Afghanistan, and southern Europe. The quince is not held in high esteem in present-day horticulture, although its fruit can be baked or made into a very tasty jelly or used to flavor cooked apples and pears.

The quince is adapted to many soil types as long as they are well drained. Relatively little pruning is needed, just enough to stimulate new shoot production and to remove dead or interfering branches. Nitrogen should be applied cautiously due to quince's high susceptibility to fire blight. Quince cultivars are propagated by T-budding on rooted 'Angers' cuttings. Rooted cuttings of the 'Angers' cultivar are used widely in pear culture as a rootstock that produces a dwarfed pear tree. Most cultivars can be propagated by vegetative cuttings, but some do root easily.

There are a number of quince cultivars including 'Pineapple,' 'Orange,' 'Van Deman,' 'Champion,' 'Smyrna,' 'Rea,' 'Meech,' and 'Portugal.' The quince is self-fruitful, so only one cultivar need be planted.

Quince fruits are ready to harvest when the skin loses its greenish color and a pronounced fragrance develops. Fruits keep well in cold storage.

Fire blight is the chief disease of the quince. Codling moth and Oriental fruit moth are major insect pests. Control requires three or four sprays per year.

Raspberry (*Rubus* spp.) ROSACEAE

Although the genus *Rubus* contains 400 to 500 species, there are only three important raspberry species, which are often grouped together as the "brambles." These are the European red raspberry (*R. idaeus* L.), which is native to many areas in Europe; the American red raspberry (*R. idaeus* var. *strigosus* Michx.), native to the eastern and northern United States and southern Canada; and the black raspberry (*R. occidentalis* L.), which is native to many areas in North America. Purple raspberries are hybrids between red raspberries and *R. occidentalis*. The red raspberry is the most important commercially.

Pliny, a Roman author who wrote just before the beginning of the Christian era, mentioned that the raspberry was originally found growing wild on Mount Ida in Greece. Linnaeus commemorated the place of discovery when he named the raspberry *R. idaeus*.

Raspberries differ from blackberries in that the fruit itself separates readily from the receptacle on which it is produced. Blackberries do not separate readily.

The largest commercial raspberry plantings are found in Russia, Yugoslavia, Poland, Scotland, the United States (Washington, Oregon, Michigan, and New York), and Canada (principally British Columbia). While world raspberry production has increased since 1900, production in the United States has decreased, probably because of virus problems and high hand-labor costs for harvesting the very tender fruits. Use of virus-free nursery stock and mechanical harvesting equipment may reverse this trend, however.

Although there are cultivar differences, the red raspberry is considered to be resistant to cold if the bushes are properly hardened before winter, but injury can occur from alternating warm and cold periods in late winter. Raspberries should be grown in areas with cool summers and on sloping sites that have good air drainage to avoid frost pockets.

Raspberries grow best in a deep, well-drained, medium-textured, slightly acid loam soil with large amounts of organic matter. Heavy fertilization with manure or inorganic fertilizers is required for strong vegetative growth and high yields. Irrigation is required in areas where rainfall is unreliable.

Raspberries are bush-type plants that fruit on biennial branched, upright canes. Primocanes grow vegetatively the first summer, initiating fruit buds in the fall. The following spring on the floricanes, short lateral shoots develop from these buds, which flower and produce fruits in early summer. After this the entire cane dies to the ground and should be pruned out to leave space for newly developing primocanes for next year's fruit. In early spring, before growth starts on the fruiting canes, they should be pruned to a height of 1.2 to 1.5 m (4 to 5 ft) to limit the crop and give larger berries of better quality.

Red raspberries produce many upright canes from the crown of the plant. After fruiting and removal of the old fruiting canes, the newly formed vegetative canes for next year's crop should be thinned out, removing weak shoots and suckers from the red raspberry plants—and all canes under about 1.3 cm (0.5 in.) in diameter from black and purple raspberries.

Primocane fruiting red raspberries are increasing in popularity. They produce a fall crop on the upper portions of current season canes followed by a spring crop on the lower portion of the same canes. Some growers are producing only the fall crop and then cutting all the canes at ground level in the dormant season. This greatly reduces pruning labor and eliminates winter-kill of canes in very cold climates. In addition, because the fall crop occurs during the "off-season," the fruit commands a premium price on the fresh market.

Some popular red raspberry cultivars are 'Latham,' 'Willamette,' 'Heritage,' 'Newburgh,' 'Fairview,' 'Meeker,' 'Taylor,' 'Chilcotin,' 'Skeena,' and 'Boyne.' Black raspberry cultivars are 'Cumberland,' 'Bristol,' 'Munger,' 'Morrison,' and 'Blackhawk.' Purple raspberries are 'Clyde,' 'Sodus,' 'Brandywine,' 'Royalty,' and 'Marion.'

Red raspberries are easily propagated by digging and transplanting in the fall the suckers arising from the roots. Or it is possible to wait until spring to dig either the 1-year suckers or new ones just starting. A piece of the old root should remain attached in either case. Red raspberries can also be propagated by root cuttings. Root pieces are cut into sections 5 or 7 cm (2 or 3 in.) long in early spring and planted about 5 cm (2 in.) deep in nursery rows.

Black and purple raspberries are propagated by tip layering. The tips of the new nonfruiting canes are pinched back when they are about 0.6 m (2 ft) tall. This causes many lateral branches to form, which arch over with their tips touching the ground. In late summer the tip of each of these laterals is buried 5 to 10 cm (2 to 4 in.) deep. Roots and a new shoot develop from the top. The following spring the cane is cut from the parent plant leaving about 15 cm (6 in.) attached to the new plant. The plant can then be dug out and set in the nursery row to grow another year or directly in place in a new planting.

Raspberry fruits are highly perishable and should be carefully picked when the berries are beginning to soften and separate easily from the cap (torus). Overripe or decaying berries should be discarded. The fruits should be chilled to 0°C (32°F) at 90 percent humidity immediately after harvest, and they can be kept for about a week. A planting should be harvested every 1 to 3 days to maintain high quality. Some raspberries are eaten fresh but more are processed for a number of products including jellies, jams, preserves, ice cream and yogurt flavoring, juices, and wine.

Raspberries are susceptible to such viruses as raspberry mosaic, leaf curl, mild streak, and ring spot. Anthracnose (*Elsinoe veneta*), powdery mildew (*Sphaerotheca humuli*), cane blight (*Leptosphaeria coniothyrium*), leaf spot (*Sphaerulina rubi*), orange rust (*Gymnoconia peckiana*), and spur blight (*Didymella applanata*) are fungal diseases that can be controlled by fungicides or cane removal and destruction. Raspberry roots are attacked by several nematode species. To avoid this problem only nematode-free nursery planting stock should be used, and the proposed planting site should be fumigated with a nematicide.

Several insect pests attack raspberries. Aphids are serious because they transmit viruses. Raspberry fruitworm feeds on buds and new foliage. Other pests are crown borers, leaf rollers, thrips, and mites. Tarnished plant bugs feed on flower parts, causing fruit to be misshapen, crumbly, and small.

Strawberry (*Fragaria* × *ananassa* Duch.)
ROSACEAE

The strawberry is a small herbaceous plant with a short central stem (the crown). From the crown grow the leaves and other structures including branch crowns, flower stalks, and runners, which are all branches originating from leaf axils of the main stem.

Various strawberry species grow wild throughout the world, but the cultivated strawberry is based on two species, *F. chiloensis* (L.) Duch., native to the west coast of North and South America, and *F. virginiana* Duch., native to the Atlantic seaboard and to the Sierra Nevada, Cascades, and Rocky Mountains of North America. Hybrids between these two species were the ancestors of all modern strawberry cultivars. A French intelligence officer, A. F. Frézier, carried *F. chiloensis* plants from near present-day Concepción, Chile, back to France in 1714. The plants were set out in European gardens near those of *F. virginiana* previously brought to Europe from the Atlantic seaboard of North America. Natural hybridization between these two species provided the basis for the large-fruited garden strawberries we know today. Since then private and government-supported plant breeders have introduced many hundreds of new cultivars. Those well adapted to one set of climatic conditions often do poorly under different conditions, so that plant breeders throughout the world have developed new cultivars for their own particular conditions. Strawberries are native to the temperate zones, but adapted cultivars are grown successfully in subtropical areas such as southern California and Florida, also.

Leading strawberry producing countries are the United States, Spain, Japan, Italy, and Poland. In the United States, California produces over 80 percent of the nation's crop, followed by Florida, Oregon, North Carolina, Washington, Michigan, and New York. In California average production is about 62.8 MT/ha (28 t/ac), with harvesting continuing from February to about mid-October. U.S. production outside California averages about 18 MT/ha but with yields of 38 to 45 MT/ha possible, especially in Florida. Canada produces substantial amounts of strawberries in Quebec, Ontario, and British Columbia.

Strawberry cultivars can be classified arbitrarily into two groups based on their fruiting habits. First, the spring cropping type are facultative short-day plants, forming fruit buds with the onset of short days in the fall. Normally, the plants then flower and fruit the next spring, after which vegetative runners develop in response to the long days of summer. Most strawberry cultivars are of this type. Popular U.S. strawberries are 'Chandler,' 'Douglas,' 'Pajaro,' 'Hood,' 'Selva,' 'Tioga,' 'Tufts,' 'Olympus,' 'Aiko,' 'Parker,' 'Linn,' 'Shuksan,' 'Apollo,' 'Atlas,' 'Redchief,' 'Tangi,' 'Honeoye,' 'Allstar,' 'Midway,' 'Surecrop,' 'Raritan,' 'Guardian,' 'Benton,' and 'Totem.'

The second group has been called "everbearing" because the plants fruit in recurrent cycles throughout the growing season, including the long days of summer. Many cultivars of this type runner very little and, because they fruit during the summer and fall months, they behave as if they were long-day plants. Cultivars of this type are 'Geneva,' 'Gem,' 'Rockhill,' 'Ogallala,' 'Arapahoe,' 'Ozark Beauty,' 'Fort Laramie,' and 'Quinalt.'

True day-neutral cultivars have been bred that are similar to the second group but with the day-neutral habit, derived from *Fragaria virginiana glauca*. They flower in continuous cycles throughout the year in mild climates such as central and coastal southern California. They differ from the everbearers in that they do not go dormant during the shortest days of the year if favorable growing temperatures prevail. The first cultivars of this type were released in California and named 'Aptos,' 'Brighton,' and 'Hecker.' Day-neutral strawberries introduced later include 'Tribute' and 'Tristar' from the USDA at Beltsville, Maryland; 'Selva' and 'Fern' from California; and 'Tillicum,' 'Sakuma,' and 'Burlington' from Washington.

As with many other temperate zone fruit species, strawberry plants develop a rest or dormancy condition in the fall that must be overcome by the chilling temperatures of winter before vigorous growth will resume in the spring. Strawberry cultivars differ in the amount of winter chilling required. Short-day cultivars originating in California, such as 'Douglas,' 'Pajaro,' and 'Chandler,' require little chilling and are suitable for regions with very mild winters. Most cultivars originating elsewhere, on the other hand, have a high winter-chilling requirement and would not be suitable for mild winter regions. 'Midway,' 'Surecrop,' and 'Raritan,' for example, have a high winter-chilling requirement.

Strawberries grow best on well-drained, slightly acidic neutral, or slightly alkaline sandy soils, although a range of soil types can be used by selecting cultivars adapted to them. Poorly drained heavy clay soils should be avoided. Sites with a gentle slope are often desirable to give good water drainage and air movement. In California, strawberries are grown on level or contoured sites to facilitate irrigation.

Irrigation with high-quality water (low in soluble salts) is essential for commercial strawberry production, particularly in semiarid regions. Irrigation is desirable even in areas where summer rainfall can be expected, as rains may not fall at critical times during the growth and development of the crop. Drip irrigation is almost universal in California and sprinklers are used only for establishment purposes. In other areas, for example, Florida and Oregon, sprinkler irrigation is used as a supplement to normal rainfall.

Strawberry blossoms are often damaged by frost in the spring in the midwestern and eastern parts of the United States, often eliminating the entire crop. The freezing point of the opened strawberry flower is near −3.1°C (26.4°F). To avoid frost damage a sloping site should be selected to allow drainage of cold air to lower levels. A northern exposure retards early spring blooming and lessens the chances of frost damages but causes the crop to mature late. Sprinkler irrigation systems are efficient and economical in protecting strawberry blossoms against frost damage. Sprinklers should be started when the air temperature in the strawberry field at plant level drops to about 1°C (34°F). Sprinklers can protect blossoms at air temperatures down to −7.2°C (19°F).

Fertilizer requirements for strawberries vary considerably with the locality, cultivar, and soil type. In California, strawberries generally respond only to nitrogen, but in Washington and Oregon phosphorus is often needed as well as potassium and sometimes sulfur, boron, and magnesium. In highly acidic soils, lime applications are helpful. In the eastern United States nitrogen is mostly used, but potassium and, less often, phosphorus are sometimes needed.

Different planting systems are used in the various strawberry-growing regions. Planting through plastic has been successful in Florida and California (Figure 26–2), but in California the polyethylene is normally applied after planting. In California the predominant method in the Central Coast is the double-row, raised-bed system with two rows [about 25 cm (10 in.) apart] of plants, spaced 20 to 35 cm (8 to 14 in.) in the rows, with the beds 129 to 132 cm (48 to 52 in.) apart center to center. In coastal California, either

Figure 26–2 Use of clear polyethylene plastic in a strawberry planting. Plants are allowed to grow through holes in plastic, which warms the soil, promotes early plant growth and fruiting, and helps keep the berries clean.

a summer or winter planting system is used, often with the plants grown as annuals under the winter planting system and biennials under the summer system. Plants grown longer than this in southern California do not receive enough winter chilling and lose vigor, and their fruits are small and inferior.

Summer plantings are established from mid-July in the Central Valley (Fresno area) to late August or early September in coastal California depending on the cultivar and location. Plants are dug from nursery fields when semidormant in January and held in cold storage at −2.2°C (28°F) packed in containers lined with thin polyethylene bags until planting time. The plants grow rapidly through the fall, producing many runners (which are removed), become semidormant in winter, and resume growth in early spring with fruit harvested from about March in southern California and late April in northern California and continuing on as late as October in some areas. The planting is then destroyed, if it is on an annual system, and the field may be cultivated, fumigated, and prepared for a winter planting.

Winter plantings are used in the coastal districts of southern and central California, which have relatively mild winters. The plants are dug from high-elevation northern California nurseries from mid-October to early November and planted in the fields immediately or after 10 to 21 days of cold storage at above-freezing temperatures. The newly set plants grow through the winter and early spring, and the short days stimulate fruit bud initiation. Harvest of the crop begins as early as January or February in southern California. The planting is destroyed in time to prepare the ground for the next planting (winter). Modifications of the winter planting system (green plants with leaves) are also used in Florida and Mexico, where California cultivars are used exclusively.

In strawberry-growing regions with cold winters, such as the northwestern, midwestern, and eastern United States, the conventional matted row system is often used and the plants are grown as perennials. Plants are set out in the spring about 0.6 m (2 ft) apart on raised beds. In many cases runners from the original plants are allowed to root until a matted row 38 to 60 cm (15 to 24 in.) wide is produced. In other cases a hill system, similar to that described for California, is used. Harvest takes place in the spring, a year after planting. Such strawberry fields may be kept for several years if weeds, pests, and diseases can be controlled. In many places where drainage is poor, the red stele disease (*Phytophthora fragaria*) is the factor limiting production.

Strawberries are easily propagated vegetatively by runner plants. If these runners encounter loose soil and adequate moisture, they form roots, grow rapidly and, in turn, give rise to additional runner plants. A critical point in planting strawberries is to have the soil level against the crown just above the root-stem junction. Plants set too low or too high may not survive. Only high-quality vigorous plants with pest- and disease-free crowns and root systems (white or straw color) should be used. Do not use plants showing moldy crowns or roots. It is best to obtain new plants from a strawberry nursery that produces plants in fumigated soil and which are free of viral, bacterial, and fungal diseases, nematodes, insects, and other pests. Preferably the plants should be certified by state agencies as being true to cultivar and disease- and pest-free.

Ripe strawberries are extremely perishable and require careful handling. They are soft with a thin tender skin, and are easily attacked by fruit decay organisms. Peak-quality berries are fresh, bright, clean, and solid red with little or no white or green showing on the surface. Fruits of some strawberry cultivars ripen slightly after they are picked but those of others do not. The fruits should be moved from the field into the shade and refrigerated immediately after harvest. Berries of some cultivars, such as 'Selva,' 'Pajaro,' and 'Chandler,' store quite well if they are handled carefully. Others, such as 'Sequoia,' do not. The ideal storage condition for strawberries is 0°C (32°F) and 90 percent relative humidity.

Much of the strawberry crop in the eastern and midwestern parts of the United States is marketed by "pick-your-own" systems, but in California the crop is moved largely through commercial fresh shipment, freezer, and preserves channels.

Strawberries are subject to many disease and insect problems. Virus susceptibility eliminated many of the early cultivars. Development by plant breeders of new virus- and red stele-resistant cultivars has been an outstanding achievement in plant science, rejuvenating an entire agricultural industry. Viruses are best controlled by planting only resistant cultivars and obtaining plants certified to be free of known viruses. *Verticillium, Rhizoctonia, Pythium,* nematodes, and weeds are best controlled by preplant soil fumigation with registered fumigants. Chloropicrin and methyl bromide combinations are used in California and Florida. Aphids, mites, and thrips also attack strawberries and, if present, must be controlled by the proper insecticides and miticides.

Walnuts (*Juglans* spp.) JUGLANDACEAE

The walnut is a large deciduous tree bearing both male catkins and the female (pistillate) flowers at different locations on the same tree. The roughened nut is enclosed in a thick husk. The edible portion is the seed inside the nut.

Fifteen species of *Juglans* are indigenous to southeastern Europe and western Asia, eastern Asia, and North and South America. The most important species, producing the Persian or English walnuts of commerce, is *Juglans regia* L. It is native to a broad area from the Carpathian Mountains eastward across Turkey, Iraq, Iran, Afghanistan, and southern Russia to northern India. It was probably moved by migrating populations from ancient Persia to Greece and later distributed throughout the Roman empire. From Italy, it spread to what is now France, Spain, Portugal, and southern Germany. There are records of it being grown in England by 1562. The early colonists from England brought seeds to America. The settlers called the resulting trees English walnuts to distinguish them from the native American black walnuts. Persian walnut is a better name than English walnut, however, since it relates more to the area where the species originated.

The black walnut (*J. nigra* L.) is the native species of eastern and central United States. It has a tasty nut but with a thick and hard-to-crack shell. It is highly prized for its lumber, which is used in making furniture, gunstocks, and cabinets. Many cultivars of this species have been selected and propagated vegetatively for nut production.

A third important species is *J. hindsii* Jeps., native to a small area in northern California. Its value is in the use of its seedlings as rootstocks for *J. regia* cultivars in the huge Persian walnut industry in California's Central Valley.

The leading countries in walnut production are China, the United States, and Turkey. However, Persian walnuts are produced commercially in several European countries including France, Italy, Yugoslavia, Poland, Germany, Czechoslovakia, and Bulgaria, as well as in Russia, and India. Most commercial U.S. production is in California with small production areas in Oregon and Washington.

Trees of the Persian walnut cultivars grown in California withstand a winter temperature of only about $-11°C$ to $-9.5°C$ (12°F to 15°F). These originated from ancestors grown in such mild-winter areas as southern France, Spain, and Iran. The most popular Persian walnut cultivars being grown in California are 'Hartley,' 'Payne,' 'Franquette,' 'Chandler,' 'Chico,' 'Sunland,' 'Ashley,' 'Vina,' 'Amigo,' and 'Howard.'

Trees of the Carpathian type of *J. regia* grown in the eastern United States are much hardier, surviving winter temperatures as low as $-37°C$ to $-42°C$ ($-35°F$ to $-45°F$) if they are fully dormant. These Carpathian walnuts originated from seeds brought to the eastern United States and Canada from *J. regia* trees growing in such cold-winter areas as Germany and the Carpathian Mountains. A limited number of trees of several cultivars of this type are in production throughout the midwestern and middle Atlantic states of the United States grafted onto *J. nigra* stock. Some Carpathian walnut cultivars are 'Broadview,' 'Hansen,' 'Metcalf,' 'Fickes,' 'Somers,' 'Jacobs,' and 'Colby.'

Trees of the mild-winter type of Persian walnuts thrive and produce heavy crops in the long, hot, dry and rainless summers of California. High summer temperatures result in sunburning of the hulls and shriveled kernels. Some damage occurs at 38°C (100°F) with severe damage at 40° to 43°C (105° to 110°F). Late spring or summer rains can increase the chances of damage by walnut blight. Spring frosts can also damage developing flowers. Persian walnuts have a winter-chilling requirement to overcome the rest or dormancy influence in the buds. With insufficient

chilling, shoot growth and bloom in the spring is delayed and abnormal.

Walnut trees grow best on a fertile, well-drained, alluvial loam soil 1.8 m (6 ft) or more in depth. Ample water sources must be available to provide irrigation throughout the summer. Sites subject to late spring or early fall frosts should be avoided. Walnut trees require annual applications of nitrogen fertilizers, and, in California, zinc deficiency often develops. The deficiency is alleviated by foliar sprays of zinc sulfate.

Walnut cultivars are self-fruitful and wind-pollinated, but often the male catkins are not shedding pollen at the time the pistillate, nut-bearing flowers are receptive (this problem is termed dichogamy). It is necessary, then, to interplant cultivars, one of which produces pollen at a time when the pistillate flowers of the principal cultivar are receptive. All walnut cultivars are cross-compatible.

In many areas of the world where Persian walnuts are grown, seedling trees are used that are propagated from seeds taken from the individual trees with good bearing characteristics and producing large high-quality nuts. This procedure, nevertheless, results in considerable variability in the resulting tree characteristics and in the nuts produced and would cause difficult problems in large-scale production and marketing. In California the cultivars are maintained by propagating all trees vegetatively, either by patch or T-budding or by whip grafting on seedling rootstocks. The Persian walnut has also been propagated by tissue culture methods.

The rootstock most widely used is *J. hindsii* seedlings or, in some cases, 'Paradox' seedlings, which are natural F_1 hybrids resulting from *J. hindsii* pistillate flowers being pollinated by *J. regia* pollen from nearby trees. To a much lesser extent, *J. regia* seedlings are also used as rootstocks.

Carpathian walnut cultivars are usually propagated by grafting onto *J. nigra* seedlings with the bark graft method on young established rootstock trees planted in place in the orchard.

Persian walnut trees are best trained to a modified central-leader system. This produces a strong, well-shaped tree with wide-angled scaffolds capable of bearing heavy crops without limb breakage. With this system four or five primary scaffolds, well spaced vertically and around the tree, are allowed to develop, with the lowest branch about 1.8 m (6 ft) from the ground.

In large commercial Persian walnut plantings the harvest is completely mechanized. Mechanical tree shakers remove the nuts from the trees in early fall just as soon as the hulls split open and separate from the nut easily and when at least 80 percent of the nuts on the trees will be removed. They fall onto smooth well-prepared soil where pickup harvesting machines immediately gather them into bins. The nuts are dried in a dehydrator, then fumigated in closed bins en route to central receiving stations.

Walnuts are subject to a number of disease and insect pests. Crown rot fungi (*Phytophthora*) can attack *J. hindsii* roots, particularly where there is considerable moisture at the soil level. Crown gall due to a bacterium attacks most types of walnut roots and is very difficult to control. Oak root fungus attacks *J. regia* roots primarily; *J. hindsii* is resistant. Walnut blight is a bacterial disease damaging the nuts. Insect pests include the walnut husk fly, navel orange worm, codling moth, aphids, and scale, but all of these can be controlled by the proper insecticide sprays. Mites and root lesion nematodes are also problems in some cases.

Walnut black line is a breakdown of the phloem tissue at the graft union of *J. regia* cultivars on *J. hindsii* or Para-dox seedling rootstocks. This generally occurs after the trees are 5 to 20 years old. The cause of this trouble is a virus carried in the Persian walnut cultivar. The virus is the walnut strain of cherry leaf roll virus. Use of infected scionwood in propagation will spread the virus. It is also spread by seed and by pollen. When the virus moves from the infected Persian walnut scion into a few cells of the *J. hindsii* or Paradox rootstock, the rootstock's cells quickly die, appearing as a black line at the union. The virus also dies within the rootstock's dead cells, but the rootstock portion beyond the black line remains virus-free.

REFERENCES AND SUPPLEMENTARY READING

CHILDERS, N. F., J. R. MORRIS, and G. S. SIBBETT. 1995. *Modern fruit science*. Tenth Edition. Gainesville, Fla.: Horticultural Publications.

HARDENBURG, R. E., A. E. WATADA, and C. Y. WANG. 1986. *The commercial storage of fruits, vegetables, and florist and nursery stocks*. USDA Agr. Handbook No. 66, Washington, D.C.

JACKSON, D. I., and N. E. LOONEY (eds.). 1999. *Temperate and subtropical fruit production*. CAB International. Wallingford, UK.

JANICK, J., and J. N. MOORE (eds.). 1996. *Fruit breeding, Vols. I, II, III*. New York: John Wiley & Sons.

ROM, R. C., and R. F. CARLSON (eds.). 1987. *Rootstocks for fruit crops*. New York: John Wiley.

SCHAFFER, B., and P. C. ANDERSEN. 1994. *Handbook of environmental physiology of fruit crops*, Vol. I, *Temperate crops*. Boca Raton, FL: CRC Press.

WESTWOOD, M. N. 1993. *Temperate zone pomology*. San Francisco: W. H. Freeman.

CHAPTER

Subtropical Fruit and Nut Crops

The fruit and nut crops considered in this chapter are grown at low altitudes in those parts of the temperate zones nearest the equator. Most grow in the true Tropics but do not produce well there, although some types, like the West Indian race of avocados, do best in strictly tropical climates. Most mature subtropical fruit and nut trees tolerate some subfreezing temperatures but are killed or severely injured below $-9.5°C$ (15°F). Certain species, like the olive, require winter cold for flower initiation. Some fruit trees, such as avocado, citrus, date, and olive are evergreen; others such as the fig, Oriental persimmon, pomegranate, and pistachio, are deciduous. They are all highly heterozygous and must be propagated by vegetative methods (rather than by seed) to maintain superior cultivars.

Avocado (*Persea americana* Mill.) LAURACEAE

The avocado, an evergreen tree, grows to a height of 4.5 to 13.6 m (15 to 45 ft) with two or more periods of shoot growth during the year. It may flower in winter or spring, and the flowering period may last as long as 6 months.

Avocados are native to southern Mexico. In Central America and the West Indies the fruit has been used as a food for centuries. Three distinguishable races are recognized based on these centers of origin: Mexican, Guatemalan, and West Indian. The three races have been designated *Persea*

americana var. *drymifolia, P. americana* var. *guatemalensis,* and *P. americana* var. *americana*. They differ in a number of traits including cold tolerance, salinity tolerance, iron chlorosis tolerance, fruit size, skin thickness, oil content, and flavor. There are no known sterility barriers among the three races or among any members of the *P. americana* complex. Wild forms of avocado are found in all these areas and trees of these races hybridize readily.

Mexico is the leading producer of avocados, followed by the United States, the Dominican Republic, Brazil, Indonesia, and Israel. Significant avocado industries exist in many other tropical and subtropical countries of the world. U.S. production is easily the highest in monetary value. California produces about 207,000 MT per year and Florida about 27,000 MT. Avocado production is limited by its sensitivity to climatic extremes, its vulnerability to root rot, and delay in developing the superior cultivars possible on the basis of available germplasm. The chief factors limiting avocado consumption worldwide are lack of familiarity and high cost.

Trees of the West Indian race produce well in tropical climates; however, those of the other two generally fail to flower or set fruit in the Tropics. On the other hand, the West Indian race sets little or no fruit in subtropical climates, such as that of southern California. In regions where minimum winter temperatures of −5.5°C to −3.5°C (22°F to 26°F) or below occur, only trees of the Mexican race can be expected to survive. If the proper race and cultivar are chosen, avocados thrive and produce well in climatic conditions from truly tropical to the warmer parts of the temperate zone.

Avocado trees grow well in a wide range of soil types provided the drainage is good. Avocado tree growth and production is best suited to acidic soils with a pH of 4.5 to 5.5. However, some West Indian race rootstocks allow production on marginal calcareous soils with an alkaline pH, such as those found in Israel and Florida. Adequate, but not excessive, soil nitrogen is conducive to good fruit production. A leaf nitrogen of about 1.8 percent is associated with the best yields of 'Fuerte' avocados.

All commercially grown avocado cultivars originated as chance seedlings. The excellent 'Fuerte' has long been the world's leading avocado cultivar but has now been replaced by 'Hass.' 'Fuerte' is a Mexican-Guatemalan hybrid, with green, pear-shaped fruits that mature over a long period of time in California (from November to May). Fruits are high in oil and weigh 227 to 454 gm (8 to 16 oz). 'Hass,' a Guatemalan-Mexican hybrid, is the chief summer-maturing cultivar grown in California. At maturity the skin of the fruit is green but gradually turns black. It weighs 168 to 336 gm (6 to 12 oz). In Florida the leading cultivars are Guatemalan-West Indian hybrids such as 'Booth,' 'Hall,' 'Lulu,' and 'Hickson.' The fruits are green and weigh from 280 to 700 gm (10 to 25 oz).

Florida avocados are generally larger than the California types but contain less oil. California harvests and ships avocados throughout the year, while the Florida season is from August through January.

Avocado flowers occur in panicles of several dozen to several hundred that develop on shoot terminals all over the tree. The number of flowers per tree is large. Avocado flowers are all perfect (having both male and female parts), but they exhibit an unusual behavior. Avocado flower behavior can be described as protogynous dichogamy with synchronous daily complementarity. A cultivar or tree is classified as type "A" if each flower is pistil receptive in the morning and sheds pollen the following day in the afternoon (a two-day cycle). Type "B" trees or cultivars are "female" (pistil-receptive) in the afternoon, then "male" (pollen-shedding) the following morning. Flowers of a given cultivar tend to behave uniformly as type A or B. Weather conditions can upset flower dichogamy and induce partial overlapping of the male and female stages, thus allowing self-pollination. While isolated trees or large blocks of one cultivar may set well, cross-pollination can markedly increase fruit set and retention. The advantage of cross-pollination has been confirmed using isozyme analysis. Pollination of flowers is dependent on large insects, principally honeybees in subtropical regions.

The avocado fruit is a berry, consisting of a leathery skin (exocarp), the fleshy, buttery mesocarp (which is eaten), and a large seed consisting mostly of two cotyledons. Avocados contain higher quantities of fiber and protein than most other fleshy fruits and are an excellent source of potassium and vitamins E, C, and betacarotene.

Avocado cultivars must be propagated vegetatively to retain their characteristics. Various methods are used: T-budding, tip grafting (whip graft method), side veneer, cleft, and bark grafting onto seedling rootstocks. A major need in California and some other regions is a rootstock resistant to *Phytophthora* root rot. In California, 'Duke 7' (Mexican) is the leading clonal rootstock. Currently, 'Bar-Duke' and 'Thomas' are two promising rootstocks being tested. West Indian race seedlings are used in Florida. Avocado trees of all ages are pruned sparingly, mainly by clipping back upright shoot tips to prevent excessive height growth. Heavy pruning encourages strong vegetative growth, reducing yields.

Avocado fruits in California are considered mature and ready to harvest only when the oil content of the flesh reaches more than 8 percent and when the seed coats within the fruits change from yellowish white to dark brown. In more tropical areas fruits are mature with less oil. Cultivars of the three horticultural races differ in oil content, the Mexican race having the highest, Guatemalan race intermediate, and West Indian the lowest. The mature fruits can be "stored" on the tree, however, for several months (Fig. 27–1). The fruits remain hard as long as they stay on the tree, softening only after harvest. Mature avocado fruits, even though firm, bruise easily and must be carefully handled. Avocados can be held for about a month in cold storage—7°C (45°F) for 'Fuerte' and 4.5°C (40°F) for 'Lulu' and 'Booth 8.' The fruit

Figure 27–1 Mature avocado fruits ready for harvest. They can be "stored" on the tree in this stage for several months. They remain hard and firm until picked, when they begin to soften. *Source:* Blue Anchor.

softens at room temperature. Mature, soft fruits can be maintained for more than 2 months at 2°C (36°F) but must be utilized shortly thereafter.

Root rots caused by *Phytophthora, Armillaria,* and *Verticillium* are the major cause of poor tree health and death. Some *Phytophthora*-resistant rootstocks are now available, but considerable research is still needed on this problem. Sunblotch virus can cause tree and fruit distortions, and is transmitted through infected seed, pruning, and grafting. Cercospora spot on fruits and leaves is the most important avocado disease in Florida. Pests include leaf roller caterpillar, greedy scale, thrips, mealybugs, and Lygus bugs.

Citrus (*Citrus* spp.) RUTACEAE

The citrus fruits, especially oranges, are one of the world's four major fruit crops, along with grapes, bananas, and apples. Commercial citrus species are native to Southeast Asia and eastern India. Most present-day cultivars have been grown for many years. The famous seedless 'Washington Navel' orange was found in Brazil in the early 1800s and propagating material was introduced to Washington, D.C. by the USDA in 1870 and from there trees were sent to Riverside, California, in 1873 where its important attributes were recognized. It is believed to have originated as a limb sport of the seedy 'Seleta' sweet orange in Brazil. 'Seleta' was introduced to Brazil by the Portuguese via the Iberian Peninsula from settlements in the Orient. Commercial citrus is produced at low elevations in the world's subtropical cli-

matic zones between 20° and 40° north and south of the equator. Citrus has comparatively little cold resistance and is not grown commercially where minimum temperatures are likely to fall below −6.5°C (20°F).

The Mediterranean area and North and Central America contain about 55 percent of the commercial plantings. The remaining 45 percent is distributed in South America (25 percent), the Far East (15 percent), and in other Southern Hemisphere countries (5 percent), including South Africa and Australia. Brazil is the largest producer of citrus fruits, mainly oranges, grapefruit, and lemons, followed by the United States, China, and Spain. About 80 percent of all citrus grown is consumed in the producing countries themselves. Oranges account for the bulk (65 percent) of the citrus produced, with mandarins, including tangerines (17 percent), lemons and limes (9 percent), grapefruit plus pommelo (5 percent), and other citrus species accounting for the rest. During the past 100 years world citrus production has increased from less than 1 million to over 100 million MT. Citrus production in the United States by crop and state is shown in Table 27–1.

Florida far overshadows all other citrus-producing states, especially in orange and grapefruit production, and is the only commercial producer of limes.

The common and scientific names and the important citrus cultivars grown in the United States are given as follows:

Common Name	Scientific Name	Important Cultivars
Grapefruit	*Citrus paradisi* Macf.	'Marsh,' 'Duncan,' 'Redblush,' 'Thompson,' 'Henderson & Ray,' 'Rio Ruby,' 'Star Ruby'
Lemon	*C. limon* (L.) Burm. f.	'Eureka,' 'Lisbon,' 'Bearss'
Lime	*C. aurantifolia* (Christm.)	'Tahiti'
Orange (sweet)	*C. sinensis* (L.) Osbeck	'Valencia,' 'Washington Navel,' 'Hamlin,' 'Pineapple,' 'Parson Brown,' 'Marrs'
Mandarins	*C. reticulata* (Blanco)	'Satsuma,' 'Dancy,'[1] 'Temple,'[1] 'Minneola,'[1] 'Orlando,'[1] 'Nova,'[1] 'Owari'

[1]Known hybrids, not pure mandarin.

Citrus species interbreed easily, and a number of interspecific and intergeneric hybrids have been developed. Some of these are listed on the following page, as well as some minor citrus species sometimes used as rootstocks.

TABLE 27–1 Average U.S. Production of Citrus by Species and State 1996–98, 1,000 Boxes

Species	Arizona	California	Florida	Texas	Total
Oranges	1,000	69,000	235,100	1,473	306,573
Grapefruit	750	8,600	52,675	5,050	67,075
Lemons	2,600	22,300			24,900
Mandarins, including tangerines	575	2,500	9,150		12,225
Limes			310		310

Common Name	Scientific Name
Calamondin	*Citrus mitis* Blanco
Citrange	*C. sinensis* × *Poncirus trifoliata*
Citrangequat	*P. trifoliata* × *C. sinensis* × *Fortunelia* spp.
Citron	*Citrus medica* L.
Cleopatra mandarin	*C. reticulate* Blanco
Kumquat	*Fortunelia* spp.
Meyer lemon	*Citrus limon* (L.) Burm. F. 'Meyer'
Pummelo (Shaddock)	*C. maxima* [Burm.] Merrill
Rough lemon	*C. jambhiri* Lush.
Sour orange	*C. aurantium* L.
Trifoliate orange	*Poncirus trifoliate* (L.) Raf.

A deep, well-drained fertile soil is optimal for growing citrus, but many kinds of soils are used. Soils with low fertility can be made more productive by adding fertilizers. Citrus generally requires enough added nitrogen fertilizers to maintain leaf nitrogen of about 2.5 percent dry weight. California soils usually have adequate phosphorus and potassium for citrus, but magnesium is occasionally deficient and may require magnesium nitrate sprays. Zinc and manganese deficiency often appears in California citrus and is corrected with zinc sulfate or manganese sulfate sprays. In alkaline or high saline soils, iron deficiency appears and the trees require iron chelate sprays. In Florida, with its less fertile soils, the fertilizer program for citrus includes nitrogen, potassium, magnesium, and possibly phosphorus as well as such micronutrient elements as boron, iron, copper, manganese, zinc, and molybdenum.

Some areas where citrus grows have long, dry periods that make irrigation mandatory. All types of irrigation are used: furrow, basin, sprinkler, and drip. Since much of the citrus crop is on the trees during the winter months, frost damage is a real hazard. In selecting the site for the grove low-lying frost pockets should be avoided. Oil burning heaters or wind machines are often installed in bearing orchards to protect against local radiation frost damage to blossoms, fruits, and trees. Usually heater fires are lit when orchard thermometers drop to a reading of −2°C or −3°C (27°F or 28°F) and are turned off at 0°C (32°F); 87 to 150 heaters per ha (35 to 60 per ac) may be required. Wind machines are operated at −1°C (30°F) and below except when there is no temperature inversion. Where water is available for flood, furrow, or sprinkler irrigation, it may be applied for heat release in cold weather.

Most commercial kinds of citrus set adequate crops without cross-pollination. A few of the citrus hybrids do, however, require cross-pollination. Bees work among citrus flowers, and hives are placed in citrus groves for collection of honey.

In subtropical regions with cool winters most citrus species bloom once a year in the spring. Flower initiation takes place in the buds about 3 months earlier. In the Tropics and warm-winter areas the flowering period may be prolonged or it may occur several times during the year. Lemons grown in the mild southern California coastal regions and in southern Florida bloom and fruit year-round.

Navel oranges from California are available in the markets from November through June, whereas Valencias are harvested from mid-March to mid-November. Florida Valencias are on the markets from March through July, while the Florida "early midseason" orange cultivars ('Pineapple,' 'Temple,' 'Hamlin,' and 'Parson Brown') are marketed from October to April. Grapefruit harvest periods are: Florida, September through July; Texas, October through April; California, December through October; and Arizona, November through August. Lemons in the United States are produced mostly in California with much smaller amounts in Arizona and Florida. California lemons are on the market year-round. Arizona lemons are available from October through February. Florida lemons are almost all used in processed forms.

Some seed-propagated citrus trees produce acceptable fruit, but most superior cultivars must be propagated by vegetative methods. In the early days of orange production in Florida, seedling trees were grown. In modern times, however, most citrus is propagated by T-budding on selected seedling rootstocks. Citrus cuttings root fairly easily, but this method of propagation is little used commercially.

Seeds of most citrus species produce nucellar embryos by apomixis in addition to the sexual embryo; the nucellar embryos are the same genotype as the female parent and thus maintain the maternal clone. These asexual nucellar seedlings can be separated from sexual embryos in the nursery by their greater vigor and uniformity of character; this permits selection for rootstock uniformity. The type of rootstock used in citrus can strongly influence fruit quality, yield, and tree size of the fruiting cultivar, as well as other horticultural characteristics such as disease resistance and soil adaptation.

Rootstocks generally used are seedlings of rough lemon, sour orange, 'Troyer' citrange, trifoliate orange,

'Cleopatra' mandarin, and 'Rangpur' lime. Because of susceptibility to tristeza (quick decline), sour orange is no longer used as a rootstock for oranges in California, South Africa, or Brazil, but still is used to some extent in Florida. Sweet orange cultivars on sour orange roots are susceptible to the tristeza virus. This combination can be used in Florida, however, because of an apparently less virulent form of the virus. The use of sweet orange, 'Cleopatra' mandarin, and trifoliate orange roots results in trees that produce high-quality fruits. However, trifoliate and trifoliate hybrid rootstocks are very susceptible to the exocortis virus and infected trees are dwarfed to some extent, depending on the virus' intensity. Trees on rough lemon roots tend to produce coarse-textured fruits low in solids and acids, but they are well adapted to light-textured soils.

All citrus cultivars and rootstocks are subject to a number of viral diseases. The best control is to use only propagating material or nursery trees produced under conditions free of such viruses. Most citrus-producing countries have set up elaborate certification programs to produce virus-free, true-to-type nursery stock. Citrus trees produce mutations frequently, and since most of these mutations are inferior, propagating material should be taken only from mother trees that have a history of producing high-quality, true-to-type fruits that are free of disease.

Pruning young citrus trees delays bearing and should be done only enough to develop a trunk and strong scaffold system. Bearing orange and grapefruit trees require only light pruning. On the other hand, bearing lemon trees need heavier pruning to facilitate orchard operations and limit their height.

Most citrus fruits store best on the tree but they can be held for a time under refrigeration. Florida oranges may be kept for about 12 weeks at 0°C (32°F) and 90 percent relative humidity. California oranges store best slightly above 4.5°C (40°C). California and Arizona grapefruit keep for 6 weeks at 15.5°C (60°F), but Texas and Florida grapefruit require 10°C (50°F), all with 90 percent relative humidity. Lower storage temperatures cause fruit deterioration in grapefruit. Fully matured and colored lemons can be held for several weeks at 0° to 4.5°C (32° to 40°F), but for prolonged storage—4 to 6 months—they should be picked dark green and held at 11°C to 14.4°C (52°F to 58°F). Limes can be stored at 9°C to 10°C (48°F to 50°F) with 85 to 90 percent relative humidity for 6 to 8 weeks. However, in the latter part of storage, the rind color turns from green to yellow.

Most California oranges are marketed fresh, but in Florida about 90 percent are processed, mostly as frozen concentrate juice. About two-thirds of Florida's grapefruit are marketed in a processed form. Production of Florida frozen concentrated orange juice has been one of the outstanding examples of the development of a new industry by modern technology. Of the total U.S. production of oranges, about 73 percent is sold in the processed form. For grapefruit, 34 percent is processed and for lemons 54 percent is sold processed.

Many insect pests attack citrus. Some important ones are citrus red mite, citrus thrips, citrus mealy bug, red scale, yellow scale, snow scale, purple scale, citrus whitefly, and aphids. In addition, certain species of nematodes attack citrus roots.

Citrus disease problems are more devastating than insect pests and must be controlled for citrus growing to be successful. Virus diseases include tristeza, exocortis, xyloporosis, vein enation, and yellow vein. All of the viruses are transmitted by budding and grafting with infected wood, and some are distributed by insect vectors. Consequently, it is best to avoid infected propagating material and susceptible rootstocks. A disease caused by a mycoplasm-like organism called stubborn is an important citrus disease in California. The only known control is disease-free propagating material. Fungal diseases attacking citrus are brown rot gummosis due to *Phytophthora*, brown rot of fruit, *Septoria* spot, and *Armillaria* root rot. Bacterial diseases are citrus blast and citrus canker, the latter devastating the Florida citrus nursery industry in the mid-1980s, when all infected trees were destroyed. A disease of unknown origin called blight is causing serious tree losses in Florida, mostly of trees on rough lemon rootstock.

Date (*Phoenix dactylifera* L.) ARECACEAE

The date palm is a tall, graceful monocotyledonous evergreen tree growing to a height of 15 m (50 ft) or more, much prized for its nutritious fruits and its ornamental value. The species is dioecious, with male and female flowers borne on separate plants.

There is evidence that the date was cultivated in Mesopotamia (present-day Iraq) as early as 3000 B.C., and it was a principal food crop for the ancient eastern Mediterranean civilizations. It later spread across North Africa. The early Spanish missionaries brought the date to the Western Hemisphere where it later became established as a crop in the hot, dry interior valleys of southern California, southern Arizona, and northern Mexico. Starting about 1900, plant explorers of the USDA visited the date-growing regions of Iraq, Egypt, Tunisia, Algeria, and Morocco and brought back offshoots of most of the best cultivars in each country. These were set out in experimental plantings in the Coachella Valley in California and the Salt River Valley in Arizona. The United States Date and Citrus Station was established by the USDA near Indio, California, in 1904 to study the cultural problems of the date in the United States.

Iran is the world's leading date-producing country, closely followed by Egypt and Iraq. Commercial quantities of dates are also grown in Algeria, Tunisia, Saudi Arabia, Morocco, Israel, and Libya, with minor production in other suitable areas from India to Mauritania. In the United States, dates are grown on about 1,942 ha (4,800 ac) in the Coachella valley of California, with some minor scattered plantings in southwestern Arizona and Texas.

The date palm withstands some winter cold, but is injured at temperatures below −7°C (20°F). To produce high-quality fruit, long, hot, dry summers with low humidity at the time of fruit maturity are required, as well as ample irrigation. The trees are grown as ornamentals in the climatic regions where the fruit produced is of poor quality.

Date palms grow on a wide range of soil types; however, a deep, well-drained, sandy loam is best. They tolerate soils more alkaline and saline than most other crop species, but do not produce the best tree growth and crops under such conditions.

The stem of the date tree has a single growing point, the terminal shoot tip, which gives the height increase. The long, 3 to 6 m (10 to 20 ft) leaves live for 3 to 7 years and are shed slowly after they die, hanging for many years around the trunk. Under cultivation dead leaves are cut off, leaving a whorl of 100 or more green leaves around the top—seven to eight leaves are required to develop each bunch of pollinated fruit.

Although the date palm is naturally wind-pollinated, pollen must be transferred artificially to the female flowers to ensure good fruit-set. The 0.6 to 1.2 m (2 to 4 ft) long male inflorescence (spadix) produces hundreds of flowers and an abundance of pollen that can be collected easily by cutting the recently opened inflorescence and drying the flowers where bees cannot rob the pollen. Most date growers keep one good male tree for every 30 to 50 females and collect pollen as required—pollen is used fresh when possible. Pollen can be dried in the shade and stored at about −18°C (0°F) for at least 1 year if necessary.

Female flowers are borne in a somewhat smaller inflorescence. When the spathe opens to expose the spadix, the female flowers are dusted either manually or mechanically with fresh pollen. In California, most of the pollination is done from late February to late April.

Fruit is thinned to improve fruit size and quality and to avoid severe alternate bearing. Usually about one-third of the spandix tip is removed at the time of pollination. When pollination is assured and the fruits are developed to perhaps 1 cm (0.5 in.) in diameter, some center strands are removed. Out of a total of 8,000 to 10,000 female flowers per spadix, 1,000 to 1,200 fruits are allowed to develop on 'Deglet Noor,' the principal California cultivar.

The developing and elongating fruit stalk is carefully bent and tied to an adjacent leaf stalk to mitigate wind damage and help support the ripening fruit. When the fruit begins to change color from green to yellow or red, bunches are usually covered with paper wrappers, open at the lower end, to protect the fruit from rain and birds.

Many hundreds of date cultivars are grown in the Middle East. Cultivars are classed by fruit type as soft, semidry, or dry. Different date-growing regions have developed and grown certain cultivars for centuries. Some of the most important, by country, are Iraq—'Zahadi,' 'Sayer,' 'Khadrawy,' 'Halawy'; Egypt—'Saidy,' 'Samani,' 'Hayami,' 'Zagloul'; Algeria—'Deglet Noor,' 'Rhars,' 'Mech Degla'; California—'Deglet Noor,' 'Medjool' (originally from Morocco), 'Zahadi,' 'Khadrawy.'

In the United States dates are usually eaten only as ripe fruit. However, in the Middle East date-growing areas several cultivars with low tannin content in the firm, colored (khalal) stage prior to softening are much appreciated as crisp, fresh fruit. A large part of the crop of 'Barhee,' 'Samany,' 'Hayany,' 'Briam,' and others is consumed fresh. Green and cull fruit at all stages of development is used as livestock feed.

Date cultivars are mainly propagated by severing rooted offshoots, which usually grow out from the main trunk near the soil level. Offshoots that form higher in the palms are more difficult to root unless girdled around the base. The monocotyledonous date plant cannot be budded or grafted or propagated by cuttings. Seed propagation is not feasible since the characteristics of the mother tree are not maintained in the female seedlings—about one-half of the seedlings are males. Tissue-culture techniques have been developed that permit large-scale clonal propagation of selected date cultivars.

The old flower stalks and leaves are removed when they become brown and interfere with pollination and harvest operations. In the United States, the thorny spines are removed from the leaf bases to facilitate pollination and other operations, but not in the Middle East.

Harvesting lasts from several weeks to several months, depending on the cultivar, as all of the fruits on a bunch do not mature at the same time and local microclimates affect ripening. Dates are usually left on the tree until they pass from the red-colored stage to the amber-dark brown color of most mature dates to allow full development of sugars and the loss of astringency. With care, fruit can be removed in the late khalal stage and ripened artificially. Keeping quality is best when the fruits have lost their watery consistency and have become pliable but not tough.

Mechanical harvesting, as practiced in the United States and Israel for the semidry cultivars 'Deglet Noor' and 'Zahidi,' requires that the fruit be left on the tree until fruit moisture is below 18 percent, then whole bunches can be removed in one or two harvests.

For home gardens, freshly harvested fruits are partially dried indoors with the drying completed in the sun. These fruits can then be stored for several years if protected from insects—storage in a home freezer is recommended. Commercially, dates are fumigated with methyl bromide immediately after harvest, cleaned by water sprays and roller brushes, and then dried by warm air currents. Some soft cultivars are only brushed or toweled. They are graded to remove culls and placed into uniform lots of ripeness, size, and appearance. If the fruits are not fully ripe, they can be placed in heated rooms at 27°C to 49°C (80°F to 120°F) for several hours or days to complete ripening. Following this dates may be dehydrated by warm, circulating air, after which they will keep for many months without refrigeration. Fresh, nondehydrated dates can be stored for several weeks in a household refrigerator.

In the United States, Bank's grass mite is a serious pest of dates. Control of this pest requires two or more applications of sulfur dust. Nitidulid beetles and certain moths are controlled with malathion dust in which a fungicide is incorporated to control fruit-rotting fungi. This is applied once or twice during the early harvest period. The paper wraps used to protect fruit bunches from rain prevent most fruit rot damage.

Fig (*Ficus carica* L.) MORACEAE

The edible fig is a deciduous tree or bush. There are almost 2,000 species in the genus *Ficus,* most of them evergreen and used as ornamentals. The fig apparently was first brought into cultivation in the southern part of the Arabian Peninsula by at least 3000 B.C. It later spread into what is now Iraq, Syria, and Turkey and on into all the Mediterranean countries. Along with the date, *vinifera* grape, and olive, the fig was an important food crop for the ancient civilizations of the eastern Mediterranean region. During the age of exploration following the discovery of America by Columbus the fig was taken to most subtropical areas of the Western Hemisphere. Turkey is now the leading fig-producing country, followed by Egypt and Greece. About 80 percent of U.S. production comes from California with small amounts grown in the Gulf States and extending to Virginia, often as home garden trees. Most of California's production is marketed as dried figs, with some canned, and a small quantity sold as fresh fruit.

Mature trees withstand winter temperatures of −12°C to −9.5°C (10°F to 15°F) depending on the cultivar, but young trees are not as cold resistant and may have to be protected by wrapping during the winter. For vigorous shoot growth to take place in the spring, buds of most *F. carica* cultivars require some winter chilling.

Fig trees grow well in a wide range of soils but do best in deep, nonalkaline clay loams. They respond to nitrogen fertilizers. In semiarid regions summer irrigation is required.

Some types of figs produce two crops a year. In the first or early summer crop, the fruits develop from latent flower buds on shoots produced the previous summer. This is termed the "breba" crop. In the second or late summer main crop, the fruits develop in the axils of leaves on shoots produced that same summer.

The edible fig is a unique "fruit." Figure 27–2 helps explain its structure. Botanically it consists of vegetative peduncle tissue called a syconium. The true fruits are the tiny

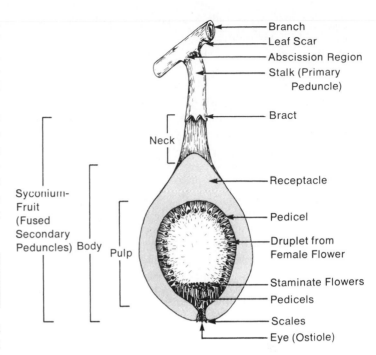

Figure 27–2 Longitudinal section through a caprifig syconium showing the various parts. The short-styled female flowers line the inner wall of the syconium. The staminate flowers, shown here, near the ostiole are typical of the male caprifig. The edible fig syconium is similar to this but has only long-styled female flowers. *Source:* Adapted from Storey, W. B. Figs. In *Advances in fruit breeding,* eds. J. Janick and J. N. Moore. West Lafayette, Ind.: Purdue University Press.

drupelets inside the cavity of the fused peduncles. Fruit growth is similar to a double sigmoid curve and shows three distinct phases. Phase I is characterized by intense cell division and differentiation with growth occurring rapidly. This is followed by a period of stasis (phase II) during which mitotic activity ceases. In phase III, rapid cell expansion procedes with a concurrent change in color and texture, and an increase in aroma.

There are four horticultural types of figs: (1) common, (2) San Pedro, (3) Smyrna—all producing only long-styled pistillate or female flowers—and (4) the inedible *caprifig* (producing both short-styled pistillate and staminate or male flowers).

In the common fig, the fruits are persistent, developing without the stimulation of pollination. Consequently, they are apparently seedless. Some cultivars of this type are 'Kadota' ('Dottato'), 'Mission,' 'Adriatic,' 'Brown Turkey,' 'Celeste,' and 'Conadria.'

The San Pedro type produces the breba crop without pollination of the flowers, but the second (late summer) crop develops only if the flowers have been pollinated as described following for the Smyrna-type fig. 'San Pedro,' 'King,' and 'Gentile' are cultivars of this type grown in California.

In the Smyrna-type fig, the breba crop is usually not produced. The second crop develops only if the pistillate flowers within the syconium are pollinated with pollen carried from the male flowers of the caprifig syconium by the tiny *Blastophaga* wasp. When this is done, hundreds of tiny fruits with viable seed and a brittle endocarp form inside the syconium of the Smyrna-type fig. The 'Calimyrna' ('Lob Injir') fig is of this type, grown on about half the 7,460 ha (18,440 ac) of figs in California. There are many more cultivars of the Smyrna type grown in the Mediterranean area. To attain pollination, fruit setting, and development of the edible syconia of the Smyrna-type fig, pollination (caprification) from the male caprifig by the *Blastophaga* wasp is necessary. Flowers of all edible figs produce no pollen. Pollen is obtained only from within the syconia of the inedible caprifigs, which are borne on caprifig trees, usually grown separately from the fruiting Smyrna-type trees. Pollen is actually transferred by the female of the small wasp (*Blastophaga psenes*), which overwinters in the pollen-producing caprifig. These caprifigs (with the wasps inside) are gathered and placed in small bags or wire baskets that are hung in the fruiting Smyrna-type trees. The emerging female wasps, covered with pollen, enter the Smyrna fruits and pollinate the long-styled pistillate flowers inside. Thus, for fruit setting of the Smyrna-type fig, caprifig trees are necessary as well as the *Blastophaga* wasp. For home gardens the common fig cultivars are preferred rather than the Smyrna-type fig cultivars with their complex pollination problems.

The fig is so easily propagated by hardwood cuttings that this is the usual method, although all types of budding and grafting as well as air layering are successful. Hardwood cuttings of 2- or 3-year wood 20 to 30 cm (8 to 12 in.) long are best taken in later winter, treated with the root-promoting indolebutyric acid, then stored in moist packing material, like sawdust, at room temperature for about 4 weeks, then planted in the nursery with just the top bud showing.

Figs can be pruned to most any system since the second (main) crop is produced in the leaf axils of the current season's shoots. Heavy dormant pruning reduces the first (breba) crop and may reduce the total crop, especially with the Smyrna type. Trees of some cultivars, if not kept pruned, can grow to a huge size with a spread of 30 m (100 ft) or more. In order to facilitate fruit harvesting, trees of others, like the 'Kadota' in California, are kept small, low, and flat-topped with a height of only 1.5 to 1.8 m (5 to 6 ft). Figs can easily be trained to an attractive espalier form against a wall.

For best flavor, figs grown in the home garden should not be picked until the fruits are soft and "wilt" at the neck, hanging down from their own weight. They are then fully mature and ripe and ready to eat. Milky latex exuding from the stem when the fruit is pulled off indicates the fruit is still immature. Figs picked when fully ripe have the best qualitative attributes (taste and aroma), but are not easily marketed as they can be badly damaged in transit. Fresh figs can be held only about a week at 0°C to 1.7°C (32°F to 35°F) with a 85 to 90 percent humidity. Fully matured figs can be either frozen or dried and still retain their flavor and color. For best

drying allow the figs to become completely mature on the tree, with maximum sugar content. Complete the drying on trays in the sun, turning frequently.

Figs are susceptible to root-knot nematodes, viruses, and endosepsis, a fruit rot in the Smyrna-type fig carried by the pollinating wasps. Other diseases affecting figs include oak root rot (*Armillaria mellea* Vahl.), anthracnose (*Colletotrichum gloeosporioides* Penz.), rust [*Kuehneola fici* (Cast.) Butl.], and leaf spot [*Alternaria alternata* (Fr.) Keissl.]. Birds feed on the developing fruits, and bees are attracted to the sweet juices from the "eye" of overmature figs.

Kiwifruit (Chinese Gooseberry) [*Actinidia deliciosa* (A. Chev.) C.F. Liang et A.R. Ferguson], ACTINIDIACEAE

The kiwifruit is a native of south central China. The plants grow naturally as deciduous fruiting vines in the hills and mountains of southern and central China. Seeds were taken to New Zealand in 1906 where the plant was named the Chinese gooseberry. It became a commercial export crop in the 1950s. It was renamed the kiwifruit after the native bird, the kiwi. Italy, New Zealand, and Chile are the leading countries in kiwifruit production. There is considerable interest in growing kiwifruit in the milder climatic parts of Europe, Japan, Australia, and Israel. In China considerable quantities of the fruits are collected in the wild each year.

The kiwifruit vine is grown on 1.8 m (6 ft) trellises, much like a grapevine. They are very vigorous and are usually planted in rows 4.5 m (15 ft) apart with the plants set 6 m (20 ft) apart in the rows. Full bearing 10-year-old vines can yield 17 to 22 MT of fruit per ha (7 to 9 MT per ac).

The kiwifruit requires frequent irrigation throughout the summer. A deep, well-drained soil is best. Mature plants need about 188 kg of nitrogen fertilizer per ha (150 lb per ac) each year. Wind protection for the plants and trellises is needed to prevent broken shoots and scarred fruit.

Flowers, fruits, and shoots of the kiwifruit plant in active growth are damaged by temperatures of −1.5°C (29°F), but dormant, leafless canes of mature vines in winter withstand temperatures as low as −12°C (10°F). Kiwifruits can generally be grown wherever citrus is cultivated in the temperate zones. Kiwifruit requires some winter cold [300 to 500 hours below 7°C (45°F)]; therefore, it does not grow well in the Tropics.

The kiwifruit is dioecious (male and female flowers borne on separate plants); about one male plant for every nine female plants—both blooming at the same time—is required for pollination. Pollen is transferred by insects, chiefly bees.

In both New Zealand and California the most popular cultivar is 'Hayward.' In California the vines bloom in May and the fruits are harvested in November. The fruit, which is a matured ovary, is about the size of a large chicken egg. It has a greenish brown rather unattractive skin densely covered with short hairs. The firm but juicy flesh is an attractive emerald green color containing many small black edible seeds. The taste is difficult to describe. It is somewhat tart but with a delicate flavor blend of citrus, gooseberry, strawberry, and pineapple.

The kiwifruit can be propagated by rooting leafy cuttings under mist after treating them with indolebutyric acid. They can also be propagated by T-budding in late summer or by whip or cleft grafting in the spring using 1-year-old kiwifruit seedlings as rootstocks.

Vines of the kiwifruit are trained to a single trunk on the trellis with one arm, cane, or cordon in each direction (bilateral cordon) on three-wire trellises. Fruiting canes are allowed to develop every 0.3 to 0.6 m (1 to 2 ft) along the side cordons (Fig. 27–3). Fruits develop from basal buds on the current season's shoots, which grew from last year's fruiting arms. Dormant pruning is required to remove old fruiting canes, leaving 1- or 2-year-old fruiting canes to produce the next year's crop. Summer pruning may be required to admit light to reduce excessive shading of fruiting wood.

In New Zealand the minimum acceptable maturity for export kiwifruit is 6.2 percent soluble solids. Kiwifruits in California are considered ready to harvest when the soluble solids reach a minimum of 6.5 percent together with a reduction in flesh firmness. For long-term storage, fruit should have a soluble solids level of 7.0 to 9.0 percent when harvested. In the past, growers were advised to cool-store fruit at 0°C as soon as possible, preferably within 24 hours of harvest. However, it was determined that rapidly cooled fruit had

a significantly higher incidence of botrytis stem-end rot (*Botrytis cinerea*) than passively cooled fruit. It is now common practice to hold the fruit at ambient temperature for several days after harvest before cool-storage. Properly protected with a fungicide, the fruits store up to 6 months in air at 0°C (32°F) with 95 percent relative humidity and ethylene maintained at < 0.01 ppm. Controlled atmospheres can increase storage life an additional 3 to 4 months (5 to 8 percent CO_2 and 1 to 2 percent O_2). Fruits are packed in trays covered with polyethylene sheets to prevent water loss and fruit shrivel during the long storage period. The fruits are usually eaten fresh but can also be preserved by canning or freezing. Kiwifruits are high in potassium and have a very high vitamin C content (105 mg/100 g), about double that of oranges.

In California there are a few serious insect and disease pests of kiwifruit, which are omniverous leaf roller, oak root fungus, and nematodes. Where nematodes occur, the soil should be fumigated before planting. In New Zealand the tortricid leaf roller caterpillar and greedy scale are the main insect pests. Botrytis and other fruit rots can be a problem in the stored fruits.

Olive (*Olea europaea* L.) OLEACEAE

The olive is an attractive evergreen tree with gray-green foliage. It was under cultivation long before the time of earliest recorded history, originating in the eastern Mediterranean area and providing ancient civilizations with one of their most valuable foods from the fruits and the oil. The olive was particularly valuable in those early days because it grew readily on a wide range of soils and tolerated the long, hot, rainless summers of the Mediterranean region. Its fruits are nourishing and the oil could be consumed or used as a fuel for lamps. Its ease of propagation by hardwood cuttings allowed perpetuation of superior types.

Olive trees live for many hundreds of years; if the tops die, a new tree often develops from the roots. They grow in all areas of the world having a Mediterranean climate—that is, moderately cold winters and long, hot, low-humidity summers. Commercial olive production lies in those areas between 30° and 45° north and south latitudes where severe winters do not occur. Spain is the world's leading olive producer, followed by Italy, Greece, Tunisia, Turkey, and Portugal. Olives in the United States are produced only in California, with an average production of about 108,864 MT (120,000 t) per year, but this crop represents less than 1 percent of the total world production.

Olive trees are killed to the ground by temperatures below −11°C (12°F). Most, but not all, cultivars require at least 2 months of winter chilling (daily fluctuating temperatures between 1.5°C and 15.5°C (35°F and 60°F) for the flowers to initiate in the buds. The trees bloom in midspring about 8 weeks after the flower buds have been initiated.

Figure 27–3 Good crop of kiwifruits on a trellised vine. Fruits are about 5 cm (2 in.) in length. *Source:* Blue Anchor.

Olive trees grow well on many different kinds of soil, ranging from rocky shallow hillsides to deep fertile valley soils, from acid soils to fairly alkaline soils. They tolerate considerable salinity and boron. They do not withstand poorly drained soils, however, and quickly die if water stands around their roots for a few weeks. To obtain satisfactory tree growth and production, moderate annual nitrogen fertilizer application is required together with irrigation through the summer months. In some soils olives respond to potassium and boron fertilizers.

All olive-producing countries have their own local cultivars, many of which have been grown for centuries. Some leading oil cultivars are 'Picual,' 'Frantoio,' 'Cornicabra,' 'Chemlaly,' 'Chetoui,' and 'Memecik.' Prominent table olive cultivars are 'Manzanillo,' 'Sevillano' ('Gordal'), 'Ascolano,' 'Conservolia,' 'Calamata,' and 'Galega.' 'Swan Hill' is a fruitless ornamental olive that is resistant to verticillium wilt and produces small amounts of nonviable pollen.

Olive flowers are produced on inflorescences of about 15 flowers each that arise in the leaf axils of shoots that grew the previous summer. Most olive flowers are perfect; that is, each flower contains the pollen-producing anthers and the fruit-producing pistil. Very often, however, the pistil aborts, resulting in nonfruiting staminate flowers. The olive is wind pollinated and trees of most cultivars set some fruit with their own pollen, but in many years fruit-set is heavier if cross-pollination with trees of other cultivars is available. In some years olives tend to set so many fruits the trees cannot mature them properly. Fruit size is small, maturity is delayed, and usually the trees will not even bloom the next year, leading to an alternate-bearing situation. In such cases fruits on overloaded trees should be thinned shortly after fruit-set in the spring. This can be done by hand, but spray thinning with an auxin-type material—naphthaleneacetic acid—is more feasible.

Most olive cultivars can be propagated by cuttings, either hardwood cuttings planted directly in the nursery row or leafy, softwood cuttings rooted in mist-propagating beds. Root initiation is greatly stimulated by applications of auxin-type chemicals. Olives can also be propagated by grafting or budding the scion cultivar onto rootstocks (seedlings or rooted cuttings). Older trees can be easily topworked by bark grafting to change cultivars.

Young olive trees should be pruned sparingly with the objective of establishing a central trunk with four or five well-spaced primary scaffold branches. Mature trees bear fruit laterally on shoots produced the previous summer, so moderate annual pruning is required to stimulate development of such fruiting shoots. Dead, diseased, broken, and interfering branches should be removed as required. As the trees become very old, they may lose vigor and productivity, but by judicious pollarding new growth can be forced out that will again yield well.

The table olive crop is harvested in autumn when the fruits are a green to straw color. Fruits left on the tree until winter, when they turn black and attain their maximum oil content, are harvested for oil extraction. Olives picked from the tree are very bitter and cannot be eaten because of a bitter glucoside in the raw fruits. This can be partially leached out with running water, or it can be neutralized by several repeat applications of a dilute alkaline solution such as 0.5 to 2.0 percent sodium hydroxide (lye), which must then be thoroughly washed out with water. If the olive fruits are exposed to the air during this process, phenolic compounds oxidize and the olives turn black. In the United States, they are then sold as "black ripe" olives. If they are kept under the solution continually, they stay green and are sold as "green ripe" olives. Hot brine is added to the fruits, then they are canned and heated in a pressurized retort to a temperature of 115°C (240°F) for 1 hour to eliminate any toxic bacteria.

The Spanish-green fermented style of processed table olives, used primarily in the Mediterranean countries, involves a different treatment. The fruits are picked green, then treated briefly with lye to remove some of the bitterness, washed thoroughly, and transferred to barrels containing a sodium chloride brine (9 to 11° Be) solution. Lactic acid fermentation starts and continues for 6 to 8 months. After fermentation, the fruits are graded according to shape, size, and color, then pasteurized and canned in brine. All types of commercial olive processing involve chemical reactions that must be carefully watched and controlled or considerable losses can occur.

Olive oil is extracted after the olives turn black on the tree in midwinter and reach their maximum oil content (20 to 30 percent). The main processing steps for olive oil extraction include washing, milling, malaxation, and centrifugation. Harvested fruits are washed, then the whole fruit (skin, flesh, and pit) is ground in crushers. The olive paste is kneaded or rolled (malaxation) after crushing. This aids in the coalescence of small oil drops into larger ones, facilitating the separation of the oil and water. The temperature should not be allowed to rise above 30°C. Higher temperatures will cause the oil to become reddish in color, have an increased acidity, and will alter the aroma constituents of the oil. After malaxation, the olive paste is diluted with water (1 liter H_2O: 1 kg paste) and separated in special centrifuges. Filtering gives a final golden clear product.

Olives are subject to several insect and disease pests. In the Mediterranean countries the olive fly (*Dacus oleae*) and the olive moth (*Pravs oleaellus*) are of most concern. These do not occur in California. Black scale (*Saissetia oleae*) can be severe on olives and is found in most olive-producing countries. Olives are attacked by a gall-producing bacteria (*Pkytomonas savastanoi*); they are also susceptible to two fungal diseases—verticillium wilt, which affects the tree's roots, and peacock spot (*Cycloconium oleaginum*), which can defoliate the trees.

Persimmon (*Diospyros* spp.) EBENACEAE

There are a number of *Diospyros* species native to various parts of the world. The most important commercially is the Oriental persimmon, *Diospyros kaki* L.f., which originated in China and was imported into Japan about 750 A.D. Many introductions of this species were made into the United States following Commodore Perry's visit to Japan in 1853. Trees of the native American species (*D. virginiana* L.) grow naturally in the southeastern and southern United States. They produce small but very tasty fruits. There are some named cultivated types, but most fruits are collected from wild, volunteer trees.

Cultivars of *D. kaki* are grown commercially and as home garden trees in mild subtropical or temperate climatic regions like the Mediterranean countries, Florida, and California. The trees do not survive winter temperatures below −18°C (0°F). China leads the world in persimmon production, followed by Japan and Korea. In the United States *D. kaki* cultivars are produced commercially in California, and many southern states have small plantings.

The Oriental persimmon is a small [6 to 9 m (20 to 30 ft)] attractive deciduous tree producing large orange-red fruits that mature in late fall. The trees grow well in a wide range of soils but do best in those that are deep and well-drained. They may require nitrogen fertilization each year to maintain vegetative growth, but excessive shoot growth—more than 30 cm (12 in.) per year—causes premature fruit drop. Irrigation is required in some regions with rainless summers.

Some astringent Oriental persimmon cultivars are 'Hachiya' and 'Tamopan,' the favored ones in California, and 'Tanenashi,' mostly grown in the southern United States. 'Fuyu' is one of the nonastringent, firm-fruited cultivars favored in Japan. It has flat, tomato-shaped fruits that can be eaten while still firm and buttery. 'Tanenashi,' 'Hachiya,' and 'Fuyu' fruits are large and heart-shaped. *D. virginiana* fruits are small, about the size of a golf ball, and a dull orange when ripe.

'Hachiya' and 'Fuyu' trees produce only pistillate (female) flowers, but the fruits set parthenocarpically (without pollination) and are generally seedless. However, if a *D. kaki* persimmon tree with male, pollen-producing flowers is nearby, some of the flowers of Hachiya or other cultivars may be pollinated and the resulting fruits will have seeds.

D. kaki cultivars are best propagated by whip grafting on seedling rootstocks, although they are easily T-budded (see Ch. 5). Seedlings of *D. lotus* have been found to be best for rootstocks in California whereas *D. virginiana* is used almost exclusively in the southern United States as a rootstock. *D. virginiana* is difficult to transplant due to a long, large taproot, and the trees tend to be short-lived (8 to 12 years). Pruning and training are done to encourage formation of a structurally strong tree with wide-angled scaffold branches because the wood of *D. kaki* trees is naturally brittle and tends to break or split under heavy fruit loads. A modified central leader form is probably best. *D. kaki* flowers are borne on the current season's growth. Flowers forming near the tips of the shoots are not as vigorous as basal flowers, but may set fruit. Sufficient annual pruning should be given to force 20 to 30 cm (8 to 12 in.) of new growth each season, but excessive pruning results in long, rank shoots that are often unfruitful.

Mature fruits of all *D. virginiana* and most *D. kaki* cultivars are very astringent until properly ripened and allowed to soften. 'Hachiya' fruits, for example, are picked when the orange-red color starts to develop but when they are still firm enough to handle for market. As the fruits ripen and soften at room temperature the astringency disappears. It is not necessary that *D. virginiana*, or *D. kaki* fruits be frosted to eliminate the astringency. Treatment of persimmon fruits with CO_2, ethylene, or vapors from alcoholic beverages promotes ripening and loss of astringency. Placing some persimmons in a plastic bag with one or two apples, ripe pears, or bananas and holding at room temperature for several days results in loss of astringency due to the ethylene released by the other fruits. Firm persimmons may be frozen whole in plastic bags and used as needed by simply thawing. When completely thawed they are soft and free of astringency. Astringency in persimmon fruits is generally due to tannin compounds. During ripening the tannins are bound into firm, insoluble particles that can no longer be tasted.

Freshly harvested, mature Oriental persimmon fruits can be stored for about 2 months at −0°C (32°F) and 90 percent relative humidity. Controlled atmosphere storage (5 to 8 percent CO_2 and 2 to 3 percent O_2) allows storage of some cultivars up to 6 months.

Persimmon trees are remarkably free of insect and disease pests; however, root-knot nematodes, crown gall, and stem pitting virus can become problems.

Pistachio (*Pistacia vera* L.) ANACARDIACEAE

Pistachio is one of the most ancient of nut crops. It was cultivated at least 3,000 years ago in the low mountain regions of central Asia, Pakistan, and India, where wild trees can still be found. Iran, the United States, and Turkey are the leading producers of pistachio nuts. In the United States pistachio nuts are produced commercially only in California where, in recent years, there have been many new plantings with 19,000 ha (46,900 ac) by 1984, primarily in the San Joaquin Valley. Another species, *P. chinensis* Bunge., which originated in China, is a much larger, beautiful shade tree widely planted in subtropical zones especially for its attractive fall leaf coloration. *P. vera* is a small-to medium-size deciduous tree that can live for several hundred years. Tree growth and production are best in regions with long, hot, dry summers and moderately cold winters. The trees survive winter temperatures of −18°C (0°F). Many subtropical and all tropical regions have insufficient winter cold to break the rest period of the buds, and, consequently, spring shoot growth is poor.

While pistachio trees survive on rocky, shallow soils in desert areas, they grow and yield best when planted in deep, well-drained sandy loams. They do well on highly calcareous soils and grow under conditions of alkalinity and salinity that would be detrimental to most other crops. Although pistachio trees stay alive under prolonged drought conditions, summer irrigation is necessary for good tree growth and productivity. The trees are resistant of cold and wind but intolerant of excessive dampness and high humidity. For vigorous tree growth and good yields nitrogen fertilization is also required.

The pistachio is dioecious (male and female flowers borne on separate trees) so that in any planting some male trees must be included to provide adequate pollination (one pollinator tree for every eight female trees). Pollen is distributed by the wind. Pistachios exhibit alternate bearing, producing a heavy crop one year followed by a light or no crop the next.

In the commercial trade pistachio nuts are classified according to the country of their origin, but each producing country has a number of local cultivars, probably selected from superior seedlings and propagated vegetatively. In California, 'Kerman' is the most widely planted female cultivar and 'Peters' is the male cultivar most used for pollination.

P. vera cultivars are somewhat difficult to propagate, the usual method being T-budding from mid-spring to fall with seedlings of *P. atlantica* or *P. terebenthis* as rootstocks if verticillium wilt is not a problem. Otherwise, use *P. integerrima* seedlings, which are resistant to verticillium wilt. The rootstock seedlings must be growing vigorously and the vegetative buds should be large, taken from dark-colored budwood on bearing trees known to produce good crops.

Pistachio trees are trained to a modified leader or vase system with well-spaced lateral branches. Pistachio flowers and fruits are borne on the previous season's growth, so pruning should be sufficient to force ample shoot growth each year. Small cuts should be made, thinning out weak, shaded branches or heading back excessively vigorous growth.

Pistachio nuts are drupes, like peaches or apricots, maturing in late summer or early fall. The outer "husk" is the exocarp and mesocarp (two outer layers of ovary wall). The hard thin, gray-white shell, which splits open just before nut maturity, is the endocarp. The part eaten is the embryo, consisting mainly of the two green cotyledons covered by a thin seed coat. Pistachio nuts are ready to harvest when the green or reddish husk becomes translucent and separates from the shell. The nuts are borne in clusters

that can be picked up by hand but, in more rapid modern operations, machine shaking of the trees drops the nuts onto catching frames. The attached husks can be removed mechanically, and empty nuts are separated by flotation. Most nuts are marketed in the split shells.

The most troublesome disease attacking pistachio trees in California is the soilborne fungus *Verticillium alboatrum*. Plantings should not be made in soil previously planted to cotton, vegetables, or strawberries as it is likely to be contaminated with the *Verticillium* organism. Trees in poorly drained wet soils may become infected with a root disease caused by *Phytophthora* spp. Since root-knot nematodes also attack pistachios, fumigation is done before the trees are planted in infested soil. There are many serious insect pests of pistachio in the Middle Eastern countries, but few have appeared so far in California plantings.

Pomegranate (*Punica granatum* L.) PUNICACEAE

This deciduous shrub or small tree is one of the most ancient of fruits. It apparently originated in the region of present-day Iran and was held in great esteem for religious ceremonies by the ancient civilizations of the Middle East. It was referred to in myths by the early Greek writers. Cultivation of pomegranates spread east to India and the Orient and west to all countries circling the Mediterranean Sea, where it became particularly popular in Spain. The city of Granada, Spain, was named after the high-quality pomegranates produced there. It is now a popular fruit in the Far East, India, and in Iraq and Iran, where certain superior cultivars have been developed. Spanish missionaries brought the pomegranate to the New World shortly after Cortés conquered Mexico in 1521, then the Franciscan padres brought it to all the missions along the California coast in the 1700s.

All commercial U.S. production of pomegranates comes from California where about 1,500 ha (3,700 ac) are grown. Pomegranates make handsome ornamentals because of their brilliant orange-red flowers in the spring, their shiny, bright-green leaves, and the large, red apple-shaped fruits ripening in the fall. Pomegranates grow well on a wide range of soils and tolerate mild alkaline conditions. It is essentially a desert plant and survives long periods of drought, but for commercial production of high-quality fruits summer irrigation is required. Some nitrogen fertilization is also needed. Local cultivars are used in the countries where the pomegranate is grown. In California the most widely planted, by far, is 'Wonderful,' followed by 'Ruby Red,' 'Early Red,' and 'Granada,' the latter two being patented. Pomegranates are self-pollinated.

Pomegranates are easily propagated by vegetative cuttings, principally hardwood cuttings planted in outdoor nurseries in early spring. Leafy cuttings root easily in mist propagating beds in the summer. Seeds germinate readily, but the seedlings are variable and do not reproduce the cultivar. The pomegranate can be allowed to grow as a bush for home garden use, but commercial growers prefer to train it as a single or multiple trunk tree. It suckers readily, so annual pruning should include sucker removal plus light thinning out of the top to encourage production of new fruiting spurs.

The pomegranate fruit is a round, apple-sized berry with a hard leathery rind that encloses many seeds. The edible part is the pulpy and juicy sac surrounding each of the numerous seeds. The inner hard and stony seed coat encloses the embryo. The pulp is removed from the fruit and the red juice extracted by pressure. It can be used for flavoring or for making juice drinks or jelly.

Fruits mature in the fall and should be picked after they become full red and the juice attains a soluble solids content of 15 percent and a crimson color, but before they split open. Often the attractive red fruits with attached calyx are just used for table decorations. Pomegranates store well and keep for up to 6 months in cold storage if the humidity is maintained at 85 to 90 percent. The control of relative humidity is critical in the storage of pomegranate fruits. At low humidity, the skin desiccates readily and the rind becomes dark and hard—the fruits become less attractive and have poor marketability. Storage at 5°C or lower resulted in chilling injury to the fruits. The severity of the symptoms increases with exposure time and at temperatures below 5°C.

Pomegranates in California have few serious insect or disease problems. Postharvest pathogens include greymold rot (*Botrytis cinerea*), heart rot (*Aspergillus niger* and *Alternaria* spp.), and penicillium rot (*Penicillium* spp.).

REFERENCES AND SUPPLEMENTARY READING

CHILDERS, N. F., J. R. MORRIS, and G. S. SIBBETT. 1995. *Modern fruit science.* Tenth Edition. Gainesville, Fla.: Horticultural Publications.

GALLETTA, G. J., and D. G. HIMELRICK (eds.). 1990. *Small fruit crop management.*

JACKSON, D. I., and N. E. LOONEY (eds.). 1999. *Temperate and subtropical fruit production.* Wallingford, UK.: CAB International.

JUNICK, J., and J. N. MOORE (eds.). 1996. *Fruit breeding,* Vols. I and II. New York: John Wiley & Sons.

MITRA, S. K. (ed.). *Postharvest physiology and storage of temperate and subtropical fruits.* Wallingford, UK.: CAB International.

SCHAFFER, B., and P. C. ANDERSEN. 1994. *Handbook of environmental physiology of fruit crops,* Vol. II, *Sub-Tropical and tropical crops.* Boca Raton, FL: CRC Press.

CHAPTER

Tropical Fruit and Nut Crops

There are a great many tropical fruit and nut species. This chapter discusses the principal ones that have gained great popularity for their adaptability to many parts of the Tropics, their high productivity and ease of culture, and their attractive, flavorful, easy-to-eat products. Some have become an indispensable part of the food supply in many households throughout the world.

There are other lesser-known tropical fruits and nuts that have for some reason not been brought into wide culture and distribution. With sufficient research to overcome their difficulties or, perhaps, by hybridization to develop new forms, fruits of some of these other lesser-known tropical species in the future will become commonplace in world markets. Some species deserving more consideration as potential tropical fruit crops, although unheard of by most people, are the durian, mangosteen, naranjilla, pejibaye, pummelo, soursop, and uvilla. The mangosteen, for example, is often described as the world's best-flavored fruit, yet it is virtually unknown outside Southeast Asia.

Banana (*Musa acuminata* Colla.) MUSACEAE

Bananas appear to have originated in Southeast Asia, spreading to India, Africa, and finally to tropical America, which now supplies the bulk of the world's commercial shipments. The first evidence of banana cultivation is found in the ancient writings (500 to 600 B.C.) of India. All bananas grown for export originated from *Musa acuminata;* whereas, some bananas favored in certain areas for domestic consumption and all plantains are hybrids between *M. acuminata* and *M. balbisiana.* Ploidy and genomic composition of the different clones are designated by A and B to represent the genomes of *M. acuminata* and *M. balbisiana,* respectively. Before the 1940s, the cultivar 'Gros Michel' (AAA) dominated the international banana trade, until it succumbed to Panama disease [*Fusarium oxysporum* f. sp. *cubense* (E.F. Sm.) Snyd. & Hans.]. Since the 1940s, the trade has adopted cultivars of the 'Cavendish' subgroup (AAA), which are resistant to Race 1 of Panama disease.

India, Ecuador, and Brazil, ranked in that order, are the leading banana-producing countries, although commercial plantings are increasing in several countries. Banana production in India averaged more than 9,000,000 MT from 1996 to 1998. The United States, Western Europe, and Japan are the major import markets. Bananas, along with grapes, oranges, and apples, are the world's most important fruit crops. Bananas, both the sweet dessert cultivars and the starchy cooking plantain types are important food items throughout the Tropics. Cooking bananas, however, seldom reach temperate zone markets. Bananas arrive at the world's markets year-round.

Bananas thrive and yield best in tropical regions where temperatures average from 27°C to 29°C (80°F to 85°F), not dropping below 15°C (60°F) nor rising above 35°C (95°F). A frost kills banana leaves, and cool temperatures—10°C to 13°C (50°F to 55°F)—impair fruit quality. Bananas require constant soil moisture, such as 10 cm (4 in.) of rain per month. If rainfall is inadequate, irrigation, preferably from high overhead sprinklers, is essential for good production. Bananas have a high nitrogen and potassium fertilizer requirement in a ratio of 1:3 to 1:5 for good yields in most soils.

The banana is a giant monocotyledonous herbaceous perennial plant with an underground, horizontal rhizome from which roots and the leafy shoots or pseudostems develop. This rhizome is the banana's true stem. The pseudostems, which reach a height of up to 6 m (20 ft) comprise many overlapping leaf sheath bases. New rolled leaves emerge through the center of previous overlapping leaf bases. Each individual leaf is about 2.4 m (8 ft) long and 0.6 m (2 ft) wide, making it one of the largest photosynthetic units known. Initiation of the inflorescence is not directly influenced by photoperiod or temperature, but occurs after the plant has produced a predetermined number of leaves. Generally, after approximately 40 ± 10 leaves have been produced, the plant enters a reproductive phase and the vegetative apex inside the leaf sheath changes to a floral apex. This grows rapidly pushing the inflorescence out the top of the pseudostem. Each inflorescence contains 5 to 13 groups of flowers ("hands") with each group enclosed in a floral bract. The hands are made up of double rows of "fingers" (Fig. 28–1).

In the commercial edible banana cultivars there are three types of flowers borne along the inflorescence stem: (1) a large group of female flowers that result in the parthenocarpic seedless fruits (fruits that develop without fertilization of the egg), (2) beyond this, a small group of hermaphroditic flowers (having both male and female parts), and (3) male flowers only. After emerging from the top of the pseudostem, the inflorescence turns and grows downward so that the developing banana fruits are at the top of the inflorescence stalk and the nonfruiting sterile portion below. The fingers are negatively geotropic and gradually turn upward during the first week or 10 days. Pulp in the parthenocarpic edible bananas develops mainly from the ovary wall and no pollination is necessary to stimulate this development. About 80 to 120 days or longer are required for the fruits to mature enough for harvest.

Banana harvesting begins about 9 months after planting, which is considerably less than for most tree fruit crops. Although each pseudostem produces fruits but once before dying, a banana planting can exist for many years as a succession of new pseudostems arise from buds along the underground, horizontal rhizome.

The seedless edible banana cultivars must be propagated vegetatively. The best propagating material for setting out new plantations is the sword suckers [0.3 to 0.9 m (1 to 3 ft) tall], which arise from the rhizome alongside the older plants. These, plus the rhizome, are cut off and planted, with new roots developing from the base of the leaf sheaths. Suckers should be taken only from plantings free of burrowing nematodes and bunchy-top virus.

Throughout the world there are about 300 named banana cultivars (many are synonyms), but the most important shipping bananas are the seedless triploid cultivars of the 'Cavendish' subgroup, including 'Williams,' 'Giant Cavendish,' and 'Robusta.' These are resistant to Panama disease, a problem that has plagued banana producers for many years. Growing these resistant, but tender, cultivars starting in the 1950s required the shipment of the bananas in protective cartons rather than as loose bunches. Before this the fusarium-susceptible 'Gros Michel' was grown. Its compact stems withstood the long journey to market as uncrated individual bunches without bruising.

Bananas are harvested when the fruits are fully mature but still hard and solid green. They do not ripen well if left on the plants, splitting open, becoming "cottony" textured, and tasteless. The bananas' "hands" are cut into clusters of 4 to 16 "fingers" for shipping to importing countries in 40-lb polyethylene-lined cartons. Upon arrival at distributing centers the cartons first go to ripening rooms. The fruit is exposed to ethylene at 1,000 μL l^{-1} (1,000 ppm) for 24 h. Temperature is varied during ripening to achieve the desired combination of peel color, pulp texture, and flavor, according to market demands. Temperatures held between 16°C and 18°C (56°F and 62°F) give a long shelf-life, while temperatures up to 25°C hasten the process, and at

Figure 28–1   *Left:* A banana fruit stalk at an early stage of development. *Right:* Cluster of mature fruits ready for harvest. The entire fruit stalk will be cut off by the sharpened chisel on the end of a pole. Note that the individual banana fruits have turned upward during their development. *Source:* United Brands Company.

temperatures above 30°C the pulp softens but the peel stays green. Humidity is held high at first (90 to 95 percent), but is reduced during the ripening process to avoid splitting of the peel. In warm weather the bananas are moved to the retail shelves just as the yellow color appears, but in cold weather the fruit is allowed to attain a more pronounced yellow before going to the markets. Good fruit storage temperatures after ripening are 13°C to 16°C (56°F to 60°F). Bananas are ready to use when a full yellow color is attained; however, the best flavor and nutritive value develop when flecks of brown appear on the skin surface. The banana is a highly nutritious food, low in sodium and fat but high in potassium and easily absorbed carbohydrates.

The most serious problem in growing bananas was Panama disease, caused by *Fusarium oxysporum* f. sp. *cubense,* Race 1, which is a soilborne organism that spreads from plant to plant, causing them to wilt and die. Resistant Cavendish-type cultivars are now grown, but reports indicate a new race of *Fusarium oxysporum* (Race 4) is killing 'Cavendish' cultivars in Australia and other banana-producing countries. Yellow Sigatoka (*Mycosphaerella musicola*) and black Sigatoka (*Mycosphaerella fijiensis*) leaf spot can result in almost complete defoliation in susceptible cultivars of

bananas and plantains. Cercospora leaf spot and bunchy top virus are also problems in some banana-growing countries. Although there are few serious insect pests, the burrowing nematode [*Radopholus similes* (Cobb) Thorne] can significantly reduce yields. Wind damage can be serious in banana plantations in windy areas. 'Dwarf Cavendish,' a small Cavendish type, is widely grown in Hawaii, South Africa, and other places where wind is a problem.

Cacao (*Theobroma cacao* L.) STERCULIACEAE

The center of genetic diversity of cacao is the Amazon Basin region of South America, originating there at least 4,000 years ago. Cacao evolved as a perennial evergreen understory tree in the tropical forest. It was naturally dispersed by foraging animals and people who ate the sweet pulp of the pods and dropped the seeds. It was first cultivated by the Mayan and Aztec civilizations of tropical Central America who used the ground beans to make a cocoa beverage. Based on archaeological information, the Mayans cultivated cacao 2,000 to 4,000 years before coming in contact with Spanish explorers. Columbus, on his fourth voyage in 1502, saw cacao beans, but it was Cortés who, a few years later, took the beans back to Spain. The beverage was improved considerably by the addition of sugar. Spain kept the secret of this new drink for about 100 years, but it gradually spread throughout Europe and became very popular. The Dutch invented chocolate candy about 1830.

Cacao production has developed into a billion dollar industry of worldwide importance in the manufacture of cocoa and chocolate. Until about 1900 most cacao plantations were in Central and South America and the Caribbean Islands. In 1879, however, cacao was introduced into West Africa where it started on small farms of 0.8 to 2.0 ha (2 to 5 ac) each. Production in this region increased phenomenally. The Côte d'Ivoire is the leading producer, followed by Ghana, Indonesia, Brazil, and Nigeria.

Cacao trees are broad-leaved evergreens that grow to a height of 5 to 8 m (16 to 26 ft) under cultivation, but may attain a height >15 m in neotropical forests. They require a truly tropical climate, with all production areas located within 20° of the equator and at altitudes below 300 m (1,000 ft). Cacao trees grow best at a mean annual temperature varying between 21°C to 27°C (70°F to 80°F). Growth slows at average temperatures < 15°C (59°F), and is minimal at average temperatures of 10°C (50°F). Production is highest in areas with high humidity and rainfall well distributed throughout the year [about 127 to 152 mm (5 to 6 in.) per month] and with little or no dry season. Winds are harmful to cacao trees by increasing water loss from the leaves and causing defoliation; thus, windbreaks are often used. Cacao trees, like some other tropical crops (coffee and tea), are usually grown in the shade of taller trees, planted especially for this purpose or retained when forests are cleared for new plantings. The optimum degree of shade depends on the fertility of the soil. On fertile or fertilized soils, no shade is needed and yields increase.

Cacao trees grow best in deep, slightly acidic soils that are high in organic matter and well drained but with a high water-holding capacity. Heavy clay soils are unsuitable. Cacao trees respond to nitrogen, phosphorus, and potassium fertilizers when shade is reduced or removed.

Cacao plantings consist largely of seedling trees that develop a wide, branching growth habit and start bearing at 2 to 6 years of age. Some high-yielding clones have been developed that are propagated vegetatively by rooting cuttings. The flowers and fruit are borne directly on the older, leafless parts of the branches and on the trunk (Fig. 28–2). Flowering and fruiting may occur throughout the year, determined largely by the rainfall pattern, but the majority of the fruit harvest occurs from September to March.

Cacao flowers are perfect (with both stamens and pistils) and may be self- or cross-pollinated for fruit setting. Pollination of flowers is by *Forcipomyia* spp. and *Euprojoannisia* spp., which are small-bodied flying insects in the Ceratopogonidae family. Lack of adequate cross-pollination may sometimes limit crop production.

Figure 28–2 Maturing cacao fruits as they develop on the older branches and trunk of the tree. *Source:* USDA.

A mature cacao tree may produce 50,000 flowers, but less than 5 percent set pods, and many of those will not reach maturity. The pod-like oval fruits are 10 to 50 cm (4 to 12 in.) long containing 30 to 50 seeds, each surrounded by a sweet, white mucilage. A good yield is 270 kg (600 lb) of dry beans (seeds) per acre, but yields of over 1,350 kg (3,000 lb) per acre have been obtained from improved cultivars. Pods usually mature in 5 to 6 months on cultivated clones and are harvested weekly when they change to a red or yellow color. They are cut open when fully ripe to remove the seeds. The mucilaginous pulp around the seeds is removed by a fermentation process, which takes 3 to 8 days and causes temperatures to rise to about 51°C (125°F). This kills the embryo in the seeds and develops certain chemical precursors that give the chocolate flavors when the beans are later sun-dried and finally roasted. Cocoa has mild stimulating properties from its theobromine and caffeine content.

Cacao trees, especially the pods, are attacked by *Phytophthora palmivora*, producing a disease called black pod, as well as by several other fungal pathogens, such as *Marasmius* spp., which cause witches'-broom disease. Recommended management to reduce the deleterious effects of witches'-broom on cacao production include the use of phytosanitation (removal of diseased plant parts), application of chemical fungicides, and the use of host resistance. A viral disease called swollen shoot has caused great damage in West Africa. Several insect pests also attack cacao.

Cashew (*Anacardium occidentale* L.) ANACARDIACEAE

The cashew originated along the northeast coast of Brazil. From Brazil, Carib natives probably took the cashew to the West Indies. However, it was not until the sixteenth century that Spanish sailors likely introduced the cashew to Central America and Panama. When the Portuguese arrived in Brazil, they soon recognized the value of the cashew apple and nut and took the crop to their colonies in the Old World. By 1590 the cashew had already been introduced to the Portuguese territories in East Africa and Goa. The delicious cashew is one of the most important nuts of commerce. It was not until the beginning of the twentieth century that the nut entered world commerce. It is believed that the cashew was adapted so well in India that from there it spread to other parts of tropical Asia and East Africa, where today in Mozambique and Tanzania cashews grow in the wild. Mozambique and Tanzania formally dominated global trade in cashew nuts, but production has declined in those countries due to political instability in Mozambique and ineffective agricultural policies in Tanzania. Currently, India and Brazil are the world's leading producers of cashew nuts. India's share is 35 percent, Brazil is second with 25 percent, and the combined East African nation's share is 25 percent. Countries throughout the tropics, including Indonesia, Vietnam, Guatemala, Costa Rica, Venezuela, Peru, Thailand, and Kenya, make up the remaining 14 percent.

The cashew is a hardy, fast-growing, evergreen perennial with a symmetrical, umbrella-like canopy. Although the tree resembles a large bush, it is a true evergreen tree and can attain heights in excess of 10 m (32 ft). Young trees are very tender to cold but when mature withstand light frosts. Roots reach vertically to a considerable depth and to a radius twice the canopy spread.

Cashew flowers are borne in terminal panicles, having about 60 percent staminate (male) and 40 percent hermaphroditic (male and female) flowers, which are insect-pollinated. Each perfect flower has one ovary, which contains one ovule. About 70 percent of the perfect flowers fail to set nuts. Flowering to nut maturity takes from 55 to 70 days. The precocious dwarf cashew tree, a selection developed in Brazilian fruit-breeding programs, has been a very important achievement because it allows hand harvest without ladders and promotes early and expanded flowering and fruit-set. With irrigation management, year-round production is possible, expanding the possibilities for both export and processing.

The cashew fruit is attached at the end of the cashew "apple," which is actually the swollen pedicel or fruit stem

Figure 28–3   Group of cashews as harvested from the tree. (A) Enlarged fleshy fruit stem or pedicel, called the cashew apple. (B) True cashew fruit containing the seed or "nut" inside. *Source:* USDA.

(Fig. 28–3). This "apple" is 7 to 10 cm (3 to 4 in.) long and about 5 cm (2 in.) wide, red or yellow in color at maturity, and is eaten as a popular "fruit" by the natives in areas where cashews are grown. The fleshy cashew apple can be eaten raw or processed into a variety of products. Cashew fruit contributes to human nutrition by supplying vitamin C, averaging 200 mg/100 g of juice, which is four times that of orange juice. In several countries, cashew apple products such as juice, wine, vinegar, soft drinks and candies are widely commercialized, further raising the aggregate value of the cashew crop. The kidney-shaped cashew nut of commerce is the seed (with the seed coats removed) produced inside the true fruit, which has a two-layered ovary wall or shell. The cashew nut shell liquid (CNSL) is located between the two layers and is extracted during processing of the nut. CNSL has high polymerizing and friction-reducing properties and is used in varnishes, lacquers, paints, and brake linings, and for waterproofing and as a preservative. There are more than 300 patented uses of cashew by-products; however, CNSL is extremely irritating to human tissues, both internal and external.

For establishment of the cashew plantation, seeds are sown directly in the field, three seeds per stake, to ensure seedling survival during the season. After about 1 year of germination, only the most vigorous seedling is left per stake. Due to the fact that most cashews are produced from seedlings, plantings are quite heterogeneous, which causes considerable variation in yield, quality, and the shape of the fruit produced. Because of the wide variations from tree to tree in seedling populations, much effort has gone into asexual propagation. Budding, grafting, layering, cuttings, and inarching have all been employed to vegetatively propagate trees with high yields. However, no simple technique has been developed that is "reliable, inexpensive, and usable on a large scale." Seedling rootstocks are difficult to grow since the delicate root system is easily damaged when transplanting rootstocks from pots to the field. The cashew industry will not develop to its fullest potential until the primitive practice of cultivating variable and unproductive trees raised from seeds has been replaced by vegetative propagation methods so that the commercial plantings will consist of uniform trees giving high predictable yields of high-quality nuts.

Seedling cashew trees bear fruit about the third year. Grafted dwarf types initiate flowering within 6 months and, when irrigated, produce up to 30 fruits by the end of the first year. Flowering lasts for 2 to 3 months and it is common to have flowers and fruits at different stages of development within the same panicle. The trees reach full production within 7 to 10 years, depending on growing conditions. Trees are productive from 20 to 30 years, with some trees bearing fruit for up to 45 years. Yields can range from 0.5 to 200 kg (1 to 220 lb) of nuts per tree per year.

Cashew apples are very bruise-sensitive, and those destined for the fresh market should be hand-harvested with the nut attached before the occurrence of natural abscission from the tree. Harvested cashew apples will keep only for about 24 hours under ambient conditions. For the processing industry, harvest may be by hand or by use of a harvest pole fitted with a collecting bag. For nut harvest and/or extraction of the shell liquid, the apples (along with the nuts) are allowed to drop to the ground to assure full maturity of the kernel. The fruits are collected weekly or daily, depending on the weather. Preparation of the cashew kernels is an intricate procedure compared with that of other nut crops. The cashew kernel is surrounded by the toxic resin, CNSL, in the mesocarp. If no precautions are taken, CNSL can contaminate the nuts and cause severe blisters on the skin of individuals who come in contact with it. Consequently, the shells are first detached from the apples and sun-dried to reduce the moisture content from 16 percent to 7 percent. At this moisture content (upon shaking the shells, the nuts rattle) the shells can be safely stored for a year or longer. To extract the nut, the shells are roasted to remove the brown, caustic CNSL, and cracked. Since the caustic oil is expelled during roasting, contact with the smoke should be avoided. The kernels are then further dried at about 68°C (155°F) to facilitate dehusking and removal of the testa (often manually), then graded and packed into appropriate storage containers. In Africa shelling is largely

mechanized in factories using machines developed in Italy, Germany, and England. Such machines have also been in use since about 1970 in northeastern Brazil and, more recently, in Central America.

Up until 1923, weevil infestations in transit practically prohibited shipments to distant countries. Now, after grading, the nuts are packed in cans that have been purged with carbon dioxide and hermetically sealed to prevent deterioration and insect damage. After roasting and salting, the kernels are vacuum-packed for retail distribution.

The cashew tree has several serious pests and diseases that vary by growing region. Insect pests include the white fly (*Aleurodicus cocois,* Curtis), a caterpillar (*Anthistarcha binoculares*), a red beetle (*Crimissa* sp.), and a thrip (*Selenothrips rubrocinctus*). The flies, *Helopeltis anacardii* and *H. schoutedeni,* cause damage by feeding on leaves, young shoots, and inflorescences. Principal diseases include anthracnose (*Colletotrichum gloeosporioides*), black mold (*Diploidium anacardeacearum*), gummosis (*Lasiodipodia theobromae*), and a fungal disease caused by *Oidium* sp. Principal pathogens in cashew nurseries include *Sclerotium rolfsii, Pythium splendens, Phytophthora* sp., and *Cylindrocladium scoparium*.

Coconut (*Cocos nucifera* L.) ARECACEAE

The coconut is by far the most important nut crop in the world. It is used for its copra, from which coconut oil is extracted. The coconut is a tall, unbranched monocotyledonous tree (palm) grown throughout all the tropical regions. The native home of the coconut is difficult to determine since over the centuries the nuts (seeds) were so easily dispersed by early ocean voyagers and by ocean currents from island to island and continent to continent. Some believe it originated in the islands of the Malaysian Archipelago, but others believe it had a Central American origin.

It is estimated that there are about 3.4 million ha (8.5 million ac) of coconuts, of which 0.98 million are in the Philippines, 0.64 in India, 0.6 in Indonesia, 0.43 in Sri Lanka, and 0.24 in various South Sea Islands. Honduras, the Dominican Republic, and Puerto Rico are the leading suppliers of coconuts for the United States. Coconuts are grown to a limited extent in Florida and Hawaii. World coconut production has been increasing at about 5 percent per year.

Although the coconut is a tropical plant with commercial production within 15° of the equator, it grows and fruits some distance from the equator as shown by plantings in southern Florida at a latitude of 26° north. Coconuts withstand some frost. Growth and production are best under high humidity with mean annual temperatures around 26.5°C (80°F) and daily fluctuations of no more than 5°C. The coconut palm is a light-requiring tree and does not grow well under shade or in very cloudy conditions.

Figure 28–4   Grove of bearing coconut palms in Hawaii. *Source:* USDA.

Coconuts need ample, well-distributed soil moisture and suffer from long, rainless periods. Annual rainfall of at least 1,520 to 1,770 mm (60 to 70 in.) is required, although irrigation can be used to supplement low rainfall. The trees grow in a wide range of soil types—beach sand, coral rock, or rich muck—provided they are at least 1.2 m (4 ft) deep. Coconuts tolerate high salt levels in the soil and salt sprays as are found along the seashore. Shallow soils with a high water table, which prevents good root development, are unsuitable since the tall palms are likely to blow over in tropical typhoons and cyclones.

Coconut palms respond to nitrogen and potassium fertilizers and, in some soils, to phosphorus, although in many naturally fertile soils no benefits have been obtained from added fertilizers. Little is known of their need for trace elements.

The coconut tree has a tall, flexible trunk marked by leaf scars and topped by a crown of leaves. Each leaf is about 6 m (20 ft) long with a central midrib and about 100 opposite leaflets. A growing point in the central crown produces new leaves and flowers (Fig. 28–4). This growing point is the only bud on the palm; hence, if it is killed, death of the entire palm

follows. These growing points, called hearts of palm, are a very nutritious and tasty food product, reminiscent of artichoke hearts. Trees of some palm species, including the coconut palm, can be grown specifically for this product, even though extracting it kills the tree.

A number of heterogeneous cultivars are recognized. Tall palms tend to be slow maturing, flowering 6 to 10 years after planting, with a life span of 80 to 100 years. Dwarf coconut palms start flowering in their third year and have a productive life of 30 to 40 years. Both male and female flowers are borne on the same many branched inflorescence arising from a leaf axil in the top of the palm. Each inflorescence can have up to 8,000 male flowers along the terminal part of the inflorescence and 1 to 30 globular female flowers near the base. Transfer of pollen from the male to the female flowers by wind, birds, or various insects is required for nuts to develop.

The coconut fruit is classified as a drupe—just as is the fruit of the peach or apricot. The hard shell commonly seen when one buys a coconut in the stores is the inner layer of the matured ovary wall of the fruit (the endocarp) (Fig. 28–5). Outside of this is the husk (the mesocarp plus

Figure 28–5 Coconut fruits. One of the husks is split open to show the nut inside. The nut consists of the inner wall (endocarp) of the fruit, covering the true seed inside, which contains the edible part of the coconut—the endosperm—in both the liquid and solid forms.

exocarp), which is usually removed when the nuts are harvested. Inside the shell (endocarp) is the true seed with a thin brown seed coat. The white "meat" of the coconut is part of the endosperm, a food storage tissue. The coconut "milk" is also endosperm in a liquid form. The tiny embryo is buried under one of the three "eyes" in the endosperm at the end where the fruit was attached to the plant. The coconut is mature and ready to eat about a year after pollination of the flower.

There are no true named clonal cultivars of the coconut. Nearly all trees are propagated by seed since no parts of the tree can be used for vegetative propagation. The coconut palm does not produce "offsets" as does the date palm. Tree characteristics of the coconut are reproduced fairly well by seed, however, so that the best strategy is to select seeds for new plantings from trees that produce large crops of high-quality nuts. There are some seed-propagated cultivars, such as 'Yellow Dwarf' and 'King,' where the seeds are gathered from isolated groves of trees that all have similar characteristics. There are also self-pollinating dwarfed selections that bear early and give high yields, flowering in 3 to 4 years from seed.

Coconuts (still enclosed in the husk) are usually germinated in a seedbed. Fully matured nuts (at least 12 months old) are planted flat on their sides and covered to their depth with soil. They must be kept moist to germinate. The embryo grows out through one of the "eyes," with roots developing at the base of the sprout and growing down through the husk. After 8 to 10 leaves have formed, the new plant plus the husk is transplanted to its permanent location with trees set about 7.5 m (25 ft) apart.

The edible part of the coconut is the white endosperm, usually eaten fresh and raw, plus the liquid endosperm (the "milk"). Much more important is the dried endosperm—known in the trade as copra. This is used principally as a source of coconut oil.

Depending on the cultivar, the average weight of fruit from tall palms ranges from 1.2 to 2 kg with nuts from 0.7 to 1.2 kg containing 0.35 to 0.6 kg endosperm and yielding 0.2 to 0.29 kg of copra. Dwarf palms bear fruits weighing 1.1 kg with nuts weighing 0.6 kg and yielding 0.2 kg of copra. Copra contains 60 to 68 percent oil of which approximately 64 percent is extractable. To prepare copra the husk is removed by knives that also split the nut in two. The "milk" is not saved. The open halves of the nuts are turned upward and stacked in trays to dry. After 4 to 6 days the "meat" is pried out with special spoons, then further dried in the sun or artificially in heated kiln dryers until the moisture content is about 5 percent. The dried product—copra—is then chopped into uniform small pieces, sacked, and delivered to large processing plants for extraction of the oil. Dried copra can be stored under refrigeration below 10°C (50°F) for several months. The copra cake left after oil extraction is ground to a meal that is a high-protein cattle feed. Fresh coconuts can be stored for 1 to 2 months at 0°C to 1.5°C (32°F to 35°F) at a relative humidity of 75 percent or less. They will keep for 2 weeks at room temperature.

The dried endosperm (copra) is an important commercial source for vegetable oil. The low content of unsaturated fatty acids present makes coconut oil resistant to oxidative rancidity. It is used in the preparation of margarines, shortenings for cooking, frying oils, and imitation dairy products. Coconut stearin is valued as a confectionery fat and as a substitute for cocoa butter. The shredded and dried copra contains 68 to 72 percent oil and < 2 percent water and is used in confectionery and bakery products. Coconut oil is also used in the manufacture of liquid and solid soaps and detergents, cosmetics, hair oil, and various lubricants.

Another use for coconut has developed recently. Fiber (coir) from the husk is being used as a replacement for peat moss in soilless potting mixes.

Coconut trees are highly susceptible to various diseases. Viruses have eliminated entire plantations. Lethal yellowing, possibly caused by a mycoplasma-like organism, is particularly severe in the Caribbean area and has killed thousands of acres of coconut palms. A similar disease has occurred in West Africa. In Florida and some other regions the common coconut is being replaced with 'Malayan Dwarf' (*Cocos* sp.), which is resistant to lethal yellowing. In the Philippines a disease called cadang-cadang has wiped out many large coconut plantations. *Phytophthora palmivora* fungus can kill the terminal growing point, leading to the death of the entire palm. Bronze leaf wilt causes death of the older leaves, eventually working its way to the younger leaves. It is a nonpathogenic problem due to unfavorable weather conditions, often appearing after a long drought.

There are insect pests that attack coconut palms but only a few are serious and these can generally be controlled by sanitary measures. A rhinoceros beetle, which attacks the terminal bud and thus kills the palm, is considered a serious pest in many coconut-growing areas.

Coffee (*Coffea* spp.) RUBIACEAE

The genus *Coffea* consists of more than 70 species, but only four species (*Coffea arabica, C. canephora, C. liberica,* and *C. dewevrei*) are involved in world coffee production. *Coffea arabica* (Aribica coffee) and *C. canephora* (Robusta coffee) account for more than 70 percent and 25 percent of the world market, respectively. *C. arabica* is native to the highlands of southwestern Ethiopia and southeastern Sudan. *C. canephora* originated in the African equatorial lowland forests from the Republic of Guinea and Liberia eastward to Uganda, Kenya, and southern Sudan, extending into central Africa. The Arabs were using coffee as a beverage as long ago as 600 A.D. and introduced it to the Mediterranean countries about 1500. It found its way into western Europe about 1615 (tea was first used there about 1610 and cocoa about 1528).

Coffee plants were brought to Brazil in 1727 and, after 40 years, it became, and still is, the world's leading coffee grower and exporter, now producing one-quarter of the world's coffee supply. Colombia ranks second in coffee production, followed by Indonesia, Vietnam, and Mexico. Latin America produces mostly *C. arabica,* whereas the African continent and Southeast Asian coffee production are dominated by *C. canephora.* This is a cheaper, less flavorful coffee, but it is now in great demand for use in blending with *C. arabica* coffees and in the manufacture of instant coffees. *C. liberica* is an important species in some areas but produces an inferior coffee. Coffee is the mainstay of the economy of the Central American countries, and it is grown to a limited extent in Puerto Rico and in Hawaii, where the famous Hawaiian Kona coffee is well-known. Young coffee seedlings are occasionally grown outdoors in the warmer parts of the United States and as indoor houseplants, making attractive ornamentals.

The typical *C. arabica* plant is a large bush to small tree with dark green leaves. It is an allotetraploid ($2n = 44$) and differs from other coffee species that are diploid ($2n = 22$). Optimum altitudes for *C. arabica* are 1,000 to 2,000 m and a temperature range of 15°C to 24°C (59°F to 75°F). *C. arabica* trees withstand temperatures near 0°C (32°F) for only a short period. Frost is one of the principal hazards in growing coffee. Cold-damaged trees that have been exposed to temperatures below freezing have considerable difficulty in recovering. Temperatures above 27°C (80°F) tend to reduce flowering and fruiting, while temperatures below 13°C (55°F) cause cessation of growth and tree stunting. *C. arabica* is self-fertile and autogamous. Fruit shape is oval and generally reaches maturity in 7 to 9 months. It is grown throughout Latin America, in Central and East Africa, in India, and to some extent in Indonesia.

C. canephora prefers altitudes of 0 to 700 m, and optimum temperatures are 24°C to 30°C. It ranges in size from a small shrub to a small tree reaching heights up to 10 m. The rounded fruits mature in 10 to 11 months and contain oval-shaped seeds, which are smaller than seeds of Arabica coffee. *C. canephora* is diploid ($2n = 22$) and self-sterile. It must be cross-pollinated and is highly heterozygous. 'Robusta' coffee is grown in West and Central Africa, throughout Southeast Asia, and to some extent in Brazil, where it is known as 'Conilon.'

Coffee trees need a continual soil moisture supply. Annual rainfall of about 1,650 mm (65 in.) is required, although supplemental irrigation can be given in drought periods. Optimum precipitation for *C. arabica* is 1,500 to 2,000 mm, while *C. canephora* requires 2,000 to 3,000 mm annually.

In many countries other than Brazil and Hawaii, coffee trees are planted in the shade of other tall-growing trees. Coffee, apparently, has a low light saturation level for photosynthesis so the trees grow adequately in the shade. However, sun-grown coffee trees give the highest yields provided all other cultural factors, particularly soil fertility, are optimal. *C. arabica* trees grown at higher elevations [900 to 1,800 m (3,000 to 6,000 ft)] where the temperatures are somewhat cooler produce a milder-type bean, from which a better drink can be prepared, than trees grown at lower and hotter elevations.

Coffee trees and their foliage are tender and are quickly injured by strong winds. Wherever such winds blow, rows of windbreak trees must be used.

The trees grow best in a slightly acid, loamy soil at least 0.9 m (3 ft) deep with good aeration and drainage. To grow and yield well coffee trees require a high level of soil fertility. They readily show symptoms of a lack of any of the essential elements. Plantations grown without added fertilizers show a steady decline in production. Coffee growers should ideally learn to recognize deficiency symptoms and apply the proper fertilizers. Deficiency symptoms of all the essential mineral elements have appeared in coffee plantations. Coffee yields vary considerably from about 2,240 kg/ha (2,000 lb/ac) of "clean" coffee (beans removed from the fruit) in Hawaii to only 400 kg/ha (360 lb/ac) in Brazil.

Most coffee trees now in bearing are seedling trees, grown in a nursery for 1 year before planting in the field. *C. arabica* is self-fertile and tends to reproduce fairly true from seed so that several seed-propagated cultivars have been developed such as 'Mondo Novo' and 'Caturra' in Brazil, 'Tico' and 'Hibrido' in Central America, 'Moko' in Saudi Arabia, the dwarf 'San Ramon' and the Jamaican 'Blue Mountain.' *C. canephora* trees are self-sterile and depend on cross-pollination; therefore, the seedlings vary considerably.

Coffee trees are planted about 2.4 m (8 ft) apart and are kept low—about 1.8 m (6 ft)—by pruning to facilitate harvesting. They start bearing full crops on the lateral branches at about 5 years and reach full production at 15 years. Clusters of 2 to 20 perfect flowers (both male and female parts) are produced in the leaf axils. The white flowers are very attractive and fragrant. Pollination is accomplished by wind or insects.

Botanically, the coffee fruit is a drupe, except that within each fruit is a mucilaginous pulp covering two thin, fibrous parchment-like "pits" (endocarp), each surrounding a single seed. The two greenish seeds (the coffee "beans") in each fruit consist mostly of a hard, thick, and folded endosperm food storage tissue covered with a thin silvery seed coat. The tiny embryo in the seed is embedded in one end of the endosperm. The fruits (called "cherries" by the growers) first are green, then become dark red when ripe.

Harvesting of coffee begins in Brazil in May at the end of the rainy season. In harvesting the coffee fruits, which are mostly picked by hand, the pickers must remove only the mature red fruits, leaving the green ones for further development. There has been some mechanical harvesting of coffee in Brazil. A coffee tree may have blossoms, green fruit, and ripe fruit all at the same time on the same branch. An average coffee tree yields about 2.25 kg (5 lb) of ripe fruits [converting to about 0.5 kg (1 lb) of roasted ground coffee]. Following harvesting the pulp must be removed from the coffee "cherries" to obtain the seeds ("beans") inside. There is both a dry and a wet process for this. In the dry process the fruits are washed, then spread out on concrete slabs in the open sun to dry, being turned several times a day. Following this they are repeatedly run through fanning and hulling machines to free the coffee beans inside. In the wet process the "cherries" are run through a depulping machine that breaks them open and squeezes the beans out of the pulpy skin. They then go into large tanks for 24 to 40 hours where the jelly-like substance surrounding the beans ferments slightly. Then the beans are thoroughly washed, spread out in the sun to dry, and continually turned and mixed. In wet weather, drying machines are used, tumbling the beans in perforated drums through which warm air is blown.

After they are well dried, the coffee beans are inspected to remove any that are defective, then they are packed into 60 kg (132 lb) bags for shipment to coffee-importing countries. Different types of coffee are often blended together by the purchasers. Roasting the green coffee beans develops the characteristic tastes and color and must be skillfully done. Coffee purchases are often based on quality as determined by taste tests by highly trained coffee tasters on sample lots of roasted and ground coffee. Millions of bags of unroasted coffee beans are shipped to the consuming countries each year for blending, roasting, grinding, and packing. Marketing of coffee produced by thousands of small farmers must necessarily be done by cooperative organizations. In Colombia, for example, the National Federation of Coffee Growers represents the growers in marketing agreements and conducts research in coffee production and processing.

Many disease and insect pests affect coffee trees. Probably the worst disease is the leaf-rust fungus (*Hemileia vas-*

tatrix). This rust first appeared in Sri Lanka in 1869 and destroyed all coffee plantations throughout the island and spread to other coffee-producing countries in that part of the world. Coffee planters in Sri Lanka dared not risk replanting coffee so they turned to tea, developing Sri Lanka into the world's second-largest tea producer (India is first). Vigorous trees well fertilized with nitrogen seem to resist coffee leaf rust, and sprays with copper compounds aid in controlling the fungus. *Coffea arabica* is susceptible, while *C. canephora* is considered to be resistant. The coffee bean borer (*Stephanoclores*) is a problem in Brazil where it is referred to as the coffee plague. The coffee leaf miner and the Mediterranean fruit fly commonly attack coffee trees in some countries.

Macadamia (*Macadamia integrifolia* Maiden & Betche)
PROTEACEAE

The macadamia is the only commercial food crop indigenous to Australia. This attractive, nut-producing evergreen tree was found along the fringes of rainforests in eastern tropical Australia. Only two of the three species of macadamia are edible: *Macadamia integrifolia* and the rough-shelled *M. tetraphylla*. *Macadamia integrifolia* was introduced into Hawaii in 1878 and was used as an ornamental until the 1920s when the first commercial plantings were made. In Hawaii, which is now the leading macadamia-producing area, it has developed into an important crop where, in 1997, 7,770 ha (19,200 ac) of trees were being grown. Australia has a similar production level as Hawaii, but has a lot of young orchards just coming into production. Together they produce over 70 percent of the world's macadamia crop. The macadamia is also grown in southern California, Arizona, southern Florida, Mexico, Guatemala, Costa Rica, Brazil, Malawi, Zimbabwe, Rhodesia, Panama, Jamaica, and South Africa. Plantings in all these areas are quite small, except for Malawi and South Africa, where sizable plantings have been established.

Commercial macadamia culture is suited only for subtropical to tropical frost-free localities, although mature trees in a dormant state have withstood temperatures as low as −6.5°C (20°F) for short periods without serious injury. The trees also tolerate temperatures up to 46°C (115°F) for short periods. Macadamia trees tolerate a wide range of soil types but are best suited to deep, well-drained soil with a high organic matter content (3 to 4 percent), and a pH of 5.0 to 5.5. The trees have a definite requirement for nitrogen, phosphorus, potassium, and magnesium fertilizers, which are best applied in small amounts several times a year. Micronutrients such as iron, zinc, manganese, and boron are also needed in Hawaiian macadamia orchards under certain conditions.

Macadamia trees require ample soil moisture throughout the year from rainfall [a minimum of about 1,524 mm (60 in.) annually] plus supplementary irrigation in dry periods to induce flowering and good fruit-set. Some of the productive orchards in Hawaii are in well-drained soils receiving as much as 3,430 mm (135 in.) of rainfall per year.

Macadamia trees have a shallow, poorly anchored root system and, in windy areas, should be protected by rows of windbreak trees. In Hawaii the Norfolk Island pine (*Araucaria heterophylla*) is commonly used in the wet areas.

The principal macadamia cultivars grown commercially in Hawaii are 'Kakea,' 'Keauhou,' 'Ikaika,' 'Keaau,' 'Waimanalo,' and 'Ka'u,' all the smooth-shell type. These are propagated by grafting in early summer, with a side wedge or a simple splice graft, onto seedlings of smooth-shell (*M. integrifolia*) or rough-shell species (*M. tetraphylla* L.) as a rootstock.

Macadamia trees usually start flowering and fruiting at 5 years. A full commercial crop is obtained after the seventh year. Flowering in Hawaii starts in November and December and continues through January, February, and March, with harvesting of the nuts starting in late July and continuing through March of the next year.

The tubular ivory-white perfect flowers of the smooth-shell types are produced in groups of three or four on pedicels with 100 to 500 groups on a whip-like terminal or axillary pendulous cluster about 15 to 25 cm (6 to 10 in.) long. Flowers of many cultivars require insects to transfer pollen from flower to flower to obtain good fruit-set. Honeybees in the orchards at bloom are believed to promote fruit-set. The ovary of each flower contains two ovules and develops into a follicle having a husk and shell. The husk, in most cultivars, splits open naturally to reveal the shell. Inside the shell is the seed ("nut"), consisting mostly of the edible cotyledons of the embryo. Double seeds are less desirable and are generally avoided.

Harvesting is done by hand in small plantings and is simply a matter of picking up the nuts as they drop. In large plantings, after the nuts drop to the ground, one special machine blows the leaves away, another sweeps the nuts into windrows, and a third pickup machine gathers the nuts for hauling to the processing plant. Harvesting is done six or seven times a year. Mechanized harvesting is under development with the tree-shaking and pickup equipment used in harvesting almonds and walnuts.

The inshell nuts are dried to 1.5 percent moisture and then are cracked. Mature nuts should contain at least 72 percent oil for optimum eating and processing quality. In the Hawaiian processing plants the cracked nuts are separated electronically from the shells. Kernels may be graded by specific gravity, which gives a good indication of oil content. Kernels with a specific gravity ≤ 1 will float in water, which indicates the oil content is 72 percent or greater. The nuts are cooked in coconut oil, cooled, inspected, salted, and vacuum-packed in cans or glass jars. Macadamia nuts, cooked and salted, are a fine specialty food with their delicate, sweet taste and crisp bite. They are expensive but much in demand by nut fanciers and gourmets.

Blossom blight diseases caused by attacks of *Botrytis* and *Phytophthora* fungi develop on macadamias and require fungicide sprays. Anthracnose sometimes causes problems. A mite attacks the flower in all stages but can be controlled by sulfur sprays and other approved miticides. Aphids and thrips damage flowers occasionally.

Mango (*Mangifera indica* L.) ANACARDIACEAE

The mango is one of the most important fruit crops of the tropical regions. It is a broad-leaved evergreen tree with a symmetrical leafy canopy producing luscious fruits of varying shapes from oblong to round. The trees can become very large—up to 30 m (100 ft) high with a 37 m (125 ft) spread (Fig. 28–6). The mango is believed to have evolved as a canopy layer species in the tropical rainforest in the Indo-Burmese region of Southeast Asia. There is evidence that the mango has been under cultivation in India for over 4,000 years. It has been associated with the economic, cultural, religious, and aesthetic life of the Indian people since prehis-

Figure 28–6 Mango trees in Hawaii. *Above:* Young tree with a good fruit crop. *Below:* Mango trees can become very large. Note the size of people in lower left corner. *Source:* Above photo courtesy W. Yee.

toric times. The Buddhist pilgrims Fa-Hien and Sung-Yun mentioned in their travel notes that the Gautama Buddha was presented a mango grove by Amradarika (500 B.C.) as a quiet place for meditation. Other travelers to India, including the Chinese Hwen T'sung (632 to 645 A.D.), the Arabs Ebn Hankal (902 to 968 A.D.) and Ebn Batuta (1325 to 1349 A.D.), and the Portuguese Lurdovei de Varthema (1503 to 1508 A.D.), all described the mango.

The mango currently ranks fifth in total production among fruit crops worldwide, only being surpassed by *Musa* (bananas and plantains), *Citrus* (all types), grapes, and apples. India is still, by far, the leading producer of mangos, harvesting 10,800,000 MT in 1997 (50 percent of the world's production). Other leading countries in mango production are China (2,107,512 MT), Thailand (1,400,000 MT), Mexico (1,190,745 MT), and Indonesia (1,000,000 MT). Mangos are also grown extensively in Pakistan, Brazil, Nigeria, the Philippines, Egypt, Congo, Haiti, Madagascar, Tanzania, Bangladesh, Dominican Republic, Vietnam, Sudan, Venezuela, Peru, Australia, Cuba, Puerto Rico, and, in small amounts, in southern coastal areas of Florida, southern California, and Hawaii.

Mangos imported into the United States come mostly from Mexico, with substantially smaller amounts from Venezuela, Guatemala, Peru, Ecuador, Haiti, and Brazil. The most popular varieties are 'Haden,' 'Keitt,' 'Kent,' and 'Tommy Atkins.'

Because mango seeds are recalcitrant and can only survive a few days, mango germplasm must have been transported as ripe fruit or seedlings when the mango was carried throughout the tropical regions by the early Spanish and Portuguese explorers. Mangos reached the Americas in the eighteenth century. It was first introduced into Hawaii from Mexico about 1810 by Don Marin, a noted Spanish horticulturist. The mango was introduced into Florida in 1861 and involved the 'No. 11' polyembryonic seedling from Cuba. Beginning in 1889, the United States Department of Agriculture (USDA) systematically introduced grafted Indian cultivars into Florida.

Although the mango is a tropical plant, mature trees have withstood temperatures as low as −4°C (25°F) for a few hours. Young trees and actively growing shoots are likely to be killed at 1°C (30°F). Flowers and small fruits are damaged if temperatures drop below 4.5°C (40°F) for a few hours. Temperatures of 24°C to 27°C (75°F to 80°F) are considered optimal for mangos during the growing season, along with high humidity. They tolerate temperatures as high as 48°C (118°F).

Temperatures affect the flowering time of the trees. A cool or dry period generally during the winter, which slows or stops growth, is beneficial in inducing flowering. The mango grows in regions of both heavy, 2,540 mm (100 in.) and light, 254 mm (10 in.) rainfall. Precipitation of 890 to 1,015 mm (35 to 40 in.) in a year, if well distributed, is best. If prolonged rainless periods occur, supplementary irrigation should be provided for good crops. In Florida, fixed overhead sprinklers

have been very successful in mango orchards because of their added benefit of frost protection during cold weather. Heavy rains during flowering can drastically reduce fruit-set.

Mango trees grow on a wide range of soil types, but the best production comes from well-drained sandy loam to loam soils at least 1.5 to 1.8 m (5 to 6 ft) deep. Little is known of the fertilizer requirements for mango. The trees are generally able to obtain sufficient mineral nutrients from the soils in which they are grown, and little experimental work has been done. Sand culture experiments have established mineral deficiency symptoms, but more field test plot work is needed to correlate mineral deficiencies with tree productivity.

There are two types, or races, of the mango that can be distinguished on the basis of their mode of reproduction and their respective centers of diversity:

1. The Indian race, grown in India, Pakistan, and Bangladesh, has highly flavored, brilliantly colored fruits well suited to commercial production. These are monoembryonic and must be propagated vegetatively. There are many cultivars of the Indian race. Some are 'Alfonso,' 'Mulgoba,' 'Haden,' and 'Tommy Atkins.' 'Haden' is a popular cultivar in Hawaii. 'Tommy Atkins' and the later ripening 'Keitt' are the principal cultivars in Florida.
2. The Indochina or Saigon race produces smaller, less attractive fruits with a yellowish-green color and nonfibrous flesh with a delicate flavor. This type is polyembryonic and generally reproduces true by seed. Cultivars of this type are generally known by the name of the district where they originated, such as 'Cambodiana,' 'Carabao,' and 'Pico.'

The mango inflorescence is a branched terminal panicle up to 0.6 m (2 ft) long, having several hundred to several thousand flowers. A tree may have from 200 to 3,000 panicles so that tremendous numbers of flowers are produced. The flowers are small, monoecious, and polygamous. Both male (the pistil aborts) and perfect flowers are formed on the same panicle. The ratio of male to perfect flowers is strongly influenced by environmental and cultural factors. Pollination is accomplished by various insects, but thought to be predominately by flies. Only a very small percentage of the flowers develop into fruits. The mango fruit is a large, fleshy drupe and is highly variable with respect to shape and size, ranging from about 85 g (3 oz) to over 2.3 kg (5 lb), depending on the cultivar. The delicious fleshy part, which is eaten, is the mesocarp of the fruit (ovary wall). Chlorophyll, carotenes, anthocyanins, and xanthophylls are all present in the fruit, but chlorophyll disappears with maturity while anthocyanins and carotenoids increase during ripening. In some cultivars masses of fibers extend from the stony pit (endocarp) into the flesh, making the fruit difficult to cut. Inside the stony pit is the true seed.

Mango cultivars did not exist in India until the late fifteenth century because there was no known method for vegetatively propagating superior selections. This changed with the establishment of Portuguese enclaves on the coast of India. Under the reign of Moghul emperor Akbar (1556 to 1605), the best seedling mangos were selected and propagated by approach grafting and large orchards were established. Most Indian mango cultivars originated in those years and have been maintained by vegetative propagation for more than 400 years. 'Alphonso,' 'Dashehari,' 'Langra,' 'Rani Pasand,' and 'Safdar Pasand' all date from that time.

Mango trees tend strongly toward biennial bearing. There may be very low numbers of perfect (male and female parts) flowers, and embryo abortion after fertilization may cause the young fruits to drop. A major problem in many mango-producing areas is getting good yields regularly and obtaining good fruit-set from the abundant bloom. The best approach is to plant cultivars with regular bearing habits like 'Tommy Atkins,' 'Keitt,' 'Kent,' and 'Pope.'

Mangos can be propagated in many ways. Polyembryonic types (which come true from seed) can be propagated by seed or by one of the following vegetative methods. Monoembryonic types, which have only a sexually produced embryo, must be propagated vegetatively if the cultivar characteristics are to be retained. Many methods are used, such as approach grafting (used in India since ancient times), whip grafting, side grafting, and veneer grafting, T-budding, chip-budding, and patch budding have also been successful. Mango is difficult to propagate by stem cuttings or layering.

The mango, like most other broad-leaved evergreen tropical and subtropical trees, requires very little pruning, although with heavy crops limbs need cutting back to prevent breakage. Young trees should be trained enough to develop a central trunk and several well-spaced primary scaffold branches to form a sturdy framework. Dead wood and crowded interfering branches should be removed from mature trees.

Mango fruits grow rapidly and are ready to harvest about 12 weeks after fruit-set. In Florida and Hawaii mango harvest is from mid-May to October, depending on the cultivar. Their fruits are best picked after the green skin color changes to yellow but the flesh is still firm. Fruits ripen 3 to 6 days after picking. If harvested too soon, the fruits of most cultivars do not develop a good flavor. Harvested fruits can be refrigerated at about 13°C (55°F) to delay ripening during shipment. Fruit ripening is best done at 21°C to 24°C (70°F to 75°F); ethylene gas treatment can be given to induce uniform ripening.

The mango is a highly esteemed tropical fruit. It is generally peeled for eating fresh, but it can also be frozen, fried, canned, or used for pies, jams, and jellies.

Mangos are subject to several disease and insect problems. Anthracnose is the most serious fungal disease causing flowers, young fruits, leaves, and twigs to turn black and drop. This is best prevented by maintaining preventive fungicide sprays on all parts of the tree. On the windward side of mountain ranges in Hawaii, which have heavy rains throughout the year, there are large mango trees that flower but never fruit because of the *Anthracnose* fungus.

Mites, mealybugs, scale insects, stem borers, and thrips can build up to sufficiently high levels on mango trees to cause severe damage and greatly limit fruit production. Mango hopper is a serious pest in India. Mediterranean and Mexican fruit flies are frequently a problem. Mangos have not developed to a greater extent commercially in Hawaii because of a quarantine against importation of the fruits into the mainland United States due to the mango seed weevil, which can be found in some mango seeds.

Papaya (*Cariaca papaya* L.) CARICACEAE

The papaya plant is a large, fast-growing, short-lived herbaceous perennial that bears clusters of delicious, mild-flavored large or small fruits varying in shape from spherical to pear-like. In some countries it is known as the "pawpaw." *C. papaya* plants have never been found in the wild, but a close relative, *C. peltata,* grows wild in southern Mexico and Central America, leading to the supposition that *C. papaya* may have originated in this region. The early Spanish and Portuguese explorers carried the papaya to most tropical countries throughout the world.

The papaya is now grown in many places in the Tropics between 32° north and south latitudes, from sea level up to about 1,500 m (5,000 ft). For heavy production of high-quality fruits, papayas must be grown in warm, 21°C to 26.5°C (70°F to 80°F), frost-free climates in full sun. Lower temperatures give poor results. Papayas grow on many soil types, but good aeration and drainage are necessary with, preferably, a soil pH of 6.5 to 7.0. Papaya trees need ample soil moisture at all times. If rainfall is lacking during parts of the year, irrigation is required to supplement the rain. For good plant growth and production the various mineral nutrients must be readily available in the soil. The kinds and amount of fertilizers required depend on the soil type. Brazil leads the world in papaya production, followed by India and Mexico.

There are different types of papaya grown in the various tropical countries—some large-fruited and some small-fruited. The main commercial cultivar being grown in Hawaii is 'Puna Solo,' which was selected from an introduction from Barbados, West Indies, in 1911. It has small fruits of about 0.5 kg (1 lb) each. Several new papaya cultivars have been developed in Hawaii, originating from controlled crosses. Some of these are 'Waimanolo,' 'Sunrise,' 'Higgins,' and 'Wilder.' Florida grows a type called 'Blue Solo,' which originated as a cross of 'Solo' with a blue-stemmed plant having large fruits. The 'Blue Solo' type is too variable, however, to be called a cultivar.

There is a wide diversity of biological types of cultivated papaya, which may be dioecious (with male and female flowers on separate plants), monoecious (male and female flowers on the same plant), or hermaphrodite (with male and female parts on the same flower). Male plants are easily determined by the flower clusters borne on long peduncles arising in the axils of the leaves. Male trees are generally cut out as soon as the sex can be determined since they are unfruitful. The flowers on female trees grow close to the stem, one on the axil of each leaf. The flower is large, bell-shaped, and the color of wax; when ready to be fertilized, it turns white and begins to open. The third type of tree has hermaphroditic (perfect) flowers, with both male and female parts. Seedlings from the 'Solo' cultivar grown in Hawaii develop into trees producing female flowers and hermaphroditic flowers on separate plants. Fruits developing from the latter type are preferred since they have a desirable pear shape, whereas those from female flowers develop into round fruits. Female 'Solo' trees are generally cut out as soon as their sex can be determined.

Papaya fruits develop in the axils of the leaves arising near the top of the tree. As the trees grow older and taller, the fruits become difficult to harvest, so they are replaced with younger plantings.

Botanically the papaya fruit is a berry. The edible part is the soft fleshy mesocarp and endocarp of the ovary wall. The thin, leathery skin (exocarp) is not eaten. The inside of the ovary wall of the fruit is lined with great numbers of seeds surrounded by a mucilaginous material developed from the seed coats. The seeds are removed before the fruits are served for eating.

Immature fruits contain papain, a protein-digesting enzyme, which is extracted from the latex of the skin. This is obtained by scoring or injuring the fruit, followed by collecting and drying the latex. The papain is used as a meat tenderizer.

Papayas are propagated by seed, planted directly in the field, in seed flats, or in individual peat pots for later transplanting. Fresh seed germinates in about 2 weeks with adequate moisture and heat. Planting distances in the field are about 2.4 m (8 ft) in the rows and 3.3 m (11 ft) between rows. Several seedlings are planted in each place and after about 5 months, when flowering starts and the trees' sex can be determined, the unwanted types are cut out.

Papaya trees produce mature fruits about 10 months after planting and then bear year-round. Individual fruits are harvested when they are at a mature-green stage and show a tinge or more of yellow at the apical end. Fruits can be picked by hand when the trees are young, but as they become taller the fruits cannot be reached. In Hawaii an interesting harvest method has been developed. A rubber cup, the so-called plumber's friend, is attached to a long pole and pushed up under the fruit and twisted, breaking the fruit loose if it is mature. The picker then catches the fruit as it falls to the ground. Tractor-mounted picking platforms designed after cherry pickers, as well as other harvest aids, have added an extra year to the production life of trees, giving it 4 years after planting. A yield of 22.4 to 56.0 thousand kg/ha (20 to 50 thousand lb/ac) of fruit can be obtained in Hawaii. One of the major problems facing papaya fruit marketing is the identification of optimum harvest maturity. Hawaii specifies minimum total soluble solids of 11.5 percent and fruit showing at least 6 percent surface coloration at the blossom end region.

Before Hawaiian papayas can be shipped to the U.S. mainland they must be treated to destroy fruit flies. A two-stage hot water dip treatment has been adopted as a quarantine procedure for papayas destined for export. The standard double-dip treatment consists of immersing the fruit in 42°C (108°F) water for 30 to 40 minutes, followed by a second immersion in 49°C (120°F) water for 20 minutes. For fruit to qualify for the double-dip method, the blossom end of the fruit should have a Hunter 'b' value of less than 27.5.

Harvested papayas can be held under refrigeration for 2 to 3 weeks at 16°C (50°F) with 1 percent O_2 and 5 percent CO_2. Fruits are ready to eat when they turn completely yellow or feel soft to a gentle squeeze.

Most papaya fruits are consumed locally because of the difficulties in long-distance transportation of the tender fruits. The practice of air shipments to the U.S. mainland and to Japan, however, has given impetus to the Hawaiian industry.

Most U.S. supplies of papaya come from Hawaii, with smaller amounts from Mexico and the Dominican Republic. Even in the warmest parts of southern California the weather is too cool in most years for papayas to ripen properly, and the industry in Florida has been hard hit by viruses and the papaya wasp.

Both fungal and viral diseases attack papayas. Anthracnose is a common fungal disease in Hawaii, attacking the fruits. A *Phytophthora* blight fungus is the most important disease and attacks aboveground portions of the plant as well as the roots. Two viral diseases, papaya mosaic and papaya ring spot, occur in Hawaii and are a threat to the industry. In Florida, in fact, viral diseases appearing from 1945 to 1950 proved catastrophic and have blocked any expansion of the industry. No good control measures have been developed.

Mites are a serious pest on papayas but can be controlled by sulfur sprays. The melon, Oriental, and Mediterranean fruit flies attack papaya fruits in Hawaii and are the cause of the strict quarantine and fumigation requirements.

Pineapple [*Ananas comosus* (L.) Merr.] BROMELIACEAE

The pineapple is a low-growing, herbaceous, perennial, monocotyledonous tropical plant producing tasty fruits that are well-known and consumed widely throughout the world, principally as canned products. The pineapple is thought to have originated in southeastern Brazil, Paraguay, and northern Argentina in central South America. Improved cultivars selected and propagated by the Indians in this region had become widely distributed throughout tropical America even before Columbus' discovery of the New World.

The botanical name, *Ananas,* was taken from the name *Nana* used by the American Indians. This is used as the common name in all European languages except English. Referring to the fruit's similarity to a pine cone, the Spanish word *pinas plus apple* led to its name in English.

The Portuguese explorers apparently disseminated the pineapple throughout the world's tropical regions in the 1500s, carrying the fruit "crowns," which could remain viable for long periods of time, in their ships. Pineapples were first brought to Hawaii in 1813 and canning started in 1892. James Dole started his pineapple plantations in 1900. Thailand is the leading country in pineapple production, followed by Brazil and the Philippines. The pineapple is an important commercial crop in many other tropical countries including the United States (Hawaii), Mexico, South Africa, Taiwan, India, China, Vietnam, and Australia. Prior to 1950 Hawaii produced about 70 percent of the world's processed pineapples, but because of ever-increasing production costs, this had decreased by the mid-1970s to less than 36 percent as other countries expanded their production. By 1990, Hawaii harvested only 5.4 percent of the total world production of pineapple.

The pineapple is a strictly tropical plant, showing severe leaf damage at freezing temperatures. Temperatures more than 32°C (90°F) can lower fruit quality, and higher daytime temperatures can sunburn exposed fruits. As the fruits start to mature, nighttime temperatures around 21°C (70°F) and dry weather are desirable for best quality.

A well-drained, slightly acidic sandy loam is best for pineapples, but they grow on a wide range of soils provided drainage and aeration are good. Preplant soil fumigation is essential to control nematodes when pineapples are grown on the same soil for many years. Pineapples need fertile soils and generally require added fertilizers, particularly nitrogen and potassium. In Hawaii, preplant ground applications of nitrogen, phosphorus, and potassium at relatively high levels are made. Subsequent foliage sprays containing nitrogen, potassium, iron, and zinc are applied by ground sprayers.

Pineapple plants need ample soil moisture throughout the year, with 50 to 100 mm (2 to 4 in.) of rain a month for good production. During rainless periods, supplementary irrigation should supply this amount of water, particularly for young plantings getting established. The pineapple plant survives very well, however, under drought conditions.

There are many pineapple cultivars, but the leading one commercially over much of the tropical region is 'Smooth Cayenne.' Although 'Smooth Cayenne' is the major

cultivar worldwide, as a fresh fruit it suffers from high acidity, low ascorbic acid, poor flavor, and a susceptibility to translucency. 'Pernambuco,' 'Queen,' 'Spanish,' and 'Mordilonus-Perolera-Maipure' are normally grown for local fresh fruit use.

Pineapples are propagated vegetatively with crowns, slips, or suckers. The crown is the vegetative shoot on top of the fruit. Slips are side shoots arising from the fruiting stem just below the fruit. Suckers are side shoots developing from the main stem above ground level. New plants from crowns require about 24 months to fruit, from slips about 20 months, and from suckers 17 months. Plantings are generally made to coincide with periods of heavy rainfall. A double-row system is used in Hawaii with plants set about 30 cm (12 in.) apart in rows 50 cm (20 in.) apart with the beds 1.5 m (5 ft) apart. Sheets of polyethylene film are often used as mulch between rows with the propagating pieces inserted through the plastic into the soil.

Each original plant set out produces one fruit at the top of the stem. This is called the plant crop. After these fruits are harvested, one or two suckers are allowed to develop from the mother plant. Each sucker, about a year later, produces one pineapple fruit ready to harvest. This is called the ratoon crop. Sometimes a second ratoon crop is grown. However, highest yields and best-quality fruits are obtained from the plant crop.

The pineapple fruit forms in a complex fashion. At the terminal vegetative growing point of the plant a central axis core develops, bearing an inflorescence of 100 or more closely attached lavender flowers arranged spirally, each subtended by a bract. This central axis core terminates in a crown or rosette of small leaves. Each of the many pineapple flowers on the inflorescence consists of a calyx with three violet-colored sepals joined at their bases, three purple petals, six stamens, and a pistil having three stigmas and three carpels. All parts of each flower, together with the bract, develop into the fleshy fruitlet tissue. As it develops, the entire inflorescence converts into the single pineapple fruit (botanically a sorosis) by the coalescence of the inner stem tissue and the fruitlets (developed ovaries and other parts of the flowers).

Fruits of commercial pineapple cultivars weigh from 1.4 to 2.7 kg (3 to 6 lb), but fruits of some types are much larger. The fibrous center tissue becomes woody and tough and is usually discarded when the fruit is eaten, as is the tough and leathery outer rind.

With solid block plantings of a single cultivar, pineapple fruits are seedless because each pineapple cultivar is self-incompatible (its own pollen does not fertilize its egg). In solid block fields where no pollen from other cultivars is present, no embryo or seed forms but, fortunately, the fruit develops anyway. Thus, it is parthenocarpic. If different cultivars are mixed in the same field so that cross-pollination can take place, the fruits can have seeds.

For orderly harvesting and marketing it is important that flowering (and subsequent fruit maturation) begin at about the same time in all plants in a field. Pineapples do not do this too well on their own. Plant scientists found in the 1930s that pineapple plants treated with ethylene or acetylene initiated flowers. A common commercial practice is to drop a few pieces of granular calcium carbide into the center of the pineapple terminal leaf cluster. In contacting the water accumulated in the center of the plant the calcium carbide reacts to release acetylene, which affects the terminal growing point in such a manner as to cause floral induction. Activity is known to be greater with applications of the calcium carbide made at night.

$$CaC_2 + 2H_2O \rightarrow H-C \equiv C-H + Ca(OH)_2$$

| calcium carbide | water | acetylene gas | calcium hydroxide |

Other chemicals, like the sodium salt of the auxin-naphthaleneacetic acid, and ethephon, an ethylene-releasing material, applied to the plants as water sprays, have also been effective in inducing floral induction in some areas. These materials appear to be more effective in cool areas than in the warm, humid Tropics.

Pineapple harvesting is semimechanized. A long endless belt attached to the harvesting machine stretches out over about 9 m (30 ft) of pineapple plants and moves slowly over the field. Workers follow along behind this belt in the rows, breaking off the fruits by a bending and twisting motion and placing them on the belt, which moves them to the waiting trucks for hauling to processing plants or shipment overseas.

The highest quality in pineapple fruits develops only when the fruits are allowed to become fully ripe on the plant. They do not become sweeter if picked at an immature stage because they have no starch reserves. A minimum reading of 12 percent total soluble solids is required for fresh fruit in Hawaii, while some countries require 14 percent before harvest. A sugar-to-acid ratio of 0.9 to 1.3 is recommended. Much of the pineapple crop is harvested fully ripe and processed and canned immediately at local canneries as slices, wedges, crushed pulp, and pineapple juice. With the advent of the large jet planes for shipment, more and more fresh pineapples are seen in local markets in Europe and the United States. European markets are largely supplied with fresh pineapples from Africa, while U.S. shipments originate mostly from Hawaii, Puerto Rico, and Mexico. Ripe, ready-to-eat fresh pineapples can be stored for 2 to 4 weeks at about 7°C (45°F) with 85 to 90 percent relative humidity. Temperatures below this are damaging. Fresh pineapple should never be frozen as freezing causes a deterioration and browning of the edible parts.

Numerous pests and diseases constantly confront pineapple growers. Mealybug wilt is best controlled through elimination by insecticides of ants that are associated with the mealybugs. *Phytophthora* fungi can cause heart rot and foot rot in the plants, particularly in poorly drained soils with high rainfall. Some diseases of the pineapple fruits are endogenous brown spot, pink disease, and fruitlet core rot. Control of nematodes by soil fumigation is necessary in some areas.

SOME MINOR TROPICAL FRUITS AND NUTS

Many tropical fruit and nut crops have found a place for themselves in certain tropical localities. For one reason or another they have not made a major impact on the world's food consumers. Flavor of the fruits may not be appealing to many, preparation for eating may be difficult, bearing may be undependable, harvesting and transportation of the crops may be difficult, or control of diseases or insect pests may be costly. Most of the plants (usually broad-leaved evergreens) are attractive and are often used in landscaping, with the fruits considered a bonus.

Breadfruit [*Artocarpus altilis* (Park) Fusb.] MORACEAE

This beautiful evergreen tree originated in the Indo-Malayan Archipelago and has been grown throughout the South Pacific. It reaches a height of 12 to 18 m (40 to 60 ft) and usually has large, deeply pinnate, lobed, glossy green leaves. There are two types: the seeded form known as breadnut and the more desirable seedless form referred to as breadfruit. The latter is the form usually grown.

The tree produces a large, smooth, or short-spined fruit 10 to 20 cm (4 to 8 in.) in diameter. As the fruits mature, greenish-brown or yellow spots appear on the green fruit surface. Inside the rind the fruit consists of a white, fibrous pulp. Breadfruit is high in carbohydrates but low in proteins and fat. The unripened but mature fruits are cooked and eaten in much the same way as other starchy tubers and roots. Today it is baked, boiled, or steamed, resembling the sweet potato in flavor. Most breadfruit is consumed locally, but there is a growing export trade from the Caribbean to Europe and North America, serving the ethnic market. However, the extreme perishability of this fruit hampers export trade. As the mature breadfruit becomes fully ripe on the tree, its starch changes to sugar and the pulp can be mixed with coconut "milk" and made into a pudding.

Breadfruit plants were being taken from Tahiti to the West Indies in a memorable voyage in 1792 by Captain Bligh and botanist David Nelson in the ship *HMS Bounty* when the epic mutiny took place. Captain James Cook had seen breadfruit used as a food in many Pacific Islands and recommended it for feeding the incoming slaves from West Africa if it could be grown in the West Indies. While breadfruit eventually was established in Jamaica, it did not live up to expectations because the slaves of that era preferred eating bananas and plantains.

Carambola (*Averrhoa carambola* L.) OXALIDACEAE

The center of origin of carambola is not clearly defined, but it is generally accepted to be Indochina and Indonesia. Carambola, or starfruit, requires tropical to warm subtropical conditions for optimum growth. It is grown in many Southeast Asian countries, as well as India, southern China, Taiwan, Australia, the United States (Hawaii and Florida), Central America, and Brazil.

There are two main groups of carambolas: the acid types, which are only consumed after cooking, and the sweet carambolas, which are ideal for fresh consumption. The sweet types are not recommended for cooked dishes since they lose their character and flavor in the process. Carambola is predominantly consumed as fresh fruit, but is also processed into dried fruit, pickles, preserves, jellies, sauces, and juice. The carambola tree is slow-growing and relatively small, rarely exceeding 9 m in height, although vigorous specimens may reach 15 m at maturity. They grow in a variety of soil types but prefer soils that are well-drained, high in organic matter, salt-free, and have a pH range of 5.5 to 7.0.

It is an evergreen tree and can flower continuously during the year under good cultural and environmental conditions. Therefore, it is possible to have flowers, green fruit, and mature yellow fruit all at the same time on a tree. The flowers of carambola are arranged in loose panicles and are born on basally branched, thin twigs 1 to 8 cm long. The reddish-pink flowers are perfect and mostly produced on the periphery of the canopy. Insects are required for good pollination, although some wind pollination can occur.

The fruit is a fleshy berry and usually weighs between 100 and 250 g. Fruits normally have five longitudinal ribs but can range from two to eight ribs. The ripe fruit is green to yellow, but can range from whitish to orange. Under tropical conditions or during the summer months of subtropical climates, the fruit is ready to harvest in 2 to 3 months after flowering, but this period can be extended from 3 to 4.5 months under typical subtropical conditions. Some of the major cultivars are 'Arken,' 'B-2,' 'B-10,' 'Cheng-Tsey,' 'Fwang Tung,' 'Golden Star,' 'Hong Hug,' 'Kaput,' 'Leng Bak,' and 'Sri Kembangan.'

Cherimoya (*Annona cherimola* Mill.) ANNONACEAE

There are several fruit-producing plants in the *Annonaceae* family including *A. squamosa* (sweetsop), *A. reticulata* (custard apple), *A. muricata* (soursop), and a hybrid, *A. squamosa* × *A. cherimola* (atemoya), but the cherimoya (*A. cherimola*) is most suited to the cooler regions of sub-tropical America and Central America. It has become an increasingly important commercial crop in the Mediterranean area, with southern Spain being the largest world producer.

Cherimoya grows fast for a few years, then slows down to form a tree about 4.5 m (15 ft) tall with a 6 m (20 ft) spread. The dull green leaves, about 15 cm (6 in.) long, drop in late spring to release the new buds buried in the base of the leaf petiole. The brown to yellowish fragrant flowers appear together with the new leaves. Pollination is a problem with different times of activity for the pistil (female) and stamens (male). Hand pollination usually increases fruit-set and improves carpel development. The growth and development of cherimoya fruit follows a double sigmoidal curve similar to the pattern for stone fruit.

The fruit is a syncarpium, or multiple ovary, with the skin being composed of overlapping scale-like carpel walls. The green fruits are smooth or bumpy, round to conical in form, and contain many seeds, each the size of a coffee bean. They range in weight from 50 to 850 g (1.75 to 30 oz) with the edible portion comprising about 60 percent of the fruit. Cherimoya fruit have a creamy white flesh with a smooth, almost custard-like texture. The flesh has a pleasant blend of sweetness and acidity with a mild flavor suggestive of banana and papaya with a tinge of pineapple. The fruit has a 3-week storage life if precooled by forced air in a cool room, then stored at 13°C with a relative humidity of 90 to 95 percent.

Commercial production of cherimoya is found in Spain, Chile, Ecuador, and Peru. Smaller plantings are located in California and in South Africa.

Feijoa (*Feijoa sellowiana* Berg) MYRTACEAE

The Feijoa is a native of South America, probably originating in southern Brazil at an altitude of 1,000 m. This evergreen bush or tree (2 to 5 m tall) with leathery oval leaves, gray succulent edible petals, and many prominent long red stamens, is grown on a limited scale in California, Uruguay, and New Zealand. The fruit has an edible, creamy flesh and gelatinous pulp. It is eaten as a dessert fruit, mixed with other fruit in a salad, stewed or canned, or used for jellies and juice. Cultivars include 'Coolidge' and 'Choiceana' from Australia, and 'Edenvale Supreme,' 'Nazemetz,' and 'Trask' from California. New Zealand released 'Apollo' and 'Gemini' in the 1980s and 'Opal Star,' 'Pounamu,' and 'Kakapo' more recently. Fruit-set in feijoa can be variable and there is evidence of self-incompatibility in some plants. Therefore, planting several cultivars is recommended even though some cultivars have been observed to set reasonable crops without the presence of other cultivars for pollination.

Seedlings may be used as rootstocks for proven scion clones, but superior rootstocks have not been selected. The main pests are Chinese wax scale, greedy scale, and leaf roller caterpillars.

Guava (*Psidium guajava* L.) MYRTACEAE

The common guava, a native of tropical America, is an evergreen shrub or small tree. It is a hardy plant from 3 to 10 m in height, and is adaptable to a wide variety of habitats. Because of its robust growth, it is found growing wild in several areas and is considered a noxious weed in some countries. Guava takes about 6 to 9 months from planting to the fruit-bearing stage and can crop practically year-round. The fruit is a berry, which is round, oval, or pear-shaped and about 5 cm (2 or 3 in.) in diameter. The thick rind, which is white, red, or yellow in color, is edible, as is the soft pulp inside, which contains numerous hard, small seeds. Guava fruits have a characteristic gritty texture due to the presence of stone cells. They may be rather acidic for a fresh dessert fruit, so they are used mostly as a component of juice drinks or for making jams and jellies. Guavas are an excellent source of vitamin C.

Another species, the red strawberry guava (*P. cattleianum* Sab.), is hardier than *P. guajava* and is grown in subtropical regions. It can withstand minimum temperatures as low as −6.5°C (20°F). The red strawberry guava is useful in landscaping as a hedge or container plant and has attractive greenish-gray bark. The leaves, about 7.6 cm (3 in.) long, are a glossy, golden-green color. The dark red fruits, which mature in the fall or winter, have a white, sweet tart flesh with several hard, small seeds.

Litchi (*Litchi chinensis* Sonn.) SAPINDACEAE

The litchi is a handsome evergreen tropical tree forming a dense symmetrical rounded form and growing to a height of 12 m (40 ft). It is indigenous to parts of southern China and is highly regarded and deeply involved in Chinese culture where it has been cultivated for thousands of years. World litchi production is currently more than 315,000 MT per year with China, India, and Taiwan dominating the market. Pro-

duction has increased in South Africa, Australia, Southeast Asia, Vietnam, Israel, and the United States. It bears fruits in clusters of 5 to 20 at the ends of shoots. The fruits have a rough, warty, hard shell that turns red when ripe and looks like a cluster of strawberries. Fruit cracking is a problem in many litchi-growing areas. Any stress that induces a check in growth of the fruit during the early period of cell division in the skin will generally suffer from cracking as the fruit rapidly increases in size before harvest.

Since litchi is a nonclimacteric fruit and does not ripen after harvest, the fruit must be harvested at optimal visual appearance and eating quality. The main postharvest disorder associated with litchi is pericarp browning, which can be caused by a wide range of stresses. Litchies that are quickly cooled to 3°C and held at a low temperature (5°C) tend to be less susceptible to moisture loss and disease. There are many cultivars that are propagated by air-layering, approaching grafting, or seed. The juicy white flesh inside the brittle shell has a gelatinous consistency. The fruits are often dried, having a sweet flavor with a raisin-like texture and flavor.

Passion Fruit (*Passiflora edulis* Sims.) PASSIFLORACEAE

This is a strong-growing, fruitful, woody perennial evergreen vine grown commercially in Australia, Brazil and other South American countries, Central America, South and East Africa, several Caribbean countries, and the United States (Florida and Hawaii). The vines are grown on high wire trellises or on lattices. While the vines require considerable heat to produce fruit, they grow but do not fruit well in cooler areas. Purple-fruited cultivars tend to fruit in spring and early summer but do not fruit well during hot weather, unlike the yellow passion fruit that thrives under lowland tropical conditions. The yellow fruits from the near relative, *P. edulis* f. *flavicarpa* Degener, which are also round and about the size of a large plum, have a delicious sprightly but acidic flavor. Passion fruits normally drop from the vine when they are mature, and the usual mode of harvest is to collect them from the ground. The extracted juice is often used commercially mixed with other fruit juices to make an attractive canned product.

REFERENCES AND SUPPLEMENTARY READING

CHAN, H. T., (ed.). 1983. *Handbook of tropical foods.* New York: Marcel Dekker.

JACKSON, D. I., and N. E. LOONEY, (eds.). 1999. *Temperate and subtropical fruit production.* Wallingford, UK: CAB International.

JUNICK, J., and J. N. MOORE, (eds.). 1996. *Fruit breeding,* Vol. I. New York: John Wiley & Sons.

MITRA, S. K. 1997. *Postharvest physiology and storage of tropical and subtropical fruits.* Wallingford, UK: CAB International.

Appendix

USEFUL WEB SITES

United Nations Food and Agriculture Organization (FAO): Global information on all aspects of agriculture.

http://www.fao.org/

United States Department of Agriculture, National Agricultural Statistics Service (NASS): Gives current and archived information for agriculture in the United States.

http://www.usda.gov/nass/

USDA Economics and Statistics System: Access to many national and international agricultural information sites.

http://usda.mannlib.cornell.edu/

Locations of all land-grant institutions in the United States:

http://www.nhq.nrcs.usda.gov/land/lgif/landgrant.html

Ohio State University's plant fact sheet database and search engine for universities and departments that teach plant science:

http://plantfacts.osu.edu

NUTRITIVE VALUES OF THE EDIBLE PARTS OF FOODS

Foods, Approximate Measures, Units, and Weight (Edible Part Unless Footnotes Indicate Otherwise)		Gm	Water Percent	Food Energy Calories	Protein Gm	Fat Gm
FRUITS AND FRUIT PRODUCTS						
Apples, raw, unpeeled, without cores:						
3 1/4-in. diam. (about 2 per lb).	1 apple	212	84	125	Trace	1
Apricots:						
Raw, without pits (about 12 per lb with pits).	3 apricots	107	85	55	1	Trace
Dried, Uncooked (28 large of 37 medium halves per cup).	1 cup	130	25	340	7	1
Avocados, raw, whole, without skins and seeds:						
California, mid- and late-winter (with skin and seed, 3 1/8-in. diam.; wt., 10 oz).	1 avocado	216	74	370	5	37
Florida, late summer and fall (with skin and seed, 3 5/8-in. diam.; wt., 1 lb).	1 avocado	304	78	390	4	33
Banana without peel (about 2.6 per lb with peel).	1 banana	119	76	100	1	Trace
Blackberries, raw	1 cup	144	85	85	2	1
Blueberries, raw	1 cup	145	83	90	1	1
Cherries, Sweet, raw, without pits and stems.	10 cherries	68	80	45	1	Trace
Cranberry sauce, sweetened, canned, strained.	1 cup	277	62	405	Trace	1
Dates:						
Whole, without pits	10 dates	80	23	220	2	Trace
Chopped	1 cup	178	23	490	4	1
Grapefruit:						
Raw, medium, 3 3/4-in. diam. (About 1 lb 1 oz):						
Pink or red	1/2 grapefruit with peel	241	89	50	1	Trace
White	1/2 grapefruit with peel	241	89	45	1	Trace
Grapes, European type (adherent skin), raw:						
Thompson Seedless.	10 grapes	50	81	35	Trace	Trace
Lemon, raw, size 165, without peel and seeds (about 4 per lb with peels and seeds).	1 lemon	74	90	20	1	Trace
Olives, pickled, canned:						
Green	4 medium or 3 extra large or 2 giant	16	78	15	Trace	2
Ripe, Mission	3 small or 2 large	10	73	15	Trace	2
Oranges, all commercial cultivars, raw:						
Whole, 2 5/8-in. diam., without peel and seeds (about 2 1/2 per lb with peel and seeds).	1 orange	131	86	65	1	Trace
Papayas, raw, 1/2-in. cubes	1 cup	140	89	55	1	Trace
Peaches:						
Raw:						
Whole, 2 1/2-in. diam., peeled, pitted (about 4 per lb with peels and pits).	1 peach	100	89	40	1	Trace
Dried:						
Uncooked	1 cup	160	25	420	5	1
Pears:						
Raw, with skin, cored:						
Bartlett, 2 1/2-in. diam. (About 2 1/2 per lb with cores and stems).	1 pear	164	83	100	1	1

Carbohydrate Gm	Calcium Mg	Phosphorus Mg	Iron Mg	Potassium Mg	Vitamin A International Units	Thiamin Mg	Riboflavin Mg	Niacin Mg	Ascorbic Acid Mg
31	15	21	.6	233	190	.06	.04	.2	8
14	18	25	.5	301	2,890	.03	.04	.6	11
86	87	140	7.2	1,273	14,170	.01	.21	4.3	16
13	22	91	1.3	1,303	630	.24	.43	3.5	30
27	30	128	1.8	1,836	880	.33	.61	4.9	43
26	10	31	.8	440	230	.06	.07	.8	12
19	46	27	1.3	245	290	.04	.06	.6	30
22	22	19	1.6	117	160	.04	.09	.7	20
12	15	13	.3	129	70	.03	.04	.3	7
104	17	11	.6	83	60	.03	.03	.1	6
58	47	50	2.4	518	40	.07	.08	1.8	0
130	105	112	5.3	1,153	90	.16	.18	3.9	0
13	20	20	.5	166	540	.05	.02	.2	44
12	19	19	.5	159	10	.05	.02	.2	44
9	6	10	.2	87	50	.03	.02	.2	2
6	19	12	.4	102	10	.03	.01	.1	39
Trace	8	2	.2	7	40	—	—	—	—
Trace	9	1	.1	2	10	Trace	Trace	—	—
16	54	26	.5	263	260	.13	.05	.5	66
14	28	22	.4	328	2,450	.06	.06	.4	78
10	9	19	.5	202	1,330	.02	.05	1.0	7
109	77	187	9.6	1,520	6,240	.02	.30	8.5	29
25	13	18	.5	213	30	.03	.07	.2	7

Continued

NUTRITIVE VALUES OF THE EDIBLE PARTS OF FOODS (continued)

Foods, Approximate Measures, Units, and Weight (Edible Part Unless Footnotes Indicate Otherwise)		Gm	Water Percent	Food Energy Calories	Protein Gm	Fat Gm
FRUITS AND FRUIT PRODUCTS *(continued)*						
Bosc, 2 1/2-in. Diam. (about 3 per lb with cores and stems).	1 pear	141	83	85	1	1
D'Anjou, 3-in. diam. (about 2 per lb with cores and stems).	1 pear	200	83	120	1	1
Pineapple: Raw, diced	1 cup	155	85	80	1	Trace
Plums:						
Raw, without pits:						
Japanese and hybrid (2 1/8-in. diam., about 6 1/2 per lb with pits).	1 plum	66	87	30	Trace	Trace
Prune-type (1 1/2-in. diam., about 15 per lb with pits).	1 plum	28	79	20	Trace	Trace
Prunes, dried, "softenized," with pits: Uncooked	4 extra large or 5 large prunes	49	28	110	1	Trace
Raisins, seedless: Cup, not pressed down	1 cup	145	18	420	4	Trace
Raspberries, red: Raw, capped, whole	1 cup	123	84	70	1	1
Strawberries: Raw, whole berries, capped	1 cup	149	90	55	1	1
Tangerine, raw, 2 3/8-in. diam., size 176, without peel	1 tangerine	86	87	40	1	Trace
LEGUMES (DRY), NUTS, SEEDS						
Almonds, shelled:						
Chopped (about 130 almonds).	1 cup	130	5	775	24	70
Beans, dry:						
Common cultivars as Great Northern, navy, and others:						
Cooked, drained:						
Great Northern	1 cup	180	69	210	14	1
Navy	1 cup	190	69	225	15	1
Red kidney	1 cup	255	76	230	15	1
Lima, cooked, drained	1 cup	190	64	260	16	1
Blackeye peas, dry, cooked (with residual cooking liquid).	1 cup	250	80	190	13	1
Brazil nuts, shelled (6-8 large kernels).	1 oz	28	5	185	4	19
Cashew nuts, roasted in oil.	1 cup	140	5	785	24	64
Coconut meat, fresh:						
Piece, about 2 by 2 by 1/2 in.	1 piece	45	51	155	2	16
Shredded or grated, not pressed down.	1 cup	80	51	275	3	28
Filberts (hazelnuts), chopped (about 80 kernels).	1 cup	115	6	730	14	72
Lentils, whole, cooked.	1 cup	200	72	210	16	Trace
Peanuts, roasted in oil, salted (whole, halves, chopped).	1 cup	144	2	840	37	72
Pecans, chopped or pieces (about 120 large halves).	1 cup	118	3	810	11	84
Walnuts:						
Black:						
Chopped or broken kernels.	1 cup	125	3	785	26	74
Persian or English, chopped (about 60 halves).	1 cup	120	4	780	18	77

Carbohydrate Gm	Calcium Mg	Phosphorus Mg	Iron Mg	Potassium Mg	Vitamin A International Units	Thiamin Mg	Riboflavin Mg	Niacin Mg	Ascorbic Acid Mg
22	11	16	.4	83	30	.03	.06	.1	6
31	16	22	.6	260	40	.04	.08	.2	8
21	26	12	.8	226	110	.14	.05	.3	26
8	8	12	.3	112	160	.02	.02	.3	4
6	3	5	.1	48	80	.01	.01	.1	1
29	22	34	1.7	298	690	.04	.07	.7	1
112	90	146	5.1	1,106	30	.16	.12	.7	1
17	27	27	1.1	207	160	.04	.11	1.1	31
13	31	31	1.5	244	90	.04	.10	.9	88
10	34	15	.3	108	360	.05	.02	.1	27
25	304	655	6.1	1,005	0	.31	1.20	4.6	Trace
38	90	266	4.9	749	0	.25	.13	1.3	0
40	95	281	5.1	790	0	.27	.13	1.3	0
42	74	278	4.6	673	10	.13	.10	1.5	—
49	55	293	5.9	1,163	—	.25	.11	1.3	—
35	43	238	3.3	573	30	.40	.10	1.0	—
3	53	196	1.0	203	Trace	.27	.03	.5	—
41	53	522	5.3	650	140	.60	.35	2.5	—
4	6	43	.8	115	0	.02	.01	.2	1
8	10	76	1.4	205	0	.04	.02	.4	2
19	240	388	3.9	810	—	.53	—	1.0	Trace
39	50	238	4.2	498	40	.14	.12	1.2	0
27	107	577	3.0	971	—	.46	.19	24.8	0
17	86	341	2.8	712	150	1.01	.15	1.1	2
19	Trace	713	7.5	575	380	.28	.14	.9	—
19	119	456	3.7	540	40	.40	.16	1.1	2

Continued

NUTRITIVE VALUES OF THE EDIBLE PARTS OF FOODS (continued)

Foods, Approximate Measures, Units, and Weight (Edible Part Unless Footnotes Indicate Otherwise)		Gm	Water Percent	Food Energy Calories	Protein Gm	Fat Gm
VEGETABLES						
Asparagus, green:						
Cooked, drained:						
Spears, 1/2-in. diam. at base:						
From raw.	4 spears	60	94	10	1	Trace
From frozen.	4 spears	60	92	15	2	Trace
Beans:						
Lima, immature seeds, frozen, cooked, drained:						
Thick-seeded types (Fordhooks).	1 cup	170	74	170	10	Trace
Thin-seeded types (baby limas).	1 cup	180	69	210	13	Trace
Snap:						
Green:						
Cooked, drained:						
From raw (cuts and French style).	1 cup	125	92	30	2	Trace
From frozen:						
Cuts	1 cup	135	92	35	2	Trace
French style	1 cup	130	92	35	2	Trace
Yellow or wax:						
Cooked, drained:						
From raw (cuts and French style).	1 cup	125	93	30	2	Trace
From frozen (cuts).	1 cup	135	92	35	2	Trace
Canned, drained solids (cuts).	1 cup	135	92	30	2	Trace
Beet greens, leaves and stems, cooked, drained.	1 cup	145	94	25	2	Trace
Beets:						
Cooked, drained, peeled:						
Whole beets, 2-in. diam.	2 beets	100	91	30	1	Trace
Canned, drained solids:						
Whole beets, small	1 cup	160	89	60	2	Trace
Broccoli, cooked, drained:						
From raw:						
Stalk, medium size.	1 stalk	180	91	45	6	1
From frozen:						
Stalk, 4 1/2 to 5 in. long.	1 stalk	30	91	10	1	Trace
Brussels sprouts, cooked, drained:						
From raw, 7-8 sprouts (1 1/4- to 1 1/2-in. diam.).	1 cup	155	88	55	7	1
From frozen.	1 cup	155	89	50	5	Trace
Cabbage:						
Common varieties:						
Raw:						
Coarsely shredded or sliced	1 cup	70	92	15	1	Trace
Finely shredded or chopped	1 cup	90	92	20	1	Trace
Cooked, drained:	1 cup	146	94	30	2	Trace
Red, raw, coarsely shredded or sliced.	1 cup	70	90	20	1	Trace
Savoy, raw, coarsely shredded or sliced.	1 cup	70	92	15	2	Trace
Cabbage, celery (also called pe-tsai or wongbok), raw, 1-in. pieces.	1 cup	75	95	10	1	Trace
Cabbage, white mustard (also called bokchoy or pakchoy), cooked, drained.	1 cup	170	95	25	2	Trace

Carbohydrate Gm	Calcium Mg	Phosphorus Mg	Iron Mg	Potassium Mg	Vitamin A International Units	Thiamin Mg	Riboflavin Mg	Niacin Mg	Ascorbic Acid Mg
2	13	30	.4	110	540	.10	.11	.8	16
2	13	40	.7	143	470	.10	.08	.7	16
32	34	153	2.9	724	390	.12	.09	1.7	29
40	63	227	4.7	709	400	.16	.09	2.2	22
7	63	46	.8	189	680	.09	.11	.6	15
8	54	43	.9	205	780	.09	.12	.5	7
8	49	39	1.2	177	690	.08	.10	.4	9
6	63	40	.0	180	290	.09	.11	.6	16
8	47	42	.9	221	140	.09	.11	.5	8
7	61	34	2.0	128	140	.04	.07	.4	7
5	144	36	2.8	481	7,400	.10	.22	.4	22
7	14	23	.5	208	20	.03	.04	.3	6
14	30	29	1.1	267	30	.02	.05	.2	5
8	158	112	1.4	481	4,500	.16	.36	1.4	162
1	12	17	.2	66	570	.02	.03	.2	22
10	50	112	1.7	423	810	.12	.22	1.2	135
10	33	95	1.2	457	880	.12	.16	.9	126
4	34	20	0.3	163	90	0.04	0.04	0.02	33
5	44	26	.4	210	120	.05	.05	.3	42
6	64	29	.4	236	190	.06	.06	.4	48
5	29	25	.6	188	30	.06	.04	.3	43
3	47	38	.6	188	140	.04	.06	.2	39
2	32	30	.5	190	110	.04	.03	.5	19
4	252	56	1.0	364	5,270	.07	.14	1.2	26

Continued

NUTRITIVE VALUES OF THE EDIBLE PARTS OF FOODS (continued)

Foods, Approximate Measures, Units, and Weight (Edible Part Unless Footnotes Indicate Otherwise)		Gm	Water Percent	Food Energy Calories	Protein Gm	Fat Gm
VEGETABLES *(continued)*						
Carrots:						
Raw, without crowns and tips, scraped:						
Whole, 7 1/2 by 1 1/8 in., or strips, 2 1/2 to 3 in. long.	1 carrot or 18 strips	72	88	30	1	Trace
Grated.	1 cup	110	88	45	1	Trace
Canned:						
Sliced, drained solids	1 cup	155	91	45	1	Trace
Cauliflower:						
Raw, chopped.	1 cup	115	91	31	3	Trace
Cooked, drained.	1 cup	125	93	30	3	Trace
Celery, Pascal type, raw:						
Stalk, large outer, 8 by 1 1/2 in., at root end.	1 stalk	40	94	5	Trace	Trace
Collards, cooked, drained:						
From raw (leaves without stems).	1 cup	190	90	65	7	1
From frozen (chopped).	1 cup	170	90	50	5	1
Corn, sweet:						
Cooked, drained:						
From raw, ear 5 by 1 3/4 in.	1 ear	140	74	70	2	1
From frozen:						
Ear, 5 in. long.	1 ear	229	73	120	4	1
Kernels.	1 cup	165	77	130	5	1
Cucumber slices, 1/8 in. thick (large, 2 1/8-in. diam.; small, 1 3/4-in. diam.):						
With peel.	6 large or 8 small slices	28	95	5	Trace	Trace
Endive, curly (including escarole), raw, small pieces.	1 cup	50	93	10	1	Trace
Kale, cooked, drained:						
From raw (leaves without stems and midribs).	1 cup	110	88	45	5	1
From frozen (leaf style).	1 cup	130	91	40	4	1
Lettuce, raw:						
Butterhead, as Boston types:						
Head, 5-in. diam.	1 head	220	95	25	2	Trace
Leaves.	1 outer or 2 inner or 3 heart leaves	15	95	Trace	Trace	Trace
Crisphead, as Iceberg:						
Head, 6-in. diam.	1 head	567	96	70	5	1
Pieces, chopped or shredded	1 cup	55	96	5	Trace	Trace
Looseleaf (bunching varieties including romaine or cos), chopped or shredded pieces.	1 cup	55	94	10	1	Trace
Muskmelons, raw, orange-fleshed (with rind and seed cavity, 5-in. diam., 2 1/3 lb).	1/2 melon with rind	477	91	80	2	Trace
Honeydew (with rind and seed cavity, 6 1/2-in. diam., 5 1/4 lb).	1/10 melon with rind	226	91	50	1	Trace
Mustard greens, without stems and midribs, cooked, drained.	1 cup	140	93	30	3	1
Okra pods, 3 by 5/8 in., cooked.	10 pods	106	91	30	2	Trace

Carbohydrate Gm	Calcium Mg	Phosphorus Mg	Iron Mg	Potassium Mg	Vitamin A International Units	Thiamin Mg	Riboflavin Mg	Niacin Mg	Ascorbic Acid Mg
7	27	26	.5	246	7,930	.04	.04	.4	6
11	41	40	.8	375	12,100	.07	.06	.7	9
10	47	34	1.1	186	23,250	.03	.05	.6	3
6	29	64	1.3	339	70	.13	.12	.8	90
5	26	53	.9	258	80	.11	.10	.8	69
2	16	11	.1	136	110	.01	.01	.1	4
10	357	99	1.5	498	14,820	.21	.38	2.3	144
10	299	87	1.7	401	11,560	.10	.24	1.0	56
16	2	69	.5	151	310	.09	.08	1.1	7
27	4	121	1.0	201	440	18	.10	2.1	9
31	5	120	1.3	304	580	.15	.10	2.5	8
1	7	8	.3	45	70	.01	.01	.1	3
2	41	27	.9	147	1,650	.04	.07	.3	5
7	206	64	1.8	243	9,130	.11	.20	1.8	102
7	157	62	1.3	251	10,660	.08	.20	.9	49
4	57	42	3.3	430	1,580	.10	.10	.5	13
Trace	5	4	.3	40	150	.01	.01	Trace	1
16	108	118	2.7	943	1,780	.32	.32	1.6	32
2	11	12	.3	96	180	.03	.03	.2	3
2	37	14	.8	145	1,050	.03	.04	.2	10
20	38	44	1.1	682	9,240	.11	.08	1.6	90
11	21	24	.6	374	60	.06	.04	.9	34
6	193	45	2.5	308	8,120	.11	.20	.8	67
6	98	43	.5	184	520	.14	.19	1.0	21

METRIC CONVERSION CHART

Into Metric			Out of Metric		
If You Know	Multiply by	To Get	If You Know	Multiply By	To Get
Length					
inches	2.54	centimeters	millimeters	0.04	inches
feet	30	centimeters	centimeters	0.4	inches
feet	0.303	meters	meters	3.3	feet
yards	0.91	meters	kilometers	0.62	miles
miles	1.6	kilometers			
Area					
sq inches	6.5	sq centimeters	sq centimeters	0.16	sq inches
sq feet	0.09	sq meters	sq meters	1.2	sq yards
sq yards	0.8	sq meters	sq kilometers	0.4	sq miles
sq miles	2.6	sq kilometers	hectares	2.47	acres
acres	0.4	hectares			
Mass (Weight)					
ounces	28	grams	grams	0.035	ounces
pounds	0.45	kilograms	kilograms	2.2	pounds
short ton	0.9	metric ton	metric tons	1.1	short tons
Volume					
teaspoons	5	milliliters	milliliters	0.03	fluid ounces
tablespoons	15	milliliters	liters	2.1	pints
fluid ounces	30	milliliters	liters	1.06	quarts
cups	0.24	liters	liters	0.26	gallons
pints	0.47	liters	cubic meters	35	cubic feet
quarts	0.95	liters	cubic meters	1.3	cubic yards
gallons	3.8	liters			
cubic feet	0.03	cubic meters			
cubic yards	0.76	cubic meters			
Pressure					
lb/in.^{2}	0.069	bars	bars	14.5	lb/in.^{2}
atmospheres	1.013	bars	bars	0.987	atmospheres
atmospheres	1.033	kg/cm^2	kg/cm^2	0.968	atmospheres
lb/in.^{2}	0.07	kg/cm^2	kg/cm^2	14.22	lb/in.^{2}
Rates					
lb/acre	1.12	kg/hectare	kg/hectare	0.882	lb/acre
tons/acre	2.24	metric tons/hectare	metric tons/hectare	0.445	tons/acre

TEMPERATURE CONVERSION TABLE

Celsius temperatures have been rounded to the nearest whole number					
F	*C*	*F*	*C*	*F*	*C*
−26	−32	19	−7	64	18
−24	−31	21	−6	66	19
−22	−30	23	−5	68	20
−20	−29	25	−4	70	21
−18	−28	27	−3	72	22
−17	−27	28	−2	73	23
−15	−26	30	−1	75	24
−13	−25	32	0	77	25
−11	−24	34	1	79	26
−9	−23	36	2	81	27
−8	−22	37	3	82	28
−6	−21	39	4	84	29
−4	−20	41	5	86	30
−2	−19	43	6	88	31
0	−18	45	7	90	32
1	−17	46	8	91	33
3	−16	48	9	93	34
5	−15	50	10	95	35
7	−14	52	11	97	36
9	−13	54	12	99	37
10	−12	55	13	100	38
12	−11	57	14	102	39
14	−10	59	15	104	40
16	−9	61	16	106	41
18	−8	63	17	108	42
				212	100

Fahrenheit to Celsius: Subtract 32 from the Fahrenheit figure, multiply by 5, and divide by 9.
Celsius to Fahrenheit: Multiply the Celsius figure by 9, divide by 5, and add 32.

LIGHT CONVERSION TABLE

Conversion of lux or footcandles (ft-c) from various light sources to watts per sq meter of photosynthetically active radiation (PAR) or to micromols of photosynthetic photon flux (PPF) per sq meter per sec.

Light Source	To Convert to:			
	Wm^{-2} PAR Divide:		$\mu mol\ m^{-2}s^{-1}$ PPF Divide:	
	lux	ft-c	lux	ft-c
Daylight[a]	by 247	by 22.9	by 54	by 5.0
Metal halide lamps	326	30.2	71	6.6
Fluorescent, cool-white lamps	340	31.5	74	7.0
Incandescent lamps	250	23.1	50	4.6
Low-pressure sodium lamps	522	48.3	106	9.8
High-pressure mercury lamps[b]	380	35.2	84	7.8

[a]Based on average spectral distribution.
[b]Different types of mercury lamps can vary $\pm$ 5 percent.

To convert measurements of light intensity in footcandles (ft-c), or illuminance in lux from various light sources into equivalent watts of photosynthetic radiation per square meter (Wm^{-2} PAR), or micromols of photosynthetic photon flux per square meter per second ($\mu mol\ m^{-2}s^{-1}$), divide the reported measurement by the appropriate constant given for the particular light source.

Example: A light intensity of 2,500 ft-c of daylight recorded in the greenhouse may be converted into equivalent Wm^{-2} PAR or $\mu mol\ m^{-2}s^{-1}$ PPF as follows:

$$\frac{2,500}{22.9} = 109\ Wm^{-2}\ PAR \quad or \quad \frac{2,500}{5.0} = 500\ \mu mol\ m^{-2}\ s^{-1}\ PPF$$

Terms Used to Describe Visible and Photosynthetic Light Energy

Lux and footcandles (ft-c) are photometric measurements based on the sensitivity of the eye to various wavelengths of electromagnetic radiation. The *lux* is the illuminance resulting from a light energy (luminous) flux of 1 lumen distributed evenly over one square meter (1 lux = 1 lumen m^{-2}). The *footcandle* is the illuminance resulting from a luminous flux of 1 lumen per square foot (1 ft-c = 1 lumen ft^{-2}).

Watts and joules are *radiometric* measurements that express energy per unit of time per unit area without respect to wavelength. Such measurements made in the PAR waveband are termed W PAR.

In electromagnetic terms, the mol is Avogadro's number of photons. Micro-mol per square meter per second ($\mu mol \cdot m^{-2} \cdot s^{-1}$) is a flux of 6.022×10^{17} photons intercepted by one square meter over one second. Each of the light sources above has its own specific distribution of wavelengths compared with daylight; each, therefore, requires its own conversion factor. Sometimes Einstein (E) is used in place of mol. The terms are equivalent. However, mol is preferred because E is not a *système international* (SI) unit.

Glossary

A horizon   The surface layer of varying thickness of a mineral soil having maximum organic matter accumulation, maximum biological activity, and/or eluviation by water of materials such as iron and aluminum oxides and silicate clays.

Abaxial   Away from the axis or central line; turned toward the base, dorsal.

Abscisic acid   A plant hormone involved in abscission, dormancy, stomatal closure, growth inhibition, and other plant responses.

Abscission zone   A layer of thin-walled cells extending across the base of a petiole or peduncle, whose breakdown separates the leaf or fruit from the stem causing the leaf or fruit to drop.

Absorption   The taking up of water by assimilation or imbibition. The taking up by capillary, osmotic, chemical, or solvent action such as the taking up of water from air, or taking up of gases by water, or taking up of mineral nutrients by plant roots.

Acclimatization   The adaptation of an individual plant to a changed climate, or the adjustment of a species or a population to a changed environment, often over several generations.

Achene   A simple, dry, one-seeded indehiscent fruit with the seed attached to the ovary wall at one point only. The so-called strawberry and sunflower "seeds" are examples.

Acid delinting   A process used to remove the short fibers (lint) from seed cotton.

Acid soil   Soil with a reaction below pH 7; more technically, a soil having a preponderance of hydrogen ions over hydroxyl ions in solution.

Action spectrum   A graph indicating the changes in response of an entity to changes in a stimulus.

Adaptation   The process of change in structure or function of an individual or population caused by environmental changes.

Adaxial   Toward the axis or center; turned toward the apex; ventral.

Adenosine diphosphate (ADP)   A nucleotide (a nitrogen-based compound) composed of adenine and ribose with two phosphate groups attached. ATP and ADP participate in metabolic reactions (both catabolic and anabolic). These molecules, through the process of being phosphorylated (accepting phosphate groups) or dephosphorylated (losing phosphate groups), transfer energy within the cells to drive metabolic processes.

Adenosine triphosphate (ATP)   ATP has three phosphoric groups attached and is the phosphorylated condition of ADP. It conveys energy needed for metabolic reactions, then loses one phosphate group to become ADP (adenosine diphosphate).

Adhesion   The molecular attraction between unlike substances such as water and sand particles.

Adsorption   The attraction of ions or molecules to the surface of a solid.

Adventitious   Refers to structures arising from an unusual place; for example, buds at places other than shoot terminals or leaf axils, or roots growing from stems or leaves.

Aeration, soil   The process by which air in the soil is replaced by air from the atmosphere.

Aerobic   An environment or condition in which oxygen is not deficient for chemical, physical, or metabolic processes.

Agar   A gelatinous substance obtained from certain species of red algae; widely used as a substrate in aseptic cultures.

Aggregate (soil)   Many primary soil particles held in a single unit as a clod, crumb, block, or prism.

Agronomy   The art and science of crop production and soil management.

Air-dry   The state of dryness at equilibrium with the moisture content in the surrounding atmosphere.

Air layer   An undetached aerial portion of a plant on which roots are caused to develop commonly as the result of wounding or other stimulation.

Air porosity   The proportion of the bulk volume of soil that is filled with air at any given time or under a given condition, such as a specified moisture tension.

Alar®   *See* Daminozide.

Albino   A plant or part of a plant lacking chlorophyll. Albinism is usually lethal in higher plants.

Aleurone   The outer layer of cells surrounding the endosperm of a cereal grain (caryopsis).

Alkali soil   *See* Sodic soil.

Alkaline soil   Any soil that has a pH greater than 7.

Allele   One of a pair or a series of factors that occur at the same locus on homologous chromosomes; one alternative form of a gene.

Allelopathy   The excretion of chemicals by plants of some species that are toxic to plants of other species.

Alluvial soil   A recently developed soil from deposited soil material that exhibits essentially no horizon development or modifications.

Alternate (Taxonomy)   A leaf arrangement with leaves placed singly at various heights on the stem; i.e., not opposite or whorled.

Alumino-silicates   Compounds containing aluminum, silicon, and oxygen as main constituents. An example is microcline, $KAlSi_3O_8$.

Ambient temperature   Air temperature at a given time and place; not radiant temperature.

Amendment (soil)   Any substance such as lime, sulfur, gypsum, or an organic material like peat moss, sawdust, or bark used to alter the properties of a soil, generally to improve its physical properties.

Amino acids   The fundamental building blocks of proteins. There are 20 common amino acids in living organisms, each having the basic formula NH_2–CHR–COOH.

Ammonification   The biochemical process whereby ammoniacal nitrogen is released from nitrogen-containing organic compounds.

Ammonium fixation   The incorporation of ammonium ions by soil fractions in such a manner that they are relatively insoluble in water and nonexchangeable by the principle of cation exchange.

Anaerobic   An environment or condition in which molecular oxygen is deficient for chemical, physical, or biological processes.

Ancymidol (A-Rest®)   A plant growth retardant effective as a spray or soil drench.

Angiosperm   One of a large group of seed-bearing plants in which the female gamete is protected within an enclosed ovary. A flowering plant.

Anion   A negatively charged particle that, during electrolysis, is attracted to positively charged surfaces.

Annual ring   The cylinder of secondary xylem added to a woody plant stem by the cambium in any one year.

Annuals   Plants living one year or less. During this time the plant grows, flowers, produces seeds, and dies.

Antagonism   Opposing action of different chemicals such that the action of one is impaired or the total effect is less than that of one acting separately.

Anther   In a flower the saclike structure of the stamen in which microspores (pollen grains) are produced; usually borne on a filament.

Anthesis   A developmental stage in flowering at which anthers rupture and pollen is shed. A state of full bloom.

Anthocyanin   A class of water-soluble pigments that account for many of the red to blue flower, leaf, and fruit colors. Anthocyanins occur in the vacuole of the cell.

Antipodal nuclei   The three or more nuclei at the end of the embryo sac opposite the egg nucleus (female gamete). They are produced by mitotic divisions of the megaspore and degenerate following sexual fertilization.

Apical dominance   The inhibition of lateral buds on a shoot due to auxins produced by the apical bud.

Apical meristem   A mass of undifferentiated cells capable of division at the tip of a root or shoot. These cells differentiate by division, allowing the plant to grow in depth or height.

Apomixis   The asexual (vegetative) production of seedlings in the usual sexual structures of the flower but without the mingling and segregation of chromosomes. Seedling characteristics are the same as those of the maternal parent.

Arable land   Land suitable for the production of crops.

A-Rest®   *See* Ancymidol.

Asexual reproduction   The production of a new plant by any vegetative means not involving meiosis and the union of gametes.

Assimilation   The transformation of organic and inorganic materials into protoplasm.

Atom   The smallest particle in which an element combines, either with itself or with other elements; the smallest quantity of matter possessing the properties of a particular element.

Autogamy   Pollination within the same flower.

Autotrophic   Plants capable of utilizing carbon dioxide or carbonates as the sole source of carbon and obtaining energy for life processes from the oxidation of inorganic elements or compounds such as iron, sulfur, hydrogen, ammonium, and nitrites, or from radiant energy. (*See* Heterotrophic.)

Available nutrient   That portion of an element or compound in the soil that can be readily absorbed, assimilated, and utilized by growing plants. (*Available* should not be confused with *exchangeable*.)

Available water   The portion of water in a soil that can be readily absorbed by plant roots. That soil moisture held in the soil between field capacity and permanent wilting percentage (available water = F.C. − P.W.P.).

Axil   The angle on the upper side of the union of a branch and main stem or of a leaf and a stem.

Axillary bud   A bud formed in the axil of a leaf.

B horizon   A soil layer of varying thickness (usually beneath the A horizon) that is characterized by an accumulation of silicate clays, iron and aluminum oxides, and humus, alone or in combination and/or a blocky or prismatic structure.

Backcross   In breeding, a cross of a hybrid with one of its parents or with a genetically equivalent organism. In genetics, a cross of a hybrid with a homozygous recessive.

Bagasse   The dry refuse of sugarcane after the juice has been expressed.

Bar   A unit of pressure equal to 1 million dynes/cm^2. Equivalent approximately to 1 atm of pressure.

Bark   The tissue of a woody stem or root from the cambium outward.

Base pair   The nitrogen bases that pair in the DNA molecule—adenine with thymine, and guanine with cytosine.

Berry   A simple fleshy fruit formed from a single ovary; the ovary wall is fleshy and includes one or more carpels and seeds. For example, fruits of the tomato and grape are botanically berries.

Biennial   A plant that completes its life cycle within two seasons. For most biennial plants the two seasons are separated by an obligate degree of cold temperature sufficient to initiate flowering and fruit formation, after which the plant dies.

Binomial   In biology each species is generally indicated by two names; first, the genus to which it belongs, and second, the species name (e.g., *Quercus suber,* cork oak).

Biocide   A combination of a bactericide and a fungicide used for cut-flower-keeping solutions, or used as a sterilant in horticultural operations.

Biodegradable   Materials readily decomposed by microorganisms such as bacteria and fungi.

Biology   The science that deals with living organisms.

Biotechnology   Coined by Karl Erecky in 1919, it refers to any product produced from raw materials with the aid of living organisms. For example, monoclonal antibodies (for immunotherapy), enzymes (lactase and rennin), human insulin, and Round-Up® resistant crops.

Biuret   A compound ($H_2NCONHCONH_2 \cdot H_2O$) formed by the thermal decomposition of urea (H_2NCONH_2) that is phytotoxic to many crops. Therefore, "biuret-free" urea is used for fertilization of crops.

Bloat   Excessive accumulation of gases in the rumen of some animals.

B-nine®   *See* Daminozide.

Bolting   Rapid production of flower stalks in some herbaceous plants after sufficient chilling or a favorable photo-period.

Botany The science of plants, their characteristics, functions, life cycles, and habits.

Brace root (anchor root) A type of adventitious root that grows from above-ground parts of the stem and serves to support some plants; for example, corn.

Bract A modified leaf, from the axil of which arises a flower or an inflorescence.

Breeder seed Seed (or vegetative propagating material; e.g., potato) increased by the originating, or sponsoring, plant breeder or institution and used as the source to increase foundation seed.

Broadcast Scattering seed or fertilizers uniformly over the soil surface rather than placing in rows.

Brushing The procedure using riverbank brush for constructing a barrier for sheltering crop plants from wind and for modifying the plant's microenvironment in order to accelerate growth.

Bud A region of meristematic tissue with the potential for developing into leaves, shoots, flowers, or combinations; generally protected by modified scale leaves.

Bud scar A scar left on a shoot when the bud or bud scales drop.

Bud sport A mutation arising in a bud and producing a genetically different shoot. Includes change due to gene mutation, somatic reduction, chromosome deletion, or polyploidy.

Budding A form of grafting in which a single vegetative bud is taken from one plant and inserted into stem tissue of another plant so that the two will grow together. The inserted bud develops into a new shoot.

Bulb A highly compressed underground stem (basal plate) to which numerous storage scales (modified leaves) are attached. Examples are lily, onion, tulip.

Bulk density (soil) The mass of a known volume (including air space) of soil. The soil volume is determined in place, then dried in an oven to constant weight at 105°C. Bulk density (D_B)= oven dry weight of soil/volume of soil.

Bunch type grass Grass that does not spread by rhizomes or stolons.

C horizon A soil layer beneath the B layer that is relatively little affected by biological activity and pedogenesis and is lacking properties diagnostic of an A or B horizon.

C_3 cycle The Calvin-Benson cycle of photosynthesis, in which the first products after CO_2 fixation are three-carbon molecules.

C_4 cycle The Hatch-Slack cycle of photosynthesis, in which the first products after CO_2 fixation are four-carbon molecules.

CO_2 compensation point The point at which the level of CO_2 causes respiration to equal photosynthesis.

Calcareous soil Soil containing sufficient calcium and/or magnesium carbonate to effervesce visibly when treated with cold 0.1 normal (0.1N) hydrochloric acid.

Calcium pectate An organic calcium compound found in the middle lamella between plant cells and serving as an intercellular cement.

Callus Mass of large, thin-walled parenchyma cells, usually developing as the result of a wound.

Calorie (gram calorie) Unit for measuring energy, defined as the heat necessary to raise the temperature of 1 g of water from 14.5°C to 15.5°C at standard pressure; 1 kilocalorie (kcal) raises the temperature of 1 kg of water 1°C. Thus 1 kcal = 1,000 cal.

Calyx The collective term for the sepals.

Cambium (vascular) A thin layer of longitudinally dividing cells between the xylem and phloem that gives rise to secondary growth.

Capillary water The water held in the "capillary" or small pores of a soil, usually with a tension greater than 60 cm of water.

Capsule (botanical) A simple, dry, dehiscent fruit, with two or more carpels.

Carbohydrate Compound of carbon, hydrogen, and oxygen in the ratio of one atom each of carbon and oxygen to two of hydrogen, as in sugar, starch, and cellulose.

Carbon dioxide fixation The addition of H^+ to CO_2 to yield a chemically stable carbohydrate. The H^+ is contributed by NADPH, the reduced (hydrogen-rich) form of $NADP^+$, produced in the noncyclic phase of the light reactions of photosynthesis. The H^+ comes originally from the photolysis of water.

Carbon-nitrogen ratio The ratio of the weight of organic carbon to the weight of total nitrogen in a soil or in organic material.

Carotene Yellow plant pigments, precursors of vitamin A. Alpha, beta, and gamma carotenes are converted into vitamin A in the animal body.

Carpel Female reproductive organ of flowering plants. In some plants one or more carpels unite to form the pistil.

Caryopsis Small, one-seeded, dry fruit with a thin pericarp surrounding and adhering to the seed; the "seed" (grain) or fruit of grasses.

Casparian strip A secondary thickening that develops on the radial and end walls of some endodermal cells.

Catalyst Any substance that accelerates a chemical reaction but does not enter into the reaction itself.

Cation A positively charged ion.

Cation exchange capacity (base-exchange capacity) A measure of the total amount of exchangeable cations that a soil can hold; expressed in meq/100 g soil at pH 7.

Catkin A type of inflorescence (a spike) generally bearing either pistillate or staminate flowers. Found on walnuts and willows, for example.

Cell The basic structure and physiological unit of plants and animals.

Cell membrane The membrane that separates the cell wall and the cytoplasm and regulates the flow of material into and out of the cell.

Cell plate The precursor of the cell wall, formed as cytokinesis starts during cell division. It develops in the region of the equatorial plate and arises from membranes in the cytoplasm.

Cell wall The outermost, cellulose limit of the plant cell; the barrier that develops between nuclei during mitosis.

Cellulose A complex carbohydrate composed of long, unbranched beta-glucose molecules, which makes up 40 to 55 percent by weight of the plant cell wall.

Cenozoic The geologic era extending from about 65 million years ago to the present time.

Center of origin A geographical area in which a species is thought to have evolved through natural selection from its ancestors.

Cereal A member of the GRAMINAE family grown primarily for its mature, dry seed.

Cereal forage Cereal crop harvested when immature for hay, silage, green chop, or pasturage.

Certified seed The progeny of foundation, registered, or certified seed, produced and handled so as to maintain satisfactory genetic identity and purity, and approved and certified by an official certifying agency.

Character   The expression of a gene in the phenotype.

Chelate (sequestering agent)   A large organic molecule that attracts and tightly holds specific cations, like a chemical claw, preventing them from taking part in inorganic reactions but at the same time allowing them to be absorbed and used by plants.

Chilling injury   Direct and/or indirect injury to plants or plant parts from exposure to low, but above-freezing temperatures (as high as 9° C or 48° F).

Chimera   A plant composed of two or more genetically different tissues. Includes *periclinal chimera,* in which one tissue lies over another as a glove fits a hand; *mericlinal chimera,* where the outer tissue does not completely cover the inner tissue; and *sectorial chimera,* in which the tissues lie side by side.

Chisel (subsoil)   A tillage implement with one or more cultivator-type shanks to which are attached knifelike units that shatter or break up hard, compact layers, usually in the subsoil.

Chlormequat (Cycocel®)   A plant growth retardant effective as a spray or a soil drench.

Chlorophyll   A complex organic molecule that traps light energy for conversion through photosynthesis into chemical energy.

Chloroplast   Chlorophyll-containing cytoplasmic body, in which important reactions of sugar or starch synthesis take place during photosynthesis.

Chlorosis   A condition in which a plant or a part of a plant is light green or greenish yellow because of poor chlorophyll development or the destruction of chlorophyll resulting from a pathogen or a mineral deficiency.

Chromosome   A specific, highly organized body in the nucleus of the cell that contains DNA.

Class (soil)   A group of soils having a definite range in a particular property such as acidity, degree of slope, texture, structure, land-use capability, degree of erosion, or drainage.

Classification   The systematic arrangement into categories on the basis of characteristics. Broad groupings are made on the basis of general characteristics and subdivisions on the basis of more detailed differences in specific properties.

Clay   (1) Soil particles less than 0.002 mm in equivalent diameter. (2) Soil material containing more than 40 percent clay, less than 45 percent sand, and less than 40 percent silt.

Claypan   A compact, slowly permeable layer of varying thickness and depth in the subsoil having a much higher clay content than the overlying material. Claypans are usually hard when dry, and plastic and sticky when wet.

Climate   Weather conditions such as wind, precipitation, and temperature that prevail in a given region.

Climacteric   The period in the development of some plant parts involving a series of biochemical changes associated with the natural respiratory rise and autocatalytic ethylene production.

Climax vegetation   Fully developed plant community in equilibrium with its environment.

Clod   A compact, coherent mass of soil produced artificially, usually by tillage operations, especially when performed on soils either too wet or too dry.

Clone   The aggregate of individual organisms originating from one sexually produced individual (or from a mutation) and maintained exclusively by asexual propagation.

Clove   One of a group of small bulbs produced, for example, by garlic and shallot plants.

Coenzyme   A substance, usually nonprotein and of low molecular weight, necessary for the action of some enzymes.

Cohesion   Holding together; a force holding a solid or liquid together, owing to attraction between like molecules.

Colchicine   An alkaloid, derived from the autumn crocus, used specifically to inhibit the spindle mechanism during cell division and thus cause a doubling of chromosome number.

Cold frame   An enclosed, unheated covered frame useful for growing and protecting young plants in early spring. The top is covered with glass or plastic, and sunlight provides heat.

Coleoptile   A transitory membrane (first leaf) covering the shoot apex in the seedlings of certain monocots. It protects the plumule as it emerges through the soil.

Coleorhiza   Sheath that surrounds the radicle of the grass embryo and through which the young developing root emerges.

Collenchyma   Elongated, parenchymatous cells with variously thickened walls, commonly at the acute angles of the cell wall.

Colloid (soil)   Organic and inorganic matter with very small particle size and a correspondingly large surface area per unit of mass.

Community   All the plant populations within a given habitat; usually the populations are considered to be interdependent.

Companion cells   Cells associated with the sieve-tubes in the phloem.

Companion crop   A crop sown with another crop and harvested separately. Small-grain cereal crops are often sown with forage crops (grasses or legumes) and harvested in the early summer, allowing the forage crop to continue to grow (e.g., oats sown as a companion crop with red clover).

Compensation point (light)   The light intensity at which the rates of photosynthesis and respiration are equal.

Complete flower   A flower that has pistils, stamens, petals, and sepals, all attached to a receptacle.

Compost   A mixture of organic residues and soil that has been piled, moistened, and allowed to decompose biologically. Mineral fertilizers are sometimes added.

Compound leaf   A leaf whose blade is divided into a number of distinct leaflets.

Cone   The woody, usually elongated seed-bearing organ of a conifer, consisting of a central stem, woody scales and bracts (often not visible), and seeds.

Contact herbicide   A chemical that kills plants on contact.

Controlled atmosphere (CA) storage   A process of fruit storage wherein the CO_2 level is raised (to 1 to 3 percent) and the O_2 level is lowered (to 2 to 3 percent).

Cork   An external, secondary tissue impermeable to water and gases produced by certain kinds of woody plants.

Cork cambium   The meristem from which cork develops.

Corm   A short, solid, vertical, enlarged underground stem in which food is stored; it contains undeveloped buds (leaf and flower). Examples are crocus, freesia, and gladiolus.

Corolla   The collective term for all petals of a flower.

Cortex   Primary tissue of a stem or root bounded externally by the epidermis and internally in the stem by the phloem and in the root by the pericycle.

Cotyledons   Leaflike structures at the first node of the seedling stem. In some dicots, cotyledons contain the stored food for the young plant not yet able to photosynthesize its own food. Often referred to as seed leaves.

Cover crop   A close-growing crop grown primarily for the purpose of protecting or improving soil between periods of regular crop production or between trees and vines in orchards and vineyards.

Crop residue   Portion of crop plants remaining after harvest.

Crop rotation   Growing crop plants in a different location in a systematic sequence to help control insects and diseases, improve the soil structure and fertility, and decrease erosion.

Cross-pollination   The transfer of pollen from a stamen to the stigma of a flower on another plant, except for clones where the plants must be two different clones.

Cross section   The surface exposed when a plant stem is cut horizontally and the majority of the cells are cut transversely.

Crown   The region at the base of the stem of cereals and forage species from which tillers or branches arise. In woody plants, the root-stem junction. In forestry, the top portions of the tree.

Crumb (soils)   A soft, porous, irregular, natural unit of structure from 1 to 5 mm in diameter.

Crust   A surface layer on soils, ranging in thickness from a few millimeters to a few centimeters. It is more compact, hard, and brittle when dry than the soil beneath it.

Cubing   Process of forming hay into high-density cubes to facilitate transportation, storage, and feeding.

Culm   Stem of grasses and bamboos; usually hollow except at the swollen nodes.

Cultivar (derived from "*culti*v*ated v*ariety")   International term denoting certain cultivated plants that are clearly distinguishable from others by any characteristic and that when reproduced (sexually or asexually) retain their distinguishing characters. In the United States *variety* is often considered synonymous with *cultivar.*

Cultivation   The growing or tending of crops.

Cure   To prepare crops for storage by drying. Dry onions, sweet potatoes, and hay crops are examples. Dehydration of fruits for storage is not considered curing.

Cuticle   An impermeable surface layer on the epidermis of plant organs.

Cutin   A clear or transparent waxy material on plant surfaces that tends to make the surface waterproof.

Cutting   A detached leaf, stem, or root that is encouraged to form new roots and shoots and develop into a new plant.

Cycocel®   *See* Chlormequat.

Cyme   A type of inflorescence which has a broad, more or less flat-topped determinate flower cluster, with the central flower opening first.

Cytochrome   A class of several electron-transport proteins serving as carriers in mitochondrial oxidation and in photosynthetic electron transport.

Cytokinesis   Division of cytoplasmic constituents at cell division.

Cytokinins   A group of plant growth hormones important in the regulation of nucleic acid and protein metabolism and in cell division, organ initiation, and delaying senescence.

Cytology   The study of cells and their components and of the relationship of cell structure to function.

Cytoplasm   The living material of the cell, exclusive of the nucleus, consisting of a complex protein matrix or gel. The part of the cell in which essential membranes and cellular organelles are found.

Daminozide (B-nine® or Alar®)   A plant growth retardant usually most effective as a spray. In some cases effective in enhancing flowering.

Damping off   A pathogenic disorder causing seedlings to die soon after seed germination.

Daylength   Number of effective hours of daylight in each 24-hour cycle.

Day-neutral plants   Those capable of flowering under either long or short daylengths.

DDT (Dichlorodiphenyltrichloroethane)   One of the earliest insecticides of the chlorinated hydrocarbon family. No longer used in some countries because of its persistence in the environment.

Deciduous   Refers to trees and shrubs that lose their leaves every fall. Distinguished from evergreens, which retain them throughout the year.

Decomposition   Degradation into simpler compounds; rotting or decaying.

Defoliant   A chemical or method of treatment that causes only the leaves of a plant to fall off or abscise.

Dehiscence   The splitting open at maturity of pods or capsules along definite lines or sutures.

Dehulled seed   Seed from which pods, glumes, or other outer covering have been removed, as sometimes with lespedeza and timothy. Also often ambiguously referred to as "hulled" seed.

Denitrification   Biological reduction of nitrate or nitrite to gaseous nitrogen or nitrogen oxides.

Deoxyribonucleic acid (DNA)   A molecule composed of repeating subunits of ribose (a sugar), phosphate, and the nitrogenous bases adenine, guanine, cytosine, and thymine. Genes, the fundamental units of inheritance on chromosomes, are sequences of DNA molecules.

Desalinization   Removal of salts from saline soil or water.

Desiccant   Substance used to accelerate drying of plant tissues.

Determinate (flowering)   The flowering of plant species uniformly within certain time limits, allowing most of the fruit to ripen about the same time.

Diatomaceous earth   A geologic deposit of fine, grayish, siliceous material composed chiefly of the remains of diatoms.

Dichogamy   Maturation of male or female flowers at different times, ensuring cross-pollination. Common in maple and walnuts.

Dicotyledonae (dicots)   The subclass of flowering plants that have two cotyledons.

Differentiation   Development from one cell to many cells, together with a modification of the new cells for the performance of particular functions.

Diffusion   The movement of molecules, and thus a substance, from a region of higher concentration of those molecules to a region of lower concentration.

Digestion   The breakdown of complex foods to simpler food materials, which are more easily utilized. Digestion requires energy.

Dihybrid cross   A cross between organisms differing in two characters.

Diluent   An essentially inert or nonreacting gas, liquid, or solid used to reduce the concentration of the active ingredient in a formulation.

Dioecious   Refers to individual plants having either staminate (male) or pistillate (female) flowers, but not both. Therefore, plants of both sexes must be grown near each other to provide pollen before fruits and seed can be produced. Examples are English holly, asparagus, ginkgo, and date palms.

Diploid (2n)   Refers to two sets of chromosomes. Germ cells have one set and are haploid; somatic cells have two sets and are diploid (except for polyploid plants).

Disease   Any change from the state of metabolism necessary for the normal development and functioning of any organism.

Disperse   (1) to break up compound particles, such as aggregates, into the individual component particles. (2) To distribute or suspend fine particles, such as clay, in or throughout a dispersion medium.

Diurnal   Recurring or repeated every day. Going through regular or routine changes daily.

Division   A propagation method in which underground stems or roots are cut into pieces and replanted.

Dolomite $CaCO_3$—$MgCO_3$; a native limestone source having a significant magnesium content.

Dominant   Referring to the gene (or the expression of the character it influences) that, when present in a hybrid with a contrasting gene, completely dominates in the development of the character. In peaches, for example, white fruit is dominant over yellow.

Dormancy   A general term denoting a lack of growth of seeds, buds, bulbs, or tubers due to unfavorable environmental conditions (external dormancy or quiescence) or to factors within the organ itself (internal dormancy or rest).

Double fertilization   The process of sexual fertilization in the angiosperms in which one nucleus from the male gametophyte fertilizes the egg nucleus to form the zygote and a second nucleus from the male gametophyte fertilizes two polar nuclei to form endosperm tissue.

Drip irrigation   A method of watering plants so that only soil in the plant's immediate vicinity is moistened. Water is supplied from a thin plastic tube at a low rate of flow. Sometimes called trickle irrigation.

Drupe   A simple, fleshy fruit derived from a single carpel, usually one-seeded, in which the exocarp is thin, the mesocarp fleshy, and the endocarp hard. Example: peach.

Dry land farming   The practice of crop production without irrigation.

Dry matter percentage   The percent of the total fresh plant material left after water is removed. The percentage is determined by weighing a sample of fresh plant material, oven-drying the sample, then reweighing the dried sample. Dry matter percentage equals dry weight divided by fresh weight times 100.

Earlywood (spring wood)   The less dense part of the growth ring. It is made up of cells with thinner walls, a greater radial diameter, and shorter length than those formed later in the year.

Ecology   The study of life in relation to its environment.

Ecosystem   A living community and all the factors in its nonliving environment.

Ecotype   Genetic variant within a species that is adapted to a particular environment yet remains interfertile with all other members of the species.

Edaphic   Pertaining to the influence of the soil on plant growth.

Electrons   Negatively charged particles that form a part of all atoms.

Element   A substance that cannot be divided or reduced by any known chemical means to a simpler substance; 92 natural elements are known.

Eluviation   The removal of soil material in suspension (or in solution) from a layer or layers of a soil. (Usually, the loss of materials in solution is described by the term *leaching*.)

Emasculate   To remove the anthers from a bud or flower before pollen is shed. Emasculation is a normal preliminary step in hybridization to prevent self-pollination.

Embryo   A miniature plant within a seed produced as a result of the union of a male and female gamete resulting in the development of a zygote.

Embryo sac   Typically, an eight-nucleate female gametophyte. The embryo sac arises from the megaspore by successive mitotic divisions.

Endemic.   Species native to a particular environment or locality.

Endocarp   Inner layer of the fruit wall (pericarp).

Endodermis   In roots, a single layer of cells at the inner edge of the cortex. The endodermis separates the cortical cells from cells of the pericycle.

Endogenous   Produced from within. Opposite of *exogenous.*

Endoplasmic reticulum   The lamellar or tubular system of the colorless cytoplasm in a cell.

Endosperm   The 3n tissue of angiospermous seeds that develops from sexual fusion of the two polar nuclei of the embryo sac and a male sperm cell. The endosperm provides nutrition for the developing embryo. A food storage tissue.

Energy (kinetic)   The capacity to do work. Examples are light, heat, chemical, electrical, or nuclear energy.

Enzyme   Any of many complex proteins produced in living cells, that, even in very low concentrations, promotes certain chemical reactions but does not enter into the reactions itself.

Epicotyl   The upper portion of the embryo axis or seedling, above the cotyledons and below the first true leaves.

Epidermis   The outer layer of cells on all parts of the primary plant body: stems, leaves, roots, flowers, fruits, and seeds. It is absent from the root cap and on apical meristems.

Epigeal germination   A type of seed germination in dicots in which the cotyledons rise above the soil surface. This occurs in beans, for example.

Epigyny   Arrangement of flower parts in which the ovary is embedded in the receptacle so that the other parts appear to arise from the top of the ovary.

Epinasty   A twisted or misshapened stem or leaf. Leaves exhibit downward curling of leaf blade due to more rapid growth on the upperside cell than the lower side.

Epiphyte   A plant that grows upon another plant yet is not parasitic.

Erosion   The wearing away of the surface soil by wind, moving water, or other means.

Ethylene   A gaseous growth hormone (C_2H_4) regulating various aspects of vegetative growth, fruit ripening, abscission of plant parts, and the senescence of flowers.

Ethephon (2-chloroethyl phosphonic acid)   A useful compound in horticultural applications which breaks down to release ethylene gas.

Etiolation   A condition involving lack of chlorophyll, increased stem elongation, and poor or absent leaf development. It occurs in plants growing under very low light intensity or complete darkness.

Evapotranspiration   The total loss of water by evaporation from the soil surface and by transpiration from plants, from a given area, and during a specified period of time.

Evergreen   Trees or shrubs that are never entirely leafless, as in pine or citrus.

Evolution   The development of a species, genus, or other larger group of plants or animals over a long time period.

Exchange capacity   The total ionic charge of the adsorption complex active in the adsorption of ions. Also called anion exchange capacity, cation exchange capacity, and base exchange capacity.

Exocarp   The outermost layer of the fruit wall (pericarp).

Exogenous Produced outside of, originating from, or because of external causes. Opposite of *endogenous*.

Explant Living tissue removed from its place in a body and placed in an artificial medium for tissue culture.

F$_1$ First filial generation in a cross between any two parents.

F$_2$ Second filial generation, obtained by crossing two members of the F$_1$ generation, or by self-pollinating plants of the F$_1$ generation.

Facultative Referring to an organism having the power to live under a variety of conditions; a facultative parasite is either parasitic or saprophytic.

Fallow Cropland left idle for one or more seasons for any number of reasons, such as to accumulate moisture, destroy weeds, and allow the decomposition of crop residue.

Family In plant taxonomy, a group of genera.

Fascicle A bundle of needle-leaves of gymnosperms such as the pines.

Fatty acid Organic compound of carbon, hydrogen, and oxygen that combines with glycerol to make a fat.

Fermentation An anaerobic chemical reaction in foods, such as the production of alcohol from sugar by yeasts.

Fertilization (floral) The union of an egg and a sperm (gametes) to form a zygote.

Fertilization (soil) Application to the soil of needed plant nutrients, such as nitrogen, phosphoric acid, potash, and others.

Fertilizer Any organic or inorganic material of natural or synthetic origin added to a soil to supply elements essential to the growth of plants.

Fibers Elongated, tapering, thick-walled strengthening cells in various parts of the plant.

Field capacity (field moisture capacity) The percentage of water remaining in a soil two or three days after having been saturated and after free drainage due to gravity has practically ceased.

Fixation (soil) The process in soil by which certain chemical elements essential for plant growth convert from a soluble or exchangeable form to a much less soluble or to a nonexchangeable form.

Fleshy fruit Any fruit formed from an ovary that has fleshy or pulpy (not dried) walls at maturity. Also, those fruits that include fleshy parts of the perianth, floral tube, or the receptacle.

Flocculate To clump together individual, tiny soil particles, especially fine clay, into small granules. Opposite of *disperse*.

Flooding A method of irrigation by which water is released from field ditches and allowed to spread over the land.

Flora A collective term for all the plant types that grow in a region.

Floral incompatibility A genetic condition in which certain normal male gametes are incapable of functioning in certain pistils.

Flower Floral leaves grouped together on a stem that, in the angiosperms, are adapted for sexual reproduction.

Fodder Coarse grasses such as corn and sorghum harvested with the seed and leaves and cured for animal feed.

Follicle A simple, dry dehiscent fruit, having one carpel and splitting along one suture. Examples are milkweeds and magnolia.

Food chain The path along which food energy is transferred within a natural plant and animal community (from producers to consumers to decomposers).

Foot-candle (ft-c) A standard measure of light (English system). The illumination of a standard light source of one candle, one foot away, falling at all points on a square foot surface. Photometric unit expressing the illuminance of 1 lumen per sq ft; 1 ft-c = 10.76 lux. *See* Lux.

Forage Vegetation used as feed for livestock, such as hay, pasture, and silage. The material is fed green or dehydrated.

Forcing A cultural manipulation used to hasten flowering or growing plants outside their natural season.

Fossil Any impression, natural or impregnated remains, or other trace of an animal or plant of past geological eras that has been preserved in the earth's crust.

Foundation seed Seed stocks increased from breeder seed, and handled to closely maintain the genetic identity and purity of a cultivar.

Friable (soil) Generally refers to a soil moisture consistency that crumbles when handled.

Fruit A mature ovary; in some plants other flower parts are commonly included as part of the fruit, e.g., the hypanthium of the apple flower surrounds the ovary.

Fumigation Control of insects, disease-causing organisms, weeds, or nematodes by gases applied in an enclosed area such as a greenhouse or under plastic laid on the soil.

Fungicide A pesticide chemical used to control plant diseases caused by fungi.

Fungus (plural, fungi) A thallus plant unable to photosynthesize its own food (exclusive of bacteria).

Furrow Small V-shaped ditch made for planting seed or for irrigating.

Furrow irrigation A method of irrigation by which the water is applied to row crops in ditches.

Gamete An haploid-generation male sperm cell or a female egg cell capable of developing into an embryo after fusion with a germ cell of the opposite sex.

Gametophyte In a seed plant the few-celled, haploid generations arising from a meiotic division and giving rise through mitosis to the male or female gametes.

Gene A group of base pairs in the DNA molecule in the chromosome that determines or conditions one or more hereditary characters.

Genetic code The sequence of nitrogen bases in a DNA molecule that codes for an amino acid or protein. In a broader sense, for example, the full sequence of events from the translation of chromosomal DNA to the final stage of the synthesis of an enzyme.

Genetic engineering The process of moving genetic material (DNA) or genes from their normal location and transferring them to another organism or the same organism in a different combination (genetic recombination) or to a different chromosome. The term *engineering* indicates human intervention.

Genetically modified organism Any organism that is the product of genetic recombination, which includes almost all organisms except those produced clonally or parthenogenically. Sometimes different species, such as plant and bacteria, can exchange genetic material; for example, wheat. Includes, but is not limited to, genetically engineered organisms. *See* Genetic engineering.

Genetics The science or study of inheritance.

Genotype The genetic makeup of a nucleus or of an individual.

Genus (plural, genera) A group of structurally or phylogenetically related species.

Geotropism Growth curvature in plants induced by gravity.

Germination (seed) Sequence of events in a viable seed starting with imbibition of water that leads to growth of the embryo and development of a seedling.

Germplasm The protoplasm of the sexual reproductive cells containing the units of heredity (chromosomes and genes).

Gibberellins A group of natural growth hormones whose most characteristic effect is to increase the elongation of cells.

Glacial till A product of glacial weathering in which rock particles varying in size from clay to boulders are deposited by the glacier on the land surface as it melts and recedes.

Glucose A simple sugar composed of 6 carbon, 12 hydrogen, and 6 oxygen atoms ($C_6H_{12}O_6$).

Graft To place a detached branch (scion) in close cambial contact with a rooted stem (rootstock) in such a manner that scion and rootstock unite to form a new plant.

Grain (caryopsis) A simple, dry, indehiscent fruit with ovary walls fused to the seed. The so-called seed of cereal or grain crops such as corn, wheat, barley, and oats is actually a fruit.

Grana Structures in the chloroplasts usually seen as stacks of parallel lamellae with the aid of an electron microscope.

Gravitational water Water that moves into, through, or out of the soil under the influence of gravity.

Green chop Forage that is chopped in the field while succulent and green and fed directly to livestock, made into silage, or dehydrated.

Green manure A crop that is plowed under while still green and growing to improve the soil.

Groundwater Water that fills all the unblocked pores of underlying material below the water table, which is the upper limit of saturation.

Group (taxonomy) A category of cultivated plants at the subspecies level that have the same botanical binomial but have characteristics different enough to warrant naming them for clarity.

Growing medium (soil mix) Soil or soil substitute prepared by combining such materials as peat moss, vermiculite, sand, or composted sawdust. Used for growing potted plants or germinating seeds.

Growth An irreversible increase in cell size and/or cell number. An increase in dry weight, regardless of cause.

Growth regulator A synthetic or natural compound that in low concentrations controls growth responses in plants.

Growth retardant A chemical that selectively interferes with normal hormonal promotion of plant growth, but without appreciable toxic effects.

Guard cells Specialized epidermal cells that contain chloroplasts and surround a stoma.

Gully erosion A process whereby water accumulates in narrow channels and, over relatively short periods, removes the soil from this narrow area to considerable depths.

Guttation Exudation of water in liquid form from plants.

Gymnosperm A seed plant with seeds not enclosed by a megasporophyll or pistil.

Gynoecium The female part of a flower or pistil formed by one or more carpels and composed of the stigma, style, and ovary.

Haploid Having only one complete set of chromosomes; referring to an individual or generation containing such a single set of chromosomes per cell.

Hardening off Adapting plants to outdoor conditions by withholding water, lowering the temperature, or nutrient supply. This conditions plants for survival when transplanted outdoors.

Hardpan A hardened soil layer, in the lower A horizon or in the upper B horizon, caused by the cementing of soil particles.

Hardy plants Plants adapted to cold temperatures or other adverse climatic conditions of an area. *Half-hardy* indicates some plants may be able to take local conditions with a certain amount of protection.

Hay Herbage of forage plants, including seed of grasses and legumes, that is harvested and dried for animal feed.

Head (botanical) A type of inflorescence, typical of the composite family, in which the individual flowers are grouped closely together on a receptacle. Example: sunflower.

Heaving The partial lifting of plants out of the ground, frequently breaking their roots, as a result of freezing and thawing of the surface soil during the winter.

Heavy soil A soil with a high content of the fine separates, particularly clay, or one with a high tractor power requirement and hence difficult to cultivate.

Heeling in Temporary storing of bare-rooted trees and shrubs by placing the roots in a trench and covering with soil or sawdust.

Helix A spiral form; a term often used in reference to the double spiral of the DNA molecule.

Herbaceous Refers to plants that do not normally develop woody tissues.

Herbarium A collection of dried and pressed plant specimens cataloged for easy reference.

Herbicide Any chemical used to kill plants; an herbicide may work against a narrow or wide range of plant species.

Herbivore Animal that subsists principally or entirely on plants or plant products.

Heredity The transmission of morphological and physiological characteristics from parents to their offspring.

Hermaphrodite flower A flower having both stamens (male) and pistils (female).

Heterosis (hybrid vigor) (1) The increased vigor, growth, size, yield, or function of a hybrid progeny over the parents that results from crossing genetically unlike organisms. (2) The increase in vigor or growth of a hybrid progeny in relation to the average of the parents.

Heterotrophic Capable of deriving energy for life processes only from the decomposition of organic compounds and incapable of using inorganic compounds as sole sources of energy or for organic synthesis. Contrast with *autotrophic*.

Heterozygous Having different genes of a Mendelian pair present in the same cell or organism; for instance, a tall pea plant with genes for both tallness (T) and dwarfness (t).

Hill Raising the soil in a slight mound for planting, or setting plants some distance apart.

Hilum The scar on a dicot seed, such as a bean seed, where it was attached to the fruit.

Histology The science that deals with the microscopic structure of plant or animal tissues.

Homologous chromosomes The two members of a chromosome pair.

Homozygous Having similar genes of a Mendelian pair present in the same cell or organism; for instance, a dwarf pea plant with genes for dwarfness (tt) only.

Horizon, soil   A layer of soil, approximately parallel to the soil surface, with distinct characteristics produced by soil-forming processes.

Hormone   A chemical substance that is produced in one part of a plant and used in minute quantities to induce a growth response in another part. For example, auxins are one type of hormone.

Hotbed   A bed of soil enclosed in a low glass or transparent plastic frame and heated with fermenting manure, electric cables, or steam pipes. Used to germinate seeds, root cuttings, and grow other plants for transplanting outside.

Hot caps   Waxpaper cones, paper sacks, or cardboard boxes with bottoms removed and placed over individual plants in spring for frost and wind protection.

Humidity, relative   The ratio of the weight of water vapor in a given quantity of air to the total weight of water vapor that quantity of air is capable of holding at a given temperature, expressed as a percentage.

Humus   The more or less stable fraction of the soil organic matter remaining after the major portion of plant and animal residues have decomposed.

Hybrid   The offspring of two plants or animals differing in one or more Mendelian characters.

Hybridization   (1) The crossing of individuals of unlike genetic constitution. (2) A method of breeding new cultivars that uses crossing to obtain genetic recombinations.

Hydrologic cycle   The movement of water from the atmosphere to the earth and its return to the atmosphere.

Hydrolysis   The chemical reaction in which water participates as a reactant and not as a solvent. Usually, the splitting of a molecule to form smaller molecules that incorporate hydrogen and hydroxyl ions, derived from water, in their structures.

Hydroponics   Growing plants in aerated water containing all the essential mineral nutrients rather than soil. Also called soilless gardening.

Hypogeal germination   In dicots, a type of seed germination in which the cotyledons remain below the soil surface; for example, peas.

Hypothesis   A proposition or supposition provisionally adopted to explain certain facts. Once proven by ultimate scientific investigation, it becomes a theory or a law.

Igneous rock   Rock formed from the cooling and solidification of magma that has not been changed appreciably since its formation.

Illuminance   The luminous flux (brightness of light) per unit area on an intercepting surface at any given point.

Imbibition   The absorption of liquids or vapors into the ultramicroscopic spaces in materials like cellulose.

Immune   Free from attack by a given pathogen; not subject to the disease.

Imperfect flower   A flower lacking either stamens or pistils.

Impervious   Resistant to penetration by fluids or by roots.

Inbred line   A pure line usually originating by self-pollination and selection.

Incompatibility (floral)   Failure to obtain fertilization and seed formation after pollination, usually because of slow pollen tube growth in the stylar tissue.

Incompatibility (graft)   Failure of two graft components (stock and scion) to unite and develop into a successfully growing plant.

Incomplete flower   A flower that is missing one or more of the following parts: sepals, petals, stamens, or pistils.

Indehiscent fruit   A fruit that does not split open naturally at maturity.

Indeterminate   Pertaining to growth of plants, the flowers of which are borne on lateral branches, the central stem continuing vegetative growth, with blooming continued for a long period. Examples are alfalfa and fuchsia.

Indigenous   Produced or living naturally in a specific environment.

Indoleacetic acid (IAA)   A natural or synthetic plant growth regulator; an auxin.

Inferior ovary   An ovary that is imbedded in the receptacle, or an ovary whose base lies below the point of attachment of the perianth.

Infiltration rate   The maximum velocity at which water can enter the soil under specified conditions, including the presence of an excess of water.

Inflorescence   An axis bearing flowers, or a flower cluster (e.g., umbel, spike, panicle).

Inheritance   The acquisition of characters or qualities by transmission from parent to offspring.

Inoculate   (1) To induce a disease in a living organism by introducing a pathogen. (2) To treat seeds of leguminous plants with bacteria to induce nitrogen-fixation in the roots.

Inorganic compound   A chemical compound that generally is not derived from life processes; compounds that do not contain carbon.

Insecticide   Any chemical (organic or inorganic) substance that kills insects.

Integrated pest management (IPM)   Using multiple methods such as cultural, biological, and chemical to control pests in a crop.

Integuments   The tissues covering or surrounding the ovule, usually consisting of an inner and outer layer; they subsequently become the seedcoats of the mature ovule.

Intercalary growth   A pattern of stem elongation typical of grasses. Elongation proceeds from the lower internodes to the upper internodes through the differentiation of meristematic tissue at the base of each internode.

Internode   The region of a stem between two successive nodes.

Interspecific cross   A cross, natural or intentional, between two species.

In vitro   Latin for "in glass." Living in test tubes; outside the organism or in an artificial environment.

In vivo   Latin for "in living." In the living organism.

Ions   Atoms, groups of atoms, or compounds that are electrically charged as a result of the loss of electrons (cations) or the gain of electrons (anions).

IPM   Integrated Pest Management.

Irradiation   In genetics and plant breeding, exposing seed, pollen, or other plant parts to X-rays or other short wave-length (gamma) radiations to increase mutation rates.

Irrigation   Applying water to the soil, other than by natural rainfall.

Isolation   The prevention of crossing among plant populations because of distance or geographic barriers (geographic isolation).

Lamina   Blade or expanded part of a leaf.

Lateral bud   A bud that grows out from the leaf axil on the side of a stem.

Latewood (summer wood)   The denser part of the growth ring produced late in the season. It is made up of xylem cells with thicker walls, smaller radial diameter, and are generally longer than those formed earlier in the growing season.

Latex   A milky secretion produced by various kinds of plants.

Lathhouse   An open structure built of wood lath or plastic screen for protecting plants from excessive sunlight or frost.

Layby application   The procedure when a material is applied with or after the last cultivation of a crop.

Layering   A form of vegetative propagation in which an intact branch develops roots as the result of contact with the soil or another rooting medium.

Leach   To remove soluble materials from soil or plant tissue with water.

Leaf mold   Partially decayed leaves useful for improving soil structure and fertility.

Leggy   Weak-stemmed and spindly plants with sparse foliage caused by too much heat, shade, crowding, or overfertilization. *See* Etiolation.

Legume   Plant member of the family LEGUMINOSAE, with the characteristic capability to fix atmospheric nitrogen in nodules on its roots if inoculated with proper bacteria.

Lenticel   An opening made up of loosely arranged cells in the periderm that permits passage of gases.

Light flux   The total light energy used by plants in the photosynthetically active region (PAR). The quantity of light (intensity) $\times$ duration of light (daylength).

Light reactions   The reactions of photosynthesis in which light energy is required: the photo (light) activation or excitement of electrons in the chlorophyll molecule, transfer of the electrons, photolysis of water, and associated reactions.

Light soil   A coarse-textured sandy soil; hence, easy to till.

Lignin   An organic substance found in secondary cell walls that gives stems strength and hardness. Wood is composed of lignified xylem cells (about 15 to 30 percent by weight).

Lime (agricultural)   A material containing the carbonates, oxides, and/or hydroxides of calcium and/or magnesium. It is used to increase soil pH and to neutralize soil acidity.

Line   Group of individuals from a common ancestry. When propagated by seed, it retains its characteristics. A type of cultivar.

Lipid   Any of a group of fats or fatlike compounds insoluble in water but soluble in certain other solvents.

Loam   A textural class for soil with prescribed amounts of sand, silt, and clay.

Locus   The fixed position of location of a gene on or in a chromosome.

Lodging   A condition in which plants are caused to bend for various reasons at or near the soil surface and fall more or less flat on the ground. Most frequently observed in cereals.

Lucerne   A name for alfalfa used in Europe, Australia, and other regions.

Lux (lx)   A standard measure of light (metric system). The illumination impinging upon a surface $1m^2$, each point of which is at a distance 1 m away from a standard light source of one candle. Photometric unit expressing the illuminance of 1 lumen per square meter; 1 lux = 0.093 ft-c. *See* Foot-candle.

Macronutrient   A chemical element, like nitrogen, phosphorus, and potassium, necessary in large amounts (usually greater than 1 ppm) for the growth of plants.

Male sterility   A condition in some plants in which pollen either is not formed or does not function normally, even though the stamens may appear normal.

Malthusian theory   Developed by T. Malthus, a British economist (1766–1834), who speculated that the world's population would grow faster than food productivity and thus world starvation would eventually occur.

Meiosis   Two successive nuclear divisions, in the course of which the diploid chromosome number is reduced to the haploid and genetic segregation occurs.

Mendel's laws   A set of three laws formulated by Gregor Mendel; each is generally true but there are numerous exceptions. The laws are (1) characters exhibit alternative inheritance, being either dominant or recessive; (2) each gamete receives one member of each pair of factors present in a mature individual; and (3) reproductive cells combine at random.

Meristem   Undifferentiated tissue whose cells can divide and differentiate to form specialized tissues, such as xylem or phloem.

Mesocarp   Middle layer of the fruit wall (pericarp).

Mesophyll   Parenchyma tissue in leaves found between the two epidermal layers.

Mesozoic   A geologic era beginning 225 million years ago and ending 65 million years ago.

Messenger RNA   Ribonucleic acid produced in the nucleus and capable of carrying parts of the message coded in chromosomal DNA. Messenger RNA moves from the nucleus to the ribosomes, where protein is synthesized in the cytoplasm.

Metabolism   The overall physiological activities of an organism.

Metamorphic rock   A rock that has been greatly altered from its previous condition through the combined action of heat and pressure.

Microclimate   Atmospheric environmental conditions in the immediate vicinity of the plant, including interchanges of energy, gases, and water between atmosphere and soil.

Micronutrient   A chemical element necessary in extremely small amounts (less than 1 ppm) for the growth of plants. Examples are boron, chlorine, copper, iron, manganese, and zinc.

Micropyle   An opening leading from the outer surface of the ovule between the edges of the two integuments inward to the surface of the nucellus.

Microspore   One of the four haploid spores that originate from the meiotic division of the microspore mother cell in the anther of the flower and that gives rise to the pollen grain.

Microspore mother cell   Diploid cell in the anther that gives rise, through meiosis, to four haploid microspores.

Middle lamella   The pectic layer lying between the primary cell walls of adjoining cells.

Millimho (mmho)   A measure of electrical conductivity, 1 mmho = 0.001 mho. The mho is the reciprocal of an ohm.

Mineral soil   A soil whose makeup and physical properties are largely those of mineral matter.

Minor element   *See* Micronutrient.

Mist propagation   Applying water in mist form to leafy cuttings in the rooting stage to reduce transpiration.

Mitochondria (singular, mitochondrion)   A minute particle in the cytoplasm associated with intracellular respiration.

Mitosis   A form of nuclear cell division in which chromosomes duplicate and divide to yield two nuclei that are identical with the original nucleus. Usually mitosis includes cellular division (cytokinesis).

Mixed bud A bud containing both rudimentary flowers and vegetative shoots.

Moisture, dry basis A basis for representing moisture content of a product. It is calculated from the net weight of water lost by drying, divided by dried weight of the material and the answer multiplied by 100 = percent.

Mol (light) Used in plant science when waveband stated is 400 to 700 nm (PAR); 1 mol = 6.022×10^{23} photons, or 1 micromol = 6.022×10^{17} photons.

Mole (M) Amount of a substance that has a weight in grams numerically equal to the molecular weight of the substance. Also called gram-molecular weight.

Molecular biology A field of biology concerned with the interaction of biochemistry and genetics in the life of an organism.

Molecule A unit of matter; the smallest portion of an element or a compound that retains chemical identity with the substance in mass. A molecule usually consists of the union of two or more atoms, and some organic molecules contain hundreds of atoms.

Monocotyledonae (monocots) The subclass of flowering plants that have only a single cotyledon at the first node of the primary stem.

Monoecious A plant with separate male and female flowers on the same plant, such as corn and walnuts.

Morphology (plant) The study or science of the form, structure, and development of plants.

Muck (soil) Highly decomposed organic material in which the original plant parts are not recognizable.

Mulch Any material such as straw, sawdust, leaves, plastic film, and loose soil that is spread upon the surface of the soil to protect the soil and plant roots from the effects of rain, soil crusting, freezing, or evaporation.

Multiple fruit A cluster of matured fused ovaries produced by separate flowers; for example, pineapple.

Mutation A sudden, heritable change appearing in an individual as the result of a change in genes or chromosomes.

Mycelium The mass of hyphae forming the body of a fungus.

Mycology A branch of botany dealing with the study of fungi.

Mycorrhiza The association, usually symbiotic, of fungi with the roots of some seed plants.

Nanometer (nm) A unit of length equal to one millionth (10^{-6}) of a millimeter or one millimicron; 1 nm equals 10 angstrom units.

Natural selection Environmental effects in channeling the genetic variation of organisms along certain pathways.

Naturalized plant A plant introduced from one environment into another in which the plant has become established and more or less adapted to a given region by growing there for many generations.

Necrosis Death associated with discoloration and dehydration of all or some parts of plant organs.

Nematodes Unsegmented roundworms abundant in many soils; important because many species of them attack plants or animals.

Neutral soil A soil in which the surface layer is neither acid nor alkaline in reaction.

Night-break lighting Low intensity lighting used in the middle of the night to change the photoperiod.

Nitrification The conversion of ammonium ions into nitrates through the activities of certain bacteria.

Nitrogen assimilation The incorporation of nitrogen into organic cell substances by living organisms.

Nitrogen fixation The conversion of atmospheric nitrogen (N_2) into oxidized forms that can be assimilated by plants. Certain blue-green algae and bacteria are capable of biochemically fixing nitrogen.

Nitrogenous base A nitrogen-containing compound found in DNA and RNA that, in sequence, specifies precise genetic information. The nitrogenous bases in DNA are adenine, thymine, cytosine, and guanine. In RNA, they are adenine, uracil, cytosine, and guanine.

Nodes Enlarged regions of stems that are generally solid where leaves are attached and buds are located. Stems have nodes but roots do not.

No-till Also known as stubble culture. A cultural system most often used with annual crops in which the new crop is seeded or planted directly in a field on which the preceding crop plants were cut down or destroyed by a nonselective herbicide rather than being removed or incorporated into the soil as is common in preparing a seed bed.

Nucellus A tissue originally making up the major part of the young ovule, in which the embryo sac develops.

Nucleic acid An acid found in all nuclei; all known nucleic acids fall into two classes, DNA and RNA.

Nucleus A dense body in the cytoplasm essential for cellular development and reproduction.

Nut A dry, indehiscent, single-seeded fruit with a hard, woody pericarp (shell), such as the walnut and pecan.

Nutrient, plant Element essential to plant growth used in the elaboration of food and tissue.

Obligate parasite An organism that must live as a parasite and cannot otherwise survive.

Obligate saprophyte An organism obliged to live only on nonliving animal or plant tissue.

Oedema Intumescence or blister formation on leaves.

Opposite An arrangement of leaves or buds on a stem. They occur in pairs on opposite sides of a single node.

Organ A part of an animal or plant body adapted by its structure for a particular function.

Organelle A specialized region in a cell, such as mitochondria, that is bound by a membrane.

Organic In chemistry, the carbon compounds, many of which are associated with living organisms.

Organic soil A soil that contains a high percentage (greater than 20 percent) of organic matter.

Osmosis The diffusion of fluids through a semipermeable or selectively permeable membrane.

Ovary The basal, generally enlarged part of the pistil in which seeds are formed. The ovary, at maturity, is a fruit. It is a characteristic organ of angiospermous plants.

Ovule A rudimentary seed, containing, before fertilization, the embryo sac, including an egg cell, all being enclosed in the nucellus and one or two integuments.

Oxidation-reduction reaction A chemical reaction in which one substance is oxidized (loses electrons, or loses hydrogen ions and their associated electrons, or combines with oxygen) and a second substance is reduced (gains electrons, or gains hydrogen ions and their associated electron, or loses oxygen).

Oxidative respiration   The chemical decomposition of foods (glucose, fats, and proteins) requiring oxygen as a terminal electron acceptor and yielding carbon dioxide, water, and energy. The energy is commonly stored in ATP.

Paclobutrazol (Bonzi®)   A plant growth retardant that inhibits gibberellic acid (GA) synthesis, which is used as a spray or soil drench.

Palatability   Term used to describe how agreeable or attractive feed stuff is to animals or how readily they consume it.

Paleozoic   A geologic era beginning about 570 million years ago and ending about 225 million years ago.

Palisade parenchyma   The cell layer in leaves immediately below the upper epidermis; packed with chloroplasts. Found in dicots, but not in monocots.

Pallet   Rectangular or square platform, usually wooden, designed for ease of mechanical handling and transportation of material (containers) placed on the platform.

Palmate   Arrangement of leaflets of a compound leaf or of the veins in a leaf. Characterized by subunits arising from a common point much as fingers arise from the palm of the hand.

Panicle   An inflorescence, common in the grass family, that has a branched central axis. An example is oats.

PAR   Abbreviation for photosynthetically active radiation; i.e., in the 400–700 nanometer (nm) spectral waveband.

Parasite   An organism obtaining its nutrients from the living body of another plant or animal.

Parenchyma   A tissue composed of thin-walled, loosely packed, unspecialized cells.

Parthenocarpy   Fruit development without sexual fertilization. Such fruits are seedless. Examples are the 'Navel' orange and some fig cultivars.

Parthenogenesis   Development of an egg into an embryo without fertilization.

Particle size (soil)   The effective diameter of a particle measured by sedimentation, sieving, or micrometric methods.

Parts per million (ppm)   Weight units of any given substance per one million equivalent weight units; the weight units of solute per million weight units of solution (i.e., 1 ppm = 1 mg/1).

Pasteurized soil   A soil that has been heated to 82°C (180°F) for at least 30 minutes to kill most of the pathogenic organisms without affecting the saprophytic soil flora.

Pasture   Area of domesticated forages, usually improved, on which animals are grazed.

Pathogen   An organism that causes disease.

Pathology   The study of diseases, their effects on plants or animals, and their treatment.

Pearl   The process of grinding off the hull, bran, aleurone, and germ of barley or rice to yield a pellet of endosperm.

Peat   Any unconsolidated soil mass of semicarbonized vegetable tissue formed by partial decomposition in water. An example is sphagnum peat moss.

Pectin   Polysaccharide from the middle lamella of the plant cell wall; jelly-forming substance found in fruit.

Ped   A unit of soil structure such as an aggregate, crumb, prism, block, or granule formed by natural processes.

Pedicel   Individual flower stalk of an inflorescence.

Pedigree   A record of ancestry.

Peduncle   Flower stalk that is borne singly; or the main stem of an inflorescence.

Percolation, soil water   The downward movement of water through soil.

Perennial   A plant that grows more or less indefinitely from year to year and usually produces seed each year.

Perfect flower   Having both stamens and pistils; a hermaphroditic flower.

Perianth   The petals and sepals of a flower, collectively.

Pericarp   The fruit wall, which develops from the ovary wall.

Pericycle   The layer of cells immediately inside the endodermis. Branch roots arise from the pericycle.

Periderm   A corky layer formed by the cork cambium at the surface of organs that are undergoing secondary growth.

Permanent wilting percentage.   *See* Wilting point.

Permeable   Referring to a membrane, cell, or cell system through which substances may diffuse.

Petal   Part of a flower, often brightly colored.

Petiole   The stalk that attaches a leaf blade to a stem.

pH (soil)   The negative logarithm of the hydrogen-ion activity of a soil.

pH (solution)   A measure of acidity or alkalinity, expressed as the negative logarithm (base 10) of the hydrogen-ion concentration. pH 7 is neutral. Values less than this indicate acidity; higher values indicate alkalinity.

Phellogen   Cork cambium, a cambium layer giving rise externally to cork and, in some plants, internally to phelloderm.

Phenology   The study of the timing of periodic phenomena such as flowering, growth initiation, or growth cessation, especially as related to seasonal changes in temperature or photoperiod.

Phenotype   The external physical appearance of an organism.

Pheromone   Any of a class of hormonal substances secreted by an individual and stimulating a physiological or behavioral response from an individual of the same species.

Phloem   A tissue through which nutritive and other materials are translocated through the plant. The phloem consists of sieve tube cells, companion cells, phloem parenchyma, and fibers.

Photomorphogenesis   Plant shape determined by light quality particularly through relative amounts of red, far-red, and blue light.

Photoperiod   That length of day or period of daily illumination required for the normal growth and sexual reproduction of some plants.

Photoperiodism   Response of a plant to the relative lengths of day and night (light and dark), particularly in respect to flower initiation and bulbing.

Photophosphorylation   The production of ATP by the addition of a phosphate group to ADP using the energy of light-excited electrons produced in the light reactions of photosynthesis. Photo = light, phosphorylation = adding phosphorus.

Photosynthesis   The process in green plants of converting water and carbon dioxide into sugar with light energy; accompanied by the production of oxygen.

Photosynthetic photon flux (PPF)   Radiometric irradiance per unit area pcr unit time at specific wavelengths (400–700 nm waveband for plants). Units micromols $m^{-2}s^{-1}$.

Phototropism   A change in the manner of growth of a plant in response to nonuniform illumination. Usually, the response, which is auxin-regulated, is a bending toward the strongest light.

Phylum   A primary division of the animal and plant kingdom.

Physiology, plant   The science of the functions and activities of living plants.

Phytochrome   A reversible protein pigment occurring in the cytoplasm of green plants. It is associated with the absorption of light that affects growth, development, and differentiation of a plant, independent of photosynthesis (e.g., in the photoperiodic response).

Piggyback   Intermodal system for carrying full trailer loads of product on a railroad flat car; also known as TOFC (trailer on flat car). The trailer is usually moved to and from the rail car by tractor truck.

Pigments   Molecules that are colored by the light they absorb. Some plant pigments are water soluble and are found mainly in the cell vacuole.

Pinching   The removal of the terminal bud or apical meristematic growth to stimulate branching.

Pistil   The seed-bearing organ in the flower, composed of the ovary, the style, and the stigma.

Pistillate flower   A female flower having pistils but no stamens.

Pith   A region in the center of some stems and roots consisting of loosely packed, thin-walled parenchyma cells.

Placenta (plural, placentae)   The tissue within the ovary to which the ovules are attached.

Plasmolysis   The separation of the cytoplasm from the cell wall because of the removal of water from the protoplast.

Plastids   The cellular organelles in which carbohydrate metabolism is localized.

Plow layer   The surface soil layer ordinarily moved in tillage.

Plow pan   A hard layer of soil formed by continual plowing at the same depth.

Plow-plant   The practice of plowing and planting a crop in one operation, with no additional seedbed preparation.

Plumule   The first bud of an embryo or that portion of the young shoot above the cotyledons.

Polar nuclei   Two centrally located nuclei in the embryo sac that unite with a second sperm cell in a triple fusion. In certain seeds the product of this fusion develops into the endosperm.

Polar transport   The directed movement within plants of compounds (usually hormones) mostly in one direction; polar transport overcomes the tendency for diffusion in all directions.

Pollen   The almost microscopic, yellow bodies that are borne within the anthers of flowers and contain the male generative (sex) cells.

Pollen mother cell   A 2n cell that divides twice (once by meiosis and once by mitosis) to form a tetrad of four pollen grains.

Pollen tube   A tubelike structure developed by the tube nucleus in the microspore that helps guide the sperm through the stigma and style of a flower to the embryo sac.

Pollination   The transfer of pollen from a stamen (or staminate cone) to a stigma (or ovulate cone).

Polyembryony   The presence of more than one embryo in a developing seed.

Polyploidy   A condition in which a plant has somatic (non-sexual) cells with more than 2n chromosomes per nucleus.

Polysaccharides   Long-chain molecules composed of units of a sugar; starch and cellulose are examples.

Pome   A simple fleshy fruit, the outer portion of which is formed by floral parts that surround the ovary (i.e., apple and pear fruits).

Pore size distribution   The volume of the various sizes of pores in a soil. It is expressed as a percentage of the bulk volume (soil plus pore space).

Porosity   That percentage of the total bulk volume of a soil not occupied by the solid particles.

Postemergence spray   A pesticide or herbicide that is applied after the crop plants have emerged from the soil.

Potting mixture (soil mix)   Combination of various ingredients such as soil, peat, sand, perlite, or vermiculite designed for starting seeds or growing plants in containers.

Pre-emergence spray   A pesticide or herbicide that is applied after planting, but before the crop plants emerge from the soil.

Primary tissue   A tissue that has differentiated from a primary meristem.

Primordium   An organ in its earliest stage of development, such as leaf primordium.

Profile (soil)   A vertical section of the soil through all its horizons and extending into the parent material.

Protein   Any of a group of nitrogen-containing compounds that yield amino acids on hydrolysis and have high molecular weights. They are essential parts of living matter and are one of the essential food substances of animals.

Proterozoic   The earliest geologic era, beginning about 4.5 to 5 billion years ago and ending 570 million years ago; also called Precambrian era.

Protoplasm   The essential, complex living substance of cells on which all vital functions of nutrition, secretion, growth, and reproduction depend.

Protoplast   The organized living unit of a single cell.

Provenance   The natural origin of a tree or group of trees. In forestry, the term is considered synonymous with geographic origin.

Pruning   A method of directing or controlling plant growth and form by removing certain portions of the plant.

Pure line   Plants in which all members have descended by self-fertilization from a single homozygous individual.

Raceme   An inflorescence in which flowers on pedicels are borne on a single, unbranched main axis.

Radial face   The wood surface exposed when a stem is cut along a radius from pith to bark, and the cut parallels the long axis of the majority of the cells.

Radicle   The part of the embryonic axis that becomes the primary root. The first part of the embryo to start growth during seed germination.

Range   Land and native vegetation that is predominantly grasses, grasslike plants, or shrubs suitable for grazing by animals.

Range management   Producing maximum sustained use of range forage without detriment to other resources or uses of land.

Raphe   Ridge on seeds, formed by the stalk of the ovule, in those seeds in which the funiculus is sharply bent at the base of the ovule.

Ray   A narrow group of cells, usually parenchyma, extending radially in the wood and bark.

Reaction (soil)   *See* pH (soil).

Receptacle   The enlarged tip of a stem on which a flower is borne.

Recessive   The condition of a gene such that it does not express itself in the presence of the contrasting (dominant) gene.

Recombination   The mixing of genotypes that results from sexual reproduction.

Reduction division   A nuclear division in which the chromosomes are reduced from the diploid to the haploid number.

Registered seed   The progeny of foundation or registered seed produced and handled so as to maintain satisfactory genetic identity and purity, and approved and certified by an official certifying agency. Registered seed is normally grown for the production of certified seed.

Replication   In cell physiology, the production of a second molecule of DNA exactly like the first molecule.

Reproduction, sexual   Development of new plants by seeds (except in apomixis).

Reproduction, vegetative (vegetative propagation)   Reproduction by other than sexually produced seed. Includes grafting, cuttings, layering, and so forth, as well as apomixis.

Respiration   The oxidation of food by plants and animals to yield energy for cellular activities.

Rest period   An endogenous physiological condition of viable seeds, buds, or bulbs that prevents growth even in the presence of otherwise favorable environmental conditions. This is referred to by some seed physiologists as *dormancy.*

Rhizobium   Genus of bacteria that live symbiotically in the roots of legumes and fix nitrogen that is used by plants.

Rhizome   An underground stem, usually horizontal and often elongated; distinguished from a root by the presence of nodes and internodes. Capable of producing new shoots.

Ribonucleic acid, RNA   A single-strand acid, formed on a DNA template, found in the protoplasm, and controlling cellular chemical activities. Whereas DNA transmits genetic information from one cell generation to the next, ribonucleic acid is an intermediate chemical translating genetic information into action.

Ribosome   A protoplasmic granule containing ribonucleic acid (RNA) and believed to be the site of protein synthesis.

Ripening   Chemical and physical changes in a fruit that follow maturation.

Rock   The material that forms the essential part of the earth's solid crust, including loose incoherent masses such as sand and gravel, as well as solid masses of granite, limestone, and others.

Root   The descending axis of a plant, usually below ground, serving to anchor the plant and absorb and conduct water and mineral nutrients.

Root cap   A mass of hard cells covering the tip of a root and protecting it from mechanical injury.

Root hair   An absorptive unicellular protuberance of the epidermal cells of the root.

Root pressure   Pressure developed in the root as the result of osmosis and causing bleeding in stem wounds of some plants.

Rooting media   Materials such as peat, sand, perlite, or vermiculite in which the basal ends of cuttings are placed vertically during the development of roots.

Rootstock (understock)   The trunk or root material to which buds or scions are inserted in grafting.

Roughage   Plant materials that are relatively high in crude fiber and low in digestible nutrients, such as straw.

Ruminant   Cud-chewing mammals such as cattle, sheep, goats, and deer, characteristically having a stomach divided into four compartments.

Runner   *See* Stolon.

Saline soil   A nonsodic soil containing sufficient soluble salts to impair plant growth.

Samara   A dry, indehiscent, simple fruit that has winglike appendages on both sides of the ovary. These appendages help carry the wind-borne fruit.

Sand   A soil particle between 0.05 and 2.0 mm in diameter.

Saprophyte   An organism deriving its nutrients from the dead body or the nonliving products of another plant or animal.

Savanna   Grassland having scattered trees, either as individuals or clumps; often a transitional type between true grassland and forest.

Scarify   To scratch, chip, or nick the seed coverings of certain species to enhance the passage of water and gases as an aid to seed germination.

Scientific method   An approach to a problem that consists of stating the problem, establishing one or more hypotheses as solutions to the problem, testing these hypotheses by experimentation or observation, and accepting or rejecting the hypotheses.

Scion   A small shoot that is inserted by grafting into a rootstock.

Scion-stock interaction   The effect of a rootstock on a scion (and vice versa) in which a scion on one kind of rootstock performs differently than it would on its own roots or on a different rootstock.

Sclerenchyma   Supporting or protective tissue in which the cells have hard lignified walls.

Scutellum   The rudimentary leaflike structure at the first node of the (embryonic) stem (culm) of a grass plant. The single cotyledon of a monocotyledonous seedling.

Secondary phloem   Phloem cells formed by activity of the vascular cambium. Secondary phloem is found in biennials and perennials, but usually not in annuals.

Secondary xylem   Xylem cells formed by activity of the vascular cambium. The development of the secondary xylem accounts for the so-called annual rings seen in most trees.

Sedimentary rock   Rock formed from material originally deposited as a sediment, then physically or chemically changed by compression and hardening while buried in the earth's crust.

Seed   The mature ovule of a flowering plant containing an embryo, an endosperm (sometimes), and a seed coat.

Seed (breeder)   Seed (or, sometimes, vegetative propagating material) directly controlled by the originator (or the sponsoring plant breeder or institution) that provides the source for the initial increases of foundation seed.

Seed (certified)   Progeny of foundation, registered, or other certified seed that is so handled as to maintain satisfactory genetic identity and/or purity and that has been approved and certified by a certifying agency.

Seed (foundation)   Seed stocks so handled as to maintain specific genetic identity and purity, such as may be designated by an agricultural experiment station. Foundation seed is the source of certified seed, either directly or through registered seed.

Seed (registered)   Progeny of foundation or other registered seed so handled as to maintain satisfactory genetic identity and purity; approved and certified by a certifying agency.

Seed potatoes   Pieces of potato tubers or whole tubers that are planted to produce new plants and subsequent commercial crops.

Seedbed   Soil that has been prepared for planting seeds or transplants.

Self-fertile   Capable of fertilization and producing viable seed after self-pollination.

Self-incompatibility Inability to produce viable seed following self-pollination. The inability is sometimes due to a pollen-borne gene that prevents pollen tube growth on a stigma with the same gene.

Self-pollination Transfer of pollen from the stamens to the stigma of either the same flower, other flowers on the same plant, or flowers on other plants of the same clone.

Self-sterility Failure to complete fertilization and obtain viable seed after self-pollination.

Semiarid Climate in which evaporation exceeds precipitation, a transition zone between a true desert and a humid climate. Usually annual precipitation is between 250 and 500 mm (10 to 20 in.).

Senescence A physiological aging process in which tissues in an organism deteriorate and finally die.

Sepals The outermost series of floral parts; usually green, leaflike structures at the base of a flower; collectively, they form the calyx.

Separation A type of propagation based upon natural breaking apart of plant segments, such as in bulbs or corms.

Sessile Used in reference to flowers, florets, leaves, leaflets, or fruits that are attached directly to a shoot and not borne on any type of a stalk.

Sexual reproduction Development of new plants by the processes of meiosis and fertilization in the flower to produce a viable embryo in a seed.

Short day plants Plants that initiate flowers only under short-day (long-night) conditions.

Sidedressing Applying fertilizer on a soil surface close enough to a plant that cultivating or watering carries the fertilizer to the plant's roots.

Silage Forage that is chemically changed and preserved in a succulent condition by partial fermentation in the preparation of food for livestock.

Silique A dry, one-seeded, dehiscent (infrequently indehiscent) fruit consisting of two carpels that form a bilocular ovary. Common in the CRUCIFERAE family.

Silo A structure for making and storing silage.

Silt A soil textural class consisting of particles between 0.05 and 0.002 mm in diameter.

Simple fruit A fruit derived from a single pistil.

Sod Top 3 to 7 cm (1 to 3 in.) of soil permeated by and held together with grass roots or grass-legume roots.

Sodic soil A soil that contains so much exchangeable sodium (more than 15 percent) that it interferes with the growth of most crop plants. Also, if the total soluble salts is more than 4 mmho per cm^2, the soil is a saline sodic soil; if it is less than 4 mmho per cm^2, the soil is a nonsaline sodic soil.

Soil The solid portion of the earth's crust in which plants grow. It is composed of mineral material, air, water, and organic matter both living and dead.

Soil air The gaseous phase of the soil; that percentage of the total volume not occupied by solid or liquid.

Soil conservation A combination of all management and land use methods that safeguard the soil against depletion or deterioration by natural or by human-induced factors.

Soil fertilization *See* Fertilization (soil).

Soil management The total tillage operations, cropping practices, fertilizing, liming, and other treatments conducted on or applied to a soil for the production of plants.

Soil organic matter The organic fraction of the soil that includes plant and animal residues at various stages of decomposition, cells and tissues of soil organisms, and substances synthesized by the soil population.

Soil pasteurization Treating the soil with heat [usually steam at 60°C to 70°C (140°F to 160°F)] to destroy most harmful pathogens, nematodes, and weed seeds. Less severe temperature treatment than soil sterilization.

Soil salinity The amount of soluble salts in a soil, expressed as parts per million, millimho/cm, or other convenient ratios.

Soil series The basic unit of soil classification; a subdivision of a family, comprising soils that are essentially alike in all major profile characteristics.

Soil solution The aqueous liquid phase of the soil and its solutes consisting of ions dissociated from the surfaces of the soil particles and of other soluble materials.

Soil sterilization Treating soil by gaseous fumigation, chemicals, heat (usually steam) at 100°C (212°F) to destroy all living organisms.

Soil structure The arrangement of primary soil particles into secondary particles, units, or peds that act as primary particles. The secondary units are characterized and classified on the basis of size, shape, and degree of distinctness.

Soil texture The relative percentages of sand, silt, and clay in a soil.

Soil tilth The physical condition of soil as related to its ease of tillage, fitness as a seedbed, and suitability for plant growth.

Soil type The lowest unit in the natural system of soil classification; a subdivision of a soil series.

Soluble salts Disassociated salts (anions and cations) in the soil that can become toxic to roots when exceeding certain levels.

Solute A substance dissolved in a solvent.

Solution A homogeneous mixture; the molecules of the dissolved substance (the solute) are dispersed among the molecules of the solvent.

Solvent A substance, usually a liquid, that can dissolve other substances (solutes).

Somaclonal variation Existing genetic variation which may not be seen until after plant cells have been through aseptic culture, or the culture may force the change.

Somatic tissue Nonreproductive, vegetative tissue. Tissue developed through mitosis that will not undergo meiosis.

Species A group of similar organisms capable of interbreeding and more or less distinctly different in geographic range and/or morphological characteristics from other species in the same genus.

Sperm A male gamete.

Spermatophyte A seed-bearing plant.

Spike An inflorescence that has a central axis on which sessile flowers are borne. Examples are some grasses, gladioli, and snapdragons.

Spongy parenchyma The cell layer in a leaf located between the palisade parenchyma and the lower epidermis; these cells have thin cell walls and are loosely packed.

Spore A reproductive cell that develops into a plant without union with other cells.

Sprigging Vegetative propagation by planting stolons or rhizomes (sprigs).

Square (botanical) An unopened flowerbud in cotton with its accompanying bracts.

Stamen The male reproductive structure of a flower. The stamen produces pollen and is composed of a filament on which is borne an anther.

Staminate (male) flower A flower having stamens but no pistils.

Starch A complex polysaccharide carbohydrate. The form of food commonly stored by plants.

Stele The vascular tissue and closely associated tissues in the axes of plants. The central cylinder of the stem.

Stem The main body of a plant, usually the ascending axis, whether above or below ground in opposition to the descending axis or root. Stems, but not roots, produce nodes and buds.

Stigma In a flower, the portion of the style to which pollen adheres.

Stock *See* Rootstock.

Stolon A slender, prostrate above-ground stem. The runners of white clover, strawberry, and bermuda grass plants are examples of stolons.

Stoma (stomate) (plural, stomata, stomates) A small opening, bordered by guard cells, in the epidermis of leaves and stems, through which gases including water vapor pass.

Stool (horticultural) Sprouts that arise from the base of the plant below ground and become rooted. They are used for vegetative propagation.

Stool or stooling (agronomic) Shoots that arise from below the soil at the base of a plant.

Stratification The practice of exposing imbibed seeds to cool 2°C to 10°C (35°F to 50°F) (sometimes warm) temperatures for a period of time prior to germination in order to break dormancy. This is a standard practice in the germination of seeds of many grass and woody species.

Stress (water) Plant(s) unable to absorb enough water to replace that lost by transpiration. Results may be wilting, cessation of growth, or death of the plant or plant parts.

Strip cropping The practice of growing crops that require different types of tillage, such as row and sod, in alternate strips along contours or across the prevailing direction of wind.

Style Slender column of tissue that arises from the top of the ovary in the flower through which the pollen tube grows toward the ovule.

Suberin A waxy, waterproofing substance in cork tissue.

Subirrigation Application of water from below the soil surface, usually from a ditch or a perforated hose or pipe, or by placing a potted plant on a constantly moist surface.

Subsoil That part of the soil below the plow layer.

Subsoiling Breaking of compact subsoils, without inverting them, with a special knifelike instrument (chisel) that is pulled through the soil at depths usually of 30 to 60 cm (12 to 24 in.) and at spacings usually of 60 to 150 cm (2 to 5 ft).

Sucrose Table sugar ($C_{12}H_{22}O_{11}$); a carbohydrate formed by chemically joining a molecule of glucose ($C_6H_{12}O_6$) with a molecule of fructose ($C_6H_{10}O_5$).

Summer fallow The tillage of uncropped land during the summer in order to control weeds and store moisture in the soil for the growth of a later crop.

Sunscald High temperature injury to plant tissue due to intense sun's rays warming the trunks of trees during winter, cracking and splitting the bark. It can be prevented by shading or whitewashing tree trunks and larger branches. Sunscald also occurs on unshaded vegetable fruits (tomatoes and melons) or to houseplants exposed to direct sunlight.

Superior ovary An ovary situated above the receptacle; all other floral parts develop below the base of the ovary.

Symbiosis An obligate relationship between two organisms of different species living together in close association for their mutual benefit. An example is the mycelium of a fungus with roots of seed plants.

Synergids The two nuclei within the embryo sac at the upper end in the ovule of the flower, which, with the third (the egg), constitute the egg apparatus.

Synergism The mutual interaction of two substances that results in an effect greater than the additional effects of the two when used independently.

Systemic A pesticide material absorbed by plants, making them toxic to feeding insects. Also, pertaining to a disease in which an infection spreads throughout the plant.

Tangential face The wood surface exposed when a cut is made at right angles to the rays and parallel to the long axis of the majority of cells.

Tannin Broad class of soluble polyphenols with a common property of condensing with protein to form a leatherlike substance that is insoluble in water.

Taproot An elongated, deeply growing primary root.

Taxon (plural, taxa) A taxonomic group of plants of any rank such as family, genus, species, and so forth.

Taxonomy The science dealing with describing, naming, and classifying plants and animals.

Tendril A slender coiling modified leaf or stem arising from stems and aiding in their support.

Tension, soil-moisture The equivalent negative pressure of water in soil. The attraction with which water is held to soil particles.

Terminal bud A bud at the distal end of a stem.

Terrace A level, usually narrow, plain bordering a river, lake, or the sea. Rivers sometimes are bordered by terraces at different levels. It also refers to a raised, more or less level strip of land usually constructed on a contour and designed to make the land suitable for tillage and prevent erosion.

Tetraploid Having four sets of chromosomes per nucleus.

Texture (soil) The relative proportions of the various-sized groups of individual soil grains in a mass of soil. It refers to the proportions of sand, silt, and clay in a given amount of soil.

Thallophytes A division of plants whose body lacks roots, stems, and leaves (e.g., mushrooms).

Thinning Removing young plants from a row to provide the remaining plants with more space to develop. Also, the removal of excess numbers of fruits from a tree so the remaining fruits will become larger.

Tile (drain) Pipe made of burned clay, concrete, or similar material, in short lengths, usually buried at the bottom of a ditch, with open joints to collect and carry excess water from the soil. The newer type polyvinyl tile is made in continuous lengths, and is perforated to allow water to enter the pipe for removal from the soil.

Tilth *See* Soil tilth.

Tissue A group of cells of similar structure that performs a special function.

Topdressing Applying materials such as fertilizer or compost to the soil surface while plants are growing.

Topsoil The upper layer of soil moved in cultivation.

Topworking (top-grafting) To change the cultivar of a tree by grafting the main scaffold branches.

Tracheid An elongated, tapering xylem cell with lignified pitted walls adapted for conduction and support.

Transfer RNA *See* Ribonucleic acid.

Translocation The transfer of food materials or products of metabolism throughout the plant.

Transpiration The loss of water vapor through the stomata of leaves.

Tuber An enlarged, fleshy, underground tip of a stem. The white (Irish) potato produces tubers as food storage organs.

Tuberous root An enlarged fleshy, underground root (e.g., sweet potato and dahlia).

Turgid Swollen, distended; referring to cells or tissues that are firm because of water uptake.

Turgor pressure The pressure within the cell resulting from the absorption of water into the vacuole and the imbibition of water by the protoplasm.

2,4-Dichlorophenoxyacetic acid (2,4-D) A selective auxin-type herbicide that kills broad-leaved plants but not grasses.

Umbel A type of inflorescence in which flowers are borne at the end of stalks that arise like the ribs of an umbrella from one point (e.g., carrot, onion).

Unavailable water Water held by the soil so strongly that the root cannot absorb it.

Understock *See* Rootstock.

Vacuole A cavity in the plant's cell bounded by a membrane in which various plant products and byproducts are stored.

Variety (botanical) A subdivision of a species with distinct morphological characters and given a Latin name according to the rules of the International Code of Botanical Nomenclature. A taxonomic variety is known by the first validly published name applied to it so that nomenclature tends to be stable.

Variety (cultivated) *See* Cultivar.

Vascular bundle A strand of tissue containing primary xylem and primary phloem and frequently enclosed by a bundle sheath of parenchyma or fibers.

Vascular cambium A meristem that produces secondary xylem and secondary phloem cells. Vascular cambium is found in biennials and perennials.

Vector A carrier; for example, an insect that carries pathogenic organisms from plant to plant.

Vegetation The plants that cover a region; it is formed of the species that make up the flora of the area.

Vegetative Referring to asexual (stem, leaf, root) development in plants in contrast to sexual (flower, seed) development.

Vein The vascular strand of xylem and phloem in a leaf.

Vernalization In reference to flowering, the process by which floral induction in some plants is promoted by exposing the plants to chilling for a certain length of time.

Vessel A series of xylem elements in the stem and root that conduct water and mineral nutrients.

Virgin soil A soil that has not been significantly disturbed from its original natural environment.

Virus A pathogen consisting of a single strand of nucleic acids encapsuled in a protein coat that can only replicate inside a living cell.

Vitamins Natural organic substances, necessary in small amounts for the normal metabolism of plants and animals.

Volatile Evaporating readily or easily dissipated in the form of a vapor.

Water potential The difference between the activity of water molecules in pure distilled water at atmospheric pressure and 3°C (standard conditions) and the activity of water molecules in any other system; the activity of these water molecules may be greater (positive) or less (negative) than the activity of the water molecules under standard conditions.

Water table The upper surface of ground water or that level below which the soil is saturated with water.

Watts (W) (light) Radiometric units expressing energy per unit of time and unit area independent of wavelength. Irradiance within the photosynthetic active region (PAR) of 400–700 nm is commonly referred to as W PAR.

Wavelength The distance between two corresponding points on any two consecutive waves. For visible light it is minute and is generally measured in nanometers.

Weathering All physical and chemical changes produced in rocks, at or near the earth's surface, by atmospheric agents.

Weed A plant not valued for use or beauty. Any plant growing where it is not wanted.

Wilting point (Permanent Wilting Point, PWP) The moisture content of soil at which plants wilt and fail to recover even when placed in a humid atmosphere.

Windbreak A planting of trees or shrubs, usually perpendicular or nearly so to the principal wind direction, to protect soil, crops, homesteads, roads, and so on, against the effects of winds.

Windrow Hay, grain, leaves, or other material swept or raked into rows to dry.

Winter hardiness The ability of a plant to tolerate severe winter conditions.

Wood Secondary nonfunctioning xylem in a perennial shrub or tree.

Xanthophyll Yellow carotenoid pigment ($C_{40}H_{56}O_2$) found along with chlorophyll in green plants.

Xylem Specialized cells through which water and minerals move upward from the soil through a plant.

Zygote A protoplast resulting from the fusion of gametes (either isogametes or heterogametes). The beginning of a new plant in sexual reproduction.

Index